APPLICATIONS OF CELL IMMOBILISATION BIOTECHNOLOGY

FOCUS ON BIOTECHNOLOGY

Volume 8B

Series Editors
MARCEL HOFMAN
Centre for Veterinary and Agrochemical Research, Tervuren, Belgium

JOZEF ANNÉ
Rega Institute, University of Leuven, Belgium

Volume Editors
VIKTOR NEDOVIĆ
*University of Belgrade,
Belgrade, Serbia and Montenegro*

RONNIE WILLAERT
*Free University of Brussels,
Brussels, Belgium*

COLOPHON

Focus on Biotechnology is an open-ended series of reference volumes produced by Springer in co-operation with the Branche Belge de la Société de Chimie Industrielle a.s.b.l.

The initiative has been taken in conjunction with the Ninth European Congress on Biotechnology. ECB9 has been supported by the Commission of the European Communities, the General Directorate for Technology, Research and Energy of the Wallonia Region, Belgium and J. Chabert, Minister for Economy of the Brussels Capital Region.

Applications of Cell Immobilisation Biotechnology

Edited by

VIKTOR NEDOVIĆ
*University of Belgrade,
Belgrade, Serbia and Montenegro*

and

RONNIE WILLAERT
*Free University of Brussels,
Brussels, Belgium*

A C.I.P. Catalogue record for this book is available from the Library of Congress.

ISBN-10 1-4020-3229-3 (HB) Springer Dordrecht, Berlin, Heidelberg, New York
ISBN-10 1-4020-3363-X (e-book) Springer Dordrecht, Berlin, Heidelberg, New York
ISBN-13 978-1-4020-3229-5 (HB) Springer Dordrecht, Berlin, Heidelberg, New York
ISBN-13 978-1-4020-3363-6 (e-book) Springer Dordrecht, Berlin, Heidelberg, New York

Published by Springer,
P.O. Box 17, 3300 AA Dordrecht, The Netherlands.

www.springeronline.com

Printed on acid-free paper

All Rights Reserved
© 2005 Springer
No part of this work may be reproduced, stored in a retrieval system, or transmitted
in any form or by any means, electronic, mechanical, photocopying, microfilming, recording
or otherwise, without written permission from the Publisher, with the exception
of any material supplied specifically for the purpose of being entered
and executed on a computer system, for exclusive use by the purchaser of the work.

Printed in the Netherlands.

EDITOR'S PREFACE

This book is the second of two volumes dedicated to cell immobilisation/encapsulation biotechnology. The first book (Focus on Biotechnology, vol. 8A) dealt with the fundamentals of cell immobilisation biotechnology, whereas the present book is focused on the applications. Actually it is an outcome of the editors' intention to gather the vast widespread information on applications of immobilisation/encapsulation biotechnology into a comprehensive reference book and to give the reader an overview of the most recent results and developments that have been realised in this domain. Extensive research in the field of cell immobilisation/ encapsulation for various processes in biotechnology has been carried out during recent years. This is clearly reflected in the voluminous publications of original research, patents, and symposia, and the development of successful commercial inventions. In addition, it is obvious that there is a rapidly growing market for such applications in many different areas including biomedicine, pharmacology, cosmetology, food technology and beverage production, agriculture, waste treatment, analytical applications and biologics production. The progress in immobilised/encapsulated cell applications has therefore become a real scientific and technological challenge in the last few years.

This book consists of 33 chapters that are arranged into 5 parts: (1) "Tissue engineering", (2) "Microencapsulation for disease treatment", (3) "Food and beverage applications", (4) "Industrial biochemical production", and (5) "Environmental and agricultural applications". In each part various related topics are presented in detail. Specifically, contents of different chapters cover a broad variety of cell types and applications: seeding and cultivation of mammalian cells on different supports for tissue engineering, encapsulation of animal cells for treatment of variety of diseases, immobilisation of microbial cells for utilisation in food industry and fermentation processes, encapsulation and stabilisation of probiotics, production of biopharmaceuticals, biologics and biofuel, waste treatment and pollution control, agricultural applications such as artificial plant seeds and artificial insemination of farm animals and development of biosensors. In addition, particular attention is given to selection, design and characterisation of cell carriers and bioreactors for different processes. 77 carefully selected experts in different fields of biotechnology and bioengineering from 46 research institutes and laboratories all over the world contributed to this volume and collectively provided a unique, rich expertise and knowledge. Today, the book presents the most comprehensive, complete, up-to-date source of information on various aspects of cell immobilisation/encapsulation applications.

This book is intended to cover needs and to be the essential resource for both academic and industrial communities interested in cell immobilisation biotechnology. An integrated approach to "biological" and "engineering/technology" aspects is pursued to reach an even wider audience such as specialists in biomedicine, pharmacy, microbiology, biology, food technology, agriculture, environmental protection, chemical, biochemical and tissue engineering who seek a broad view on cell immobilisation/encapsulation applications. Due to the in depth review of various applications of cell immobilisation, this volume, in combination with the first volume, can be used as a "handbook" of "Cell Immobilisation Biotechnology".

We express our gratitude and appreciation to our many colleagues, who as experts in their fields, have contributed to this volume. We would also like to thank the series editor, Marcel Hofman, and the publisher, Kluwer/Springer, for their excellent support in assuring the high quality of this publication.

Viktor A. Nedović and Ronnie Willaert

Belgrade/Brussels, September 2004

TABLE OF CONTENTS

EDITOR'S PREFACE v
TABLE OF CONTENTS 1
PART 1
TISSUE ENGINEERING
Bio-artificial organs 17
The example of artificial pancreas 17
Claudio Nastruzzi, Giovanni Luca, Giuseppe Basta and Riccardo Calafiore 17
   1. Introduction to bio-artificial organs 17
   2. Diabetes mellitus 17
   3. Application of bio-artificial pancreas 18
   4. General characteristic of bio-artificial pancreas 20
   5. Devices for bio-artificial pancreas 21
      5.1. Macrodevices 21
      5.2. Microcapsule 22
         5.2.1. General considerations 22
            5.2.1.1. Uncoated Ca-alginate microcapsules. 22
            5.2.1.2. Barium alginate microcapsules. 23
            5.2.1.3. Agarose microcapsules. 23
            5.2.1.4. Poly-ethylene glycol microcapsules. 24
            5.2.1.5. Other polymers. 24
            5.2.1.6. Conformal alginate microcapsules. 25
            5.2.1.7. Polyaminoacid coated alginate microcapsules. 25
            5.2.1.8. Scaling-up protocols for alginate microcapsules. 26
   6. Microcapsule immunobarrier capacity 28
   7. Site of transplantation 28
   8. Intracapsular environmental conditions 28
   9. Co-encapsulation of drug delivery devices 29
      9.1. Effects of anti-oxidants on β-cell function 29
      9.2. Effects of vitamin D_3 on insulin secretion 30
   10. Conclusions 33
   References 35
Bioartificial pancreas: an update 39
Ales Prokop and Jeffrey M. Davidson 39
   1. Introduction 39
   2. Islet immobilisation techniques: towards a perfect capsule 40
      2.1. Drop generation and capsule size 41
      2.2. Process and reactor design, product performance 42
      2.3. Biologically relevant problems 43
   3. Islet sourcing 44
      3.1. Embryonic stem cells (ESC) 48
      3.2. Adult pluripotent stem cells (APSC) 48
      3.3. Embryonic vs. adult SC 50

References ... 51
Bioartificial skin ... 55
Barbara Zavan, Roberta Cortivo, Paola Brun, Carolin Tonello
and Giovanni Abatangelo ... 55
   1. Introduction ... 55
   2. Requirements ... 56
      2.1. Anatomic ... 56
      2.2. Surgical ... 57
      3. Tissue engineering of human skin ... 58
      3.1. Scaffolds ... 58
      3.2. Cells ... 59
         3.2.1. Epidermis ... 59
         3.2.2. Dermis ... 61
   4. Conclusions ... 65
   References ... 66
Bioartificial liver ... 69
Clare Selden ... 69
   1. Introduction ... 69
   2. What is the best cell for a bioartificial liver, and how many are required? ... 69
   3. Bioreactor design – initial clinical experience ... 70
   4. Bioreactor designs – experimental models ... 73
   5. Cell encapsulation ... 74
   6. Alginate hydrogels ... 75
   7. Alternative approach ... 80
   8. Conclusion ... 80
   References ... 80
Tissue-engineered blood vessels and the future of tissue substitutes ... 85
Lucie Germain, Karina Laflamme and François A. Auger ... 85
   1. Introduction ... 85
   2. TEBV as a conceptual continuum ... 85
   3. Reconstructing small diameter blood vessel by the self-assembly approach ... 87
   4. Methodology for TEBV reconstruction by the self-assembly approach ... 88
   5. Histological and phenotypic characteristics of the reconstructed TEBV ... 90
   6. Functional characteristics of the reconstructed TEBV ... 90
   7. Vascularisation of the tissue construct by the addition of capillaries *in vitro* ... 93
   8. Conclusion ... 95
   Acknowledgements ... 95
   References ... 95
Tissue engineering of cartilage and myocardium ... 99
Bojana Obradovic, Milica Radisic and Gordana Vunjak-Novakovic ... 99
   1. Introduction ... 99
      1.1. Tissue engineering requirements ... 100
      1.2. Tissue engineering model system ... 100
   2. Cartilage tissue engineering ... 103
      2.1. Articular cartilage ... 103
      2.2. Clinical need ... 103

2.3. Tissue engineering 103
2.4. Cell sources 104
2.5. Scaffolds 104
2.6. Bioreactor hydrodynamics 105
2.7. Growth factors 106
2.8. Duration of culture 108
2.9. Spatial and temporal patterns of chondrogenesis 109
2.10. Mathematical model of cartilage development 110
    2.10.1. Mathematical model of GAG accumulation
          in cultured cartilage explants 110
    2.10.2. Mathematical model of GAG accumulation in engineered cartilage
          constructs 113
3. Cardiac tissue engineering 116
3.1. Myocardium (cardiac muscle) 116
3.2. Clinical need 116
3.3. Tissue engineering 117
3.4. Cells 117
3.5. Scaffolds 118
3.6. Bioreactor hydrodynamics 118
    3.6.1. Static dishes 120
    3.6.2. Interstitial flow 121
3.7. Mathematical model of oxygen distribution in a tissue construct 123
3.8. Mechanical stimulation 126
3.9. Electrical stimulation of construct contractions 127
4. Summary 129
References 129

Tissue engineered heart 135
Kristyn S. Masters and Brenda K. Mann 135
1. Introduction 135
2. Heart valves 135
2.1. Valve biology 136
2.2. Valve substitutes 137
    2.2.1. Mechanical valves 137
    2.2.2. Tissue valves 138
2.3. Tissue engineered heart valves 139
    2.3.1. Cell immobilisation in acellular valves 139
    2.3.2. Cell immobilisation in porous matrices 140
    2.3.3. Cell immobilisation in hydrogels 142
2.4. Critical considerations for tissue engineered heart valves 143
    2.4.1. Cell source 143
    2.4.2. Material properties 144
2.5. The ideal tissue engineered valve 144
3. Cardiac muscle 145
3.1. Biological considerations 145
3.2. Cell source 145
3.3. Scaffold materials 146

   3.4. Cultivation conditions 148
   4. Conclusions 149
   References 149
Bone tissue engineering 153
Pankaj Sharma, Sarah Cartmell and Alicia J. El Haj 153
   1. Introduction 153
   2. Mesenchymal stem cells 154
   3. Carrier scaffolds 155
   4. *Ex vivo* conditioning of constructs 157
   5. Animal studies 159
   6. Human studies 161
   References 163
Stem cells – potential for tissue engineering 167
M. Minhaj Siddiqui and Anthony Atala 167
   1. Introduction 167
   2. What is a stem cell? 168
      2.1. Adult stem cells 169
      2.2. Foetal stem cells 170
      2.3. Embryonic stem cell 170
      2.4. Somatic cell nuclear transfer: therapeutic cloning 171
   3. Potential and how cells are differentiated 171
      3.1. Non-specific differentiation and selection 172
      3.2. Gene transduction induced differentiation 172
      3.3. Growth factor and media formulation induced differentiation 173
   4. Advantages in tissue engineering 173
      4.1. Self-renewal 174
      4.2. Multipotency 174
   5. Research directions 174
      5.1. Skin 175
      5.2. Bone 175
      5.3. Cartilage 175
      5.4. Renal 176
      5.5. Skeletal and cardiac muscle 177
   6. Ethical and political considerations in stem cell biology 178
   References 178
PART 2
MICROENCAPSULATION FOR DISEASE TREATMENT
Challenges in cell encapsulation 185
Gorka Orive, Rosa Mª Hernández, Alicia R. Gascón and José Luis Pedraz 185
   1. Introduction 185
   2. Immunoisolation approaches 186
   3. Potential advantages of cell encapsulation technology 187
   4. Materials used in cell encapsulation 187
   5. Cell lines 190
   6. Therapeutic applications 191
   7. Perspectives and concluding remarks 193

References 193
Protein therapeutic delivery using encapsulated cell platform 197
Marcelle Machluf 197
    1. Introduction 197
    2. Anti-angiogenic protein therapy 198
    3. Anti-angiogenic gene delivery 198
    4. Genetically engineered cells delivering therapeutics 199
    5. Cell encapsulation – a platform for delivering therapeutics 200
    6. Cell encapsulation – delivering anti-angiogenic therapeutics 201
    7. Future perspective 205
    References 205
Cell encapsulation therapy for malignant gliomas 211
Anne Mari Rokstad, Rolf Bjerkvig, Terje Espevik and Morten Lund-Johansen 211
    1. Introduction 211
    2. Glioma growth and invasiveness 212
    3. Glioma treatment 213
        3.1. Surgery 213
        3.2. Irradiation 213
        3.3. Chemotherapy 213
    4. Glioma treatment with alginate bioreactors 213
        4.1. Optimising the alginate bioreactor 214
        4.2. Improvement of the alginate microcapsules for proliferating cells 217
        4.3. Choosing cell-lines suited for encapsulation 218
        4.4. The biocompatibility of foreign materials in the brain 219
        4.5. The biocompatibility of the alginate bioreactors 220
    5. Conclusion 222
    References 222
Gene therapy using encapsulated cells 229
Gonzalo Hortelano 229
    1. Introduction 229
    2. Gene therapy 230
    3. Genetic engineering of cells 231
        3.1. Viral vectors 232
        3.2. Retrovirus 232
        3.3. Lentivirus 233
        3.4. Adenovirus 233
        3.5. Adeno-associated virus 234
        3.6. Herpes 234
        3.7. Non-viral vectors 234
    4. Selection of cells for encapsulation 235
        4.1. Established cell lines/primary cells 235
        4.2. Proliferative/quiescent cells 236
        4.3. Allogeneic/xenogeneic cells 237
    5. Applications of encapsulated cells in gene therapy 237
        5.1. Cancer 237
        5.2. Neurological conditions 238

    5.3. Erythropoietin 238
    5.4. Encapsulated cells to treat metabolic diseases 239
    5.5. Gene therapy of haemophilia B 240
    5.6. Gene therapy of haemophilia A 242
    6. Concluding remarks 242
    References 243

Artificial cells for blood substitutes, enzyme therapy, cell therapy and drug delivery 249
Thomas Ming Swi Chang 249
    Abstract 249
    1. Introduction 249
    2. Artificial cells containing enzymes for inborn errors of metabolism and other conditions 250
    3. Artificial cells for cell therapy 250
    4. Red blood cell substitutes 251
        4.1. Polyhemoglobin as blood substitutes 251
        4.2. Polyhemoglobin containing catalase and superoxide disumutase 251
        4.3. Recombinant human hemoglobin 252
        4.4. Other new generations of modified hemoglobin blood substitutes 252
    5. General 252
    Acknowledgements 252
    References 253

PART 3
FOOD AND BEVERAGE APPLICATIONS

Beer production using immobilised cells 259
Viktor Nedović, Ronnie Willaert, Ida Leskošek-Čukalović, Bojana Obradović and Branko Bugarski 259
    1. Introduction 259
    2. Carrier selection and design 261
    3. Reactor design 263
    4. ICT applications for the brewing industry 264
        4.1. Flavour maturation of green beer 264
        4.2. Production of alcohol-free or low-alcohol beer 265
        4.3. Production of acidified wort using immobilised lactic acid bacteria 266
        4.4. Continuous main fermentation 266
    5. Summary 268
    6. References 269

Application of immobilisation technology to cider production: a review 275
Alain Durieux, Xavier Nicolay and Jean-Paul Simon 275
    Abstract 275
    1. Introduction 275
    2. Cell immobilisation in cider production 277
        2.1. Immobilisation of yeast for apple juice fermentation 278
        2.2. Immobilisation of *O. oeni* for malolactic fermentation in cider 280
    3. Conclusions 282
    References 282

Wine production by immobilised cell systems 285
Charles Divies and Remy Cachon 285
   1. Introduction 285
   2. Immobilised cell technology and heterogeneous bioreactors 285
   3. Potential of immobilised cell systems for applications in oenology 287
      3.1. Alcoholic fermentation 287
      3.2. Bottle-fermented sparkling wines ("méthode champenoise") 288
      3.3. Production of sparkling wines in closed reactors 289
      3.4. The malolactic fermentation 290
   4. Conclusion 290
   References 291

Immobilised cell technologies for the dairy industry 295
Christophe Lacroix, Franck Grattepanche, Yann Doleyres and Dirk Bergmaier 295
   1. Introduction 295
   2. Immobilisation techniques 296
      2.1. Entrapment within polymeric networks 296
      2.2. Adsorption to a preformed carrier 296
      2.3. Membrane entrapment 297
      2.4. Microencapsulation 297
   3. Microbial growth in gel beads 299
      3.1. Biomass distribution in gel beads 299
      3.2. Cell release from beads 300
      3.3. Cross-contamination phenomenon 300
   4. Changes of culture characteristics during immobilised-cell fermentations 301
   5. Biological stability 303
      5.1. Psychrotroph contaminants 303
      5.2. Bacteriophage contaminants 303
      5.3. Plasmid stability 305
   6. Applications of immobilised cell technology 305
      6.1. Biomass production 305
         6.1.1. Starter production 306
            6.1.1.1. Lactic acid bacteria starter 306
            6.1.1.2. Probiotic cultures 307
         6.1.2. Prefermentation of milk 308
      6.2. Metabolite production 310
         6.2.1. Lactic acid production 310
         6.2.2. Exopolysaccharide production 311
         6.2.3. Bacteriocin production 312
      6.3. Cell protection 313
   7. Conclusions 314
   References 314

Food bioconversions and metabolite production 321
P. Heather Pilkington 321
   1. Introduction 321
   2. Food bioconversions 322
      2.1. Sake production 322

2.2. Soya sauce production 323
2.3. Mead production 324
2.4. Removal of malic acid from coffee beans 324
2.5. De-bittering of citrus juice 324
2.6. "Ugba" food snack production 325
2.7. Removal of simple sugars 325
    2.7.1. Purification of food-grade oligosaccharides 325
    2.7.2. Glucose removal from egg 326
2.8. Sugar conversions 326
2.9. Hydrolysis of triglycerides and proteins in milk 327
2.10. Vitamins 328
3. Metabolite production 328
   3.1. Amino acids 328
   3.2. Organic acids 329
   3.3. Alcohols 330
   3.4. Enzymes 331
   3.5. Bacteriocins 331
References 331

Immobilised-cell technology and meat processing 337
Linda Saucier and Claude P. Champagne 337
Introduction 337
1. Historical use of meat fermentation 338
2. New approach to meat preservation 339
3. Probiotic cultures and health 341
4. Immobilisation/encapsulation and starter production 341
   4.1. Technological cultures for meat fermentation 342
   4.2. Probiotic cultures for meats 345
5. Applications of ICT cultures in meat 345
   5.1. Meat fermentation using immobilised cells 345
   5.2. Improving the use of protective cultures *via* ICT 347
   5.3. Other applications of ICT in meat 348
6. Conclusion 349
References 350

Bioflavouring of foods and beverages 355
Ronnie Willaert, Hubert Verachtert, Karen Van Den Bremt, Freddy Delvaux and Guy
   Derdelinckx 355
1. Introduction 355
2. Definition of bioflavouring 356
3. Processes for flavour production 356
4. Bioproduction of natural flavours 358
   4.1. Microorganisms 358
       4.1.1. De novo biosynthesis 359
           4.1.1.1. Biosynthesis of lactones. 359
           4.1.1.2. In situ bioflavouring. 359
           4.1.1.3. Production of flavour metabolite organic acids. 360
       4.1.2. Bioconversion by microorganisms 362

  4.1.2.1. Biosynthesis of vanillin. 362
  4.1.2.2. Biosynthesis of flavours starting from fatty acids. 363
  4.1.2.3. Biosynthesis of aldehydes. 365
  4.1.2.4. Bioflavouring of beer. 365
 References 368

PART 4
INDUSTRIAL BIOCHEMICAL PRODUCTION

Production of ethanol using immobilised cell bioreactor systems 375
Argyrios Margaritis and Peter M. Kilonzo 375
 1. Introduction 375
 2. Immobilised cell systems 376
  2.1. Basic principles of cell immobilisation 378
  2.2. Cell immobilisation by adsorption 379
  2.3. Cell immobilisation by covalent bonding 380
  2.4. Cell immobilisation by physical entrapment within porous matrices 381
   2.4.1. Calcium alginate matrix 381
   2.4.2. Carrageenan matrix 384
   2.4.3. Polyacrylamide gel matrix 386
   2.4.4. Epoxy resin matrix 386
   2.4.5. Gelatin polymer matrix 387
  2.5. Cell immobilisation by containment behind a membrane barrier 388
  2.6. Cell immobilisation by self-aggregation 389
 3. Immobilised cell bioreactor types 391
 4. Ethanol production from non-conventional feedstock using immobilised cell systems 395
  4.1. The production of ethanol from cheese whey 396
  4.2. The production of ethanol from jerusalem artichokes 396
  4.3. The production of ethanol from cellulose and cellobiose 397
  4.4. Production of ethanol from xylose 399
 Acknowledgement 400
 References 400

Production of biopharmaceuticals through microbial cell immobilisation 407
Tajalli Keshavarz 407
 1. Introduction 407
 2. The use of immobilised microorganisms in production of antibiotics 408
  2.1. Penicillin production 408
  2.2. Cephalosporin-c production 413
  2.3. Other antibiotics 414
 3. The use of immobilised microorganisms in biotransformation in two liquid phase systems 417
 4. Concluding remarks 419
 References 420

Production of biologics from animal cell cultures 423
James Warnock and Mohamed Al-Rubeai 423
 1. Introduction 423
 2. Cell retention 424

   3. Cell encapsulation 426
   4. Cell entrapment 427
       4.1. Stirred tank bioreactors 429
       4.2. Fluidised-bed bioreactors 431
       4.3. Packed-bed bioreactors 433
   5. Conclusions 435
   References 436

Stabilisation of probiotic microorganisms 439
Helmut Viernstein, Josef Raffalt and Diether Polheim 439
   1. Introduction 439
   2. Standard pathways for production of microorganism and related stabilisation
      techniques 442
   3. Stabilisation methods 444
       3.1. Capsule forming processes 444
          LBC ME10 445
          Bifina® 445
          BiActon® 445
       3.2. Specific solutions 446
          LifeTop™ Cap 446
          LifeTop™ Straw 446
       3.3. Extrusion 446
       3.4. Hard (gelatine) capsules 447
          Bioflorin® capsules 447
          Probio-Tec®QUATRO-cap-4 447
          Omniflora® N 447
          BioTura® 447
       3.5. Liquid formulations 447
       3.6. Techniques for solid particle coating and aggregation 448
       3.7. Pastes and ointments 448
          Lactiferm® Ironpaste 449
       3.8. Powder formulations 449
          Antibiophilus sachets and capsules 449
          Effidigest® sachets 449
          Lactiferm® 449
          Effervescent formulation 450
       3.9. Tablet processing 450
          Bion® 3 and Multibionta® 450
          Gynoflor® 451
          Paidoflor® 451
          Lacto 451
   4. Summary 451
   References 452

Growth of insect and plant cells immobilised using electrified liquid jets 455
Mattheus F. A. Goosen 455
   1. Introduction 455
   2. Droplet generation using an electrified liquid jet 456

    2.1. Production of alginate beads using electrified liquid jets 456
    2.2. Effect of electrostatic field on cell viability 458
    2.3. Forces acting on droplet in an electric field 458
  3. Culture of encapsulated insect cells 462
    3.1. Growth of cells and virus in microcapsules 463
    3.2. Biocompatibility of encapsulation solutions 464
  4. Encapsulation and growth of plant tissue in alginate 466
  5. References 467

Plant cell immobilisation applications 469
Ryan Soderquist and James M. Lee 469
  1. Introduction 469
  2. Significance of plant cell immobilisation 470
  3. Methods of plant cell immobilisation 471
  4. Important considerations for plant cell immobilisation 473
  5. Potential benefits of plant cell immobilisation 474
  6. Conclusions 477
  References 477

PART 5
ENVIRONMENTAL AND AGRICULTURAL APPLICATIONS

Wastewater treatment by immobilised cell systems 481
Hiroaki Uemoto 481
  1. Introduction 481
  2. Immobilisation techniques 481
    2.1. Adsorption 482
    2.2. Entrapment 482
    2.3. Carrierless immobilisation 483
  3. Applications of immobilised cells 483
    3.1. Nitrification 483
    3.2. Simultaneous nitrification and denitrification 486
    3.3. Denitirification 489
    3.4. Metal accumulation 490
  4. Conclusions 491
  References 491

Immobilised cell strategies for the treatment of soil and groundwaters 495
Ludo Diels 495
  1. Introduction 495
  2. *In situ* treatment 496
  3. Soil and aquifer biofilms 498
  4. Viruses in soil and groundwater: do they play a role? 499
  5. Crossing the bridge between groundwater and surface water 499
  6. Pump and treat 500
  7. *In situ* bioprecipitation 503
  8. Bacterial contamination of aquifers 503
  9. Conclusions 504
  References 504

Application of immobilised cells for air pollution control 507
Marc A. Deshusses 507
  1. Introduction 507
  2. Microbiology of gas phase bioreactors 511
    2.1. Microflora 511
    2.2. Biofilm architecture 513
    2.3. Secondary degraders and predators 514
    2.4. Biodegradation and growth kinetics 515
  3. Biofilter and biotrickling filter applications 517
    3.1. Definitions and performance reporting 517
    3.2. Examples of applications 519
  4. Current research, emerging topics 521
  References 524

Artificial seeds 527
Eugene Khor and Chiang Shiong Loh 527
  1. Introduction 527
  2. Definitions of artificial seed 527
  3. Plant materials used for encapsulation 528
    3.1. Somatic embryos and microspores-derived embryos 528
    3.2. Shoot buds and shoot-tips 529
    3.3. Seeds 529
    3.4. Orchid protocorms and protocorm-like bodies 530
    3.5. Mycorrhizal fungi 530
    3.6. Hairy roots-derived materials 531
  4. Matrix material selection for artificial seeds 532
    4.1. Fluid drilling 532
    4.2. Polymeric coating 532
    4.3. Hydrogels 533
    4.4. Two-coat system 533
    4.5. Taking stock 535
  5. Outlook 536
  References 536

Spermatozoal microencapsulation for use in artificial insemination of farm animals 539
Raymond L. Nebel and Richard G. Saacke 539
  1. Introduction 539
  2. Procedures for microencapsulation of sperm 540
  3. Capsule size and maintenance of sperm viability *in vitro* 541
  4. Microcapsule behaviour in the female tract 542
  5. Evaluation of encapsulation technology in cattle artificial insemination – field trials 545
  6. Conclusions 547
  References 547

Biosensors with immobilised microbial cells using amperometric and thermal detection principles 549
Ján Tkáč, Vladimír Štefuca and Peter Gemeiner 549

1. Biosensors 549
2. Microbial biosensors 549
3. Immobilisation strategies 550
4. Improvement of microbial biosensor characteristics 551
5. Amperometric and thermal microbial biosensors 551
    5.1. Amperometric biosensors with the use of *Gluconobacter oxydans* 552
        5.1.1. Gluconobacter oxydans as a prospective biocatalyst 552
        5.1.2. Whole cell Gluconobacter oxydans biosensors 553
            5.1.2.1. Oxygen-based G. oxydans biosensors 553
            5.1.2.2. Mediated Gluconobacter sp. biosensors 555
        5.1.3. Conclusion and future perspectives of G. oxydans biosensors 556
    5.2. Thermal biosensors 557
        5.2.1. Analytical applications 558
        5.2.2. Investigation of properties of immobilised cells 559
        5.2.3. Conclusions and perspectives 563
    References 563
Index **567**

PART 1

TISSUE ENGINEERING

BIO-ARTIFICIAL ORGANS

The example of artificial pancreas

CLAUDIO NASTRUZZI[1], GIOVANNI LUCA[1], GIUSEPPE BASTA[2] AND RICCARDO CALAFIORE[2]
[1]Department of Medicinal Chemistry and Pharmaceutics, and [2]Department Internal Medicine, Section of Internal Medicine and Endocrine and Metabolic Sciences, School of Medicine, University of Perugia, I-06100 Perugia, Italy – Fax: +39-075-5847469 – Email: nas@unipg.it

1. Introduction to bio-artificial organs

The design and production of bio-artificial organs is one of the most challenging applications of a relatively new science: tissue engineering. Tissue engineering has been defined as "an interdisciplinary field which applies the principles of engineering and life sciences towards the development of biological substitutes that aim to maintain, restore or improve tissue function" [1].

In general, every organ that can be broken apart into single cells or cell clusters, without disrupting the original function is potentially suitable for generating a bio artificial organ. However, it is generally difficult to preserve the functionality of cellular units placed in environmental conditions that usually are quite different from their native site. In addition, only few materials enable creation of a tissue/material interface that is fully biocompatible towards both the immobilised cells and the host's tissue. In particular, special care must be taken not only to assess the material's physical-chemical properties and biocompatibility, but also to select the tissue's sources, in compliance with physiological competence and safety principles. Finally, the overall suitability of the newly developed biohybrid devices for clinical application must also be carefully assessed.

2. Diabetes mellitus

Diabetes mellitus (DM) is the most common endocrine disorder; the pathology is associated to metabolic abnormalities such as elevated blood glucose levels and their biochemical consequences that may provoke acute illness or result during the time course of the disease into secondary chronic complications [2]. These mainly affect eyes, kidney, nerves and blood vessels. Although it is not an immediate life-threatening disease anymore, in the majority of instances, provided that the patients are promptly and adequately treated, DM still represents a potentially lethal and certainly highly disabling disease. It is estimated that over 100 million people in the world actually suffer for either type 1 or type 2 DM. At the clinical onset of type 1 Diabetes mellitus (T1DM) that affects mainly, but not exclusively, adolescent/young individuals, the majority of islet β-cells has been completely destroyed by autoimmunity.

Since T1DM patients will require life-long, multiple daily insulin injections, they unfortunately represent a heavy burden to society. In fact, if these individuals have seen their life expectancy to gradually improve from the introduction of insulin therapy, their increased longevity has not resulted in elimination of the risk for developing chronic, often very serious complications of the disease, such as premature blindness, terminal renal insufficiency, vascular disease and disabling neuropathy. These complications may occur in spite of the selected insulin therapy regimens because it is very difficult to mimic performance of the glucose sensing apparatus incorporated in the normal β-cells. Consequently, the physiological stimulus-coupled insulin secretory response is far from being reproduced by subcutaneous exogenous insulin injections.

3. Application of bio-artificial pancreas

Bio-Artificial Pancreas (BAP), composed of insulin producing cells that are protected from the host's immune reaction by biocompatible, selective permeable and chemically stable artificial membranes, would ideally apply to the potential cure of T1DM by transplantation. In fact, BAP that contains insulin-producing cells would provide for continuous insulin delivery under strict regulation by the extra-cellular glucose levels. The goal of such a "totally automatic treatment" for T1DM could be also theoretically accomplished by creating an "artificial pancreas". In this instance, a glucose sensor [3] implanted subcutaneously, would continuously record extra-cellular glucose levels and convert the chemical information into an electronic signal, by a mini-computerised system. This would regulate, in turn, release of pre-stored insulin by a mini-pump, in order to maintain normoglycemia. This machinery, while able to reverse hyperglycemia, has been preliminarily proven to function, so far, for only very limited periods of time, due to still pending technical problems.

The most actual and effective way to prevent the onset of secondary complications in patients with T1DM is based on strict blood glucose control. This goal may be achieved by injecting either short- or long-acting insulin molecules, as frequently as 4 times a day. Unfortunately, this therapeutic option is associated with major drawbacks: (a) patients are exposed to high risk for developing severe hypoglycemia, (b) risk for developing secondary complications of the disease is attenuated, but not eliminated [4] and (c) patients' compliance with intensive insulin therapy may be quite poor, because of both, strict blood glucose monitoring that involve multiple finger pricks, and insulin injections. The imperfect blood glucose control achieved by subcutaneously-injected exogenous insulin lies on a fundamental pitfall consisting of the physical distance between the insulin injection site and the liver, that represents the first physiological site of insulin action.

In consideration of the above reviewed concerns, the "T1DM problem" could be solved only if the original physiological model, namely viable and functional insulin producing β-cells, was fully reconstituted. Such a goal might be accomplished by transplanting the endocrine pancreas as a whole organ or as isolated human/nonhuman, primary or engineered islet cells. BAP could offer the opportunity to graft the islet cells within artificial membranes so as to prevent immune destruction of the transplanted tissue with no recipient's general immunosuppression. The actual transplantation protocols comprise: (a) whole pancreas transplantation and (b) islet cell transplantation.

The isolated islet transplantation would be incomparably less invasive, in comparison with whole pancreatic graft, since the islets, that represent only 1-2% of the total pancreatic mass, are usually injected into the porta vein by slow beg mediated infusion [5], directly into the liver, under local anaesthesia. However, in spite of the potential privileges associated with islets over whole organs, the involved immune problems are similar.

Human islet allografts in generally immunosuppressed patients with T1DM, according to the "Edmonton protocol", University of Alberta, Edmonton, Canada, have resulted in successful reversal of hyperglycemia, and discontinuation of the daily exogenous insulin therapy, in 70% of the original cases, throughout 2 years of post-transplant follow-up [5]. The principle that islet transplantation may provide a final cure for T1DM has been proved, at least two major issues remain unfulfilled: (a) the restricted availability of cadaveric human donor pancreata, in addition to the fact that two organs per recipient are needed according to the Edmonton protocol, strictly hampers progress of islet allograft into widespread clinical trials; (b) although the "steroid-free" immunosuppressive regimen adopted by Edmonton, comprising Sirolimus and Daclizumab, in combination with low-dose Tacrolimus, seems to be less toxic and more effective than the conventionally employed immunosuppressive agents, its potential long-term adverse side effects are unknown. In our opinion, to comply with safety requirements, general immunosuppression in patients receiving islet allografts should be obviated.

A potential solution could be to use alternate islet sources to both avoid tissue shortage, while preventing recurrent autoimmunity, and in parallel develop alternate strategies to protect islet grafts from the immune destruction, without general immunosuppression. Either adult [6] or neonatal [7,8] porcine islets (see Figure 1) might successfully replace human donor tissue, provided that they were properly immunoprotected. On one hand, pork insulin would be acceptable to humans, and on the other hand it would be possible to create specific pathogen free (SPF) pig breeding stocks that comply with standard safety requirements.

With regard to alternatives to general immunosuppression, two main strategies seem to be feasible: (a) the use of properly assembled, physical immunobarriers or (b) the possibility of inducing a state of donor's immune unresponsiveness. This second approach would consist of employing new agents in conjunction with the induction of microchimerism, the latter being obtained by grafting homologous bone marrow cells, before islets.

4. General characteristic of bio-artificial pancreas

The basic principles for the development of a safe and reliable prototype of BAP should respond to the general characteristics summarised in Table 1. In addition, it should be considered that islets cells constitute a highly differentiated cellular system, equipped with a sophisticated glucose sensing apparatus that is very difficult to reproduce.

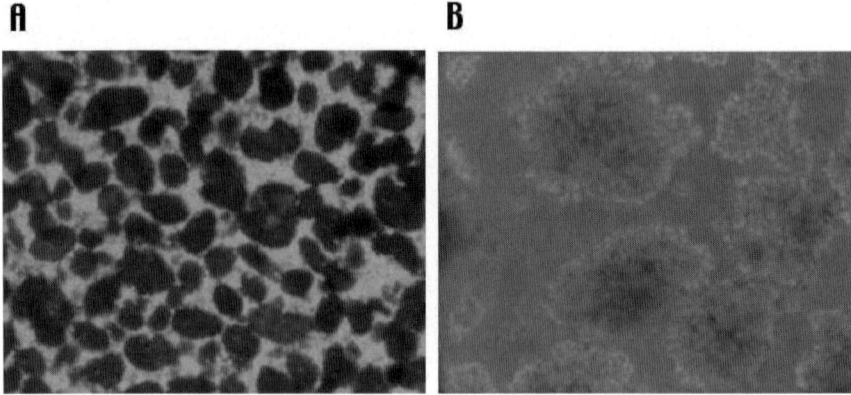

Figure 1. Optical photomicrographs of freshly isolated adult pig islets (A) and neonatal porcine cell clusters (NPCC's) (B) (50x). Islets were stained with dithizone.

Table 1. General characteristics of bio-artificial pancreas.

- Unlimited access to islet tissue sources (both from primary or engineered human/nonhuman islet cells).
- Availability of artificial materials (polymers) that can be properly engineered.
- Absence of any harmful effects on both, the transplanted tissue and the host.
- Elimination of any kind of recipient's general immunosuppression.
- Islets should be able to survive in new microenvironment, while retaining the ability to secrete insulin, according to physiological algorithms, in order to avoid delays of insulin action.
- The artificial component should respond to specific characteristics, such as membrane's physical-chemical properties that are compatible with membrane's selective porosity, filtration and permeability.
- The membrane's selective cut off properties should strictly interdict access to cellular as well as humoral mediators of the islet graft-directed immune destruction.

Moreover, adult islet β-cells do not replicate (mitotic rate 1-2%) [9], and there is consequently no hope that any eventual loss in the transplanted islet mass may be replaced by islet cell proliferation.

5. Devices for bio-artificial pancreas

Different prototypes of BAP have been designed and developed over the past few decades [10]. In general, it is still valid the classification that keeps macrodevices and microcapsules as distinct systems.

5.1. MACRODEVICES

A wide range of biomaterials has been considered for islet graft immunoisolation. The most common of them (Table 2) are still in use, with special regard to PAN-PVC, although many once innovative concepts are progressively fading away. For instance, hollow fibres have been extensively employed for subcutaneous [11] or intraperitoneal islet grafts [12]. The fibres were made of a selective permeable macro-membrane (molecular weight cut-off 50-70 kD) (Figure 2) that while blocking the inlet to immune cells or antibodies, was supposed to be highly biocompatible and allow for biochemical exchange. While successful in a few rodent trials, these devices have shown to correct diabetes in higher mammalians only sporadically. Major limitations derived from the low islet survival rate inside the chambers, possibly due to restricted nutrient supply and fibroblast overgrowth of the chambers that resulted in membrane's pore clogging.

Table 2. Biomaterials for macrodevice production.

- Polyacrylonitrile-polyvynilchloride (PAN-PVC)
- Acrylates
- Cellulose acetate
- Cellulose mixed esters
- Polytetrafluoroethylene (PTFE)
- Alginates
- Polyethylene glycol (PEG)

It was commonly observed that when loaded "free" in the device, the islets were associated with quite a limited survival rate. This specific problem was overcome using a gel matrix to embed the islets prior to the fibre's loading process. The matrix, usually an alginate gel, greatly improved the islet viability. Unfortunately, the trade-off was that the matrix-embedded islet loading capacity was very low, thereby imposing the use of larger-size devices to obtain sufficient metabolic results in the grafted diabetic animal models.

Another possible strategy to use macrodevices for islet transplantation has consisted of seeding the islets in a special compartment chamber directly anastomosed to blood vessels, usually as arterio-vein shunts. In this way, the islets were continuously perfused by blood ultrafiltrate, which seemingly facilitated biochemical exchange. The membrane, at contact with the blood stream, was associated with an appropriate MWCO (commonly below 100 kD) so as to avoid that immune cells or antibodies would cross the islet containing chamber's wall. The vascular approach, while brilliant in principles, and preliminarily associated with results that were convincingly positive in diabetic dogs [13], although not in humans, where only partial and transient remission of hyperglycemia was achieved [14].

5.2. MICROCAPSULE

5.2.1. General considerations

Microcapsules (usually fabricated with calcium alginate and further coated with a ionically complexed polyaminoacid layer) have represented the most widely known and studied kind of microimmunobarrier for islet cell transplantation (Figure 3). Microcapsules are typically fabricated by suspending the islets in sodium alginate (AG) and subsequently by passing the alginate-islet mixture through a microdroplet generator (Figure 3). The islet containing microdroplets are collected in a calcium chloride bath, which immediately turned them into gel microspheres. The calcium alginate gel microbeads are then subsequently coated with aminoacidic polycations, such as poly-L-lysine (PLL), or uniquely in our laboratory, poly-L-ornithine (PLO). The outer coat was comprised of highly purified sodium alginate that provided the multi-layered, selective permeable membrane with additional biocompatibility (Figure 4). These microcapsules have undergone several adjustments, in our and a few other laboratories [15,16].

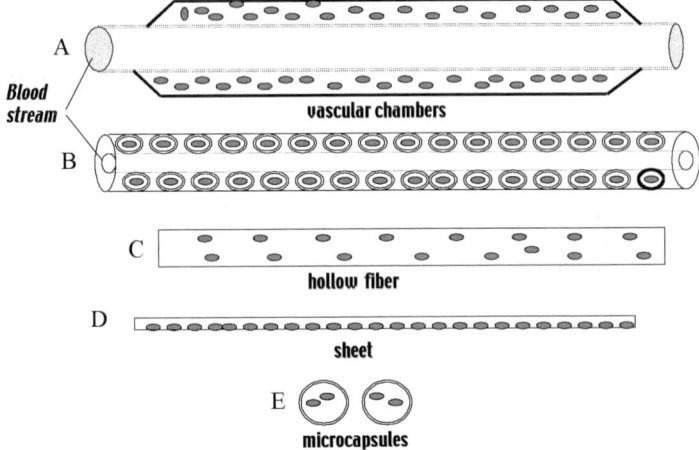

Figure 2. Schemes of different devices for islet immunoisolation. Vascular chamber containing matrix-embedded islets (A); vascular prosthesis containing alginate/polyornithine encapsulated cells (B); hollow fiber containing matrix embeded cells (C); laminar alginate thin sheet containing monolayered islets (D) and alginate based microcapsules (E).

5.2.1.1. Uncoated Ca-alginate microcapsules. Because of the very well known ability of the alginates to form highly biocompatible hydrogels, whether these are prevalently comprised of guluronic or mannuronic acid derivatives, some groups have prepared alginate beads that were uncoated with poly-L-lysine or other poly-aminoacids [17]. Usually big Ca-AG microcapsules, measuring up to 3 mm in diameter, were fabricated and employed to successfully immunoprotect either allo- [18] or xenogeneic islet grafts in diabetic dogs or rodents, respectively [19]. Although the capsules were proven to protect the islet allografts from acute rejection, a treatment course with general immunosuppressants was scheduled for the recipients in order to achieve sustained remission of hyperglycemia. The uncoated AG beads have posed a special problem that still remains open. In fact, MWCO associated with uncoated AG-capsules should amount several hundred of kD, thereby virtually allowing for either immunoglobulins or large-size complement fractions to cross the membrane.

5.2.1.2. Barium alginate microcapsules. Based on the principle that AG can be complexed with different divalent cations to form hydrogels, some authors have employed barium ions to generate Ba-AG based microbeads in a one-step procedure [20]. Although simplicity of the method is attractive, and further refinements have permitted to initiate the study of smaller-size microcapsules [21]. The major issues with this microencapsulation procedure are: (a) the toxicity associated with Ba ions (Ba is toxic to live cells), since Ba ions release cannot be precisely quantified and (b) the questionable "immunoisolation" properties of the microcapsules that have been only partially elucidated. In particular, Ba-AG microcapsules do not seem able to provide islet xenografts with sufficient immunoprotection [22].

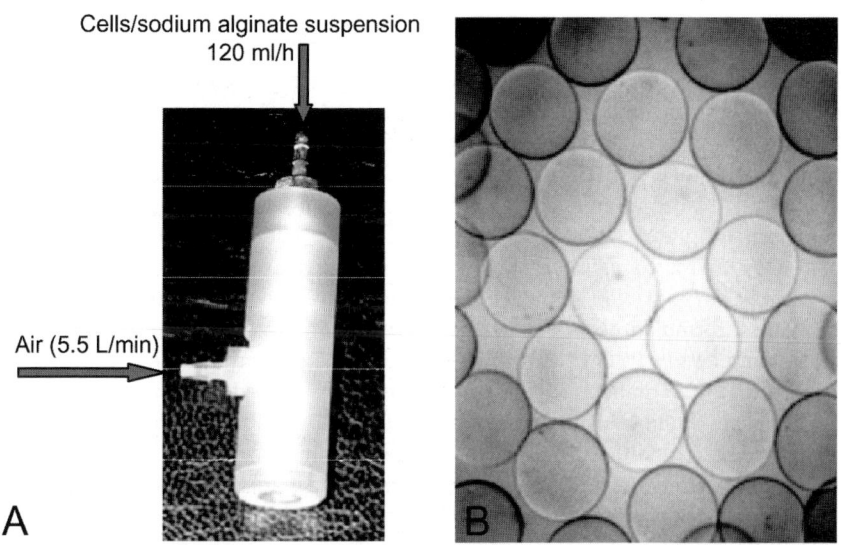

Figure 3. A. Picture of a air-driven droplet generator device used for the production of alginate microcapsules. B. Photomicrograph of alginate microcapsules.

5.2.1.3. Agarose microcapsules. Agarose, like alginate, is a gelling agent extracted from seaweeds, and it is composed of repeating units of alternating β-D-galactopyranosyl and 3,6-anhydro-α-L-galactopyranosyl. Depending upon temperature conditions, agarose forms thermally reversible gels, with gelling occurring at temperatures that are far below the gel fusion point [23]. Iwata et al. have developed several technical procedures to form agarose microcapsules to immunoprotect transplanted islets from the host's immune attack. Typically, these authors have employed 5% agarose gel solutions to formulate their capsules, with no major problems adversely affecting the enveloped islet cell viability, as assessed by in vitro studies. However, a potential technical problem lied on the capsules polydispersity, with variable amounts of empty microspheres burdening the overall graft volume. In terms of immunobarrier competence, agarose capsules have performed satisfactorily, within an islet allograft system (typically mouse islets into allogeneic mice with streptozotocin-induced diabetes) upon intraperitoneal transplantation [24]. Further advances to cope with immune problems, specially associated with islet xenografts, have consisted of coating the agarose microcapsule's gel core with outer layers. These were composed of sulfated polyanions, complement activators (i.e., poly-styrene sulfonic acid-polybrene polyion), and carboxymethylcellulose, the latter being added to improve the capsule's biocompatibility [24]. Although these specially engineered agarose microcapsules have apparently prolonged the islet xenograft survival (hamster islets into C57BL/6 diabetic mice), there is no compelling evidence that blocking complement factors would per se provide the final solution to the multiple-task problems posed by encapsulated islet xenografts in nonimmunosuppressed diabetic recipients.

5.2.1.4. Poly-ethylene glycol microcapsules. The most appealing property of poly-ethylene glycol (PEG) for fabrication of immunoprotective microcapsules has consisted of its capability to form a protein-repellent surface. In fact, protein adhesion has been indicated to initiate the cell aggregation process, on the capsule's surface, which could impair the membrane's diffusion properties [25]. PEG has been used to either coat alginate-poly-L-lysine gel microbeads [26] or create conformal coatings [27] that tightly envelop each individual islet. The latter specifically addresses the capsular volume issue, already indicated as a potential cause for the encapsulated islet graft failure. However, PEG has so far gained only limited confidence as a basic material for microcapsules formulation. Aside of in vitro and in vivo biocompatibility studies, conducted with empty PEG capsules, still limited data on graft of islet containing PEG microcapsules into diabetic recipients have been reported. Furthermore, while chemically elegant and sharp, the fabrication procedure of PEG microcaspules requires a laser-induced photopolymerisation process that is initiated by eosin. No sufficient data on the process safety have been provided to endorse this approach as a potential substitute for alginates, agarose and other biomaterials. This being the actual state of the art with PEG microcapsules, in vivo studies, targeting both, biocompatibility and immunobarrier competence of these membranes in transplant lower and higher animal models of diabetes are warranted.

5.2.1.5. Other polymers. Chitosan has been proposed as a biomaterial that may be suitable for immobilisation of cells that retain viability and function, using a semi-automated process involving use of alginates within the microcapsule's chemical formulation [28]. The multiple functional and therapeutic properties (i.e., wound healing, etc.) of chitosan require further investigation to assess whether the systematic use of this polymer, either alone or in combination with other molecules, would result in improving the microcapsules physical-chemical structure. This being the case, the environmental conditions to which the islets are exposed, would consistently improve.

Hydroxyethyl-methacrylate-methyl-methacrylate (HEMA-MMA) is a polyacrylate copolymer [29] prepared by solution polymerisation. Although the polymer is to some extent hydrophilic (25-30% water uptake), it also associates with mechanical strength, elasticity and durability over time. This polymer has proven suitable to fabricate small size capsules that might represent the ultimate answer to the final graft volume-related encapsulated islet problems. A possible concern eventually lies on the use of potential harmful technical steps, during the capsule fabrication process, such as exposure to shear forces and organic solvents that might damage living cells.

5.2.1.6. Conformal alginate microcapsules. Several methods have been described to envelop the isolated islets within conformal coatings. The rationale behind this procedure is to create immunoisolatory membranes that tightly envelop each individual islet thereby eliminating any idle dead space between islet and outer capsule's membrane. Among a few reported procedures to fabricate CM, our own consists of suspending the islets in a two-phase aqueous system, comprising AG, dextran and PEG that are emulsified extemporaneously. The microemulsion droplets engulf each islet, during brief incubation on a rocking plate, prior to immediate gelling upon reaction on Ca-chloride. The gel microbeads are sequentially coated with PLO, at particular molar ratios, in order to avoid any membrane's shrinkage, and finally with diluted AG [30].

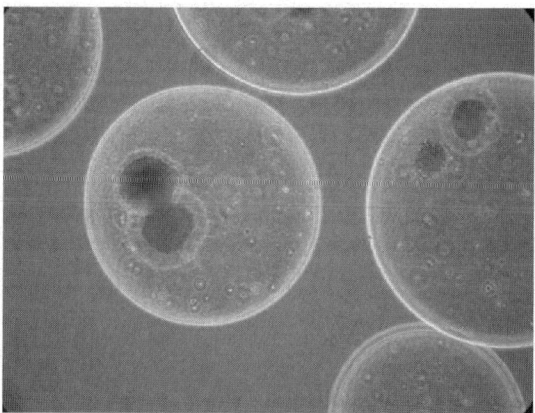

Figure 4. Freshly prepared alginate/polyornithine medium size microcapsules (100x).

5.2.1.7. Polyaminoacid coated alginate microcapsules. We actually believe that "medium-size microcapsules" (Figure 4 and 5), measuring an average 300-400 µm in diameter, represent a good compromise between volume and functional/immunoisolation performance. In order to reduce the capsule's size to 350-400 µm, and thereby obtain smaller but not conformal microcapsules, we have modified physical-chemical variables of our original method for microcapsules fabrication [30]. In particular, we have adjusted the parameters reported in Table 3. By making technical modifications, we have been able to obtain microcapsules that measured approximately half the size of the original microcapsules. Canine islet allografts, immunoprotected in MSM, into two spontaneously diabetic, insulin-dependent dogs, was associated with full reversal of hyperglycemia and discontinuation of exogenous insulin that lasted in one animal for as long as 600 days of transplantation (Figure 6).

5.2.1.8. Scaling-up protocols for alginate microcapsules. In our laboratory, the upscaling of   the alginate microcapsules production have been performed using an automated encapsulator (Figure 7). The encapsulator's technology is based on the principle that a laminar liquid jet is broken into equally sized droplets by a superimposed vibration. The optimal vibration parameters can be determined in the light of the stroboscope incorporated into the instrument.

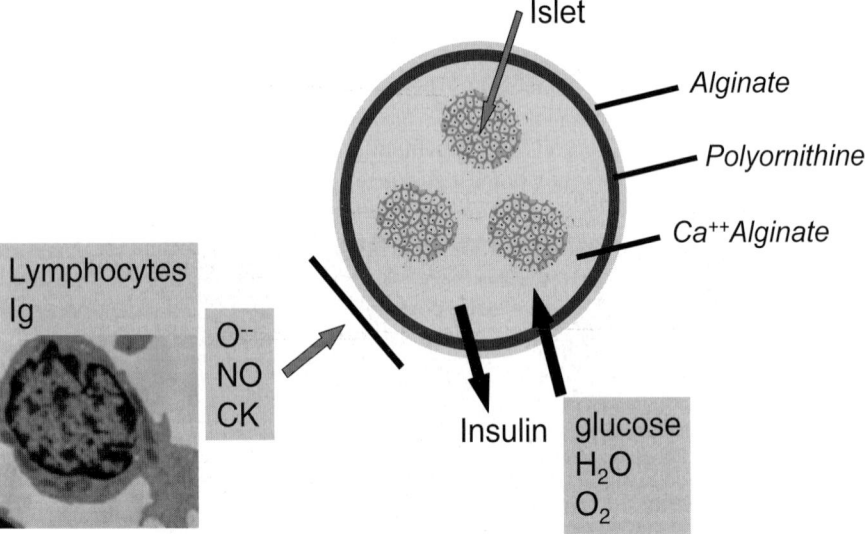

Figure 5. Scheme depicting a alginate/polyornithine microcapsule and its immunobarrier properties.

Table 3. Technical modifications applied to increase the in vivo performances of alginate/polyornithine microcapsules.

- AG viscosity (increased)
- AG/islet suspension extrusion flow rate (increased)
- Air flow (increased)
- Temperature of the suspension (decreased)

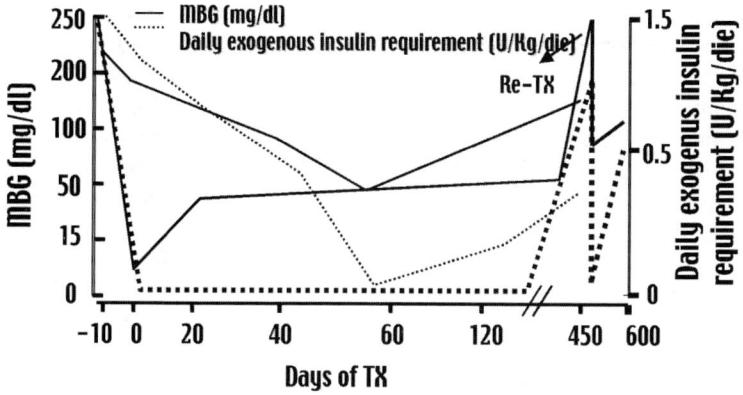

Figure 6. Long-term remission of hyperglycemia and discontinuation of exogenous insulin administration in two dogs with spontaneous insulin-dependent diabetes.

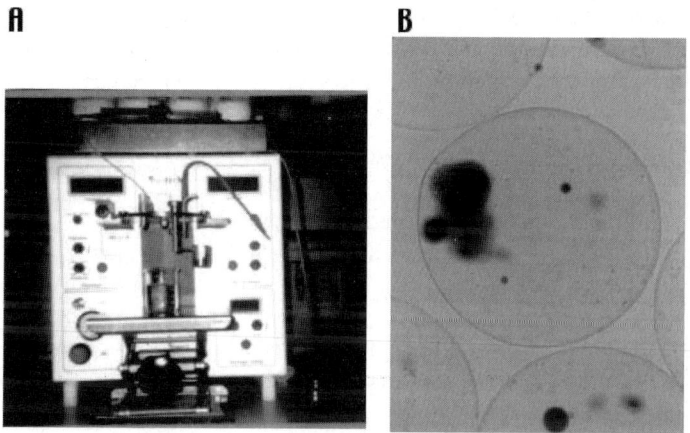

Figure 7. A. Automatic encapsulator used for the scaling-up of alginate microcapsule production, B. Alginate microcapsules produced with the automatic encapsulator (50x).

All parts of the instrument, which come in direct contact with the beads, can be sterilised by autoclaving. The syringe, or the pressurisable recipient – filled with the alginate solution and the cells – is connected to the reaction vessel, via a luerlock connection. This step is performed inside a sterile hood. Thereafter, the remaining encapsulation procedures can be performed on the laboratory bench. The polymer-cell mixture is forced into the pulsation chamber by either a syringe pump or by air pressure. The liquid then passes through a precisely drilled sapphire-nozzle and separates into equal size droplets on exiting the nozzle. The uniformity of the bead chain is visible in the stroboscope light allowing for the immediate manipulation and determination of optimal bead formation parameters. Washing solutions, or other reaction solutions, will be added aseptically through a membrane filter and drained off through the filtration disk in the bottom plate.

The entire scaling-up process has been evaluated with the aim to find the optimal parameters in order to obtain: uniform microbeads with spherical shape and narrow size distribution, selectable bead size - bead diameters from 300 µm to bigger than 600 µm. sterile production, high productivity, from 400 to 3000 beads per second depending on the bead size and finally high encapsulation efficiency, low loss of product material and full viability of encapsulated cells.

6. Microcapsule immunobarrier capacity

Encapsulated islet allografts seem to be better protected than xenografts from the host's immune response. However, it is not clear whether it may solely relate to substantial differences between allo- and xenogeneic immune reactivity. Initial reports have suggested that the microcapsules may be unable to prevent permeation of low molecular weight humoral molecules, released by xenogeneic islets. These could trigger, in turn, the host's immune attack, through indirect immune pathways. Furthermore, the eventual "helper" role of not immunospecific environmental factors in promoting the islet xenograft destruction may not be excluded.

7. Site of transplantation

The vast majority of the encapsulated islet transplantation (TX) trials has been conducted using the peritoneal cavity as an elective TX site, mainly due to the large final capsule's TX volumes (over 100 ml/dog or 180 ml/patient for standard capsules measuring > 600 µm in diameter), employed in these studies. However, the smaller-size capsules could suit other sites within the mesenteric area, such as artificially constructed omental pouches or mesentery flaps. Since microencapsulated, unlike free islets cannot be grafted intrahepatically, employment of alternate sites that are as close as possible to the liver should be pursued, in order to enhance their metabolic performance.

8. Intracapsular environmental conditions

The islets are comprised of both endocrine and nonendocrine cells, the latter including for instance fibroblasts that could thrive at expense of the β-cells, in terms of oxygen and nutrient consumption. Moreover, if intrahepatically allografted free islets seem to survive for extraordinary long periods of time, it should be noted that free, unlike microencapsulated islets are re-innervated and re-vascularised at TX site. The encapsulated islets will only rely on passive nutrient diffusion. Consequently, life-span of the encapsulated islets likely is to be finite. Finally, neonatal islets could require the presence of factors that are known to implement their time-related maturation and acquisition of functional competence. On this purpose, Sertoli's cells (SC) that originally are situated in the testis have been reported to possibly serve as a "nursing" cellular as well as immunomodulatory system by the expression of immune-related molecules (*i.e.*, FasL, TGFα, *etc.*) for the islets [31]. SC, by releasing a number of growth factors could mimic effects of the extra-cellular pancreatic matrix from which the islets have physically been disconnected. These factors have been shown to enhance survival of foetal/neonatal as well as adult islets.

We have recently undertaken the task of developing composite bio-artificial pancreas to implement islet TX survival and function within the special immunoisolatory micro-environment provided by the AG/PLO microcapsules. The composite bio-artificial pancreas represents a hybrid entity where the isolated either allo- or xenogeneic islets are co-enveloped within AG/PLO microcapsules, together with "nursing" cellular systems or possibly pharmaco-active substances that could prolong the islet cell's both survival as well as retention of their functional competence. In parallel, these systems could also attenuate the impact of TX-directed either immune or not immunospecific destruction mechanisms.

In particular, we have observed, at day 10, significant increase in endogenous insulin output from the islets that had been co-cultured with Sertoli cells, in comparison with either islets alone. The enhanced insulin secretion, deriving from the islet-Sertoli cells co-cultivation, was significantly higher than islet alone. Interestingly, although insulin released from I+SC co-cultures under static glucose incubation increased on day 10, the statistical significance was reached only when the I/SC ratio was 1:20000 ($p < 0.05$). Data were analysed by ANOVA test (Figure 8).

In vivo reversal of hyperglycemia was fully achieved in both experimental groups and sustained in 100% of the recipients, at 45 days post-TX, when some animals of the (I) but not (I+SC) group started failing. 50% of the animals belonging to the (I+SC) group was associated with full graft survival throughout 100 days post-TX. At that time no animals with (I) grafted, but two animals with (I+SC) were still euglycemic (Figure 9).

Moreover the tissue pellet, at day 12, undertook fixation and double staining (BrdU as mitosis marker and anti-insulin Ab as a β-cell marker) to detect any eventual β-cell proliferating activities, under laser confocal microscopy (LCM) and image analysis examination (IA). Therefore, SC seems to promote islet β-cell proliferation as well as insulin secretion and, furthermore, improve the functional performance of encapsulated homologous rat islet grafts in non-immunosuppressed diabetic mice.

9. Co-encapsulation of drug delivery devices

9.1. EFFECTS OF ANTI-OXIDANTS ON β-CELL FUNCTION

It is known that pancreatic islets may be vulnerable to adhesion of activated leukocytes, cytokines, and free oxygen radicals released by macrophages that may induce cell damage. In an attempt to solve or alleviate these unwanted effects, some laboratories have recently proposed the use of vitamins as antioxidants to prolong β-cell function in both patients or islet graft recipients with T1DM [14-16]. For instance, it has been reported that dietary supplement of vitamin E can help prevention of rejection of pancreatic xenografts [32]. The xenografted animals were fed with vitamin E at low (150 mg/kg) or high (8000 mg/kg) doses. These experiments demonstrated that orally administered vitamin E, was able to attenuate leukocyte-endothelial cell interactions, thereby preserving the islet microvasculature. Also recently, the effects of vitamins E and C and N-acetyl-L-cysteine on β-cell function in diabetic C57BL/KsJ-db/db mice were studied. The findings suggested that antioxidants could be beneficial to diabetes, evidencing, on the contrary, the negative impact of oxidation on β-cell function. In addition, we have recently demonstrated that vitamin D_3 (cholecalciferol) and E (D-alpha-tocopherol) indeed improve the *in vitro* viability and islet performance of free neonatal porcine cell clusters (NPCC's) [33], possibly confirming beneficial effects of anti-oxidants on cell apoptosis.

9.2. EFFECTS OF VITAMIN D_3 ON INSULIN SECRETION

In recent experiments, we have addressed to examine the effects of anti-oxidising agents such as vitamins D_3 on RPI's *in vitro* function. Our results seem to demonstrate that the vitamins may improve the *in vitro* islet cell performance, possibly confirming the beneficial effects of anti-oxidants on cell apoptosis. While is still premature to speculate on the impact of these findings on islet transplantation which will, per se, warrant "ad hoc" study, introduction of vitamins D_3 in the *in vitro* standard islet culture maintenance protocols appears to be meaningful. *In vivo* graft trials with vitamin pre-incubated, free or microencapsulated islets, will unfold whether these anti-oxidising agents may significantly prolong the islet transplant longevity.

Figure 10 shows both a schematic representation of multifunctional microcapsules for islet encapsulation and optical microphotographs of multifunctional alginate microcapsules containing both islets and Vitamin D_3 containing cellulose acetate (CA) microspheres.

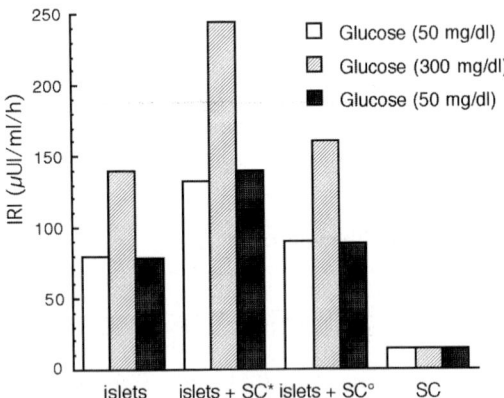

Figure 8. Endogenous insulin output, under static glucose incubation, from islets co-cultured in the presence of Sertoli's cells versus islets alone, at day 10 from isolation. *20,000 SC/islets; °10,000 SC/islets.

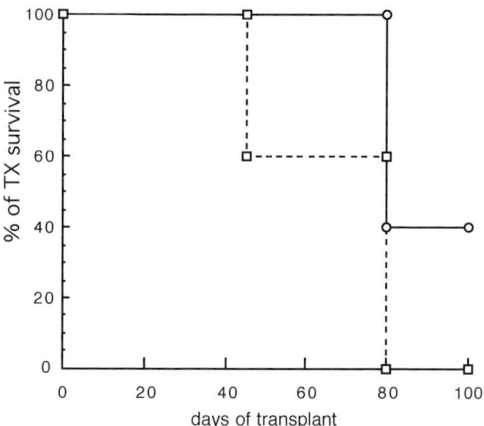

Figure. 9. Comparison of glycemic levels in a diabetic mice transplanted with islets in the presence of co-encapsulated Sertoli's cells (circles) versus islets alone (squares).

Data reported by Figure 11 showed that the amount of insulin secreted by islets co-cultured with free vitamin D_3 and, particularly, with vitamin D_3 entrapped in CA microspheres, was significantly higher as compared to the islets alone (Panel A).

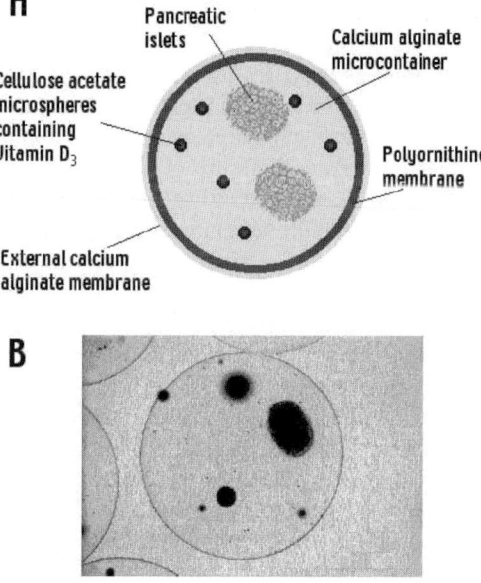

Figure. 10. Schematic representation of multifunctional microcapsules for islet encapsulation (panel A) and optical microphotographs of multifunctional alginate microcapsules containing both islets and Vitamin D_3 containing cellulose acetate (CA) microspheres (panel B). Bar correspond to 165 µm.

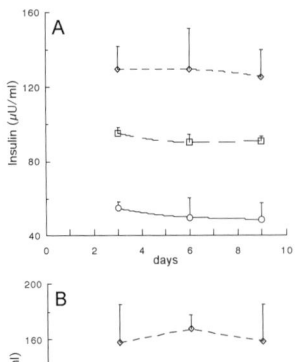

Condition	islets	Islets + free Vit D₃	Islets + Vit D₃ microspheres
islets			
Islets + free Vit D3	7.9 <0.01		
Islets + Vit D3 microspheres	7.8 <0.01	4.04 <0.05	
Islets in NAG/PLO	1.43 n.s.		
Islets in NAG/PLO + free Vit D3	12.67 <0.001	14.45 0.001	
Islets in NAG/PLO + Vit D3 multicopartimental	6.9 <0.01		1.96 n.s.

t-student test for kinetic of insulin release

Panel A: Control untreated RPI (circles); RPI cultivated in the presence of 2 mM free vitamin D_3 (squares); RPI cultivated in the presence of encapsulated vitamin D_3 (20 mM) (diamonds).

Panel B: in vitro insulin release by NAG/PLO microencapsulated islets. Control untreated NAG/PLO microencapsulated RPI (circles); NAG/PLO microencapsulated RPI cultivated in the presence of encapsulated vitamin D_3 (20 mM) (squares); co-microencapsulated RPI + Vit. D_3 entrapped in cellulose acetate microsphere (diamonds).

Figure 11. In vitro *insulin release by free (panel A) and microencapsulated (panel B) RPI. Data represent the mean of 5 independent cells batches, each insulin determination was performed in triplicate ± SD.*

Moreover, *in vitro* insulin release from NAG/PLO microencapsulated islets plus vitamin D_3 entrapped in CA microspheres and, particularly, from islets co-microencapsulated with Vitamin D_3 in CA microspheres, was significantly higher in comparison with microencapsulated islets alone (Panel B).

Figure 12 (Panel A) shows *in vitro* insulin release before and after islets lysis at day 9 of culture of islets alone, islets plus free vitamin D_3 and islets plus vitamin D_3 entrapped in CA microspheres; Panel B shows *in vitro* insulin release, before and after islets lysis at day 9 of culture, of NAG/PLO microencapsulated islets, NAG/PLO microencapsulated islets plus Vitamin. D_3 entrapped in cellulose acetate microspheres, and co-microencapsulated islets plus Vitamin. D_3 entrapped in cellulose acetate microspheres.

RPI, after 6 days of culture with or without vitamins, were compared for the amount of released insulin at low (50 mg/dl/h) or high (300 mg/dl/h) glucose concentrations. The results of this experiment are reported in Figure 13.

As expected, incubation with high glucose, increased the RPI's insulin secretory rate; this effect was further magnified by the addition of islets plus free vitamin D_3 and, particularly, islets plus vitamin D_3 entrapped in cellulose acetate microspheres (Panel A); panel B shows that *in vitro* insulin release of NAG/PLO microencapsulated islets plus vitamin D_3 entrapped in cellulose acetate microspheres and, particularly, when the islets were co-microencapsulated islets with vitamin D_3 entrapped in CA microspheres, was significantly higher as compared to microencapsulated islets alone. This final data suggests that vitamin D_3 could be eventually used in order to obtain a beneficial, possible synergistic effect on insulin output of the adult RPI.

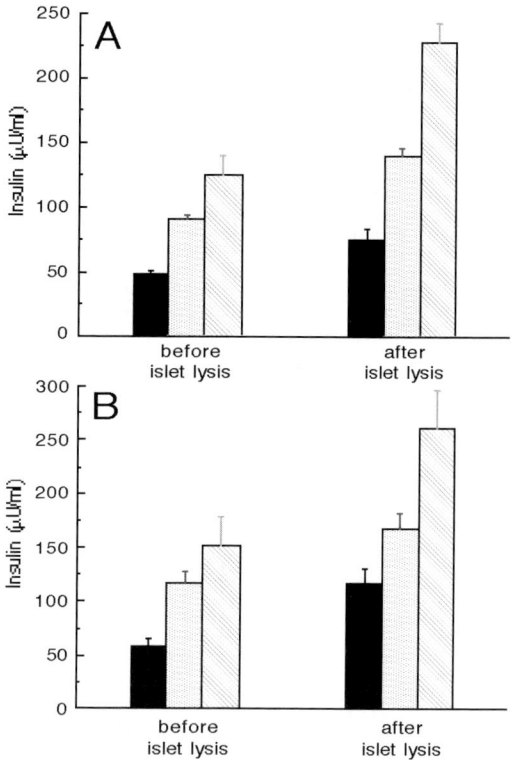

Figure 12. In vitro *insulin release from RPI before and after islets cell lysis at day 9 of culture. Panel A: insulin release by free RPI. Control untreated RPI (filled bars); RPI cultivated in the presence of 2 mM free vitamin D_3 (dotted bars); RPI cultivated in the presence of encapsulated vitamin D_3 (20 mM) (striped bars). Panel B: insulin release by NAG/PLO microencapsulated islets. Control untreated NAG/PLO microencapsulated RPI (filled bars); NAG/PLO microencapsulated RPI cultivated in the presence of encapsulated vitamin D_3 (20 mM) (dotted bars); co-microencapsulated RPI + Vit. D_3 entrapped in cellulose acetate microsphere (striped bars). Data represent the mean of 5 independent cells batches, each insulin determination was performed in triplicate ± SD.*

10. Conclusions

BAP seems to offer the practical opportunity to transplant either human or nonhuman islets with no recipient's general immunosuppression. While the former is clearly limited by the restricted availability of cadaveric human donor pancreata, the latter could rely on the improved performance of methods for separation and purification of adult or neonatal islets from selected, SPF pig breeding stocks. Should the problems of potential transmission of adventitious microbial agents be overcome, pig islets could represent an ideal tissue resource for human transplants, for all the above-mentioned reasons.

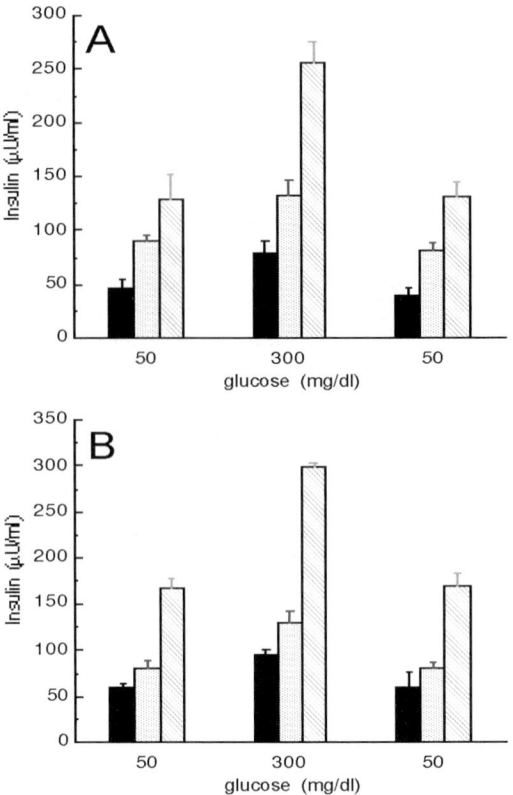

Figure 13. In vitro *insulin release from RPI, under static glucose incubation at the indicated concentrations, at day 6 of cell culture. Panel A: insulin release by free RPI. Control untreated RPI (filled bars); RPI cultivated in the presence of 2 mM free vitamin D_3 (dotted bars); RPI cultivated in the presence of encapsulated vitamin D_3 (20 mM) (striped bars). Panel B: insulin release by NAG/PLO microencapsulated islets. Control untreated NAG/PLO microencapsulated RPI (filed bars); NAG/PLO microencapsulated RPI cultivated in the presence of encapsulated vitamin D_3 (20 mM) (dotted bars); co-microencapsulated RPI + vitamin D_3 entrapped in cellulose acetate microsphere (striped bars). Data represent the mean of 5 independent cells batches, each insulin determination was performed in triplicate.*

Two alternatives are possibly ready to take over the pig islet source in the more or less far future: first, engineered insulin-producing cells; second, *in vitro* regenerated islet cells. This could rely on methods that are more and more refined for generating cellular systems that are genetically engineered not only to secrete insulin, but lately also to reproduce the glucose sensing apparatus that is uniquely associated with β-cells [34]. Regulation of insulin secretion, appears to be an indispensable feature of any insulin producing tissue. Caution is still required prior to endorsing the use of artificial tissue that is derived from tumour cell lines, for transplant of diabetes patients. A body of additional study, in this area, is warranted for the next years to come, although this cell source seems to be the most appealing solution to fulfil not only islet mass problems but also immune identity problems. Immune determinants of the artificial β-cells could be, in fact, engineered in a way that the cells would not be recognized by the immune system [34]. Recently, methods to engineer endogenous cells that would be "educated" synthesize and release insulin upon pharmacologic stimulation have been described [35]. If successful, this approach could even obviate the need for transplantation. This ingenious strategy that has been validated, at the moment and very preliminarily, only in rodents and requires further safety and efficacy studies, remains a very interesting, potential strategy for the final cure of IDDM. Recently, attempts to regenerate islets from pancreatic duct stem cells have preliminarily succeeded [36]. This procedure could make a virtually unlimited β-cell mass available *in vitro*, to fulfil the demand for cell transplant in diabetes. The actual system's performance, that seems to require long-term *in vitro* islet cell growing protocols, as well as safety restrictions need to be further elucidated in the near future.

An important common feature of these new engineered cells strategies is that with a few exceptions, all of them will require an immunoprotection and most likely, an immunoisolation membrane. Therefore, BAP likely represents the best approach to put the engineered or otherwise artificially generated insulin producing cells at work.

References

[1] Goldstein, S.A. (2002) Tissue engineering: functional assessment and clinical outcome. Ann. NY Acad. Sci. 961: 183-192.
[2] Wilson, J.D.; Foster, D.V.; Kronenberg, H.M. and Larsen, P.R. (1998) Williams Textbook of Endocrinology. 9th edition, Saunders: Diabetes Mellitus; pp. 973-1060.
[3] Daniloff, G.Y. (1999) Continuous glucose monitoring: long-term implantable sensor approach. Diabetes Technol. Ther. 1(3): 261-266.
[4] U.K. Prospective Diabetes Study Group (1999) Quality of life in type 2 diabetic patients is affected by complications but not by intensive policies to improve blood glucose or blood pressure control (UKPDS 37). Diabetes Care 22(7): 1125-1136.
[5] Shapiro, A.M.; Lakey, J.R.; Ryan, E.A.; Korbutt, G.S.; Toth, E.; Warnock, G.L.; Kneteman, N.M. and Rajotte, R.V. (1999) Islet transplantation in seven patients with type 1 diabetes mellitus using a glucocorticoid-free immunosuppressive regimen. N. Engl. J. Med. 343(4): 230-238.
[6] Basta, G.; Falorni, A.; Osticioli, L.; Brunetti, P. and Calafiore R. (1995) Method for mass retrieval, morphologic and functional characterization of porcine islets of Langerhans: a potential xenogeneic tissue resource for transplantation in IDDM. J. Invest. Med. 43(6): 555-566.
[7] Korbutt, G.S.; Elliott, J.F.; Ao, Z.; Smith, D.K.; Warnock, G.L. and Rajotte R.V. (1996) Large scale isolation, growth, and function of porcine neonatal islet cells. J. Clin. Invest. 97(9): 2119-2129.
[8] Yoon, K.H.; Quickel, R.R.; Tatarkiewicz, K.; Ulrich, T.R.; Hollister-Lock, J.; Trivedi, N.; Bonner-Weir, S. and Weir, G. (1999) Differentiation and expansion of Beta-cell mass in porcine neonatal pancreatic cell clusters transplanted into nude mice. Cell Transpl. 8: 673-689.
[9] Swenne, I. (1983) Effect of aging on the regenerative capacity of the pancreatic □-cell of the rat. Diabetes 32: 14-19.
[10] Lanza, R.P.; Sullivan, J.S. and Chick, W.L. (1990) Islet transplantation with immunoisolation. Diabetes 41: 1503-1510.
[11] Lacy, P.E.; Hegre, O.D.; Gerasimidi-Vazeou, A.; Gentile, F.T. and Dionne, K.E. (1991) Maintenance of normoglycemia in diabetic mice by subcutaneous xenografts of encapsulated islets. Science 254: 1782-1784.

[12] Lanza, R.P.; Beyer, A.M.; Staruk, J.E. and Chick, W.L. (1993) Biohybrid artificial pancreas: long-term function of discordant islet xenografts in streptozotocin-diabetic rats. Transplant. 56: 1067-1072.
[13] Maki, T.; Otsu, J.; O'Neill, J.J.; Dunleavy, K.; Mullon, C.J.P.; Solomon, B.A. and Monaco, A.P. (1996) Treatment of diabetes by xenogeneic islets without immunosuppression: use of a vascularized bioartificial pancreas. Diabetes 45: 342-347.
[14] Calafiore, R. (1992) Transplantation of microencapsulated pancreatic human islets for the therapy of diabetes mellitus: a preliminary report. ASAIO J. 38(1): 34-37.
[15] Brunetti, P.; Basta, G.; Falorni, A.; Calcinaro, F.; Pietropaolo, M. and Calafiore, R. (1991) Immunoprotection of pancreatic islet grafts within artificial microcapsules. Int. J. Artif. Organs 14(12): 789-791.
[16] Lanza, R.P.; Kuhtreiber, W.M. and Chick, W.L. (1995) Encapsulation technologies. Tissue Eng. 1: 181-196.
[17] Lanza, R.P.; Jackson, R.; Sullivan, A.; Ringeling, J.; MacGrath, C.; Kuhtreiber, W. and Chick, W.L. (1999) Xenotransplantation of cells using biodegradable microcapsules. Transplant. 67(8): 1105-1111.
[18] Lanza, R.P. and Chick, W.L. (1997) Transplantation of pancreatic islets. Ann. NY Acad. Sci. 831: 323-331.
[19] Lanza, R.P.; Ecker, D.; Kuhtreiber, W.M.; Staruk, J.E.; Marsh, J. and Chick, W.L. (1995) A simple method for transplanting discordant islets into rats using alginate gel spheres, Transplant. 59(10): 1485-1487.
[20] Zekorn, T.D.C.; Horcher, A.; Siebers, U.; Schnettler, R.; Hering, B.; Zimmermann, U.; Bretzel, R.G. and Federlin, K. (1992) Barium-cross-linked alginate beads: A simple, one-step-method for successful immuno isolated transplantation of islets of Langerhans. Acta Diabetol. 29: 99-106.
[21] Zekorn, T.; Horcher, A.; Siebers, U.; Schnetter, R.; Hering, B.; Zimmermann, U.; Bretzel, R.G. and Federlin, K. (1992) Alginate coating of islets of islets of Langerhans: in vitro studies on a new method of immunoisolated transplantation. Acta Diabetol. 29: 41-45.
[22] Zekorn, T. and Bretzel R.G. (1999) Immunoprotection of islets of Langerhans by microencapsulation in barium alginate beads. In: Kuhtreiber, W.M.; Lanza, R.P. and Chick W.L. (Eds.) Cell encapsulation technology and therapeutics. Birkhauser, Boston; pp. 90-96.
[23] Dumitriu, S. (2002) Polysaccharides as biomaterials. In: Dumitriu, S. (Ed.) Polymeric Biomaterials, Marcel Dekker, New York; pp. 1-61.
[24] Iwata, H and Ikada, H. (1999) Agarose. In: Kuhtreiber, W.M.; Lanza, R.P. and Chick W.L. (Eds.) Cell encapsulation technology and therapeutics. Birkhauser, Boston; pp. 97-107.
[25] Sawhney, A.S. (1999) Poly(ethylene glycol). In: Kuhtreiber, W.M.; Lanza, R.P. and Chick W.L. (Eds.) Cell encapsulation technology and therapeutics. Birkhauser, Boston; pp. 108-116.
[26] Sawhney, A.S. and Hubbell J.A. (1992) Poly(ethylene oxide)-graft-poly(l-lysine) copolymers to enhance the biocompatibility of poly(l-lysine)-alginate microcapsule membrane. Biomaterials 13: 863-870.
[27] Sawhney, A.S.; Pathak, C.P. and Hubbell J.A. (1994) Modification of Langerhans islet surfaces with immunoprotective poly(ethylene glycol) coatings. Biotechnol. Bioeng. 44: 383-386.
[28] Kim, S.K.; Choi, J.; Balmaceda, E.A. and Rha, C. (1999) Chitosan. In: Kuhtreiber, W.M.; Lanza, R.P. and Chick W.L. (Eds.) Cell encapsulation technology and therapeutics. Birkhauser, Boston; pp. 151-172.
[29] Weber, C.J.; Kapp, J.A.; Hagler, M.K.; Safley, S.; Chryssochoos, J.T. and Chaikof E.L. (1999) Long-term survival of poly-L-lysine-alginate microencapsulated islet xenografts in spontaneously diabetic NOD mice. In: Kuhtreiber, W.M.; Lanza, R.P. and Chick W.L. (Eds.) Cell encapsulation technology and therapeutics. Birkhauser, Boston; pp. 117-137.
[30] Calafiore, R., and Basta, G. (1999) Alginate/poly-L-ornithine microcapsules for pancreatic islet cell immunoprotection. In: Kuhtreiber, W.M.; Lanza, R.P. and Chick W.L. (Eds.) Cell encapsulation technology and therapeutics. Birkhauser, Boston; pp. 138-150.
[31] Luca, G.; Calvitti, M.; Becchetti, E.; Basta, G.; Angeletti, G.; Santeusanio, F.; Brunetti, P. and Calafiore, R. (1998) Method for separation morphological and functional characterization of Sertoli's cells from the rat pre-pubertal testis : a potential nursing cell system for pancreatic islets. Diab. Nutr. Metabol. 11: 307-315.
[32] Vajkoczy, P.; Lehr, H.A.; Hybner, C.; Arfors, K.E. and Menger, M.D. (1997) Prevention of pancreatic islet xenograft reaction by dietary vitamin E. Am. J. Pathol. 150: 1487-1495.
[33] Luca, G.; Nastruzzi, C.; Basta, G.; Brozzetti, A.; Saturni, A.; Mughetti, D.; Ricci, M.; Rossi, C.; Brunetti, P. and Calafiore, R. (2000) Effects of anti-oxidizing vitamins on in vitro cultured porcine neonatal pancreatic islet cells. Diabetes Nutr. Metab. 13(6): 301-307.

[34] Efrat, S. (1999) Genetically engineered pancreatic β-cell lines for cell therapy of diabetes. In: Hunkeler, D.; Prokop, A.; Cherrington, A.; Rajotte, R. and Sefton M. (Eds.) Bioartificial Organs II, Technology, Medicine & Materials. Ann. NY Acad. Sci. 875: 286-293.

[35] Rivera, V.M.; Wang, X.; Wardwell, S.; Courage, N.L.; Volchuk, A.; Keenan, T.; Holt, D.A.; Gilman, M.; Orci, L.; Cerasoli, F.; Rothman, J.E. and Clackson, T. (2000) Regulation of protein secretion in the endoplasmic reticulum. Science 287: 826-830.

[36] Ramiya, V.K.; Maraist, M.; Arfors, K.E.; Schatz, D.A.; Peck, A.B. and Cornelius, J.G. (2000) Reversal of insulin-dependent diabetes using islets generated in vitro from pancreatic stem cells. Nature Medicine 6(3): 278-282.

BIOARTIFICIAL PANCREAS: AN UPDATE

ALES PROKOP[1] AND JEFFREY M. DAVIDSON[2,3]

[1]107 Olin Hall, Chemical Engineering, 24th Avenue South & Garland Avenue, Vanderbilt University, Nashville, TN 37235-1604, USA –
Fax: 615-343-7951 – Email: Ales.Prokop@vanderbilt.edu
[2]Department of Pathology, Vanderbilt University School of Medicine, Nashville, TN 37212-2561, USA – Email: jeff.davidson@vanderbilt.edu
[3]Research Service, Department of Veterans Affairs Medical Center, Nashville, TN 37212-2637, USA

1. Introduction

Regenerative medicine is a fast-emerging field that seeks to develop new treatments, repair and function enhancement, or replace diseased or damaged tissue and organs using techniques such as tissue engineering, cellular therapies, and the development of artificial and bioartificial organs. Replacing damaged or diseased body parts and restoring their function could be achieved through surgical transfer of natural tissue, either of the patient's or donor's. With the introduction of new healthy cells into the body, curable diseases will include those that require the replacement of non-functional or functionally-inadequate tissues: Alzheimer's disease, Parkinson's disease, kidney diseases, liver diseases, diabetes, spinal cord injury and others.

The increasing incidence of diabetes has created a strong desire to develop a reliable, safe, biological source of insulin production. Conventional transplantation cannot possibly meet present or projected future demands, although the whole pancreas or islet transplantation in a patient with severe diabetes can effectively restore insulin production. A lack of availability of donor pancreata requires the development of alternative sources of islets. The success of this technology depends upon identifying an appropriate source of graft material and developing a biocompatible platform for prolonged survival of transplant material. Thus the bioartificial pancreas has become an important bioengineering objective. Allograft or xenograft islet sources require sophisticated encapsulation to ensure adequate blood supply, to avoid immune surveillance/rejection, and to limit fibrous capsule formation. The ultimate source of new pancreatic tissue will probably come from advances in tissue regeneration. In the meantime, a promising strategy is the identification and propagation of islet progenitor cells/tissue from somatic or embryonic stem cell sources, such as *ex vivo* expansion of stem/progenitor cells in culture and their differentiation into functional islets. An overview of alternative cell sourcing is presented in Table 1. We will discuss additional points in the update to our recent review [1].

Table 1. Overview of sourcing.

Embryonic Stem Cells	Fetal Stem Cells	Adult Stem Cells	Adult Cells
Unlimited self-renewal and differentiation	Intermediate committed progenitors	Indermediate between fetal and adult	Terminally differentiated progeny
Capable of multilineage development	Restricted differentiation potential	Intermediate beween fetal and adult cells	No proliferation (or very limited) capacity

Table 2. Attributes of ideal capsule (immunoisolation device).

Attribute	Problems/Comments
Capsule size uniformity	Satellite droplets; wide size distribution
Capsule sphericity	Assymetry
Capsule surface uniformity	Surface roughness; fibrotic reactions
Membrane uniformity	Requires uniform permselective membrane with no defects
Suitable capsule size to minimise implant volume	Nutrient and oxygen supply, diffusion limited; large capsules and beads controversial
Gelled vs. chelated capsule interior	Not yet resolved; solid signal could be beneficial
Conformal coating of implant/thin film	Permselectivity not yet defined; immunoprotection questioned
Mechanical stability	Most capsules currently used not stable
Permselectivity	Methods to be refined
Reject empty capsules	Minimise implant volume; not yet standard procedure
Low toxicity	Availability of purified polymers; presence of pyrogens, mitogens, heavy metals
Implant loading	Nutrient and oxygen supply
Implant centering	Implant protruding beyond the wall
Proper implant source	Limited for pancreas/islets; cell stem technology could provide a solution
Polymer conducive to vascularisation	Problem with implant engulfment
Implant entrapped in viscous polymeric solution/gel	Permselectivity and immunoprotective barrier questioned

2. Islet immobilisation techniques: towards a perfect capsule

This part will define an ideal capsule and how this goal has been pursued (Table 2). It will include discussion on: (i) drop size and uniformity optimisation, (ii) reactor design configurations and product performance, and (iii) material selection, biocompatibility and minimisation of fibrotic growth. For the sake of completeness, some older references are also included.

2.1. DROP GENERATION AND CAPSULE SIZE

The size of microcapsules is an important parameter in the microencapsulation of mammalian cells (and islets). The use of smaller capsules decreases the total implant volume and improves nutrient and oxygen supply.

The practice of primary drop production is based on the theory and practice of the Rayleigh instability of liquid jets [2-3], extended to non-Newtonian fluids [4]. A cylindrical column of liquid spontaneously breaks into drops due to a disturbance from either the jet itself or from external excitation. A superposition of a small amplitude shear perturbation (due to gravity and pressure) on a thin liquid jet results in an oscillatory elongational flow [5]. After an initial pinch-off to form the primary drops, additional breaks occur in the pinched liquid strands, resulting in the formation of small secondary satellite drops, with a broad distribution of sizes [6].

Better size uniformity is obtained when an external disturbance is applied, be it mechanical, electro-mechanical (piezoelectric) or electrostatic in nature. When a jet is mechanically excited at a suitable single high frequency, drops of uniform size are formed. Satellite droplets, however, also result from such mechanically excited jet [7].

A standard double nozzle jet (fluid/air, air stripping) is another example of a device based on a mechanical disturbance of a liquid jet. For most conditions, any mechanical excitation will result in generation of satellite droplets. Under very controlled conditions, however, double-jet generator will provides uniform-sized drops, with minimal satellite drop production [8].

Electrostatic droplet generation (electrospray) results from an application of net charge on the surface of drops, reducing the binding force due to surface tension that holds the drop together. If the surface charge density is sufficiently high, the electrostatic force overcomes surface tension and the drop disintegrates into fine droplets [9-10]. A high voltage electrostatic pulse is typically applied between a positively charged needle and grounded plate containing the receiving bath solution (*e.g.*, in case of calcium alginate beads, calcium chloride hardening solution). This way, the primary droplets can be easily spatially separated from the shroud (secondary droplets) [11-12]. Besides the size uniformity, perfect sphericity and surface uniformity is another requirement. The surface roughness may affect cell adhesion at microcapsule transplantation sites. Smooth external surface yields implants with minimal fibrotic reaction over long period of time whereas rough surfaces elicit significant fibrotic growth [13]. Depending on composition of receiving bath and anti-gelling and gelling cations (NaCl) used at the capsule preparation, surface roughness can be adjusted using feedback from such measurements [14-15].

Size is an adjustable parameter for droplet generators, discussed above, within a limited range. Majority of investigators in the bioartificial pancreas field have used microcapsules of 500-800 μm size range. The capsule size is of critical importance in terms of supply of nutrients and oxygen.

Oxygen supply limitations can have deleterious effects on viability and functionality of encapsulated islets. This can be a contributing or primary cause for poor performance or for failure of an implant.

Theoretical predictions based on oxygen diffusion and consumption models agree with experimental data for size of nonviable core in single islet culture, and for loss of viability in high-density cultures. Insulin secretory capacity may increase little or even decrease as islet loading increases. Similar crowding effects can occur with encapsulated islets in the peritoneal cavity or with naked islets in the liver where prior to vascularisation, locally high concentration of islets can occur. In a planar device, islets cultivated too close together develop non-viable regions. The density of cultured islets affects viability. As density increases, functionality declines quickly. Maximum performance occurs at relatively low density. Consequently, using a large number of islets may not be better than a small number [16-17].

The physical state of the capsule interior is often a neglected parameter. Most investigators in the cell encapsulation business employ a gelled capsule interior for cell immobilisation. On the other hand, for APA capsule (alginate-PLL-alginate coat) used in a "standard" bioartificial pancreas scenario, the gelled capsule core is typically dissolved by chelating out small cations (*e.g.*, calcium by incubation in a citrate solution).

The gelled interior was demonstrated to be somewhat inferior for islet function (diminished biphasic insulin release to glucose stimulation)[18]. This issue remains highly unexplored and open to further investigations. It is known that cells taken from their milieu within the tissue network lack access to any "solid" signals experienced in a normal state. This is particularly true for extracellular matrix components, often known to produce interpenetrating systems (gels) themselves or with other components (*e.g.*, collagens, laminin, elastin, proteoglycans, *etc.*).

In any case, the capsule interior, be it chelated or unchelated, exhibits a rather high viscosity. Such environment seems to be beneficial to the survival of the implant. It has been clearly documented that the islet physiology and implant survival is improved for encapsulated islets, as compared to naked islet transplant [19].

2.2. PROCESS AND REACTOR DESIGN, PRODUCT PERFORMANCE

The APA capsule production involves a sequential batch processing, including preparation of gelled beads, coating (to develop a permselective membrane) and chelating of the droplet interior. An alterative processing may involve an internal gelling step as suggested by Poncelet [20], resulting in a more homogeneous capsule in terms of calcium profile.

The APA capsule, although frequently used, lacks sufficient mechanical strength. A new capsule design, featuring a very distinct capsule wall of visible thickness with adjustable mechanical properties and permeability control was introduced [21,14,22]. Such capsules, generated from a multicomponent polymeric polyelectrolyte mixture (alginate-cellulose sulfate-PMCG-calcium chloride; PMCG is poly(ethylene-co-guanidine) hydrochloride, a synthetic oligomer similar to cationic polyamino acids) in a continuous system [23] possess a very high stability and uniformity. Several other multicomponent capsule membrane systems have been described [24-26].

The requirement for a semipermeable wall may be relaxed for two other types of immunoisolation devices: macrocapsules and conformal coating. In both cases, polymer itself may provide enough "immunoprotection", a necessary permeability barrier for encapsulated islets or cells, without applying a special permselective membrane in the form of capsule wall.

The polymeric macrocapsule evolved from agarose microbeads for xenografts, proposed by Iwata [27-28], employs a temperature-induced sol-gel transition of low-gelling temperature agarose. Macrobead, made of agarose and collagen (8 mm in diameter) was introduced by Jain *et al.* [29] for xenografs. Weir and his collaborators [30-31] developed calcium or barium alginate beads (about 1-3 mm in size) for islets. Barium alginate microbeads (2-3 mm) and calcium-alginate-PLL macrobeads, although achieving normoglycemia, demonstrated significantly impaired insulin secretion, perhaps due to the considerable diffusional time lag associated with the peritoneal site and their large size. Allogeneic or syngeneic mouse islets in 1 mm barium-alginate beads normalized glycemia in diabetic BALB/c and NOD mice for longer than 250 days. The retrieved capsules were remarkably free of cellular overgrowth and islets responded to glucose stimulation. Thus, the polymeric macrocapsule provides new avenue for exploration. The reasons for such performance are not clear. Perhaps, the tortuous and long paths within the macrobeads provide sufficient permselectivity and subsequent protection of islets. The permselectivity of such beads remains to be evaluated quantitatively.

Conformal coating: Hubbell [32] and Sefton [33] introduced a device consisting of a very thin film layer of polymeric membrane that conforms to the shape of the cell aggregate and minimises the polymer's contribution to the total implant volume. A variation of Hubbell's photopolymerisable PEG film using a protein-reactive PEG-isocyanate treatment of islets exists [34].

The mechanical instability of the APA capsules is also an issue for microcapsules with coatings, in spite of the application of multiple layer coatings [35-36]. Beads are not mechanically stable. Methods for evaluation of mechanical stability were briefly summarised by Rehor *et al.* [37]. They applied a uniaxial compression to microcapsules and evaluated their deformation and force at bursting. Multicomponent capsules are many times sturdier compared to standard APA capsules and an important attribute for intraperitoneal implantations [14].

2.3. BIOLOGICALLY RELEVANT PROBLEMS

Little work has been done for microcapsules with permselective membranes. Recently, Brissova *et al.* [38] described a rigorous test for evaluation of capsule diffusion properties. This was based on inverse size exclusion chromatography with dextran standards using microcapsules packed in a column. In another paper, Brissova *et al.* [39] described an influx/efflux method using a radioactive label to follow the fate of biologically relevant molecules (IgG and IL-1β) that is applicable to other encapsulation systems. The uptake of labeled IgG was followed *via* radioactivity of IgG bound to protein A Sepharose beads (PAS) entrapped within the capsules as well as that of IL-1β through IL-1β IgG polyclonal antibody bound to PAS. Although this method demonstrates ingress of these molecules into the microcapsule, diffusion is driven by an association constant of a given substance to PAS, not necessarily representing an *in vivo* situation of uptake by cells.

The centering of an implant within the microcapsule and avoidance of its capturing at the capsule wall are issues very relevant to an attainment of perfect implant immunoprotection [40]. The reduction of the capsule size increases the possibility of islets contacting the capsule wall and protruding beyond the capsule surface, compromising implant acceptance (rejection).

Continuous processing of uniform capsules has been adequately addressed by the loop reactor for production of multicomponent polymeric capsules by Anilkumar *et al.* [23] and Lacik [41]. The looped geometry of the reactor keeps the capsules suspended in the polymeric fluid, thereby avoiding their sedimentation. The process of islet centering during anion drop pinch-off, and oscillations that follow, have been examined by Anilkumar *et al.* [40]. The processing of the anion drops in the looped reactor, with uniform exposure to the cationic liquid and continuous tumbling of the drops, prevents the sedimentation of the islets within the capsules.

Imperfect and empty capsules should be optimally rejected during capsule preparation. Development of diagnostics based on laser-enabled electro-optical methods will enable rapid online interrogation of each capsule. Blank capsules and capsules not meeting the requirements can be thereby sorted out. The processing is performed while encapsulated islets are in media under incubator conditions to minimise stress [42]. The rejection of empty capsules is an important issue, directed towards the minimisation of the implant volume, particularly for peritoneal applications.

Proper material chemistry and purity of polymers represents another requirement for perfect capsule. There are a number of references on purification of polymers used for fabrication of capsules, particularly those based on alginate (*e.g.*, [43]). Among the remaining problems are the removal of pyrogens, mitogens and heavy metals.

Pronova Biomedical (Norway) is filling a need for marketing of pure polysaccharides for microencapsulation (alginate and chitosan). The elimination of contaminants is of extreme importance for securing a long-term implant function and improving biocompatibility.

Biomaterial vascularisation should include (1) selection of materials that support angiogenesis without leading to endothelial cell activation are sought [44]; and (2) promotion of vascularisation by allowing slow release of angiogenic growth factors (VEGF, bFGF) from hydrogel matrix [45-46].

3. Islet sourcing

Regenerative medicine assists the patient's own body mechanisms in growing healthy tissues and cells to take the place of damaged ones. Cells are typically expanded *ex vivo*, incorporated into a device, which is then integrated into the human body. The issues of integration include tissue-biomaterial interface (such as polymeric scaffolds), the response to foreign substances, and basic transplantation/immunologic issues (allo- *versus* xenotransplantation).

This section will address the use of stem cells (SC) as an enabling technology with enormous potential for facilitating regenerative medicine. Pluripotent cells derived from early embryos or fetal tissues appear capable of indefinite expansion and have the capacity to differentiate into many tissue types. Multipotent stem cells derived from adult tissues are being found to have considerably broader differentiation capacity than originally thought. Stem cells have potential to revolutionize regenerative (replacement) medicine. Unlike traditional mammalian cell culture, which involves robust, plentiful, well-characterised cells (and their environment), stem cells are delicate, scarce, poorly characterised and difficult to manipulate. The success of stem cell based therapies will depend on interplay of developmental biology, genetics, genomics, pharmacology, and engineering approaches.

Soluble factors from endodermal SC affect mesodermal SC development
Endodermal SC → Mesodermal SC

Steps involved:	Pancreas	Blood
Proliferation	Liver	Brain
Commitment	Gut (intestine)	Bone
Lineage progression	Thymus	Cartilage
Differentiation		Cardiac
& Maturation		Neural
		Tendon
		Adipose

Figure 1. Stem cell development/organogenesis occurs from endodermal and mesodermal parts of embryo.

In the early embryo, the undifferentiated inner cell mass give rise to an embryo with three primordial germ layers. The middle layer of cells (known as the mesoderm) and certain portions of the inner endoderm give rise to groups of cells that differentiate into many different organs (Figure 1). The process of tissue (re)generation from stem cells in adults follows a similar sequence of events, such as proliferation, commitment, lineage progression and differentiation and maturation, reaching its final form. However, the abundance and environment for regeneration is often compromised by other responses to injury.

The early endoderm, which is derived from the embryonic endoderm after gastrulation (Figure 1), gives origin to the digestive and respiratory tracts and their derivatives, thyroid gland, liver, intestine and pancreas. An interaction from the mesoderm seems to be critical for their development, supplying soluble factors and ECM molecules. These mesenchymal signals function as stimulants or repressors depending on the stage of organ development [47]. Embryologically, the pancreas arises from the interaction of two early germ layers: the endoderm (primitive gut) and the mesoderm (primordial connective tissue). A series of inductive interactions between these two germ layers promotes the morphogenesis of glandular structures such as the thyroid gland, lung, salivary glands, and pancreas. During morphogenesis, populations of exocrine pancreatic epithelial cells and endocrine islet cells differentiate from progenitor populations. Part of the specification of epithelial (endodermal) differentiation depends on signals from and cross talk with the adjacent, supporting mesenchyme. Pancreatic endodermal stem cells have not yet been isolated, but many epithelia, such as the gut, contain resident stem cells.

Cell growth and differentiation of endocrine progenitors in the pancreas provides a model of programmed epithelial morphogenesis. It is generally thought that pancreatic endocrine cells originate from a subpopulation of undifferentiated progenitors residing within the ductal epithelium of the fetal pancreas [48]. This process involves cell budding from the duct, migration into the surrounding mesenchyme, differentiation and clustering into highly organised islet of Langerhans, in addition to vasculogenesis and generation of neural network. As with mesodermal tissues, adult pancreas can presumably be also regenerated in part from adult progenitors present in such organs (in very low frequency) [49].

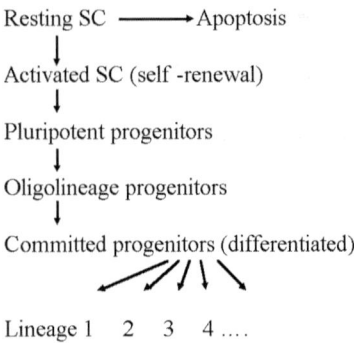

Figure 2. *Stem cell progeny is regulated by programmed cell death (apoptosis). During embryonic and fetal development the self-renewal (proliferation) decreases as the lineage commitment increases.*

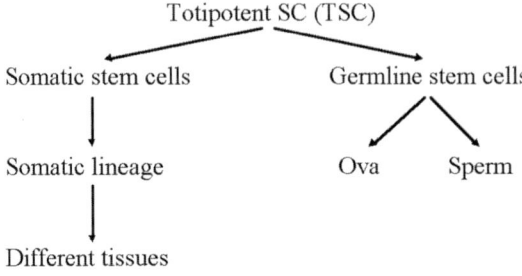

Figure 3. *Totipotent stem cells diverge into somatic and germline cells.*

Stem cells are defined as clonogenic cells capable of both self-renewal and multilineage differentiation [50]. They have low rates of proliferation, but they give rise to daughter cells that have rapid growth rates. The earliest stem cells in the embryo are totipotent (the capability to differentiate into any of the cells of the body) which give rise to non-self renewing oligolineage progenitors, which in turn give rise to progeny that are more restricted in their differentiation potential and finally mature into functional cells with different lineage (Figure 2). At the same time, totipotent stem cells in the embryo can diverge into germ and somatic stem cells (Figure 3).

Although stem cells can be identified by their morphology or location, panels of molecular markers can often define the position of the stem cells. The existence of a common precursor cell for all four-islet cell types has been proposed (Figure 4) by many and experimentally proven by few (*e.g.* [52]). The endocrine, exocrine (acini) and duct cells seem to rise from common progenitor cells in the dorsal and ventral regions of the embryonic endoderm [53-54]. Endoderm proximal to the cardiogenic mesoderm develops into liver, whereas endoderm distal to cardiac mesoderm develops into the pancreas. Transcription factors Ptf1a+/Pdx1+ determine whether intestinal or pancreatic lineage arises [55]. The pancreatic lineage is further specified into exocrine, endocrine and duct progenitors (Figure 4), by means of action of numerous gene products (transcription factors). It should be noted that alternative models for pancreatic cell lineage commitment also exist [56-57]. Generally, a combination of lineage tracing (labeling of precursor cells) and gene inactivation (targeted gene inactivation using Cre recombinase [58]) has been used. Further examination will undoubtedly lead to more details.

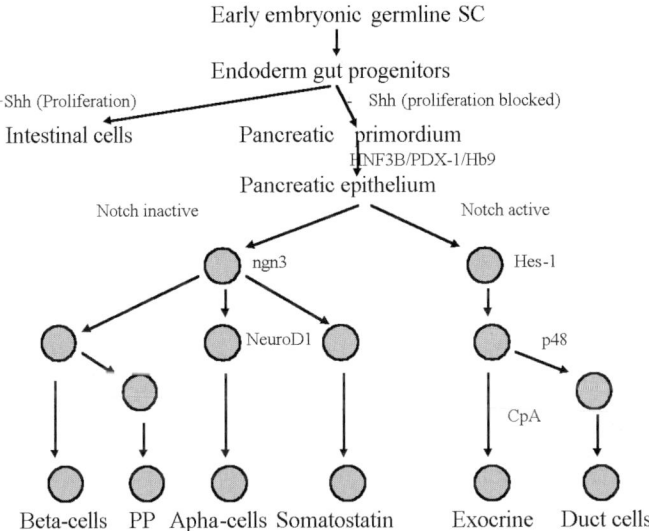

Figure 4. Lineage specification in embryonic pancreas involves multiple steps, regulated by transcription factors (genes)(adapted from Doherty [51]).

Besides the lineage specification, the assembly of functional beta-cells involves morphogenesis, responsible for trapping cell populations together for inductive interactions and for building complex, three-dimensional structures of beta-cells and other cell types [59]. Particularly, branching (budding) morphogenesis, a process in which a tree-like organ is regenerated by a reiterated combination of outgrowth, elongation of a stem, and subdivision of terminal buds, is of importance for pancreas. Underlying mechanism of budding involves interactions between two cell populations (mesenchymal and epithelial) in a reactive-diffusion mode, including coordinated positive and negative feedback loops.

3.1. EMBRYONIC STEM CELLS (ESC)

The embryonic stem cells are derived from outgrowth of the totipotent cells of the pre-implantation embryo. The employment of *in vitro* differentiation methods for ESC to force them into the required differentiation and lineage appears to be quite difficult [60]. Strategies to induce *in vitro* differentiation of ESC include [61]:

- Changes in nutrient environment, *e.g.*, reduction in glucose and serum concentrations, in order to facilitate increase in proportion of islet precursors;
- Employment of feeder cell lines, enabling removal of toxic factors from the medium or addition of growth factors by providing secondary cell line (feeder layer of cells are mitotically inactive but metabolically active). Recently, feeder-free growth of undifferentiated human SC has been reported, employing conditioned medium and extracellular cell matrix (ECM) proteins, which are required for adhesion of cells to the culture dish and for the binding of cellular growth factors [62];
- Addition of ECM components and adhesion molecules;
- Stimulation of embryoid body (EB) formation in culture to facilitate multicellular interactions;
- Addition of growth factors which modify proliferation and gene expression;
- Induction of differentiation by certain agents (DMSO, retinoid acid, *etc.*);
- Engineering expression or repression of certain genes (transcription factors).

Differentiation often requires that the steps have to be employed in an ordered sequence. After a progenitor population is developed, the next step usually requires selection in order to increase the purity of the required phenotype. Soria [61] has employed cell-trapping method (see also Table 2) to encode markers suitable for later selection and identification. The following step, maturation, is the least explored one. In ESC-derived insulin expressing cells, low insulin content and low responsiveness to secretagogue-induced insulin release is still a problem. Two-week nicotinamide culture treatment has led to considerable improvement of these properties [61]. The final step of the ESC culture protocol is testing of cells for proper karyotype and lack of cellular transformation. In the clonogenic assay, cells are grown in methylcellulose, a semisolid culture medium, enabling formation of individual colonies. Overall, it should be understood that the diverse aspects of development of embryonic germ cells are regulated by multiple and temporarily distinct signals.

3.2. ADULT PLURIPOTENT STEM CELLS (APSC)

The potential to form new differentiated cells of given specification is not limited to embryonic cells, since many (epithelial) tissues contain self-renewing progenitor cell populations. It is known that the formation of pancreatic endocrine tissue is not restricted to the pancreatic period of embryonic development [63]. Many investigators are convinced that the islet neogenesis in adults is probably similar in its intrinsic mechanism to that of the embryonic development [48,57]. The process of neogenesis is defined as differentiation of new beta-cells (BC) from either replication of pre-existing BC or from precursor cells located within the pancreatic ducts. The process of neogenesis is most studied in animal models, where regeneration of the pancreas has been induced (by physical or chemical injury). It also occurs in the NOD mouse, where increased islet neogenesis attempts to compensate for the islet cell death. As the process of neogenesis seems to recapitulate the early events of pancreatic development, it lends to support the existence of ductal SC progenitors that contribute to BC regeneration.

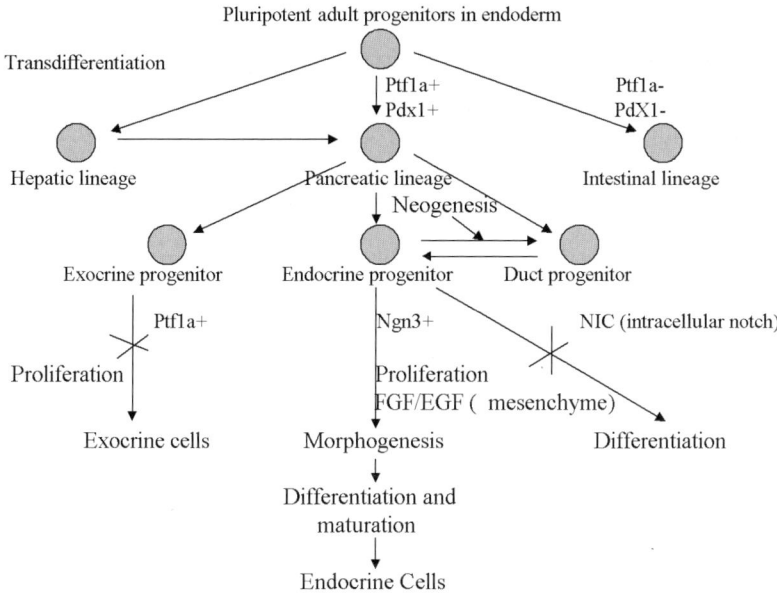

Figure 5. Adult stem cells: repetition of embryonic development? Initially, cell division is necessary for bud formation in the embryonic pancreas. Growth factors required normally come from the mesenchyme. Continued encouragement of proliferation only leads to exocrine lineage expansion, however. Endocrine cells can only be generated from undifferentiated progenitors when differentiation is blocked at an intermediate stage (NIC) and proliferation curbed at later stages.

Efforts have been made to identify pluripotent adult progenitors and to develop an *in vitro* model of BC development for transplantation in diabetes. Although the existence of adult endocrine progenitor cells is elusive, there are new indications that these stem cells are most likely embedded within the ductal cells of adult pancreas [63]. The manipulation of such endogenous stem cells to form BC and intact islets are our ultimate goal. We will be able to do so as more is known about lineage specification genes, transcription factors, and media and growth factor requirements. Initially, cell division is required for bud formation in the embryonic pancreas, and the growth factors required normally come from the mesenchyme. Continued encouragement of proliferation only leads to exocrine lineage expansion, however (Figure 5). Thus, controlled proliferation and differentiation is a necessity. The successful generation of functional mouse islets

from adult SC (ductal) includes several steps of a combined *in vitro/in vivo* procedure [64]:
- Ductal cell isolation in serum-free media;
- Induction of proliferation of SC by serum addition to allow for expansion with minimal differentiation. The limited expansion capability of adult SC is restricted;
- Enhancement of proliferation and differentiation of partially specified progenitors by addition of glucose;
- *In vivo* maturation upon implantation.

Such cells were reported to be only partially responsive to glucose, indicating that the process of islet maturation has not been fully accomplished. No human adult SC studies have been reported. It is obvious that the generation of functional islets from APSC is far from being solved.

Table 3. Stem-cell based approaches to diabetes treatment (adapted from Doherty [51], 2001).

Embryonic Stem Cells (ESC)
Cultured pancreatic ductal epithelium [67]
Cultured human islet buds derived from acinar/ductal tissue [68]
Engineered neonatal islets [69]
Islet Progenitor Cells
Cell-trapping system to select ES cells followed by embryonic bodies formation [61]
Nestin-positive EB [70]
EB formation by human SC [71]
Adult Pluripotent Stem Cells (APSC)
Cultured and transdifferentiated cells (not yet available)
In vivo transdifferentiation of liver cells [72]

3.3. EMBRYONIC VS. ADULT SC

A summary of stem cell based approaches to diabetes treatment is presented in Table 3. It is generally accepted that stem cells collected from tissues of adults, fetal tissue or older embryos are typically more restricted in their developmental potency [50]. However, recent studies have demonstrated a much broader capacity for adult stem cell differentiation than previously recognised. While embryonic SC, proliferate indefinitely in culture and retain their potential to differentiate into virtually any cell type, they, in principle, might serve as a source of large quantities of any derived cell for transplantation into patient. As more is known about lineage specification genes and defined medium and growth factor requirements of ES cells, we will be able to more precisely control their expansion and cell type specification. Because ES cells are derived prior to implantation (into utero), certain immuno-related cell surface proteins might not be expressed. ESC are not, however, fully immunologically compatible with most patients who require cell transplantation. Customised, a patient-by-patient approach might be needed ('therapeutic cloning'). There is also a danger of undifferentiated ES cells being turned into teratomas (benign tumors containing a mixture of different tissue types) after being transplanted. Considering federal and ethical restrictions (US) imposed on use of embryonal stem cells, it is clear that the debate on relative merits will continue.

One approach to employ adult organ-specific SC is to activate endogenous pancreatic precursor cells to form BC and even intact islets to replace defective ones. Compared to embryonic SC, adult SC provide an autologous transplantation mode, without risk of rejection, as adult cells are isolated from the patient and their progeny is transplanted back [73]. Until we efficiently and reproducibly ensure that stable, fully functional BC can be generated *in vitro*, this attempt will remain hypothetical.

References

[1] Prokop, A. (2001) Bioartificial pancreas: Materials, devices, function, and limitations. Diabetes Technol. Therapeut. 3: 431-449.
[2] Lord Rayleigh (1878) Proc. London Math. Soc. 10: 4.
[3] Rayleigh, J.W.S. (1945) The Theory of Sound. Vol.2, 2nd edition. Dover Publishing, New York.
[4] Weber, C. (1931). ZAMM 11: 136.
[5] Crane, L.; Birch, S. and McCormack, P.D. (1964) The effect of mechanical vibration on the break-up of a cylindrical water jet in air. Brit. J. Appl. Phys. 15: 743-750.
[6] Chan Man Fong, C.F. and Kee, D.D. (1993) A material function from the vibrating jet. J. Non-Newt. Fluid Mech. 46: 111-119.
[7] Henderson, D.M.; Pritchard, W.G. and Smolka, L.B. (1997) On the pinch-off of a pendant drop of viscous fluid. Phys. Fluids 9: 3188-3194.
[8] Matsumoto, S.; Kobayashi, H. and Takashima, Y. (1986) Production of monodispersed capsules. J. Microencapsul. 3: 25-31.
[9] Lord Rayleigh (1882) On the equilibrium of liquid conducting masses charged with electricity. Phil. Mag. 14: 184.
[10] Bugarski, B.; Li, Q.; Goosen, M.F.A.; Poncelet, D. and Neufeld, R.J. (1994) Electrostatic droplet generation: Mechanism of polymer droplet formation. AIChE J. 40: 1026-1031.
[11] Gomez, A. and Tang, K. (1994) Charge and fission of droplets in electrostatic sprays. Phys. Fluids 6: 404-414.
[12] Halle, J.-P.; Leblond, F.A.; Pariseau, J.-F.; Jutras, P.; Brabant, M.-J. and Lepage, Y. (1994) Studies on small (<300 µm) microcapsules: II – Parameters governing the production of alginate beads by high voltage electrostatic pulses. Cell Transplant. 3: 365-372.
[13] Xu, K.; Hercules, D.M.; Lacik, I. and Wang, T.G. (1998) Atomic force microscopy used for the surface characterization of microcapsule immobilization devices. J. Biomed. Mater. Res. 41: 461-467.
[14] Lacik, I.; Brissova, M.; Anilkumar, A.V.; Powers, A.C. and Wang, T. (1998) New capsule with tailored properties for the encapsulation of living cells. J. Biomed. Mater. Res. 39: 52-60.
[15] Lacik, I.; Anilkumar, A.V. and Wang, T.G. (2001) A two-step process for controlling the surface smoothness of polyelectrolyte-based microcapsules. J. Microencapsul. 18: 479-490.
[16] Colton, C.K. (2002) Oxygen requirements of islets. In: Encapsulation & Immunoprotective Strategies of Islet Cells, Workshop Proc. Report, December 6-7, 2001, Washington, D.C.; RTI International.
[17] Weir, G. (2002) Encapsulation of islets: How difficult are the problems? In: Encapsulation & Immunoprotective Strategies of Islet Cells, Workshop Proc Report, December 6-7, 2001, Washington, D.C.; RTI International.
[18] Garfinkel, M.R.; Harland, R.C. and Opara, E.C. (1998) Optimization of the microencapsulated islet for transplantation. J. Surg. Res. 76: 7-10.
[19] Rajotte, R. (2000) Advances in islet isolation and transplantation – Past, present, and future. In: Bioartificial Organs III. Tissue Sourcing, Immunoisolation and Clinical Trials, 7-12 October 2000, Davos, Switzerland, discussion.
[20] Poncelet, D. (2001) Production of alginate beads by emulsification/internal gelation. Ann. NY Acad. Sci. 944: 74-82.

[21] Wang, T.; Lacik, I.; Brissova, M.; Anilkumar, A.V.; Prokop, A.; Hunkeler, D.; Green, R.; Shahrokhi, K. and Powers, A.C. (1997) An encapsulation system for the immunoisolation of pancreatic islets. Nature Biotechnol. 15: 358-362.
[22] Prokop, A.; Hunkeler, D.; Powers, A.C.; Whitesell, R.R. and Wang, T.G. (1998) Water soluble polymers for immunoisolation II: Evaluation of multicomponent microencapsulation systems. Adv. Polymer Sci .136: 53-73.
[23] Anilkumar, A.V.; Lacik, I. and Wang, T.G. (2001) A novel reactor for making uniform capsules. Biotechnol. Bioeng. 75: 581-589.
[24] Schneider, S.; Feilen, P.J.; Slotty, V.; Kampfner, D.; Preuss, S.; Berger, S.; Beyer, J. and Pommersheim, R. (2001) Multilayer capsules: a promising microencapsulation system for transplantation of pancreatic islets. Biomat. 22: 1961-1970.
[25] Tatarkiewics, K.; Sitarek, E.; Fiedor, P.; Sabat, M. and Orlowski, T. (1994) In vitro and in vivo evaluation of protamine-heparin membrane for microencapsulation of rat Langerhans islets. Artif. Organs 18: 736-739.
[26] Tun, T.; Inoue, T.; Hayashi, H.; Aung, T.; Gu, Y.J.; Doi, R.; Kaji, H.; Echigo, Y.; Wang, W.J.; Setoyama, H.; Imamura, M.; Maetani, S.; Morikawa, N.; Iwata, H. and Ikada, Y. (1996) A newly developed three-layer agarose microcapsule for a promising biohybrid aritificial pancreas: rat to mouse xenotransplantation. Cell Transplant. 5 (Suppl. 1): S59-S63.
[27] Iwata, H.; Takagi, T.; Amemiya, H.; Shimizu, H.; Yamashita, K.; Kobayashi, K. and Akutsu, T. (1992) Agarose for a bioartificial pancreas. J. Biomed. Mater. Res. 26: 967-977.
[28] Iwata, H.; Kobayashi, K.; Takagi, T.; Oka, T.; Yang, H.; Amemiya, H.; Tsuji, T. and Ito, F. (1994) Feasibility of agarose microbeads with xenogeneic islets as a bioartificial pancreas. J. Biomed. Mater. Res. 28: 1003-1011.
[29] Jain, K.; Yang, H.; Cai, B.-R.; Haque, B.; Hurvitz, A.I.; Diehl, C.; Miyata, T.; Smith, B.H.; Stenzel, K.; Suthanthiran, M. and Rubin, A.L. (1995) Retrievable, replaceable, macroencapsulated pancreatic islet xenograft. Transplant. 59: 319-324.
[30] Duvivier-Kali, V.F.; Omer, A.; Parent, R.J.; O'Neil, J.J. and Weir, G.C. (2001) Complete protection of islets against allorejection and autoimmunity by a simple barium-alginate membrane. Diabetes 50: 1698-1705.
[31] Trivedi, N.; Keegan, M.; Steil, G.M.; Hollister-Lock, J.; Hasenkamp, W.M.; Colton, C.K.; Bonner-Weir, S. and Weir, G.C. (2001) Islets in alginate macrobeads reverse diabetes despite minimal acute insulin secretory responses. Transplant. 71: 203-211.
[32] Hill, R.S.; Cruise, G.M.; Hager, S.R.; Lamberti, F.V.; Yu, X.; Garufis, C.L.; Yu, Y. and Mundwiller, K.E. (1997) Immunoisolation of adult porcine islets for the treatment of diabetes mellitus. The use of photopolymerizable polyethylene glycol in the conformal coating of mass-isolated porcine islets. Ann. NY Acad. Sci. 831: 332-343.
[33] May, M.H. and Sefton, M.V. (1999) Conformal coating of small particles and cell aggregates at a liquid-liquid interface. Ann. NY Acad. Sci. 875: 126-134.
[34] Panza, J.L.; Wagner, W.L.; Rilo, H.L.; Rao, R.H.; Beckman, E.J. and Russell, A.J. (2000) Treatment of rat pancreatic islets with reactive PEG. Biomat. 21: 1155-1164.
[35] Weber, C.J; Norton, J.E. and Reemtskma, K. (1993) Method for microencapsulation of cells or tissue. US Patent 5,227,298.
[36] Chang, S.J.; Lee, C.H.; Hsu, C.Y. and Wang, Y.J. (2002) Biocompatible microcapsules with enhanced mechanical strength. J. Biomed. Mater. Res. 59: 118-126.
[37] Rehor, A.; Canaple, L.; Zhang, Z.B. and Hunkeler, D. (2001) The compressive deformation of multicomponent microcapsules: Influence of size, membrane thickness, and compression speed. J. Biomat. Sci. - Polymer Edn. 12: 157-170.
[38] Brissova, M.; Petro, M.; Lacik, I.; Powers, A.C. and Wang, T. (1996) Evaluation of microcapsule permeability via inverse size exclusion chromatography. Analyt. Biochem. 242: 104-111.
[39] Brissova, M.; Lacik, I.; Powers, A.C.; Anilkumar, A.V. and Wang, T. (1998) Control and measurement of permeability for design of microcapsule cell delivery system. J. Biomed. Mater. Res. 39: 61-70.
[40] Anilkumar, A.V.; Hmelo, A.B. and Wang, T.G. (2001) Core centering of immiscible compound drops in capillary oscillations: Experimental observation. J. Coll. Interface Sci. 242: 465-469.
[41] Lacik, I. (2003) Polyelectrolyte complexes for microcapsule formation. In: Nedovic, V. and Willaert, R. (Eds.) Fundamentals of Cell Immobilisation Biotechnology, Kluwer Academic Publishers, the Netherlands; pp.103-120.

[42] Bachalo, W. (2002) Factors affecting the transplant efficacy of islets microencapsulated with purified alginate. In: Encapsulation & Immunoprotective Strategies of Islet Cells, Workshop Proc Report, December 6-7, 2001, Washington, D.C.; RT International.
[43] Prokop, A. and Wang, T.G. (1997) Purification of polymers used for fabrication of an immunoisolation barrier. Ann. NY Acad. Sci. 831: 223-231.
[44] Risbud, M.V.; Bhonde, M.R. and Bhonde, R.R. (2001) Effect of chitosan-polyvinyl pyrrolidone hydrogel on proliferation and cytokine expression of endothelial cells: Implications in islet immunoisolation. J. Biomed. Mat. Res. 57: 300-305.
[45] Trivedi, N.; Steil, G.M.; Colton, C.K.; Bonner-Weir, S. and Weir, G.C. (2000) Improved vascularization of planar membrane diffusion devices following continuous infusion of vascular endothelial growth factor. Cell Transplant. 9: 115-124.
[46] Prokop, A.; Kozlov, E.; Non, S.N.; Dikov, M.M.; Sephel, G.C.; Whitsitt, J.S. and Davidson, J.M. (2001) Towards retrievable vascularized bioartificial pancreas (VBAP): Induction and long-lasting stability of polymeric mesh implant vascularized with help of fibroblast growth factors a and b and hydrogel coating. Diabetes Technol. Therapeut. 3: 245-262.
[47] Gittes, G.K.; Galante, P.E.; Hanahan, D. Rutter, W.J. and Debas, H.T. (1996) Lineage-specific morphogenesis in the developing pancreas: Role of mesenchymal factors. Development 122: 430-447.
[48] Edlund, H. (2002). Pancreatic organogenesis – Developmental mechanisms and implications for therapy. Nature Rev. Genetics 3: 524-532.
[49] Bouwens, L and Pipeleers, D.G. (1998) Extra-insular cells associated with ductals are frequent in adult human pancreas. Diabetol. 41: 629-633.
[50] Weisman, I.L. (2000) Stem cells: Units of development, units of regeneration, and units in evolution. Cell 100: 157-168.
[51] Doherty, K. (2001) Growth and development of the islets of Langerhans: Implications for the treatment of diabetes mellitus. Curr. Opinion Pharmacol. 1: 641-650.
[52] Percival, A.C. and Slack, J.M.W. (1999) Analysis of pancreatic development using a cell lineage label. Exp. Cell Res. 247: 123-132.
[53] Detsch, G.; Jung, J.; Zheng, J.; Lora, J. and Zaret, K.S. (2002) A bipotential precursor population for pancreas and liver within the embryonic endoderm. Development 128: 871-881.
[54] Bort, R. and Zaret, K. (2002) Paths to the pancreas. Nature Gen. 32: 8585.
[55] Kawaguchi, Y.; Cooper, B.; Gannon, M.; Ray, M.; MacDonald, R.J. and Wright, C.V.E. (2002) The role of the transcriptional regulator Ptf1a in converting intestinal to pancreatic progenitors. Nature Gen. 32: 128-134.
[56] Gu, G.; Dunauskaite, J. and Melton, D.A. (2002) Direct evidence for the pancreatic lineage: NGN3+ cells are islet progenitors and are distinct from duct progenitors. Development 120: 2447-2457.
[57] Herrera, P.L. (2002) Defining the cell lineages of the islets of Langerhans using transgenic mice. Int. J. Dev. Biol. 46: 97-103.
[58] Kuhn, R and Torres, R.M. (2002) Cre/loxP recombination system and gene targeting. Math. Mol. Biol. 180: 175-204.
[59] Hogan, B.L.M. (1999) Morphogenesis. Cell 96: 225-233.
[60] Schwitzgebel, V.M. (2001) Programming of the pancreas. Mol. Cell Endocrinol. 185: 99-108.
[61] Soria, B. (2001) In vitro differentiation of pancreatic beta-cells. Differentiation 68: 205-219.
[62] Xu, C.; Inokuma, M.S.; Denham, J.; Golds, K.; Kundu, P.; Gold, J.D. and Carpenter, M.K. (2001) Feeder-free growth of undifferentiated human embryonic stem cells. Nature Biotechnol. 19: 971-974.
[63] Lipsett, M. and Finegard, D.T. (2002) Beta-cell neogenesis during prolonged hypeglycemia in rats. Diabetes 51: 1834-1841.
[64] Soria, B.; Roche, E.; Berna, G.; Leon-Quinto, T.; Reig, J.A. and Martin, F. (2000) Insulin secreting cells derived from embryonic stem cells normalize glycemia in streptozotocin-induced diabetic mice. Diabetes 49: 157-162.
[65] Slack, J.M.W. (1995) Developmental biology of the pancreas. Development 121: 1589-1580.
[66] Peck, A.B.; Chaudhari, M.; Cornelius, J.G. and Ramiya, V.K. (2001) Pancreatic stem cells: Building blocks for a better surrogate islet to treat type 1 diabetes. Ann. Med. 33: 186-192.
[67] Ramiya, V.K.; Maraist, M.; Arfors, K.E.; Schatz, D.A.; Peck, A.B. and Cornelius, J.G. (2000) Reversal of insulin-dependent diabetes using islets generated in vitro from pancreatic stem cells. Nature Med. 6: 278-282.

[68] Bonner-Weir, S.; Taneja, M.; Weir, G.C.; Taterkiewics, K.; Song, K.H.; Sharma, A. and O'Neil, J.J. (2000) In vitro cultivation of human islets from expanded ductal tissue. Proc. Natl. Acad. Sci. USA 97: 7999-8004.
[69] MacFarlane, W.M.; Chapman, J.C.; Shepherd, R.M.; Hashmi, M.N.; Kamimura, N.; Cosgrove, K.E.; O'Brien, R.E.; Barnes, P.D.; Hart, A.W. and Doherty, H.M. (1999) Engineering a glucose-responsive human insulin-secreting cell line from islets for Langerhans isolated from a patient with persistent hyperinsulinemic hypoglycemia of infancy. J. Biol. Chem. 274: 34059-34066.
[70] Lumelsky, N.; Blondel, O.; Laeng, P.; Velasco, I.; Ravin, R. and McKay, R. (2001) Differentiation of embryonic stem cells to insulin-secreting structures similar to pancreatic islets. Science 292: 1389-1394.
[71] Assady, S.; Masor, H.; Amit, M.; Itskovitz-Eldor, J.; Skorecki, K.L. and Tzukerman, M. (2001) Insulin production by human embryonic stem cells. Diabetes 50: 1691-1697.
[72] Ferber, S.; Halkin, A.; Cohen, H.; Ber, I.; Einav, Y.; Goldberg, I.; Barshack, I.; Seijffers, R.; Kopolovic, J. and Kaiser, N. (2000) Pancreatic and duodenal homeobox gene 1 induces expression-induced hyperglycemia. Nature Med. 6: 568-572.
[73] Orkin, S.H. and Morrison, S.J. (2002) Stem-cell competition. Nature 418: 25-27.

BIOARTIFICIAL SKIN

BARBARA ZAVAN, ROBERTA CORTIVO, PAOLA BRUN,
CAROLIN TONELLO AND GIOVANNI ABATANGELO
Department of Histology, Microbiology and Medical Biotechnologies –
University of Padova, Viale G. Colombo 3, I-35121, Padova, Italy –
Fax: ++39-49-8276079 – Email: abatange@civ.bio.unipd.it

1. Introduction

Tissue engineering is an emerging interdisciplinary field that applies the principles of biology and engineering to the development of viable substitutes that restore, maintain, or improve the function of human tissues or organs. This form of therapy differs from standard therapies in that the engineered tissue becomes integrated with the patient's tissue, affording a potentially permanent and specific cure of the disease state. Although cells have been cultured, or grown, outside the body for many years, the possibility of growing complex, three-dimensional tissues, literally replicating the design and function of human tissue, is a recent development. The intricacies of this process require input from many fields, including engineering, hence the term, tissue engineering.

Skin was the first human organ to be reconstructed *in vitro* using tissue engineering technology. Skin models emerged as early as 1975 to 1980, and contributed to our understanding of human skin physiology, pharmacology, and pathophysiology. At the same time, cell culture techniques were developed for the replacement of large skin defects and the enhancement of wound healing in humans. The main objective of skin substitution is restoration of the anatomy and physiology of uninjured skin. With this new technique, it is possible to produce large quantities of biological material suitable for the treatment of specific pathologies (*e.g.* burns, scars, cutaneous ulcers, congenital anomalies). With such a treatment, reduction of morbidity and mortality from full-thickness skin wounds is possible. Morbidity from grafting of autologous, split-thickness skin [1] occurs at both the treatment and donor site [2]. Acute wounds that require grafting include excised burns, burn scars, and congenital cutaneous anomalies (*i.e.* giant nevus). Autologous full-thickness skin grafts, free flaps and pedicle flaps [3,4] are the gold standard of therapy used to restore all structures and functions of uninjured skin. In these cases, skin donor sites and treatment sites must be equal in size. Tissue expansion is a new surgical technique that allows stretching of the skin by approximately a factor of 2, but some complications can arise, including rupture or infection of the expander, and necrosis of expanded skin before transplantation. Skin substitutes that contain cultured cells can provide large quantities of grafts for wound treatment, but can restore only some anatomic structures and physiologic functions of skin. The full potential of skin substitute engineering will not be realised until the complex structure of native skin is ideally reproduced.

2. Requirements

2.1. ANATOMIC

Skin, also known as the cutis or integument, has a surface area of 1.5–2.0 m^2 [5]. This largest and most complex organ in the human body is also the most frequently injured. It is well known that skin prevents dehydration through evaporative water loss, provides thermal regulation, and acts as a barrier against chemical and infectious insult [6]. It is comprised of four tissues: epidermis, dermis, basement membrane and subcutis (Fig. 1). The upper layer, epidermis, is the tissue in direct contact with the external environment and consists of epithelial cells, known as keratinocytes, forming a 10 layer-thick fragile sheet (around 0.1 mm). Epidermis is divided into a number of strata formed in part by the dynamic processes of cell proliferation, maturation and death, and thus, reflecting the different stages of keratinocyte maturation. The epidermis and dermis are separated by the basement membrane, a 20 nm thick multi-layered dynamic structure that mechanically stabilises the tight junctions between the upper epidermis and underlying connective tissue via hemidesmosomes. The thicker dermis layer is comprised of a loose array of fibroblasts (mesenchymal cells) and vasculature (endothelial cells), forming a 2-5 mm thick connective tissue that sustains the overlying epidermis. The dermal matrix provides considerable strength to skin by virtue of the arrangement of collagen fibers. This collagenous meshwork is interwoven with varying quantities of elastin fibers, proteoglycans, glycosaminoglycans (GAGs, *i.e.*, hyaluronic acid, dermatan sulfate, chondroitin-6-sulphate and heparin sulphate), fibronectin and other components. Finally, the subcutis, found underneath the dermis, is a 0.4-4 mm thick tissue composed primarily of fat cells (adipocytes). Skin is also composed of a variety of appendages known as adnexa, such as hair follicles, sweat glands and sebaceous glands, all generally embedded in the dermis.

Because the skin serves as a protective barrier against the outside world, any break in it must be rapidly and efficiently mended. The physiologic mechanism that coordinates the ordered repair of cutaneous tissues is generally referred to as wound healing. It can be defined as a well- orchestrated series of temporary overlapping actions involving different cells, matrix components, and biochemical reactions. Different types of cells respond to environmental signals in a specific manner in order to carry out their genetically programmed role in proliferation, differentiation and function of tissues. Cells synthesise proteins necessary for their proliferation and migration during the specific phases of wound healing: inflammation, granulation tissue formation, re-epithelialisation, matrix production and remodelling [7,8]. Thus, each layer of skin has a pivotal role to play in the normal functioning of skin. When one of these becomes dysfunctional, skin disorders arise. Healing of injured tissues involves both regeneration of the epidermis and repair of the dermis resulting in scar tissue [9].

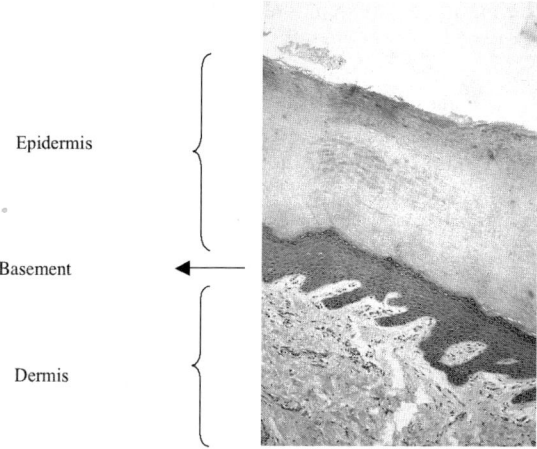

Figure 1. Skin structure: epidermis, dermis, basement membrane.

2.2. SURGICAL

Cutaneous lesions can be divided into two broad categories: acute and chronic. The former refers to a group of pathologies that has been provoked by a trauma such as burns or accidents, while the latter refers to permanent, and often unresolved, lesions such as ulcers or genetic pathologies. In both cases, an impairment of the healing process is at the origin of a poor clinical result.

From a surgical point-of-view, these pathologies are still a challenge for the clinician who has to deal, in almost all cases, with the problem of tissue loss. Historically, the treatment of skin loss has focused on the design of a temporary wound closure. Attempts to cover wounds and severe burns have been reported from historical sources as far back as 1500 B.C. Since then, a large number of temporary wound dressings have evolved, including membranes or sheets made from natural and synthetic polymers, skin grafts from human cadavers (homografts, or allografts), and skin grafts from animals (heterografts, or xenografts).

Acellular coverage affords only a temporary dressing, since it does not integrate with cutaneous tissues. Either surgical removal or natural rejection occurs within a short period. Drawbacks such as wound contamination and delay in natural remodeling have been associated with the use of synthetic polymers. Similarly, when using natural or naturally derived coverage (*i.e.*, devitalised homologous or heterologous skin), the immune response of the patient is a major limiting factor.

It is now evident that the ideal wound coverage for cutaneous lesions is an autologous graft that provides both full immuno-compatibility and permanent functionality since it is completely integrated over time. Although possible, the technique is not always available to the surgeon, for example in the case of large deep burns. In addition, problems such as donor site morbidity may add unwanted complications to the pathology treated. To resolve this dilemma, combinations of self and non-self devices have been developed.

In 1952, the discovery that keratinocytes obtained from skin specimens and treated with trypsin maintained viability dramatically changed skin substitution research. The culture of human keratinocytes culminated in the work of Rheinwald and Green (1975) who determined the fundamental procedure for keratinocyte growth *in vitro*. Other significant advances followed this first successful experiment, related mostly to perfecting the best culture conditions. Subsequently, the most important advancement was the transfer of this laboratory technique to the operating room [10,11].

Prerequisites for successful skin grafts are: absence of antigenicity and toxicity; adherence to wounds; histocompatibility; control of fluid loss and infection; mechanical stability and compliance; cost effectiveness; and availability [12,13]. Anatomic deficiencies in engineered epithelial sheets decrease the probability that these requirements can be met. Consequently, complications associated with the use of cultured epithelial autografts (CEA) *versus* split-thickness skin grafts, and which may increase rather than decrease risks to patient recovery, include: reduced rates of engraftment [14], increased microbial contamination [15,16]; mechanical fragility [17], increased time to healing [18], increased regrafting, and very high cost [19] (Table 2 [20]). These shortcomings have made it clear that engineered substitutes should be made of the skin's two main components, the epidermis and the underlying supporting dermis, to fulfil biological properties such as mechanical strength, good handling and wound adherence. Complications associated with early models of skin substitutes do not preclude their long-term potential for medical advantages in wound care.

Table 2. Clinical limitations and considerations for use of engineered skin [20].

Limitation	Consideration
Mechanical fragility	Special dressings and nursing care
Susceptibility to microbial contamination	Non-cytotoxic topical antimicrobial agents
Decreased rates of engraftment	Increased regrafting
Increased time to heal	Delay of recovery
Very high cost	Resource allocation

3. Tissue engineering of human skin

3.1. SCAFFOLDS

Improved understanding of biological processes has facilitated the development of new classes of biomaterials, polymers, and diagnostic and analytical reagents. This emerging technology can be used to introduce better functioning materials that can be processed from naturally occurring or synthetic resources, or a combination of these. Cellular and other biologic components may also be added.

These material components are usually called matrices or scaffolds. Their function is to direct the growth of cells migrating from surrounding tissue or of cells seeded within these porous structures. They provide a suitable substrate for cell attachment, proliferation, differentiated function, and cell migration. These matrices/scaffolds can be permanent or biodegradable (temporary). Although there are many biocompatible materials that could potentially be used to construct matrices/scaffolds, a biodegradable material is desirable, since the role of these matrices/scaffolds is usually temporary. Many natural or synthetic biodegradable polymers have been developed (*e.g.*, collagen, hyaluronic acid derivatives, poly alpha-hydroxyesters, and polyanhydrides), and many more are currently under development.

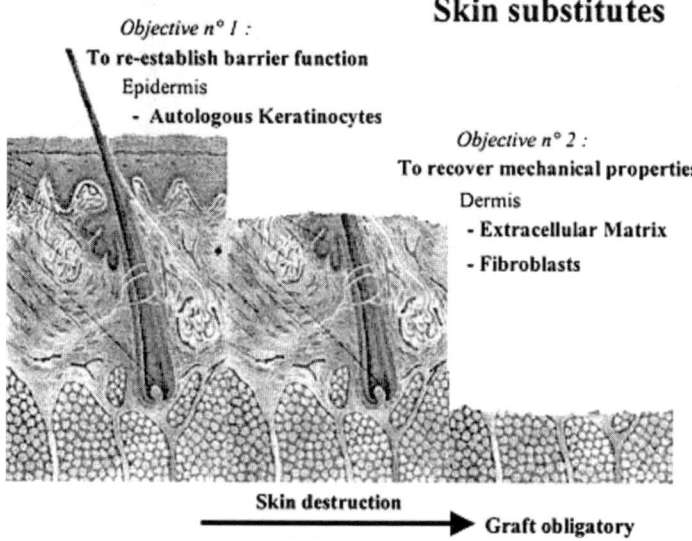

Figure 2. The depth of tissue destruction and necessity to replace multiple functions will determine the type of skin substitute used. [20]

3.2. CELLS

For extensive skin defects such as burns or giant nevi, approaches that provide only an epidermis and those that also provide a dermis (Figure 2) must be distinguished since each of these tissues has specific functions. Commercial products and experimental models for epidermal and/or dermal repair have been configured from individual and combined materials as shown in Table 3 [20]. Skin constructs that are presently commercially available are based on the following techniques:
- Epidermis: with only keratinocytes;
- Dermis: with fibroblasts;
- Co-cultures of fibroblasts and keratinocytes.

3.2.1. Epidermis

The minimum requirement for a viable skin construct: to re-establish a barrier function and avoid infection and water loss, is fulfilled by the horny layer of the epidermis, a product of terminal keratinocyte differentiation. After the breakthroughs of Rheinwald and Green in the mid seventies, the technology of cultured epithelial autograft (CEA) is now available to a large number of hospitals worldwide. In addition, cell culture parameters, harvest procedures and usage indications have been strictly defined and CEA production is being performed on an industrial basis.

Table 3. Engineered skin substitutes* [21].

Trademark name	Source	"Dermis"	"Epidermis"
EpiCel™	Genzyme Tissue Repair	Allodermis	Cultured auto HK
Integra™	Integra Life Sciences	Collagen-GAG^a & silicone	Thin autograft
AlloDerm™	LifeCell Corporation	Acellular dermal matrix	Thin autograft
DermaGraft™	Advanced Tissue Sciences	PGA^b/PLA^c + allo HK	Thin autograft
N/a	Univ. Cincinatti/Shriners	Collagen-GAG + auto HF	Cultured auto HK
LaserSkin™	Fidia Bioplymers (Italy)	Hyaluronic acid	Cultured auto HK
PolyActive™	HC Implants (The Netherlands)	PEO^d/PBT^e + auto HF	Cultured auto HK
ApliGraft™	Organogenesis, Inc.	Collagen gel + allo HF	Cultured allo HK
ORCEL™	Ortec International, Inc.	Collagen + allo HF	Cultured allo HK
TransCyte™	Advanced Tissue Sciences	Allo HF	BioBrane™

* The above list of products is not all-inclusive, exclusive, or an endorsement of one particular brand.
^{a} GAG = glycoaminoglycan; ^{b} PGA = polyglycolic acid; ^{c} PLA = polylactic acid; ^{d} PEO = polyethylene oxide; ^{e} PBT = polybutylene terephthalate.

Table 4. Components of engineered skin [20].

Dermal substitutes	Epidermal substitutes
Autologous cultured fibroblasts	Autologous cultured keratinocytes
Allegeneic cultured fibroblasts	Allogeneic cultured keratinocytes
Collagen-GAG^a, collagen gel	Thin epidermal graft
Acellular cadaveric skin matrix	Epidermal suction blisters
PLA^b/PGA^c, PEO^d/PBT^e	Epidermal cell suspensions

^{a} GAG = glycoaminoglycan; ^{b} PLA = polylactic acid; ^{c} PGA = polyglycolic acid; ^{d} PEO = polyethylene oxide; ^{e} PBT = polybutylene terephthalate

Cumulative clinical data on the clinical application of CEAs, mostly for extensive burns treatment, has revealed several limitations of this technique. Firstly, time requirements for the production of CEA are high, as keratinocytes in standard conditions form epithelial sheets at least 3 weeks after initiation of culture. Secondly, graft take-rates and the number of implantation procedures are often unpredictable. Furthermore, the use of enzymes to detach CEA from plastic culture dishes and the contraction of epithelial sheets are additional problems [22]. Previous attempts of using cultured epidermal grafts alone did not yield satisfactory results because their fragility and mechanical instability resulted in blistering of the epidermis. The grafted skin was reported to also be more prone to contractures, yielding poor cosmetic results. The time, cost and facilities involved in culturing these epidermal sheets are also factors for consideration.

Research in this area has led to the transplantation of epidermal cells using a delivery system that also serves as a substrate for cell growth. Most of the non-biological surfaces act as passive physical supports that do not interact with cells. For the second generation of CEAs, better delivery vehicles, *i.e.*, biomaterials, have been developed to overcome these limitations [23]. The ideal delivery vehicle should possess all relevant properties in terms of its biocompatibility and tolerability, it should be totally degradable in order to avoid device removal, and, finally, should guarantee effective epithelial sheet stability once grafted onto the patient. A large number of biomaterials have been tested so far, both in pre-clinical and clinical experimental conditions [22]. Among these, one of the most promising backing material for the culture of human autologous keratinocytes is a hyaluronic acid-based microperforated membrane. The rationale for the use of hyaluronan-derived products in skin repair lies in the key role of hyaluronic acid in the extracellular matrix of human skin and in the wound healing process [24]. One popular and commercially viable biological delivery system is derived from Laser skin. Laser skin, manufactured by FIDIA Advanced Biopolymer, Italy, is a biodegradable delivery system for keratinocyte grafting. It is a thin transparent membrane, composed of a semi-synthetic derivative of hyaluronic acid completely esterified to yield a benzyl derivative. Microholes are drilled with a laser across the membrane and the seeded keratinocytes populate these pores and grow out as colonies both above and beneath the membrane. The membrane can be peeled from the dish without 'dispase' and can be sterilised by gamma radiation [25].

3.2.2. Dermis

An advancement in the design of the ideal skin equivalent evolved from basic research and clinical observations that revealed the CEA take to be strongly dependent on wound bed condition, particularly that of the dermal layer [26,22]. Several studies have shown that the presence of a dermal layer is of great importance for the regulation of growth and differentiation of cultured keratinocytes. Indeed, fibroblasts embedded in their extracellular matrix (ECM) form a supportive and regulatory layer for those cells. Although the barrier function depends on the epidermis (through keratinocyte differentiation), there is a need to improve CEA grafting by incorporating dermal tissue and promoting the overall functionality of the engrafted zone. The immediate benefit (*i.e.*, promotion of cultured epidermis graft take), in which the extracellular matrix has a key role, should be distinguished from the medium- and long-term benefits, in which fibroblasts become more important given their involvement in extracellular matrix synthesis and remodeling, and in keratinocyte growth and differentiation. The first attempts to stimulate take rates of CEAs involved improving wound bed preparation with meticulous "débridement" and/or bio-interactive dressings. Subsequently, to resolve the problem of unacceptable graft *versus* host reactions, methodology was developed for the application of acellular dermal substitutes such as cadaver or devitalised heterologous dermis.

At present, allogenic, cryopreserved cadaver skin is used to initially cover the patient's wounds (*i.e.*, wound closure) while the CEA is in preparation, and to prepare the graft bed (*i.e.*, vascularisation). Once the CEA is ready, the epidermis of the cadaver skin is removed using a dermatome, and the cultured epidermis is grafted. This wound bed preparation technique [31] facilitates CEA take and improves results in fold regions such as the elbow. A functional neodermis is obtained after 3 to 5years [32,33]. Nevertheless, cryopreserved cadaver skin is often difficult to obtain.

Given that fibroblasts are the main constituent of the dermis and are normally used to grow keratinocytes, the concept of co-culture was introduced in order to produce *in vitro* a composite skin replacement composed of an epithelial layer overlaid onto a dermal substitute. The presence of fibroblasts within the grafted region provides many advantages. Several lines of research are being pursued to define this methodology, including studies of allogenic fibroblast persistence in humans. While fibroblasts have been shown to persist in animals, [34] and allogenic fibroblasts have been reported to give similar results to autologous fibroblasts, their ultimate fate in humans is unknown [35]. Definition of fibroblast populations within the dermis is ongoing. There is clear evidence that, depending on the anatomical region or the depth within the dermis, fibroblasts can be classified with functional criteria. Cell markers are needed to identify and select which cells would be most efficient within dermal substitutes. At present, there is still no commercially available fibroblast marker despite active research in this field [36].

Lastly, the harvesting of mesenchymal stem cells, which seem to participate in wound healing, from bone marrow [37] or peripheral blood [38] is being evaluated. In addition to skin substitutes for permanent replacement, other devices can help to rapidly cover patients' wounds (despite eventual rejection of allogenic cells) and accelerate healing. It is already known that allogenic epidermal cell cultures promote external ulcer healing, [39] probably by secreting factors (matrix metalloproteinases, growth factors, *etc.*) that promote wound cleansing and stimulate the activity of cells present at the wound site.

The importance of the scaffolding material for the creation of a dermal-like tissue cannot be overstated. In fact, the design of a dermal-like tissue critically depends on the use of an ideal scaffold that allows the dermis to develop with a three-dimensional architecture. One of the most important features of the scaffold is its porosity, which has to fit the neo-vessel ingrowth from the host tissue (angiogenesis) during the wound healing process. In addition, the biomaterial should be fully biocompatible and totally degradable since it will be substituted by a fully functional ECM in the long term Finally, the material should be resistant to mechanical forces and easy to apply on the wound area.

The following guidelines have emerged from clinical experience:
- Homogeneity of the cultured epidermal sheet is crucial for a good cosmetic result.
- One-step grafting of a co-cultured dermis and epidermis is not yet possible, and would be too lengthy for emergency use (e.g., burns).
- Preliminary grafting of a collagen matrix is haemostatic, reduces pain, and prepares the graft bed.
- Most importantly, the presence of living fibroblasts accelerates the reappearance of a functional neodermis (less than 1 year instead of 3 to 5 years when the graft bed is prepared with only cryopreserved cadaver skin).

Biomaterial technology has produced a series of scaffolds tested for fibroblast culture and synthesis of a neo-extracellular matrix, which plays a key role in CEA attachment *via* the basement membrane [27]. In order to obtain a dermal-like tissue, fibroblasts have been cultured into various kinds of scaffolds, such as PGL meshes, acemannan polymers, collagen lattices, *etc*. Recently, an allogenic cultured human skin equivalent was generated for the treatment of venous ulcers [28,29]. This skin substitute consists of a dermal equivalent composed of type I bovine collagen that contains living human dermal fibroblasts with an overlying cornified epidermal layer of living human keratinocytes. A different approach is used with autologous fibroblasts, which when cultured into the appropriate scaffolds, can proliferate and deposit the main extracellular components, giving rise to a living autologous dermal equivalent.

The Genzyme Tissue Repair's Epicel$^{(TM)}$ is a cultured epidermal autograft used as a permanent skin replacement for patients with burn injury. It requires a stamp-sized biopsy from the patient's skin to culture epidermal sheets in approximately 16 days. The cultured epidermal autograft functions as a permanent cover for the patient and is not rejected. Though cultured autografts are permanent and life saving, the length of time required for their preparation limits their popularity for use in burns.

An alternative to autografts is offered by the development of a cultured epithelium from an allogenic donor. This technique is recommended for burn and leg ulcer treatment, and split thickness skin graft donor sites [40-45]. Keratinocytes isolated from older donors were shown to exhibit slow growth in culture and to have reduced culture life span and colony forming ability. In contrast, keratinocytes derived from neonatal foreskin provided a potent stimulus to healing in a variety of wounds. The application of an allograft is an outpatient procedure that causes little discomfort to the patient. Furthermore, in contrast to autografting, donor site injury is not inflicted. The use of cultured allogenic cells also permits immediate graft availability and the possibility of bulk manufacture and preservation in a suitable condition for future use. Due to the accelerated proliferation of newborn epidermal cells, the preparation of allografts is faster and easier. Allografts provide a temporary wound covering that releases multiple cytokines and promotes permanent re-epithelialisation by quiescent host keratinocytes.

Apligraf® (Graft skin) is a two-layered living skin with appearance and handling characteristics similar to normal skin [46]. It has been used without adverse immune responses in animal studies [47], and has been found to be safe and effective for providing overlay coverage of widely meshed autografts. An epidermal layer is formed from human neonatal foreskin-derived keratinocytes, which are organised as they would be in human skin with a differentiated stratum corneum. The dermis has dermal fibroblasts, also derived from neonatal foreskin, in a type-I bovine collagen lattice. Collagen assembles into a gel in which these human fibroblasts are interspersed, contracting the network of collagen fibers. A suspension of epidermal cells is added to the surface of the collagen/fibroblast layer. After several days of growth submerged in tissue culture medium, the surface of the skin equivalent is exposed to air to promote epidermal differentiation. After 7-10 days of incubation under these conditions, a matured cornified epidermis develops at the air liquid interface [48]. Apligraf is supplied as either a circular disk 65-75 mm in diameter or as a 4-8 inch rectangle with a thickness of 0.5-0.75 mm in a sterile plastic carrier intended for a single use. Apligraf is readily available, uniform in composition and regulated by the FDA to assure good manufacturing practices (GMP), quality materials, and definitive documentation of safety and efficacy. Apligraf is indicated for non-healing ulcers [28] along with standard compression therapy. It is also used to treat acute wound types such as donor sites and cancer excision sites [50]. In severe ulcers that do not heal, Apligraf has been used over skin grafting, while in the case of burn wounds, Apligraf meshed with an autograft has shown promise. Organogenesis manufactures Apligraf and Novartis Pharma AG retains the global marketing rights.

Integra® artificial skin is a two layered dermal and epidermal equivalent. The dermal replacement layer is made of a porous matrix of fibers of cross-linked bovine tendon collagen and a glycosaminoglycan (chondroitin-6-sulfate). The appropriate pore size of 20–50 µm is critical to allow the optimal ingrowth of native fibroblasts and endothelial cells [51]. The temporary outer layer or epidermal analog is medical grade 100 µm thick silicone, to be replaced after 10–14 days with an ultra thin auto-epidermal graft with a wide mesh [51]. This collagen-GAG-dermal architecture approaches that of normal dermis and completely biodegrades after 30 days, thus encouraging the ingrowth of native fibroblasts and endothelial cells, facilitating the formation of a completely biologic, native neodermis that histologically resembles and functions like normal dermis [52,53]. Upon adequate vascularisation of the dermal layer and availability of donor autograft tissue, the temporary silicone layer is removed and a thin meshed layer of epidermal autograft is placed over the neodermis. Cells from the epidermal autograft grow and form confluent stratum corneum, thereby closing the wound and reconstituting a functional dermis and epidermis [23]. In an effort to improve the quality of healed skin, dermal replacements can be used beneath the cultured epithelial grafts. This approach is consistent with various reports addressing the importance of the survival and maturation of keratinocytes.

Dermagraft® consists of human dermal fibroblasts from neonatal foreskin seeded into a three-dimensional bioabsorbable scaffold: Biobrane. Biobrane is a synthetic membrane bound to one surface of a nylon mesh and coated with porcine collagen. The Silastic membrane serves as an epidermis. Foreskin fibroblasts proliferate on the nylon mesh, secrete collagen, fibronectin, growth factors, GAGs and Tenashin. Dermagraft® is a non-reactive dermal tissue that induces less granulation tissue formation and less bleeding on removal [54,55]. It is a temporary skin substitute and has to be replaced with an autologous skin graft [56].

Recently, a biocompatible, acellular connective tissue material (AlloDerm™) made from human dermis has been introduced as an alternative to conventional autogenous tissue. This material eliminates the need for donor sites, and minimises postoperative discomfort and complications. It also integrates well with the recipient site, and provides an excellent colour match with the surrounding tissues, making it aesthetically pleasing [57].

Composite cultured skin (CCS Ortec International Inc, NY) consists of neonatal keratinocytes and fibroblasts cultured in distinct layers within a bovine type-I collagen scaffold. FDA approved clinical trials are ongoing in burn patients and in patients with epidermolysis bullosa [58].

Another biopolymer used as scaffolding for fibroblasts and keratinocytes is Polyactive™. The skin substitute obtained has a dense top layer that serves as a substrate for keratinocyte culture, and a porous under layer that is critical for wound adhesion and serves as a template for dermal regeneration. Tensile properties of these two layered matrices are to a considerable extent dependent on top layer and under layer composition and thickness. Elasticity moduli are thought to be in the range of those previously reported for human skin [59].

A new biologic wound covering for the treatment of partial-thickness burns is TransCyte (Advanced Tissue Sciences, La Jolla, California, formerly marketed as Dermagraft-Transitional Covering). This material is composed of human newborn fibroblasts that are cultured on the nylon mesh of Biobrane (Dow B. Hickam, Inc, Sugarland, Tex). The thin silicone membrane bonded to the mesh provides a moisture vapour barrier for the wound [60].

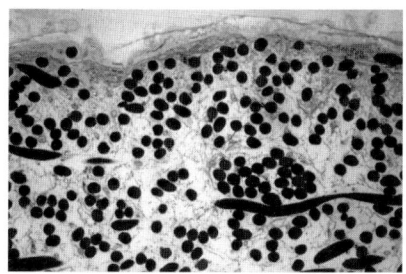

Figure 3. Fibroblasts on Hyaff 11® non-woven meshes.

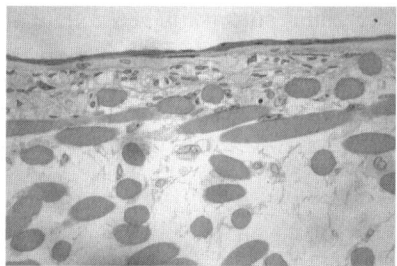

Figure 4. Collagen type I on Hyaff 11® non-woven meshes.

Naturally occurring polysaccharides have also been shown to be excellent bioactive materials, modulating cell-cell interactions, cell-substrate interactions and cell proliferation [61]. Recently, hyaluronic acid (HA) has been extensively studied in the biomaterials field. New classes of insoluble polymers have been developed using a variety of esterifications of the carboxyl group of HA with different types of alcohols. In particular, the benzylic ester of hyaluronan, referred to as HYAFF11®, produces some promising biomaterials (Fidia Advanced Biomaterial, Abano Terme PD) that have been utilised in the preparation of highly biocompatible and biodegradable devices, such as membranes and non-woven tissues. When implanted in body tissues, these devices are well tolerated and degraded. A dermal-like support has been generated by first seeding human fibroblasts onto three-dimensional scaffolds comprised of HYAFF (Hyalograft 3D® Fidia Advanced Biomaterial, Abano Terme PD), and then subsequently using this as a living support for the culture of keratinocyte laminae. Studies have shown that keratinocytes have differentiated underneath a spontaneous "physiological" re-organisation of the dermal-epidermal junction. On the non-woven matrix, a multi-stratified epithelium can be observed lying on a "dermal like" three-dimensional structure formed by fibroblast cells that actively secrete ECM molecules such as different collagen types (I, II, IV), laminin and fibronectin (Figure 3 and 4). These living dermal equivalents can be used in clinical practice for the treatment of different skin defects such as full thickness surgical wounds or chronic ulcers [30].

4. Conclusions

In conclusion, the current state of the art for the creation of a complete cutaneous substitute is far from being wholly realised. The many studies conducted thus far reveal that tissue engineering research is continually breaking new ground. Burn patients were the first targets for such tissue substitutes, after which chronic diseases such as venous ulcers have followed. The more experience gained by the surgeon, the more feedback is available to the basic scientist for the improvement of the product and the broadening of its clinical indications. One example of this is the revelation that it is necessary with certain wounds to apply first a dermal substitute and then cover this with an epidermal substitute. In other cases, a single, one step procedure such as grafting a dermal-epidermal equivalent may be more indicated.

Presently, progress in cell culture and biomedical material technologies have added two important tools to the surgeon's armamentarium: the epidermis and dermis. These can be reconstituted in the laboratory from small biopsies of the same recipient. Other skin-related tissues will follow in the next few years, culminating with the ultimate goal of the creation of a fully transplantable replica of the skin with adnexa and vasculature.

References

[1] Robson, M.C.; Barnett, R.A.; Leitch, I.O. and Hayward, P.G. (1992) Prevention and treatment of postburn scars and contracture. World J. Surg. 16: 87–96.
[2] McHugh, A.A.; Fowlkes, B.J.; Maevsky, E.I.; Smith, D.J. Jr.; Rodriguez, J.L. and Garner, W.L. (1997) Biomechanical alterations in normal skin and hypertrophic scar after thermal injury. J. Burn Care Rehabil. 18: 104–108.
[3] Mast, B.A and Newton, E.D. (1996) Aggressive use of free flaps in children for burn scar contractures and other soft-tissue deficits. Ann. Plast. Surg. 36: 569–575.
[4] Isenberg, J.S. and Price, G. (1996) Longitudinal trapezius fasciocutaneous flap for the treatment of mentosternal burn scar contractures. Burns 22: 76–79.
[5] Asmursen, P.D. (1986) Einfurung and Grundlage: Rohstoffe, Die Hant, Klebetchnologie, Hamburg Ag. Compendium Medical, Berlin: Hamburg, Biersdorf, vol. 1.
[6] Mast, B.A. (1992) The skin. In: Cohen, I.K.; Diegelmann, R.F. and Lindbald, W.J. (Eds.) Wound healing: biochemical and clinical aspects, W.B. Saunders, Philadelphia, USA; pp. 344–355.
[7] Clark, R.A.F. and Henson, P.M. (1988) The molecular and cellular biology of wound repair. Plenum Press, New York, USA.
[8] Jennings, R.W. and Hung, T.K. (1992) In: Adzick, N.S. and Longacker, M.T. (Eds.) Fetal wound healing. Elsevier, New York, USA; pp. 25.
[9] Calvin, M. (1998) Cutaneous wound repair Wounds. A Compendium of Clinical Research and Practice 12, pp. 32.
[10] Gallico, G.G. 3rd.; O'Connor, N.E.; Compton, C.C.; Kehinde, O. and Green, H. (1984) Permanent coverage of large burn wounds with autologous cultured human epithelium. N. Engl. J. Med. 311(7): 448-51.
[11] Limova, M. and Mauro, T. (1995) Treatment of leg ulcers with cultured epithelial autografts: clinical study and case reports. Ostomy Wound Manage 41(8): 48-50, 52, 54-60.
[12] Pruitt, B.A. Jr. and Levine, S. (1984) Characteristics and uses of biologic dressings and skin substitutes. Arch. Surg. 199: 312–322.
[13] Hansbrough, J.F. (1992) Wound coverage with biologic dressings and cultured skin substitutes. Landes, Austin, USA.
[14] Odessey, R. (1992) Addendum: Multicenter experience with cultured epithelial autografts for treatment of burns. J. Burn Care Rehabil. 13: 174–180.
[15] Pittelkow, M.R. and Scott, R.E. (1986) New techniques for the in vitro culture of human skin keratinocytes and perspectives on their use for grafting of patients with extensive burns. Mayo Clin. Proc. 61: 771–777.
[16] Boyce, S.T.; Greenhalgh, D.G.; Kagan, R.J.; Housinger, T.; Sorrell, J.M.; Childress, C.P.; Rieman, M. and Warden, G.D. (1993) Skin anatomy and antigen expression after burn wound closure with composite grafts of cultured skin cells and biopolymers. Plast. Reconstr. Surg. 91: 632–641.

[17] Desai, M.H.; Mlakar, J.M.; McCauley, R.L.; Abdullah, K.M.; Rutan, R.L.; Waymack, J.P.; Robson, M.C. and Herndon, D.N. (1991) Lack of long term durability of cultured keratinocyte burn wound coverage: a case report. J. Burn Care Rehabil. 12: 540–545.
[18] Williamson, J.S.; Snelling, C.F.; Clugston, P.; Macdonald, I.B. and Germann, E. (1995) Cultured epithelial autograft: Five years of clinical experience with twenty-eight patients. J. Trauma 39: 309–319.
[19] Rue, L.W.; Cioffi, W.G.; McManus, W.F. and Pruitt, B.A. Jr. (1993) Wound closure and outcome in extensively burned patients treated with cultured autologous keratinocytes. J. Trauma 34: 662–667.
[20] Boyce, S.T. and Warden G.D. (2002) Principles and practices for treatment of cutaneous wounds with cultured skin substitutes. Am. J. Surg. 183: 445-456.
[21] Coulomb, B. and Dubertret, L. (2002) Skin cell culture and wound healing. Wound Rep. Regen. 10: 109.
[22] Rennekampff, H.O.; Kiessig, V. and Hansbrough, J.F. (1996) Current concepts in the development of cultured skin replacements. J. Surg. Res. 62(2): 288-295.
[23] Rennekampff, H.O.; Kiessig, V.; Griffey, S.; Greenleaf, G. and Hansbrough, J.F. (1997) Acellular human dermis promotes cultured keratinocyte engraftment. J. Burn Care Rehabil. 18(6): 535-544.
[24] Oksala, O.; Salo, T.; Tammi, R.; Hakkinen, L.; Jalkanen, M.; Inki, P. and Larjava, H. (1995) Expression of proteoglycans and hyaluronan during wound healing. J. Histochem. Cytochem. 43(2): 125-135.
[25] Zacchi, V.; Soranzo, C.; Cortivo, R.; Radice, M.; Brun, P. and Abatangelo, G. (1998) *In vitro* engineering of human skin-like tissue. J. Biomed. Mater. Res. 40(2): 187-194.
[26] Fusenig, N.E.; Limat, A.; Stark, H.J. and Breitkreutz, D. (1994) Modulation of the differentiated phenotype of keratinocytes of the hair follicle and from epidermis. J. Dermatol. Sci. Jul. 7 suppl. S142-151.
[27] Lamme, E.N.; de Vries, H.J.; van Veen, H.; Gabbiani, G.; Westerhof, W. and Middelkoop, E. (1996) Extracellular matrix characterization during healing of full-thickness wounds treated with a collagen/elastin dermal substitute shows improved skin regeneration in pigs. J. Histochem. Cytochem. 44(11): 1311-1322.
[28] Falanga, V. (1998) Apligraf treatment of venous ulcers and other chronic wounds. J. Dermatol. 25(12): 812-817.
[29] Falanga, V.; Margolis, D.; Alvarez, O.; Auletta, M.; Maggiacomo, F.; Altman, M.; Jensen, J.; Sabolinski, M. and Hardin-Young, J. (1998) Rapid healing of venous ulcers and lack of clinical rejection with an allogeneic cultured human skin equivalent. Human Skin Equivalent Investigators Group. Arch. Dermatol. 134: 293–300.
[30] Galassi, G.; Brun, P.; Radice, M.; Cortivo, R.; Zanon, G.F.; Genovese, P. and Abatangelo, G. (2000) *In vitro* reconstructed dermis implanted in human wounds: degradation studies of the HA-based supporting scaffold. Biomaterials 21(21): 2183-91.
[31] Cuono, C.B.; Langdon, R.; Birchall, N.; Barttelbort, S. and McGuire, J. (1987) Composite autologous-allogeneic skin replacement: development and clinical application. Plast. Reconstr. Surgery 80: 626-35.
[32] Compton, C.C.; Gill, J.F.; Bradford, D.A.; Regauer, S.; Gallico, G.G. and O'Connor, N.E. (1989) Skin regenerated from cultured epithelial autografts on full-thickness burn wounds from 6 days to 5 years after grafting. Lab. Invest. 60: 600-612.
[33] Neveux, Y.; Rives, J.M.; Lebreton, C.; Gentilhomme, E.; Saint-Blancar, P. and Carsin, M. (1995) Clinical interest of cutaneous models reproduced in vitro for severeburn treatment: histopathologic and ultrastructural study. Cell Biol. Toxicol. 11: 173-178.
[34] Sher, S.E.; Hull, B.E.; Rosen, S.; Church, D.; Friedman, L. and Bell, E. (1983) Acceptance of allogeneic fibroblasts in skin equivalent transplants. Transplant. 36: 5052-5057.
[36] Sorrell, J.M.; Carrino, D.A.; Baber, M.A.; Asselinean, D. and Caplan, A.I. (1999) A monoclonal antibody recognizes a glycosaminoglycan epitope in both dermatan sulfate and chondroitin sulfate proteoglycans of human skin. Histochem. J. 31: 549-558.
[37] Prockop, D. (1997) Marrow stromal cells as stem cells for nonhematopoietic tissues. Science 276: 71-74.
[38] Bucala, R.; Spiegel, L.; Chesney, J.; Hogan, M. and Cerami, A. (1994) Circulating fibrocytes define a new leukocyte subpopulation that mediates tissue repair. Mol. Med. 1: 71-81.
[39] Leigh, I.M.; Purkis, P.E.; Navsaria, H.A.; Phillips, T.J. (1987) Treatment of chronic venous ulcers with sheets of cultured allogenic keratinocytes. Br. J. Dermatol. 117: 591-597.
[40] De Luca, M.; D'Anna, F.; Bondanza, S.; Franzi, A.T. and Cancedda, R. (1992) Treatment of leg ulcers with cryopreserved allogeneic cultured epithelium – a multicenter study. Arch. Dermatol. 28: 633–638.
[41] Mol, M.A.; Nanninga, P.B.; van Eendenburg, J.P.; Westerhof, W.; Mekkes, J.R. and van Ginkel, C.J. (1991) Grafting of venous leg ulcers. An intraindividual comparison between cultured skin equivalents and full-thickness skin punch grafts. J. Am. Acad. Dermatol. 24: 77–82.

[42] Teepe, R.G.; Koebrugge, E.J.; Ponec, M. and Vermeer, B.J. (1990) Fresh *versus* cryopreserved cultured allografts for the treatment of chronic skin ulcers. Br. J. Dermatol. 122: 81-89.
[43] Teepe, R.G.; Roseeuw, D.I.; Hermans, J.; Koebrugge, E.J.; Altena, T.; de Coninck, A.; Ponec, M. and Vermeer, B.J. (1993) Randomized trial comparing cryopreserved cultured epidermal allografts with hydrocolloid dressings in healing chronic venous ulcers. J. Am. Acad. Dermatol. 29: 982-988.
[44] Phillips, T.J.; Kehinde, O.; Green, H. and Gilchrest, B.A. (1989) Treatment of skin ulcers with cultured epidermal allografts. J. Am. Acad. Dermatol. 21: 191-199.
[45] Gilchrest, B.A. (1983) *In vitro* assessment of keratinocyte aging. J. Invest. Dermatol. 81: 184s-189s.
[46] Trent, J.F. and Kirsner, R.S. (1998) Tissue engineered skin: Apligraf, a bi-layered living skin equivalent. Int. J. Clin. Pract. 52: 408-413.
[47] Parenteau, N.L. (1998) Skin equivalents, In: Leigh, I.M. and Watt, F.M. (Eds.) Keratinocytes methods. Cambridge University Press, London; pp. 44-55.
[48] Eaglstein, W.H. and Falanga, V. (1998) Tissue engineering and the development of Apligraft a human skin equivalent. Adv. Wound Care 11: 1-8.
[50] Eaglstein, W.H.; Iriondo, M. and Laszlo, K. (1995) A composite skin substitute (graftskin) for surgical wounds. A clinical experience. Dermatol. Surg. 10: 839-843.
[51] Yannas, I.V.; Burke, J.F.; Gordon, P.L.; Huang, C. and Rubenstein, R.H. (1980) Design of an artificial skin. II. Control of chemical composition. J. Biomed. Mater. Res. 14: 107-132.
[52] Burke, J.F. (1987) Observations on the development and clinical use of artificial skin-an attempt to employ regeneration rather than scar formation in wound healing. Jpn. J. Surg. 17: 431-438.
[53] Bruke, J.F. (1984) The effects of the configuration of an artificial extracellular matrix on the development of a functional dermis. In: Trelstad, R. (Ed.) The role of extracellular matrix in development, Liss, New York; pp. 351-355.
[54] Purdue, G.F.; Hunt, J.L.; Still, J.M. Jr.; Law, E.J.; Herndon, D.N.; Goldfarb, I.W.; Schiller, W.R.; Hansbrough, J.F.; Hickerson, W.L.; Himel, H.N.; Kealey, G.P.; Twomey, J.; Missavage, A.E.; Solem, L.D.; Davis, M.; Totoritis, M. and Gentzkow, G.D. (1997) A multicenter clinical trial of a biosynthetic skin replacement, Dermagraft-TC, compared with cryopreserved human cadaver skin for temporary coverage of excised burn wounds. J. Burn Care Rehabil. 18: 52-57.
[55] Hansbrough, J.F.; Mozingo, D.W.; Kealey, G.P.; Davis, M.; Gidner, A. and Gentzkow, G.D. (1997) Clinical trials of a biosynthetic temporary skin replacement, Dermagraft Transitional Covering, compared with cryopreservedhuman cadaver skin for temporary coverage of excised burn wounds. J. Burn Care Rehabil. 18: 43-51.
[56] Purdue, G.F. (1996) Dermagraft-TC pivotal safety and efficacy study. J. Burn Care Rehabil. 18: 513-514.
[57] Haeri, A.; Clay, J. and Finely, J.M. (1999) Compend Contin Educ The use of an acellular dermal skin graft to gain keratinized tissue. Dent. 1999 20(3): 233-234, 239-242 (quiz 244).
[58] Schwartz, S. (1997) In: A new composite cultured skin product for treatment of burns and other deep dermal injuries. Presented at the Bioengineering of Skin Substitutes Conference. Boston, Mass. (USA); pp. 19.
[59] Beumer, G.J.; van Blitterswijk, C.A.; Bakker, D. and Ponec, M. (1993) A new biodegradable matrix as part of a cell seeded skin substitute for the treatment of deep skin defects: a physico-chemical characterisation. Clin. Mater. 14(1): 21-27.
[60] Noordenbos, J.; Dore, C. and Hansbrough, J.F. (1999) Safety and efficacy of TransCyte for the treatment of partial-thickness burns. J. Burn Care Rehabil. 20(4): 275-281.
[61] Chen, W.Y.J. and Abatangelo, G. (1999) The functions of Hyaluronan in wound repair. Wound Rep. Reg. 7: 79-89.

BIOARTIFICIAL LIVER

CLARE SELDEN
*Centre for Hepatology, Royal Free and University College Medical School Royal Free Campus, Rowland Hill Street, Hampstead, London, NW3 2PF, UK – Fax: +44 (0) 207 433 2852 –
Email: c.selden@rfc.ucl.ac.uk*

1. Introduction

Acute liver failure in man is catastrophic, usually striking out of the blue, often in the previously fit and young, and is associated with a high mortality. Currently, the only available treatment is a whole organ transplant, which, although reasonably successful, is associated with life-long immunosuppression, and is often unavailable for those patients who need it most. Between 20-30% of transplant patients die whilst on the waiting list to receive an organ, in the West alone. Moreover, acute liver failure arising from hepatitis affects millions of people worldwide, but especially in the less developed countries.

Against this background, there is, however, hope for a better treatment if one can exploit the liver's ability to repair and regenerate after damage. Experimentally, there is complete restoration of liver mass in rodents after a 70% removal of the liver, within seven to ten days. Fortunately, this is also the case in man, when a hepatic resection is carried out for removal of a tumour, or other surgical trauma. Indeed, in those patients who have not survived acute hepatic failure, post-mortem analysis shows the liver has been attempting to regrow, with several regenerative nodules, but such regeneration that takes place has not provided sufficient functional liver mass rapidly enough for patient survival. It is this remarkable feature of the liver that a bioartificial liver machine aims to exploit. The rationale is to provide sufficient liver function to "buy time" ideally for the liver to repair and regenerate, but in the first instance to tide the patient over until a suitable organ becomes available for transplant and the patient is well enough for the operation.

2. What is the best cell for a bioartificial liver, and how many are required?

The complexity of the functions of the liver, which encompass synthetic capacity, metabolic activity and detoxification potential, make it a considerable challenge to mimic in a machine. It is generally agreed that a "biological" component is required, which begs the question of which cells are needed; this latter question is not within the remit of this review, but both animal and human cells have been utilised in a bioartificial liver, and will be alluded to in this chapter. Most systems have utilised either primary hepatocytes or hepatocyte cell lines. The liver is composed of both parenchymal and non-parenchymal cells; the former provide most of the metabolic and synthetic functions, but it is clear that the non-parenchymal cell population influences cell function *in vivo*. So far co-cultures have not been utilised in bioartificial liver systems currently being tested clinically. Since conventional monolayer cell culture has failed to provide cell performance which mimics *in vivo* liver function, redesign of culture conditions has been necessary. Experimental studies have repeatedly shown that hepatocytes in culture have higher expression of liver specific function when grown as 3-dimensional cuboidal cultures [1-3].

The question of number of cells required to provide sufficient liver function is controversial. A 70 kg man has a liver of 1-1.5 kg, containing in the region of $2 \cdot 10^{11}$ liver cells. Whilst it is known that man can survive with only 30% liver mass, this assumes that the 30% is fully functional and healthy. In acute liver failure that is not the case, and moreover the toxins found in liver failure plasma will have a detrimental effect not only on the patients' liver but also on the cells in the bioreactor. Cell numbers of 10^9 -10^{10} have been utilised in clinical trials. Potentially as many as $2 \cdot 10^{11}$ may be required. These impacts on the choice of cell immobilisation and the design of the bioreactor to hold the cultured cells.

3. Bioreactor design – initial clinical experience

Several different cell immobilisation techniques have been used in the current bioartificial devices. Mammalian epithelial cells are fastidious cells, usually requiring some form of adherent substrate and a high nutrient and oxygen supply [4]. The definition of success lies in provision of comparable cell performance *in vitro*, as is found *in vivo* [5].

In man factors in the circulation provide much of the nutrients required by the liver *via* the portal vein. The blood flow through the liver *in vivo* is approximately 1.5 l/min, of which 1-1.2 l/min is *via* the portal vein carrying nutrients and oxygen. Experimental studies have shown that *in vivo* function of the liver is maintained by this high blood flow, and that *in vitro*, cells in a bioreactor perform better when perfused at high flow rates. Design of the bioreactor to work efficiently at high flow rates adds a level of complexity and cell immobilisation techniques need to be developed with that long term view in mind. Recently there have been reports of nanotechnology approaches being used to design cell immobilisation techniques suitable for a hepatic bioreactor [6], which may well be effective on a small scale, but the "scaling up" problems remain considerable.

The nature of disease progression in acute liver failure also dictates the requirements of a bioartificial liver machine; a patient deteriorates rapidly and needs "extracorporeal" liver support within 24-48 hours of admission to hospital. This requires a machine to be available "off the shelf" and to be of sufficient mass to completely support the patient from the start of therapy. For this the biological component of the liver must be either cryopreservable and fully utilisable within 24 hours, or constantly available. Some systems are being designed with cryopreservability in mind [7-12]; others, for example from the Gerlach laboratory [13] have established maintenance of function for about two weeks and intend to have a stand-by circuit continuously available. Pragmatically the ability to cryopreserve and reconstitute with 24 hours would seem a more suitable solution given the geographical distribution of fulminant hepatic failure.

Cell immobilisation, as utilised in the bioartificial liver field, encompasses cell encapsulation and artificial substrates encouraging a three-dimensional phenotype. Most systems currently in clinical trial use some form of hollow fibre cartridges, however, these are not ideal, and experimentally, other systems using radial flow, flat bed, and fluidized-bed reactors to house cells as spheroids or encapsulated are being explored.

The first capillary membrane liver cell bioreactor was designed and built by Knazek et al. in 1972 [14], using the semi-permeable membranes for both cell adhesion and metabolite exchange [14,15]. Thereafter development of bioreactors for liver cells has included reactors for suspension culture without cell immobilisation [16,17], some form of cell immobilisation, and reactors which provide both cell immobilisation and mass transfer functions [18]. Previous reactor technology had focused on non-anchorage dependent cells; hepatocytes are very anchorage dependent, at least in monolayer culture, and initial results of those reactors used for liver cells grown in suspension culture illustrated the rapid loss of viability within a few hours of cell isolation [19].

Two of the four bioartificial machines which have been tested clinically, and are currently in clinical trial, rely on the use of hollow fibre technology. This technology has some inherent disadvantages for liver cells since it relies on a semi-permeable membrane with fixed molecular weight cut-offs, however, much of the clinical applications of kidney dialysis cartridges could be used, thereby by-passing much of the required characterisation legislation for use in therapeutic devices. In principle, the cells providing the liver specific function are cultured in the extracapillary space, whilst blood or plasma is perfused through the lumen of the hollow fibres. There is fairly rapid membrane fouling of the pores due to the high protein content of blood and/or plasma. Moreover the nutrients, metabolites and toxins pass through the membranes by simple diffusion and mass transfer is compromised rapidly for cells furthest from the fibres. A high flow rate is used, partly to overcome some of the mass transfer problems, but the disadvantage is that there is probably increased membrane fouling. Reid et al., working on a rat hepatocyte bioreactor on a small scale have attempted to overcome the problem of membrane fouling using large pore hydrophobic fibres and cross-flow to reduce resistance [20], but scale up to human size remains a problem. If low molecular weight cut-offs are used the small toxins and nutrients can cross the membrane but several liver toxins are albumin bound hence necessitating cut-offs of $\geq$ 70kDa. With that cut-off there is a risk that immunoglobulins can also cross the barrier. This is clearly a problem when using xenogeneic hepatocytes e.g. porcine or canine. Some systems use very large pores (> 1 million Da); these are less sensitive to technical failure, but have the added risk that small fragments of cells and cell debris, including DNA could cross the membrane into the patients' circulation.

Nonetheless there is some positive data arising from the use of hollow fibre cartridges using both porcine primary hepatocytes and human liver tumour cell lines. 1987 saw the initial use of a bioartificial liver clinically, in a single patient with acute liver failure. A kidney dialysis cartridge was filled with previously cryopreserved rabbit hepatocytes. The patient's symptoms improved, with reduced serum bilirubin levels and improved neurological function. The patient was discharged from hospital providing the first evidence that there was some value in developing this approach [21].

The next development of the technology for clinically applicable BALs came simultaneously from two groups; Sussman *et al.* utilised the C3A human hepatoblastoma cell line [22] and Demetriou utilised porcine primary hepatocytes [23]. Both groups utilised hollow fibre cartridges, but Demetriou also used microcarriers as a cell immobilisation technique seeding those into the extracellular space, whilst Sussman simply inoculated cells into the extracellular space. Using these devices their safety during extracorporeal liver support has been established, but their efficacy so far has not. The first clinical trial of 24 patients (12 treated with BAL and 12 controls) showed feasibility and safety but no overall clinical benefit above the controls [24]. A larger trial of more than 170 patients using the Demetriou porcine system, who were treated at several centres in both USA and Europe, has recently reported preliminary results which appear encouraging, but a full analysis is still awaited [25]. There are sporadic reports of clinical use of hollow fibre bioreactors for individual patients; one such is a recent description of a patient with fulminant hepatic failure treated with 70-100 g porcine hepatocytes in the Excorp Medical Bioartficial Liver Support System (BLSS) [26]. Although slight changes in biochemistries were noted post BLSS the patient exhibited transient hypotension and thrombocytopaenia and and a lowered glucose level at the start of the procedure. Reports such as these emphasise the importance of multi-centre controlled clinical trials with sufficient patients in each arm of the trial to make a true comparison.

The architecture of the liver (Figure 1) is such that hepatocytes lie as single cell plates bounded on each side by a sinusoidal space through which blood flows, carrying nutrients and oxygen in *via* the portal vein, and toxins out either *via* the bile duct system or the hepatic artery and vein. This achieves low metabolite gradients *in vivo*. As well as the high blood flow alluded to above there is a high oxygen tension in the liver, required since hepatocytes have high oxygen consumption rates. The bioreactor designed by Gerlach and colleagues has most nearly tried to reproduce the liver architecture artificially utilising separate ports for metabolite exchange and oxygenation, and also using different membrane chemistries optimal for the different functions *e.g.* cell attachment, nutrient supply – toxin removal, oxygenation and carbon dioxide removal. The resultant bioreactor is composed of many discrete bundles of fibres each supporting only a few hepatocytes. Initially used with porcine hepatocytes alone this system is now used with a mix of parenchymal and non-parenchymal liver cells, and cells spontaneously aggregated to form 3-dimensional structures with some appropriate ultrastructure. There have been many experimental studies carried out with this system [27-30]. Some patients have also been treated with this machine, but no controlled clinical trial has been completed so far. Anecdotal reports suggest some initial successes. Recent modifications have included the use of human hepatocytes taken from discarded donor livers although these are rarely available; this development has come about presumably to overcome the current ethical questions arising form the use of animal cells in human therapies such as zoonoses.

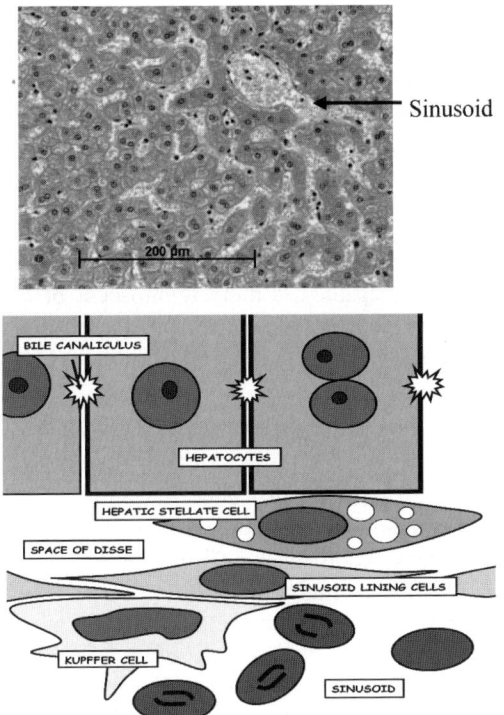

Figure 1. (Top) Histological section of normal liver stained with Haematoxylin and Eosin. Note single cell thick plates. (Bottom) schematic showing arrangement of hepatocytes as single cell thick plates and non-parenchymal cells – after Standish and Scheuer.

The fourth bioreactor currently in clinical trial comes from Flendrig et al. [31] in Holland. Their design utilises non-woven polyester membranes rolled to allow cell growth between the interstices of the sheets. A carefully controlled study in pigs induced to ischaemic liver failure provided evidence that blood ammonia and bilirubin levels fell, and survival time improved in animals treated with the Dutch design of bioreactor. There is no control trial data in patients available to date.

4. Bioreactor designs – experimental models

Although not tested in patients there is considerable investigation at the experimental level of new designs of bioreactors. Bader has instigated two types of reactor based on different principles. The first utilises a clinically available hollow fibre oxygenator used in open heart surgery, with homogeneous interfibre distances of 200 µm. This small distance was helpful in allowing cells to aggregate into relatively small spheroids and achieved a cell density of 2.5×10^7 /ml. Biochemical comparisons of this system filled with porcine hepatocytes compared favourably with collagen-gel cultures of the same cells over a period of 22 days in culture [32]. In a second design porcine hepatocytes and non-parenchymal cells were grown between flat sheets on an oxygen permeable surface. Similar biochemical analyses were performed over 18 days, and microscopy demonstrated cellular organisation reminiscent of hepatocyte plates *in vivo* [33].

To improve contact between the perfusing blood or plasma and the liver cells in the bioreactor a radial flow reactor has been developed, in which the perfusing fluid crosses the hepatocyte filled space from centre to periphery. Measurement of biochemical parameters of function *e.g.* detoxification as demonstrated by MEGX production from Lidocaine, oxygen consumption and carbon dioxide release, showed improvements in both performance and longevity of function in this bioreactor design compared with conventional hollow fibre cartridges [34]. Moreover structural analyses suggested a lobule-like arrangement of hepatocytes more reminiscent of the liver *in vivo* [35]. In addition, a higher mass of cells could be maintained utilising the radial flow reactor improving the problems associated with scale up of reactors for therapeutic use.

5. Cell encapsulation

Encapsulation techniques tried with liver cells can be divided into two main classes: incorporation into a hydrogel and encapsulation within a semi-permeable membrane. Substrates used for encapsulation must be resistant to degradation and mechanical stress, both of which are likely at high flow rates.

If animal cells are to be used for therapeutic purposes in man, there must be some immunoisolation, preventing the free passage of immunoglobulins and complement proteins. This presents an immediate conflict as the liver produces and secretes many useful proteins which are of high molecular weight and would also be excluded. One example is the clotting factor fibrinogen which is 340 kDa. Any membrane excluding immunoglobulins at 160 kDa would also exclude fibrinogen. Since it has not been clearly established exactly what functions are required to counteract hepatic failure, the current goal is to mimic as many of the liver specific functions as possible.

Mass transfer capabilities also influence cell immobilisation design, both from a metabolic perspective, providing good nutrient access and toxin removal [36-38], but also in the large mass considerations. Since the patient is linked up to an "extracorporeal" device *via* venous access the overall volume must be kept to a minimum.

The choice of encapsulating media is between naturally occurring and synthetic materials. Bio-compatibility is an important consideration when a patient's blood or plasma will be in direct contact with the encapsulating material.

Encapsulation regimes have utilised relatively inert substances such as alginate, agarose and cellulose on the one hand and extremely bioactive substances such as collagen gels and matrigel derived from the Englebreth Holm Swarm sarcoma on the other. The latter provide excellent support for liver cells in culture, supplying both integrin binding capacity and growth factors [39]. The disadvantages however, for therapeutic use in man are several. Firstly, the animal nature of these substrates and their bio-interactions are notable; for example collagen is known to cause clotting when exposed to blood or plasma, and the tumour origin of matrigel are likely to preclude its use in man. Attempts at recreating exactly the components of matrigel recombinantly have not proven successful. Secondly, they offer no immunoisolation, an absolute requirement if xenogeneic cells are to be utilised. Thirdly, they offer limited mechanical resistance and may be subject to sheer stress in a high flow system.

A recent report has however, attempted to modify the collagen gel system to improve this last parameter, by combining the biological advantages of collagen with the more bio-inert capabilities of sol-gel SiO_2 coated as a superficial microlayer over the collagen gel. Metabolic capabilities of SiO_2 coated collagen gels were comparable with collagen gel alone but mechanical stability was increased more than 10-fold [40].

Some attempts to recreate, recombinantly, specific properties of the binding capacities of these bioactive gels have been more successful. Kobayashi, using an immortalised human cell line NKNT-3, also developed cellulose microspheres containing the RGD peptide. Cells were immobilised within 24 hours and exhibited differentiated function as exemplified by ammonia clearance and at the ultrastructural level showed glycogen granules, mitochondria and extensive endoplasmic reticulum [41].

6. Alginate hydrogels

Our own work has focused on alginate as an encapsulation medium for human hepatocyte cell lines [2,5,7,42-44]. Unlike most other reports of alginate encapsulated cells, we have not used a poly-L-lysine coating, nor have we solubilised the alginate matrix. Our system therefore is an hydrogel with a final alginate concentration of 1%. Having compared low, medium and high viscosity, and high manuronic (M) *versus* high guluronic (G) acid alginates we have shown hepatocyte specific function to be best supported by high M medium viscosity alginate. Cell performance for a range of liver specific activities is markedly increased when cells are cultured in alginate compared with conventional monolayer culture. Figure 2 shows the synthesis and secretion of five liver specific proteins covering a variety of cell functions; albumin acting as a carrier protein and maintaining the oncotic pressure, alpha-1-antitrypsin and alpha-1-acid glycoprotein, being two acute phase reactants, and prothrombin and fibrinogen being clotting factors. Each specific activity is significantly upregulated compared with monolayer culture. Similar increases were noted with cytochrome P450 enzyme activities (Figure 3) and androstene-dione metabolism (Figure 4), reflecting a general improvement of liver specific function. Of particular note was the provision of urea synthesis in alginate encapsulated cells (7 µmol/million nuclei/48h) at normal *in vivo* levels (6 µmol/million nuclei/48h), compared with entirely undetectable levels in monolayer culture. Cells seeded in alginate are initially single cells which proliferate *in situ* to form multicellular spheroids. Transmission electron microscopy of such spheroids indicates remarkable cell-to-cell contact, gap junctions and junctional complexes and multiple microvilli (Figure 5). Moreover there is evidence of highly transcriptionally active cell cytoplasm with considerable endoplasmic reticulum. We have shown increased levels of extracellular matrix production by spheroidal cultures which may contribute to the overall improved function.

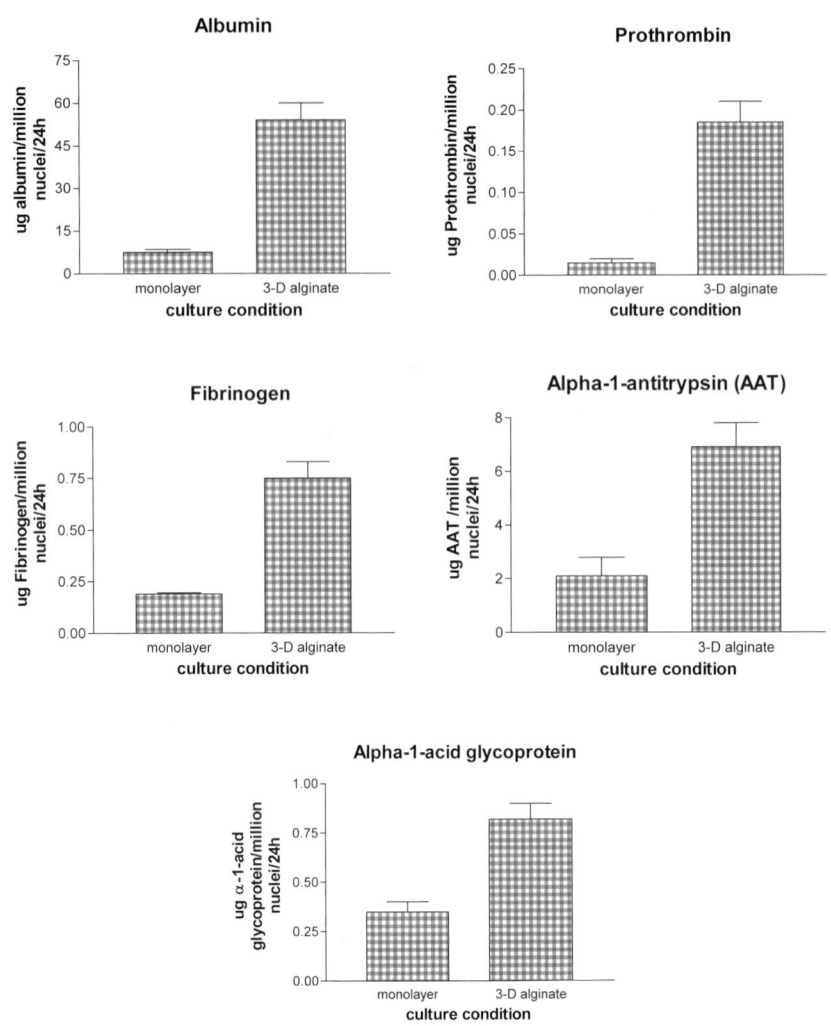

Figure 2. Protein secretion by monolayer and alginate encapsulated human hepatocyte cell lines (HepG2). Results are expressed as mean ± SD(n-1), n=12 for monolayer, n= 24 for alginate.

This increased cell performance is limited to about 10 days in spheroidal culture. Thereafter, in spite of continued viability and proliferation there is a decrease in per cell specific activities of all the parameters we routinely measure. For scale up to a human therapeutic system we need to understand the mechanisms involved in both the initial upregulation and subsequent down regulation of function. Microarray analyses are shedding some light on mechanisms involved and suggest involvement of stress-related proteins and mitochondrial function changes.

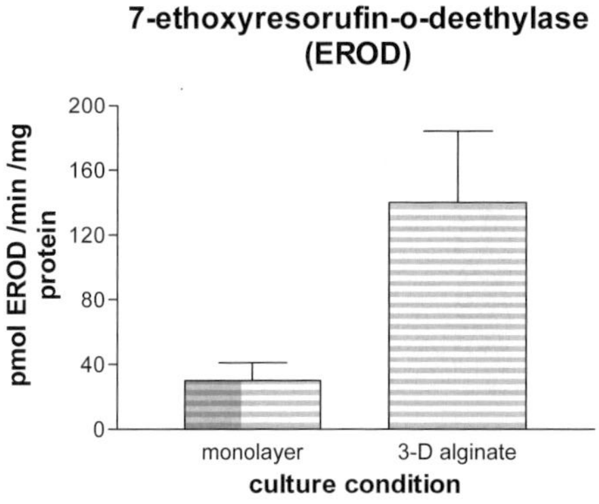

7-ethoxyresorufin-o-deethylase (EROD)

Figure 3. Cytochrome P450 1A1 function, as exemplified by ethoxyresorufin deethylase activity in alginate encapsulated Hep G2 cultures compared with monolayer culture. (mean ± SD(n-1), n=16).

Using encapsulated cells the design of a bioreactor is still an important aspect when developing a bioartificial liver. Two groups have used porcine hepatocytes in calcium-alginate beads in a fluidised bed reactor to treat experimental liver failure induced by devascularisation in the pig [45,46]. Using 3-8 10^9 hepatocytes some improvement in ICP, ammonia removal and urea production was achieved; compared with monolayer cultures as controls metabolic activities were considerably improved but it is well known that metabolic activity in conventional monolayer culture are only a few per cent of those achieved *in vivo*, so that a better comparison would be with *in vivo* performance levels. These particular designs allowed only a rather low flow-rate of 20-90 ml/min compared with most hollow fibre based designs of several hundred millilitres per minute. I have already alluded to the importance of perfusion flow rate above. Nonetheless the data support the idea that encapsulated cells can be utilised in a bioartificial liver, and there is the advantage that both manipulation and cryopreservation of the system is relatively simple.

Achieving sufficient cells per unit volume remains a considerable challenge for a bioartificial liver suitable for therapeutic intervention, since there is a major volume limitation set by the patient's blood volume. There are two aspects influencing the final cell density in a BAL. The first is the efficiency of cell seeding within any scaffold and the second is the effect of cell performance at high cell density. Using fixed reactor material be it polymer based, non-woven, resinous, sponges or films, the efficiency attained by simply inoculating cells into the reactor is rarely more than 30%; this can be improved somewhat by utilising a low speed centrifugation procedure combined with intermittent resuspension to impregnate the porous scaffold with cells [47], however the limitation is still a maximum of less than 50% seeding. Primary cells do not proliferate therefore the starting point has at least a 50% void volume. Systems utilising cell lines may overcome that limitation, but the second aspect then comes into play – that of inhibition of cell function at high cell density [48].

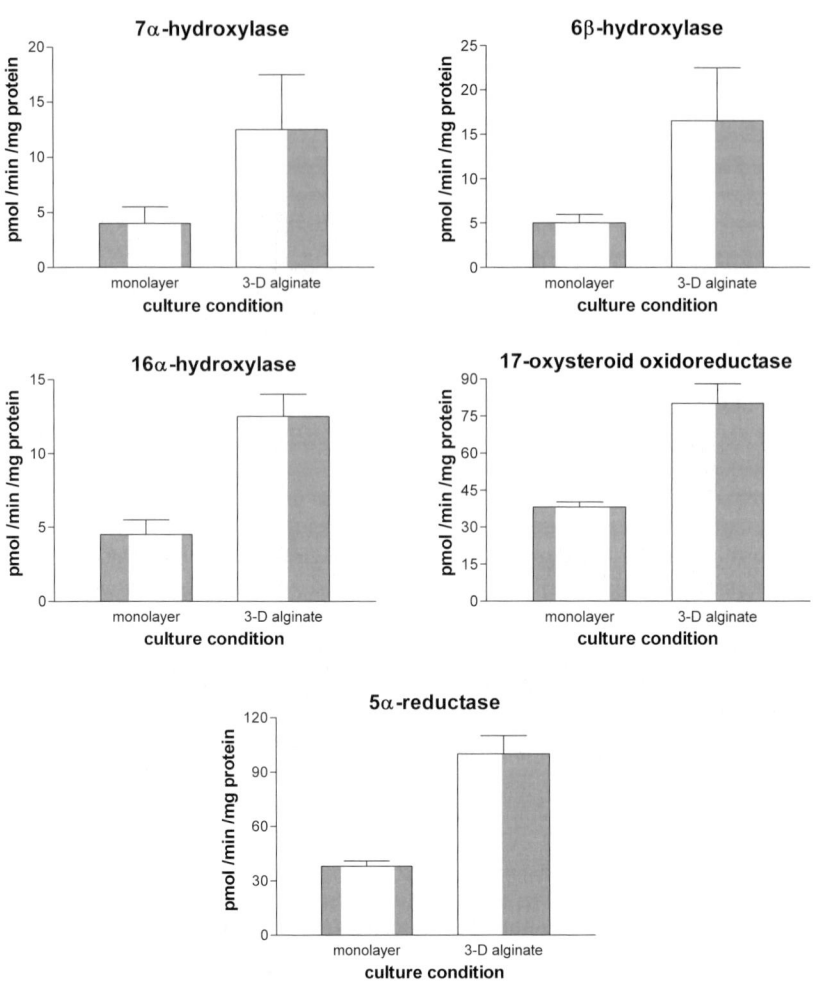

Figure 4. Steroid metabolism as exemplified by androstene dione metabolism in alginate encapsulated cells as compared with monolayer culture (mean ± SD(n-1), n=4).

Most cell scaffolds have been derived from synthetic polymers such as polyether sulphone or cellulose acetate /cellulose nitrates. Some more recent non-woven scaffolds have been fabricated from polymers of hyaluronic esters, which since it is naturally occurring and biodegradable may have some advantages in supporting hepatocyte growth *ex vivo*. Hyaluronic acid is a natural glycosaminoglycan present in connective tissue which influences a myriad of cell behaviours such as cell adhesion, motility and proliferation. It can be made water insoluble by esterification with different alcohols forming benzyl or ethyl esters. The nature and extent of esterification influences cell behaviours with different cell types [49,50]. Rat hepatocytes seeded onto both ethyl and benzyl hyal esters maintained differentiated function better when the ester polymers were presented as non-woven fabrics than as films, presumably due to the better 3-dimensional cellular organisation. The ethyl ester supported hepatocyte function better than the benzyl ester irrespective of the geometric configuration [51].

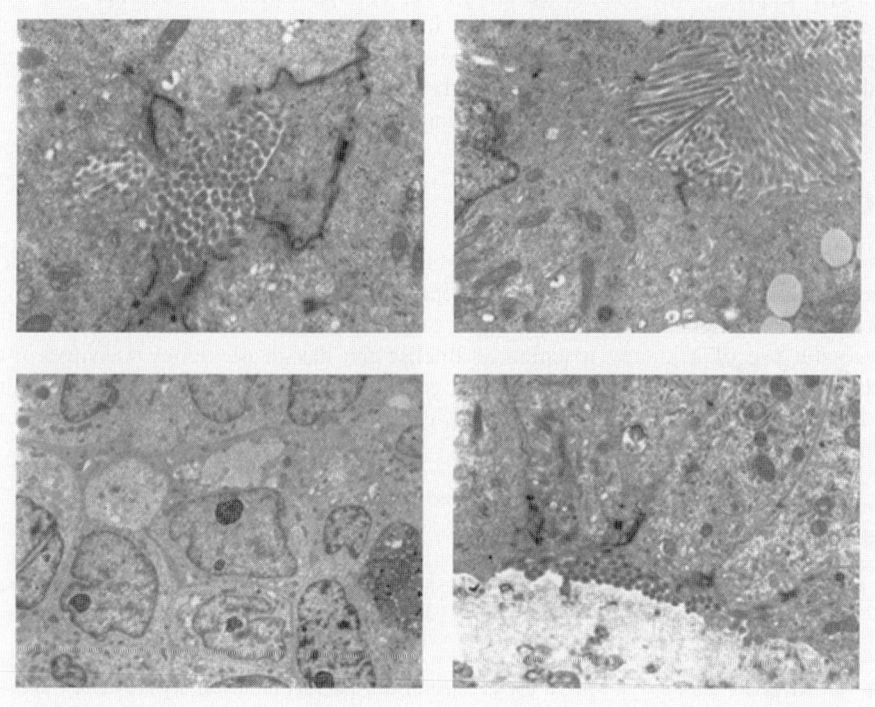

Figure 5. Transmission electron microscopy of alginate encapsulated HepG2 cells. Note transcriptionally active cytoplasm, numerous microvilli, desmosomes and junctional complexes.

Porous microcarriers fabricated from poly D,L-lactide-co-glycotides, when collagen coated, support the growth of human hepatocyte cell lines and rat primary liver cells [52]. These microcarriers are both biocompatible and biodegradable by the cells, a feature which might help to overcome the void volume problem if the spaces could then be occupied by proliferating cells. It is possible to envisage their use with cell lines but as primary liver cells do not multiply the system may not lend itself to a bioreactor for primary cell cultures.

Xenogeneic hepatocytes are most often encapsulated in beads surrounded by a semipermeable membrane. Poly-L-lysine is the polymer tested with hepatocytes to date, using rabbit, goat and porcine hepatocytes. Using polymers with a the molecular weight cut-off of 70 kDa most studies have demonstrated some degree of protection from antibody mediated cell lysis, when exposed to human serum, but metabolic function is limited to detoxification of small molecules *e.g.* ammonia [53,54].

7. Alternative approach

An alternative to both hollow fibre cartridges and encapsulated cells is the use of foams as a cell support, which encourage the growth of hepatocytes as spheroids. One such system used a multicapillary polyurethane foam packed-bed module to treat pigs with experimental liver failure. Some success was reported in reducing ammonia levels and maintaining glucose homeostasis, but no data is provided on the synthetic capacity of the system in, for example, producing clotting factor proteins [55].

8. Conclusion

In summary, the age of the bioartificial liver, whilst already here in respect of clinical trials for machines utilising hollow fibre technology, is still in its infancy with respect to optimal cell immobilisation. However, it took about thirty-five years from the first successful use of a kidney machine to finalise the design of kidney machines used today. It is likely that the more input that goes into optimising the initial cell seeding and culture, the more functionally successful the biological component of the artificial liver will be.

References

[1] Gomez-Lechon, M.J.; Jover, R.; Donato, T.; Ponsoda, X.; Rodriguez, C.; Stenzel, K.G.; Klocke, R.; Paul, D.; Guillen, I.; Bort, R. and Castell, J.V. (1998) Long-term expression of differentiated functions in hepatocytes cultured in three-dimensional collagen matrix. J. Cell Physiol. 177(4): 553-562.

[2] Selden, C.; Shariat, A.; McCloskey, P.; Ryder, T.; Roberts, E. and Hodgson, H. (1999) Three-dimensional *in vitro* cell culture leads to a marked upregulation of cell function in human hepatocyte cell lines--an important tool for the development of a bioartificial liver machine. Ann. N.Y. Acad. Sci. 875: 353-363.

[3] Berthiaume, F.; Moghe, P.V.; Toner, M. and Yarmush, M.L. (1996) Effect of extracellular matrix topology on cell structure, function, and physiological responsiveness: hepatocytes cultured in a sandwich configuration. FASEB J. 10(13): 1471-1484.

[4] Freshney, R.I. (1992) Culture of epithelial cells. New York, Wiley-Liss.

[5] Selden, C.; Roberts, E.; Shariat, A.; Tootle, R. and Hodgson, H.J.F. (1997) A human hepatocyte line, when grown in a 3-dimensional culture configuration, synthesises urea at levels commensurate with in vivo production. Hepatology 26: 196A.

[6] Amirpour, M.L.; Ghosh, P.; Lackowski, W.M.; Crooks, R.M. and Pishko, M.V. (2001) Mammalian cell cultures on micropatterned surfaces of weak-acid, polyelectrolyte hyperbranched thin films on gold. Anal. Chem. 73(7): 1560-1566.

[7] Khalil, M.; Shariat-Panahi, A.; Tootle, R.; Ryder, T.; McCloskey, P.; Roberts, E.; Hodgson, H. and Selden, C. (2001) Human hepatocyte cell lines proliferating as cohesive spheroid colonies in alginate markedly upregulate both synthetic and detoxificatory liver function. J. Hepatol. 34(1): 68-77.

[8] Wu, L.; Sun, J.; Wang, L.; Wang, C.; Woodman, K.; Koutalistras, N.; Horvat, M. and Sheil, A.G. (2000) Cryopreservation of primary porcine hepatocytes for use in bioartificial liver support systems. Transplant. Proc. 32(7): 2271-2272.

[9] Naik, S.; Santangini, H.A.; Trenkler, D.M.; Mullon, C.J.; Solomon, B.A.; Pan, J. and Jauregui, H.O. (1997) Functional recovery of porcine hepatocytes after hypothermic or cryogenic preservation for liver support systems. Cell Transplantation 6(5): 447-454.

[10] Dixit, V.; Darvasi, R.; Arthur, M.; Lewin, K. and Gitnick, G. (1993) Cryopreserved microencapsulated hepatocytes--transplantation studies in Gunn rats. Transplantation 55: 616-622.
[11] Guyomard, C.; Rialland, L.; Fremond, B.; Chesne, C. and Guillouzo, A. (1996) Influence of alginate gel entrapment and cryopreservation on survival and xenobiotic metabolism capacity of rat hepatocytes. Toxicol. Appl. Pharmacol. 141(2): 349-356.
[12] Canaple, L.; Nurdin, N.; Angelova, N.; Saugy, D.; Hunkeler, D. and Desvergne, B. (2001) Maintenance of primary murine hepatocyte functions in multicomponent polymer capsules--in vitro cryopreservation studies. [see comments]. J. Hepatol. 34(1): 11-18.
[13] Obermayer, N.; Busse, B.; Grunwald, A.; Monch, E.; Muller, C.; Neuhaus, P. and Gerlach, J.C. (2001) Biochemical characterization of bioreactors for hybrid liver support: serum-free liver cell coculture of nonparenchymal and parenchymal cells. Transplant. Proc. 33(1-2): 1930-1931.
[14] Knazek, R.A.; Gullino, P.M.; Kohler, P.O. and Dedrick, R.L. (1972) Cell culture on artificial capillaries: an approach to tissue growth in vitro. Science 178(56): 65-66.
[15] Knazek, R.A.; Kohler, P.O. and Gullino, P.M. (1974) Hormone production by cells grown in vitro on artificial capillaries. Exp. Cell Res. 84(1): 251-254.
[16] Margulis, M.S.; Erukhimov, E.A.; Andreiman, L.A.; Kuznetsov, K.A.; Viksna, L.M.; Kuznetsov, A.I. and Deviatov, V.V. (1992) The use of hemoperfusion through a suspension of cryopreserved hepatocytes in acute liver failure. Vestn. Khir. Im. I. I. Grek. 148(1): 83-87.
[17] Margulis, M.S.; Erukhimov, E.A.; Andreiman, L.A. and Viksna, L.M. (1989) Temporary organ substitution by hemoperfusion through suspension of active donor hepatocytes in a total complex of intensive therapy in patients with acute hepatic insufficiency. Resuscitation 18(1): 85-94.
[18] Olumide, F.; Eliashiv, A.; Kralios, N.; Norton, L. and Eiseman, B. (1977) Hepatic support with hepatocyte suspensions in a permeable membrane dialyzer. Surgery 82(5): 599-606.
[19] Gerlach, J.; Kloppel, K.; Schauwecker, H.H.; Tauber, R.; Muller, C. and Bucherl, E.S. (1989) Use of hepatocytes in adhesion and suspension cultures for liver support bioreactors. Int. J. Artif. Organs 12(12): 788-792.
[20] Macdonald, J.M.; Wolfe, S.P.; Roy-Chowdhury, I.; Kubota, H. and Reid, L.M. (2001) Effect of flow configuration and membrane characteristics on membrane fouling in a novel multicoaxial hollow-fiber bioartificial liver. Ann. N.Y. Acad. Sci. 944: 334-343.
[21] Matsumura, K.N.; Guevara, G.R.; Huston, H.; Hamilton, W.L.; Rikimaru, M.; Yamasaki, G. and Matsumura, M.S. (1987) Hybrid bioartificial liver in hepatic failure: preliminary clinical report. Surgery 101(1): 99-103.
[22] Sussman, N.L.; Gislason, G.T. and Kelly, J.H. (1994) Extracorporeal liver support. Application to fulminant hepatic failure. J. Clin. Gastroenterol. 18(4): 320-324.
[23] Watanabe, F.D.; Mullon, C.J.; Hewitt, W.R.; Arkadopoulos, N.; Kahaku, E.; Eguchi, S.; Khalili, T.; Arnaout, W.; Shackleton, C.R.; Rozga, J.; Solomon, D. and Demetriou, A.A. (1997) Clinical experience with a bioartificial liver in the treatment of severe liver failure. A phase I clinical trial. Ann. Surg. 225(5): 484-491.
[24] Ellis, A.J.; Hughes, R.D.; Wendon, J.A.; Dunne, J.; Langley, P.G.; Kelly, J.H.; Gislason, G.T.; Sussman, N.L. and Williams, R. (1996) Pilot-controlled trial of the extracorporeal liver assist device in acute liver failure. Hepatology 24(6): 1446-1451.
[25] Stevens, A.C. (2001) An interim analysis of a phase II/III prospective randomized, multicenter, controlled trial of the HEPATASSIST bioartificial liver support system for the treatment of fulminant hepatic failure. Hepatology 34(4): 299a. Ref. Type: Abstract.
[26] Mazariegos, G.V.; Patzer, J.F.; Lopez, R.C.; Giraldo, M.; Devera, M.E.; Grogan, T.A.; Zhu, Y.; Fulmer, M.L.; Amiot, B.P. and Kramer, D.J. (2002) First clinical use of a novel bioartificial liver support system (BLSS). Am. J. Transplant. 2(3): 260-266.
[27] Gerlach, J.C. and Neuhaus, P. (1994) Culture model for primary hepatocytes. In vitro Cell Dev. Biol. Anim. 30A(10): 640-642.
[28] Gerlach, J.C.; Encke, J.; Hole, O.; Muller, C.; Ryan, C.J. and Neuhaus, P. (1994) Bioreactor for a larger scale hepatocyte in vitro perfusion. Transplantation 58(9): 984-988.
[29] Gerlach, J.C.; Encke, J.; Hole, O.; Muller, C.; Courtney, J.M. and Neuhaus, P. (1994) Hepatocyte culture between three dimensionally arranged biomatrix-coated independent artificial capillary systems and sinusoidal endothelial cell co-culture compartments. Int. J. Artif. Organs 17(5): 301-306.

[30] Gerlach, J.C.; Fuchs, M.; Smith, M.D.; Bornemann, R.; Encke, J.; Neuhaus, P. and Riedel, E. (1996) Is a clinical application of hybrid liver support systems limited by an initial disorder in cellular amino acid and alpha-keto acid metabolism, rather than by later gradual loss of primary hepatocyte function? Transplantation 62(2): 224-228.

[31] Flendrig, L.M.; Calise, F.; Di Florio, E.; Mancini, A.; Ceriello, A.; Santaniello, W.; Mezza, E.; Sicoli, F.; Belleza, G.; Bracco, A.; Cozzolino, S.; Scala, D.; Mazzone, M.; Fattore, M.; Gonzales, E. and Chamuleau, R.A. (1999) Significantly improved survival time in pigs with complete liver ischemia treated with a novel bioartificial liver. Int. J. Artif. Organs 22(10): 701-709.

[32] Jasmund, I.; Langsch, A.; Simmoteit, R. and Bader, A. (2002) Cultivation of Primary Porcine Hepatocytes in an OXY-HFB for Use as a Bioartificial Liver Device. Biotechnol. Prog. 18(4): 839-846.

[33] De Bartolo, L. and Bader A. (2001) Review of a flat membrane bioreactor as a bioartificial liver. Ann. Transplant. 6(3): 40-46.

[34] Morsiani, E.; Galavotti, D.; Puviani, A.C.; Valieri, L.; Brogli, M.; Tosatti, S.; Pazzi, P. and Azzena, G. (2000) Radial flow bioreactor outperforms hollow-fiber modules as a perfusing culture system for primary porcine hepatocytes. Transplant. Proc. 32(8): 2715-2718.

[35] Puviani, A.C.; Lodi, A.; Tassinari, B.; Ottolenghi, C.; Ganzerli, S.; Ricci, D.; Pazzi, P. and Morsiani, E. (1999) Morphological and functional evaluation of isolated rat hepatocytes in three dimensional culture systems. Int. J. Artif. Organs 22(11): 778-785.

[36] Catapano, G. (1996) Mass transfer limitations to the performance of membrane bioartificial liver support devices. Int. J. Artif. Organs 19(1): 18-35.

[37] Giorgio, T.D.; Moscioni, A.D.; Rozga, J. and Demetriou, A.A. (1993) Mass transfer in a hollow fiber device used as a bioartificial liver. ASAIO J. 39(4): 886-892.

[38] Goosen, M.F. (1999) Physico-chemical and mass transfer considerations in microencapsulation. Ann. N.Y. Acad. Sci. 875: 84-104.

[39] Nagaki, M.; Miki, K.; Kim, Y.I.; Ishiyama, H.; Hirahara, I.; Takahashi, H.; Sugiyama, A.; Muto, Y. and Moriwaki, H. (2001) Development and characterization of a hybrid bioartificial liver using primary hepatocytes entrapped in a basement membrane matrix. Dig. Dis. Sci. 46(5): 1046-1056.

[40] Muraca, M.; Vilei, M.T.; Zanusso, G.E.; Ferraresso, C.; Boninsegna, S.; Dal Monte, R.; Carraro P. and Carturan, G. (2002) SiO(2) entrapment of animal cells: liver-specific metabolic activities in silicaoOverlaid hepatocytes. Artif. Organs 26(8): 664-669.

[41] Kobayashi, N.; Taguchi, T.; Noguchi, H.; Okitsu, T.; Totsugawa, T.; Watanabe, T.; Matsumura, T.; Fujiwara, T.; Urata, H.; Kishimoto, N.; Hayashi, N.; Nakaji, S.; Murakami, T. and Tanaka, N. (2001) Rapidly functional immobilization of immortalized human hepatocytes using cell adhesive GRGDS peptide-carrying cellulose microspheres. Cell Transplant. 10(4-5): 387-392.

[42] Selden, C; Khalil, M. and Hodgson H. (2000) Three dimensional culture upregulates extracellular matrix protein expression in human liver cell lines--a step towards mimicking the liver in vivo? Int. J. Artif. Organs 23(11): 774-781.

[43] Selden, C.; Leiper, K.; Ryder, T.; Roberts, E.A.; Kono, Y.; Parker, K.; Davis, P. and Hodgson, H.J.F. (1996) Human liver cell lines proliferate freely and maintain their differentiated phenotype secreting high levels of liver specific proteins when grown in 3-dimensional culture for over 20 days. Hepatology 24(4): 134A.

[44] Selden, C.; Roberts, E.; Stamp, G.; Parker, K.; Winlove, P.; Ryder, T.; Platt, H. and Hodgson, H. (1998) Comparison of three solid phase supports for promoting three- dimensional growth and function of human liver cell lines. Artif. Organs 22(4): 308-319.

[45] Hwang, Y.J.; Kim, Y.I.; Lee, J.G.; Lee, J.W.; Kim, J.W. and Chung, J.M. (2000) Development of bioartificial liver system using a fluidized-bed bioreactor. Transplant. Proc. 32(7): 2349-2351.

[46] Desille, M.; Fremond, B.; Mahler, S.; Malledant, Y.; Seguin, P.; Bouix, A.; Lebreton, Y.; Desbois, J.; Campion, J.P. and Clement, B. (2001) Improvement of the neurological status of pigs with acute liver failure by hepatocytes immobilized in alginate gel beads inoculated in an extracorporeal bioartificial liver. Transplant. Proc. 33(1-2): 1932-1934.

[47] Yang, T.H.; Miyoshi, H. and Ohshima, N. (2001) Novel cell immobilization method utilizing centrifugal force to achieve high-density hepatocyte culture in porous scaffold. J. Biomed. Mater. Res. 55(3): 379-386.

[48] Khalil, M.; Shariat-Panahi, A.; Tootle, R.; Ryder, T.; McCloskey, P.; Roberts, E.; Hodgson, H. and Selden, C. (2001) Human hepatocyte cell lines proliferating as cohesive spheroid colonies in alginate markedly upregulate both synthetic and detoxificatory liver function. J. Hepatol. 34(1): 68-77.

[49] Williams, D.F. (1976) Biomaterials and biocompatibility. Med. Prog. Technol. 4(1-2): 31-42.

[50] Williams, D.F. (1989) A model for biocompatibility and its evaluation. J. Biomed. Eng. 11(3): 185-191.
[51] Catapano, G.; De Bartolo, L.; Vico, V. and Ambrosio, L. (2001) Morphology and metabolism of hepatocytes cultured in Petri dishes on films and in non-woven fabrics of hyaluronic acid esters. Biomaterials 22(7): 659-665.
[52] Xu, A.S. and Reid, L.M. (2001) Soft, porous poly(D,L-lactide-co-glycotide) microcarriers designed for ex vivo studies and for transplantation of adherent cell types including progenitors. Ann. N.Y. Acad. Sci. 944: 144-159.
[53] Khan, A.A.; Capoor, A.K.; Parveen, N.; Naseem, S.; Venkatesan, V. and Habibullah, C.M. (2002) In vitro studies on a bioreactor module containing encapsulated goat hepatocytes for the development of bioartificial liver. Indian J. Gastroenterol. 21(2): 55-58.
[54]Desille, M.; Mahler, S.; Seguin, P.; Malledant, Y.; Fremond, B.; Sebille, V.; Bouix, A.; Desjardins, J.F.; Joly, A.; Desbois, J.; Lebreton, Y.; Campion, J.P. and Clement, B. (2002) Reduced encephalopathy in pigs with ischemia-induced acute hepatic failure treated with a bioartificial liver containing alginate-entrapped hepatocytes. Crit. Care Med. 30(3): 658-663.
[55] Nakazawa, K.; Ijima, H.; Fukuda, J.; Sakiyama, R.; Yamashita, Y.; Shimada, M.; Shirabe, K.; Tsujita, E.; Sugimachi, K. and Funatsu, K. (2002) Development of a hybrid artificial liver using polyurethane foam/hepatocyte spheroid culture in a preclinical pig experiment. Int. J. Artif. Organs 25(1): 51-60.

TISSUE-ENGINEERED BLOOD VESSELS AND THE FUTURE OF TISSUE SUBSTITUTES

The self-assembly as a novel approach to tissue-engineering

LUCIE GERMAIN, KARINA LAFLAMME AND FRANÇOIS A. AUGER
*Experimental Organogenesis Laboratory (LOEX), Saint-Sacrement Hospital of the CHA, Québec, Qc, Canada G1S 4L8 and Department of Surgery, Laval University, Sainte-Foy, Qc, Canada – Fax: 418-682-8000
Email: lucie.germain@chg.ulaval.ca*

1. Introduction

It is seldom acknowledged that the very future of tissue engineering hinges upon the successful reproduction of the cardiovascular system. Since the tissue engineering of the heart is discussed elsewhere in this volume, the focus of the present chapter will be on the vascular system.

The previous axiom has been clearly extolled in many physiology textbook, but has unfortunately received only recently all the scrutiny it deserves from tissue engineers.

Thus, if the brain is the command centre, the human body must have an extremely efficient import/export system for the nutrition of all its tissues with the accompanying disposal of waste or toxic by-products of the normal metabolism. Our own group (LOEX) has been made keenly aware of the paramount role of vasculature by our clinicians. This has been so because every new project we put in place within our research team involve: biomedical biologists, bioengineers, and physicians specialised in the targeted organ.

Thus the importance of reproducing functional vascular substitutes was frequently raised in many projects. These discussions were the impetus for our ongoing interest in tissue engineered blood vessels (TEBV) since 1989.

2. TEBV as a conceptual continuum

The first goal of our endeavour in vascular tissue engineering was the creation of a blood vessel substitute. This was a daunting task since such a goal required the *in vitro* creation of not only a three-dimensional tubular structure but also the combination of three tissue layers. One must also remember that a different cell type produces each of these layers: *adventitia* with perivascular fibroblasts, *media* with smooth muscle cells and *intima* from endothelial cells.

A testimony to the complexity of this tissue engineering project is that very few articles followed the seminal work of Weinberg and Bell in 1986 [1-13]. Even though these authors had reproduced successfully a vascular tubular structure applying the cell seeded collagen gel approach. The final model necessitated the addition of a Dacron® mesh to attain an acceptable level of mechanical resistance [1]. Furthermore, all these experiments were carried out with animal cells when it is known that human vascular cells are much more fidgety in culture.

In this context, our publication, in 1993 of a totally human TEBV [14], was an article of particular significance that rekindled the interest of the scientific community for such a project. However this substitute still lacked some of the critical characteristics of a vascular substitute as seen in table 1.

Table 1. Ideal characteristics of a vascular substitute.

Biocompatibility and haemocompatibility
Mechanical properties: resistance and endurance
Good handling and suturability
Long-term patency

Our second generation of TEBV was presented a few years later. Such a TEBV was obtained through the self-assembly approach that was first proposed by our LOEX group [4]. This TEBV was a much more sophisticated vascular substitute presenting excellent mechanical characteristics and also very valuable histological and functional characteristics.

Moreover, we then extended the spectrum of our vascular tissue engineering effort to encompass the creation of capillaries [15]. This new endeavour was prompted by two reasons. First, if we were to recreate complex TEBV the presence of *vasa vasorum*, a specialised type of capillaries, were of utmost importance. Second, nearly all human tissues demand an excellent blood irrigation system that is provided at the microscopic level by the capillary vessels.

Thus, the LOEX did set about to create a capillary bed in a tissue engineered skin substitute as first attempt in organ vascularisation. The ensuing positive results did not have a trivial significance since they allowed not only better skin construct for burned patient therapy, but can also be regarded as paradigm shift in the manner tissue engineers can envision capillarisation of large organs.

Evidently, there are other approaches to the vascularisation of organs and tissues [16-18], but in the following sections we shall present our own culture systems for obtaining macroscopic and microscopic blood vessels by newly developed tissue engineering methodologies.

Thus our group's interest in the vascular system spawns a full spectrum of tissue engineered substitutes allowing for a fascinating cross fertilisation process along the way.

3. Reconstructing small diameter blood vessel by the self-assembly approach

The aim of our tissue-engineering program is to develop living tissue constructs as close as possible to natural organs. There is a pressing clinical need in the field of vascular grafting. The number of bypass grafts a patient could necessitate during his lifetime combined with the limited availability of autologous arteries and saphenous veins renders imperative the development of new substitutes for small diameter blood vessels (less than 5 mm in diameter). However, the particular haemodynamic conditions present in these small vessels constitute a challenge when they have to be replaced. A functional endothelial layer is evidently the very best lining, the gold standard, for the inner coverage of any vascular graft. The addition of endothelial cells to synthetic prosthesis is currently used to improve their patency [19-21].

The first model that we developed was based on the inclusion of human fibroblasts and smooth muscle cells in a collagen gel [14]. The resistance of this construct was not sufficient to sustain even normal blood pressure. Previously, Weinberg and Bell [1] had presented a bovine construct that had to be reinforced with a Dacron® mesh in order to attain a supraphysiologic burst pressure for an eventual transplantation. The decrease in the resistance of the construct observed with time in culture was attributed to metalloproteinase secretion by the cells included in the gel. Obtaining a resistant natural substitute from this collagen gel methodology proved to be difficult, some teams have now managed to get a supraphysiologic resistance [6,9].

We have designed a second generation of tissue engineered construct that was based on the observation that mesenchymal cells can reconstitute their own extracellular matrix when they are provided with adequate conditions.

The self-assembly approach is a new concept aiming at the reconstruction of an organ in a fashion resembling its formation *in vivo*. The starting material is a solely isolated cell that will secrete *de novo* and organise a relatively thick extracellular matrix thus surrounding them in a tissue-like fashion. Then, this living tissue sheet is rolled over a mandrel and assembled with the other tissue layers in a three-dimensional construct (Figure 1). Both biochemical and mechanical stimulations have to be controlled adequately to reconstruct a tissue displaying the appropriate histological and functional properties. Both cell-cell and cell-extracellular matrix interactions are taking place in reconstructed tissue that also respond to mechanical stimuli [14,22,23]. The stability of the living tissue produced is particularly important since there is no addition of exogenous extracellular matrix or synthetic material that would bring mechanical support if the endogenous extracellular matrix is degraded. Therefore, if the enzymatic degradation is over stimulated, the resistance of the tissue may be decreased.

Thus, our method was based on the exclusive use of human cells and their culture in the absence of any exogenous collagen or synthetic material. All three layers of the blood vessel, the *media*, the *adventitia* and the *intima* were sequentially added to an acellular inner membrane to form a living tubular structure, leading to a true tissue engineered blood vessel.

4. Methodology for TEBV reconstruction by the self-assembly approach

The first step in living tissue reconstruction is the isolation and characterisation of each cell type of interest. Since such a step is so crucial, the determination of purity and quality of each cell culture must be carefully scrutinised [14,24]. The second step is the production of living tissue sheets. They are produced from cultures of mesenchymal cells in the presence of serum and sodium ascorbate [4]. Finally, the third step is the assembly and maturation of all the tissue layers in a tubular form.

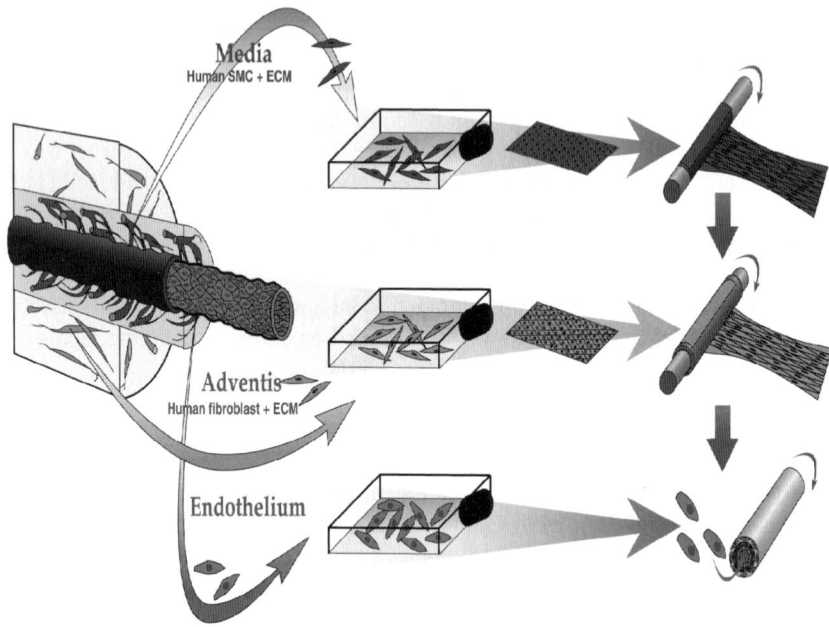

Figure 1. Schematic representation of the production steps for the tissue-engineered blood vessel. Different cell types of the TEBV are extracted from the umbilical vein. Endothelial cells (EC) are obtained enzymatically. Smooth muscle cells and fibroblasts are isolated by the explant method. To induce extracellular matrix formation, SMC and fibroblasts are cultured in medium supplemented with serum and sodium ascorbate. A SMC sheet is wrapped around an inner membrane over a mandrel to produce a cylinder composed of concentric sheet layers that will constitute the media. After a week of maturation, a sheet of fibroblasts is rolled around the vascular media to provide the adventitia. After a maturation period of 6 weeks, the inner tubular mandrel is removed and the TEBV is cannulated at both ends for luminal endothelial cell seeding.

The endothelial cells and the smooth muscle cells were isolated from human umbilical cords by the method of Jaffe [25] and Ross [26], respectively. The fibroblasts were cultured from the dermal portion obtained by thermolysin digestion of skin biopsies [27,28] and used in passage 4 to 10. Fibroblasts and smooth muscle cells were seeded and cultured in DME, supplemented with 50 µg/ml of sodium ascorbate (Sigma), 10% foetal calf serum (Gibco BRL, Burlington, Canada) and antibiotics (penicillin and gentamicin 100U/ml 25µg/ml). Twenty-eight to 35 days later, the living sheets, comprising fibroblasts and the matrix they synthesised, were peeled off from the flasks. A shorter time was necessary for the production of living tissue sheets from smooth muscle cells.

A tubular construct was then obtained in the following fashion: the sheets were detached from the plastic flasks and sequentially wrapped around a tubular mandrel. A smooth muscle cell sheet was first rolled over an acellular inner membrane (dehydrated tubular tissue formed from a fibroblast sheet) to reconstruct the *media*. One week later, a fibroblast sheet was added in order to mature, during seven weeks in culture, into a living reconstructed *adventitia*. Then, the endothelial cells were seeded in the lumen that was created after the removal of the mandrel. These cells then attached to the inner membrane to form a confluent endothelium.

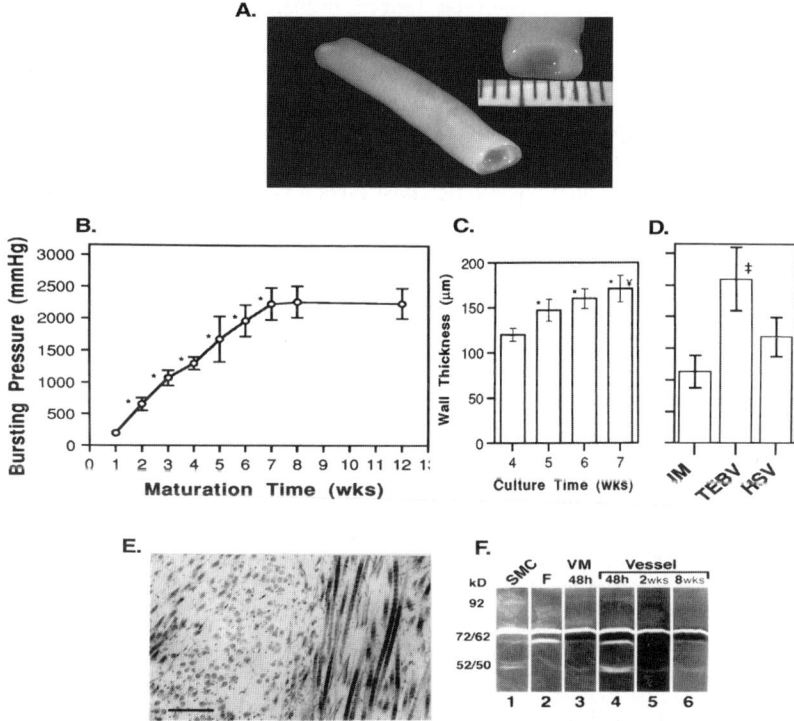

Figure 2. Characterisation of the TEBV. A) Macroscopic view of a mature TEBV. The vessel is self-supporting when removed from culture medium (open lumen 3 mm). B) Burst strength of tissue-engineered adventitia as a function of maturation time. C) Wall thickness of adventitia as a function of culture time after sheet detachment and rolling. D) Burst strength of rehydrated inner membrane alone, matured TEBV and human saphenous veins (HSV). E) Transmission electron micrograph of the adventitial matrix. Uranyl acetate and lead citrate stain. Scale bar 500 nm. F) Gelatine zymogram showing gelatinase activity in conditioned culture medium. Lane 1 and 2: SMC and fibroblast sheets prior to rolling. Lane 3: vascular media 48 h after rolling. Lane 4 to 6: addition of the fibroblast sheet over the media. Reprinted by permission from L'Heureux et al. [4]

5. Histological and phenotypic characteristics of the reconstructed TEBV

The macroscopic aspect of the reconstructed tissue engineered blood vessel did resemble its native counterpart. It had a tubular form with an open lumen (Figure 2A). Histologic and scanning electron microscopic observations revealed a confluent endothelium for the inner surface (Figures 3A and B). The endothelial cells expressed von Willebrand factor and incorporated acetylated low-density lipoproteins indicating that they were functional. Smooth muscle cells and a dense extracellular matrix comprising collagen and glycosaminoglycans respectively, surrounded fibroblasts, in the *media* and *adventitia*. Elastin was detected in the *adventitia*. This characteristic is unique since it was not observed in any other vascular equivalents produced previously *in vitro* with various methods even when they were cultured under pulsatile conditions [1,5].

Interestingly, this three-dimensional organisation was beneficial for the cells since they reacquired their quiescent phenotype. Indeed, human vascular smooth muscle cells are known to loose their differentiated phenotype and stop their expression of proteins such as desmin when they are cultured as monolayers on plastic substrates, they then display a proliferative phenotype. In the *media* of our TEBV, the reexpression of desmin was observed. This intermediate filament, specific for smooth muscle cells, was noted solely in the *media*, not in the *intima* or *adventitia* [4]. The presence of smooth muscle cells presenting a differentiated phenotype constitutes a distinct advantage of the present construct since it is known that the dedifferentiation of smooth muscle cells is associated with diseases such as atherosclerosis [29].

6. Functional characteristics of the reconstructed TEBV

The first aim in the reconstruction of a tissue is to achieve histological and phenotypical characteristics close to that of the native tissue. However, the ultimate goal remains to produce a tissue displaying the functions of its native counterpart. Therefore, representative functions of each layer constituting the blood vessel were analysed: the haemocompatibility of the endothelium, the resistance of the *adventitia*, the vasocontractile properties of the *media*. Furthermore we evaluated the suturability and did an *in vivo* assessment of the TEBV.

The addition of endothelial cells into the lumen of the TEBV construct led to the formation of a complete endothelium after a 7-day maturation period (Figure 3B). This endothelium exhibited an elegant cobblestone morphology. The Weibel-Palade bodies were observed in the cytoplasm of the cells confirming that they were indeed endothelial cells. The absence of platelet aggregates following heparinised human whole blood circulation when the endothelial cells were present at the inner surface of the construct indicates that the endothelial cells inhibited platelet aggregation (Figure 3B). As expected in the absence of endothelium, platelets adhered to the inner membrane confirming that extracellular matrix promoted platelet adhesion and aggregation (Figure 3A). Therefore, this result was very encouraging for the transplantation experiments since a confluent endothelium with an antithrombotic phenotype is necessary at the time of grafting for the success of small-diameter vascular conduits replacement [19]. The lack haemocompatibility was the main limitation to the use of synthetic biomaterials and was responsible for their unduly low long-term patency rate [30].

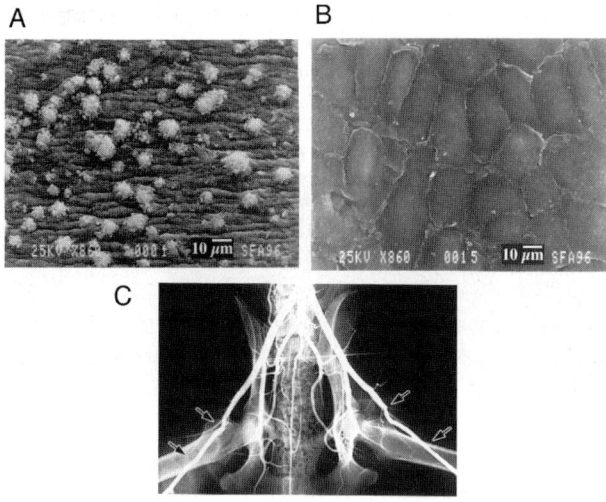

Figure 3. Inhibition of platelet adhesion by the endothelium and in vivo grafting of the TEBV. Scanning electron micrographs of unendothelialised inner membrane (A) promoted platelet adhesion and activation whereas endothelialised inner membrane (B) almost completely inhibited the process. Angiogram of the lower limbs 7 days after implantation of 2 unendothelialised TEBV (C). Two patents TEBV are visible (arrows) providing normal blood flow in both legs. Reprinted by permission from L'Heureux et al. [4].

One of the crucial properties of a blood vessel is its capacity to sustain blood pressure. Therefore, the mechanical resistance of a vascular prosthesis must be sufficient at the time of transplantation. The level of mechanical resistance of the *adventitia* was assessed *in vitro* by testing its burst strength. The mechanical resistance of the *adventitia* regularly increased during the maturation in culture. After 2 weeks of maturation the *adventitia* already presented a bursting pressure over 500 mmHg. This value continued to increase regularly reaching a plateau of over 2000 mmHg (Figure 2B). The thickness of the *adventitia* also increased during the culture period (Figure 2C). The resistance of the complete TEBV was about 20 times higher than the mean systolic blood pressure and was above the burst level of comparable saphenous veins (Figure 2D). This latter vessel is a traditional choice for bypass surgery. Such an elevated burst strength as displayed by the TEBV has been attributed to the particularly well-organised extracellular matrix of the *adventitia* (Figure 2E) presenting many characteristics of a mature collagen scaffold [31,32] comprising collagen type I and III, fibronectin, *etc.* The fibril bundles were closely packed and oriented in perpendicular directions to one another. Furthermore, a network of elastin-associated microfibers parallel to the collagen fibrils was also observed (Figure 2E; [33]). Moreover, the gelatinase activity was transiently induced in the first days of maturation of the *adventitia* but was generally down-regulated thereafter (Figure 2F). Such a constant level of collagenase activity is contributing to the long-term mechanical stability of our TEBV. An hypothesis for the difference between our data and the results obtained by Weinberg and Bell [1] with the collagen gel-based model where the resistance diminished by about 50% between the 4th and 12th week of maturation, is that an extracellular matrix produced *de novo* by the cells induces less metalloproteinase expression and/or more tissue inhibitors of metalloproteinase (TIMP) than a biochemically extracted and reassembled matrix in which cells have been seeded.

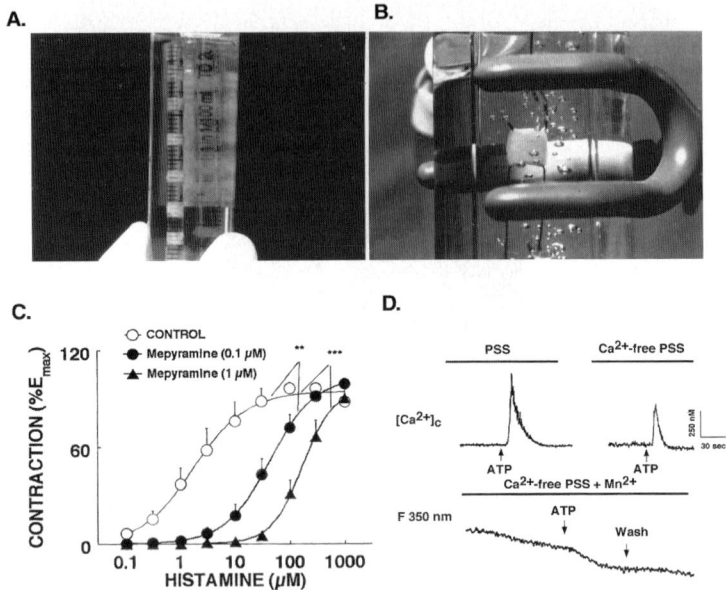

Figure 4. Pharmacological studies of the human tissue-engineered vascular media (TEVM). A) Rings of the TEVM on a 4.5-mm tubular support during maturation. B) A TEVM ring mounted on a force transducer in an organ bath. C) Concentration-effect curves of histamine producing contraction (% maximum response, Emax obtained with 3 mM ATP) of human TEVM in the absence and presence of selective antagonists. D) Effects of ATP on $[Ca^{2+}]c$ and Ca^{2+} influx in human TEVM layer. Top: representative traces showing $[Ca^{2+}]c$ in the presence (normal physiological salt solution, PSS) and absence (Ca^{2+}-free PSS) of extracellular Ca^{2+}. Bottom: manganese quench (at 350nm) produced by ATP in a Fura-2 loaded preparation. Reprinted by permission from L'Heureux et al. [34].

This high mechanical strength allowed the grafting of this living TEBV without addition of any synthetic material. The implantation of human reconstructed blood vessels into dogs demonstrated that they could be handled and sutured by conventional surgical techniques. The patency rate of 50% obtained was very significant considering that the endothelium was not added to avoid the hyperacute rejection in the intrinsic xenogeneic situation of the experiment (Figure 3C). The blood was then in contact with the thrombogenic collagen matrix in the absence of endothelium (Figure 2A). The suturability and handling characteristics of the TEBV graft were evaluated as "tissue-like" by an experienced vascular surgeon. The absence of exogenous extracellular matrix in the TEBV should provide the advantage of an accelerated remodelling after grafting because of a simplified healing phenomenon since in this model the phase of biomaterials resorption that at times leads to a slow inflammatory reaction has been bypassed.

The *media* layer of the blood vessel, through its ability to contract, is responsible for the control of blood flow in response to physiological stimuli. The cohesiveness and mechanical strength of the reconstructed human *media* allow its mounting on force transducers in organ bath *in vitro* to perform standard pharmacological experiments (Figures 4A and B). The vasoactive contractile/relaxation responses of our reconstructed *media* were shown to be comparable to those of normal blood vessels [34]. It contracted under stimuli such as histamine (Figure 4C). The sustained contraction to vasoconstrictor agonist such as ATP was associated with transient increases in cytosolic calcium concentration (Figure 4D). Therefore, in addition to its value as a *media* layer in the vascular prosthesis for bypass surgery, this reconstructed *media* provides also a new model with applications in pharmacological research. Indeed, the human nature of this construct, combined to its completely natural composition, is attractive for pharmacologic experimentation where the response of animal tissue may be different and the supply of normal human blood vessel can be very limited.

7. Vascularisation of the tissue construct by the addition of capillaries *in vitro*

In blood vessels as well as in all other tissues, the nutritional aspect is very important. Thus, the production of a large or thick construct must be accompanied of an adequate blood supply to ensure its viability. After transplantation, the revascularisation process must be efficient to prevent necrosis of the tissue. The pre-existing capillaries in the transplanted organ have been shown to play a deciding role in its survival after implantation *in vivo*. Indeed, the revascularisation of cadaver skin within a few days after grafting on burn patients results from the inosculation between pre-existing capillary plexus of the cadaver skin and the wound vasculature [35]. Therefore, the presence of capillaries within *in vitro* tissue engineered living constructs should lead to a significant advantage for their survival rate.

According to this goal, we took advantage of our experience in tissue-engineered human blood vessels to demonstrate, for the first time, the feasibility of reconstructing a human capillary supply in a living tissue construct produced *in vitro*.

Culturing keratinocytes over dermal fibroblasts seeded in a collagen-chitosan sponge was chosen as the model of skin production by culturing keratinocytes over dermal fibroblasts seeded in a collagen-chitosan sponge [36]. In this biopolymer, fibroblasts secreted and organised an extracellular matrix within the pores and over the sponge. In contrast, the seeding of endothelial cells from human umbilical veins in the sponge, without fibroblasts, did not lead to the formation of particular structures and a very limited extracellular matrix production was observed (Figure 5A). However, when endothelial cells were added with the fibroblasts, these cells spontaneously reorganised within the dense extracellular matrix synthesised by the fibroblasts into capillary-like structures with a lumen (Figures 5B, C) and deposited type IV collagen (Figure 5D) on their outer side. The ultrastructural characterization of the reconstructed tissue confirmed that their cells surrounding the lumens possessed Weibel-Palade bodies indicating that they were truly endothelial cells. Moreover, these cells were able to spontaneously reform cell-cell junctions (Figures 5E, F).

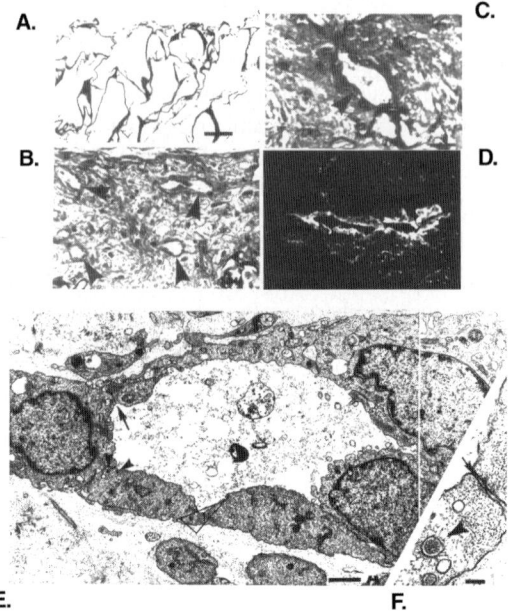

Figure 5. Histological and immunohistochemical staining and ultrastructural characterisation of capillary-like tubular structures present in the mesenchymal portion of in vitro tissue-engineered endothelialised skin equivalent. Biopolymers were seeded with HUVEC (A) HUVEC and fibroblasts (B, C) HUVEC, fibroblasts and keratinocytes (D-F) and cultured for 31 days. Note that fibroblasts were numerous and filled the pores with newly synthesised extracellular matrix when cultured with HUVEC (B, C). The formation of capillary-like tubular structures in the newly synthesised extracellular matrix was observed in these cocultures of fibroblasts and HUVEC (B, C). Immunostained frozen sections shows that human type IV collagen is secreted (D). Transmission electron microscopy of a capillary-like structure shows a closed tube formed by seven cells with intercellular junctions (arrow)(E). Weibel-Palade bodies were also seen (arrowheads) (F). Note that the lumen was filled with cellular debris but did not contain dense extracellular matrix components (E,. F). (F) is an enlargement of triangular panel in E. Bars indicate 60 μm (A); 10 μm (C); 30 μm (D); 1 μm (E) and 0.1 μm (F). Reprinted by permission from Black et al. [15].

Taken together, these results suggest that the extracellular matrix organised by the fibroblasts as well as the growth factors they secreted did provide excellent conditions for the formation of capillaries within this mesenchymal tissue. In the absence of such cell-cell and cell-extracellular matrix interactions, there was no capillary-tube formation. The stability of the *in vitro* reconstructed human capillary-like structures is another feature of our model especially since the addition of tumour promoting agents like phorbol 12-myristate 13-acetate (PMA), or specific external growth factors were not necessary for the induction of angiogenesis *in vitro*. This contrasts with other human angiogenic models that are limited by a very short-term survival period [37,38].

The application of this methodology for obtaining capillary-tube formation to other mesenchymal tissues will likely improve the nutritional supply of the engineered constructs after grafting.

8. Conclusion

In conclusion, this TEBV reconstructed blood vessel is a promising substitute for clinical applications in the vascular surgery of small diameter arteries (*e.g.* coronary heart vessels). Other *in vitro* applications comprise its use as a model to understand the development of various pathologies such as atherosclerosis or to shed new light on specific pharmacological pathways.

The addition of capillary-like structures to mesenchymal tissues is a step towards the creation of more complex organs that rely on a microvascular system for their survival. Thus the production of more sophisticated vascularised organs by tissue engineering brings about a novel approach to replace wounded or diseased tissues with living substitutes. These substitutes in turn will restore and provide the function and healing power of the native tissue. The continuous improvement of methods for culturing cells and reconstructing three-dimensional structures has led to the engineering of an increasing number of tissues such as skin, cornea, bronchi, ligaments, bladder [39-45]. These new therapeutic horizons should benefit many patients in the future.

Acknowledgements

The authors are grateful to their collaborators on the development of reconstructed living substitutes by tissue engineering Drs. Raymond Labbé, (LOEX), Jean-Claude Stoclet and Ramaroson Andriantsitohaina (Faculté de Pharmacie, Université Louis Pasteur de Strasbourg) and all members of the LOEX laboratory for their kind help, advice and technical assistance in relation to the work presented in this review.

References

[1] Weinberg, C.B. and Bell, E. (1986) A blood vessel model constructed from collagen and cultured vascular cells. Science 231: 397-400.
[2] Matsuda, T.; Akutsu, T.; Kira, K. and Matsumoto, H. (1989) Development of hybrid compliant graft: rapid preparative method for reconstruction of a vascular wall. ASAIO Trans. 35: 553-555.
[3] Tranquillo, R.T.; Girton, T.S.; Bromberek, B.A.; Triebes, T.G. and Mooradian, D.L. (1996) Magnetically orientated tissue-equivalent tubes: application to a circumferentially orientated media-equivalent. Biomaterials 17: 349-357.
[4] L'Heureux, N.; Paquet, S.; Labbe, R.; Germain, L. and Auger, F.A. (1998) A completely biological tissue-engineered human blood vessel. FASEB J. 12: 47-56.
[5] Niklason, L.E.; Gao, J.; Abbott, W.M.; Hirschi, K.K.; Houser, S.; Marini, R. and Langer, R. (1999) Functional arteries grown *in vitro*. Science 284: 489-493.
[6] Seliktar, D.; Black, R.A.; Vito, R.P. and Nerem, R.M. (2000) Dynamic mechanical conditioning of collagen-gel blood vessel constructs induces remodeling *in vitro*. Ann. Biomed. Eng. 28: 351-362.

[7] Niklason, L.E.; Abbott, W.; Gao, J.; Klagges, B.; Hirschi, K.K.; Ulubayram, K.; Conroy, N.; Jones, R.; Vasanawala, A.; Sanzgiri, S. and Langer, R. (2001) Morphologic and mechanical characteristics of engineered bovine arteries. J. Vasc. Surg. 33: 628-638.
[8] Nerem, R.M. and Seliktar, D. (2001) Vascular tissue engineering. Ann. Rev. Biomed. Eng. 3: 225-243.
[9] Neidert, M.R.; Lee, E.S.; Oegema, T.R. and Tranquillo, R.T. (2002) Enhanced fibrin remodeling *in vitro* with TGF-beta1, insulin and plasmin for improved tissue-equivalents. Biomaterials 23: 3717-3731.
[10] Grassl, E.D.; Oegema, T.R. and Tranquillo, R.T. (2002) Fibrin as an alternative biopolymer to type-I collagen for the fabrication of a media equivalent. J. Biomed. Mater. Res. 60: 607-612.
[11] Girton, T.S.; Barocas, V.H. and Tranquillo, R.T. (2002) Confined compression of a tissue-equivalent: collagen fibril and cell alignment in response to anisotropic strain. J. Biomech. Eng. 124: 568-575.
[12] Lai, J.Y.; Yoon, C.Y.; Yoo, J.J.; Wulf, T. and Atala, A. (2002) Phenotypic and functional characterization of *in vivo* tissue engineered smooth muscle from normal and pathological bladders. J. Urol. 168: 1853-1857; discussion 1858.
[13] Campbell, J.H.; Efendy, J.L.; Han, C.L. and Campbell, G.R. (2000) Blood vessels from bone marrow. Ann. N.Y. Acad. Sci. 902: 224-229.
[14] L'Heureux, N.; Germain, L.; Labbe, R. and Auger, F.A. (1993) *In vitro* construction of a human blood vessel from cultured vascular cells: a morphologic study. J. Vasc. Surg. 17: 499-509.
[15] Black, A.F.; Berthod, F.; L'Heureux, N.; Germain, L. and Auger, F.A. (1998) *In vitro* reconstruction of a human capillary-like network in a tissue- engineered skin equivalent. FASEB J. 12: 1331-1340.
[16] Supp, D.M.; Wilson-Landy, K. and Boyce, S.T. (2002) Human dermal microvascular endothelial cells form vascular analogs in cultured skin substitutes after grafting to athymic mice. FASEB J. 16: 797-804.
[17] Burg, K.J.; Holder, W.D. Jr.; Culberson, C.R.; Beiler, R.J.; Greene, K.G.; Loebsack, A.B.; Roland, W.D.; Eiselt, P.; Mooney, D.J. and Halberstadt, C.R. (2000) Comparative study of seeding methods for three-dimensional polymeric scaffolds. J. Biomed. Mater. Res. 51: 642-649.
[18] Eiselt, P.; Kim, B.S.; Chacko, B.; Isenberg, B.; Peters, M.C.; Greene, K.G.; Roland, W.D.; Loebsack, A.B.; Burg, K.J.; Culberson, C.; Halberstadt, C.R.; Holder, W.D. and Mooney, D.J. (1998) Development of technologies aiding large-tissue engineering. Biotechnol. Prog. 14: 134-140.
[19] Zilla, P.; von Oppell, U. and Deutsch, M. (1993) The endothelium: a key to the future. J. Card. Surg. 8: 32-60.
[20] Cines, D.B.; Pollak, E.S.; Buck, C.A.; Loscalzo, J.; Zimmerman, G.A.; McEver, R.P.; Pober, J.S.; Wick, T.M.; Konkle, B.A.; Schwartz, B.S.; Barnathan, E.S.; McCrae, K.R.; Hug, B.A.; Schmidt, A.M. and Stern, D.M. (1998) Endothelial cells in physiology and in the pathophysiology of vascular disorders. Blood 91: 3527-3561.
[21] Zilla, P.; Deutsch, M. and Meinhart, J. (1999) Endothelial cell transplantation. Semin. Vasc. Surg. 12: 52-63.
[22] Goulet, F.; Poitras, A.; Rouabhia, M.; Cusson, D.; Germain, L. and Auger, F.A. (1996) Stimulation of human keratinocyte proliferation through growth factor exchanges with dermal fibroblasts *in vitro*. Burns 22: 107-112.
[23] Lopez Valle, C.A.; Auger, F.A.; Rompre, P.; Bouvard, V. and Germain, L. (1992) Peripheral anchorage of dermal equivalents. Br. J. Dermatol. 127: 365-371.
[24] Germain, L.; Remy-Zolghadri, M. and Auger, F. (2000) Tissue engineering of the vascular system: from capillaries to larger blood vessels. Med. Biol. Eng. Comput. 38: 232-240.
[25] Jaffe, E.A.; Nachman, R.L.; Becker, C.G. and Minick, C.R. (1973) Culture of human endothelial cells derived from umbilical veins. Identification by morphologic and immunologic criteria. J. Clin. Invest. 52: 2745-2756.
[26] Ross, R. (1971) The smooth muscle cell. II. Growth of smooth muscle in culture and formation of elastic fibers. J. Cell Biol. 50: 172-186.
[27] Germain, L.; Rouabhia, M.; Guignard, R.; Carrier, L.; Bouvard, V. and Auger, F.A. (1993) Improvement of human keratinocyte isolation and culturing using thermolysin. Burns 19: 99-104.
[28] Berthod, F.; Germain, L.; Guignard, R.; Lethias, C.; Garrone, R.; Damour, O.; van der Rest, M. and Auger, F.A. (1997) Differential expression of collagen XII and XIV in human skin and in reconstructed skin. J. Invest. Dermatol. 108: 737-742.
[29] Ross, R. (1999) Atherosclerosis--an inflammatory disease. N. Engl. J. Med. 340: 115-126.
[30] Greenwald, S.E. and Berry, C.I. (2000) Improving vascular grafts: The importance of mechanical and haemodynamic properties. J. Pathobiol. 190: 292-299.

[31] Birk, D.E. and Linsenmayer, T.F. (1994) Collagen fibril assembly, deposition, and organization into tissue-specific matrices. In: Yurchenco, P.D.; Birk, D.E. and Mecham, P.R. (Eds.) Extracellular Matrix Assembly and Structure. Academic Press, San Diego; pp. 91-128.
[32] Fleischmajer, R.; Contard, P.; Schwartz, E.; MacDonald, D.; Jacobs, L. and Sakai L.Y. (1991) Elastin-associated microfibrils (10nm) in a three-dimensional fibroblast culture. J. Invest. Dermatol. 97: 638-643.
[33] Pouliot, R.; Larouche, D.; Auger, F.A.; Juhasz, J.; Xu, W.; Li, H. and Germain, L. (2002) Reconstructed human skin produced *in vitro* and grafted on athymic mice. Transplantation. 73: 1751-1757.
[34] L'Heureux, N.; Stoclet, J.C.; Auger, F.A.; Lagaud, G.J.; Germain, L. and Andriantsitohaina, R. (2001) A human tissue-engineered vascular media: a new model for pharmacological studies of contractile responses. FASEB J. 15: 515-524.
[35] Young, D.M.; Greulich, K.M. and Weier, H.G. (1996) Species-specific in situ hybridization with fluorochrome-labeled DNA probes to study vascularization of human skin grafts on athymic mice. J. Burn Care Rehabil. 17: 305-310.
[36] Black, A.; Berthod, F.; L'Heureux, N.; Germain, L. and Auger, F.A. (1998) *In vitro* reconstruction of a human capillary-like network in a tissue-engineered skin equivalent. FASEB J. 12: 1331-1340.
[37] Troyanovsky, B.; Levchenko, T.; Mänsson, G.; Matvijenko, O. and Holmgren, L. (2001) Angiomotin: an angiostatin binding protein that regulates endothelial cell migration and tube formation. J. Cell Biol. 152: 1247-1254.
[38] Bayless, K.J.; Salazar, R. and Davis, G.E. (2000) RGD-dependent vacuolation and lumen formation observed during endothelial cell morphogenesis in three-dimensional fibrin matrices involves the alpha(v)beta(3) and alpha(5)beta(1) integrins. Am. J. Pathol. 156: 1673-1683.
[39] Germain, L.; Carrier, P.; Auger, F.A.; Salesse, C. and Guerin, S.L. (2000) Can we produce a human corneal equivalent by tissue engineering? Prog. Retin Eye Res. 19: 497-527.
[40] Germain, L.; Auger, F.A.; Grandbois, E.; Guignard, R.; Giasson, M.; Boisjoly, H. and Guerin, S.L. (1999) Reconstructed human cornea produced *in vitro* by tissue engineering. Pathobiology 67: 140-147.
[41] Goulet, F.; Germain, L.; Caron, C.; Rancourt, D.; Normand, A. and Auger, F. (1997) Tissue-engineered ligament. In: Yahia, L. (Ed.) Ligaments and ligamentoplasties. Springer-Verlag, Berlin; pp. 367-377.
[42] Goulet, F.; Germain, L.; Rancourt, D.; Caron, C.; Normand, A. and Auger, F.A. (1997) Tendons and ligaments. In: Lanza, R.; Langer, R. and Chick, W. (Eds.) Textbook of Tissue Engineering. Academic Press Ltd., San Diego; pp. 633-644.
[43] Paquette, J.-S.; Goulet, F.; Boulet, L.-P.; Laviolette, M.; Tremblay, N.; Chakir, J.; Germain, L. and Auger, F.A. (1998) Three-dimensional production of bronchi *in vitro*. Can. Resp. J. 5: 43.
[44] Laplante, A.; Germain, L.; Auger, F. and Moulin, V. (2001) Mechanisms of wound reepithelialization: hints from a tissue-engineered reconstructed skin to long-standing questions. FASEB J. 15: 2377-2389.
[45] Atala, A. (2001) Bladder regeneration by tissue engineering. BJU Int. 88: 765-770.

TISSUE ENGINEERING OF CARTILAGE AND MYOCARDIUM

BOJANA OBRADOVIC[1], MILICA RADISIC[2] AND GORDANA VUNJAK-NOVAKOVIC[2]
[1]*Department of Chemical Engineering, Faculty of Technology and Metallurgy, University of Belgrade, Karnegijeva 4, PO Box 3503, 1120 Belgrade, Serbia and Montenegro – Fax: 381-11-3370387 – Email: bojana@tmf.bg.ac.yu*; [2]*Harvard – MIT Division of Health Sciences and Technology, Massachusetts Institute of Technology, E25-330, 45 Carleton Street, Cambridge MA 02139, USA – Fax: 1-617-2588827 – Email: milica@mit.edu; gordana@mit.edu*

1. Introduction

A variety of exciting new strategies has emerged over the past decade to address the clinical problem of tissue failure. Tissue engineering is particularly significant because it can provide biological substitutes of compromised native tissues. As compared to the transplantation of cells alone, engineered tissues have the potential advantage of immediate functionality. As compared to transplantation of native tissues, engineered tissues can alleviate the scarcity of suitable tissue transplants, as well as donor-recipient compatibility and disease transmission (for allografts), and donor site morbidity (for autografts). Engineered tissues can also serve as physiologically relevant models for controlled studies of cells and tissues under normal and pathological conditions. Ideally, a lost or damaged tissue could be replaced by an engineered graft that can re-establish appropriate structure, composition, cell signalling and function of the native tissue. In light of this paradigm, the clinical utility of tissue engineering will likely depend on our ability to replicate the site-specific properties of the particular tissue across different size scales. In engineered constructs, the cells should conform to a specific differentiated phenotype, while the composition and architectural organisation of the extracellular matrix (ECM) should provide the necessary functional properties inherent to the tissue being replaced. Ideally, an engineered graft should provide regeneration, rather than repair, and undergo remodelling in response to environmental factors [1-3]:

- Repair is rapid replacement of the damaged, defective or lost tissue with functional new tissue that resembles, but does not replicate the structure, composition and function of the native tissue.
- Regeneration is slow restoration of all components of the repair tissue to their original condition such that the new tissue is indistinguishable from normal tissue with respect to structure, composition and functional properties.

- Remodelling is the change in tissue structure and composition in response to the local and systemic environmental factors that alters the functional tissue properties.

1.1. TISSUE ENGINEERING REQUIREMENTS

In general, the tissue engineering requirements can be summarised as follows. To begin with, it is necessary to generate a graft of a desired size and shape to repair a specific defect. Next, the grafts should have the biochemical composition, histomorphology and ultrastructure mimicking those of the native tissue being replaced. Furthermore, it is often necessary for a graft to provide immediate functionality at some minimal level. For load-bearing tissues for example, mechanical competence of the engineered tissue can critically determine if the graft will survive implantation. An engineered graft should also have the capacity to fully integrate (structurally and functionally) with the adjacent host tissues. Additional requirements include specific structural and functional properties such as compressive stiffness for cartilage, contractile function for myocardium, and vascularisation for most tissues.

Cells, biomaterial scaffolds, biochemical and physical regulatory signals have been utilised in a variety of ways to engineer tissues, *in vitro* and *in vivo*. Tissue engineering generally involves the presence of reparative cells, a structural template, facilitated transport of nutrients and metabolites, and a provision of molecular and mechanical regulatory factors. Rather than providing a comprehensive review of tissue engineering, we focus here on one approach based on bioreactor cultivations of dissociated cells on biodegradable scaffolds. Two distinctly different tissues: articular cartilage (load-bearing skeletal tissue) and myocardium (contractile heart tissue) that perform functions vital for health and survival, are used as paradigms of functional tissues that are of great clinical interest. The incidence of diseases including osteoarthritis and heart failure is constantly increasing and, in the absence of curative interventions, has resulted in substantial human suffering and medical expense. We explore here feasibilities of the tissue engineering model system regarding specific requirements for *in vitro* functional assembly of these two engineered tissues.

1.2. TISSUE ENGINEERING MODEL SYSTEM

One envisioned scenario of clinically relevant tissue engineering involves the use of autologous cells, a biodegradable scaffold (designed to serve as a structural and logistic template for tissue development), and a bioreactor (designed to enable environmental control and support cell differentiation and functional assembly into an engineered tissue) (Figure 1). Cells are generally isolated from a small tissue sample, expanded in culture under conditions selected to yield sufficient number for seeding a clinically sized scaffold and in some cases transfected to (over-)express a gene of interest. Scaffolds should be made of biocompatible materials, preferentially those already approved for clinical use. Scaffold structure determines the transport of nutrients, metabolites and regulatory molecules to and from the cells, whereas the scaffold chemistry may have an important role in cell attachment and differentiation. The scaffold should biodegrade at the same rate as the rate of tissue assembly and without toxic or inhibitory products.

Mechanical properties of the scaffold should ideally match those of the native tissue being replaced, and the mechanical integrity should be maintained as long as necessary for the new tissue to mature and integrate.

In this approach, a bioreactor should ideally provide all necessary conditions in *in vitro* environment for rapid and orderly tissue development by cells cultured on a scaffold. In general, a bioreactor is designed to perform one or more of the following functions: establish a desired spatially uniform cell concentration within the scaffold during cell seeding, maintain controlled conditions in culture medium (*e.g.*, temperature, pH, osmolality, levels of oxygen, nutrients, metabolites, regulatory molecules), facilitate mass transfer, and provide physiologically relevant physical signals (*e.g.*, interstitial fluid flow, shear, pressure, compression) during cultivation of cell-polymer constructs.

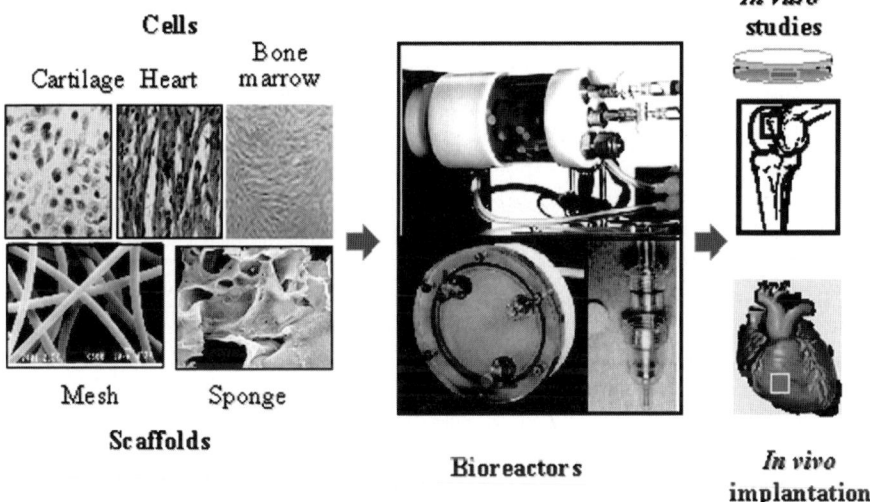

Figure 1. Tissue engineering based on cell cultivation on biomaterial scaffolds in bioreactors. Cells (e.g., from cartilage, heart or bone marrow) are cultured on a scaffold (e.g., highly porous, biodegradable mesh or a sponge) in a bioreactor (e.g., rotating bioreactor or perfused cartridge [4]). The resulting constructs are used for controlled in vitro studies or implanted in vivo (e.g., to repair an osteochondral defect [5] or injured myocardium [6-8]).

Three representative culture vessels that are frequently used for tissue engineering are compared in Figure 2. All culture vessels are operated in an incubator (to maintain the temperature and pH) with continuous gas exchange and periodic medium replacement. Flasks contain constructs that are fixed in place by threading onto needles and cultured either statically or with magnetic stirring, with gas exchange through loosened side arm caps. Rotating vessels contain constructs that are freely suspended in culture medium between two concentric cylinders, the inner of which serves as a gas exchange membrane. The vessel rotation rate is adjusted to maintain each construct settling at a stationary point within the vessel. This experimental set-up thus enables the evaluation of the effects of flow and mass transfer on engineered tissues and the selection of suitable culture environments to be further explored and optimised.

a)

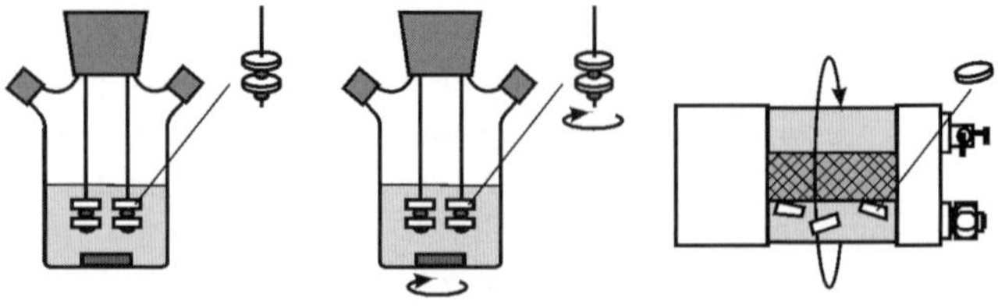

b) Cultivation parameter	Static flask	Mixed flask	Rotating vessel
Vessel diameter (cm)	6.5	6.5	14.6/5.1
Medium volume (cm^3)	120	120	110
Tissue construct or explant[1]	Fixed in place n = 12 per vessel	Fixed in place n = 12 per vessel	Freely settling n = 12 per vessel
Medium exchange (3 cm^3 per tissue per day)	Batch-wise	Batch-wise	Batch-wise
Gas exchange	Continuous via surface aeration	Continuous via surface aeration	Continuous via an internal membrane
Stirring/rotation rate (s^{-1})	0	0.83 – 1.25	0.25 – 0.67
Flow conditions	Static fluid	Turbulent[2]	Laminar[3]
Mixing mechanism	None	Magnetic stirring	Settling in rotational flow
Mass transfer in bulk medium	Molecular diffusion	Convection (due to medium stirring)	Convection (due to tissue settling)
Fluid shear at tissue surfaces	None	Steady, turbulent	Dynamic, laminar
Reference	9-13	9-13	9–11, 14-19

[1] 5 mm diameter by 2 mm thick discs
[2] The smallest turbulent eddies had a diameter of 250 µm and velocity of 0.4 cm/s [20,9]
[3] Tissues were settling in a laminar tumble-slide regimen in a rotational field [14,21,22]

Figure 2. Representative bioreactors. a) Schematic presentation of tissue cultivation in static flasks, mixed flasks and rotating vessels; b) Overview of the operating conditions for each vessel type (based on [23-25]).

In this chapter, the bioreactor hydrodynamics are discussed in light of their effects on tissue development, and rationalised by mathematical models accounting for the biosynthesis and transport rates within engineered cartilage and myocardium.

2. Cartilage tissue engineering

2.1. ARTICULAR CARTILAGE

Articular cartilage is an avascular tissue containing only one cell type, the chondrocyte, which generates and maintains an extracellular matrix (ECM) consisting of a fibrous network of collagen type II and glycosaminoglycan (GAG)-rich proteoglycans [26]. The main function of articular cartilage is to allow joint mobility while transferring compressive and shear forces. The biomechanical behaviour of cartilage is determined by the balance between the swelling pressure of proteoglycan gel carrying fixed charge density and the restraining properties of the collagen network [27]. The equilibrium modulus, a measure of cartilage stiffness under compressive loading, increases with increasing wet weight fraction of proteoglycan [28–30] and collagen [31] and decreasing wet weight fraction of water [28-30]. The hydraulic permeability, a measure of cartilage resistance to fluid flow during compression, has been inversely correlated with the wet weight fraction of GAGs [28] and the compressive strain [32]. In developing cartilage, the increase in confined-compression modulus and decrease in hydraulic permeability were associated with increases in fractions of ECM components [33].

2.2. CLINICAL NEED

Once damaged, cartilage has minimal capacity for self-repair [34]. The clinical demand for cartilage repair is large, in particular in older patients. Osteoarthritis (OA), the hallmark of which is progressive cartilage degeneration, affects approximately 20 million individuals per year in the United States only [35]. Most of these patients developed degenerative joint disease either from wear and tear on their joints, or subsequent to trauma. Joint degeneration can range from localised defects to the complete loss of large surface areas of cartilage. The increasing incidence of OA (due to an aging population and better health care) challenges our current technology and fuels the need for development of tissue engineering solutions. Therapeutic interventions, such as analgesics, physical therapy, and surgery have mixed results [36,34]. For clinically relevant tissue engineering, engineered grafts need to provide an equal or better alternative to the existing treatment options and standards of care.

2.3. TISSUE ENGINEERING

Functional restoration of articular cartilage remains a challenge, and none of the existing treatment regimens gives a consistently good outcome [1]. Orthopaedic tissue engineering has a potential to provide orderly and mechanically competent regeneration of compromised cartilage. Clearly, the main objectives are symptom relief and the reestablishment of normal load-bearing function of the articular surface. One approach to functional tissue engineering of cartilage involves the *in vitro* cultivation of cartilaginous constructs that would have a capacity to develop site- and scale-specific structural and biomechanical properties of native articular cartilage, and integrate firmly and completely to the adjacent host tissues. It is thought that an engineered graft should mediate matrix remodelling in a fashion similar to that present in immature tissue [25]. The cell – scaffold – bioreactor system (Figure 1) has been designed to utilise some of the factors known to enhance chondrogenesis *in vitro* (*e.g.*, [4]) and *in vivo* (*e.g.*, [5]). It involves an integrated use of chondrogenic cells, biodegradable scaffolds and bioreactors towards the *in vitro* engineering of functional tissue constructs.

2.4. CELL SOURCES

The cells used thus far to engineer cartilage have varied with respect to donor age (embryonic, neonatal, immature or adult), differentiation state (precursor or phenotypically mature), and the method of preparation (selection, expansion, gene transfer) [14,25,37]. High and spatially uniform cell density was associated with rapid chondrogenesis in engineered constructs based on differentiated chondrocytes and bone marrow derived progenitor cells [4]. Growth factors supplemented sequentially to culture medium markedly and significantly improved the compositions and mechanical properties of cartilage engineered starting from chondrocytes [38,39] and bone marrow derived chondrogenic cells [40]. Gene transfer of human IGF-I (an anabolic factor of cartilage development) into bovine articular chondrocytes also improved construct properties [41].

2.5. SCAFFOLDS

Most studies suggest that cell cultivation on a three-dimensional scaffold is essential for promoting orderly regeneration of cartilage, *in vivo* and *in vitro*. Scaffolds investigated to date vary with respect to material chemistry (*e.g.* collagen, agarose, synthetic polymers), geometry (*e.g.* gels, fibrous meshes, porous sponges), structure (*e.g.* porosity, distribution, orientation and connectivity of the pores), mechanical properties (*e.g.* compressive stiffness, elasticity), and degradation (see *e.g.*, [4] for a review).

One extensively used, representative scaffold for cartilage tissue engineering is the highly porous mesh made of biodegradable fibrous polyglycolic acid (PGA), a material used for decades to make absorbable surgical sutures [12,13]. When PGA scaffolds were seeded at sufficiently high densities of bovine chondrocytes (typically $5 \cdot 10^6$ cells per disc-shaped scaffold, 5 mm diameter by 2 mm thick) and cultivated *in vitro*, polymer degradation rate matched the rate of ECM accumulation [16]. PGA scaffolds also supported cartilaginous differentiation of chick bone marrow stromal cells (BMSCs) [42]. However, mammalian BMSCs required a more mechanically stable scaffold [43]. Macroporous polymer foams made of an 80:20 blend of poly(lactic-co-glycolic acid)/poly(ethylene glycol) [44] enabled the control of scaffold structure, degradation and mechanical properties, and supported the growth and differentiation of bovine calf BMSCs [40]. Other materials successfully used for cartilage tissue engineering include agarose gel [2,45-49], fibrous and porous polylactic and polylactic-glycolic acid [50,20,51-53], fibrous and porous benzylated hylaruonic acid [54]. The beneficial effects of scaffolds were amplified in hydrodynamically active environments [54].

2.6. BIOREACTOR HYDRODYNAMICS

Hydrodynamic factors present during culture can modulate chondrogenesis in at least two ways: via associated effects on mass transport between the developing tissue and culture medium (*e.g.* oxygen, nutrients, growth factors), and by physical stimulation of the cells (*e.g.* shear, pressure). *In vivo*, mass transfer within articular cartilage involves diffusion in conjunction with fluid flow that accompanies tissue loading and unloading. *In vitro*, mass transfer has been shown to determine the size and composition of engineered constructs and cultured cartilage explants [11,19,55]. Explants were sectioned from middle sections of full thickness plugs of bovine articular cartilage in form of 5 mm diameter by 2 mm thick discs. Constructs were prepared by seeding chondrocytes isolated from full thickness cartilage into 5 mm diameter by 2 mm thick scaffolds made of fibrous PGA. Constructs and freshly harvested cartilage explants were subdivided into three groups each and cultured for up to 6 weeks in static flasks, mixed flasks or rotating bioreactors (Fig. 2).

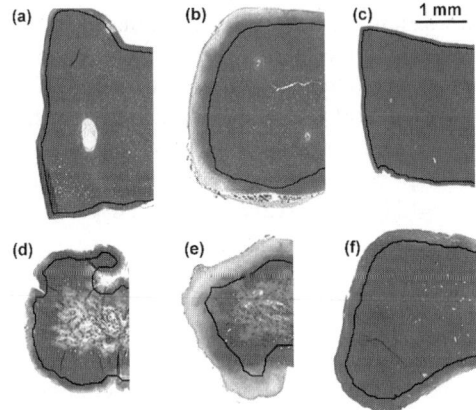

Figure 3. Effects of hydrodynamic factors on tissue morphology. Histological sections of cartilage explants (a, b, c) and engineered cartilage constructs (d, e, f) cultured for 6 weeks in static flasks (a, d), mixed flasks (b, e) and rotating bioreactors (c, f) (see Figure 2 for details of each hydrodynamic environment). Safranin-O stained cross-sections were bisected and one representative half is shown for each group. (Reproduced with permission from Vunjak-Novakovic et al. [55]).

In static construct cultures, GAG accumulated mostly at the periphery (Fig. 3d), presumably due to diffusionally constrained mass transfer. Likewise, in static explant cultures, the presence of GAG was better maintained at the periphery than in the central zone (Fig. 3a). In mixed flasks, mixing enhanced mass transport throughout culture medium and at the tissue surfaces, but the associated turbulent shear was related to the formation of an outer fibrous capsule at construct and explant surfaces (Fig. 3b, e). Only in rotating vessels were GAG concentrations in cultured constructs and explants high and spatially uniform (Fig. 3c, f). All tissues contained an external region with elongated cells and low GAG content (Fig. 3, solid lines). The thickness of this region was 70 - 265 µm for constructs and explants cultured in static flasks and rotating vessels and ~ 450 µm for constructs and explants cultured in mixed flasks [56]. The gradients in GAG level at the tissue surfaces were consistent with the measured GAG loss into the culture medium [19]. Over 6 weeks of culture, the fractional release of newly synthesised GAG was 10 – 30% in static flasks and rotating vessels, as compared to 40 – 60% in mixed flasks.

The effects of bioreactor hydrodynamics on biochemical compositions of cultured tissues were substantial, and comparable for engineered constructs and cartilage explants (Table 1). After 6 weeks of cultivation, constructs and explants from rotating vessels had significantly higher wet weights and accumulated significantly higher total amounts of GAG and total collagen, as compared to either static or mixed flasks. Importantly, the wet weight fractions of GAG and collagen were also markedly and significantly higher in constructs and explants cultured in rotating vessels than in those cultured in static and mixed flasks. The total amount of all quantified tissue components (cells, GAG and collagen) in cultured constructs and explants increased from static flasks to mixed flasks and rotating vessels. Also, cartilage-specific components represented most of the sample dry weight for tissues cultured in rotating vessels, whereas tissues from static and mixed flasks contained substantial fractions (30 – 40%) of unspecified components. Taken together, these data strongly suggest that bioreactors with dynamic laminar flow support chondrogenesis in native and engineered cartilage much better than either static or turbulent flow conditions. In addition, the same effects observed for native and engineered cartilage with respect to all measured parameters suggest that engineered cartilage represents an accurate yet controllable tissue model for basic *in vitro* studies.

2.7. GROWTH FACTORS

Growth factors supplemented sequentially to culture medium (TGF-β/FGF-2 early, IGF-I later) markedly and significantly improved the compositions and mechanical properties of engineered cartilage [39]. After 4 weeks of culture, constructs contained up to 4.5% ww GAG, up to 4.5% ww collagen and had equilibrium moduli of up to 400 kPa [39]. Gene transfer of IGF-I also resulted in markedly larger amounts of GAGs and collagen, both total and per unit DNA, and had 4-fold higher equilibrium moduli after 4 weeks of culture, as compared to non-transfected or lacZ constructs [41]. The observed enhancement of chondrogenesis by spatially defined over-expression of human IGF-I suggested that cartilage tissue engineering based on genetically modified cells may be advantageous compared to either gene transfer or tissue engineering alone. Beneficial effects of growth factors can be amplified by dynamic mechanical loading. TGF-β and IGF-I interacted with dynamic loading applied during culture in a synergetic manner and improved the compositions and mechanical properties of cultured constructs to the extent greater than the sum of effects of either stimulus applied alone [57-59]. Likewise, hydrodynamically active environment present in rotating bioreactors amplified the beneficial effects of polymer scaffolds on construct compositions and mechanical properties and yielded engineered cartilage that had equilibrium moduli of 400 – 540 kPa after only 4 weeks of bioreactor cultivation [54].

Table 1: Effects of bioreactor hydrodynamics on biochemical compositions of engineered constructs and cartilage explants cultured for 6 weeks (# significantly different from mixed flasks; * significantly different from static flasks; n = 3-4).

Parameter	Initial 3 days	6-week constructs		
		Static flask	Mixed flask	Rotating vessel
Wet weight (mg per sample)	78 ± 14	191 ± 9	175 ± 12	237 ± 21 *,#
Cells (millions per sample)	6.6 ± 1.0	9.7 ± 0.1	16.6 ± 3.3	14.4 ± 2.0 *
GAG (mg per sample)	0.56 ± 0.10	5.33 ± 0.62	3.82 ± 0.10	11.20 ± 1.27 *,#
Total collagen (mg per sample)	0.38 ± 0.12	2.78 ± 0.24	4.79 ± 0.06	8.15 ± 0.73 *,#
Cells (% wet weight)	0.81 ± 0.02	0.49 ± 0.02	0.94 ± 0.16	0.61 ± 0.03 *,#
GAG (% wet weight)	0.71 ± 0.03	2.73 ± 0.20	2.19 ± 0.17	4.71 ± 0.41 *,#
Total collagen (% wet weight)	0.48 ± 0.08	1.41 ± 0.08	2.74 ± 0.16	3.79 ± 0.05 *,#
Type II collagen (% total)	67.5 ± 4.3	75.4 ± 12.4	54.8 ± 6.3	83.6 ± 3.9 #
Cells + GAG + collagen (% wet weight)	2.00 ± 0.03	4.63 ± 0.13	5.88 ± 0.33	9.03 ± 0.53 *,#
Cells + GAG + collagen (% dry weight)	26.4 ± 4.7	62.4 ± 3.6	63.1 ± 3.6	86.9 ± 5.1 *,#
Parameter	Initial (native cartilage)	6-week explants		
		Static flask	Mixed flask	Rotating vessel
Wet weight (mg per sample)	54 ± 8	167 ± 17	189 ± 18	225 ± 32 *,#
Cells (millions per sample)	4.8 ± 1	8.9 ± 1.6	13.9 ± 4.5	14.3 ± 1.2 *
GAG (mg per sample)	3.2 ± 0.7	4.2 ± 0.3	6.0 ± 0.3	15.3 ± 0.5 *,#
Total collagen (mg per sample)	3.9 ± 1.4	8.2 ± 0.8	13.1 ± 2.9	16.7 ± 4.9 *
Cells (% wet weight)	0.90 ± 0.10	0.53 ± 0.04	0.73 ± 0.03	0.67 ± 0.24
GAG (% wet weight)	6.10 ± 0.03	2.55 ± 0.15	3.41 ± 1.10	6.88 ± 0.41 *,#
Total collagen (% wet weight)	8.80 ± 1.30	4.97 ± 0.56	7.12 ± 1.10	7.35 ± 0.98 *
Type II collagen (% total)	100.4 ± 5.4	85.6 ± 2.7	74.0 ± 6.4	84.2 ± 5.2 #
Cells + GAG + collagen (% wet weight)	15.93 ± 2.45	8.05 ± 0.68	11.26 ± 2.07	14.89 ± 0.71 *
Cells + GAG + collagen (% dry weight)	100.6 ± 12.4	69.2 ± 4.5	68.5 ± 15.8	95.0 ± 5.5 *,#

Taken together, these studies demonstrated that the application of growth factors in concert with physical stimuli can help reduce the time needed to engineer immature (foetal-like) but functional cartilage to within approximately 4 weeks. Synergetic effects between the hydrodynamic environment, scaffolds and growth factors will likely determine the optimal conditions for functional tissue engineering of cartilage.

2.8. DURATION OF CULTURE

With increasing cultivation time, the constructs more closely approximated articular cartilage both structurally and functionally (Table 2) [15,11]. The changes in biomechanical construct properties correlated with the respective changes in biochemical construct compositions. The fraction of GAG increased progressively from very low at 3 days to significantly higher than physiological at 7 months, whereas the fraction of total collagen increased during the first 6 weeks but remained at this level for the duration of culture. Biomechanical construct properties correlated with construct compositions. Six-week constructs had 75% as much GAG and 40% as much collagen per unit wet weight, equilibrium moduli of approximately 0.175 MPa, and hydraulic permeabilities that were 4-fold higher than those for native cartilage. In 7-month constructs, both the equilibrium modulus and the hydraulic permeability became comparable with those measured for native cartilage (Table 2) [15,11]. Importantly, the wet weight fraction of collagen, which was subnormal when compared to adult articular cartilage, corresponded to that measured for foetal cartilage, both at 6 weeks and 7 months of cultivation. Likewise, the equilibrium modulus and hydraulic permeability of 6-week constructs were in the range of values measured for foetal cartilage [25]. These studies suggest that foetal-like cartilage can be engineered over a period of 6 weeks *in vitro*, and that prolonged cultivation does not necessarily result in improved construct structures, in particular with respect to collagen contents.

Table 2. Effects of cultivation time on biochemical compositions and mechanical properties of constructs cultured in rotating vessels (*significantly different from 6-week data; # significantly different from native cartilage; n = 3-4).

Parameter	Construct cultivation time			Native cartilage (freshly explanted)
	3 days	6 weeks	7 months	
GAG (% wet weight)	0.71 ± 0.03	4.71 ± 0.41	8.83 ± 0.93*,#	6.81 ± 1.12
Total collagen (% wet weight)	0.48 ± 0.08	3.79 ± 0.05	3.68 ± 0.27 #	9.69 ± 1.68
Equilibrium modulus (MPa)	~ 0	0.172 ± 0.035	0.932 ± 0.049 *	0.939 ± 0.026
Hydraulic permeability ($\times 10^{15}$ m^4/Ns)	~ ∞	10.4 ± 5.0	3.7 ± 0.2	2.4 ± 0.6

Prolonged cultivation can also decrease the capacity of engineered cartilage for integration with native cartilage. In immature constructs, integration involved cell proliferation and the formation of strong cartilaginous tissue bond at the interface with native cartilage, whereas more mature constructs that were cultured for longer time, integrated less well, as evaluated both histologically and biomechanically [60]. Bioreactor studies of construct integration with native cartilage suggested that the duration of cultivation is likely to be determined by a proper balance between the compressive stiffness and integrative potential at the time of implantation. These studies also implied the need to provide additional physiologically relevant regulatory signals during bioreactor cultivation to achieve more rapid chondrogenesis. When implanted in large osteochondral defects in adult rabbit knees, engineered constructs cultured for 4-6 weeks and sutured to an osteoconductive support, remodelled over 6 months into cartilage with normal architectural features (tidemark, columnar arrangement of chondrocytes), composition and mechanical properties, and into new subchondral bone. These results suggest that engineered cartilage has the capacity to further develop following implantation, and to remodel into physiologically thick and mechanically stable osteochondral tissue.

2.9. SPATIAL AND TEMPORAL PATTERNS OF CHONDROGENESIS

The progression of chondrogenesis in engineered constructs based on bovine calf chondrocytes cultured on fibrous PGA scaffolds in rotating bioreactors has been associated with temporal and spatial changes in local concentrations of the cells and matrix shown in Figure 4. Cells at the construct periphery proliferated more rapidly during the first 4 days of culture and initiated the matrix deposition in this same region (Fig. 4d). Over time, chondrogenesis progressed both inward towards the construct centre and outward from its surface. Cell density gradually decreased and became more uniform, as the cells separated themselves by newly synthesised matrix and the construct size increased (Fig. 4d, e, f). After 10 days of culture, cartilaginous tissue was formed at the construct periphery (Fig. 4b).

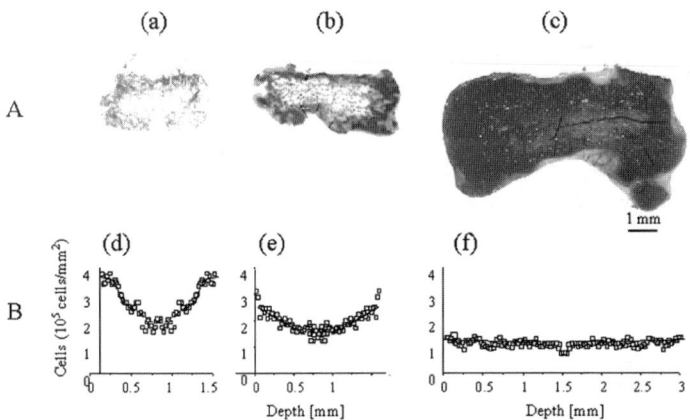

Figure 4. In vitro chondrogenesis. (A) Full cross-sections of tissue constructs after (a) 3 days, (b) 10 days and (c) 6 weeks of culture. Stain: safranin-O/fast green. Scale bar: 1 mm. (B) Spatial profiles of cell distribution after (d) 3 days, (e) 10 days and (f) 6 weeks of culture (measured by image processing) (based on data reported in Obradovic et al. [18]).

By 6 weeks of culture, self-regulated cell proliferation and deposition of cartilaginous matrix yielded constructs that had physiological cell density and spatially uniform distributions of matrix components (Fig. 4c). Notably, the patterns of chondrogenesis were quite different for cells embedded in gels where cell clusters formed and accumulated matrix over time but remained separated from each other after 6 weeks of culture [61]. Construct compositions and mechanical properties were generally better for fibrous meshes [11,62,39,37] than for agarose gels [61], which may be due to the spatial continuity of the cartilaginous matrix formed in constructs based on fibrous meshes. In contrast, mechanical stimulation improved construct compositions only for chondrocytes cultured in agarose gel [57-59], which may be due to the enhanced signal transduction and fluid flow through the gel between the cell clusters.

2.10. MATHEMATICAL MODEL OF CARTILAGE DEVELOPMENT

To facilitate data interpretation, mathematical models were developed, which yielded GAG concentrations as a function of time and position within the cultured tissues [18]. Production of GAGs was taken as a marker of chondrogenesis in light of prior association of GAG deposition with that of collagen type II, the other major component of cartilage tissue matrix [16]. To experimentally verify the models, we developed a high-resolution (40 μm) image analysis method that enabled the measurement of local GAG concentrations in histological tissue sections [56].

2.10.1. Mathematical model of GAG accumulation in cultured cartilage explants

The basic model was applied to the *in vitro* cultures of native cartilage explants in order to determine the effects of the conditions and duration of cultivation on an already formed tissue. Specifically, the model of chondrogenesis in cultured cartilage explants helped interpret the histological appearances of native cartilage explants cultivated for 6 weeks in static flasks, mixed flasks and rotating bioreactors (Fig. 3a, b, c) [63].

GAG concentration profiles within native cartilage explants were analysed as a function of GAG synthesis by the cells, diffusion of newly synthesised and not yet collagen-immobilised GAG and the resulting accumulation within the tissue matrix. In healthy cartilage *in vivo*, GAG concentration is maintained by a balance between GAG synthesis and catabolism [64]. Accordingly, local GAG kinetics were formulated as product inhibited with C_l as the maximum GAG concentration. The temporal changes in local GAG concentration (C_G) within a disc-shaped cartilage explant can be expressed as:

$$\frac{\partial C_G}{\partial t} = D_G \left(\frac{\partial^2 C_G}{\partial r^2} + \frac{1}{r}\frac{\partial C_G}{\partial r} + \frac{\partial^2 C_G}{\partial z^2} \right) + \rho \cdot k \cdot \left(1 - \frac{C_G}{C_l}\right) \qquad (1)$$

where r and z are cylindrical coordinates (GAG concentration is independent of the third coordinate due to symmetry), t is time of cultivation, D_G is the coefficient of GAG diffusion within the tissue, ρ is the cell density, and k is the apparent synthesis rate constant. Initial and boundary conditions for a disc-shaped explant are listed in Table 3.

Table 3. Initial and boundary conditions for an explant (disc of a diameter d, thickness h).

Initial conditions	$t = 0$	$0 \leq r < d/2; 0 \leq z < h/2$	$C_G = C_l$	GAG concentration within explants is equal to the limiting GAG concentration
Boundary conditions	$t \geq 0$	$r = d/2; 0 \leq z \leq h/2$ $z = h/2; 0 \leq r \leq d/2$	$C_G = 0$	GAG concentration at the explant surfaces is equal to that measured in the bulk medium (~ 0)
	$t > 0$	$r = 0; 0 \leq z \leq h/2$	$\dfrac{\partial C_G}{\partial r} = 0$	symmetry conditions along explant axes
		$z = 0; 0 \leq r \leq d/2$	$\dfrac{\partial C_G}{\partial z} = 0$	

Based on the experimental measurements of GAG concentration profiles, we assumed that the inner tissue phase of cultured explants was not significantly affected by external hydrodynamic conditions, and that a constant GAG concentration C_l was maintained. This constant GAG concentration and the corresponding cell density were determined by image analysis of histological cross-sections. GAG diffusion coefficient, D_G, was estimated from measured rates of GAG release into the culture medium and the corresponding concentration gradients at the explant surfaces, using the following transport equation:

$$V_{med}\frac{dC_{Gmed}}{dt} = D_G S \frac{\partial C_G}{\partial \delta}\bigg|_{\delta=0} \qquad (2)$$

where V_{med} is the volume of medium, C_{Gmed} is GAG concentration in medium, and S is the surface area of the explant. All parameters affecting GAG synthesis rate were lumped in the apparent synthesis rate constant, k, which was the only adjustable model parameter. Estimated apparent GAG diffusion coefficients, D_G, (Eq. 2) and apparent GAG synthesis rate constants, k, determined by least-squares fits to experimental GAG distributions are presented in Table 4.

Table 4. Model parameters for cultures of native cartilage explants.

	k [% ww GAG day^{-1} (10^5 cell/mm^3)$^{-1}$]	D_G [cm^2/s]
Static flask	4.8	$5.31 \pm 0.05 \cdot 10^{-11}$
Mixed flask	0.2	$5.31 \cdot 10^{-11} - 5.00 \cdot 10^{-10}$*
Rotating bioreactor	6.9	$5.31 \pm 0.05 \cdot 10^{-11}$

* value of D_G in this culture was predicted to increase as a 2nd order polynomial function of cultivation time

GAG concentration profiles in cartilage explants cultivated for 6 weeks in static flasks, mixed flasks and rotating vessels were calculated using the numerical model [63]. Very good agreements of model predictions with experimental data for GAG concentrations were obtained for all culture conditions (average SD = ± 0.1%, Fig. 5).

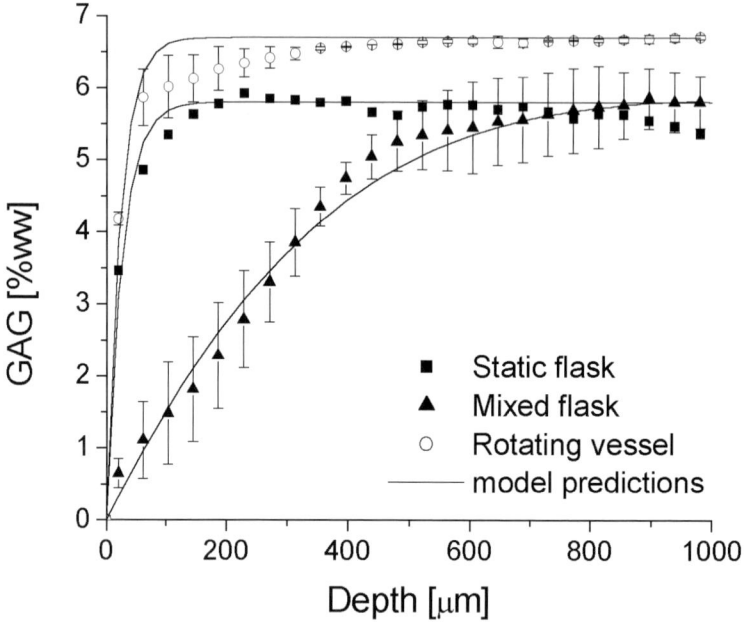

Figure 5. GAG distributions in cartilage explants cultured for 6 weeks in different hydrodynamic environments; experimental data (symbols, average ± SD, n = 2-4) and model predictions (lines) [63].

Modelling of GAG distribution in native cartilage explants quantified the effects of hydrodynamic conditions on GAG synthesis and diffusion rates. As expected, diffusion coefficients of GAG were comparable for explants cultured in static flasks and rotating bioreactors, whereas dynamic laminar flow in rotating vessels had strong positive effect on the kinetic constants of GAG synthesis. For explants cultured in mixed flasks, GAG distribution profiles could be reconciled only if an order of magnitude increase in D_G over 6 weeks of cultivation was incorporated in the model (Table 4), possibly due to the formation of an external capsule that became progressively depleted of GAG. In addition, GAG synthesis rate in mixed flasks was markedly lower than in either static flasks or rotating vessels, consistent with the presence of the outer thick fibrous capsule containing high concentration of elongated fibroblast-like cells (Fig. 3b). In overall, these findings suggest that the flow regime (preferably dynamic laminar than steady turbulent) rather than the magnitude of shear stress at tissue surfaces (similar for mixed flasks and rotating vessels) determined the synthesis and accumulation of tissue matrix.

2.10.2. Mathematical model of GAG accumulation in engineered cartilage constructs

The same modelling approach was extended to cultures of tissue engineered cartilage in rotating bioreactors with the aim to rationalise the temporal and spatial progression of GAG accumulation in constructs and quantify the effects of oxygen. The choice of culture system was dictated by structural and mechanical properties of engineered constructs, which were superior in rotating vessels as compared to other culture systems studied (Table 1). Oxygen concentration in culture medium was shown to significantly affect the *in vitro* chondrogenesis such that low oxygen concentrations (42.7 ± 4.5 mmHg as compared to 86.5 ± 7.3 mmHg) suppressed the growth of both native and engineered cartilage [17]. The model predicted the concentrations of oxygen and GAG as functions of spatial position within the tissue and time of culture, and incorporated the dependence of GAG synthesis rate on local oxygen concentration.

Table 5. *Initial and boundary conditions for a construct of a diameter d and thickness h.*

Initial conditions	$t = 0$	$0 \leq r < d/2;\ 0 \leq z < h/2$	$C_G = 0$ $C_{O_2} = C^o_{O_2}$	GAG concentration within the construct is negligible; oxygen concentration is equal to that in the bulk medium
Boundary conditions	$t \geq 0$	$r = d/2;\ 0 \leq z \leq h/2$ $z = h/2;\ 0 \leq r \leq d/2$	$C_G = 0$ $C_{O_2} = C^o_{O_2}$	GAG and oxygen concentrations at construct surfaces are equal to those measured in the bulk medium
	$t > 0$	$r = 0;\ 0 \leq z \leq h/2$	$\dfrac{\partial C_G}{\partial r} = 0$	symmetry condition
			$\dfrac{\partial C_{O_2}}{\partial r} = 0$	
		$z = 0;\ 0 \leq r \leq d/2$	$\dfrac{\partial C_G}{\partial z} = 0$	symmetry condition
			$\dfrac{\partial C_{O_2}}{\partial z} = 0$	

The central hypothesis was that the rate of GAG synthesis depended on local oxygen concentration (C_{O2}) according to the first order kinetics. The model equation (1) is then slightly modified:

$$\frac{\partial C_G}{\partial t} = D_G \left(\frac{\partial^2 C_G}{\partial r^2} + \frac{1}{r}\frac{\partial C_G}{\partial r} + \frac{\partial^2 C_G}{\partial z^2} \right) + \rho \cdot k \cdot \left(1 - \frac{C_G}{C_l}\right) \cdot C_{O_2} \qquad (3)$$

where the same notation as in 2.10.1. is retained, and a disc-shaped construct is assumed.

Temporal changes in local oxygen concentration were assumed to be governed by oxygen transport from culture medium to the cells (molecular diffusion) and cellular consumption (Michaelis-Menten kinetics), that is:

$$\frac{\partial C_{O_2}}{\partial t} = D_{O_2}\left(\frac{\partial^2 C_{O_2}}{\partial r^2} + \frac{1}{r}\frac{\partial C_{O_2}}{\partial r} + \frac{\partial^2 C_{O_2}}{\partial z^2}\right) - \rho \cdot \frac{Q_m C_{O_2}}{C_m + C_{O_2}} \quad (4)$$

where D_{O_2} is the oxygen diffusion coefficient in constructs, Q_m is the maximum rate of oxygen consumption, and C_m is the C_{O_2} at half-maximum consumption rate. The initial and boundary conditions are listed in Table 5.

In order to incorporate appreciable tissue growth over time (Fig. 4a, b, c), construct diameter (d) and thickness (h) were forced to vary according to experimental observations. The resulting changes in cell density (ρ) were experimentally determined by image analysis of histological cross-sections of samples taken at different time points (Fig. 4d, e, f) and interpolated as a function of time. The limiting GAG concentration, C_l, and the coefficient of GAG diffusion, D_G, were estimated as described above (see 2.10.1.). The maximum rate of oxygen consumption was experimentally determined, while C_m and oxygen diffusion coefficient, D_{O2}, were set to values reported in literature. All other parameters affecting GAG synthesis rate were lumped into the synthesis rate kinetic constant, k, which was the only adjustable model parameter. Based on experimental evidence [16], the change of k with time t was incorporated into the model:

$$k = k_o[1 + A(t - t_o)] \quad (5)$$

where k_o is the apparent kinetic rate of GAG synthesis in the initial culture period, and A is the experimentally determined parameter. The increase in k was not observed at low oxygen tensions.

The model predicted a gradual decrease of oxygen concentration from the construct surface towards its centre, as a result of oxygen consumption by the cells. Due to the higher total number of cells and larger diffusional distances, the decrease in oxygen concentration was markedly higher in 6-week as compared to 10-day constructs (Fig. 6a, b).

Model predictions for concentration profiles of GAG (Fig. 6c, d, lines) were consistent with those measured via image processing of tissue samples (Fig. 6c, d, data points). The coefficient of GAG diffusion in engineered constructs was estimated to 7 10^{-11} cm^2/s and the only adjustable parameter, the initial value of apparent GAG synthesis rate constant, k_o, was determined by a least-squares fit to the GAG distribution measured in constructs cultured for 10 days. As shown in Figure 6, the qualitative and quantitative agreements of model predictions with experimental data for both time points were excellent (average SD = ± 0.2 % wet weight GAG).

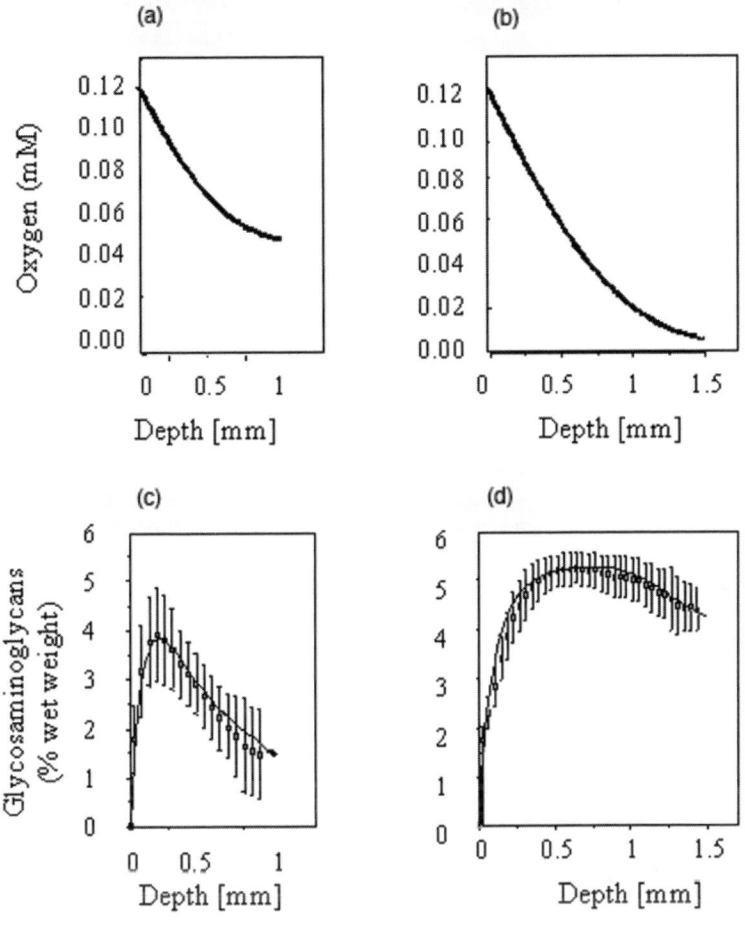

Figure 6. In vitro *chondrogenesis. Spatial profiles of oxygen distribution after (a) 10 days and (b) 6 weeks of culture (model predictions). Spatial profiles of glycosaminoglycan distribution after (c) 10 days and (d) 6 weeks of culture (data points: measured by image processing; lines: model predictions) (based on data reported in [18]).*

Mathematical models presented here, although only grossly approximating the processes of GAG accumulation in cartilaginous tissues under *in vitro* conditions, indicated some of the mechanisms and parameters affecting tissue regeneration. As information is accumulated about biochemical pathways and the regeneration of other tissue components, in particular type II collagen, the model may be refined and extended and, ultimately, provide a tool for understanding and optimising the cultivation of functional equivalents of native tissues for a variety of clinical applications.

3. Cardiac tissue engineering

3.1. MYOCARDIUM (CARDIAC MUSCLE)

The myocardium (cardiac muscle) is a highly differentiated tissue composed of cardiac myocytes and fibroblasts with a dense supporting vasculature and collagen-based extracellular matrix. The myocytes form a three-dimensional syncytium that enables propagation of electrical signals across specialized intracellular junctions to produce coordinated mechanical contractions that pump blood forward. Only 20-40% of the cells in the heart are cardiac myocytes but they occupy 80-90% of the heart volume. The average cell density in the native rat myocardium is on the order of $5 \cdot 10^8$ cells/cm^3. Morphologically, intact cardiac myocytes have an elongated, rod shaped appearance. Contractile apparatus of cardiac myocytes consists of sarcomeres arranged in parallel myofibrils. High metabolic activity is supported by the high density of mitochondria and electrical signal propagation is provided by specialised intercellular connections, gap junctions [65,66].

The control of heart contractions is almost entirely self-contained. Groups of specialised cardiac myocytes (pace makers), fastest of which are located in the sinoatrial node, drive periodic contractions of the heart. Majority of the cells in the myocardium are non-pace maker cells and they respond to the electrical stimuli generated by pace maker cells. Excitation of each cardiac myocyte is followed by the increase in the amount of cytoplasmic calcium, which triggers mechanical contraction. The propagation of the electrical excitation through the tissue by ion currents in the extracellular and in the intercellular space results in synchronous contraction that enables expulsion of the blood from the heart.

3.2. CLINICAL NEED

Cardiovascular disease is responsible for a preponderance of health problems in the developed countries, as well as in many developing countries. Heart disease and stroke, the principal components of cardiovascular disease, are the first and the third leading cause of death in the U.S., accounting for nearly 40% of all deaths. Congenital heart defects, which occur in nearly 14 of every 1000 newborn children [67], are the most common congenital defects and the leading cause of death in the first year of life [68,69]. Cardiovascular diseases result in substantial disability and loss of productivity, and largely contribute to the escalating costs of health care. About 61 million Americans (almost one-fourth of the population) live with cardiovascular diseases, such as coronary heart disease, congenital cardiovascular defects, and congestive heart failure, and 298.2 billion dollars were spent in 2001 to treat these diseases [70]. The economic impact of cardiovascular disease on the U.S. health care system is expected to grow further as the population ages.

Once damaged, the heart is unable to regenerate. Heart failure affects over five million Americans [71], and is the leading cause of morbidity and mortality in developed countries [72]. Currently, the only definitive treatment for end stage heart failure is cardiac transplantation. However, the limited availability of organs for transplantation has led to prolonged waiting periods that are often not survivable [73]. Repair of myocardial injuries has been attempted by injection of myogenic cells into scarred myocardium [74-76] and the replacement of scarred tissue with engineered grafts [6,7,77].

3.3. TISSUE ENGINEERING

Tissue engineering has emerged over the last decade as an interdisciplinary field with tremendous potential. Tissue engineering offers a possibility of creating tissue constructs to be used for repair of larger injuries or congenital malformations. In addition cardiac tissue constructs may be utilized for studies of normal and pathological tissue function *in vitro*. Substantial progress has been made in areas of biopolymers [78], cell-material interactions [79] and bio-mimetic culture devices [80]. In addition, functional tissues have been developed and implanted *in vivo,* including (but not limited to) cartilage [16], bone [81], bladder [82] and blood vessels [83,84]. However, fundamental and all-encompassing problems remain. One of the most important ones is mass transfer into the tissues that are greater than 100-200 μm in thickness, both during the *in vitro* cultivation and following implantation *in vivo*. This explains why tissue engineering has been most successful with tissues that are either thin (*e.g.*, bladder) or have low oxygen requirements (*e.g.*, cartilage). There are multiple challenges to be overcome to produce viable, functional cardiac tissue:

- Development of appropriate biodegradable and biocompatible scaffolds that can provide adequate mass transfer, vascularisation, and transduction of mechanical and electrical signals.
- Development of bio-mimetic culture systems that promote differentiated function and excitation - contraction coupling of cardiac cells.
- Development of functional vascular networks embedded into the cardiac muscle tissue to promote nutrient and oxygen transfer, angiogenesis and integration with the host vasculature.

In current approaches, foetal or neonatal rat cardiomyocytes are seeded onto scaffolds (collagen sponges, polyglycolic acid meshes) or cast in collagen gels, and cultivated immersed in the culture medium in static or mixed dishes, spinner flasks or rotating vessels [85,86]. The metabolism and viability of the resulting constructs are assessed during culture by monitoring levels of glucose consumed, lactate produced and the release of lactate dehydrogenase in the samples of culture medium. Cell distribution, morphology and construct structure are assed by histology. Expression of cardiac specific markers is assessed by immunohistochemistry and confirmed by Western blots. Gene expression is assessed by RT-PCR. Cardiac specific ultrastructural features are detected by transmission electron microscopy. Functional assessment of engineered constructs has been based on electrophysiological studies [87,88], monitoring of synchronous contractions in response to electrical stimuli [89] and measurement of force of contraction in paced [77] or spontaneously contracting constructs [90].

3.4. CELLS

Three dimensional cardiac tissue constructs were successfully cultivated in dishes using variety of scaffolds and cell sources. Foetal rat ventricular cardiac myocytes were expanded after isolation, inoculated into collagen sponges and cultivated in static dishes for up to 4 weeks [6]. The cells proliferated with time in culture and expressed multiple sarcomeres. Adult human ventricular cells were used in a similar system, although they exhibited no proliferation [91]. Foetal cardiac cells were also cultivated on alginate scaffolds in static 96-well plates. After 4 days in culture the cells formed spontaneously beating aggregates in the scaffold pores [7]. Cell seeding densities of the order of 10^8 cells/cm^3 were achieved in the alginate scaffolds using centrifugal forces during seeding [92]. Neonatal rat cardiac myocytes formed spontaneously contracting constructs when inoculated in collagen sponges within 36 hr [90] and maintained their activity for up to 12 weeks. The contractile force increased upon addition of Ca^{2+} and epinephrine.

Two-week constructs based on neonatal rat cardomyocytes exhibited spontaneously beating areas, whereas constructs based on embryonic chick myocytes exhibited no contractions and reduced in size by 60%. Immunohistochemistry, revealed presence of large number of non-myocytes in the constructs based on embryonic chick hear cells, while constructs based on neonatal rat cells consisted mostly of elongated cardiomyocytes [85].

Constructs based on the cardiomyocytes enriched by preplating exhibited lower excitation threshold (ET), higher conduction velocity, higher maximum capture rate (MCR), and higher maximum and average amplitude [87].

3.5. SCAFFOLDS

The scaffolds utilised for cardiac tissue engineering include collagen fibres [93], collagen sponges [6,91,90] and polyglycolic acid meshes [85,87,86]. The main advantage of a synthetic scaffold such as PGA is that it provides mechanical stability, while scaffolds based on natural cell polymers such as collagen enable rapid cell attachment.

The scaffold free approaches include casting the cells in collagen gels followed by mechanical stimulation [94-96,77] and stacking of confluent cardiac cell monolayers [8]. The main advantage of scaffold free approaches is higher active force generated by such tissues. However, the main disadvantage remains tailoring the shape and dimensions of the scaffold free engineered tissues.

As an alternative, gels (Matrigel®) were combined with scaffolds (collagen sponge) to achieve rapid cell inoculation and attachment along with the possibility of tailoring tissue shape and dimensions through the use of scaffolds [89].

3.6. BIOREACTOR HYDRODYNAMICS

The representative bioreactors utilised for tissue engineering of the myocardium include static or mixed dishes, static or mixed flasks and rotating vessels (Fig 2). These bioreactors offer three distinct flow conditions (static, turbulent, and laminar) and therefore differ significantly in the rate of oxygen supply to the surface of the tissue construct. Oxygen transport is a key factor for myocardial tissue engineering due to the high cell density, very limited cell proliferation and low tolerance of cardiac myocytes for hypoxia. In all configurations oxygen is supplied only by diffusion from the surface to the interior of the tissue construct.

Table 6. Overview of cardiac tissue engineering studies

Bioreactor type	Cell source	Scaffold type	Construct size	Initial cell number	Ref.
Static dish (96 well plate)	Foetal rat CM	Alginate	6 mm diameter x 1 mm thick	$3 \cdot 10^5$	[7]
Static dish (5 ml culture medium)	Foetal rat CM	Bovine collagen (Gelfoam)	5 x 5 x 1 mm	$\sim 10^4$	[6]
Static dish (4 ml culture medium)	Neonatal rat CM	Bovine collagen (Tissue Fleece)	20 x 15 x 2.5 mm	$2 \cdot 10^6$	[90]
Static dish	Neonatal rat CM	Bovine collagen	5 x 5 x 3 mm	$0.5 \cdot 10^6$	[99]
Mixed dish (20rpm, 4ml culture medium)	Neonatal rat CM	Bovine collagen (Ultrafoam)	10 mm diameter x 1.5 mm thick	$6\text{-}12 \cdot 10^6$	[98]
Spinner flask (0, 50, 90 rpm, 120 ml culture medium)	Embryonic chick CM Neonatal rat CM	PGA	5 mm diameter x 2 mm thick	$1.3\text{-}8 \cdot 10^6$	[85]
Spinner flask (50 rpm, 120 ml culture medium)	Neonatal rat CM	PGA, sPGA	5 mm diameter x 2 mm thick	$8 \cdot 10^6$	[87,86]
Rotating vessels (11 rpm, 100 ml culture medium)	Neonatal rat CM	PGA, sPGA, lPGA	5 mm diameter x 2 mm thick	$8 \cdot 10^6$	[85,87,86]
Rotating vessel	Neonatal rat CM	Bovine collagen	5 x 5 x 3 mm	$0.5 \cdot 10^6$	[99]
Perfused cartridge (0.2-3 ml/min)	Neonatal rat CM	PGA	11 mm diameter x 2 mm thick	$24 \cdot 10^6$	[97,100]
Perfused cartridge (0.2-3 ml/min)	Neonatal rat CM	Bovine collagen (Ultrafoam)	10 mm diameter x 1.5 mm thick	$6\text{-}12 \cdot 10^6$	[98]
Cyclic stretch (1.5-2 Hz, 1-20 % strain)	Embryonic chick CM Neonatal rat CM	Collagen gel		$1\text{-}2.5 \cdot 10^6$	[95,77]
Cyclic stretch (1.3 Hz)	Human heart cells	Bovine collagen (Gelfoam)	30 x 20 x 3 mm 20 x 20 x 3 mm	$3\text{-}30 \cdot 10^6$	[101]
Electrical stimulation (square biphasic pulses, 2 ms, 1 Hz, 5 V)	Neonatal rat CM	Bovine collagen (Ultrafoam)	8 x 6 x 1.5 mm	$5 \cdot 10^6$	*

*Radisic et al. (unpublished)

3.6.1. Static dishes

Static dishes remain the most widely used set-up for cardiac tissue engineering (Table 6). In static dishes, oxygen and nutrients are supplied mainly by diffusion (Fig. 2), which is capable of satisfying oxygen demand of only ~100 μm thick surface layer of compact tissue, whereas the construct interior remained mostly acellular [97]. In contrast, cartilage has been successfully grown in static dishes to millimetre thicknesses. In orbitally mixed dishes, the rate of delivery of oxygen and nutrients to construct surfaces can be increased, but diffusion remains the main mechanism of mass transport within the tissue. Diffusional transport of oxygen to cardiac myocytes within dish-grown constructs resulted in prevalently anaerobic glucose metabolism [98]. Additional limitation of the culture in static or mixed dishes is that the bottom surface of the construct (the surface closest to the bottom of the dish) often lacks proper oxygenation and nutrients yielding asymmetric cell distribution with compact tissue mostly on the top surface.

To improve cell survival and assembly on all surfaces of the engineered tissue, cardiac constructs were cultivated suspended in the spinner flasks (Table 6). Cultivation in spinner flasks (at stirring rates of up to 90 rpm) improved construct properties [85], presumably due to enhanced mass transport at construct surfaces (Fig. 2). After 2 weeks of culture, constructs from mixed flasks had significantly higher cellularity index (~ 20 μg DNA/construct) and metabolic activity (~ 150 MTTunits/mg DNA) than those from static flasks (~ 5 μg DNA/construct and ~ 50 MTT units/mg DNA). Mixing maintained medium gas and pH levels within the physiological range yielding a more aerobic glucose metabolism (L/G ~ 1.5) in mixed flasks as compared to static flaks (L/G>2) [85]. Constructs contained a peripheral tissue-like region (50-70 μm thick) in which cells stained positive for tropomyosin and organised in multiple layers in a 3-D configuration [87] Electrophysiological studies conducted using a linear array of extracellular electrodes showed that the peripheral layer of the constructs sustained macroscopically continuous impulse propagation on a centimetre-size scale [87]. However, construct interiors remained empty due to the diffusional limitations of the oxygen transport within the bulk tissue, and the density of viable myocytes was orders of magnitude lower than that in the neonatal rat ventricles [87]. Additional drawback of the cultivation in spinner flasks is that turbulent flow conditions may induce cell damage and dedifferentiation and result in the formation of a fibrous capsule at construct surfaces.

Laminar conditions of flow in rotating vessels (Table 6) enabled the maintenance of oxygen concentration in medium and pH within the physiological range, and resulted in mostly aerobic cell metabolism (L/G ~ 1) [85]. The metabolic activity of cells within constructs increased into the range of values measured for neonatal rat ventricles (~ 250 units MTT/mg DNA) and was significantly higher than in mixed flasks (~ 150 units MTT/mg DNA). The index of cell hyperthrophy was also comparable to the neonatal tissue (18 mg protein/mg DNA). However, the construct cellularity remained 2-6 times lower than in native heart ventricles. In the best experimental group (heart cells enriched for cardiac myocytes by pre-plating, laminin-coated PGA scaffolds, low serum concentration) the outer layer of viable tissue was up to 160 μm thick [86]. Cells expressed cardiac-specific markers (*e.g.*, tropomyosin, gap junction protein connexin, creatin kinase-MM, sarcomeric myosin heavy chain) at levels that were lower than in neonatal rat ventricles but higher than in constructs cultured in spinner flasks [86]. Electrophysiological properties were also improved, as evidenced by the prolonged action potential duration (APD, a measure of electrophysiological functionality of cell membrane), higher maximum capture rates (a measure of construct response to electrical pacing), and more physiological response to drugs. In particular, pharmacological studies done with 4-aminopyridine indicated that a decrease in transient outward potassium current may be responsible for the observed differences in APD and MCR [88]. In overall, dynamic laminar flow in rotating bioreactors improved properties of the peripheral tissue layer, but the limitations of the diffusional transport of oxygen to the construct interior were not overcome and constructs remained largely acellular.

3.6.2. Interstitial flow

In an attempt to enhance mass transport within cultured constructs, we developed a perfusion bioreactor that provides interstitial medium flow through the cultured construct at velocities similar to those found in native myocardium (~ 400–500 µm/s, [102]). In such a system oxygen and nutrients are supplied to the construct interior by both diffusion and convection.

In early studies, constructs were prepared by seeding cardiac myocytes onto PGA scaffolds in mixed flasks, a method that has been successfully used to seed chondrocytes. After 3 days, constructs were transferred into a perfusion cartridge and pulsatile flow of medium through the construct was provided by a peristaltic pump (0.2 - 3 ml/min) (Table 6). Gas exchange between culture medium and incubator air occurred in an external coil of silicone tubing within the medium recirculation loop. Perfusion during construct cultivation improved cell distribution, viability and differentiation. However, the overall cell density remained low due to the limitations of oxygen transport to the cells inside constructs during scaffold seeding in mixed flasks [97,100].

In subsequent studies, we developed a system that enabled the maintenance of oxygen supply to the cells at all times during cell seeding and construct cultivation, in order to achieve physiologic density of viable cells (Fig. 7, Table 6). The establishment of physiologic cell density is an essential requirement for functional tissue engineering because cardiac myocytes have no capacity for proliferation. The rate, yield and uniformity of cell seeding were significantly improved by combining (1) rapid cell inoculation into the scaffold, using Matrigel® as a delivery vehicle, and (2) immediate establishment of interstitial medium flow (Fig. 7A, [89]). Neonatal rat cardiomyocytes suspended in Matrigel® were inoculated into collagen sponges (10 mm diameter by 1.5 mm thick discs) at a physiologic cell density (1.35 10^8 cells/cm^3) and subjected to medium perfusion (0.5 ml/min, corresponding to ~ 400 µm/s). During the initial 1.5 h, the direction of medium flow was alternated to enhance cell attachment and prevent the washout of loosely bound cells; unidirectional medium flow was applied for the remained of cultivation (Fig 7B).

Medium perfusion markedly improved construct properties according to all tested parameters [98]. After 7 days, the viability of cells in perfused constructs was indistinguishable from that of freshly isolated cells (~ 85%), as compared to rather low viability of dish-grown constructs (~ 45%). The molar ratio of lactate produced to glucose consumed in perfused constructs (L/G ~ 1) indicated aerobic cell metabolism, in contrast to anaerobically metabolising (L/G ~ 2) dish-grown constructs. Perfused constructs and native ventricles had more cells in the S phase than in the G2/M phases, whereas the cells from dish-grown constructs appeared unable to complete the cell cycle and accumulated in the G2/M phase. Cells expressing cardiac-specific differentiation markers (sarcomeric α-actin, cardiac troponin I, sarcomeric tropomyosin) were present throughout the perfused constructs, while the dish-grown constructs exhibited spatially non-uniform cell distributions, with most cells located within a 100-300 μm thick surface layer, around an empty interior.

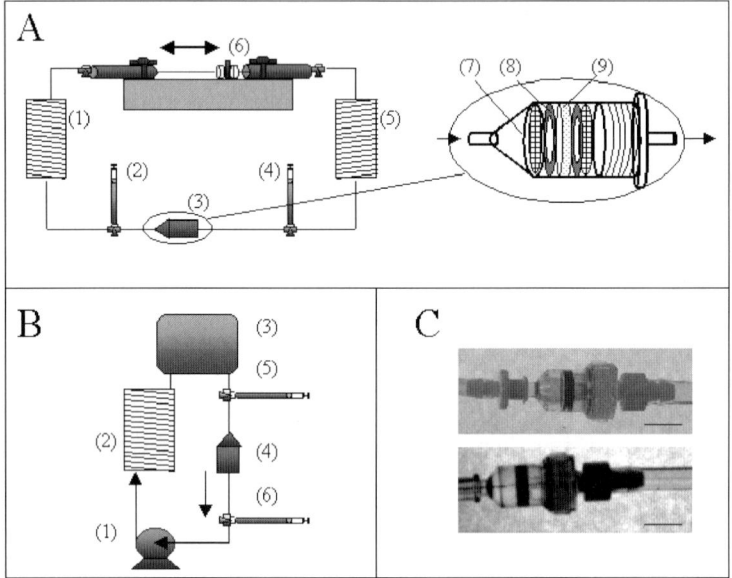

Figure 7. Perfused cartridges with interstitial flow of culture medium. (A) Scaffold seeding. Gel-cell inoculated scaffolds were placed between two stainless steel screens (7) and two silicone gaskets (8), and transferred into cartridges (3, one scaffold per cartridge). The screens (85% open area) provided mechanical support during perfusion and the gaskets (1 mm thick, 10 mm OD, 5 mm ID) routed the culture medium directly through the construct. Air bubbles were displaced by culture medium injected from the de-bubbling syringe (4) into the downstream syringe (2). Each cartridge was placed in a perfusion loop consisting of a push-pull pump (6), two gas exchangers (1, 5) and two syringes (2, 4). The total volume of medium was 8 ml. For a period of 1.5 h, the pump was programmed to the flow rate of 0.5 ml/min, with the reversal of flow direction after 2.5 ml was perfused in a given direction. (B) Construct cultivation. Each cartridge with a seeded construct (4) was transferred into a perfusion loop consisting of one channel of a multi-channel peristaltic pump (1), gas exchanger (a coil of thin silicone tubing, 3m long) (2), reservoir bag (3) and two syringes (5, 6). The total volume of medium in each loop was 32 ml. The flow rate was set at 0.5 ml/min and maintained throughout the 7 day cultivation. Constructs seeded and cultivated in orbitally mixed dishes (35 mm well, 8 ml medium, 25 rpm) served as controls. (C) Flow visualization. top: bulk flow through the construct centre, bottom: stream of tracer dye. Scale bar: 10 mm.

glucose consumed in perfused constructs (L/G ~ 1) indicated aerobic cell metabolism, in contrast to anaerobically metabolising (L/G ~ 2) dish-grown constructs. Perfused constructs and native ventricles had more cells in the S phase than in the G2/M phases, whereas the cells from dish-grown constructs appeared unable to complete the cell cycle and accumulated in the G2/M phase. Cells expressing cardiac-specific differentiation markers (sarcomeric α-actin, cardiac troponin I, sarcomeric tropomyosin) were present throughout the perfused constructs, while the dish-grown constructs exhibited spatially non-uniform cell distributions, with most cells located within a 100-300 μm thick surface layer, around an empty interior.

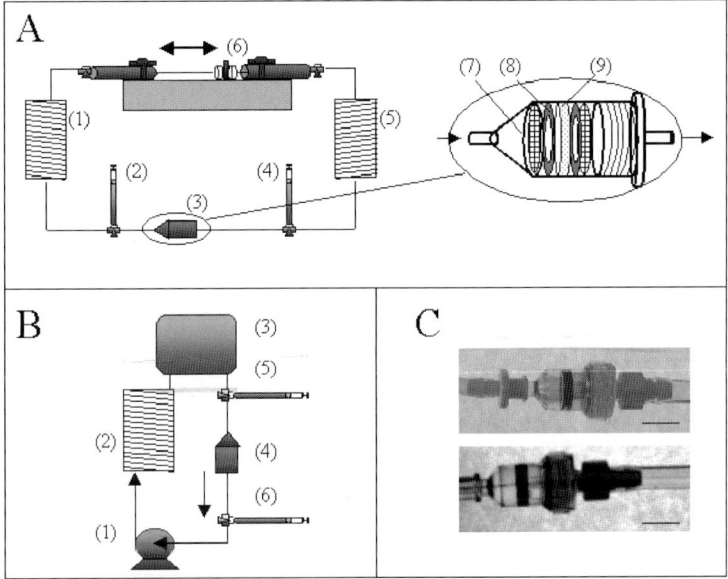

Figure 7. Perfused cartridges with interstitial flow of culture medium. (A) Scaffold seeding. Gel-cell inoculated scaffolds were placed between two stainless steel screens (7) and two silicone gaskets (8), and transferred into cartridges (3, one scaffold per cartridge). The screens (85% open area) provided mechanical support during perfusion and the gaskets (1 mm thick, 10 mm OD, 5 mm ID) routed the culture medium directly through the construct. Air bubbles were displaced by culture medium injected from the de-bubbling syringe (4) into the downstream syringe (2). Each cartridge was placed in a perfusion loop consisting of a push-pull pump (6), two gas exchangers (1, 5) and two syringes (2, 4). The total volume of medium was 8 ml. For a period of 1.5 h, the pump was programmed to the flow rate of 0.5 ml/min, with the reversal of flow direction after 2.5 ml was perfused in a given direction. (B) Construct cultivation. Each cartridge with a seeded construct (4) was transferred into a perfusion loop consisting of one channel of a multi-channel peristaltic pump (1), gas exchanger (a coil of thin silicone tubing, 3m long) (2), reservoir bag (3) and two syringes (5, 6). The total volume of medium in each loop was 32 ml. The flow rate was set at 0.5 ml/min and maintained throughout the 7 day cultivation. Constructs seeded and cultivated in orbitally mixed dishes (35 mm well, 8 ml medium, 25 rpm) served as controls. (C) Flow visualization. top: bulk flow through the construct centre, bottom: stream of tracer dye. Scale bar: 10 mm.

creatin kinase-MM, sarcomeric myosin heavy chain) at levels that were lower than in neonatal rat ventricles but higher than in constructs cultured in spinner flasks [86]. Electrophysiological properties were also improved, as evidenced by the prolonged action potential duration (APD, a measure of electrophysiological functionality of cell membrane), higher maximum capture rates (a measure of construct response to electrical pacing), and more physiological response to drugs. In particular, pharmacological studies done with 4-aminopyridine indicated that a decrease in transient outward potassium current may be responsible for the observed differences in APD and MCR [88]. In overall, dynamic laminar flow in rotating bioreactors improved properties of the peripheral tissue layer, but the limitations of the diffusional transport of oxygen to the construct interior were not overcome and constructs remained largely acellular.

3.6.2. Interstitial flow

In an attempt to enhance mass transport within cultured constructs, we developed a perfusion bioreactor that provides interstitial medium flow through the cultured construct at velocities similar to those found in native myocardium ($\sim$ 400–500 µm/s, [102]). In such a system oxygen and nutrients are supplied to the construct interior by both diffusion and convection.

In early studies, constructs were prepared by seeding cardiac myocytes onto PGA scaffolds in mixed flasks, a method that has been successfully used to seed chondrocytes. After 3 days, constructs were transferred into a perfusion cartridge and pulsatile flow of medium through the construct was provided by a peristaltic pump (0.2 - 3 ml/min) (Table 6). Gas exchange between culture medium and incubator air occurred in an external coil of silicone tubing within the medium recirculation loop. Perfusion during construct cultivation improved cell distribution, viability and differentiation. However, the overall cell density remained low due to the limitations of oxygen transport to the cells inside constructs during scaffold seeding in mixed flasks [97,100].

In subsequent studies, we developed a system that enabled the maintenance of oxygen supply to the cells at all times during cell seeding and construct cultivation, in order to achieve physiologic density of viable cells (Fig. 7, Table 6). The establishment of physiologic cell density is an essential requirement for functional tissue engineering because cardiac myocytes have no capacity for proliferation. The rate, yield and uniformity of cell seeding were significantly improved by combining (1) rapid cell inoculation into the scaffold, using Matrigel® as a delivery vehicle, and (2) immediate establishment of interstitial medium flow (Fig. 7A, [89]). Neonatal rat cardiomyocytes suspended in Matrigel® were inoculated into collagen sponges (10 mm diameter by 1.5 mm thick discs) at a physiologic cell density (1.35 10^8 cells/cm^3) and subjected to medium perfusion (0.5 ml/min, corresponding to $\sim$ 400 µm/s). During the initial 1.5 h, the direction of medium flow was alternated to enhance cell attachment and prevent the washout of loosely bound cells; unidirectional medium flow was applied for the remained of cultivation (Fig 7B).

Medium perfusion markedly improved construct properties according to all tested parameters [98]. After 7 days, the viability of cells in perfused constructs was indistinguishable from that of freshly isolated cells ($\sim$ 85%), as compared to rather low viability of dish-grown constructs ($\sim$ 45%). The molar ratio of lactate produced to

In response to electrical stimulation, perfused constructs contracted synchronously, had lower excitation thresholds (ET, the minimum voltage needed to induce sustained construct contractility, a measure of cell coupling within the construct), and recovered their baseline levels of ET and MCR following treatment with a gap junction blocker. In contrast, dish-grown constructs exhibited arrhythmic contractile patterns and failed to recover their baseline MCR levels. ET and MCR of both perfusion and dish grown constructs were inferior compared to the native tissue.

Taken together, these studies suggested that the immediate establishment of interstitial medium flow markedly enhanced the control of oxygen supply to the cells throughout constructs and thereby enabled engineering of constructs that were compact and consisted of viable aerobically metabolising cells. However, most cells in perfused constructs were round and mononucleated, indicating that some of the regulatory signals – either molecular or physical – were not present in the culture environment.

3.7. MATHEMATICAL MODEL OF OXYGEN DISTRIBUTION IN A TISSUE CONSTRUCT

In order to rationalise experimental data for oxygen transport and consumption in engineered cardiac constructs with an array of channels, we developed a mathematical model. The construct is divided into an array of cylindrical domains, each representing a channel with medium flow surrounded with a tissue space, like in a standard Krogh cylinder model. To simplify the equations of continuity, we assumed no radial velocity in the channel lumen, and no convective flow in the tissue region. Both axial and radial diffusion are taken into account in the tissue region. The consumption of oxygen occurs by Michaelis-Menten kinetics. The model yields concentration profiles of oxygen and cells based on numerical simulation of the diffusive-convective oxygen transport and its utilization by the cells.

Tissue construct is divided into an array of parallel domains consisting of a channel lumen and tissue space around it. Oxygen distribution within and around one channel can be modelled according to the Krogh tissue cylinder model. Schematic of the tissue construct is presented in Figure 8.

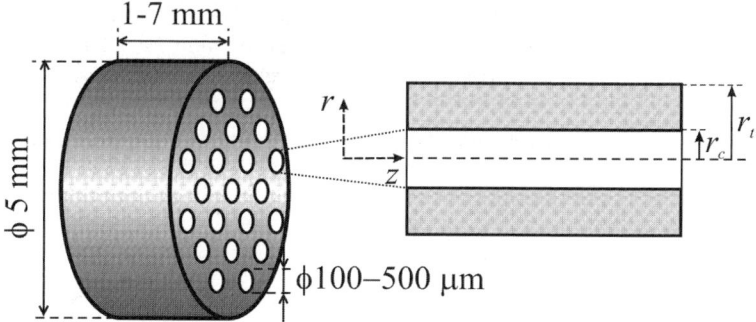

Figure 8. Schematics of the tissue construct with a parallel channel array (r_c channel radius, r_t-one half of the center-to-center spacing between the channels).

Table 7. Boundary conditions.

Channel	Tissue annulus	
$C^{O2}_{aq}(r,0) = C_{in}$	$C^{O2}_{t}(r,0) = C_{in}$	O_2 concentration at the construct entrance is equal to that in the inlet medium
$C^{O2}_{aq}(r,L) = C_{out}$	$C^{O2}_{t}(r,L) = C_{out}$	O_2 concentration at the construct outlet is equal to that in the outlet medium
$\partial C_{aq}/\partial r(0,z) = 0$	$\partial C^{aq}_{t}/\partial r(r_t,z) = 0$	symmetry conditions
$D_{aq}\dfrac{\partial C^{O2}_{aq}}{\partial r}(r_c,z) = D_t\dfrac{\partial C^{O2}_{t}}{\partial r}(r_c,z)$	$C^{O2}_{aq}(r_c,z) = C^{O2}_{t}(r_c,z)$	equal flux of O_2 at the interface, concentrations of O_2 in the medium and the tissue are the same

To simplify the equations of continuity for oxygen it can be assumed that both density and diffusivity are constant in both regions. Due to the tight packing of cells seeded onto the scaffold at high density and the presence of gel as a delivery vehicle, radial and axial velocities in the tissue annulus can be neglected. Similarly, the radial velocity component in the channel lumen is assumed to be negligible. At a steady state, oxygen transport can be described according to:

$$V(r)\frac{\partial C^{O2}_{aq}}{\partial z} = D_{aq}\left[\frac{1}{r}\frac{\partial}{\partial r}\left(r\frac{\partial C^{O2}_{aq}}{\partial r}\right) + \frac{\partial^2 C^{O2}_{aq}}{\partial z^2}\right] \quad (6)$$

where $V(r)$ is the velocity profile, D_{aq} is the diffusion coefficient in the aqueous phase and C^{O2}_{aq} is the local oxygen concentration. Because of high Peclet numbers (Pe, a function of medium flow rate, channel geometry and spacing) it was possible to neglect axial diffusion.

Governing equation for oxygen distribution in the tissue region can be expressed as:

$$0 = D_t\left[\frac{1}{r}\frac{\partial}{\partial r}\left(r\frac{\partial C^{O2}_{t}}{\partial r}\right) + \frac{\partial^2 C^{O2}_{t}}{\partial z^2}\right] + R^{O2}_{t} \quad (7)$$

where D_t is diffusion coefficient of oxygen in tissue space, R^{O2}_{t} is oxygen consumption rate and C^{O2}_{t} is local oxygen concentration in the tissue space.

The consumption of oxygen can be described according to Michaelis-Menten kinetics:

$$R_t^{O2} = -\frac{Q_m C_t^{O2}}{C_m + C_t^{O2}} \qquad (8)$$

where Q_m is the maximum rate of oxygen consumption, and C_m is the C_t^{O2} at the half-maximum consumption rate. The equations are subject to boundary conditions listed in Table 7. In model equations, L is the length of tissue construct, r_c is the channel radius and r_t is the half distance between centres of two channels.

The model can be solved for any set of experimental parameters. One representative set is listed in Table 8.

Table 8. Model parameters. Channel and tissue space radiuses, construct length and the inlet oxygen concentration can be varied.

Parameter	Value
Channel radius, r_c	90 μm
Tissue space radius, r_t	250 μm
Construct length, L	3.5 mm
Average medium velocity in channels, U_m	1.0 cm/s
Maximum O_2 consumption rate, Q_m	38 μM/s
Michaelis-Menten constant, C_m	13 μM
O_2 diffusion coefficient in medium, D_{aq}	2.4 10^{-5} cm^2/s
O_2 diffusion coefficient in tissue, D_t	2.0 10^{-5} cm^2/s
Inlet O_2 concentration, C_{in}	220 μM
Outlet O_2 concentration, C_{out}	190 μM

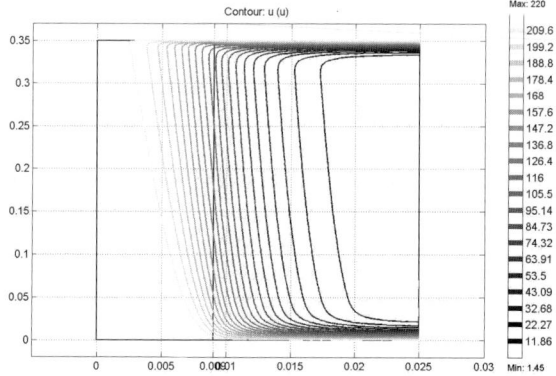

Figure 9. Oxygen distribution in and around a channel in the tissue construct. Tissue construct length [cm] vs. channel and tissue space radius [cm] (half-channel shown).

The model was solved using finite element method and a commercial software FEMLAB, to obtain oxygen concentration profiles shown in Figure 9.

According to the model, even at high velocities (~ 1 cm/s) oxygen would almost completely be depleted in the region of ~ 150 μm around the channel wall, indicating that the minimal wall-to-wall spacing between the channels should be ~ 300 μm. At the given velocity, enough oxygen was supplied to the cells for the entire length of the channel. However, if the velocity is decreased in order to decrease shear rate (*e.g.* to prevent washout of cells immediately after inoculation) the oxygen supply would dramatically decrease and supplementation of the culture medium with the oxygen carrier (*e.g.*, perfluorocarbon emulsion) may be necessary.

3.8. MECHANICAL STIMULATION

To provide appropriate mechanical stimulation neonatal rat cardiac myocytes were reconstituted in collagen gel and cultivated in the presence of cyclic stretch (Table 6). In one set-up neonatal rat ventricular myocytes were suspended in a gel consisting of collagen I and Matrigel® [95]. For each piece of tissue, 0.7 ml of the cell/gel mix was poured into a well (11 x 17 x 4 mm) made of silicone rubber containing one set of Velcro coated silicone tubes (7 mm length, 3 mm OD 2 mm ID). The mixture was allowed to gel at 37°C for 60 min before culture medium was added. After 4 days in culture, tissues were transferred for an additional 6 days into a motorised stretching device that applied either unidirectional or cyclic stretch (1.5 Hz, strain rate of up to 20%). Mechanical stimulation enhanced the alignment of cardiac myocytes, and resulted in higher mitochondrial density and longer myofilaments. As compared to the non-stimulated controls, stimulation markedly increased the RNA/DNA ratio (by 100%) and protein/cell ratio (by 50%). The force of contraction was also higher in stretched constructs, both under basal conditions and after stimulation with isoprenaline [95].

In an improved set-up, neonatal rat cardiac cells were suspended in the collagen/Matrigel® mix and cast into circular molds [77]. After 7 days in culture, the strips of cardiac tissue were placed around two rods each fixed to a stretching bar of a custom made mechanical stretcher and subjected to unidirectional and cyclic stretch at 10% strain rate and 2 Hz. Mechanical stimulation improved the formation of interconnected and aligned cardiac muscle bundles with morphological features resembling adult rather than immature native tissue. Fibroblasts and macrophages were found through the constructs, and the capillary structures positive for CD31 were also noted. Cardiomyocytes exhibited well developed ultrastructural features: sarcomeres arranged in myofibrils, with well developed Z, I, A H and M bands, gap and adherence junctions, T tubules, and well developed basement membrane. The constructs exhibited contractile properties similar to the native tissue with high ratio of twitch to resting tension and strong β-adrenegenic response. Action potentials characteristic of rat ventricular myocytes were recorded.

Using another system for mechanical stimulation, cyclic mechanical stretch (1.33 Hz) was applied to the constructs based on collagen scaffold and human heart cells (isolated from children undergoing repair of Tetralogy of Fallot) [101]. A rectangular piece of tissue was fixed at one end to the bottom of a square dish; the other end is attached to a steel rod the cyclic movement of which is induced by dynamically changing magnetic filed. Constructs subjected to chronic stretch had improved cell distribution and collagen matrix formation.

3.9. ELECTRICAL STIMULATION OF CONSTRUCT CONTRACTIONS

In native myocardium, synchronous contractions of electro-mechanically coupled cardiomyocytes are induced by electrical signals. We explored the application of electrical stimuli on cultured constructs, in order to induce contractions by physiologically relevant mechanisms, in an attempt to further enhance cell differentiation and functional assembly (Table 6). The culture chamber was designed by fitting a Petri dish with two stimulating electrodes connected to a commercial cardiac stimulator (Nihon Kohden) (Fig. 10A). Collagen sponges (6 mm x 8 mm x 1.5 mm) were seeded with neonatal rat ventricular cells ($5 \cdot 10^6$) using Matrigel®. Constructs were stimulated using supra-threshold square biphasic pulses (2 ms duration, 1 Hz, 5 V), and the stimulation was applied for up to 8 days. Immunohistochemical staining and contractile responses measured in these feasibility studies indicated that the stimulation should be initiated 3 days after seeding, a period that appeared to be sufficient for the cells to recover from isolation and attach to the scaffolds.

Electrical stimulation significantly enhanced contractile responses of constructs to electrical pacing as compared to non-stimulated controls. The measured amplitude of synchronous contractions was on average 7 times higher in stimulated than non-stimulated group (Fig. 10D), and it progressively increased with time in culture (Fig. 10E). Excitation threshold (ET), defined as the minimum amplitude of stimulation at 60 bpm necessary to observe synchronous beating of the entire construct, was lower for stimulated than non-stimulated constructs (Fig. 10F). Consistently, the maximum capture rate was significantly higher for stimulated than for non-stimulated constructs (Fig. 10G). Excitation – contraction coupling of cardiac myocytes in stimulated constructs was also evidenced by recording the electrical activity using a platinum electrode positioned ~ 2 mm away from the pair of stimulating electrodes (carbon rods). The shape, amplitude (~ 100 mV) and duration (~ 200 ms) of electrical potentials recorded for cells in constructs that were electrically stimulated during culture (Fig. 10H) were similar to the action potentials reported previously for cells from mechanically stimulated constructs. Importantly, the improved contractile properties were not the result of increased cellularity.

Stimulated constructs exhibited higher levels of α-myosin heavy chain (α-MHC), connexin 43, creatin kinase-MM and cardiac troponin I expression as assessed from Western blots and compared to unstimulated constructs. The ratio of α-MHC (adult) and β-MHC (neonatal) isoforms (~ 1.5 at Day 3), decreased to 1.4 in non-stimulated group and increased to 1.8 in the stimulated group by the end of cultivation period, indicating enhanced differentiation in the stimulated group. Stimulated constructs had thick aligned myofibers within the ~ 200 μm surface layer that expressed sarcomeric α-actin, troponin I, α-MHC and β-MHC resembling myofibers in the native heart and elongated nuclei. In contrast, early and non-stimulated constructs generally exhibited lower level of tissue organisation, less cell alignment and contained round nuclei. In addition, stimulated constructs exhibited stronger staining for connexin-43 than either initial (day 3) or non-stimulated constructs (Fig. 10I).

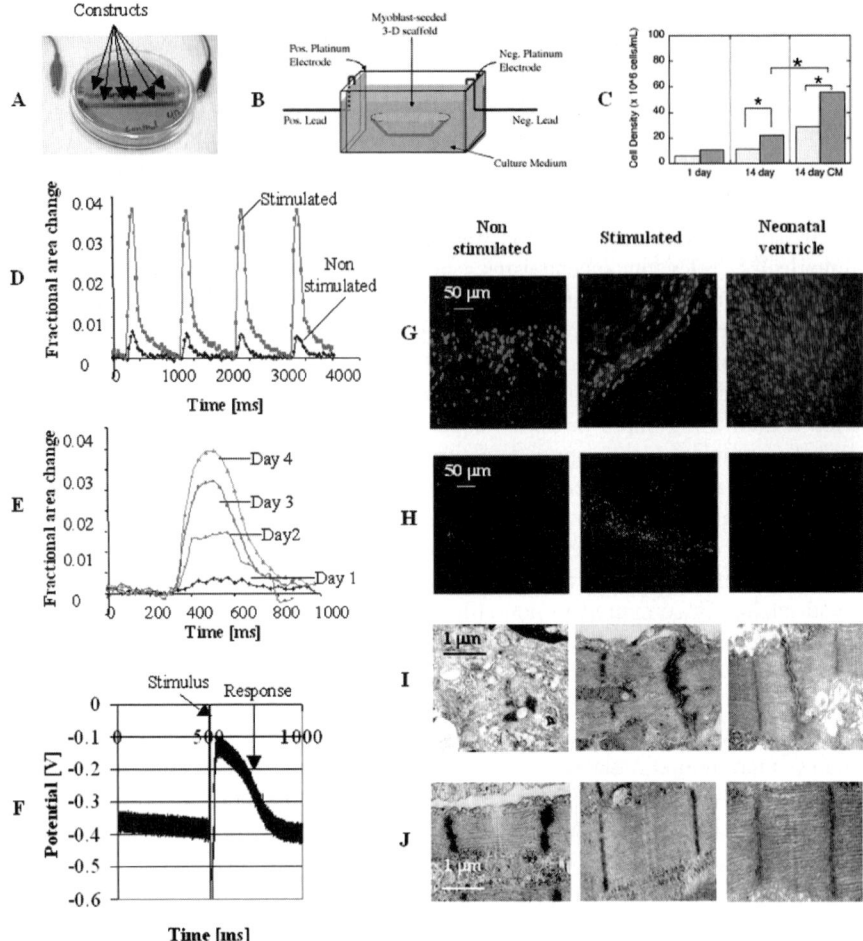

Figure 10. Electrical stimulation in situ *promoted cell differentiation and functionality of engineered constructs. A. A Petri dish with two stimulating electrodes; supra-threshold amplitude; results D-J B. A chamber with two platinum electrodes; sub-threshold amplitude; result C C. Enhanced cell proliferation in stimulated (dark grey bars) as compared to non-stimulated constructs (light grey bars). * significantly different groups. D Contraction amplitude was 7 times higher in stimulated than non-stimulated constructs. E Contraction amplitude progressively increased with time of stimulation. F. Contractile response was associated with electrical activity. Expression of α-MHC (G) and gap junction protein connexin-43; immunostains (H) were markedly higher in stimulated group. Stimulated constructs exhibited well developed myofibrils with parallel sarcomeres and intercalated discs placed symmetrically between the Z lines. In contrast, non-stimulated constructs exhibited poorly developed sarcomeres and intercalated discs. I Intercalated discs, J. Sarcomeres. (from A & C-J Radisic et al. unpublished; B Niklason et al. unpublished).*

For colour illustration please go to www.springeronline.com and enter ISBN: -14020-3229- 3, click on the link "Figure 10 colour illustration".

As an alternative contact please see the first named author of this chapter.

At the ultrastructural level, cells in stimulated constructs were elongated, contained elongated, centrally positioned nuclei and exhibited specialised cytoplasmic features, characteristic for native myocardium. Gap junctions, intercalated discs, and microtubules were all more frequent in stimulated group compared to the non-stimulated group and qualitatively more similar to those of neonatal rat heart (Fig. 10J). Cells in stimulated constructs contained aligned myofibrils, and well developed sarcomeres with clearly visible M, Z lines, H, I and A bands. In most cells, Z lines were aligned, and the intercalated discs were positioned between two Z lines. Mitochondria (between myofibrils) and abundant glycogen were detected. In contrast, non-stimulated constructs had poorly developed cardiac-specific organelles and poor organisation of ultrastructural features.

These studies suggest that electrical stimulation during *in vitro* tissue culture progressively enhanced the excitation-contraction coupling and improved the properties of engineered myocardium at the cellular, ultrastructural and tissue levels.

4. Summary

Tissue engineering can provide functional cell-based grafts to restore normal function of a compromised native tissue and to serve as physiologically relevant models for biological research. One approach to functional tissue engineering involves the *in vitro* cultivation of immature but functional tissue constructs by using: (i) cells isolated from a small tissue harvest and expanded *in vitro*, (ii) a biodegradable scaffold designed to serve as a structural and logistic template for tissue development, and (iii) a bioreactor designed to provide environmental conditions necessary for the cells to regenerate a functional tissue structure. Overall, the conditions are designed to mimic those present in a developing tissue. Two distinctly different tissues that are of high clinical and scientific interest - articular cartilage and myocardium – were discussed in light of this approach. Within the scope of this chapter, the state of the art of tissue engineering can be described by our ability to engineer immature yet functional engineered cartilage and myocardium, and the need to evaluate functional improvement *in vivo* (in rigorous animal models) following the implantation of engineered tissue grafts.

References

[1] Buckwalter, J.A. and Mankin, H.J. (1998) Articular cartilage: degeneration and osteoarthritis, repair, regeneration, and transplantation. Instr. Course Lect. 47: 487-504.
[2] Einhorn, T.A. (1998) The cell and molecular biology of fracture healing. Clin. Orthop. 355S: S7-S21.
[3] O'Driscoll, S.W. (2001) Preclinical cartilage repair: current status and future perspectives. Clin. Orthop. 391 Suppl: S397-S401.
[4] Freed, L.E. and Vunjak-Novakovic, G. (2000) Tissue engineering bioreactors. In: Lanza, R.P.; Langer R. and Vacanti, J. (eds), Principles of Tissue Engineering. Academic Press, pp. 143-156.
[5] Schaefer, D.; Martin, I.; Jundt, G.; Seidel, J.; Heberer, M.; Grodzinsky, A.J.; Bergin, I.; Vunjak-Novakovic, G. and Freed, L.E. (2002) Tissue engineered composites for the repair of large osteochondral defects. Arthritis Rheum. 46: 2524-2534.
[6] Li, R.-K.; Jia, Z.Q.; Weisel, R.D.; Mickle, D.A.G.; Choi, A. and Yau, T.M. (1999) Survival and function of bioengineered cardiac grafts. Circulation 100: II63-II69.

[7] Leor, J.; Aboulafia-Etzion, S.; Dar, A.; Shapiro, L.; Barbash, I.M.; Battler, A.; Granot, Y. and Cohen, S. (2000) Bioengineerred cardiac grafts: A new approach to repair the infarcted myocardium? Circulation 102: III56-III61.
[8] Shimizu, T.; Yamato, M.; Isoi, Y.; Akutsu, T.; Setomaru, T.; Abe, K.; Kikuchi, A.; Umezu, M. and Okano, T. (2002) Fabrication of pulsatile cardiac tissue grafts using a novel 3- dimensional cell sheet manipulation technique and temperature- responsive cell culture surfaces. Circ. Res. 90: e40-e48.
[9] Vunjak-Novakovic, G.; Freed, L.E.; Biron, R.J. and Langer, R. (1996) Effects of mixing on the composition and morphology of tissue-engineered cartilage. AIChE J. 42: 850-860.
[10] Vunjak-Novakovic, G.; Obradovic, B.; Martin, I.; Bursac P; Langer, R. and Freed, L.E. (1998) Dynamic cell seeding of polymer scaffolds for cartilage tissue engineering. Biotechnol Prog. 14:193-202.
[11] Vunjak-Novakovic, G.; Martin, I.; Obradovic, B.; Treppo, S.; Grodzinsky, A.J.; Langer, R. and Freed, L.E. (1999) Bioreactor cultivation conditions modulate the composition and mechanical properties of tissue engineered cartilage. J. Orthop. Res. 17: 130-138.
[12] Freed, L.E.; Marquis, J.C.; Vunjak-Novakovic, G.; Emmanual, J. and Langer, R. (1994) Composition of cell-polymer cartilage implants. Biotechnol. Bioeng. 43: 605-614.
[13] Freed, L.E.; Vunjak-Novakovic, G.; Biron, R.; Eagles, D.; Lesnoy, D.; Barlow, S. and Langer, R. (1994) Biodegradable polymer scaffolds for tissue engineering. Bio/Technology 12: 689-693.
[14] Freed, L.E. and Vunjak-Novakovic, G. (1995) Tissue engineering of cartilage. In: Bronzino, J.D. (ed), Biomedical Engineering Handbook. CRC Press, pp. 1788-1807.
[15] Freed, L.E.; Langer, R.; Martin, I.; Pellis, N. and Vunjak-Novakovic, G. (1997) Tissue engineering of cartilage in space. Proc. Natl. Acad. Sci. USA 94: 13885-13890.
[16] Freed, L.E.; Hollander, A.P.; Martin, I.; Barry, J.R.; Langer, R. and Vunjak-Novakovic, G. (1998) Chondrogenesis in a cell-polymer-bioreactor system. Exp. Cell Res. 240: 58-65.
[17] Obradovic, B.; Carrier, R.L.; Vunjak-Novakovic, G. and Freed, L.E. (1999) Gas exchange is essential for bioreactor cultivation of tissue engineered cartilage. Biotechnol. Bioeng. 63: 197-205.
[18] Obradovic, B.; Meldon, J.H.; Freed, L.E. and Vunjak-Novakovic, G. (2000) Glycosaminoglycan deposition in engineered cartilage: experiments and mathematical model. AIChE J. 46: 1860-1871.
[19] Obradovic, B.; Martin, I.; Freed, L.E. and Vunjak-Novakovic, G. (2001) Bioreactor studies of natural and tissue engineered cartilage. Ortopedia Traumatologia Rehabilitacja 3: 181-189.
[20] Cherry, R.S. and Papoutsakis, T. (1988) Physical mechanisms of cell damage in microcarrier cell culture bioreactors. Biotechnol. Bioeng. 32: 1001-1014.
[21] Neitzel, G.P.; Nerem, R.M.; Sambanis, A.; Smith, M.K.; Wick, T.M; Brown, J.B.; Hunter, C.; Jovanovic, I.P.; Malaviya, P.; Saini, S. and Tan S. (1998) Cell function and tissue growth in bioreactors: fluid, mechanical and chemical environments. J. Japan Soc. Microgr. Appl. 15(suppl. II): 602-607.
[22] Clift, R.; Grace, J.R. and Weber, M.E. (1978) Bubbles, drops and particles. Academic Press, New York, pp. 142-168
[23] Freed, L.E.; Martin, I. and Vunjak-Novakovic, G. (1999) Frontiers in tissue engineering: in vitro modulation of chondrogenesis. Clin. Orthop. 367S: S46-S58.
[24] Freed, L.E. and Vunjak-Novakovic, G. (2000) Tissue engineering of cartilage. In: Bronzino, J.D. (Ed.) The Biomedical Engineering Handbook, CRC Press, pp. 124-1-124-26.
[25] Vunjak-Novakovic, G. (2003) Fundamentals of tissue engineering: scaffolds and bioreactors. In: Caplan, A.I. (Ed.) Tissue Engineering of Cartilage and Bone, John Wiley, 34-51.
[26] Buckwalter, J.A. and Mankin, H.J. (1997) Articular cartilage, part I: Tissue design and chondrocyte-matrix interactions. J. Bone Joint Surg. Am. 79A: 600-611.
[27] Maroudas, A.I. (1976) Balance between swelling pressure and collagen tension in normal and degenerate cartilage. Nature 260: 808-809.
[28] Maroudas, A. (1979) Physiochemical properties of articular cartilage. In: Freeman, M.A.R. (Ed.) Adult articular cartilage. Pitman Medical, pp. 215-290.
[29] Armstrong, C.G. and Mow, V.C. (1982) Biomechanics of normal and osteoarthritic articular cartilage. In: Wilson, P.D. and Straub, L.R. (Eds.) Clinical Trends in Orthopaedics. Thieme-Stratton, pp. 189-197.
[30] Mow, V.C. and Ratcliffe, A. (1997) Structure and function of articular cartilage and meniscus. In: Mow, V.C. and Hayes, W.C. (Eds.) Basic Orthopaedic Biomechanics. Lippincott-Raven, pp. 113-177.
[31] Sah, R.L.; Trippel, S.B. and Grodzinsky, A.J. (1996) Differential effects of serum, insulin-like growth factor-I, and fibroblast growth factor-2 on the maintenance of cartilage physical properties during long-term culture. J. Orthop. Res. 14: 44-52.
[32] Lai, W.M.; Mow, V.C. and Roth, V. (1981) Effects of nonlinear strain-dependent permeability and rate of compression on the stress behavior of articular cartilage. J. Biomech. Eng. 103: 61-66.

[33] Williamson, A.K.; Chen, A.C. and Sah, R.L. (2001) Compressive properties and function-composition relationships of developing bovine articular cartilage. J. Orthop. Res. 19: 1113-1121.
[34] Buckwalter, J.A. and Mankin, H.J. (1997) Articular cartilage, part II: Degeneration and osteoarthrosis, repair, regeneration, and transplantation. J. Bone Joint Surg. Am. 79A: 612-632.
[35] Praemer, A.; Furner, S. and Rice, D.P. (1999) Musculoskeletal Conditions in the United States. American Academy of Orthopaedic Surgeons. Rosemont, IL.
[36] Lim, K.; Shahid, M. and Sharif, M. (1996) Recent advances in osteoarthritis. Singapore Med. J. 37: 189-193.
[37] Vunjak-Novakovic, G. and Goldstein, S.A. (2003, in press). Biomechanical principles of cartilage and bone tissue engineering. In: Mow, V.C. and Huiskes, R. (Eds.) Basic Orthopaedic Biomechanics and Mechanobiology, Lippincott-Williams and Wilkens.
[38] Blunk, T.; Sieminski, A.L.; Gooch, K.J.; Courter, D.L.; Hollander, A.P.; Nahir, A.M.; Langer, R.; Vunjak-Novakovic, G. and Freed, L.E. (2002) Differential effects of growth factors on tissue-engineered cartilage. Tissue Eng. 8: 73-84.
[39] Pei, M.; Seidel, J.; Vunjak-Novakovic, G. and Freed, L.E. (2002) Growth factors for sequential cellular de- and re-differentiation in tissue engineering. Biochem. Bioph. Res. Co. 294: 149-154.
[40] Martin, I.; Shastri, V.P.; Padera, R.F.; Yang, J.; Mackay, A.J.; Langer, R.; Vunjak-Novakovic, G. and Freed, L.E. (2001) Selective differentiation of mammalian bone marrow stromal cells cultured on three-dimensional polymer foams. J. Biomed. Mater. Res. 55: 229-235.
[41] Madry, H.; Padera, R.; Seidel, J.; Langer, R.; Freed, L.E.; Trippel, S.B. and Vunjak-Novakovic, G. (2002) Gene transfer of a human insulin-like growth factor I cDNA enhances tissue engineering of cartilage. Hum. Gene Ther. 13: 1621-1630.
[42] Martin, I.; Padera, R.F.; Vunjak-Novakovic, G. and Freed, L.E. (1998) In vitro differentiation of chick embryo bone marrow stromal cells into cartilaginous and bone-like tissues. J. Orthop. Res. 16: 181-189.
[43] Martin, I.; Shastri, V.; Padera, R.F.; Langer, R.; Vunjak-Novakovic, G. and Freed, L.E. (1999) Bone marrow stromal cell differentiation on porous polymer scaffolds. Trans. Orthop. Res. Soc. 24: 57.
[44] Shastri, V.P.; Martin, I. and Langer, R. (2000) Macroporous polymer foams by hydrocarbon templating. Proc. Natl. Acad. Sci. USA 97: 1970-1975.
[45] Lipshitz, H.; Etheredge, R. 3rd and Glimcher, M.J. (1976) Changes in the hexosamine content and swelling ratio of articular cartilage as functions of depth from the surface. J. Bone Joint Surg. Am. 58: 1149-1153.
[46] Frost, H.M. (1989) The biology of fracture healing. An overview for clinicians. Part II. Clin. Orthop. 248: 294-309.
[47] Kempson, G.E. (1991) Age-related changes in the tensile properties of human articular cartilage: a comparative study between the femoral head of the hip joint and the talus of the ankle joint. Biochim. Biophys. Acta 1075: 223-230.
[48] Einhorn, T.A. (1994) Enhancement of fracture healing by molecular or physical means: An overview. In: Brighton, C.T.; Friedlaender G. and Lane, J.M. (Eds.) Bone formation and repair. American Academy of Orthopaedic Surgeons, pp. 223-238.
[49] Baltzer, A.W.A.; Lattermann, C.; Whalen, J.D.; Wooley, P.; Weiss, K.; Grimm, M.; Ghivizzani, S.C.; Robbins, P.D. and Evans, C.H. (2000) Genetic enhancement of fracture repair: Healing of an experimental segmental defect by adenoviral transfer of the BMP-2 gene. Gene Ther. 7: 734-739.
[50] McKibbin, B. (1978) The biology of fracture healing in long bones. J. Bone Joint Surg. Br. 60B: 150-162.
[51] Kawamura, S.; Wakitani, S.; Kimura, T.; Maeda, A.; Caplan, A.I.; Shino, K. and Ochi, T. (1998).Articular cartilage repair-rabbit experiments with a collagen gel-biomatrix and chondrocytes cultured in it. Acta Orthop. Scand. 69: 56-62.
[52] Wakitani, S.; Goto, T.; Young, R.G.; Mansour, J.M.; Goldberg, V.M. and Caplan, A.I. (1998) Repair of large full-thickness articular cartilage defects with allograft articular chondrocytes embedded in a collagen gel. Tissue Eng. 4: 429-444.
[53] Elder, S.H.; Kimura, J.H.; Soslowsky, L.J.; Lavagnino, M. and Goldstein, S.A. (2000) Effect of compressive loading on chondrocyte differentiation in agarose cultures of chick limb-bud cells. J. Orthop. Res. 18: 78-86.
[54] Pei, M.; Solchaga, L.A.; Seidel, J.; Zeng, L.; Vunjak-Novakovic, G.; Caplan, A.I. and Freed, L.E. (2002) Bioreactors mediate the effectiveness of tissue engineering scaffolds. FASEB J. 16: 1691-1694.
[55] Vunjak-Novakovic, G.; Obradovic, B.; Martin, I. and Freed, L.E. (2002) Bioreactor studies of native and tissue engineered cartilage. Biorheology 39: 259-268.

[56] Martin, I., Obradovic, B., Freed, L.E. and Vunjak-Novakovic, G. (1999) A method for quantitative analysis of glycosaminoglycan distribution in cultured natural and engineered cartilage. Ann. Biomed. Eng. 27: 656-662.
[57] Mauck, R.L.; Soltz, M.A.; Wang, C.C.B.; Wong, D.D.; Chao, P.G.; Valhmu, W.B.; Hung, C.T. and Ateshian, G.A. (2000) Functional tissue engineering of articular cartilage through dynamic loading of chondrocyte-seeded agarose gels. J. Biomech. Eng. 122: 252-260.
[58] Mauck, R.L.; Seyhan, S.L.; Ateshian, G.A. and Hung, C.T. (2002) Influence of seeding density and dynamic deformational loading on the developing structure/function relationships of chondrocyte-seeded agarose hydrogels. Ann. Biomed. Eng. 30: 1046-1056.
[59] Mauck, R.L.; Nicoll, S.B.; Seyhan, S.L.; Ateshian, G.A. and Hung, C.T. (2003) Synergistic action of growth factors and dynamic loading for articular cartilage tissue engineering. Tissue Eng. 9: 597-611.
[60] Obradovic, B.; Martin, I.; Padera, R.F.; Treppo, S.; Freed, L.E. and Vunjak-Novakovic, G. (2001) Integration of engineered cartilage. J. Orthop. Res. 19: 1089-1097.
[61] Buschmann, M.D.; Gluzband, Y.A.; Grodzinsky, A.J.; Kimura, J.H. and Hunziker, E.B. (1992) Chondrocytes in agarose culture synthesize a mechanically functional extracellular matrix. J. Orthop. Res. 10: 745-758.
[62] Gooch, K.J.; Blunk, T.; Courter, D.L.; Sieminski, A.L.; Bursac, P.M.; Vunjak-Novakovic, G. and Freed, L.E. (2001) IGF-I and mechanical environment interact to modulate engineered cartilage development. Biochem. Bioph. Res. Co. 286: 909-915.
[63] Obradovic, B.; Bugarski, D.; Petakov, M.; Bugarski, B.; Meinel, L. and Vunjak-Novakovic, G. (2003) Effects of bioreactor hydrodynamics on native and tissue engineered cartilage. Proceedings of the 1st International Congress on Bioreactor Technology in Cell, Tissue Culture and Biomedical Applications, S. Sorvari (Ed.) July 14-18, 2003, Tampere, Finland, pp. 61-70.
[64] Hascall, V.C.; Sandy, J.D. and Handley, C.J. (1999) Regulation of proteoglycan metabolism in articular cartilage. In: Archer, C.W. (Ed.) Biology of the synovial joint. Harwood Academic Publishers, Chapter 7.
[65] MacKenna, D.A.; Omens, J.H.; McCulloch, A.D. and Covell, J.W. (1994) Contribution of collagen matrix to passive left ventricular mechanics in isolated rat heart. Am. J. Physiol. 266: H1007-H1018.
[66] Brilla, C.G.; Maisch, B.; Rupp, H.; Sunck, R.; Zhou, G. and Weber, K.T. (1995) Pharmacological modulation of cardiac fibroblast function. Herz 20: 127-135.
[67] Gillum, R.F. (1994). Epidemiology of congenital heart disease in the United States. Am. Heart J. 127: 919-927.
[68] Hoffman, J.I. (1995) Incidence of congenital heart disease: I. Postnatal incidence. Pediatr. Cardiol. 16: 103-113.
[69] Hoffman, J.I. (1995) Incidence of congenital heart disease: II. Prenatal incidence. Pediatr. Cardiol. 16: 155-165.
[70] Lysaght, M.J. and Reyes, J. (2001) The growth of tissue engineering. Tissue Eng. 7: 485-493.
[71] Rich, M. (1997) Epidemiology, pathophysiology, and etiology of congestive heart failure in older adults. J. Am. Geriatr. Soc. 45: 968-974.
[72] Dominguez, L.; Parriaello, G.; Amato, P. and Licata, G. (1999) Trends of congestive heart failure: epidemiology contrast with clinical trial results. Cardiologia 44: 801-808.
[73] Evans, R.W. (2000) Economic impact of mechanical cardiac assistance. Prog. Cardiovasc. Dis. 43: 81-94.
[74] Soonpaa, M.H.; Koh, G.Y.; Klug, M.G. and Field, L.J. (1994) Formation of nascent intercalated disks between grafted fetal cardiomyocytes and host myocardium. Science 264: 98-101.
[75] Scorsin, M.; Marotte, F.; Sabri, A.; Le Dref, O.; Demirag, M.; Samuel, J.-L.; Rappaport, L. and Measche, P. (1996) Can grafted cardiomyocytes colonize peri-infarct myocardial areas? Circulation 94: II337-II340.
[76] Connold, A.L.; Frischknecht, R.; Dimitrakos, M. and Vrbova, G. (1997) The survival of embryonic cardiomyocytes transplanted into damaged host myocardium. J. Muscle Res. Cell Motil. 18: 63-70.
[77] Zimmermann, W.H.; Schneiderbanger, K.; Schubert, P.; Didie, M.; Munzel, F.; Heubach, J.F.; Kostin, S.; Nehuber, W.L. and Eschenhagen, T. (2002) Tissue engineering of a differentiated cardiac muscle construct. Circ. Res. 90: 223-230.
[78] Wang, Y.; Ameer, G.A.; Sheppard, B.J. and Langer, R. (2002) A tough biodegradable elastomer. Nat. Biotechnol. 20: 602-606.
[79] Hubbell, J.A. (1999) Bioactive biomaterials. Curr. Opin. Biotechnol. 10: 123-129.
[80] Niklason, L.E. (1999) Replacement arteries made to order. Science 286: 1493-1494.
[81] Niklason, L.E. (2000) Engineering of bone grafts. Nat. Biotechnol. 18: 929-930.

[82] Oberpenning, F.; Meng, J.; Yoo, J.J. and Atala, A. (1999) De novo reconstitution of a functional mammalian urinary bladder by tissue engineering. Nat. Biotechnol. 17: 149-155.
[83] L'Heureux, N.; Paquet, S.; Labbe, R.; Germain, L. and Auger, F.A. (1998) A completely biological tissue-engineered human blood vessel. FASEB J. 12: 47-56.
[84] Niklason, L.E.; Gao, J.; Abbott, W.M.; Hirschi, K.K.; Houser, S.; Marini, R. and Langer, R. (1999) Functional arteries grown *in vitro*. Science 284: 489-493.
[85] Carrier, R.L.; Papadaki, M.; Rupnick, M.; Schoen, F.J.; Bursac, N.; Langer, R.; Freed, L.E. and Vunjak-Novakovic, G. (1999) Cardiac tissue engineering: cell seeding, cultivation parameters and tissue construct characterization. Biotechnol. Bioeng. 64: 580-589.
[86] Papadaki, M.; Bursac, N.; Langer, R.; Merok, J.; Vunjak-Novakovic, G. and Freed, L.E. (2001) Tissue engineering of functional cardiac muscle: molecular, structural and electrophysiological studies. Am. J. Physiol. Heart Circ. Physiol. 280: H168-H178.
[87] Bursac, N.; Papadaki, M.; Cohen, R.J.; Schoen, F.J.; Eisenberg, S.R.; Carrier, R.; Vunjak-Novakovic, G. and Freed, L.E. (1999) Cardiac muscle tissue engineering: toward an in vitro model for electrophysiological studies. Am. J. Physiol. Heart Circ. Physiol. 277: H433-H444.
[88] Bursac, N.; Papadaki, M.; White, J.A.; Eisenberg, S.R.; Vunjak-Novakovic, G. and Freed, L.E. (2003) Cultivation in rotating bioreactors promotes maintenance of cardiac myocyte electrophysiology and molecular properties. Tissue Eng. 9: 1243-1253.
[89] Radisic, M.; Euloth, M.; Yang, L.; Langer, R.; Freed, L.E. and Vunjak-Novakovic, G. (2003) High density seeding of myocyte cells for tissue engineering. Biotechnol. Bioeng. 82: 403-414.
[90] Kofidis, T.; Akhyari, P.; Boublik, J.; Theodorou, P.; Martin, U.; Ruhparwar, A.; Fischer, S.; Eschenhagen, T.; Kubis, H.P.; Kraft, T.; Leyh, R. and Haverich, A. (2002) In vitro engineering of heart muscle: Artificial myocardial tissue. J. Thorac. Cardiovasc. Surg. 124: 63-69.
[91] Li, R.-K.; Yau, T.M.; Weisel, R.D.; Mickle, D.A.G.; Sakai, T.; Choi, A. and Jia, Z.Q. (2000) Construction of a bioengineered cardiac graft. J. Thorac. Cardiovasc. Surg. 119: 368-375.
[92] Dar, A.; Shachar, M.; Leor, J. and Cohen, S. (2002) Cardiac tissue engineering Optimization of cardiac cell seeding and distribution in 3D porous alginate scaffolds. Biotechnol. Bioeng. 80: 305-312.
[93] Akins, RE.; Boyce, R.A.; Madonna, M.L.; Schroedl, N.A.; Gonda, S.R.; McLaughlin, T.A.; Hartzell, C.R.; (1999) Cardiac organogenesis in vitro: reestalishment of three-dimensional tissue architecture by dissociated neonatal rat ventricular cells. Tissue Eng. 5:103-18.
[94] Eschenhagen, T.; Fink, C.; Remmers, U.; Scholz, H.; Wattchow, J.; Weil, J.; Zimmermann, W.; Dohmen, H.H.; Schafer, H.; Bishopric, N.; Wakatsuki, T.; Elson, E.L. (1997) Three-dimensional reconstitution of embryonic cardiomyocytes in a collagen matrix: a new heart muscle model system. FASEB J. 11:683-94.
[95] Fink, C.; Ergun, S.; Kralisch, D.; Remmers, U.; Weil, J. and Eschenhagen, T. (2000) Chronic stretch of engineered heart tissue induces hypertrophy and functional improvement. FASEB J. 14: 669-679.
[96] Zimmermann, W.H.; Fink, C.; Kralisch, D.; Remmers, U.; Weil, J.; Eschenhagen, T. (2000) Three-dimensional engineered heart tissue from neonatal rat cardiac myocytes. Biotechnol Bioeng. 68:106-14.
[97] Carrier, R.L.; Rupnick, M.; Langer, R.; Schoen, F.J.; Freed, L.E. and Vunjak-Novakovic, G. (2002) Perfusion improves tissue architecture of engineered cardiac muscle. Tissue Eng. 8: 175-188.
[98] Radisic, M.; Yang, L.; Boublik, J.; Cohen, R.J.; Langer, R.; Freed, L.E. and Vunjak-Novakovic, G. (2004) Medium perfusion enables engineering of compact and contractile cardiac tissue. Am. J. Physiol. Heart Circ. Physiol. 286: H507-H516.
[99] van Luyn, M.J.; Tio, R.A.; Gallego, Y.; van Seijen, X.J.; Plantinga, J.A.; de Leij, L.F.; DeJongste, M.J. and van Wachem, P.B. (2002) Cardiac tissue engineering: characteristics of in unison contracting two- and three-dimensional neonatal rat ventricle cell (co)-cultures. Biomaterials. 24: 4793-801.
[100] Carrier, R.L.; Rupnick, M.; Langer, R.; Schoen, F.J.; Freed, L.E. and Vunjak-Novakovic, G. (2002) Effects of oxygen on engineered cardiac muscle. Biotechnol. Bioeng. 78: 617-625.
[101] Akhyari, P.; Fedak, P.W.M.; Weisel, R.D.; Lee, T.Y.J.; Verma, S.; Mickle, D.A.G. and Li, R.K. (2002) Mechanical stretch regimen enhances the formation of bioengineered autologous cardiac muscle grafts. Circulation 106: I137-I142.
[102] Fournier, R.L. (1998) Basic Transport Phenomena in Biomedical Engineering. Philadelphia: Taylor & Francis; pp. 24.

TISSUE ENGINEERED HEART

KRISTYN S. MASTERS[1] AND BRENDA K. MANN[2]
[1]Department of Chemical Engineering, University of Colorado, ECCH 111, Box 424, Boulder, CO 80309 and [2]Keck Graduate Institute, 535 Watson Drive, Claremont, CA 91711, USA – Fax: 909-607-9826 – Email: bmann@kgi.edu

1. Introduction

Cardiovascular disease is a significant cause of morbidity and mortality in the U.S. and developed countries. Successful treatment has often been limited by the poor performance of synthetic materials utilised for tissue replacement. A lofty goal for cardiovascular tissue engineering is the development of a completely tissue engineered heart. Progress towards this goal will likely be made through the parallel development of effective tissue engineered components of the cardiovascular system. These individual components include blood vessels, heart valves, and cardiac muscle. The components would later be synthesised into a larger organ structure, namely a heart. Such an ambitious goal will require substantial improvements in cell immobilisation technologies, as well in cell source and *in vitro* conditioning prior to implantation. Achievements made during the development of these individual tissue components will provide therapeutic advances and enhance our understanding of fundamental issues involved in cardiovascular tissue engineering. As tissue engineered blood vessels have been covered in another chapter, issues concerning cell immobilisation for producing tissue engineered heart valves and cardiac muscle will be discussed in this chapter.

2. Heart valves

Over 80,000 people in the U.S. receive heart valve replacements each year, and this number has been steadily rising over the last decade [1]. Valve substitutes currently available have enabled these patients to experience an enhanced quality of life and have extended patient survival. Yet, in 50-60% of patients with substitute valves, complications associated with these valve replacements necessitate re-operation or cause death within 10 years postoperatively [2].

Present heart valve substitutes consist of either mechanical prosthetic valves or tissue valves that are derived from either human or animal tissue. While significant progress has been made over the last 40 years toward the development of more efficient and durable options for valve replacements, existing technologies have still not facilitated the creation of valve substitutes that are long-lasting and capable of growth and repair. Valve failure can occur *via* several mechanisms which are discussed later in the chapter.

The field of tissue engineering has the potential to make significant contributions toward the development of an improved heart valve replacement. Great advances have recently been made toward the creation of a functional tissue engineered heart valve *via* the immobilisation of cells within a variety of natural and synthetic matrices [3-5]. This chapter will evaluate the success of this research as well as identify current obstacles in heart valve tissue engineering.

2.1. VALVE BIOLOGY

The human heart contains four heart valves, each with slightly different characteristics and each experiencing a different haemodynamic environment. The mechanical burden placed on these valves is significant, as they endure bidirectional flexing some 40 million times a year [6]. Two of these valves, the mitral and tricuspid, are atrio-ventricular valves; they regulate blood flow between the heart atria and ventricles. The remaining two valves are the pulmonary valve, located at the junction of the pulmonary artery and the right ventricle, and the aortic valve, which is situated at the junction between the aorta and left ventricle. The aortic valve, shown in Figure 1a, is the most extensively studied, most frequently diseased, and most widely transplanted valve in the heart [7]. Aortic valves withstand harsh haemodynamic conditions and endure large cyclical deformations, with changes in area as high as 50% [7].

While heart valves are relatively small, thin structures (on the order of a few hundred microns thick), their composition is surprisingly complex. There are two general cell populations within valves; these are endothelial cells, which provide a non-thrombogenic external lining, and valvular interstitial cells, which comprise the bulk of the valve cell population [8]. Valvular interstitial cells (VICs) are a heterogeneous cell population, possessing properties of both fibroblasts and smooth muscle cells. For this reason, they have not been well characterised, although they have recently started receiving more attention due to their unique properties. VICs are responsible for synthesising, remodelling, and repairing the matrix components of the valve [8]. The organisation and relative proportions of the valve matrix is paramount to valve function, thus emphasising the importance of these interstitial cells.

As mentioned above, the arrangement of the valve extracellular matrix (ECM) is crucial to valve function. Heart valves are anisotropic, meaning that their mechanical properties are not the same in all directions. This behaviour stems from the unique distribution and alignment of ECM within the valves. Valves are composed of three distinct layers: the ventricularis, the spongiosa, and the fibrosa (Figure 1b). On the outflow surface, the fibrosa consists of bundled collagen fibres aligned circumferentially. The ventricularis is located on the inflow surface, and is comprised of collagen with radially aligned elastin fibres. The spongiosa is sandwiched between the fibrosa and ventricularis; it consists primarily of glycosaminoglycans (GAGs), which acts as a hydrophilic "buffer zone" between the two other layers in order to dissipate shear stresses caused by differential movement of the layers.

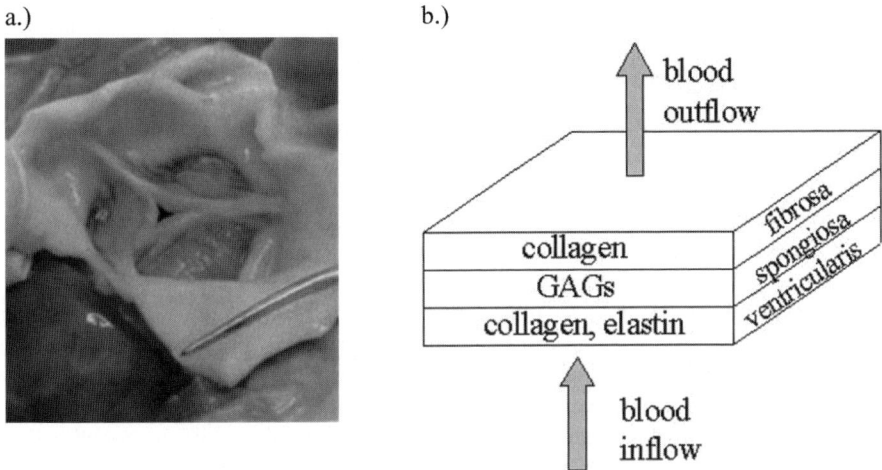

Figure 1. a.) Picture of a porcine aortic heart valve. b.) Heart valves possess a complex distribution of extracellular matrix proteins. These matrix elements are arranged into three distinct layers which impart upon the valve the necessary mechanical properties.

Valvular dysfunction may occur through a variety of mechanisms. Its causes include congenital heart defects, damage due to diseases such as rheumatic fever and syphilis or infections such as endocarditis, and onset of calcific stenosis as a result of aging [9]. Valve dysfunction may lead to regurgitation of blood into the ventricle or incomplete ejection of blood into the aorta. These events result from incomplete coaptation of valve leaflets, severe stenosis of the leaflets that prevents them from being flexible enough to allow outflow of blood, or tears in the valves. While surgical repair of a native valve is always preferable to its complete replacement, such repair is frequently not feasible, thus necessitating use of a valve substitute.

2.2. VALVE SUBSTITUTES

2.2.1. Mechanical valves

Mechanical valves are made entirely from synthetic materials, including metal alloys, carbon, or various polymers such as Dacron®. These prostheses have enjoyed overall success in the treatment of valvular heart disease, and approximately 60% of patients undergoing valve replacement surgery will receive some form of mechanical valve [7]. Several design conformations of mechanical valves have been used (*e.g.* ball-in-cage, tilting disc, and bileaflet disc), and the evolution of these designs reflects improvements in both the durability and haemodynamics of these valves. Yet, while mechanical valves have been favoured because of their durability, they are limited by the thrombogenic nature of the materials used to construct the valve. The resulting thrombus formation may not only occlude the valve, but could also lead to thromboembolism or stroke (Table 1). As a result, all patients receiving mechanical valve replacements must remain on anticoagulant therapy. However, not only is the use of anticoagulants associated with further complications such as haemorrhaging, but it also precludes any patients who have conditions not compatible with anti-coagulant use (*e.g.* bleeding problems or ulcers) from receiving a mechanical valve replacement [7]. Furthermore, mechanical valves are noisy, induce turbulent flow profiles, and are prone to sudden and catastrophic failure [6].

2.2.2. Tissue valves

The category of tissue valve substitutes includes heterografts, homografts, and autografts, as well as combinations of these tissues with synthetic components (bioprostheses). Heterografts are most often derived from porcine aortic valves or bovine pericardium. Homografts can be obtained from cadavers or organ donor tissue, while autografts include the transfer of valve tissue from one site to another (*e.g.* pulmonary valve to aortic valve setting), or the fashioning of a valve from autologous pericardial tissue. Use of a tissue valve replacement averts several of the complications associated with mechanical valves. Namely, anticoagulant therapy is not necessary following implantation of a tissue valve. However, a significant drawback of tissue valves is their limited durability (Table 1). Within 12-15 years of implantation, more than 50% of porcine aortic valve bioprostheses fail due to tissue degeneration [10]. Valve failure is highly age-dependent, with younger patients experiencing significantly higher failure rates than patients over 65 years old [7].

Table 1. Main categories of valve-related complications.

Category	Description	Predominant Valve Type
Thrombotic events	Thromboembolism, thrombosis, anticoagulation-related haemorrhage	Mechanical
Endocarditis	Inflammation and infection of the valve	Mechanical and Tissue
Structural dysfunction	Failure or dysfunction of prosthesis biomaterials	Mechanical: Acute Tissue: Gradual
Non-structural dysfunction	Tissue overgrowth, paravalvular leak, haemolysis	Mechanical and Tissue

There are several factors that contribute to limited tissue valve durability. First, in the case of hetero- and homografts, the tissues are non-living; prior to implantation, these tissues have been chemically cross-linked by a fixative such as glutaraldehyde. Glutaraldehyde treatment is necessary in order to enhance material stability and diminish antigenicity while preserving thromboresistance and sterility [11]. However, this chemical cross-linking also deleteriously affects the valve tissue, as the valve becomes acellular and less compliant. Without the presence of cells, the valve is incapable of remodelling and repairing itself, which is essential for proper long-term valve operation. The main culprit in valve tissue deterioration, however, is calcification. Calcification occurs gradually over time, and may be stimulated by cyclic mechanical stresses and potentiated by glutaraldehyde cross-linking [11]. As the process of calcification continues, there is an associated loss of tissue tensile strength. Tissue valves eventually fail due to stenosis and regurgitation induced by excessive stiffening of the tissue, or due to primary tissue failure with tearing. Non-calcific valve deterioration may also occur, as tearing at stressed regions may result from abnormal valve motion or the loss of the valve's ability to remodel [7].

2.3. TISSUE ENGINEERED HEART VALVES

The shortcomings of present valve replacement options are numerous, and the disadvantages of these valves become even more pronounced in the cases of children and adolescents. Children with congenital defects account for a significant portion of patients requiring valve replacement surgeries. Yet, failure of current tissue valve replacements occurs rapidly in young patients, and available valve substitutes do not allow for growth and repair, which are essential for these patients. Fabrication of a tissue engineered valve comprised of autologous cells grown upon a biodegradable scaffold would theoretically enable the creation of a functional valve tissue capable of growth and remodelling in response to changes in its physiological environment. Several approaches have been used in the development of tissue engineered valve constructs, and these are summarised in Figure 2.

2.3.1. Cell immobilisation in acellular valves

One approach to create a tissue engineered (TE) valve replacement is the repopulation of decellularized valves with autologous cells. The primary advantage of this strategy is that the appropriate matrix composition of the valve is already present and does not have to be reconstructed. Decellularization of porcine aortic valves can be accomplished by various methods and ultimately results in complete removal of cells and soluble matrix proteins with retention of insoluble fibrillar collagen and elastin structures and minimal alteration to overall 3D matrix arrangement [12,13]. These valve scaffolds can then be recellularized, with the goal of restoring the valve's regenerative capacity. Seeding acellular valves with endothelial cells has resulted in the formation of confluent endothelial cell cultures upon the valve and their maintenance *in vitro* for up to 3 days [13], demonstrating the ability of the acellular matrix to support cell attachment and growth.

Human neonatal dermal fibroblasts have also been used to repopulate these scaffolds [12]. Under either static or dynamic flow culture conditions, dermal fibroblasts attached to and migrated into acellular porcine valve constructs. These cells were capable of both proliferating and producing extracellular matrix during an 8-week culture upon the acellular valve scaffolds. Decellularised valves were also seeded with myofibroblasts and found to maintain normal valve function after 3 months implantation *in vivo* in the pulmonary valve position of sheep [14]. Not only did these valves show evidence of matrix protein synthesis and remodelling, but a confluent layer of endothelial cells was also observed on the valve surfaces. The ability of cells in these studies to efficiently repopulate the valve and to supplement the existing valve matrix with further production of ECM demonstrates the potential of this system for the creation of a viable valve capable of growth and remodelling.

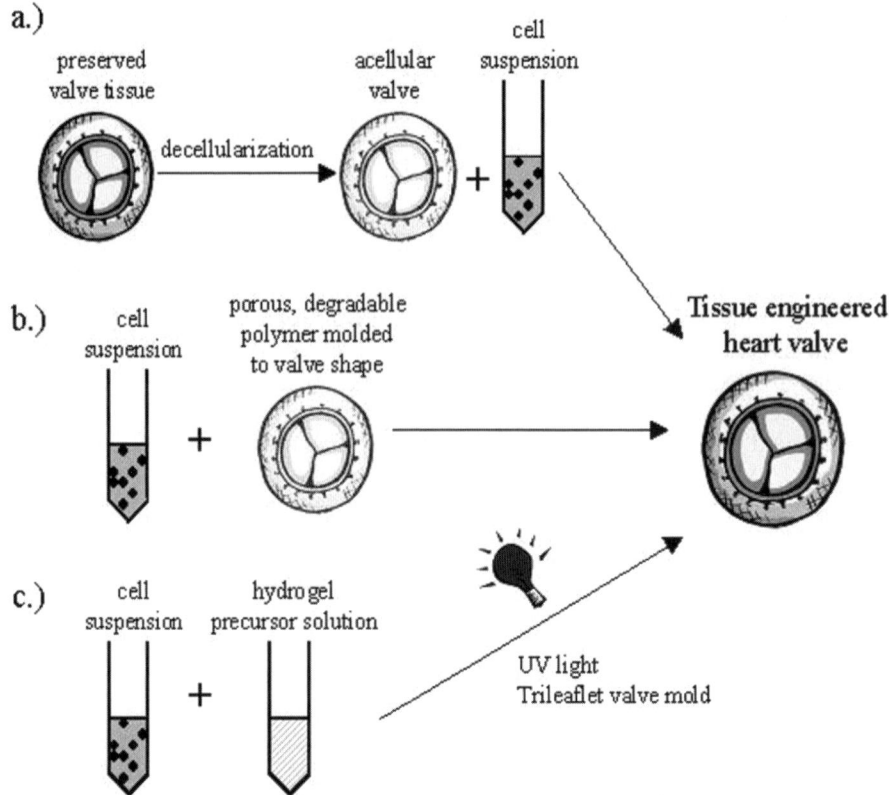

Figure 2. Cells may be immobilized in a variety of materials in order to create a tissue engineered valve. In a.), a decellularized valve is repopulated with autologous cells, thus making use of the existing matrix structure of the original valve. The most common method of forming a tissue engineered valve is shown in b.), where cells are cultured with porous, biodegradable materials, whereupon cells attach to and infiltrate the scaffolds. Lastly, c.) demonstrates the use of photopolymerization to encapsulate cells within a scaffolding material.

2.3.2. Cell immobilisation in porous matrices

The majority of tissue engineered valve research has involved the use of porous, hydrolytically degradable matrices as scaffolds for cell adhesion and growth. Polyglycolic acid (PGA) and polylactic acid (PLA) are highly biocompatible materials that have been explored for a wide range of tissue engineering applications [15]. Single valve leaflets consisting of ovine endothelial cells and myofibroblasts seeded onto highly porous PGA or PGA/PLA scaffolds were cultured under static conditions *in vitro* and implanted into the right posterior pulmonary valve position in lambs [3,16]. At 11 weeks following implantation, these valves exhibited no evidence of valvular stenosis and displayed only slight regurgitation. Additionally, the valve replacements showed signs of matrix organisation and remodelling, indicating the ability of the engineered valve tissue to respond and adapt to its physiological environment, thus mimicking the behaviour of native valves. *De novo* tissue formation was not observed in controls consisting of unseeded PGA scaffolds.

A significant limitation of the aforementioned PGA valve constructs, however, was the nonpliability of the scaffolds. Because of the PGA inflexibility, only the single leaflet design could be achieved; it was not possible to fabricate a trileaflet heart valve [17], which is ultimately necessary for proper valve function. To overcome this obstacle, subsequent studies have examined the combination of PGA with mouldable, thermoplastic polymers such as polyhydroxyalkanoates [18], which are biodegradable polyesters. PGA has been sandwiched between moulded external layers of a polyhydroxyalkanoate (PHA) to form trileaflet valves, which were able to open and close synchronously in a pulsatile flow bioreactor. This composite material construct supported vascular cell attachment and proliferation under pulsatile flow conditions [18]. In addition, cells demonstrated increasing collagen production and orientation in the direction of flow after 4 days culture in the bioreactor.

In order to avoid the fabrication of a multi-layered polymer construct, the use of PHA alone has also been explored. Porous scaffolds of PHA were created *via* a salt-leaching technique [19]. Trileaflet valves moulded from PHA were able to open and close synchronously in a pulsatile flow bioreactor, as well as support the attachment and growth of vascular cells [17,20]. After 8 days of culture under flow conditions, staining for collagen and glycosaminoglycans was positive, although elastin was not detected. These constructs were reportedly also more pliable than previous PGA-containing scaffolds. Following *in vitro* cultivation with ovine vascular cells and subsequent seeding with endothelial cells, trileaflet PHA scaffolds were also implanted *in vivo* in the pulmonary valve position in lambs for up to 17 weeks [4]. Echocardiography of the tissue engineered valves revealed mild stenosis and regurgitation in all animals. All constructs were covered by tissue at the time of explantation, although a confluent endothelial cell layer was not observed. The matrix composition of the TE valves consisted primarily of GAGs and collagen, with collagen content reaching 116% of that present in native valves after 17 weeks *in vivo*. Upon tissue explantation at 17 weeks, it was also found that the PHA polymer scaffold had experienced a molecular weight loss of just 30%.

The prolonged persistence of PHA *in vivo* may in fact be detrimental to the development of a functional valve by delaying the ingrowth of autologous tissue. An ideal scaffold will possess a degradation profile that is consistent with the rate of tissue formation, such that the material provides a transient mechanical scaffold until the cells have produced sufficient amounts of their own matrix proteins. Because the degradation time of PHA may be unsuitable for valve tissue formation, a system with faster degradation kinetics has been examined [5,21]. Nonwoven PGA was coated with a thin layer of poly-4-hydroxybutyrate (P4HB), a rapidly absorbed and thermoplastic biopolymer. These composite constructs were seeded with arterial wall myofibroblasts and endothelial cells and cultured *in vitro* in a pulsatile flow bioreactor for time periods up to 28 days. At 14 days, there was evidence of significant collagen and GAG synthesis, while elastin was not detected in any construct up to 28 days in *in vitro* culture. Following 14 days of *in vitro* culture, scaffolds were also implanted in the pulmonary valve position in lambs. At 20 weeks following implantation, leaflets exhibited no evidence of thrombus or stenosis, although mild to moderate regurgitation was observed. Histological analysis of explanted valves revealed a layered matrix structure approximating that of native valves, with evidence of elastin formation and partial coverage by endothelial cells. DNA content of valves increased steadily throughout the study, indicating constant cell proliferation. Collagen levels increased to 180% that of native valve tissue, while GAG content was 140% that of native valves after 20 weeks *in vivo*. Changes in matrix composition of the valves over time may indicate that the cells are actively remodelling and restructuring the matrices. Evolution of cell phenotype was also observed, as cells progressively showed greater resemblance to those found in native valves [5]. While constructs initially possessed a higher tensile strength than native tissue, this decreased over time in conjunction with scaffold degradation, with the final tissue displaying a tensile strength similar to that of native valves. These results indicate that the degradation kinetics of the PGA/P4HB material may be more appropriate for valve tissue development than the previously used PHA.

2.3.3. Cell immobilisation in hydrogels

An emerging area in heart valve tissue engineering is the use of hydrogels for cell attachment and encapsulation. Specifically, photopolymerisable hydrogels have been used to encapsulate a variety of cell types, including chondrocytes [22], osteoblasts [23], and vascular smooth muscle cells [24]. Photopolymerisation offers many advantages over other conventional polymerisation techniques. Namely, the photopolymerisation process is mild, allowing the encapsulation of cells within a hydrogel while maintaining cell viability. Photopolymerisation also enables spatial control of gel formation, translating to the ability to easily form complex structures, such as trileaflet heart valves. Furthermore, these matrices can be readily modified to contain covalently attached peptide sequences [24] or growth factors [25] that influence cellular differentiation and tissue growth. The degradation profiles of non-degradable photopolymers such as those derived from poly(vinyl alcohol) and poly(ethylene glycol) may also be modified to contain enzymatically degradable peptide sequences such that hydrogel degradation is tailored to match the rate of tissue ingrowth [24]. Hydrogels for TE valves may also be constructed from methods other than photopolymerisation. Fibrin gels, for instance, have shown promise for use as scaffolds in cardiovascular tissue engineering [26]. These gels are enzymatically degradable and are formed by the reaction of fibrinogen with thrombin to form fibrin.

Hydrogel materials currently being investigated for heart valve tissue engineering include modified polymers based on poly(vinyl alcohol) [27], poly(ethylene glycol) [28], and hyaluronic acid [28,29]. The use of hyaluronic acid (HA) is particularly exciting, as recent discoveries have demonstrated that this polysaccharide is required for cardiac morphogenesis and native heart valve formation [30]. In fact, disruption of HA synthase during embryogenesis fatally alters heart valve development *in vivo* [31]. Initial results for valvular interstitial cells encapsulated within photopolymerised HA are positive [28] and indicate potential for this material as a scaffold for tissue engineered heart valves.

2.4. CRITICAL CONSIDERATIONS FOR TISSUE ENGINEERED HEART VALVES

2.4.1. Cell source

As described earlier, the two predominant cell types in native valves are endothelial and valvular interstitial cells. While endothelial cells play a critical role in maintaining the non-thrombogenicity of the valve, their presence alone is not sufficient for the construction and maintenance of a healthy valve. The adaptive, complex, and dynamic structure of heart valves can be primarily attributed to the interstitial cells, as they constantly remodel and repair the valve [8]. Yet, as already noted, these cells are highly uncharacterised, and have thus far rarely been used in tissue engineered valve models. A variety of alternative cell types have been incorporated into TE valves, including dermal fibroblasts [12] and myofibroblasts [5,21]. Myofibroblasts and vascular smooth muscle cells exhibit many similarities to VICs, and are thus logical alternatives. Dermal fibroblasts, however, are not naturally present in vascular structures and exhibit marked phenotypic differences from native valve cells. The motivations for exploring this cell type have been the ease with which the cells are obtained and their rapid proliferation in *in vitro* cultures. Additionally, some cell types have been shown to redifferentiate and reorganize in response to different physiological environments, suggesting the possibility that the dermal fibroblasts may assume a more vascular phenotype when placed in a heart valve setting [32,33].

The optimal choice for cell type in the context of creating a TE valve remains unclear. A preliminary study investigated the *in vivo* performance of TE valves constructed from either autologous or allogeneic tissue [34]. Animals that received valves containing autologous cells were free of major complications, while three of four animals in the allogeneic group developed infectious complications, and the implanted allogeneic leaflets showed evidence of shrinkage, deterioration, and inflammatory response. These results reveal that allogeneic tissue may not be suitable for the creation of a TE valve. The effects of cell origin and phenotype upon the development and quality of tissue engineered valve constructs have also been examined [35]. In this study, TE valves were created by seeding either dermal fibroblasts or arterial wall myofibroblasts onto PGA scaffolds and implanting these scaffolds into the lamb pulmonary valve position for 10 weeks. The results indicated significant differences between valves created using the two different cell types. Leaflets seeded with dermal cells were thicker, more contracted, and less organised than the leaflets of arterial wall origin. Elastic fibres were also more abundant in valves derived from the arterial wall myofibroblasts, although overall mechanical strength and collagen content were not altered. Thus, while a viable valve was maintained regardless of cell origin, only the arterial wall myofibroblasts demonstrated the ability to adequately remodel and maintain proper valve function. These conclusions clearly illustrate the importance of cell type selection. With ongoing research in the area of VIC characterisation, it is likely that this cell type will soon begin to play a more prominent role in valve tissue engineering. Furthermore, use of multipotential mesenchymal stem cells has great potential, as this may obviate the issue of selecting a single cell phenotype.

2.4.2. Material properties

Materials employed in tissue engineering applications must possess suitable characteristics and physical properties for their intended use. As evidenced by the studies by Sodian *et al.* described earlier [4,21], scaffold degradation time must be tailored to match the growth of the new valve tissue. Prolonged persistence of the polymer beyond what is needed may delay tissue formation. Yet, premature degradation of the scaffold is also detrimental, as the immature valve tissue alone does not possess sufficient mechanical strength to function and continue to grow. Materials can be either hydrolytically or enzymatically degradable. Hydrolytic degradation kinetics may be altered *via* changes in material chemistry or by blending with other materials. This type of degradation occurs independently of cellular ingrowth into the scaffold. Enzymatic degradation is regulated by cell-secreted enzymes, and is therefore dependent upon the migration of cells into the scaffold, leading to degradation kinetics that are determined by the rate of tissue growth. Numerous other factors affect the design and selection of appropriate biomaterials, including scaffold porosity, mechanical properties, biocompatibility, and ability to support cell adhesion. Cell phenotype is highly dependent upon cell-material interactions, and cell behaviour can be significantly altered *via* changes in scaffold characteristics [36], underscoring the importance of material design. Lastly, advances in stereolithography combined with x-ray computed tomography have recently enabled better reproductions of the complex anatomic structure of valve leaflets [37], demonstrating how the evolvement of material fabrication techniques can impact scaffold design.

2.5. THE IDEAL TISSUE ENGINEERED VALVE

Despite the significant progress made toward the creation of a tissue engineered heart valve, there remain numerous obstacles to overcome before this technology can be applied clinically. Performance of tissue engineered valves thus far has been moderately successful when implanted in the pulmonary valve position of lambs. However, aortic and mitral valve dysfunction and disease are the most prevalent reasons for valve replacement in humans. The mechanical and haemodynamic stresses endured by both of these valves are much greater than those found in the pulmonary valve setting, meaning that valve replacements must withstand much harsher conditions than those experienced in previous *in vivo* studies. Complete endothelialisation of valve blood-contacting surfaces is also an unmet goal of TE valves. This endothelium is essential in maintaining a nonthrombogenic valve surface. Endothelial cells have been successfully seeded upon many materials to create confluent cultures, although exposure to *in vivo* shear stress conditions frequently causes a loss of endothelial cell coverage. Success of TE valves must also mean that they are resistant to calcification and tissue overgrowth and exhibit stable mechanical properties. As events such as calcification occur over a prolonged period of time, analysis of these phenomena requires long-term studies which have not yet been performed on tissue engineered valves. Lastly, the ability of the cells within the valve scaffolds to secrete appropriate proportions of matrix components and actively engage in valve remodelling and repair is paramount to the formation of a functional, durable valve. As described earlier, implantation of cell-seeded valve constructs in lambs resulted in a layered valve structure [5], indicating significant progress toward the formation of valves with appropriate matrix distributions. However, the effects of even slight deviations in matrix composition and arrangement from that of native valves are unknown, yet could be substantial.

3. Cardiac muscle

3.1. BIOLOGICAL CONSIDERATIONS

Cardiomyocytes become terminally differentiated shortly after birth, causing these cells to lose their ability to divide. As a result, cardiac muscle will not regenerate following injury, such as that caused by myocardial infarction. Further, the *in vivo* environment of cardiac muscle is highly dynamic, with the tissue exposed to stresses due primarily to cyclic stretch. These issues place significant restraints on the efforts to develop tissue engineered cardiac muscle. Therefore, aspects such as cell source, directed differentiation of cells to become cardiomyocytes, and providing the proper mechanical environment during *in vitro* culture will all be important in developing tissue engineered cardiac muscle. Such engineered tissue could serve not only as a component of a completely tissue engineered heart, but also as a treatment for damaged myocardium in congestive heart failure and myocardial infarction, as well as a model *in vitro* system for examining cardiomyocyte behaviour in response to chemical and mechanical stimulation.

Another biological consideration is that any engineered myocardial tissue will need to have similar architecture and function as native tissue. Natural cardiac muscle fibres consist of highly aligned cardiomyocytes containing myofibrils oriented parallel to the fibre axis, and adjacent cardiomyocytes are interconnected at their ends through intercalated disks (specialised junctional complexes). Cell culture studies of cardiac myofibrils and intercalated disks are complicated by the fact that cardiomyocytes become extremely flattened and exhibit disorganised myofibrils and diffuse intercellular junctions with neighbouring cells. In addition, all muscle cells lose their ability to contract effectively when cultured *in vitro*. The suitability of engineered myocardial tissue will depend on maintaining proper contractile function and electrophysiological properties.

3.2. CELL SOURCE

Since adult cardiomyocytes are terminally differentiated, researchers have begun looking for other sources of cells for cardiac muscle replacement. Whatever the cell source, it must be readily available, it should not trigger the immune system, and there should be ample supply of the cells. One option may be to use embryonic stem (ES) cells. ES cells are characterised by their capacity to proliferate for extended periods of time in culture without differentiating, yet they retain the capacity to differentiate into every tissue type in the body. Studies have shown that human ES cells form embryoid bodies (EBs) containing derivatives of all three germ layers when allowed to spontaneously differentiate [38]. Kehat *et al.* [39] found that when human ES cells were plated, 8.1% of the resulting EBs had spontaneously contracting areas. Staining of cells within these contracting demonstrated the presence of cardiac myosin heavy chain, α-actinin, desmin, and cardiac troponin I, indicating that the cells may have functional properties consistent with cardiomyocytes. Further, electron microscopy showed some myofibrillar organisation, demonstrating structural properties consistent with early-stage cardiac tissue. While such a study only examined spontaneous differentiation, the addition of specific biochemical factors, such as growth factors, to the culture medium may help to steer the cells into a particular lineage [40], such as cardiomyocytes. Additionally, seeding ES cells into scaffolds may provide further signalling for such differentiation as they would be in a more physiologic three-dimensional environment.

As there are important ethical issues to consider with ES cells, another alternative would be to use undifferentiated myoblasts, termed satellite cells, from skeletal muscle. One advantage of satellite cells is the possibility of using autologous cells, eliminating the need for immunosuppression. Autologous satellite cells, obtained from skeletal muscle and cultured *in vitro*, have been implanted into damaged myocardium in a canine model [41]. These cells survived and formed new tissue that resembled cardiac muscle at the site of injury. Dorfman *et al.* [42] isolated rat skeletal muscle satellite cells and labelled them with DAPI, a fluorescent DNA-binding dye. These cells were then implanted into the left ventricular wall of isogenic rats. This effectively marked the implanted cells so they could be distinguished from the native tissue. Over a period of four weeks, the implanted myoblasts displayed a progressive differentiation into fully developed striated muscle fibres. Other cell types that have been used include embryonic cardiac myocytes [43], myoblast cell lines [44], and cardiomyocytes, smooth muscle cells, and fibroblasts seeded onto scaffolds [45-47]. While various researchers have demonstrated some degree of success using different cell types, it is not yet clear which cell source may be optimal. Again, seeding cells onto three-dimensional scaffolds may provide signalling to aid in differentiation, particularly if exogenous factors shown to affect differentiation are added to the scaffolds.

3.3. SCAFFOLD MATERIALS

While research has shown that transplanting isolated cells into damaged myocardium can enhance cardiac function, using an engineering approach may offer better control over the development of new tissue. Such an approach would include seeding the cells onto some type of scaffold, which could then be implanted immediately or cultured *in vitro* prior to implantation. Different types of scaffold materials have been used, including natural materials, such as an extracellular matrix protein, and synthetic polymers.

As with many other tissues, collagen gels have been used as a scaffold for engineering cardiac muscle. Embryonic chick cardiac myocytes have been seeded in collagen gels [48]. Following culture, these myocytes displayed characteristic physiological responses to both physical and pharmacological stimuli. Neonatal rat cardiomyocytes have also been seeded into collagen gels, and cultured up to 13 weeks [49]. These engineered myocardial tissues showed continuous, rhythmic, and synchronised contractions over the culture period, and electrocardiograms showed physiologic patterns. When the tissues were stimulated with Ca^{2+}, epinephrine, electrical stimulation, or the tissues were stretched, an enhanced force development was observed.

Another natural scaffold material that has been used for seeding cardiomyocytes is alginate [50]. Three-dimensional porous alginate scaffolds with a pore diameter of 100 µm were seeded with neonatal rat cardiomyocytes and cultured *in vitro* for four days. These cell-scaffold constructs were then implanted into rats that had undergone myocardial infarction. Nine weeks after implantation, myofibers embedded in collagen fibres and intense neovascularisation was found in the constructs, with little of the original scaffold material remaining.

Various synthetic polymers have also been used as a scaffold material to engineer cardiac muscle. Polystyrene microcarrier beads have been seeded with rat ventricular cardiomyocytes [51]. Following culture in a bioreactor, the cardiomyocytes formed three-dimensional aggregates that were spontaneously contractile. Other polymers that have been used include PGA (plain, surface hydrolysed, or coated with laminin) and poly(lactic-co-glycolic acid) (PLGA) [47,52-55]. While the number of polymers that have been used to develop cardiac muscle is small at this point compared to the number used for other tissues, such as bone, skin, and blood vessel, this number is likely to increase over the next decade as the push to develop a completely tissue engineered heart increases.

One difficulty in trying to engineer tissues is in developing a tissue with an overall structure and organisation that resembles native tissue. In order to achieve such resemblance, the organisation of the cells may need to be directed by the scaffold itself. A scaffold that contains a micropatterned surface or mimics the desired final structure may be able to provide direction for the cells. Cardiomyocytes have been cultured on micropatterned lanes of laminin or PLGA [55]. The cells were elongated, conformed to the spatial constraints of the pattern, and their myofibrils aligned parallel to the lanes. The cells demonstrated a bipolar localisation of N-cadherin and connexin43, cell-cell junction molecules, resembling the ultrastructure of intercalated disks. Widely spaced lanes resulted in each lane of cardiomyocytes beating independently, while narrowly spaced lanes had cells bridging the lanes, resulting in aligned fields of synchronously beating cardiomyocytes.

An alternative to using a scaffold for engineering tissue is to culture cells on a surface until a cell sheet is formed. These cell sheets can then be removed from the surface and used to form the tissue. Chick embryonic cardiomyocytes have been cultured into sheets on poly(N-isopropylacrylamide) surfaces [56,57]. An advantage of these surfaces is that a simple change in temperature will cause the cell sheet to detach from the surface without damage. These cardiomyocyte sheets were then layered into stacks. The layered cell sheets rapidly adhered to one another, establishing cell-cell junctions typically found in cardiac muscle, and pulsed spontaneously and synchronously. Layered sheets that were implanted subcutaneously in rats displayed spontaneous beating and neovascularisation within the contractile tissue up to three weeks after implantation.

While all of these studies show promise, optimisation of scaffold parameters, such as pore size and porosity, size of the construct, and combination of cell type with scaffold material, needs to be done in order to achieve fully functional engineered cardiac muscle. It is likely that many more scaffold materials will be examined in the future for developing tissue engineered cardiac muscle. For example, hydrogels such as PEG, PVA, and HA that show promise in engineering other cardiovascular tissue [24,27-29] may also prove promising for engineering cardiac muscle. It is also likely that optimisation of the scaffold itself will ultimately need to be combined with optimisation of *in vitro* culture conditions, which can provide both chemical and mechanical signals for the cells.

3.4. CULTIVATION CONDITIONS

Mass transport limitations are often present in cell-scaffold constructs, which result in tissue constructs having an interior with very little to no cellularity surrounded by a peripheral tissue-like region with dense cells. In order to provide the cells in the middle of the construct with sufficient nutrients and oxygen during long culture periods, direct perfusion of the constructs could be used. Carrier *et al.* [53] cultured neonatal rat cardiac myocytes seeded on fibrous PGA scaffolds in spinner flasks or directly perfused with medium. Perfused constructs contained more uniformly distributed cells with enhanced differentiation. Further, perfusion has been found to affect oxygen concentration, which in turn affects DNA and protein contents within the constructs [52]. Increased DNA and protein content correlated with improved structure and function of the engineered cardiac tissue.

While direct perfusion of cell-scaffold constructs has been shown to be beneficial, culturing the constructs under laminar flow rather than turbulent flow has also been found to enhance engineered cardiac muscle [47,54]. Laminar flow conditions can be provided by using a rotating bioreactor rather than a typical spinner flask. Laminar flow during culture of rat neonatal cardiac myocytes seeded on PGA scaffolds resulted in increased cell viability and differentiation, and improved electrical coupling of cardiac myocytes.

Another approach for *in vitro* culture of cardiac muscle is to subject cell-scaffold constructs to mechanical stimuli, such as pulsatile flow and cyclic stretch, as would be experienced *in vivo*. Neonatal rat cardiac myocytes seeded in collagen gels have been subjected to phasic mechanical stretch [58]. Cells in these engineered tissues are interconnected and longitudinally oriented. The cardiac muscle bundles that develop have morphological features that resemble native differentiated myocardium, including organised sarcomeres, adherens junctions and gap junctions, and a basement membrane surrounding the myocytes. Further, the engineered tissues displayed contractile characteristics of native myocardium. Sodian *et al.* [59] have developed a closed-loop, perfused bioreactor that combines continuous, pulsatile perfusion. With the pulsatile nature of the flow, the constructs are also subjected to periodic stretching. Such a reactor can then provide both biochemical and biomechanical signals to the cells in the constructs.

While mechanical signals have been shown to influence the development of cardiac tissue, biochemical signals have also been shown to affect cells seeded within scaffolds. Serum content of culture medium has been found to influence the resulting engineered cardiac muscle of cardiomyocytes seeded on PGA scaffolds [54]. Ascorbate and growth factors have been found to affect the behaviour (proliferation, extracellular matrix production) of other vascular cell types seeded within scaffolds [25,60], and it is likely that such biochemical signals will also impact cells being cultured to produce cardiac muscle, no matter what cell type is being used. Combining mechanical stimulation with biochemical signalling through the use of scaffolds and *in vitro* culture may result in improved cardiac muscle that will provide tissue to repair injured myocardium and eventually lead to a completely tissue engineered heart.

4. Conclusions

The progress towards development of individual cardiac components to date has shown promise, although significant advances must be made before these engineered tissues will achieve routine clinical use. Many of the advances that will be required will be in cell immobilisation. The scaffolds themselves used for cell seeding will need improved biocompatibility and mechanical properties, better control over scaffold ultrastructures, development of novel bioactive biomaterials that control cell behaviour, and materials that can sense and respond to their biological environment. Other advances will occur from research in vascular cell biology. Better understanding of the interactions between cells and their environment as well as mechanisms of vascular wound healing will greatly facilitate tissue engineering efforts. Mechanical conditioning has been shown to influence the behaviour of cells in the cardiovascular system, and elucidating the mechanisms behind these responses will aid in designing systems for immobilising cells and culturing them *ex vivo*. Advances in stem cell technologies and gene therapy will also likely aid in engineering each of the components making up a tissue engineered heart. Interdisciplinary cooperation between engineers, scientists, and physicians will allow innovations and discoveries to be rapidly applied to development of clinically useful, tissue engineered heart valves and cardiac muscle, and ultimately to development of a tissue engineered heart.

References

[1] National Centre for Health Statistics, ICD-9-CM Code 35.2.
[2] Hammermeister, K.; Sethi, G.; Henderson, W.; Oprian, C.; Kim, T. and Rahimtoola, S. (1993) A comparison of outcomes in men 11 years after heart-valve replacement with mechanical valve or bioprosthesis. N. Engl. J. Med. 328: 1289-1296.
[3] Shinoka, T.; Ma, P.; Shum-Tim, D.; Breuer, C.; Cusick, R.; Zund, G.; Langer, R.; Vacanti, J. and Mayer, J. (1996) Tissue-engineered heart valves. Autologous valve leaflet replacement study in a lamb model. Circulation 94, Suppl. II: 164-168.
[4] Sodian, R.; Hoerstrup, S.; Sperling, J.; Daebritz, S.; Martin, D.; Moran, A.; Kim, B.; Schoen, F.; Vacanti, J. and Mayer, J. (2000) Early *in vivo* experience with tissue-engineered trileaflet heart valves. Circulation 102, Suppl. III: 22-29.
[5] Rabkin, E.; Hoerstrup, S.; Aikawa, M.; Mayer, J. and Schoen, F. (2002) Evolution of cell phenotype and extracellular matrix in tissue-engineered heart valves during in-vitro maturation and in-vivo remodeling. J. Heart Valve Dis. 11: 308-314.
[6] Love, J. (1997) Cardiac prostheses. In: Lanza, R.; Langer, R. and Chick, W. (Eds.) Principles of tissue engineering. R.G. Landes Company, Georgetown, TX; pp. 365-379.
[7] Schoen, F. (1999) Future directions in tissue heart valves: impact of recent insights from biology and pathology. J. Heart Valve Dis. 8: 350-358.
[8] Mulholland, D. and Gotlieb, A. (1996) Cell biology of valvular interstitial cells. Can. J. Cardiol. 12: 231-236.
[9] Bender, J. (1992) Heart valve disease. In: Zaret, B.; Moser, M. and Cohen, L. (Eds.) Yale University School of Medicine Heart Book. Hearst Books, New York; pp. 167-175.
[10] Jamieson, W.; Munro, A.; Miyagishima, R.; Allen, P.; Burr, L. and Tyers, G. (1995) Carpentier-Edwards standard porcine bioprosthesis: clinical performance to seventeen years. Ann. Thorac. Surg. 60: 999-1006.

[11] Schoen, F. and Levy, R. (1999) Tissue heart valves: current challenges and future research perspectives. J. Biomed. Mater. Res. 47: 439-465.
[12] Zeltinger, J.; Landeen, L.; Alexander, H.; Kidd, I. and Sibanda, B. (2001) Development and characterization of tissue-engineered aortic valves. Tissue Eng. 7, 9-22.
[13] Bader, A.; Schilling, T.; Teebken, O.; Brandes, G.; Herden, T.; Steinhoff, G. and Haverich, A. (1998) Tissue engineering of heart valves - human endothelial cell seeding of detergent acellularized porcine valves. Eur. J. Cardiothorac. Surg. 14: 279-284.
[14] Steinhoff, G.; Stock, U.; Karim, N.; Mertsching, H.; Timke, A.; Meliss, R.; Pethig, K.; Haverich, A. and Bader, A. (2000) Tissue engineering of pulmonary heart valves on allogenic acellular matrix conduits. Circulation 102, Suppl. III: 50-55.
[15] Kohn, J. and Langer, R. (1996). Bioresorbable and bioerodible materials. In: Ratner, B.; Hoffman, A.; Schoen, F. and Lemons, J. (Eds.) Biomaterials Science. Academic Press, San Diego, CA; pp. 64-72.
[16] Zund, G.; Breuer, C.; Shinoka, T.; Ma, P.; Langer, R.; Mayer, J. and Vacanti, J. (1997) The *in vitro* construction of a tissue engineered bioprosthetic heart valve. Eur. J. Cardiothorac. Surg. 11: 493-497.
[17] Sodian, R.; Sperling, J.; Martin, D.; Egozy, A.; Stock, U.; Mayer, J. and Vacanti, J. (2000) Fabrication of a trileaflet heart valve scaffold from a polyhydroxyalkanoate biopolyester for use in tissue engineering. Tissue Eng. 6: 183-188.
[18] Sodian, R.; Sperling, J.; Martin, D.; Stock, U.; Mayer, J. and Vacanti, J. (1999) Tissue engineering of a trileaflet heart valve - Early *in vitro* experiences with a combined polymer. Tissue Eng. 5: 489-493.
[19] Mooney, D.; Breuer, C.; McNamara, K.; Vacanti, J. and Langer, R. (1995) Fabricating tubular devices from polymers of lactic and glycolic acid for tissue engineering. Tissue Eng. 1: 107-118.
[20] Sodian, R.; Hoerstrup, S.; Sperling, J.; Daebritz, S.; Martin, D.; Schoen, F.; Vacanti, J. and Mayer, J. (2000) Tissue engineering of heart valves: *in vitro* experiences. Ann. Thorac. Surg. 70: 140-144.
[21] Hoerstrup, S.; Sodian, R.; Daebritz, S.; Wang, J.; Bacha, E.; Martin, D.; Moran, A.; Gulesarian, K.; Sperling, J.; Kaushal, S.; Vacanti, J.; Schoen, F. and Mayer, J. (2000) Functional living trileaflet valves grown *in vitro*. Circulation 102, Suppl. III: 44-49.
[22] Bryant, S. and Anseth, K. (2002) Hydrogel properties influence ECM production by chondrocytes photoencapsulated in poly(ethylene glycol) hydrogels. J. Biomed. Mater. Res. 59: 63-72.
[23] Burdick, J. and Anseth, K. (2002) Photoencapsulation of osteoblasts in injectable RGD-modified PEG hydrogels for bone tissue engineering. Biomaterials 23: 4315-4323.
[24] Mann, B.K.; Gobin, A.G.; Tsai, A.T.; Schmedlen, R.H. and West, J.L. (2001) Smooth muscle cell growth in photopolymerized hydrogels with cell adhesive and proteolytically degradable domains: synthetic ECM analogs for tissue engineering. Biomaterials 22: 3045-3051.
[25] Mann, B.K.; Schmedlen, R.H. and West, J.L. (2001) Tethered-TGF-β increases extracellular matrix production of vascular smooth muscle cells. Biomaterials 22: 439-444.
[26] Ye, Q.; Zund, G.; Benedikt, P.; Jockenhoevel, S.; Hoerstrup, S.; Sakiyama, S.; Hubbell, J. and Turina, M. (2000) Fibrin gel as a three dimensional matrix in cardiovascular tissue engineering. Eur. J. Cardiothorac. Surg. 17: 587-591.
[27] Nuttelman, C.; Henry, S. and Anseth, K. (2002) Synthesis and characterization of photocrosslinkable, degradable poly(vinyl alcohol)-based tissue engineering scaffolds. Biomaterials 23: 3617-3626.
[28] Masters, K.; Shah, D.; Davis, K. and Anseth, K. (2002) Designing scaffolds for valvular interstitial cells. In: Second Joint EMBS-BMES Conference; Houston, TX; pp. 860-861.
[29] Ramamurthi, A. and Vesely, I. (2002) *In-vitro* synthesis of elastin sheets on crosslinked hyaluronan gels for tissue engineering of aortic valves. In: Second Joint EMBS-BMES Conference; Houston, TX; pp. 854-855.
[30] Camenisch, T.; Schroeder, J.; Bradley, J.; Klewer, S. and McDonald, J. (2002) Heart-valve mesenchyme formation is dependent on hyaluronan-augmented activation of ErbB2-ErbB3 receptors. Nat. Med. 8: 850-855.
[31] Camenisch, T.; Spicer, A.; Brehm-Gibson, T.; Biesterfeldt, J.; Augustine, M.; Calabro, A.; Kubalak, S.; Klewer, S. and McDonald, J. (2000) Disruption of hyaluronan synthase-2 abrogates normal cardiac morphogenesis and hyaluronan-mediated transformation of epithelium to mesenchyme. J. Clin. Invest. 106: 349-360.
[32] Eghbali, M.; Tomek, R.; Woods, C. and Bhambi, B. (1991) Cardiac fibroblasts are predisposed to convert into myocyte phenotype: specific effect of transforming growth factor beta. Proc. Natl. Acad. Sci. USA 88: 795-799.

[33] Nishiyama, T.; Tsunenaga, M. and Akutsu, M. (1993) Dissociation of actin microfilament organization from acquisition and maintenance of elongated shape of human dermal fibroblasts in three-dimensional collagen gel. Matrix 13: 447-455.
[34] Shinoka, T.; Breuer, C.; Tanel, R.; Zund, G.; Miura, T.; Ma, P.; Langer, R.; Vacanti, J. and Mayer, J. (1995) Tissue engineering heart valves: valve leaflet replacement study in a lamb model. Ann. Thorac. Surg. 60, Suppl.: S513-S516.
[35] Shinoka, T.; Shum-Tim, D.; Ma, P.; Tanel, R.; Langer, R.; Vacanti, J. and Mayer, J. (1997) Tissue-engineered heart valve leaflets: does cell origin affect outcome? Circulation 96, Suppl. II: 102-107.
[36] Chen, C.; Mrksich, M.; Huang, S.; Whitesides, G. and Ingber, D. (1997) Geometric control of cell life and death. Science 276: 1425-1428.
[37] Sodian, R.; Loebe, M.; Hein, A.; Martin, D.; Hoerstrup, S.; Potapov, E.; Hausmann, H.; Lueth, T. and Hetzer, R. (2002) Application of stereolithography for scaffold fabrication for tissue engineered heart valves. ASAIO J. 48: 12-16.
[38] Itskovitz-Eldor, J.; Schuldiner, M.; Karsenti, D.; Eden, A.; Yanuka, O.; Amit, M.; Soreq, H. and Benvenisty, N. (2000) Differentiation of human embryonic stem cells into embryoid bodies comprising the three embryonic germ layers. Mol. Med. 6: 88-95.
[39] Kehat, I.; Kenyagin-Karsenti, D.; Snir, M.; Segev, H.; Amit, M.; Gepstein, A.; Livne, E.; Binah, O.; Itskovitz-Eldor, J. and Gepstein, L. (2001) Human embryonic stem cells can differentiate into myocytes with structural and functional properties of cardiomyocytes. J. Clin. Invest. 108: 407-414.
[40] Schuldiner, M.; Yanuka, O.; Itskovitz-Eldor, J.; Melton, D.A. and Benvenisty, N. (2000) Effects of eight growth factors on the differentiation of cells derived from human embryonic stem cells. Proc. Natl. Acad. Sci. USA 97: 11307-11312.
[41] Chiu, R.C.; Zibaitis, A. and Kao, R.L. (1995) Cellular cardiomyoplasty: Myocardial regeneration with satellite cell implantation. Ann. Thorac. Surg. 60: 12-18.
[42] Dorfman, J.; Duong, M.; Zibaitis, A.; Pelletier, M.P.; Shum-Tim, D.; Li, C. and Chiu, R.C. (1998) Myocardial tissue engineering with autologous myoblast implantation. J. Thorac. Cardiovasc. Surg. 116: 744-751.
[43] Soonpaa, M.H.; Koh, G.Y.; Klug, M.J. and Field, L.J. (1994) Formation of nascent intercalated disks between grafted fetal cardiomyocytes and host myocardium. Science 264: 98-101.
[44] Koh, G.Y.; Klug, M.J.; Soonpaa, M.H. and Field, L.J. (1993) Differentiation and long-term survival of C2C12 myoblast grafts in heart. J. Clin. Invest. 92: 1548-1554.
[45] Li, R.K.; Jia, Z.Q.; Weisel, R.D.; Mickle, D.A.; Choi, A. and Yau, T.M. (1999) Survival and function of bioengineered cardiac grafts. Circulation 100: 1163-1169.
[46] Li, R.K.; Yau, T.M.; Weisel, R.D.; Mickle, D.A.; Sakai, T.; Choi, A. and Jia, Z.Q. (2000) Construction of a bioengineered cardiac graft. J. Thorac. Cardiovasc. Surg. 119: 368-375.
[47] Carrier, R.L.; Papadaki, M.; Rupnick, M.; Schoen, F.J.; Bursac, N.; Langer, R.; Freed, L.E. and Vunjak-Novakovic, G. (1999) Cardiac tissue engineering: cell seeding, cultivation parameters, and tissue construct characterization. Biotechnol. Bioeng. 64: 580-589.
[48] Eschenhagen, T.; Fink, C.; Remmers, U.; Scholz, H.; Wattchow, J.; Weil, J.; Zimmermann, W.; Dohmen, H.H.; Schafer, H.; Bishopric, N.; Wakatsuki, T. and Elson, E.L. (1997) Three-dimensional reconstitution of embryonic cardiomyocytes in a collagen matrix: a new heart muscle model system. FASEB J. 11: 683–694.
[49] Kofidis, T.; Akhyari, P.; Boublik, J.; Theodorou, P.; Martin, U.; Ruhparwar, A.; Fischer, S.; Eschenhagen, T.; Kubis, H.P.; Kraft, T.; Leyh, R. and Haverich, A. (2002) In vitro engineering of heart muscle: Artificial myocardial tissue. J. Thorac. Cardiovasc. Surg. 124: 63-69.
[50] Leor, J.; Aboulafia-Etzion, S.; Dar, A.; Shapiro, L.; Barbash, I.M.; Battler, A.; Granot, Y. and Cohen, S. (2000) Bioengineered cardiac grafts: A new approach to repair the infarcted myocardium? Circulation 102: III56-III61.
[51] Akins, R.; Boyce, R.; Madonna, M.; Schroedl, N.; Gonda, S.; McLaughlin, T. and Hartzell, C. (1999) Cardiac organogenesis in vitro: reestablishment of three-dimensional tissue architecture by dissociated neonatal rat ventricular cells. Tissue Eng. 5: 103–118.
[52] Carrier, R.L.; Rupnick, M.; Langer, R.; Schoen, F.J.; Freed, L.E. and Vunjak-Novakovic, G. (2002) Effects of oxygen on engineered cardiac muscle. Biotechnol. Bioeng. 78: 617-625.
[53] Carrier, R.L.; Rupnick, M.; Langer, R.; Schoen, F.J.; Freed, L.E. and Vunjak-Novakovic, G. (2002) Perfusion improves tissue architecture of engineered cardiac muscle. Tissue Eng. 8: 175-188.
[54] Papadaki, M.; Bursac, N.; Langer, R.; Merok, J.; Vunjak-Novakovic, G. and Freed, L.E. (2001) Tissue engineering of functional cardiac muscle: molecular, structural, and electrophysiological studies. Am. J. Physiol. Heart Circ. Physiol. 280: H168–H178.

[55] McDevitt, T.C.; Angello, J.C.; Whitney, M.L.; Reinecke, H.; Hauschka, S.D.; Murry, C.E. and Stayton, P.S. (2002) *In vitro* generation of differentiated cardiac myofibers on micropatterned laminin surfaces. J. Biomed. Mater. Res. 60: 472-479.

[56] Shimizu, T.; Yamato, M.; Akutsu, T.; Shibata, T.; Isoi, Y.; Kikuchi, A.; Umezu, M. and Okano, T. (2002) Electrically communicating three-dimensional cardiac tissue mimic fabricated by layered cultured cardiomyocyte sheets. J. Biomed. Mater. Res. 60: 110-117.

[57] Shimizu, T.; Yamato, M.; Isoi, Y.; Akutsu, T.; Setomaru, T.; Abe, K.; Kikuchi, A.; Umezu, M. and Okano, T. (2002) Fabrication of pulsatile cardiac tissue grafts using a novel 3-dimensional cell sheet manipulation technique and temperature-responsive cell culture surfaces. Circ. Res. 90: e40-e48.

[58] Zimmermann, W.-H.; Schneiderbanger, K.; Schubert, P.; Didie, M.; Munzel, F.; Heubach, J.F.; Kostin, S.; Neuhuber, W.L. and Eschenhagen, T. (2002) Tissue engineering of a differentiated cardiac muscle construct. Circ. Res. 90: 223-230.

[59] Sodian, R.; Lemke, T.; Loebe, M.; Hoerstrup, S.P.; Potapov, E.V.; Hausmann, H.; Meyer, R. and Hetzer, R. (2001) New pulsatile bioreactor for fabrication of tissue-engineered patches. J. Biomed. Mater. Res. (Appl. Biomater.) 58: 401-405.

[60] Hoerstrup, S.P.; Zund, G.; Schnell, A.M.; Kolb, S.A.; Visjager, J.F.; Schoeberlein, A. and Turina, M. (2000) Optimized growth conditions for tissue engineering of human cardiovascular structures. Internatl. J. Artif. Organs 23: 817-823.

BONE TISSUE ENGINEERING

PANKAJ SHARMA, SARAH CARTMELL AND ALICIA J. EL HAJ

Centre for Science and Technology in Medicine, North Staffordshire Hospital, University of Keele School of Medicine, United Kingdom – Fax: 01782 717079 – Email: bea17@keele.ac.uk

1. Introduction

In the USA, over 1 million orthopaedic operations involve bone repair for replacement surgery, trauma, abnormal development or skeletal deficiency [1]. Approximately 6.5 million fractures occur annually in the USA, of which 15% heal with difficulty [2]. Currently, there are few effective methods of treating fractures, which progress to non-union. The most common method involves placing a bone graft, which can be derived from the patient (autograft bone marrow and bone matrix) or from a registered bone bank (allograft bone matrix without cells) into the defect site. Over 250,000 bone grafts are performed annually in the United States [2]. Bone grafting generates new bone *via* transplantation of either the patients untreated bone marrow which has not been purified for bone specific stem cells or donor bone matrix which has been derived from other patients and stored in sterile frozen conditions in a bone bank prior to use [3].

Bone grafts are avascular and rely on diffusion for survival. The size of the defect and viability of the host bed can therefore limit their use. In large defects the bone graft can often be resorbed before osteogenesis is complete [4-5]. Harvesting autologous bone for grafting adds to the overall time required for surgery, and is associated with donor site morbidity in the way of infection, pain, and haematoma formation [6-7]. Autologous bone grafts are also limited in supply. On the other hand allografts may induce cell mediated immune responses to alloantigens, and may also transmit pathogens such as HIV [8]. Technically, it can be difficult to shape bone grafts to fit the defect well. For these reasons tissue engineering offers great potential for the construction of new musculoskeletal tissues.

Tissue engineering is an interdisciplinary field that relies on cooperation between physicians, scientists and engineers. It can be defined as the application of scientific principles to the design and construction of living tissues. Bone regeneration *via* tissue engineering techniques requires a number of components: stem cells such as bone marrow derived osteoblasts capable of differentiation into mature bone cells, a suitable carrier which aids in filling large sites for repair, can deliver cells to specific sites and then function as a scaffold for growth, and in some cases, a viable well vascularised host bed [9-10]. In addition, a morphogenetic signal can be provided by bioactive compounds that are coated onto the scaffold or by the scaffold itself which may be osteoinductive. Endogenous host cells and delivered stem cells can then respond to this osteoinductive signal and initiate bone repair.

The key challenge for bone tissue engineering is to provide a more successful method of clinical treatment than existing treatment options. This establishes the goals for a tissue-engineered product quite clearly. The first major aim is to provide a treatment solution which does not require any further revision surgery. With metal implants such as hip replacements, particulate wear debris can result in bone resorption and aseptic loosening, necessitating revision surgery [11]. A biological substitute must ensure that such complications are avoided. Secondly, the approach will need to be tailored to meet individual requirements. For small defects, the use of stem cell therapies may be sufficient to stimulate repair whereas in the case of larger repair sites there is a requirement for an off the shelf product. This product must be created rapidly *ex vivo* for use in surgery to ensure adequate trade-off between costs of production *versus* successful repair. In addition, the treatment must eliminate the problems of donor site scarcity, immune rejection and pathogen transfer which are all potential weaknesses of current methodologies.

In this chapter, we will address the major components of bone tissue engineering alongside the developing complexities of treatments. Animal studies have gone someway towards proving efficacy but as yet few clinical trials are underway.

2. Mesenchymal stem cells

Mesenchymal stem cells (MSCs) are capable of undergoing differentiation into a variety of specialised mesenchymal tissues, including bone, tendon, cartilage, muscle, ligament, fat and marrow stroma [12]. During human development MSCs can be found in various tissues. In adults they are prevalent in bone marrow, however they can also be found in muscle, fat, skin, and around blood vessels [13]. It is difficult to accurately quantify the number of MSCs within the human body, however it has been estimated that there is 1 mesenchymal stem cell per 100,000 nucleated cells in bone marrow [14]. There is also evidence to suggest that the number of MSCs decreases with age and infirmity [15-16]. It is possible that this observation may influence the outcome of reparative processes of skeletal tissue.

The technique for isolating and purifying MSCs from marrow specimens, of individuals of any age, has been perfected and techniques have also been developed to mitotically expand the MSCs *in vitro* [17-18]. Human MSCs can be expanded over 1 billion-fold without any loss in their osteogenic potential [14]. This property is invaluable for the purposes of tissue engineering, and allows an individual patient's MSCs to be harvested and expanded *in vitro* prior to being re-implanted into their body. For the purposes of musculoskeletal tissue engineering, MSCs have been seeded onto carrier scaffolds prior to being re-implanted into animal models. However, MSCs have been administered intravenously, as an adjuvant in the infusion regimen of bone marrow transplantation following high dose chemotherapy [19-20].

Inductive agents that cause MSCs to differentiate and progress along individual lineage pathways have been identified, however the molecular details that govern regulation of each lineage pathway are currently areas of active research [21-23]. Dexamethasone is known to induce osteogenic differentiation of human MSCs *in vitro*, however it induces mouse MSCs to differentiate into adipocytes [21,24]. Recombinant human Bone Morphogenetic Protein (rhBMP) induces osteogenic differentiation of mouse MSCs in low doses, but much higher doses are required to elicit the same response in human MSCs [25].

Another approach is to use the patients' own differentiated cells for treatment. The technique of autologous chondrocyte implantation (ACI) has been used extensively for cartilage repair, and has the advantage of not causing immune rejection [26-27]. This treatment involves deriving cells from a healthy site in the body, proliferating them *ex vivo*, and then returning them to sites of cartilage loss. The problem with this methodology is that it relies upon the patient's cells being healthy and suitable for treatment, which may not be the case in rheumatoid or osteoarthritic patients. Also, iatrogenic injury may be caused to the healthy donor site at the time of chondrocyte harvesting. In bone repair, there have been few attempts to mimic this approach using periosteal derived cells from long bones or equivalent. This is mainly due to the well described stromal cell differentiation pathway, which proves to be more appropriate for consideration as marrow cells are easily accessible and available in large numbers.

3. Carrier scaffolds

A scaffold functions as a carrier to deliver cells or osteogenic substances to the site of bone loss. In addition, it functions as a template for the process of bone regeneration. In the initial stages the scaffold provides mechanical stability and cell anchorage sites. There are several important attributes that a tissue-engineered bone construct should possess. The scaffold material should promote bone in-growth, and it should degrade in a predictable manner. It should be possible to fashion the construct into a desired shape, allowing bony defects to be filled. The material used for the scaffold should be biocompatible, and must retain its properties after being sterilised. It should also allow uniform loading and retention of mesenchymal stem cells, and should promote rapid vascular in-growth. Ideally, the constructs should be available to surgeons at short notice.

Carrier scaffolds can have the properties of osteoinduction and osteoconduction. An osteoinductive material allows bone repair to occur in a location that would not heal if left untreated [28]. An osteoconductive material guides repair in a location where normal healing would occur if left untreated.

Various types of material have been used for the construction of scaffolds in tissue engineering (Figure 1). These materials can be naturally occurring or synthetic, and have varying degradation times. Some materials form a gel, allowing them to be used in an injectable form, however others form a more rigid structure. Commonly used materials include ceramics, polymers and composites [29]. Ceramics and polymers can be either absorbable or nonabsorbable, and polymers are either naturally occurring or synthetic. Typically ceramic materials such as hydroxyapatite have long degradation times, often a year or more.

Although many materials such as demineralised bone matrix, collagen composites, fibrin, calcium phosphate and hydroxyapatite have been used as carrier scaffolds, there has been much recent interest in synthetic absorbable polymers. The more commonly used polymers are polyglycolic acid, polylactic acid, and polyethylene glycol. Bulk degradation is a key feature of the poly(alpha-hydroxyacids) group [29]. These materials have a shorter degradation time ranging from days to months depending on the type of polymer. Polymers can differ in their molecular weight, polydispersity, crystallinity, and thermal transitions. The relative hydrophobicity and percent crystallinity can affect cellular phenotype. Variation in the pore size of the scaffold can have a profound effect on the attachment and long term survival of cells, for a specific cell type there is an optimal pore topography and size [30-32].

Figure 1. Poly-L-lactide scaffold with a pore size of 250-350 μm seeded with mesenchymal stem cells (Picture included with kind permission of Karen Hampson, Centre for science and technology in medicine, North Staffordshire Hospital, University of Keele School of Medicine).

One problem with the use of polymers is that when they lose mass following exposure to an aqueous media, this loss is accompanied by a release gradient of by-products which are acidic in nature [16,19-21,29]. An important aspect of the application of these polymers to bone tissue engineering is the need for sufficient vascularisation and metabolic activity to ensure the removal of these waste products and avoid local disturbances [21-24].

Current research also features the development of polymers either as composites, potentially with additional bioactive agents or with surface modifications to improve cell adhesion or proliferation. Hybrid materials incorporate tricalcium phosphate, HA and basic salts into polymer matrices. The aim of this is to improve biocompatibility, cell spreading and adhesion on a less hydrophobic surface and to buffer the acidic by products of the poly-matrices [16,25,28]. In addition, growth factors, small molecular weight peptides and other chemical agents can be added within or to the surface of these scaffolds [33-34]. These strategies involve providing cellular cues to the 3-dimensional environment that encourage cell adhesion, osteogensis or in the case of the mechanoactive scaffolds improves production of mechanically appropriate tissue.

Osteoblast proliferation is sensitive to surface topography, strain or other mechanical stimuli. As a result, particle size, shape, and surface roughness affect cellular adhesion, proliferation and phenotype. In the future, scaffolds will exploit these characteristics and instead of merely holding cells in place they shall function as bioactive matrices designed to encourage cellular attachment. Polymers incorporating an integrin polypeptide sequence or consisting entirely of repeating polypeptide sequences have been designed [35-37]. The scaffold will mimic the extracellular matrix allowing cells to attach to the incorporated cell surface adhesion proteins. This feature may allow proliferating cells to respond to mechanical stimuli in a more physiological manner.

4. *Ex vivo* conditioning of constructs

Static culture is simple in design and operation but there are nutrient diffusion limitations with large constructs. Cell proliferation at the exterior of a large construct is apparent in static culture, however the centres of these constructs often have poor cell viability and activity due to limited nutrient diffusion. An example of this can be seen in Figure 2, where a porous demineralised human trabecular bone scaffold has been seeded with primary rat stromal cells and grown in static *in vitro* culture for seven weeks. Micro-computed tomography shows the original scaffold morphology and the matrix mineralisation location on the scaffold after seven weeks. The matrix mineralisation is located mainly at the periphery of the constructs.

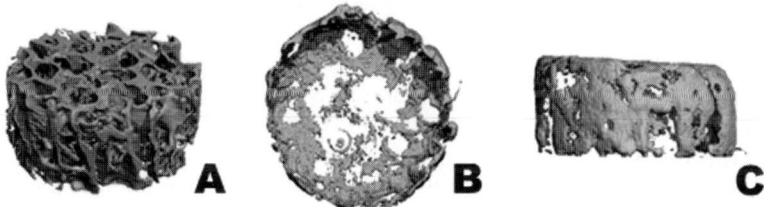

Figure 2. Micro-computed tomography images of A) Original scaffold (demineralised human trabecular bone) morphology prior to seeding with primary rat stromal cells and culturing in static in vitro conditions for seven weeks B) Top view of matrix mineralisation on scaffold after seven weeks in vitro static culture C) side view of mineralised matrix located at periphery of scaffold. (Figure included with kind permission of Dr Robert Guldberg, Georgia Institute of Technology).

Bioreactors control the biochemical and biomechanical environment for a tissue-engineered construct. The biochemical environment is controlled by allowing transport of nutrients (such as glucose, dissolved oxygen) to the cells and degradation products from the cells seeded throughout the construct. Other environmental factors such as pH, growth factors and other cell signalling molecules available to the construct can also be controlled. Biomechanical stimuli can be provided in many forms. Shear stress in varying degrees is applied to the cells due to the flowing culture media in the bioreactor. Other mechanical stimuli applied to tissue-engineered bone constructs include axial compression of the cell seeded scaffold or tensile forces.

Current bioreactor designs include spinner flasks, rotating wall vessels and perfusion systems [38-44]. Spinner flasks consist of a container full of culture media where several cell seeded constructs are suspended *via* a vertical wire from the top of the flask. A magnetic stirrer bar is included at the bottom of the flask and is rotated typically at a speed such as 50 rpm. Although this environment provides improved nutrient diffusion and promotes cell proliferation throughout the constructs, the shear forces that act on the constructs are not homogeneous.

Rotating wall vessels include the bioreactors used by NASA to study antigravity effects on bone cell activity. The tissue-engineered constructs are all grown together in the same media and 'tumble' continuously in a rotating vessel. The shear force experienced by the constructs is lower in the rotating wall vessel than in the spinner flask, as the culture media is cyclically perfused through the centre of the vessel *via* a tube. This tube is porous and the culture media is gradually introduced to the vessel, without causing high shear forces to the constructs. Similar bioreactors to this are manufactured by Synthecon Inc, where various types of rotating vessels are available [45-49].

Perfusion bioreactors have been studied by a variety of research groups for bone tissue engineering purposes [40-42,50-51]. Typically, this type of bioreactor consists of a scaffold with an individual culture media reservoir specifically for that one construct. This allows independent control over the biochemical environment to each construct. A porous scaffold is seeded with cells and the culture media is perfused through the construct either transversely or axially. The flow rate of the culture media is obviously important as is the morphology and porosity of the construct. Research into varying the rate of the perfusing media has shown that for an 80% porous construct seeded with 2 million osteoblasts, optimum flow rates were below 0.1 ml/min [52]. This flow rate produced the highest proliferation of cells. However, a flow rate of 2ml/min, although produced constructs with significantly less cell proliferation, up-regulated production of bone-related genes such as osteopontin and osteonectin. It may be that the increased shear stress applied to the cells from the higher flow rate induced this up-regulation.

As well as perfusing media through the porous constructs, it is possible to apply mechanical forces such as axial compression to the individual constructs. In this manner, axial compression of 5000 microstrain to periosteal cells suspended in porous microcarrier beads has been shown to up-regulate production of total RNA synthesis by twofold, whereas DNA synthesis remained constant after 24 hours [44].

Tensile forces have also been applied to osteoblast seeded scaffolds cultured *in-vitro* [34]. A tensile strain of 1000 microstrain was applied to primary human osteoblast seeded porous PLLA scaffolds for 30 minutes daily at 1 Hz. These conditions showed an up-regulation in alkaline phosphatase production by the cellular constructs in comparison to the controls with no load. Addition of a calcium channel agonist incorporated into the PLLA scaffold further elevated the alkaline phosphatase production with the load present.

Novel ways of trying to deliver mechanical forces to cells in a bioreactor are still currently being researched. An example of this includes the development of magnetic particle technology for applying forces directly to the cells whilst cultivating them in a perfusion bioreactor [53]. This method of applying mechanical forces uses magnetic particles, of varying sizes, from nanometres to microns in diameter. These particles are coated with a variety of proteins such as RGD (Arg-Gly-Asp) and attached *via* these proteins to the cell membrane. When an oscillating magnetic field is applied to the cell/particle, the particle undergoes a translational and rotational movement in response to the field. This in turn, applies a stretch and torque directly to the cell, in the order of 10 piconewtons. Translational stretches and torques have been applied to individual cells in this manner in the short-term (less than 24 hours) and an up-regulation of Ca^{2+} influx into cells and significant alterations in the cytoskeletal network such as actin filament stiffening has been reported as a result of such strains [54-56]. Long-term studies utilising this technique have been performed for bone tissue engineering purposes [53]. A 21-day culture of primary human bone cells in monolayer, with RGD coated 4.5-micron chromium dioxide particles attached, underwent a 1hz cylical magnetic field for 1 hour daily. These conditions showed an up-regulation in osteopontin gene production and early production of mineralised matrix in the experimental groups in comparison to the controls employed [53]. The use of magnetic particle technology is now being applied in 3D by cultivating the cell/magnetic particles on porous PLLA scaffolds in a perfusion bioreactor. Preliminary studies, in the authors' laboratory, have shown an up-regulation in osteocalcin and osteopontin gene production after 7 days, in cells with particles attached that were exposed to a magnetic field, compared to the control group which had no magnetic field exposure.

5. Animal studies

Using animal studies, comparisons can be made in the use of varying types of scaffolds at differing states of maturity with and without seeded stem cells. Synthetic polymers can be used without cells to improve fracture healing in experimental animal models. They essentially function as a scaffold allowing in-growth of the host's own cells, in addition they also avoid soft tissue interposition between the bone ends. A poly-L-lactide membrane, 250 μm thick with a 15-20 μm pore size, was used to cover 1 cm defects in the radii of rabbits. After 18–24 months, the defects had grossly healed and histologically cortical bone had been regenerated [57]. Implantation of poly-L-lactide and poly-L-*co*-D,L-lactide membranes synthesised with calcium carbonate, resulted in rapid new bone formation in critical defects of 25% of the length of the Yucatan pig radius [58]. However absorbable polymeric membranes were unable to induce bone formation when used to cover 4 cm defects in sheep tibiae [59]. Although when the same synthetic membranes were used in conjunction with cancellous bone graft a significant amount of bone healing was induced. This demonstrates how combined approaches may be required to cope with larger defects.

Several experimental animal studies have assessed bioresorbable scaffolds seeded with different sources of cells [60-61]. Porous hydroxyapatite scaffolds seeded with bone marrow derived osteoblasts, have also been evaluated in animal models. The construct seeded with cells, and a control hydroxyapatite scaffold were implanted subcutaneously into syngeneic rats and harvested at set intervals. One week after implantation, alkaline phosphatase and osteocalcin levels were much higher in the osteoblast seeded group than the control group. Alkaline phosphatase and osteocalcin have been shown to be useful chemical markers of osteogenesis, and osteoinduction can be measured by increased levels [60-61]. Two weeks after implantation a mineralised, collagenous extracellular matrix was noted in the osteoblast group, and after four weeks mature bone was seen by light microscopy.

A synthetic polymer mesh, fashioned into a tubular construct and seeded with osteoprogenitor cells, has shown promise for healing defects in rat femora [62]. Ueda and colleagues [63] harvested bone marrow from the fibulae of rabbits, and expanded it *in vitro*. The marrow mesenchymal stem cells were seeded onto a β-tricalcium phosphate scaffold and then implanted into the right maxillary sinus floor of Japanese white rabbits. A β-tricalcium phosphate scaffold soaked only in medium was implanted into the left maxillary sinus floor. More bone formation was noted on the mesenchymal stem cell side. Newly formed bone was found to have a lamellar structure, and the result resembled an autogenous bone graft. This technique could be used for maxillary sinus floor augmentation, allowing longer implants to be inserted during maxillo-facial surgery.

It may be possible to enhance the activity of cells seeded onto carrier scaffolds. Uemara and colleagues [64] used a hydroxyapatite scaffold coated with osteopontin, and seeded this scaffold with bone marrow derived osteoblasts. They implanted this construct and non-coated constructs (seeded with marrow cells only) into rats. They harvested the constructs at 18, 32 and 46 days after implantation. The ALP activity of marrow cells was measured, and although later values were similar, at 18 days the osteopontin coated group showed 40% greater ALP activity.

Some groups have attempted to reconstruct whole bones and joints by tissue engineering methods. Weng and colleagues constructed a mandibular condyle by using bovine periosteal cells and articular chondrocytes [65]. These cells were seeded onto a polyglycolic and ploy-L-lactic acid scaffold, which had been moulded into the shape of a human mandibular condyle. An alginate hydrogel was used to secure the cells in place on the scaffold. The constructs were then implanted into subcutaneous pockets created in athymic mice. The constructs were harvested 12 weeks after implantation, and were found to have retained a good shape. There was a clear interface between bone and cartilage, and histology showed that the hyaline cartilage was organised in 3 distinct layers, closely resembling articular cartilage. These findings were all similar to the features of a native mandibular condyle.

Similar techniques were used to construct *de novo* phalanges and a distal interphalangeal joint (DIPJ) [66]. As well as bovine periosteal cells and chondrocytes, tenocytes were also used to create a capsule around the joint. Constructs were harvested form the athymic mice 20 and 40 weeks after implantation. At 20 weeks a good shape had been retained and bone trabeculae had been formed. By 40 weeks, an intact junction between articular cartilage and subchondral bone was present with a proteoglycan matrix around the articular cartilage. Cortical bone formation was seen with cancellous and lamellar bone showing the presence of haversian systems and vascularisation. Fibrocartilage developed between the articular cartilage and tendon capsule, suggesting that cells could re-orientate into structures to retain the architecture observed *in vivo*. By 40 weeks only 5% of the polymeric content of the scaffold was present, the rest having been degraded.

The use of human MSCs has been investigated in athymic rat models, to define potential differences between species in osteogenic potential. Human MSCs, loaded onto a hydroxyapatite/tricalcium phosphate scaffold, have been implanted into athymic rats to evaluate their ability to initiate bone repair [67]. Segmental defects measuring 8 mm were created in the femur of the rats, the MSC loaded scaffold was implanted into one side and a cell free scaffold was implanted into the contralateral femur. Animals were sacrificed at regular intervals to assess healing of the defect. By 8 weeks radiographic and histological evidence of bone formation was apparent. Histomorphometric analysis revealed increasing bone formation by 12 weeks, angiogenesis was also noted within the scaffold pores and some evidence of remodeling was noted. Immunohistochemical analysis using antibodies, showed that the bone formed in the early stages of repair was derived from human MSCs. Biomechanical analysis found that torsional strength and stiffness were approximately 40 % that of intact control limbs, which was twice as much as seen in the cell free scaffold limb. All the MSC loaded scaffolds contained a contiguous column of bone from one end of the scaffold to the other. However, bone formation was only noted at the interface regions in the cell free scaffolds, where osteoconduction into the ceramic occurred. These studies confirm that human MSCs can initiate osteogenesis *in vivo*.

6. Human studies

Collagraft, which is a combination of 65% hydroxyapatite and 35% tricalcium phosphate, is a bone graft substitute that has been used for a number of years. Randomised controlled studies have shown that Collagraft functions as well as autologous bone grafting for the treatment of long bone fractures [68-69].

Several studies have reported on the use of demineralised bone matrix (DBM) in humans with and without cells or growth factor additives. Tiedeman and colleagues [70] used DBM and autogenous bone marrow to achieve a 77% union rate (30 of 39 patients) in a series of patients with non-unions, arthrodeses, acute fractures with bone loss, and osseous defects. An extensive review by Russell and Block [71] evaluated the use of DBM, in orthopaedic patients, in situations that would have normally warranted the use of bone graft. Regardless of the degree of complexity 80% of authors reported favourable results with the use of DBM.

In one study, 25 patients with established fracture non-unions, that included partial and complete segmental defects, were managed with a composite of allogeneic bone and human BMP [67]. The fractures were resistant to other forms of treatment, and 22 of the 25 had not responded to electromagnetic stimulation whilst 23 of the 25 had undergone an average of 3 surgical procedures which failed to achieve union. In 15 cases the bone/BMP composite was applied as an onlay graft and as an inlay graft in 10 cases. The fractures united at an average of 6 months (range 3 – 14 months), with 19 patients achieving an excellent or good result. Twenty fractures united after the original surgery, and 4 of the remaining 5 united after the second operation. Persistent infection was the cause of the single failure of union seen in this study. It should however be noted that supplementary cancellous bone graft was used in 7 cases, and it is difficult to estimate the effect of this. Furthermore, the authors believe that better surgical technique was implemented during the most recent surgery. The good results achieved may have been solely attributable to this, and therefore the effect of BMP in this study is merely circumstantial. Even so the above two studies demonstrate that combined approaches such as using rhBMP along with biodegradable scaffolds, once this technology is fully developed may offer an important treatment option for managing difficult non-unions and structural problems.

There are few reports in the literature of a tissue-engineered construct being used in human patients. A *de novo* tissue-engineered distal phalanx was implanted into the thumb of a 36-year-old patient [72]. Periosteal cells were harvested from the patient and expanded *in vitro* for 9 weeks. A porous coral scaffold was surgically implanted into the patient's thumb, and the cultured cell suspension was injected into the scaffold. An alginate hydrogel was used to secure the cells in place. Six weeks after implantation MRI studies demonstrated significant vascular perfusion of the construct. Three months after surgery, the patient was able to return to work as a landscaper, and 1 year after surgery a Greenleaf evaluation demonstrated only 10% impairment of hand function (average 22% in such injuries treated normally). Twenty-eight months after surgery, his thumb was of normal length, and he was able to perform most activities of daily living. However he had no active range of motion at the interphalangeal joint, and this joint had become encapsulated in fibrous tissue. Ten months after surgery the mineral density of the construct was greater than the contralateral phalanx (0.481 g/c^3 compared to 0.382 g/c^3). However, quantitative histomorphometric analysis revealed that 30% of the volume of the implant was coral, 5% was lamellar bone and ossified endochondral tissue, and the rest was soft tissue and blood vessels.

Quarto and colleagues used a cell based tissue engineering approach to treat 3 patients with large bone defects ranging from 4-7 cm in the tibia, ulna and humerus [73]. Osteoprogenitor cells were derived from bone marrow and expanded *ex vivo*. Hydroxyapatite scaffolds, of appropriate size and shape were implanted at the lesion sites, seeded with the expanded osteoprogenitor cells. In all 3 patients, clear evidence of abundant callus formation along the implants and good integration at the interfaces was noted 2 months after surgery. External fixation, which was provided for initial mechanical stability, was subsequently removed 6-13 months after surgery.

Whilst these cases serve to highlight the great potential of bone tissue engineering, this technology is still very much in the early stages of clinical application. As a result, at the present moment in time there are no orthopaedic devices incorporating living cells that are approved by the Food and Drug Administration (FDA) for tissue engineering. However, there are several bone void filler materials, which do not incorporate any tissue derived material, such as Collagraft, which are approved by the FDA. These products are formally classified as tissue engineering materials.

Future studies need to assess the potential for neovascularisation and nerve regeneration into implanted tissue-engineered constructs. The benefit of *ex vivo* conditioning on tissue-engineered bone constructs also needs to be further evaluated, to determine whether any significant advantage is gained compared to standard unconditioned stem cell therapy. Hopefully over time, with better understanding and further research, we will be able to realise the full potential of bone tissue engineering.

References

[1] Chaput, C.; Selani, A. and Riard, A.H. (1996) Artificial scaffolding materials for tissue extracellular matrix repair. Curr. Opin. Orthop. 7: 62-68.
[2] Braddock, M.; Houston, P.; Campbell, C. and Ashcroft, P. (2001) Born again bone: tissue engineering for bone repair. News Physiol. Sci. 16: 208-213.
[3] Vacanti, C.A. and Vacanti, J.P. (2000) The science of tissue engineering. Orthop. Clin. North Am. 31: 351-355.
[4] Brown, K.L.B. and Cruess, R.L. (1982) Bone and cartilage transplantation surgery. J. Bone Jt. Surg. Am. 64-A: 270-279.
[5] Enneking, W.F.; Eady, J.L. and Burchardt, H. (1980) Autogenous cortical bone grafts in the reconstruction of segmental skeletal defects. J. Bone Jt. Surg. Am. 62-A: 1039-1058.
[6] Summers, B.N. and Eisenstein, S.M. (1989) Donor site pain from the ilium. J. Bone Jt. Surg. 71-B: 677-680.
[7] Younger, E.M. and Chapman, M.W. (1989) Morbidity at bone graft site. J. Orthop. Trauma 3: 192-195.
[8] Oreffo, R.O. and Triffitt, J.T. (1999) Future potentials for using osteogenic stem cells and biomaterials in orthopaedics. Bone 25(2): 5S-9S.
[9] Croteau, S.; Rauch, F.; Silvestri, A. and Hamdy, R.C. (1999) Bone morphogenetic proteins in orthopaedics: from basic science to clinical practice. Orthopaedics 22: 686-95.
[10] Harakas, N.K. (1984) Dimineralized bone-matrix-induced osteogenesis. Clin. Orthop. Rel. Res. 239-251.
[11] Willert, H.G.; Bertram, H. and Buchhorn, G.H. (1990) Osteolysis in alloarthroplasty of the hip. The role of ultra high molecular weight polyethylene wear particles. Clin. Orthop. Rel. Res. 258: 95-107.
[12] Caplan, A.I. (1994) The mesengenic process. Clin. Plastic Surg. 21: 429-435.
[13] Caplan, A.I. and Bruder, S.P. (2001) Mesenchymal stem cells: building blocks for molecular medicine in the 21st century. Trends Mol. Med. 7(6): 259-264.
[14] Bruder, P.; Jaiswal, N. and Haynesworth, S.E (1997) Growth kinetics, self-renewal and the osteogenic potential of purified human mesenchymal stem cells during extensive subcultivation and following cryopreservation. J..Cell Biochem. 64: 278-94
[15] Inoue, K.; Ohgushi, H.; Yoshikawa, T.; Okumura, M.; Sempuku, T.; Tamai, S. and Dohi, Y. (1997) The effect of aging on bone formation in porous hydroxyapatite: biochemical and histological analysis. J. Bone Miner. Res. 12: 989-994.
[16] Kahn, A.; Gibbons, R.; Perkins, S. and Gazit, D. (1995) Age-related bone loss: a hypothesis and initial assessment in mice. Clin. Orthop. Rel. Res. 313: 69-75.
[17] Kadiyala, S.; Young, R.G.; Thiede, M.A. and Bruder, S.P. (1997) Culture expanded canine mesenchymal stem cells possess osteochondrogenic potential in vivo and in vitro. Cell Transplant. 6: 125-134.
[18] Fortier, L.A.; Nixon, A.J.; Williams, J. and Cable, C.S. (1998) Isolation and Chondrocytic Differation of Equine Bone Marrow – Derived Mesenchymal Stem Cells. Am. J. Vet. Res. 59: 1182-7118.
[19] Lazarus, H.M.; Haynesworth, S.E.; Gerson, S.L.; Rosenthal, N.S. and Caplan, A.I. (1995) *Ex vivo* expansion and subsequent infusion on human bone-marrow derived stromal progenitor cells (mesenchymal progenitor cells): implications for therapeutic use. Bone Marrow Transplant. 16: 557-564.
[20] Koc, N.; Gerson, S.L.; Cooper, B.W.; Dyhouse, S.M.; Haynesworth, S.E.; Caplan, A.I. and Lazarus, H.M. (2000) Rapid hematopoietic recovery after co-infusion of autologous blood stem cells and culture expanded marrow mesenchymal stem cells in advanced breast cancer patients receiving high dose chemotherapy. J. Clin. Oncol. 18: 307-316.
[21] Jaiswal, N.; Haynesworth, S.E.; Caplan, A.I. and Bruder, S.P. (1997) Osteogenic differentiation of purified, culture-expanded human mesenchymal stem cells *in vitro*. J. Cell Biochem. 64: 295-312.
[22] Johnstone, B.; Hering, T.M.; Caplan, A.I.; Goldberg, V.M. and Yoo. J.U. (1998) *In vitro* chondrogenesis of bone-marrow derived mesenchymal progenitor cells. Exp. Cell Res. 238: 265-272
[23] Wakitani, S.; Saito, T. and Caplan, A.I. (1995) Myogenic cells derived from rat bone marrow mesenchymal stem cells exposed to 5-azacytidine. Muscle Nerve 18: 1417-1426.

[24] Dennis, J.E. and Caplan, A.I. (1996) Differentiation potential of conditionally immortalized mesenchymal progenitor cells form adult marrow of a H-2K^b-tsA58 transgenic mouse. J. Cell Physiol. 167: 523-538.
[25] Dennis, J.E.; Merriam, A.; Awadallah, A.; Yoo, J.U.; Johnstone, B. and Caplan, A.I. (1999) A quadripotential mesenchymal progenitor cell isolated from the marrow of an adult mouse. J. Bone Miner. Res. 14: 1-10.
[26] Brittberg, M.; Tallheden, T.; Sjogren-Jansson, B.; Lindahl, A. and Peterson, L. (2001) Autologous chondrocytes used for articular cartilage repair: an update. Clin. Orthop. 391S: 337-348.
[27] Richardson, J.B.; Caterson, B.; Evans, E.H.; Ashton, B.A. and Roberts, S. (1999) Repair of human articular cartilage after implantation of autologous chondrocytes. J. Bone Joint. Surg. Br. 81(6): 1064-1068.
[28] Bostrom, R.D. and Mikos, A.G. (1997) Tissue engineering of bone. In: Atala, A.; Mooney, D.; Vacanti, J.P. and Langer, R. (Eds.) Synthetic biodegradable polymer scaffolds. Birkhauser, Boston, USA; pp. 215-234.
[29] Hutmacher, D.W. (2000) Scaffolds in tissue engineering bone and cartilage. Biomaterials 21(24): 2529-2543.
[30] Holy, C.E.; Dang, S.M.; Davies, J.E. and Shoichet, M.S. (1999) In vitro degradation of a novel poly(lactide-co-glycolide) 75/25 foam. Biomaterials 20: 1177-1185.
[31] Tsuruga, E.; Takita, H.; Itoh, H.; Wakisaka, Y. and Kuboki, Y. (1997) Pore size of porous hydroxyapatite as the cell-substratum controls BMP-induced osteogenesis. J. Biochem. 121: 1317-1324.
[32] Pineda, L.M.; Busing, M.; Meinig, R.P. and Gogolewski, S. (1996) Bone regeneration with resorbable polymeric membranes. III. Effect of poly(L-lactide) membrane pore size on the bone healing process in large defects. J. Biomed. Mater. Res. 31: 385-394.
[33] Yang, X.; Tare, R.S.; Partridge, K.A.; Roach, H.I.; Clarke, N.M.; Howdle, S.M.; Shakesheff, K.M. and Oreffo, R.O. (2003) Induction of human osteoprogenitor chemotaxis, proliferation, differentiation, and bone formation by osteoblast stimulating factor-1/pleiotrophin: osteoconductive biomimetic scaffolds for tissue engineering. J. Bone Miner. Res. 18(1): 47-57.
[34] Yang, Y.; Magnay, J.L.; Cooling, L. and El Haj, A. (2002) Development of a 'mechano-active' scaffold for tissue engineering. Biomaterials 23: 2119-2126.
[35] Hern, D.L. and Hubbell, J.A. (1998) Incorporation of adhesion peptides into nonadhesive hydrogels useful for tissue resurfacing. J. Biomed. Mater. Res. 39: 266-76.
[36] Harrison, D.; Johnson, R.; Tucci, M.; Puckett, A.; Tsao, A.; Hughes, J. and Benghuzzi, H. (1997) Interaction of cells with UHMWPE impregnated with the bioactive peptides RGD, RGE, or Poly-L-lysine. Biomed. Sci. Instrum. 34: 41-46.
[37] Shakesheff, K.; Cannizzaro, S. and Langer, R. (1998) Creating biomimetic microenvironments with synthetic polymer-peptide hybrid molecules. J. Biomater. Sci. Polym. Ed. 9: 507-518.
[38] Sikavitsas, V.I.; Bancroft, G.N. and Mikos, A.G. (2002) Formation of three-dimensional cell/polymer constructs for bone tissue engineering in a spinner flask and a rotating wall vessel bioreactor. J. Biomed. Mater. Res. 62(1): 136-148.
[39] Gooch, K.J.; Kwon, J.H.; Blunk, T.; Langer, R.; Freed, L.E. and Vunjak-Novakovic, G. (2001) Effects of mixing intensity on tissue-engineered cartilage. Biotechnol. Bioeng. 72: 402-407.
[40] Rucci, N.; Migliaccio, S.; Zani, B.M.; Taranta, A. and Teti, A. (2002) Characterization of the osteoblast-like cell phenotype under microgravity conditions in the NASA-approved Rotating Wall Vessel bioreactor (RWV). J. Cell Biochem. 85: 167-179.
[41] Qiu, Q.; Ducheyne, P.; Gao, H. and Ayyaswamy, P. (1998) Formation and differentiation of three-dimensional rat marrow stromal cell culture on microcarriers in a rotating-wall vessel. Tissue Eng. 4: 19-34.
[42] Goldstein, A.S.; Juarez, T.M.; Helmke, C.D.; Gustin, M.C. and Mikos, A.G. (2001) Effect of convection on osteoblastic cell growth and function in biodegradable polymer foam scaffolds. Biomaterials 22: 1279-1288.
[43] Van Den Dolder, J.; Bancroft, G.N.; Sikavitsas, V.I.; Spauwen, P.H.; Jansen, J.A. and Mikos, A.G. (2003) Flow perfusion culture of marrow stromal osteoblasts in titanium fiber mesh. J. Biomed. Mater. Res. 64A(2): 235-241.
[44] Shelton, R.M. and El Haj, A.J. (1992) A novel microcarrier bead model to investigate bone cell responses to mechanical compression *in vitro*. J. Bone Mineral. Res. 7(2): S403-405.
[45] Unsworth, B.R. and Lelkes, P.I. (1998) Growing Tissues in Microgravity. Nat. Med. 4: 901-907.

[46] Duray, P.H.; Hatfill, S.J. and Pellis, N.R. (1997) Tissue Culture in Microgravity. Science and Medicine May/June: 45-55.
[47] Qiu, Q.Q.; Ducheyne, P. and Ayyaswamy, P.S. (2001) 3D bone tissue engineered with bioactive microspheres in simulated microgravity. In Vitro Cell Dev. Biol. Anim. 37: 157-165.
[48] Qiu, Q.Q.; Ducheyne, P. and Ayyasawamy, P.S. (1999) Fabrication, characterization and evaluation of bioceramic hollow microspheres used as microcarriers for 3D bone tissue formation in rotating bioreactors. Biomaterials 20: 989-1001.
[49] Granet, C.; Laroche, N.; Vico, L.; Alexandre, C. and Lafage Profust, M.H. (1998) Rotating-wall vessels, promising bioreactors for osteoblastic cell culture; comparison with other 3D conditions. Med. Biol. Eng. Comput. 36: 513-519.
[50] Bancroft, G.N.; Sikavitsas, V.I.; Van Den Dolder, J.; Sheffield, T.L.; Ambrose, C.G.; Jansen, J.A. and Mikos, A.G. (2002) Fluid flow increases mineralized matrix deposition in 3D perfusion culture of marrow stromal osteoblasts in a dose-dependent manner. Proc. Natl. Acad. Sci. U.S.A. 99(20): 12600-12605.
[51] Porter, B.D.; Zauel, R.; Cartmell, S.H.; Stockman, H.W.; Fyhrie, D. and Guldberg, R. (2003) 3D computational modeling of media flow through scaffolds in a perfusion bioreactor. In: Transactions of 49th Annual Meeting of the Orthopaedic Research Society, 2nd-5th February New Orleans, LA (USA): Paper #0274.
[52] Porter, B.; Cartmell, S. and Guldberg, R. (2001) Design of a 3D perfused cell culture system to evaluate bone regeneration technologies. In: Transactions of the 47th Annual Meeting of the Orthopaedic Research Society, 25th-28th February San Fransisco, CA (USA): Paper #0052.
[53] Cartmell, S.H.; Dobson, J.; Verschueren, S.B. and El Haj, A.J. (2002) Development of magnetic particle techniques for long term culture of bone cells with intermittent mechanical activation. IEEE Transactions on NanoBioscience. 1: 92-97.
[54] Bausch, A.R.; Hellerer, U.; Essler, M.; Aepfelbacher, M. and Sackmann, E. (2001) Rapid stiffening of integrin receptor-actin linkages in endothelial cells stimulated with thrombin: a magnetic bead microrheology study. Biophys. J. 80: 2649-2657.
[55] Glogauer, M.; Arora, P.; Yao, G.; Sokholov, I.; Ferrier, J. and McCulloch, C.A. (1997) Calcium ions and tyrosine phosphorylation interact coordinately with actin to regulate cytoprotective responses to stretching. J. Cell Sci. 110: 11-21.
[56] Wu, Z.; Wong, K.; Glogauer, M.; Ellen, R.P. and McCulloch, C.A. (1999) Regulation of stretch-activated intracellular calcium transients by actin filaments. Biochem. Biophys. Res. Commun. 261: 419-425.
[57] Meinig, R.P.; Rahn, B.; Perren, S.M. and Gogolewski, S. (1996) Bone regeneration with resorbable polymeric membranes: treatment of diaphyseal bone defects in the rabbit radius with poly(l-lactide) membrane. A pilot study. J. Orthop. Trauma 10: 178-190.
[58] Meinig, R.P.; Buesing, C.M.; Helm, J. and Gogolewski, S. (1997) Regeneration of diaphyseal bone defects using resorbable poly(L/DL-lactide) and poly(D-lactide) membranes in the Yucatan pig model. J. Orthop. Trauma 11: 551-558.
[59] Gugala, Z. and Gogolewski, S. (1999) Regeneration of segmental diaphyseal defects in sheep usinf resorbable polymeric membranes: a preliminary study. J. Orthop. Trauma 13: 187-195.
[60] Price, P.A.; Lothringer, S.A.; Baukol, A.H. and Reddi, A.H. (1981) Developmental appearance of the vitamin K-dependent protein of bone during calcification: analysis mineralizing tissue in human, calf and rat. J. Biol. Chem. 256: 3781-3784.
[61] Weinreb, M.; Shinar, D. and Rodan, G.A. (1990) Different pattern of alkaline phosphatase, osteopontin and osteocalcin expression in developing rat bone visualized by in situ hybridization. J. Bone Miner. Res. 5: 831-842.
[62] Puelacher, W.C.; Vacanti, J.P.; Ferraro, N.F.; Schloo, B. and Vacanti, C.A. (1996) Femoral shaft reconstruction using tissue-engineered growth of bone. Int. J. Oral Maxilofac. Surg. 25: 223-228.
[63] Ueda, M.; Sumi, Y.; Mizumo, H.; Honda, M.; Oda, T.; Wada, K.; Boo, J.S. and Hata, K.I. (2000) Tissue engineering: applications for maxillofacial surgery. Mater. Sci. Eng. C13: 7-14.
[64] Uemara, T.; Nemoto, A.; Liu, Y.; Kojima, H.; Dong, J.; Yabe, T.; Yoshikawa, T.; Ohgushi, H.; Ushida, T. and Tateishi, T. (2001) Osteopontin involvement in bone remodeling and its effect on in vivo osteogenic potential of bone marrow-derived osteoblasts/ porous hydroxyapatite constructs. Mater. Sci. Eng. C17: 33-36.
[65] Weng. Y.; Cao, Y.; Silva, C.A.; Vacanti, M.P.; Vacanti, C.A. (2001) Tissue-engineered composites of bone for mandible condylar reconstruction. J. Oral Maxilofac. Surg. 59: 185-190.

[66] Isogai, N.; Landis, W.; Kim, T.H.; Gerstenfeld, L.C.; Uptom, J. and Vacanti, J.P. (1999) Formation of phalanges and small joints by tissue engineering. J. Bone Jt. Surg. Am. 81-A: 306-316.
[67] Johnson, E.E.; Urist, M.R. and Finerman, G.A. (1992) Resistant nonunions and partial or complete segmental defects of long bones. Treatment with implants of a composite of human bone morphogenetic protein (BMP) and autolyzed, antigen-extracted allogeneic (AAA) bone. Clin. Orthop. Rel. Res. 229-237.
[68] Cornell, C.N.; Lane, J.M.; Chapman, M.; Merkow, R.; Seligson, D.; Henry, S.; Gustilo, R. and Vincent, K. (1991) Multicenter trail of Collagraft as bone graft substitute. J. Orthop. Trauma 5(1): 1-8.
[69] Chapman, M.W.; Bucholz, R. and Cornell, C. (1997) Treatment of acute fractures with a collagen-calcium phosphate graft material. A randomized clinical trial. J. Bone Jt. Surg. Am. 79(4): 495-502.
[70] Tiedeman, J.J.; Garvin, K.L.; Kile, T.A. and Connolly, J.F. (1995) The role of a composite, demineralized bone matrix and bone marrow in the treatment of osseous defects. Orthopaedics 18: 1153-1158.
[71] Russell, J.L. and Block, J.E. (1999) Clinical utility of demineralized bone matrix for osseous defects, arthrodesis, and reconstruction: impact of processing techniques and study methodology. Orthopedics 22(5): 524-531.
[72] Vacanti, C.A.; Bonassar, L.J.; Vacanti, M.P. and Shufflebarger J. (2001) Replacement of N avulsed phalanx with tissue-engineered bone. N. Eng. J. Med. 344: 1511-1514.
[73] Quarto, R, Mastrogiacomo, M.; Cancedda R.; Kutepov, S.M.; Mukhachev, V.; Lavroukov, A.; Kon, E. and Marcacci, M. (2001) Repair of large bone defects with the use of autologous bone marrow stromal cells. N. Eng. J. Med. 344(5): 385-386.

STEM CELLS – POTENTIAL FOR TISSUE ENGINEERING

M. MINHAJ SIDDIQUI[1] AND ANTHONY ATALA[2]

[1]*Laboratory for Tissue Engineering and Cellular Therapeutics, Children's Hospital and Harvard Medical School, 300 Longwood Avenue, Boston, MA-02115, USA – Fax: 617-232-3692;*
[2]*Institute for Regenerative Medicine, Dept. of Urology, Medical Center Blvd., Winston-Salem, NA-27157, USA – Fax: 336-716-5701 – Email: aatala@wfubmc*

1. Introduction

The integration of stem cells and tissue-engineered scaffolds has the potential to revolutionise the field of regenerative medicine. It promises great things including the ability to grow organs composed of multiple cell types and complex structures, therapies for the correction of congenital genetic diseases, and the promise of readily obtainable immunocompatible tissues [1-6]. Yet, the research of both tissue-engineering and stem cell biology, as we know it in 2003, is in its infancy. Much work remains to be done before the true clinical promises of these fields are to be realised.

Of great interest nonetheless is the fact that even though stem cell biology is in its infancy with few experiments that have truly demonstrated its ability to cure disease, stem cells have still generated an incredible amount of public fascination. Much of this is due to the controversy in the method by which embryonic stem cells are derived, but a great deal of the interest is balanced by the ability that many claim of the cells to cure today's incurable diseases including diabetes, ischemic heart disease, liver failure, renal failure, Parkinson's disease, and spinal cord injuries to name a few [7-9].

Such a surge in attention since the initiation of the field of embryonic stem cells in only 1998 is visible not only in the public interest, but in academic and industrial laboratories as well. Many labs and even universities in the last five years have incorporated some aspect of stem cell biology within their umbrella of interests [7]. Such a trend is both extremely beneficial and very dangerous to the field in that with more minds comes more creative thought and novel ideas, but also in a field that is still so poorly defined and understood, more voices saying conflicting things can mean ever more confusion as to what the true potential of stem cells as future therapies really is [2,5,10]. The need for concrete definitions and clarity in discussion, as well as a great deal of basic science research characterising the current sources and biochemical processes of stem cells is sorely needed to aid in the maturation of this science. Further investigation into other potential sources of stem cells is also of immense interest as we are sure to discover various other sources that are ideal for different applications. In this chapter, we will try to allude to the current state as well as the future directions of stem cell biology especially as they relate in conjunction with tissue-engineering to regenerative medical therapies.

2. What is a stem cell?

Hence we arrive at the most fundamental question of any discussion on stem cells: what exactly is a stem cell? What is the common thread that links embryonic stem cells, haematopoietic stem cells, mesenchymal stem cells, oval cells (liver stem cells), satellite cells (muscle stem cells), neurospheres (neural stem cells), and the like? What is it about these cells that separate them from cells such as erythrocytes, hepatocytes, or neurons? The question is troubling in that as fundamental as it is to the remainder of the discussion, there has been no clear consensus reached as to the absolute definition of a "stem cell" [1,2,11]. However, two themes are consistently presented as properties a certain set of cells ought to present with if they are to be indeed characterised as stem cells [1,2,5,12]. Those concepts include:
- Ability to renew through division to maintain a population of cells that possess the same properties as the original cells.
- A capacity to differentiate into multiple cell types that are unique both from the parent cell in terms of gene expression and that are unique from each other.

These basic concepts of self-renewal and differentiation potential are the tenets from which stem cells derive their strength. Self-renewal is crucial especially in the examples of stem cells that are found in the body (such as oval cells, satellite cells, haematopoietic stem cells) in that such cells are often the source of renewal for the multiple cell types that are found in individual organs, and hence must have the ability to self-renew to maintain the existence of the population itself throughout the lifespan of the host [13,14]. This trait does function to separate stem cells from many, but not all, mature somatic cells which tend to have limited ability to self-renew [2,14]. Similarly, the trait of potential to differentiate into multiple cell types is argued to be important simply by definition of stem cells. It is important to note the difference between those cells that can give rise to multiple cell lineages and those that are only capable of transforming into another cell type. The former cell types are argued to be immature or progenitor cells undergoing a process of maturation into more specialised cells rather then truly differentiating stem cells [15,16]. A stem cell should be able to give rise to more stem cells and cells of other types. There is still some disagreement on this point however.

Such traits of self-renewal and multipotency are of great benefit in tissue engineering because they overcome an obstacle regarding the inability of certain cell types, such as neurons or chondrocytes, to expand and hence establish functional constructs of significant useable therapeutic volume. The nature of the self-renewal ability and the degree of differentiation potential of different stem cells are linked to the source of the cells [6,17-19]. We will briefly examine some of the main sources of stem cells and then move on to discuss some areas of interesting research using stem cells in conjunction with tissue engineering for regenerative medicine.

It is important to note however that even with a definition in mind of what a stem cell is, identifying a certain cell population as a stem cell population can still be quite a challenge [1,2]. While it is a requirement of a stem cell to be self-renewing, discovering media conditions which allow a certain cell to thrive in long term culture is quite difficult and has not yet been found for adult stem cells [1-3,5,14] (adult stem cells are therefore characterised stem cells because they are present throughout the life of an individual rather than directly observed behaviour *in vitro*). Similarly, while it is simple to state that a cell should have potential to differentiate into two or more different cell types, it is very difficult to actually find the exact media conditions that lead to the desired differentiation and sometimes even difficult to ascertain if two sets of differentiations are truly unique populations [2]. Additionally, to definitively show that a cell population is indeed a population of stem cells and not simply a group of cells that are collectively potent for multiple cell lineages, *in vitro* experiments must be done using a clonally derived stem cell population rather than heterogeneous groups of candidate stem cells. A great deal of resources and effort can be spent attempting to prove that a population of potential stem cells really are stem cells.

2.1. ADULT STEM CELLS

The field of adult stem cells is, especially in the specific area of haematopoietic stem cells, better understood and researched than any other aspect of stem cell biology [11,20]. Much work had been done on tissue specific stem cells, such as stem cells of the gastrointestinal tract, long before the existence of pluripotential embryonic stem cells was even known. Adult stem cells tend to be tissue specific self-renewing populations of cells that can differentiate into various cell types often associated with a certain organ system [14,21,22]. They are quite rare and found in very low numbers on the order of 1 in 10,000 cells within the tissue of interest [2]. Current known niches of such stem cells include bone marrow, brain, liver, skin, skeletal muscle, gastrointestinal tract, pancreas, the eye, blood, and dental pulp [14,17,23]. Of these, the most studied are CD34+ haematopoietic stem cells isolated from bone marrow that are capable of producing cells of the lymphoid and myeloid lineages in blood. Such cells are the only currently available therapeutic application of stem cells and are used for a variety of purposes usually entailing the replacement or reestablishment of the immune system of a host after a disease or toxic therapy. Great difficulty has been encountered in maintaining adult stem cells in culture and to date, there is no known efficient means of maintaining and expanding a long-term culture of any adult stem cell in large numbers. Isolation has also proven to be quite problematic as these cells are present in extremely low ratios in the adult tissue. Such cells are often selected utilising Fluorescent Activated Cell Sorting (FACS) or Magnetic Activated Cell Sorting (MACS) against surface markers specific to the stem cell of interest [3,11].

A notable exception to the tissue specific potential of stem cells is the mesenchymal stem cell or what is more recently called the multipotent adult progenitor cell derived from bone marrow stroma [2, 24-26]. Such a cell has been shown to differentiate *in vitro* into numerous tissue types and similarly differentiate developmentally in blastocyst injection into multiple tissues including neuronal, adipose, muscle, liver, lungs, spleen, and gut but notably not bone marrow or gonads [26]. While current use of adult stem cells is quite limited, there is great potential in future utilisation of such cells for the use of tissue specific regenerative therapies. These cells benefit from a great advantage in that they can be used in autologous therapies hence avoiding any immune rejection complications [2,5].

2.2. FOETAL STEM CELLS

Populations of stem cells have been found that are prominent in the foetus but sharply dwindle with the development of the baby. Such cells are interesting not only because of their renewal and differentiation potential, but also because such progenitor stem cells tend to be easily isolated based on anatomic location of dissection. Induction into mature tissue utilising these cells tends to be more straightforward using these cells as opposed to using ES cells [27]. Especially notable in research amongst such cell types are neurospheres which are neuronal stem cells and are thought to give rise to multiple neuron cell types in the developing foetal brain [28]. John Gearhart's group also discovered a pluripotential source of stem cells on the gonadal ridge of 5-9 week old aborted foetal tissue [29]. While such cells hold great promise with their ability to self-renew and readily differentiate *in vitro*, their source of extraction severely limits their future clinical potential.

2.3. EMBRYONIC STEM CELL

Embryonic stem (ES) cells are cells that are derived from the inner cell mass of a blastocyst. They are in essence what the public speaks of when it says "stem cells". These cells are also currently the most pluripotent cells known in *in vitro* experimentation. ES cells have a peculiar feature that they differentiate spontaneously into structures termed embryoid bodies if removed from the presence of a differentiation inhibitory growth factor called Leukaemia Inhibitory Factor (LIF). Also, when injected into animals, these cells tend to form tumours termed teretomas that are composed of a multitude of cell types. These cell types encompass endoderm, mesoderm, and ectoderm germ layers and have been noticed to contain tissue such as neural tissue, cardiac, cartilage, bone, and in some cases even hair and teeth structures [30].

The history of ES cell research dates back about forty years encompassing cells isolated from teretoma tumours that were present in newborn humans and mice (embryonal carcinoma cells) [1]. However, the current state of interest was sparked mainly in 1998 with the discovery by James Thomson's lab of a protocol for extraction and maintenance *in vitro* of embryonic stem cells derived from the inner cell mass of 5 day discarded blastocysts from an *in vitro* fertilisation clinic [30]. Since then, numerous advances have been made including discoveries of various protocols to differentiate the cells *in vitro* into myogenic [30], adipocyte [31], osteogenic [32], neural [33], and hepatic lineages [34] to name a few. Blastocyst injection experiments have also shown that these cells are in fact capable of producing all the tissues of a mouse and are hence in fact totipotent [35], although the validity of such experiments because of the possibility of fusion induced pluripotentiality is still under discussion [2].

Current research is focused on understanding the pathways of both maintenance of "stemness" character as well as of differentiation of the cells into various cell types [36]. Many techniques are being developed such as positive selection of the cell type of interest out of non-specifically differentiated embryoid bodies, transfection with genetic material encoding transcription factors that induce differentiation, protocols utilising growth factors to induce differentiation, and other media formulations that favour the induction a certain cell types [15, 16, 37]. The protocols that currently exist for differentiation of stem cells into most tissues are still at best inefficient and much work remains to be done to understand how the various pathways of lineage induction can be initiated.

There are many advantages supporting the use of ES cells including their relative ease of extraction, high expansion capabilities, and high degree of pluripotency. However, there are also multiple potential disadvantages to consider especially when thinking about clinical applications. One consideration is that for the use of transplantation, these foreign cells will induce immune rejection. Also, by the nature of their growth, undifferentiated ES cells have a propensity to form teretomas that would be of concern in any clinical application [38-42]. Finally, there are also significant ethical considerations that play very prominent determinants as to the availability of this technology for clinical and research use [43].

2.4. SOMATIC CELL NUCLEAR TRANSFER: THERAPEUTIC CLONING

Cells derived from somatic cell nuclear transfer are in essence the same in properties as ES cells and come with the same set of advantages, disadvantages as those cells except for a few notable exceptions. These cells are derived by transfer of a nucleus from an adult somatic cell into an enucleated oocyte followed by extraction of the inner cell mass from the resultant blastocyst. The isolated cells are ES cells and can be cultured and manipulated utilising ES cell protocols [2]. An advantage to such a cell line is that these cells are genetically identical to the somatic cell donor. Hence, if a therapy is developed utilising these cells, the threat of rejection on transplantation is non-existent [27]. A significant concern with the use of these cells is the additional ethical dilemma that although in therapeutic cloning no attempt is made to actually clone a human being, the process involves steps that generate embryos which if implanted, could feasibly under idealised conditions lead to the growth of a clone [8].

3. Potential and how cells are differentiated

The truly exciting aspect of stem cells from any source is not that they simply exist, but that they can be differentiated into various tissues. The exact protocol necessary to carry out such a differentiation is a matter of intense research and the area of most concentration amongst researchers working in the field of stem cell biology today [44,45]. A combination of basic science study of developmental biology, gene therapy, and cell culture biology influence researchers as they strive to develop protocols that can reliably and efficiently transform a stem cell into a desired differentiated cell [15,16,37]. Each protocol comes with its own set of advantages and disadvantages not only experimentally, but also within considerations as integrated with future clinical therapy. We will examine some common approaches that are being utilised to induce differentiation of adult, foetal, and embryonic stem cells into various lineages.

3.1. NON-SPECIFIC DIFFERENTIATION AND SELECTION

The approach of allowing embryonic stem cells to differentiate uncontrolled when removed from the presence of the cytokine LIF has perhaps been one of the quickest ways to assess the differentiation potential of embryonic stem cells. ES cells must be grown in the presence of LIF to maintain the "stemness" characteristic. Once removed, they aggregate into masses termed embryoid bodies and spontaneously differentiate into multiple cell types including bone, adipocyte, endothelial, skeletal, smooth, and cardiac muscle, hepatic, pancreatic, gut, neuronal to name a few [30]. Experiments that allow ES cells to form embryonic bodies and then probe cross-sections from these bodies with antibodies specific to proteins traditionally expressed by a certain cell type (such as insulin for pancreatic beta cells or albumin for hepatocytes) allows for a quick experiment that verifies the ability of those stem cells to differentiate into the cell type of interest. Some researchers have been able to find media formulations that favour the survival of one cell type over other cell types [2,15,16]. Hence, they allow embryonic bodies to form and then positively select cells of interest by placing the bodies in media that favours the growth of the cell of interest and does not support the survival or growth of other cell types. This protocol does not particularly induce differentiation itself, but rather selects for a desired cell type amongst a mixed population of variously differentiated cells [1,2,30]. Positive selection can feasibly be performed in multiple other ways including cell sorting using FACS or MACS against a surface marker generally only expressed on the cells of interest [46]. This method has the advantage in that it is simple, but suffers from the shortcoming that it usually does not yield high numbers of cells. Additionally, this method annuls a key advantage of stem cells in tissue engineering in that cell differentiation must occur *in vitro* on a plate and cannot take place in a scaffold [6]. Cells obtained in this manner of isolations from embryoid bodies are hence not much different than somatic cells isolated from tissue. Thus, the possibility that a multipotent stem cell could be seeded on a scaffold and allowed to differentiate into the two or three cell types of interest while in the three dimensional environment of the scaffold does not exist when utilising embryoid bodies [47,48].

3.2. GENE TRANSDUCTION INDUCED DIFFERENTIATION

Gene therapy *via* transduction of cells with plasmids encoding transcription factors known to be key "gatekeeper" factors in the development of certain tissues can be a very powerful way of effecting directed differentiation of ES and many alternate source stem cells [49]. An advantage to this method is that it is efficient in influencing stem cells to differentiate along the desired pathway in a directed manner. It is also effective in some cases at inducing differentiation of some fairly complicated cell types such as pancreatic beta cells for which few other effective protocols exist. Additionally, if the right transcription factor is chosen, a family of cells rather than only one cell type can be induced [50]. Such a process of inducing a family of cells can be of great benefit when trying to tissue-engineer a complex tissue composed of a mixture of multiple cell types [2,51]. Cell populations can also be purified to consist only of transfected cells by including a resistance protein to some cytotoxic chemical and then treating the whole cell population to that toxin hence killing only non-transfected cells [37]. Two key disadvantages to this method are that the process of introduction of genetic material into cells through various vectors such as viral or electroshock based come with their own set of risk factors that need to be assessed when considering clinical therapy potential. Also, the transcription factor expression tends to be constitutive which can result in cells being driven down a differentiation pathway excessively in a certain direction (such as a predominance of myeloid cell differentiation in cells infected with the HoxB4 gene) [52]. This skewed differentiation profile may not be ideal for cells being considered as long term therapies. Some work is being done with controlled expression systems that may solve this problem [53].

3.3. GROWTH FACTOR AND MEDIA FORMULATION INDUCED DIFFERENTIATION

A combination of growth factors and other signalling factors in the growth medium can lead to the differentiation of cells into desired cell types. Often, the decision of which growth factors or chemicals to use can be decided through review of developmental biology research and examination of the extracellular milieu of cells that develop into the tissue of interest, for example liver bud cells in an embryo [1,2,54]. Of special interest are the growth factors secreted by local surrounding cells. This information, when coupled with cell media formulations that have been discovered to be ideal for the growth of the desired cell types, can lead to efficient induction of a population of stem cells into the desired lineage [37]. The greatest impetus to the rapid growth of this method is that it is simply a very difficult task to discover a combination of growth factors and chemicals that are effective in inducing differentiation of certain stem cells into the cell type of choice [15,16]. Most progress relies on educated trial and error experiments. However, these protocols also suffer from a shortcoming, there is great variability in the effectiveness of the different lineage differentiations. Some protocols such as a protocol to induce myogenic differentiation utilising 5-aza-2'deoxycytidine work very reliably and can transform a vast majority of a plate of cells into the desired type within a week [55]. Some protocols, such as Hepatocyte Growth Factor (HGF) to induce hepatic differentiation, can take upwards of 45 days *in vitro* and still result in a very low proportion of cells that are differentiated [54]. This method of extracellular signalling induced differentiation, if optimised, is ideal for inducing differentiation within the context of use with tissue-engineered scaffolds as it is a one time treatment which can be circulated throughout a scaffold and has no lingering effects.

Most major advancements in the field of stem cell biology in the next few years will surely mean an increase in sophistication of available protocols and effectiveness of the protocols to induce differentiation of various stem cell types into differentiated cells of interest.

4. Advantages in tissue engineering

Discussing the advantages of stem cells in tissue engineering is a tricky business in that while the many potential benefits are readily understood, very few of them have been experimentally demonstrated. We will attempt to highlight a few of the successes that have been shown and are being investigated as therapies as well as further discuss the potential impact that the synthesis of stem cell biology and tissue engineering can have on regenerative medicine. Stem cells, whether they are of adult, foetal, or embryonic origin, possess two key advantages over regular differentiated cells to the ultimate performance of a tissue-engineered construct as a viable long term regenerative therapy: self-renewal and multipotency [6,17,47].

4.1. SELF-RENEWAL

Tissue engineering differs in its goal from the general field of bioprosthetics in that by integrating cells into the constructs, it hopes to establish an implant which will not only compensate for the current deficit, but can maintain itself for long-term use and can even adapt to future environments [47,48]. Fully committed somatic cells often suffer from the shortfall that they are not self-renewing. Thus, as a long term implant, a construct seeded only with end stage differentiated cells leaves something to be desired in that such tissue, unless self-renewing stem cells migrate in from surrounding tissue, will not maintain itself long-term and adapt with variable renewal in response to different environmental cues in the body. If a progenitor stem cell population can be established in the tissue-engineered construct, such cells could not only help organise and generate the current tissue to be transplanted, but could remain present and maintain not only their own survival but the renewal of the whole implant through the life of the patient [5,41,42,47,48].

4.2. MULTIPOTENCY

Virtually every tissue in the body is a complex mix of multiple cell types interacting and interlaced at the individual cellular level. While tissue engineering can build scaffolds that dictate the superstructure of the tissue being reconstructed, the field has not become mature enough to place individual cells in various arrangements in a practical manner. Stem cells that differentiate into a few relevant cell types in the tissue of interest can be extremely powerful as such cells seeded on a scaffold could be differentiated into the desired cell types [16,37]. Such a mixture of cells could process local cues to correctly structure themselves into highly complex formations in the scaffold much as they do during development. In such a way, structures even as complex as glomeruli can be generated in a tissue-engineered scaffold (see renal example under Research directions) [27].

When these advantages are combined, the possibility of realizing a goal of tissue engineering to generate fully functional, long-term tissue becomes ever closer to reality.

5. Research directions

Much work is being done to combine stem cells with scaffolds to develop therapies. We will discuss skin, bone, cartilage, renal, skeletal, and cardiac muscle research as a few relevant examples.

5.1. SKIN

Engineered skin grafts are established therapies which, although do not set out to use stem cells, because of the way in which they are prepared incorporate adult stem cells and require stem cells for their success. Typically a tissue-engineered skin graft consists of a scaffold to support keratinocyte growth and a population of seeded keratinocytes obtained from a skin sample that was plated and expanded *in vitro*. It has been noted that the success of these skin grafts can be negatively affected by certain culture conditions and scaffold compositions. In some cases, there is even consistent encouraging initial engraftment followed by poor long-term prognosis [47,56]. Current research has raised the possibility that such outcomes may be due to culture or scaffold induced depletion of the stem cell population found in the seeding cell population. These skin stem cells are called holoclones and are the focus of improving future skin graft therapies. Such therapies are considering using scaffolds and culture conditions that would increase the proportion of holoclones present in the seeding population of epidermal cells as well as considering utilising holoclone specific surface markers to help enrich the proportion of these cells in the seeding population [57,58].

5.2. BONE

There has been success in utilising mesenchymal stem cells (also known as skeletal stem cells) in combination with appropriate osteoconductive hydroxyappatatite scaffolds and growth factors to repair critical bone gap lengths that could not be fixed with scaffold and growth factors alone [47]. In this protocol, mesenchymal stem cells, which are in the stromal cells of the bone marrow, are isolated and expanded *in vitro* so that they can be seeded on the scaffold that will be implanted. This technique has been found to be very successful and the protocol is currently in clinical trials as a therapy for critical gap lengths that do not heal spontaneously in impaired individuals such as the elderly [59,60].

5.3. CARTILAGE

Cartilage regeneration suffers from a critical obstacle that chondrocytes are fairly dormant cells and do not expand or regenerate much matrix after isolation. Stem cells of various sources that have potential to differentiate into chondrocytes after an initial stage of expansion are being considered to help overcome this problem [61]. Currently, the greatest challenge lies in devising a reliable protocol for differentiation of ES cells or bone marrow derived mesenchymal stem cells into chondrocytes [2]. Such cells would then be seeded on appropriate scaffolds consisting of polylactic acid and polyglycolic acid copolymers possibly embedded with gels such as hyaluronic acid [61]. This is a high need application with a very large potential market and is currently under heavy investigation.

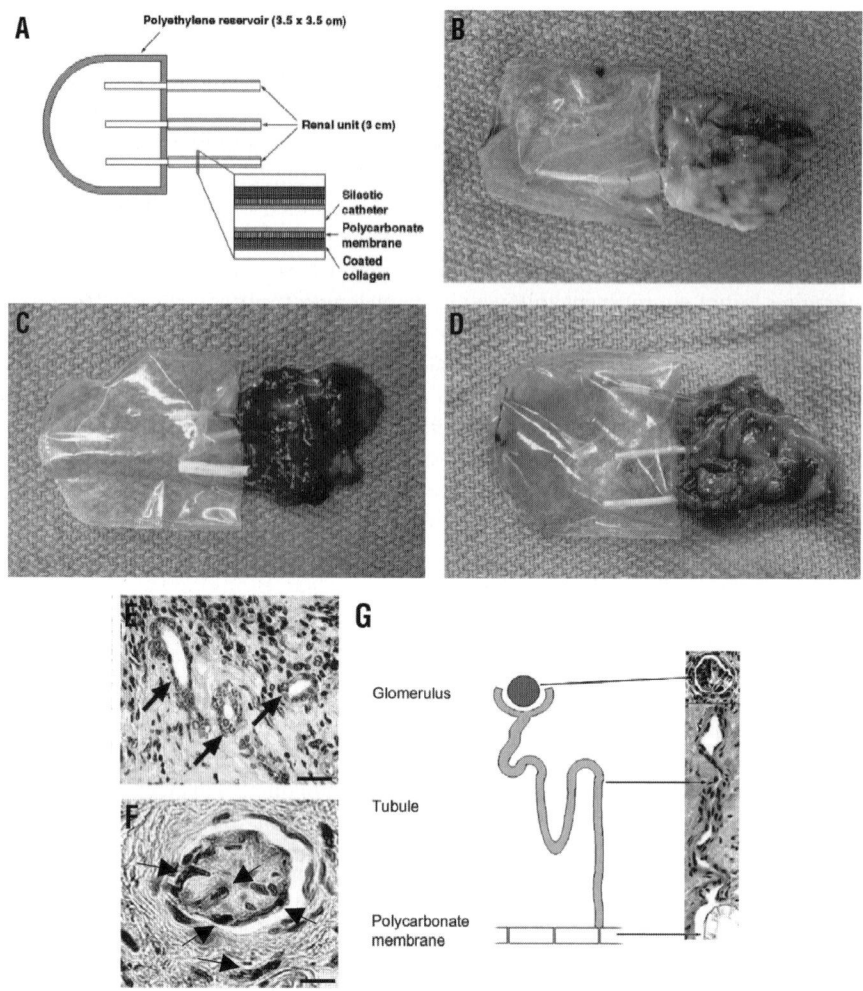

Figure 1. Tissue-engineered renal units. A. Illustration of renal unit and units retrieved 3 months after implantation. B. Unseeded control. C. Seeded with allogeneic control cells. D. Seeded with SCNT derived foetal cells showing accumulation of urine-like fluid. E. Organised tubules (arrows) were shown in the retrieved explant. F. Immunohistochemical analysis using Factor VII antibodies (arrows) identified the vascular structures. G. Development of glomerular structure in continuity via tubules with the polycarbonate membrane. (bar = 200 micrometers in E and F).

5.4. RENAL

Current advances in renal devices utilising foetal derived stem cells demonstrate the potential that combining stem cells with tissue engineering can have in terms of regenerating very complex structures at a cellular level. Tubular constructs consisting of layers of collagen, polycarbonate membrane, and silicone were seeded with cells derived from bovine fêtes metanephros cells that had been expanded *in vitro*. Such implants demonstrated production of dilute urine 12 weeks after implantation in the cows and on histology showed development of glomeruli and tubules (Figure 1) [27]. Although the foetal source of cells makes it difficult to transition such a technology into human use, such finding do demonstrate what the potential is of combining other cell sources such as adult or ES cells with correctly designs scaffolds.

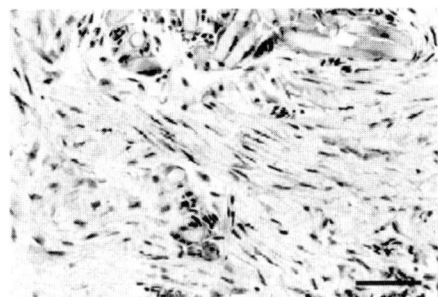 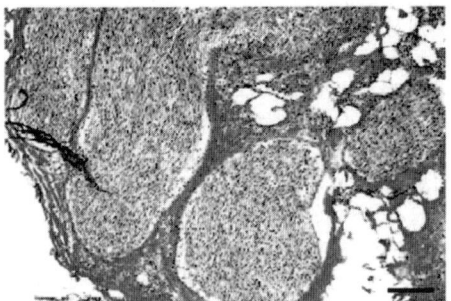

Figure 2. Left: *Cardiac muscle tissue was differentiated after isolation of foetal stem cells from a somatic cell nuclear transfer (SCNT) bovine foetus. Retrieved cloned cardiac tissue shows a well-organized cellular orientation six weeks after implantation (bar = 100 micrometers).* Right: *Similarly differentiated skeletal muscles derived from SCNT cells showed well-organized bundle formation at twelve weeks implantation (bar = 800 micrometers).*

5.5. SKELETAL AND CARDIAC MUSCLE

Exciting work is taking place in the areas of cardiac and skeletal muscle regeneration. Skeletal muscle regeneration work is mainly focused as a potential therapy for neuromuscular disorders. It is envisioned that stem cells could be altered utilising gene therapy to correct genetic disorders and such cells could be expanded and both disseminated systemically as well as implanted using constructs into specific areas were some gain of function could offer a significant improvement upon the quality of life. Much work is currently taking place in animal models and there have been many cases of successful generation of tissue (Figure 2) [62]. However, the efficiency of regeneration is not yet high enough to transition into clinical trials. Cardiac muscle work is mainly concerned with usage of stem cells as a potential therapy for post-ischemic heart disease. Multiple avenues are being investigated as to how cardiac tissue can be regenerated *via* direct injection or tissue engineered construct facilitated regeneration. There has been some success in animal models of differentiating stem cells into cardiac tissue. Clinical trials are being conducted utilising injection of stem cells but none utilising *in vitro* differentiated cells [63].

Further work is being conducted in various tissues including blood vessels, cornea, teeth, liver, pancreas, and neuronal cells and the list is constantly growing [47].

6. Ethical and political considerations in stem cell biology

No discussion on stem cells could be complete without discussing the ethical dilemmas that accompany the science. Such discussion tends to focus around the source of acquisition of embryonic stem and foetal stem cells that are human embryo blastocysts, either discarded from *in vitro* fertilisation clinics or specifically created for research, and discarded human foetuses from abortions respectively. Adult stem cells enter the discussion in the capacity as options that are being considered as alternatives to ES cells and foetal cells.

The crux of the dialogue between the two sides especially concerning ES cells comes down to when each side believes life is initiated. Perhaps oversimplifying but for the sake of a short overview, the opinions can be broken into two sides [8-10,43]. On one hand there is the group of people who believe that life begins at conception or some close variant of that concept, and to such people the destruction of an embryo at any stage is a violation to the sanctity of life. On the other hand are people who believe life begins with implantation or some time after implantation. Their argument often insists that the natural course of the embryo in its current *in vitro* environment is death, and hence it cannot be considered a living being with the same rights as such, especially when within that embryo is a source of cells which could be of immense therapeutic potential on currently living adults [8].

Many scientific pursuits can often engender deep ethical discussions, but few deal with the idea of preservation of life in both extremes as directly as this. Hence, stem cell biology is unique in that the future of the field is in many ways not only links to the development of the science, but the decisions that society makes about the ethical, and hence legal, permissibility of the science. Such a discussion has occurred and is ongoing around the world and has resulted in a current status in the United States by the order of the President that federally funded human embryonic stem cell research is to be conducted only on already established lines and that no new human ES lines are to be generated utilising federal funding. Foetal cadaver derived cell research is allowed with some regulation. As the science develops and the true potential of various cell sources is further understood, ethical discussions are sure to mature and result in adaptations of the legal stances on these technologies [2,8-10,43].

References

[1] Marshak, D.R.; Gardner, R.L. and Gottlieb, D. (2001) Stem Cell Biology. Cold Spring Harbor Laboratory Press. NY, USA.
[2] Stem Cells: Scientific Progress and Future Research Directions. Department of Health and Human Services. June 2001. http://www.nih.gov/news/stemcell/scireport.htm
[3] Potten, CS. (1997) Stem Cells. Academic Press. London, UK.
[4] Mattson, M.P. and Zant, G.V. (2002) Stem Cells: A Cellular Fountain of Youth. Elsevier. Amsterdam, The Netherlands.

[5] Stem Cells and the Future of Regenerative Medicine. Committee on the Biological and Biomedical Applications of Stem Cell Research; Board on Live Sciences, National Research Council; Board on Neuroscience and Behavioral Health, Institute of Medicine. National Academy Press. Washington, D. C., 2002.
[6] Haverich, A. and Graf, H. (2002) Stem Cell Transplantation and Tissue Engineering. Springer. Berlin, Germany.
[7] Marshall, E. (2000) The business of stem cells. Science 287(5457): 1419-1421.
[8] Ethical Issues in Human Stem Cell Research: Volume I. National Bioethics Advisory Commission. Rockville, Maryland, September 1999. http://www.georgetown.edu/research/nrcbl/nbac/pubs.html
[9] McLaren, A. (2001) Ethical and social considerations of stem cell research. Nature 414(6859): 129-131.
[10] The Use of Embryonic Stem Cells in Therapeutic Research: Report of the Intrnational Bioethics Committee on the Ethical Aspects of Human Embryonic Stem Cell Research. Division of Human Sciences, Philosophy, and the Ethics of Sciences and Technology. Paris, April 6, 2001.
[11] Quesenberry, P.J.; Stein, G.S.; Forget, B. and Weissman, S. (1998) Stem Cell Biology and Gene Therapy. Wiley-Liss. New York, USA.
[12] Watt, F.M. and Hogan, B.L. (2000) Out of Eden: stem cells and their niches. Science 287(5457): 1427-1430.
[13] Preston, S.L.; Alison, M.R.; Forbes, S.J.; Direkze, N.C.; Poulsom, R. and Wright, N.A. (2003) The new stem cell biology: something for everyone. Mol. Pathol. 56(2): 86-96.
[14] Presnell, S.C.; Petersen, B. and Heidaran, M. (2002) Stem cells in adult tissues. Semin. Cell Dev. Biol. 13(5): 369-376.
[15] Moody, S.A. (1999) Cell Lineage and Fate Determination. Academic Press. San Diego, USA.
[16] Maclean, N. and Hall, B.K. (1987) Cell Commitment and Differentiation. Cambridge University Press. Cambridge, UK.
[17] Al-Rubeai, M. (1999) Cell Engineering. Kluwer Academic Publishers, Dordrecht, The Netherlands.
[18] Bell, E. (1993) Tissue Engineering: Current Perspectives. Birkhauser.
[19] Oka, M.S. and Rupp, R.G. (1992) Cell Biology and Biotechnology: Novel Approaches to Increased Cellular Productivity. Springer-Verlag.
[20] Ballas, C.B.; Zielske, S.P. and Gerson, S.L. (2002) Adult bone marrow stem cells for cell and gene therapies: implications for greater use. J. Cell Biochem. Suppl. 38: 20-28.
[21] Young, H.E.; Duplaa, C.; Young, T.M.; Floyd, J.A.; Reeves, M.L.; Davis, K.H.; Mancini, G.J.; Eaton, M.E.; Hill, J.D.; Thomas, K.; Austin, T.; Edwards, C.; Cuzzourt, J.; Parikh, A.; Groom, J.; Hudson, J. and Black, A.C. Jr. (2001) Clonogenic analysis reveals reserve stem cells in postnatal mammals: I. Pluripotent mesenchymal stem cells. Anat. Rec. 263(4): 350-360.
[22] Vogel, G. (2000) Can old cells learn new tricks? Science 287(5457): 1418-1419.
[23] Spradling, A.; Drummond-Barbosa, D. and Kai, T. (2001) Stem cells find their niche. Nature 414(6859): 98-104.
[24] Devine, S.M. (2002) Mesenchymal stem cells: will they have a role in the clinic? J. Cell Biochem. (Suppl.) 38: 73-79.
[25] Jackson, K.A.; Majka, S.M.; Wulf, G.G. and Goodell, M.A. (2002) Stem cells: a minireview. J. Cell Biochem. Suppl. 38:1-6.
[26] Jiang, Y.; Jahagirdar, B.N.; Reinhardt, R.L.; Schwartz, R.E.; Keene, C.D.; Ortiz-Gonzalez, X.R.; Reyes, M.; Lenvik, T.; Lund, T.; Blackstad, M.; Du, J.; Aldrich, S.; Lisberg, A.; Low W.C.; Largaespada, D.A. and Verfaillie, C.M. (2002) Pluripotency of mesenchymal stem cells derived from adult marrow. Nature 418(6893): 41-49.
[27] Lanza, R.P.; Chung, H.Y.; Yoo, J.J.; Wettstein, P.J.; Blackwell, C.; Borson, N.; Hofmeister, E.; Schuch, G.; Soker, S.; Moraes, C.T.; West, M.D. and Atala, A. (2002) Generation of histocompatible tissues using nuclear transplantation. Nat. Biotechnol. 20(7): 689-696.
[28] Laywell E.D.; Rakic, P.; Kukekov, V.G.; Holland, E.C. and Steindler, D.A. (2000) Identification of a multipotent astrocytic stem cell in the immature and adult mouse brain. Proc. Natl. Acad. Sci. USA 97(25): 13883-13888.
[29] Shamblott, M.J.; Axelman, J.; Wang, S.; Bugg, E.M.; Littlefield, J.W.; Donovan, P.J.; Blumenthal, P.D.; Huggins, G.R. and Gearhart, J.D. (1998) Derivation of pluripotent stem cells from cultured human primordial germ cells. Proc. Natl. Acad. Sci. USA 95(23): 13726-137231.
[30] Thomson, J.A.; Itskovitz-Eldor, J.; Shapiro, S.S.; Waknitz, M.A.; Swiergiel, J.J.; Marshall, V.S. and Jones, J.M. (1998) Embryonic stem cell lines derived from human blastocysts. Science 282(5391): 1145-1147.

[31] Dani, C. (1999) Embryonic stem cell-derived adipogenesis. Cells Tissues Organs 165(3-4): 173-180.
[32] Zur Nieden, NI.; Kempka, G. and Ahr, H.J. (2003) In vitro differentiation of embryonic stem cells into mineralized osteoblasts." Differentiation 71(1): 18-27.
[33] Li, M.; Pevny, L.; Lovell-Badge, R. and Smith, A. (1998) Generation of purified neural precursors from embryonic stem cells by lineage selection. Curr. Biol. 8(17): 9 71-74.
[34] Miyashita, H.; Suzuki, A.; Fukao K.; Nakauchi, H. and Taniguchi, H. (2002) Evidence for hepatocyte differentiation from embryonic stem cells *in vitro*. Cell Transplant. 11(5): 429-34.
[35] Nagy, A.; Rossant, J.; Nagy, R.; Abramow-Newerly, W. and Roder, J.C. (1993) Derivation of completely cell culture-derived mice from early-passage embryonic stem cells. Proc. Natl. Acad. Sci. USA 90(18): 8424-8428.
[36] Ramalho-Santos, M.; Yoon, S.; Matsuzaki, Y.; Mulligan, R.C. and Melton, D.A. (2002) "Stemness": transcriptional profiling of embryonic and adult stem cells. Science 298(5593): 597-600.
[37] O'Shea, K.S. (2001) Directed differentiation of embryonic stem cells: genetic and epigenetic methods. Wound Repair Regen. 9(6): 443-459.
[38] Lovell-Badge, R. (2001) The future for stem cell research. Nature 414(6859): 88-91.
[39] Weissman, I.L. (2000) Translating stem and progenitor cell biology to the clinic: barriers and opportunities. Science 287(5457): 1442-1446.
[40] Gee, A.P. (2002) Regulatory issues in cellular therapies. J. Cell Biochem. Suppl. 38: 104-112.
[41] Stocum, D.L. (2001) Stem cells in regenerative biology and medicine. Wound Repair Regen. 9(6): 429-442.
[42] Lindblad, W.J. (2001) Stem cells in mammalian repair and regeneration. Wound Repair Regen. 9(6): 423-424.
[43] Green, R.M. (2001) Four moral questions for human embryonic stem cell research. Wound Repair Regen. 9(6): 425-428.
[44] Triffitt, J.T. (2002) Stem cells and the philosopher's stone. J. Cell Biochem. Suppl. 38: 13-19.
[45] Donovan, P.J. and Gearhart, J. (2001) The end of the beginning for pluripotent stem cells. Nature 414(6859): 92-97.
[46] Ying, Q.L.; Stavridis, M.; Griffiths, D.; Li, M. and Smith, A. (2003) Conversion of embryonic stem cells into neuroectodermal precursors in adherent monoculture. Nat. Biotechnol. 21(2): 183-186.
[47] Bianco, P. and Robey, P.G. (2001) Stem cells in tissue engineering. Nature 414(6859): 118-121.
[48] Alsberg, E.; Anderson, K.W.; Albeiruti, A.; Rowley, J.A. and Mooney, D.J. (2002) Engineering growing tissues. Proc. Natl. Acad. Sci. USA 99(19): 12025-12030.
[49] Ballas, C.B.; Zielske, S.P. and Gerson, S.L. (2002) Adult bone marrow stem cells for cell and gene therapies: implications for greater use. J. Cell Biochem. Suppl. 38: 20-28.
[50] Blyszczuk, P.; Czyz, J.; Kania, G.; Wagner, M.; Roll, U.; St-Onge, L. and Wobus, A.M. (2003) Expression of Pax4 in embryonic stem cells promotes differentiation of nestin-positive progenitor and insulin-producing cells. Proc. Natl. Acad. Sci. USA 100(3): 998-1003.
[51] Griffith, L.G. and Naughton, G. (2002) Tissue engineering--current challenges and expanding opportunities. Science 295(5557): 1009-1014.
[52] Kyba, M.; Perlingeiro, R.C. and Daley, G.Q. (2002) HoxB4 confers definitive lymphoid-myeloid engraftment potential on embryonic stem cell and yolk sac hematopoietic progenitors. Cell 109(1): 29-37.
[53] Liu, P.; Jenkins, N.A. and Copeland, N.G. (2002) Efficient Cre-loxP-induced mitotic recombination in mouse embryonic stem cells. Nat. Genet. 30(1): 66-72.
[54] Schwartz, R.E.; Reyes, M.; Koodie, L.; Jiang, Y.; Blackstad, M.; Lund, T.; Lenvik, T.; Johnson, S.; Hu, W.S. and Verfaillie, C.M. (2002) Multipotent adult progenitor cells from bone marrow differentiate into functional hepatocyte-like cells. J. Clin. Invest. 109(10): 1291-1302.
[55] Bittira, B.; Kuang, J.Q.; Al-Khaldi, A.; Shum-Tim, D. and Chiu, R.C. (2002) *In vitro* preprogramming of marrow stromal cells for myocardial regeneration. Ann. Thorac. Surg. 74(4): 1154-1159; discussion 1159-1160.
[56] Ruszczak, Z. and Schwartz, R.A. (2000) Modern aspects of wound healing: An update. Dermatol. Surg. 26(3): 219-229.
[57] Pellegrini, G.; Dellambra, E.; Golisano, O.; Martinelli, E.; Fantozzi, I.; Bondanza, S.; Ponzin, D.; McKeon, F. and De Luca, M. (2001) p63 identifies keratinocyte stem cells. Proc. Natl. Acad. Sci. USA 98(6): 3156-3161.
[58] Mathor, M.B.; Ferrari, G.; Dellambra, E.; Cilli, M.; Mavilio, F.; Cancedda, R. and De Luca, M. (1996) Clonal analysis of stably transduced human epidermal stem cells in culture. Proc. Natl. Acad. Sci. USA 93(19): 10371-10376.

[59] Yamada, Y.; Seong Boo, J.; Ozawa, R; Nagasaka, T.; Okazaki, Y.; Hata, K. and Ueda, M. (2003) Bone regeneration following injection of mesenchymal stem cells and fibrin glue with a biodegradable scaffold. J. Craniomaxillofac. Surg. 31(1): 27-33.
[60] Rose, F.R. and Oreffo, R.O. (2002) Bone tissue engineering: hope *vs* hype. Biochem. Biophys. Res. Commun. 292(1): 1-7.
[61] Risbud, M.V. and Sittinger, M. (2002) Tissue engineering: advances in *in vitro* cartilage generation. Trends Biotechnol. 20(8): 351-356.
[62] Partridge, T.A. Stem cell route to neuromuscular therapies. Muscle Nerve 27(2): 133-141.
[63] Stamm, C.; Westphal, B.; Kleine, H.D.; Petzsch, M.; Kittner, C.; Klinge, H.; Schumichen, C.; Nienaber, C.A.; Freund, M. and Steinhoff, G. (2002) Autologous bone-marrow stem-cell transplantation for myocardial regeneration. Lancet 361(9351): 45-46.

PART 2

MICROENCAPSULATION FOR DISEASE TREATMENT

CHALLENGES IN CELL ENCAPSULATION

GORKA ORIVE, ROSA Mª HERNÁNDEZ, ALICIA R. GASCÓN AND JOSÉ LUIS PEDRAZ

Department of Pharmacy & Pharmaceutical Technology, Faculty of Pharmacy, University of the Basque Country, Vitoria, Spain – Fax: +34 945013040 – Email: knppemuj@vc.ehu.es

1. Introduction

In the last few decades, the field of controlled drug delivery is developing rapidly. By controlling the exact dose and the continuous drug delivery in the body, side-effects are minimised and repeated shots can be substantially reduced, thereby improving the life quality of the patients. Many drug delivery approaches are being developed and optimised for the long-term secretion of therapeutic products, especially in the field of biotechnology, where most of the drugs used are proteins or peptides. Among all these new technologies, cell immobilisation approaches represent an alternative strategy in which cells working as drug-factories are immobilised and immunoprotected within polymeric and biocompatible devices. Based on this concept, a wide spectrum of cells may be encapsulated, avoiding or at least reducing the administration of immunosuppressive drugs and the implementation of strict and tedious immunosuppressive protocols.

Scientists have been exploring the concept of immunoisolation of cells and tissues since the beginning of the last century. The initial attempts of Bisceglie to enclose tumour cells in polymer membranes [1] and the development of "artificial cells" by T.M.S. Chang [2] challenged the ingenuity of the scientists world-wide. Indeed, the potential impact of this technology conjures up visions of optimised cell immobilisation devices which could fulfil the exigent requirements applicable to any other pharmaceutical drug, including performance, biosecurity, tolerance, retrievability, scale-up and cost.

The long path of cell immobilisation technology has been full of succeeds and defeats. On the one hand, its proof of principle as a method for continuous therapeutic peptide delivery has been demonstrated successfully in animal models of several diseases such as hormone-based deficiencies, haemophilia, central nervous system (CNS) diseases or cancer. Furthermore, the functional applicability of cell encapsulation in humans has also been reported in several clinical trials [3,4]. However, the general feeling is that the field has not lived up to expectations [5]. In fact, although much efforts have been focused on the field, the reality is that, to date, no product is on the market [6]. Some possible explanations for this are the lack of reproducible results in animal models, the requirement of a standardised technology and the urgent need for reproducible and biocompatible materials that provide stable and immunocompatible devices. In addition, the long-term secretion of the therapeutic products by the enclosed cells represents also another important consideration if the promise of cell encapsulation technology is to be realised.

All these pitfalls have changed the attitude of many research groups preferring small steps with a more complete understanding of the system. Due to this step-wise approach of the main research questions, more insight into the parameters delimiting success of the field have been gained. Not surprisingly, this has brought the whole technology much closer to a realistic clinical proposal.

2. Immunoisolation approaches

A number of immunoisolating devices have been refined during the past several years. These include cell adhesion or aggregation systems, cell entrapment macrodevices and microcapsules. In the former, immobilisation results usually from attachment between cells and the support due either to the production of adhesive polymers by the cells or to covalent or ionic cross-linkers. Cells can also be included in a defined volume by use of encapsulation devices with a size ranging from 100 or 200 micrometers to a few millimetres. In this approach, cells are enclosed within immobilisation systems which allow oxygen, nutrients, waste products and therapeutic products to diffuse through the membranes, yet prevent antibodies and/or other immune cells from entering and destroying the immobilised cells.

Macroencapsulation describes selectively permeable polymers shaped in hollow fibers and filled with cell suspensions. Hollow fibers constitute the technological basis for the extracorporeal bioartificial livers. Furthermore, some degree of success has been shown after intrastriatal implants of polymer-macroencapsulated cells in monkeys [7]. In microencapsulation, cells are included within polymeric matrices coated usually by polications which increase the mechanical resistance of the capsules while controlling the permeability properties. In general, it is assumed that microcapsules have a better surface/volume ratio than the tube geometry of the hollow fibers, which offers the major advantage of better permeability. Additionally, small microcapsules can be implanted in close contact to the blood stream, which could be beneficial for the long-term functionality of the enclosed cells. In contrast, macrocapsules are much easier to retrieve once implanted, offering additional safety advantages.

Researchers are trying to come up with various solutions to help the microencapsulation and macroencapsulation devices surpass their main limitations. For example, solid-support systems of expanded polytetrafluoroethylene have been designed to facilitate the retrievability of the microencapsulated cells [8]. Moreover, photolithographic technology has been employed to engineer reproducible capsules. The latter are coated by an immunoisolation membrane made of silicon that has been surface micromachined to form several thousand pores each as small as 10 nm in diameter [9]. As a result of these and other advantages the rejection of the capsules by the body's immune system will be prevented and if problems arise during the treatment phase the retrievability of the beads may be feasible.

To address the problem of permeability, the synthetic semi-permeable membrane of the hollow fibers is being carefully studied. Recently, researchers have observed that a low initial cell load significantly improves the survival of the graft, probably due to the progressive adaptation of the enclosed cells to their environment [10]. Similarly, the inclusion of solid matrix such as polyvinylalcohol (PVA) within the macroencapsulation devices facilitates cell adhesion and expansion [10].

3. Potential advantages of cell encapsulation technology

Cell encapsulation is an interdisciplinary field of biotechnology, which combines the principles of pharmaceutical technology, biology and genetics towards the development of a cell-based therapeutic product to restore or improve native tissue function without graft rejection.

One potential advantage of transplanting the cells within polymeric devices compared with the classical peptide encapsulation is that therapeutic products will be produced "*de novo*" at a constant rate, giving rise to a more physiological concentration of the biologics. Furthermore, if the encapsulation device breaks down, the toxicity caused by a quick delivery of a high concentration of the drug to the bloodstream will be avoided. Another advantage compared with "*in vivo*" gene therapy approaches is that encapsulated cells do not modify or alter the genome of the host, which represents another important safety concern. In fact, if the viral vector stitches itself into a cell's genes, it can cause the cell to mutate and become cancerous. This was clearly observed when two children who has gene therapy for severe combined immunodeficiency disease (SCID) developed leukaemia [11].

Interestingly, cell encapsulation technology allows the targeting of the therapeutic approach since capsules could be implanted near to the therapeutic target. This would enable more of the drug getting to the specific site where is needed, minimising the possible side-effects. For example, in the treatment of CNS diseases such as Parkinson's or Huntington's disease, encapsulated cells are implanted proximal to the damaged neurons, exerting neuroprotective effects and preventing cell loss and its associated behavioural abnormalities in these disorders [12]. Similarly, in the case of cancer treatment, immobilised cells expressing cytochrome P450 were delivered by supraselective angiography to the intra-arterial placement, allowing the local activation of ifosfamide to its active toxic components, phosphomustard and acrolein [13].

Finally, the encapsulation of primary cells *i.e.* islets of Langerhans, allows not only the immunoisolation of the transplanted islets but the development of intelligent medicines with the ability to control the needs of the patient by releasing insulin according to the demand and thereby avoiding the frequent complications associated with hyperglycaemia.

4. Materials used in cell encapsulation

Researchers have understood that selection of suitable materials is fundamental for the future therapeutic success of the transplanted implant. In fact, graft failure is interpreted to be a consequence of insufficient biocompatibility of the microcapsule materials, resulting in progressive fibrotic overgrowth all around the capsule and the subsequent necrosis of the immobilised cells. Totally biocompatible materials that neither interfere with cell function nor trigger an immune response of the patient must be employed to ensure the long-term function of the enclosed cells. Indeed, impurities may contribute to device failures and to undesirable host responses for many implants. Additionally, the intrinsic properties of the materials and the configuration of the encapsulation devices will determine the mechanical stability of the final therapeutic system. This is particularly important since many of the graft failures observed, specially with microcapsules, are due to the fragility of the devices.

Materials have also major implications in the biosecurity of the therapeutic approach. Based on this concept, non-toxic materials with a high batch-to-batch reproducibility should be selected if the approval of the main regulatory issues is to be assumed. Additionally, the easy scale-up of the manufacturing process need also to be considered.

Polymers employed as biomaterials can be naturally derived, synthetic or a mixture of both. Natural occurring biomaterials are usually more biocompatible than synthetic ones, while the latter have a more reproducible composition [14]. The overwhelming majority of the scientific literature has employed alginate as a naturally derived polymer for microcapsule elaboration. This is in part due to its excellent gel-forming properties and biocompatibility. Alginates are composed of variable regions of D-mannuronic acid and L-guluronic acid interspaced with regions of alternating blocks. They have hydrogel-forming capacity with many divalent or multivalent cations, but most current transplantation work is done with calcium (Ca^{+2}). Barium (Ba^{+2}) has also been employed for the elaboration of alginate microcapsules. The cross-linkage of alginate with Ba^{+2} cations produces a much more mechanically resistant matrix than with Ca^{+2} ions, avoiding the use of alternative polycation coatings, which could reduce the biocompatibility of the system. Barium alginate microcapsules have been successfully used for the entrapment of islets of Langerhans, prolonging islet survival without immunosuppression and providing complete protection against allorejection and the recurrence of autoimmune diabetes [15]. However, the higher toxicity of Ba^{+2} in relation to Ca^{+2} will render more difficult its clinical development.

Although alginate biocompatibility has been extensively investigated, there is a disagreement in the literature. Induction of foreign body reaction and fibrosis have been reported for most commercial alginates [16,17], whilst other reports show little or no immunoresponse around alginate implants [18]. Moreover, the immunogenicity of the uronic composition of the alginates is also controversial. In fact, some research groups observed a cellular overgrowth of 90% of the capsules when high-M alginate was used [19], whereas others found guluronic acids to be associated with more severe fibrotic overgrowth [20]. Overall, studies have shown that commercial alginates are contaminated with varying amounts of different mitogens such as heavy metals or polyphenols, endotoxins and pyrogens or immunogenic materials as proteins, that are difficult to remove and may provoke an inflammatory reaction. Consequently, commercial available alginates have been purified by free-flow electrophoresis, resulting in alginate preparations that do not provoke foreign-body reactions for at least three weeks after implantation in the peritoneal cavity of rodents [21]. Furthermore, when microcapsules prepared with non-purified and purified alginates of different compositions were implanted in the peritoneal cavity of Balb/c mice in order to evaluate their antigenic activity, it was observed that non-purified capsules produced a detectable antibody response when compared with their preimmune serum. However, this antigenic response was not seen when purified microcapsules were transplanted (Figure 1) [22]. Non purified alginates also produce a significant secretion of the proinflammatory cytokine, tumour necrosis factor (TNF) and an intense activation and proliferation of lymphocytes when compared with purified alginates (unpublished data).

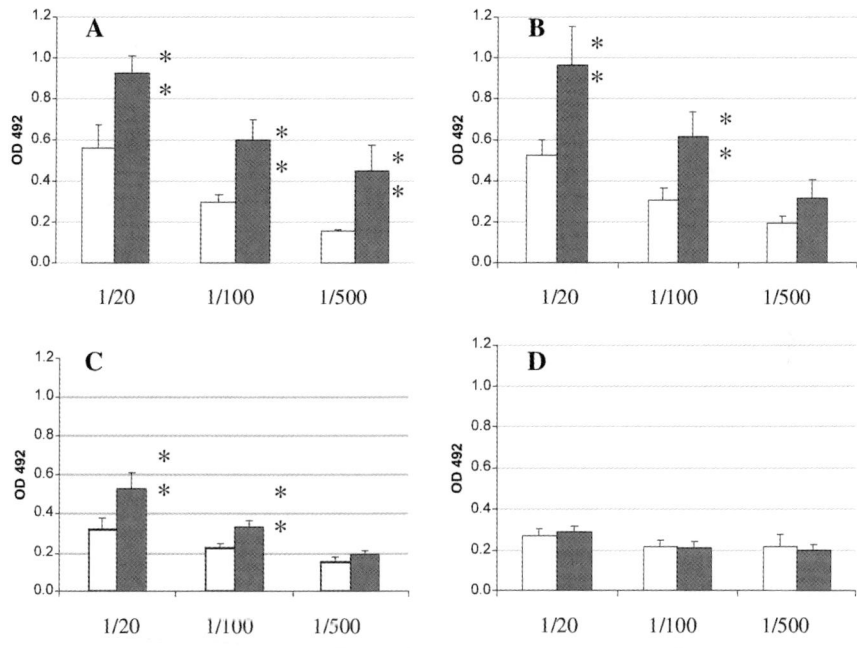

Figure 1. (A) Antibodies in serum from positive control animals; (B) antibodies against non-purified M-rich alginate capsules; (C) antibodies against non-purified G-rich alginate capsules and (D) antibodies against purified alginate capsules. **$P<0.05$ versus preimmune serum. Reprinted from [22], Copyright (2002), with permission from Elsevier.

All these results highlight the idea that the purity of individual alginate preparation, rather than their chemical composition, is probably of greater importance in determining microcapsule biocompatibility. Nonetheless, graft survival is not only influenced by the purity of biomaterials used in capsule elaboration. In fact, fibrotic overgrowth is always found in a small portion of the capsules, in spite of using purified alginates [23]. Therefore, other factors such as the microgeometry and individual imperfections of the immobilisation devices as well as the transplantation site should be carefully considered.

The discovery and optimisation of suitable immune-compatible polycations represent another area of research. Lim and Sum proposed more than 20 years ago to coat the alginate capsules with polycations such as poly-L-lysine (PLL), inducing the formation of complexes at the capsule surface [24]. The presence of these complexes increase the mechanical stability of the capsule, modulate its porosity and stabilise the gel resulting in a reduction of osmotic swelling [25]. However, since PLL induces fibrosis on alginate capsules *via* the induction of cytokines [26], it has been concluded that pericapsular reactions might be prevented by keeping the amount of polycation from the capsules at minimum [27] or neutralising the PLL membrane with an outer purified alginate coating.

In the last few years, researchers have been studying modified alginates and alternative membrane chemistries to improve the biocompatibility and mechanical resistance of the alginate-PLL capsules. For example, a novel enzymatically tailored alginate with improved biocompatibility and increased resistance to osmotic swelling has been reported [28,29]. Recently, it has also been demonstrated that the viability of the encapsulated cells is highly dependent on the molecular weight of the alginates. Thus, proper tailoring of alginates could help to improve the viability of the immobilised cells [30]. Moreover, RGD-containing (R: arginine, G: glycine, D: aspartic acid) peptide sequences that regulate cell behaviour can be grafted onto the alginates, enabling the control of proliferation and differentiation of the entrapped cells [31].

Other approaches study replacing PLL by other polycations such poly-L-ornithine (PLO) [32], poly(methylene-co-guanidine) (PMCG) [33] or photopolymerised poly(thylene glycol) diacrylate [34]. Coating beads with several layers of polyanions and polycations [35] and even the use of agarose or cellulose as polymer matrix is also under research [36,37].

5. Cell lines

Encapsulation requires suitable cell sources that produce therapeutic products for prolonged periods of time. This is particularly important if this technology is to be implemented for the treatment of chronic diseases. The intrinsic properties and nature of the cell sources should be taken into account before encapsulation. For example, the use of primary cells *versus* established cells and/or allogeneic *versus* xenogeneic cells could have important physical and immunological considerations. In this regard, although xenogeneic are more widely available, their current use is controversial due to the possible transmission of animal viruses to humans [38].

Optimal adaptation of the immobilisation device to the selected cells and *vice versa* is also an important consideration. In a recent study, it was observed that hybridoma cells presented a better adaptation to the liquefied core microcapsules, whereas myoblast and fibroblast cells grew better in solid matrices as they resembled the natural culture conditions of the cell line [39]. In addition, to avoid areas of necrosis within the devices, cells that do not proliferate after encapsulation should be selected.

Cell lines should also be tested extensively for viruses and tumourigenicity. In recent years, gene therapy has contributed to the characterisation of modified cells with an improved biosecurity level. Gene expression can be regulated through the use of enhancer and repressor elements positioned near or within the region of a promoter. These gene regulatory systems could control the expression of a gene by exogenously administered drugs. In a recent paper, the group of Patrick Aebischer demonstrated the feasibility of a doxycycline-based system to modulate the long-term secretion of erythropoietin by encapsulated C_2C_{12} myoblasts both *in vitro* and *in vivo*. Results show that hematocrits of DBA/2J mice varied between basal levels (40-50%) and elevated levels (70-90%) due to the presence or absence of doxycycline, suggesting that this artificial transgene regulatory system could become a biosecurity element to control the long-term delivery of therapeutic products *in vivo* [40].

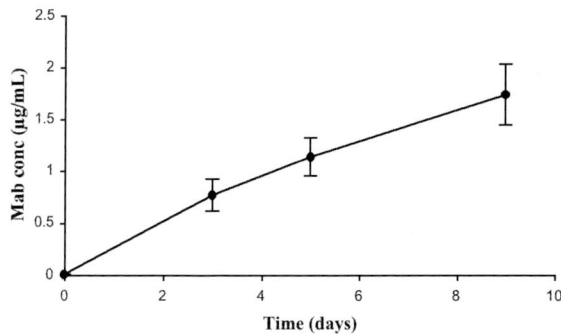

Figure 2. Cumulative production of antibody by the immobilised cells during 9 days of culture.

6. Therapeutic applications

The technology of cell encapsulation has been applied in the treatment of multiple diseases ranging from classical mendelian disorders caused by a genetic or enzymatic dysfunction to CNS diseases. Furthermore, this therapeutic strategy also has major implications in the development of bioartificial organs such as pancreas or liver. In fact, organ or tissue failure is a common and dramatic concern of health care systems and, in the United States, spending for organ replacement therapies represents approximately 1% of the GDP.

Several recent studies have focused on the eradication of cancer by using inhibitors of tumour angiogenesis. In fact, several lines of direct evidence show that angiogenesis is essential for the growth and persistence of solid tumours and their metastases [41-43]. Based on this concept, engineered cells secreting both angiostatin [44] and endostatin [45,46] have been encapsulated and their efficacy demonstrated in different tumour models.

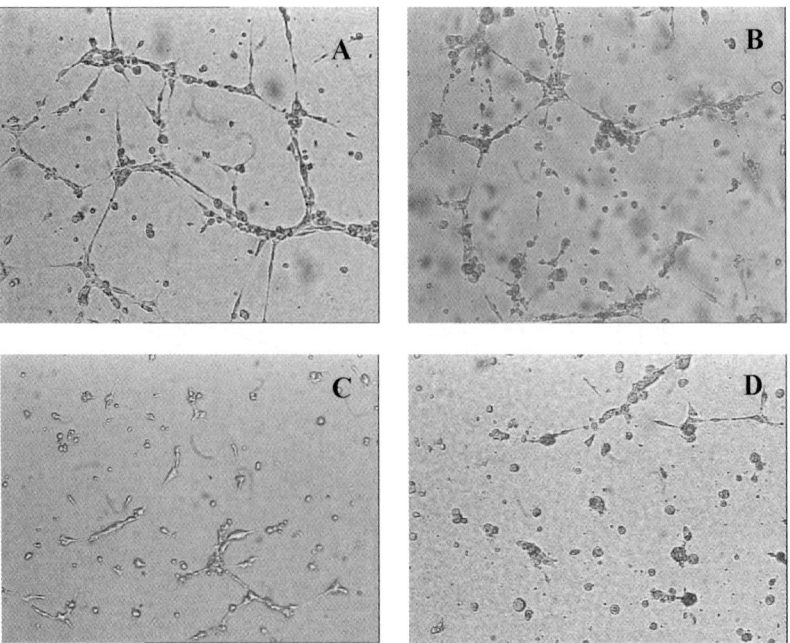

Figure 3. Photographs of the in vitro *angiogenesis, matrigel assay. (A) Control sample (endothelial cells without antibody); (B) antibody concentration: 0.7 µg/mL; (C) antibody concentration: 1.1 µg/mL; (D) antibody concentration: 1.7 µg/mL.*

The long-term secretion of monoclonal antibodies represents another tool for treating a wide range of diseases including cancer. In a recent study, anti (VE)-cadherin antibody secreting hybridoma cells were microencapsulated and their *in vitro* antibody production and antiangiogenic capacity was tested [47]. Anti (VE)-cadherin antibodies have been reported to impede the formation of microtubules and consequently to inhibit the final step of angiogenesis [48].

Immobilised cells secreted a cumulative concentration of 1.7 µg/mL during the 9 days of study (Figure 2). Furthermore, a clear relation between the total cumulative antibody concentrations and the antiangiogenic effects in the *in vitro* matrigel assay was found. Indeed, an antibody concentration of 0.7 µg/mL induced a capillary tube inhibition of 30.6%, whereas a concentration of 1.1 µg/mL resulting in an inhibition of 70.4%. Finally, an 87.8% inhibition was obtained in microtubule formation for a total antibody concentration of 1.7 µg/mL (Figure 3).

Delivery of a variety of other effectors such as inducible nitric oxide synthase, cytochrome P-450 enzyme and IL-2, have also been applied successfully for the treatment of various types of cancer [49-51].

Table 1. *The past years have seen several firsts, including more accurate assays and advances in regulatory and ethical issues. Modified from Trends in Pharmacological Sciences, Vol. 24, G. Orive et al., Cell microencapsulation technology for biomedical purposes: novel insights and challenges, 207-210 [64]. Copyright (2003), with permission from Elsevier.*

Recent applications and firsts	Description	Ref.
Atomic force microscopy	3-D view of the capsules to analyse imperfections on the surface which might reduce biocompatibility	[52]
X-ray photoelectron spectroscopy	Protein adsorption on capsules and surface analysis after transplantation	[53]
Analysis of the capsules by other techniques	Confocal laser scanning microscopy (CLSM), advanced nuclear magnetic resonance (NMR), scanning acoustic microscopy (SAM), induction of apoptosis by Jurkat cells	[54-56]
Stability studies	Osmotic pressure test, Compression resistance study	[57]
Kinetic characterisation studies	Mass transfer coefficient assay, protein ingress and egress, etc.	[58,59]
Delivery of retroviral vectors	Encapsulated cells producing retroviral vector particles for *in vivo* gene therapy.	[60]
Transient immunosuppression	Transient immunosuppression improves graft survival	[10,61]
Large-scale microcapsule production	Inotech system (www.inotechintl.com) Jet cutter system (http://www.geniaLab.com) Electrostatic droplet generator with holder for multiple needles	-
Co-microencapsulation	Co-encapsulation of Sertoli's cells to improve functional performance and activity of the immobilised cells	[32]
Smaller microcapsules	Novel microcapsules of 100-200 μm for implantation in the CNS.	[62]
2002 US Pharmacopeia and National Formulary	Cell-based products have been approved as a new category of therapy product.	[63]

7. Perspectives and concluding remarks

Cell encapsulation is an alternative non-viral approach, which allows the long-term delivery of therapeutic products and the control of the drug's pharmacokinetics and the effect, thereby modulating the efficacy of the therapeutic strategy.

In recent years there have been some interesting research developments that have renewed the potential impact of this technology (Table 1). Although many challenges remain, many experimental therapies are now easing their way into the clinic and several new clinical trials are expected in the next years. In summary, the hope that encapsulated cells may be used as therapeutics seems increasingly likely to be realised.

References

[1] Bisceglie, V. (1933) Uber die antineoplastische immunitat; heterologe Einpflnzung von Tumoren in Huhner-embryonen. Ztschr. Krebsforsch. 40: 122-140.

[2] Chang, T.M.S. (1964) Semipermeable microcapsules. Science 146: 524-525.
[3] Soon-Shiong, P.; Heintz, R.; Merideth, N.; Yao Quiang, X.; Yao, Z. and Zheng, T. (1994) Insulin independence in a type I diabetic patient after encapsulated islet transplantation. Lancet 343: 950-951.
[4] Hasse, C.; Klock, G.; Schlosser, A.; Zimmermann, U. and Rothmund, M. (1997) Parathyroid allotransplantation without immunosuppression. Lancet 351: 1296-1297.
[5] Orive, G.; Hernández, R.M.; Gascón, A.R.; Calafiore, R.; Chang, T.M.S.; De Vos, P.; Hortelano, G.; Hunkeler, D.; Lacík, I.; Shapiro, A.M.J. and Pedraz, J.L. (2003) Cell encapsulation: promise and progress. Nat. Med. 9: 104-107.
[6] Orive, G.; Hernández, R.M.; Gascón, A.R.; Igartua, M. and Pedraz, J.L. (2002) Controversies over stem cell research. Trends Biotechnol. 20: 382-387.
[7] Aebischer, P.; Schluep, M.; Déglon, N.; Joseph, J.M.; Hirt, L.; Heyd, B.; Goddard, M.; Hammang, J.P.; Zurn, A.D.; Kato, A.C.; Regli, F. and Baetge, E.E. (1996) Intrathecal delivery of CNTF using encapsulated genetically modified xenogeneic cells in amyotrophic lateral sclerosis patients. Nat. Med. 2: 696-699.
[8] De Vos, P. and Marchetti, P. (2002) Encapsulation of pancreatic islets for transplantation in diabetes: the untouchable islets. Trends Mol. Med. 8: 363-366.
[9] Leoni, L. and Desai, T.A. (2001) Nanoporous biocapsules for the encapsulation of insulinoma cells: biotransport and biocompatibility. IEEE Trans. Biomed. Eng. 48: 1335-1341.
[10] Scheneider, B.L.; Schwenter, F.; Pralong, W.F. and Aebischer, P. (2003) Prevention of the initial host immuno-inflammatory response determines the long-term survival of encapsulated myoblasts genetically engineered for erythropoietin delivery. Mol. Ther. 7(4): 506-514.
[11] Check, E. (2002) Regulators split on gene therapy as patient shows signs of cancer. Nature 419: 545-546.
[12] Emerich, D.F.; Winn, S.R.; Hantraye, P.M.; Peschanski, M.; Chen, E.Y.; Chu, Y.; McDermott, P.; Baetge, E.E. and Kordower, J.H. (1997) Protective effect of encapsulated cells producing neurotrophic factor CNTF in a monkey model of Huntington´s disease. Nature 386:395-399.
[13] Lörh, M.; Hoffmeyer, A.; Kröger, J.C.; Freund, M.; Hain, J.; Holle, A.; Karle, P.; Knöfel, W.T.; Liebe, S.; Müller, P.; Nizze, H.; Renner, M.; Saller, R.M.; Wagner, T.; Hauenstein, K.; Günzburg, W.H. and Salmons B. (2001) Microencapsulated cell-mediated treatment of inoperable pancreatic carcinoma. Lancet 2001 357: 1591-1592.
[14] Angelova, N. and Hunkeler, D. (1999) Rationalizing the design of polymeric biomaterials. Trends Biotechnol. 17: 409-421.
[15] Dauvivier-Kali, V.; Omer, A.; Parent, R.J.; O´Neil, J. and Weir, G.C. (2001) Complete protection of islets against allorejection and autoimmunity by a simple barium-alginate membrane. Diabetes 50: 1698-1705.
[16] Espevik, T.; Otterlei, M.; Skjåk-Bræk, G.; Ryan, L.; Wright, S.D. and Sundan, A. (1993) The involvement of CD14 in stimulation of cytokine production by uronic acid polymers. Eur. J. Immunol. 23: 255-261.
[17] Cole, D.R.; Waterfall, M.; McIntyre, M.; Baird, J.D. (1992) Microencapsulated islet grafts in the BB/E rat: a possible role for cytokines in graft failure. Diabetologia 35: 231-237.
[18] De Vos, P.; De Haan, B.J.; Pater, J. and Van Schilfgaarde, R. (1996) Association between capsule diameter, adequacy of encapsulation, and survival of microencapsulated rat islet allografts. Transplantation 62: 893-899.
[19] Soon-Shiong, P.; Otterlei, M.; Skjåk-Bræk, G.; Smidsrød, O.; Heintz, R.; Lanza, P. and Espevik, T. (1991) An immunology basis for the fibrotic reaction to implanted microcapsules. Transplant. Proc. 23: 758-759.
[20] Clayton, H.A.; London, N.J.; Colloby, P.S.; Bell, P.R. and James, R.F. (1991) The effect of capsule composition on the biocompatibility of alginate-poly-l-lysine capsules. J. Microencapsul. 8: 221-233.
[21] Zimmermann, U.; Klöck, G.; Federlin, K.; Haning, K.; Kowaslski, M.; Bretzel, R.G.; Horcher, A.; Entenmann, H.; Siebers, U. and Zekorn, T. (1992) Production of mitogen contamination free alginates with variable rations of mannuronic to guluronic acid by free flow electrophoresis. Electrophoresis 13: 269-274.
[22] Orive, G.; Ponce, S.; Hernández, R.M.; Gascón, A.R.; Igartua, M. and Pedraz, J.L. (2002) Biocompatibility of microcapsules for cell immobilization elaborated with different type of alginates. Biomaterials 23: 3825-3831.

[23] De Vos, P.; De Haan, B.J.; Wolters, G.H.J.; Stubbe, J.H. and Van Schilfgaarde, R. (1997) Improved biocompatibility but limited graft survival after purification of alginate for microencapsulation of pancreatic islets. Diabetologia 40: 262-270.
[24] Lim, F. and Sun, A.M. (1980) Microencapsulated islets as bioartificial endocrine pancreas. Science 210: 908-909.
[25] Thu, B.; Bruheim, P.; Espevik, T.; Smidsrød, O.; Soon-Shiong, P. and Skjåk-Bræk, G. (1996) Alginate polycation microcapsules. I. Interaction between alginate and polycation. Biomaterials 17: 1031-1040.
[26] Strand, B.L.; Ryan, L.; Veld, P.I.; Kulseng, B.; Rokstad, A.M.; Skjåk-Bræk, G. and Espevik, T. (2001) Poly-L-lysine induces fibrosis on alginate microcapsules via the induction of cytokines. Cell Transplant. 10: 263-275.
[27] King, A.; Sandler, S. and Andersson, A. (2001) The effect of host factors and capsule composition on the cellular overgrowth on implanted alginate capsules. J. Biomed. Mat. Res. 57: 374-383.
[28] King, A; Strand, B; Rokstad, AM; Kulseng, B; Andersson A; Skjåk-Bræk, G and Sandler S (2003). Improvement of the biocompatibility of alginate/poly-L-lysine/alginate microcapsules by the use of epimerized alginate as a coating. J. Biomed. Mat. Res. 64:533-539.
[29] Strand, B.L.; Mørch, Y.A.; Syvertsen, K.R., Espevik, T. and Skjåk-Bræk, G. (2003) Microcapsules made by enzymatically tailored alginate. J. Biomed. Mat. Res. 64: 540-550.
[30] Kong, H.J.; Smith, M.K. and Mooney, D.J. (2003) Designing alginate hydrogels to maintain viability of immobilised cells. Biomaterials 24: 4023-4029.
[31] Rowley, J.A. and Mooney, D.J. (2002) Alginate type and RGD density control myoblast phenotype. J. Biomed. Mat. Res. 60: 217-223.
[32] Luca, G.; Calafiore, R.; Basta, G.; Ricci, M.; Calvitti, M.; Neri, L.; Nastruzzi, C.; Becchetti, E.; Capitani, S.; Brunetti, P. and Rossi, C. (2001) Improved function of rat islets upon co-microencapsulation with Sertoly cells in alginate/poly-L-ornithine. AAPS Pharmasci. Tech. 2(3): 1-7.
[33] Orive, G.; Hernández, R.M.; Gascón, A.R.; Igartua, M. and Pedraz, J.L. (2003) Development and optimisation of alginate-PMCG-alginate microcapsules for cell immobilisation. Int. J. Pharm. 259: 57-68.
[34] Cruise, G.M.; Hegre, O.D.; Lamberti, F.V.; Hager, S.R.; Hill, R.; Scharp, D.S. and Hubbell, J.A. (1999) In vitro and in vivo performance of porcine islets encapsulated in interfacially photopolymerized poly(ethylene glycol) diacrylate membranes. Cell Transplant. 8: 293-306.
[35] Schneider, S.; Feilin, P.J.; Slotty, V.; Kampfner, D.; Preuss, S.; Berger, S.; Beyer, J. and Pommersheim, R. (2001) Multilayer capsules: a promising microencapsulation system for transplantation of pancreatic islets. Biomaterials 22: 1961-1970.
[36] Risbud, M.V.; Bhargava, S. and Bhonde, R.R. (2003) In vivo biocompatibility evaluation of cellulose macrocapsules for islet immunoisolation: implications of low molecular weight cut-off. J. Biomed. Mat. Res. 66: 86-92.
[37] Kobayashi, T.; Aomatsu, Y.; Iwata, H.; Kin, T.; Kanehiro, H.; Hisanaga, M.; Ko, S., Nagao, M. and Nakajima, Y. (2003) Indefinite islet protection from autoimmune destruction in nonobese diabetic mice by agarose microencapsulation without immunosuppression. Transplantation 75(5): 619-625.
[38] Murphy, F.A. (1996) The public health risk of animal organ and tissue transplantation into humans. Science 273: 746–747.
[39] Orive, G.; Hernández, R.M.; Gascón, A.R.; Igartua, M. and Pedraz, J.L. (2003) Survival of different cell lines in alginate-agarose microcapsules. Eur. J. Pharm. Sci. 18:23-30.
[40] Sommer, B.; Rinsch, C.; Payen, E.; Dalle, B.; Schneider, B.; Déglon, N.; Henri, A.; Beuzard, Y. and Aebischer, P. (2002) Long-term doxycycline-regulated secretion of erythropoietin by encapsulated myoblasts. Mol. Ther. 6(2): 155-161.
[41] Folkman, J. (1989) What is the evidence that tumors are angiogenesis dependent? J. Natl. Cancer Inst. 82: 4-6.
[42] Kim, K.J.; Li, B.; Winer, J.; Armanini, M.; Gillet, N.; Phillips, H.S. and Ferrara, N. (1993) Inhibition of vascular endothelial growth factor-induced angiogenesis suppresses tumor growth in vivo. Nature 362: 841-844.
[43] Millauer, B.; Schwner, L.K.; Plate, K.H.; Risau, W. and Ullrich, A. (1994) Glioblastoma growth inhibited in vivo by a dominant-negative Flk-1 mutant. Nature 367: 576-579.
[44] Cirone; P.; Bourgeois, M. and Chang, P.L. (2003) Antiangiogenic cancer therapy with microencapsulated cells. Hum. Gene Ther. 14: 1065-1077.

[45] Read, T.A.; Sorensen, D.R.; Mahesparan, R.; Enger, P.Ø.; Timpl, R.; Olsen, B.R.; Hjelstuen, H.B.; Haraldseth, O. and Bjerkvig, R. (2001) Local endostatin treatment of gliomas administered by microencapsulated producer cells. Nat. Biotechnol. 19: 29-34.
[46] Joki, T.; Machluf, M.; Atala, A.; Zhu, J.; Seyfried, N.T.; Dunn, I.F.; Abe, T.; Carroll, R.S.; Black, P.McL. (2001) Continuous release of endostatin from microencapsulated engineered cells for tumor therapy. Nat. Biotechnol. 19: 35-39.
[47] Orive, G.; Hernández, R.M.; Gascón, A.R.; Igartua, M.; Rojas, A. and Pedraz, J.L. (2001) Microencapsulation of an anti VE-cadherin antibody secreting 1B5 hybridoma cells. Biotechnol. Bioeng. 76: 285-294.
[48] Bach, T.L.; Barsigian, C.; Chalupowich, D.G.; Busler, D.; Yaen, C.H.; Grant, D.S. and Martinez, J. (1998) VE-cadherin mediates endothelial cell capillary tube formation in fibrin and collagen gels. Exp. Cell Res. 238: 324-334.
[49] Xu, W.; Liu, L. and Charles, I.G. (2002) Microencapsulated iNOS-expressing cells cause tumor suppression in mice. FASEB J. 16: 213-215.
[50] Lörh, M.; Hummel, F.; Faulmann, G.; Ringel, J.; Saller, R.; Hain, J.; Günzburg, W.H. and Salmons, B. (2002) Microencapsulated, CYP2B1-transfected cells activating ifosfamide at the site of the tumor: the magic bullets of the 21st century. Cancer Chemother. Pharmacol. 49(1): 21-24.
[51] Cirone, P.; Bourgeois, M.; Austin, R.C. and Chang, P.L. (2002) A novel approach to tumor suppression with microencapsulated recombinant cells. Hum. Gene Ther. 13: 1157-1166.
[52] Hillgärtner, M.; Zimmermann, H.; Mimietz, S.; Jork, A.; Thürmer, F.; Scheneider, H.; Nöth, U. and Hasse, C. (1999) Immunoisolation of transplants by entrapment in 19F-labelled alginic gels: production, biocompatibility, stability, and long-term monitoring of functional integrity. Mat. Wiss. U. Werkstofftech. 30: 783-792.
[53] De Vos, P.; Van Hoogmoed, C.G.; De Haan, B.J. and Busscher, H.J. (2002) Tissue responses against immunoisolating alginate-PLL-alginate capsules in the immediate posttransplant period. J. Biomed. Mat. Res. 62:430-437.
[54] Zimmermann, H.; Hillgärtner, M.; Manz, B.; Feilen, P.; Brunnenmeineier, F.; Leinfelder, U.; Weber, M.; Cramer, H.; Schneider, S.; Hendrich, C.; Volke, F. and Zimmermann, U. (2003) Fabrication of homogeously cross-linked, functional alginate microcapsules validated by NMR-, CLSM- and AFM- imaging. Biomaterials 24(12): 2083-2096.
[55] Leinfelder, U.; Brunnenmeier, F.; Cramer, H.; Schiller, J.; Arnold, K.; Vásquez, J.A. and Zimmermann, U. (2003) A highly sensitive cell assay for validation of purification regimes of alginates. Biomaterials 24(23): 4161-4172.
[56] Klemenz, A.; Schwinger, C.; Brandt, J. and Kressler, J. (2003) Investigation of elasto-mechanical properties of alginate microcapsules by scanning acoustic microscopy. J. Biomed. Mat. Res. 65: 237-243.
[57] Van Raamsdonk, J.M. and Chang, P.L. (2000) Osmotic pressure test: A simple, quantitative method to asses the mechanical stability of alginate microcapsules. J. Biomed. Mater. Res. 54: 264-271.
[58] Lewinska, D.; Rosinski, S.; Hunkeler, D.; Poncelet, D. and Werynski, A. (2002) Mass transfer coefficient in characterization of gel beads and microcapsules. J. Memb. Sci. 5409: 1-8.
[59] Bartkowiak, A. and Hunkeler, D. (1999) New microcapsules based on oligoelectrolyte complexation. Ann. NY Acad. Sci. 875: 36-45.
[60] Saller, R.M.; Indraccolo, S.; Coppola, V.; Esposito, G.; Stange, J.; Mitzner, S.; Heinzmann, U.; Amadori, A.; Salmons, B. and Günzburg, W.H. (2002) Encapsulated cells producing retroviral vectors for *in vivo* gene therapy. J. Gene Med. 4: 150-160.
[61] Peduto, G.; Rinsch, C.; Schneider, B.L.; Rolland, E. and Aebischer, P. (2000) Long-term unresponsiveness to encapsulated xenogeneic myoblasts after transient immunosuppression. Transplantation 70: 78-85.
[62] Ross, C.J.D. and Chang, P.L. (2002) Development of small alginate microcapsules for recombinant gene product delivery to the brain. J. Biomat. Sci. Polym .Edit. 13: 953-962.
[63] (2002) U.S.Pharmacopeia and National Formulary 1046: 2762-2790.
[64] Orive, G.; Hernández, R.M.; Gascón, A.R.; Igartua, M. and Pedraz, J.L. (2003) Cell microcapsulation technology for biomedical purposes: novel insights and challenges. Trends Pharm. Sci. 24(5): 207-210.

PROTEIN THERAPEUTIC DELIVERY USING ENCAPSULATED CELL PLATFORM

MARCELLE MACHLUF
The Laboratory of Mammalian Cell Technology, Dep. of Food Engineering and Biotechnology, The Technion, Haifa, Israel – Fax: 972-4-8293399 – Email: machlufm@tx.technion.ac.il.

1. Introduction

Proteins play an essential role in all biological processes and reactions, thus serving as potential therapeutic agents. Delivering proteins and peptides for cancer therapeutics, has been the major focus of the pharmaceutical and biotechnology industry. The development of proteins as therapeutic agents poses many challenges. Proteins are naturally synthesised in small amounts, consumed quickly and are stabilised by different intra and extracellular components. However, for therapeutic applications there are substantially different conditions from those in which proteins naturally act. To be successful as a therapeutic, particularly for cancer therapy, a protein must be highly purified in large amounts and be able to have a shelf life of one to two years. Most importantly it must be effective in the *in vivo* environment and be able to reach target cell or tissue while it is biologically active. Most therapeutic proteins are large molecules with complex structure, two components that makes them unstable. In addition proteins have low oral and transdermal bioavailabilities therefore need to be administered frequently by injection or infusion. Taking all of these together increases the demand for improved methods for the delivery of protein pharmaceuticals thus, results in the development of numerous technologies and delivery methods. Looking at the literature, for the past 30 years, numerous protein delivery methods and systems have been described and studied for protein therapy while focusing mainly on cancer and vascular diseases. The drug delivery systems used for these studies includes injectable depots composed of polymeric matrix (degradable and non degradable) or vesicles such as liposomes, transdermal patches and implantable or orally administered minipumps [1]. Nevertheless, most of these methods and systems have failed in animal studies and only few have reached human clinical trials. This can be explained by the obstacles facing the delivery of therapeutics such as unsuitable drug carrier (immunogenicity and stability), high manufacturing costs, biodegradability and instability of the agents, dosage capability, inappropriate distribution and site specific targeting. Furthermore, when delivering therapeutics systemically, the administration of proteins and the amount of protein delivered to the circulation is dependent upon the efficiency of the delivery method as well as the inherent biology of the delivery route.

An alternative method, which is extensively studied for the delivery of therapeutics, is a "cell based-bioreactor". This long-term delivery platform is based on the entrapment of cells, which may be genetically engineered to produce the molecules of interest, in a polymeric capsule rather then the protein itself. Once implanted the encapsulated cells produce and secrete the protein of interest in a constant rate for prolonged time thus supplying the physiological and effective concentration of the therapeutic protein. Furthermore, this system bypass obstacles concerned with biosafety such as uncontrolled cell growth and genomic alteration that may occur when using for example only gene therapy approaches. This chapter will discuss the advantageous use of cell encapsulation platform for the delivery of anti-angiogenic cancer therapeutics *versus* gene therapy and cell-based delivery approaches. Although cancer therapeutics are used as an example here, the concepts discussed in this chapter are also applicable for other diseases as described in other chapters of this book.

2. Anti-angiogenic protein therapy

Angiogenesis, the recruitment and continuing formation of new blood vessels by tumours, is a complex process with multiple, sequential and interdependent steps [2-4]. Judah Folkman [5] hypothesised, that solid tumours could not grow beyond 1–2 mm without developing their own blood supply. This observation has led to the development of a new therapeutic pathway, anti-angiogenesis. Angiogenesis inhibitors, include tissue inhibitors of matrix metalloproteinases [6-8], chemokines [9-11], tyrosine kinase inhibitors [12-15], interleukins [16,17], and naturally occurring proteolytic fragments of large precursor molecules such as endostatin, vasostatin, canstatin, angiostatin and others [18-24]. These anti-angiogenic molecules exert their inhibitory functions by multiple mechanisms on endothelial cell proliferation, migration, protease activity, as well as the induction of apoptosis. To date, more then 30 angiogenic inhibitors are in clinical trials and many new ones are being studied *in vitro* and in *in vivo* in animal models (for review see Hagedorn *et al.* [25]).

Animal experiments and human clinical trials have shown that high and steady state levels of the anti-angiogenic protein must be administrated in order to achieve and maintain anti-tumour effects [18,19,26-29]. Due to there relatively short half-lives, anti-angiogenic inhibitors have to be administrated on a long-term basis and in large, constant quantities, in order to maintain tumour inhibition. Collectively, these obstacles encourage evaluation of cell and gene therapy approaches for the delivery of protein based anti-angiogenic therapeutics.

3. Anti-angiogenic gene delivery

Gene therapy using angiogenic inhibitors is one approach being used to bring anti-angiogenic therapeutics into the clinic. To date, there are close to 200 gene therapy clinical trials for the treatment of various diseases, and more than 50% of these are designed for the treatment of cancer [30-33]. Anti-angiogenic *in vivo* gene therapy studies have demonstrated the effectiveness of this strategy by demonstrating a reduction in tumour growth in animal models [30,33-35]. Unlike protein therapy, anti-angiogenic gene therapy does not require high doses of DNA such as of protein for injection. Gene therapy approach can localised the delivery and sustained expression of the anti-angiogenic agent, and has the ability to inhibit multiple angiogenic pathways by delivering more than one gene. Using gene therapy generates properly folded inhibitor molecules and the potential for decreased cost [33,36]. Therefore, the use of the gene therapy approach for anti-angiogenic therapy can avoid repeated injections, instability of the protein and the high cost of purified protein production. It is also much easier and less expensive to produce high quantities of cDNA vector then proteins.

Nevertheless, clinical trials using gene therapy have failed to reach phase III due to several hurdles particularly concerning the type of viruses used [30,37,38]. Retrovirus can integrate into the cellular genome and inactivate host tumour suppression gene or activate proto-oncogenes *in vivo*. Retrovirus can also promote gene transfection of limited size gene (7 KB) and infect only dividing cells whereas most mammalian tissues consist primarily of non-dividing cells. Adenovirus, the second most commonly used viral vector, has the disadvantage of infecting all tissues including germ cells, when delivered *in vivo* [39,40]. Additional challenges for *in vivo* anti-angiogenic gene therapy are to understand how angiogenic inhibitors function, how tumour vessels differ from normal blood vessels, and how to target tumour vessels with appropriate therapies. Other obstacles, which need to be addressed, are the heterogeneity and genetically instability of tumour cell populations, diffusion limitation of gene vehicles, type of vehicle used to transfect the cells and possible drug resistant of tumour cells to the expressed protein.

4. Genetically engineered cells delivering therapeutics

Genetically engineered cells that over-express the protein of interest is an alternative strategy for direct gene delivery. The engineered cells can be considered 'a living drug delivery system' which provides an unlimited protein source. As long as the cells are viable and functional, they are able to release the desired products in a physiological manner. This therapeutic mode eliminates most of the complex preparation/formulation processes used in the more traditional protein/peptide delivery systems. Additionally, transplanted cells may be able to replace an entire cell type, acting as an artificial organ and providing needed metabolic functions. Genetically modified cells have been used to secrete a wide therapy of transgene products including, Factor IX [41-43], growth hormone [44], erythropoietin [45], cilliary neurotrophic factor (CNTF) [46], dopamine [47], endostatin [48] and Neuropilin-1 [79].

In general, two approaches have been used to deliver cells for therapeutic purposes. The simplest involves the direct injection of a bolus of cells in the form of a cell suspension either into the blood stream, a body cavity, or directly into the parenchyma of a particular tissue. Alternatively, the cells may be implanted in association with one or more biomaterials that serve as a vehicle for their delivery. The biomaterial may consist of a simple extracellular matrix, or may be made of synthetic materials, or a combination of the two. This approach has been used extensively in the field of tissue engineering where autologous or allogeneic cells are seeded onto a polymeric scaffold and then implanted *in vivo* for the replacement or regeneration of organs [49-53]. Among the natural polymers used for such applications are the collagens, glycosaminoglycan, starch, chitin, chitosan, and alginate [52,54,55]. These have been used to repair nerves, skin, cartilage, liver, bone and blood vessels.

Many synthetic restorable polymers, such as poly (α-hydroxyl esters), polyanhydrides, polyorthoesters, and polyphosphazens, have also been developed for cell transplantation. By far, the family of PLA (poly Lactic Acid) is the most commonly used synthetic biomaterial for cell scaffolding. However, the major challenge when delivering allogenic cells (whether engineered or not) by injection or by a polymeric scaffold is the need to circumvent the host immune reaction to the implanted cells. The seeded cells are implanted in an opened matrix and are exposed to the host immune system.

5. Cell encapsulation – a platform for delivering therapeutics

The encapsulation system consists of viable cells surrounded by a non-degradable, selectively permeable barrier that physically isolates the transplanted cells from host tissue and the immune system. This cell bioreactor relies on host homeostatic mechanisms for the control of pH, metabolic waste removal, electrolytes and nutrients. Numerous encapsulation techniques have been developed over the years. One of the most studied cell microencapsulation methods has been based upon alginate. Alginates are polysaccharides extracted from various species of brown algae (seaweed) and purified to a white powder. The alginates have different characteristics of viscosity and reactivity based on the specific algal source and the ions in solution. The alginates are linear unbranched polymers containing α (1-4)-linked D-mannuronic acid (M) and α(1-4)-linked L-guluronic acid (G) residues. Although these residues are epimers (D-mannuronic acid residues being enzymatically converted to L-guluronic after polymerization) and only differ at C5, they possess very different conformations; D-mannuronic acid being $^4C_1$ with diequatorial links between them and L-guluronic acid being $^1C_4$ with diaxial links between them. Hydrocolloids like alginate can play a significant role in the design of a controlled-release product. At low pH, hydration of alginic acid leads to the formation of a high-viscosity "acid gel." Alginate is also easily gelled in the presence of a divalent cation as the calcium ion. The ability of alginate to form two types of gel depends on pH, *i.e.*, an acid gel and an ionotropic gel, gives the polymer unique properties compared to neutral macromolecules. So far more than 200 different alginate grades and a number of alginate salts have been manufactured. The potential use of the various qualities as pharmaceutical agents has not been evaluated fully, but alginate is likely to make an important contribution in the development of polymeric delivery systems.

Using a hydrogel such as alginate has several advantages, for example the soft and pliable features of the gel reduce the mechanical irritation to the surrounding tissue. Furthermore, alginate has hydrophilic properties, which minimises protein absorbance and cell adhesion, thus exhibiting a high degree of biocompatibility. The complexation between the polyanionic alginate and a polycation Poly-L (Lysine) (PLL) has been the first utilised for cell encapsulation [56]. This complex forms a semi-permeable membrane, which allows the controlled delivery of different bioactive substances *in vivo* while preventing the diffusion of antibodies and other components of the immune system. The alginate-poly-L-lysine membranes allow the free exchange of nutrients and oxygen between the implanted cells and the host while preventing the escape and elimination of encapsulated cells. Alginate microcapsules have been used for various applications particularly for the encapsulation of pancreatic islet cells for insulin delivery [57-59]. A large number of studies have shown that intraperitoneal xenograft of alginate-PLL encapsulated rat, dog, pig or human islets into diabetic mice, dogs or human can regulate blood glucose levels. This method has also been used for the encapsulation of cells, which release cytokines and hormones [60-62]. Broad application of encapsulation cells delivery have also been demonstrated for the delivery of neuroactive agents in the treatment of different conditions such as age-related degeneration [63,64], Alzheimer's disease [65-67], amyotrophic lateral sclerosis [68], neuroprotection [69], Huntington's disease [70], and Parkinson's disease [71-73].

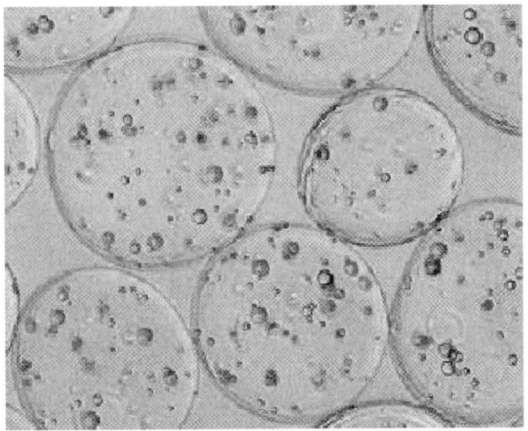

Figure 1. Phase-contrast microscopy of cell-loaded microcapsules shows the uniform size and cell distribution.

6. Cell encapsulation – delivering anti-angiogenic therapeutics

Encapsulation of non-autologous cells, which are genetically engineered to secrete anti-angiogenic proteins, is a unique strategy that can be used to combat tumours. Such a system facilitates the continuous release of biologically active anti-angiogenic protein thus, overcoming obstacles such as the short half-life of the proteins, long-term administration, high doses, and cost. Two reports, one coming from our laboratory and one from the group of Read *et al.* represents the first application of the encapsulation technology to angiogenesis inhibition [48,74]. Our study demonstrated the use of alginate encapsulated Baby Hamster Kidney cells (BHK-21), which have been engineered to continuously secrete high levels of human endostatin for the treatment of a malignant human brain tumour, glioblastoma (Figure 1). Glioblastomas are poorly understood lethal tumours with a 12-18 month median survival despite aggressive treatment. Mitosis, invasion, and angiogenesis are three cardinal features of their behaviour. There are presently no effective therapeutic agents in the clinic that successfully block any one of these attributes effectively. Endostatin, a 20-kd fragment of collagen XVIII with demonstrated anti-angiogenic activity, is being tested in phase I clinical trials as a protein infusion for various cancers [75,76]. The recombinant, biologically active protein is difficult to produce, and is rapidly cleared from the blood. Endostatin has been tested in murine tumour models, with varying success, by gene therapy delivery vectors, including adenoviral vectors, adeno associated viral vectors, in vitro transfections, polymerised plasmids, and DNA cationic liposomes [32,35,77,78]. To achieve significant tumour regression, 2.5-mg/kg recombinant endostatin was administrated once daily for 16 days in a Lewis lung carcinoma model [19]. The quantities of protein needed for this therapy, the purification procedure for large-scale production, and the attendant costs of these processes has greatly hindered current efforts to improve its efficacy *in vivo*.

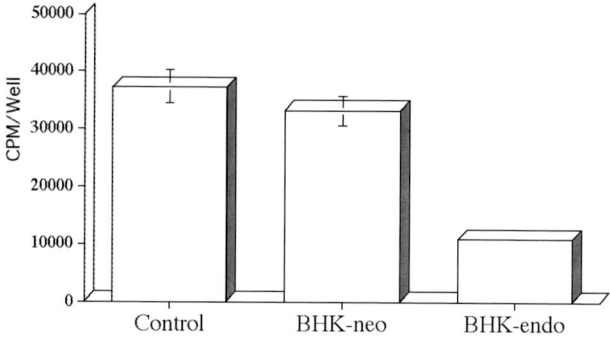

Figure 2. The effect of BHK-producing endostatin and BHK-neo (control) cells on the proliferation of BCE cells in culture.

Therefore, it seemed logical to investigate the ability of an endostatin-encapsulated system for inhibiting the growth of a human glioblastoma. An *in vitro* proliferation assay performed on bovine capillary endothelial cells (BCE cells) using encapsulated baby hamster kidney (BHK)-endo cells resulted in 67.2% inhibition in proliferation over a three day time period (Figure 2). By comparison, previous proliferation experiments had shown a 25% inhibition when the same amount of endostatin was added daily to cultured cells [19]. The potent inhibition of BCE proliferation may be explained by the continuous release of endostatin during the course of the experiment, rather than being added exogenously once at the beginning of the assay. A semi *in vivo* tube formation assay using porcine aortic endothelial cells transfected with KDR (PAE/KDR) was employed to confirm the biological activity of the endostatin released from the capsules. Conditioned media from U-87 cells stimulated tube formation (10 tubes/well), indicating the potential antigenic activity of factors released from U-87 cells. The anti-angiogenic activity of the released endostatin was confirmed by the observation that no tube formation was detected for the cells treated with U-87MG media. A series of *in vivo* studies performed in nude mice bearing subcutaneous, human glioma tumours showed that a single administration of microencapsulated-engineered BHK cells (adjacent to the tumour), which continuously secrete biologically active endostatin significantly, inhibited the tumours. Animals treated with encapsulated BHK-endo cells exhibit a 72% inhibition in tumour growth 21 days post microcapsule injection (Figure 3). The same amount of the implanted encapsulated BHK-endo cells has been shown to release an average of 150.8 ng/ml/week *in vitro*, which may give an indication for the concentration of endostatin released *in vivo*.

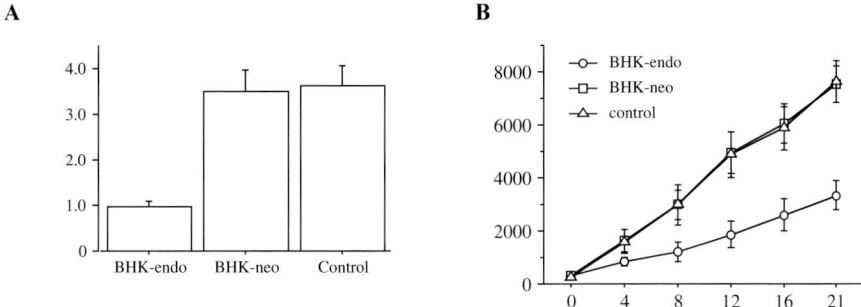

*Figure 3. Inhibition of subcutaneous U87MG human glioma cell xenograft: **A**. Mice bearing U87MG tumours were treated with encapsulated BHK-neo, BHK-endo cells, or control. * p < 0.01; **B**. 21 days after treatment, U87MG tumours were harvested and weighed. Errors are standard error of the mean (SEM). p < 0.001.*

In another study, we have used the same concept of encapsulating engineered cells to study how exogenously supplied vascular endothelial growth factor (VEGF) affects the progression of acute myeloid leukaemia (AML) *in vivo* and conversely, if VEGF antagonists repress this process [79]. VEGF is a potent angiogenic factor *in vivo* and has been implicated in the growth of a wide variety of solid tumours. Neuropilin-1 (NRP-1) acts as a high-affinity receptor for VEGF on endothelial and tumour-derived cells [80] but it does not have a kinas enzymatic activity and acts as a co receptor for KDR by enhancing VEGF binding to KDR. A naturally occurring soluble NRP-1 protein (sNRP-1) is produced by tumour cells and it can inhibit VEGF receptor binding *in vitro* and tumour growth *in vivo* when over expressed in prostate cancer cells [81]. We have engineered a VEGF antagonist, based on the sequence of the extracellular portion of NRP-1. This region contains the VEGF-binding site of NRP-1 [82] and was shown to specifically block VEGF165 activity [83]. Since many tumour cells express NRP-1 and bind VEGF via this receptor, sNRP-1 represents a novel VEGF antagonist that can bind circulating VEGF and prevent it from interacting with tumour endothelial cells and displace tumour cell NRP-1–bound VEGF.

In our leukaemia model induced subcutaneously, mice injected with encapsulated cells secreting VEGF showed an accelerated subcutaneous tumour growth compared with mice injected with control cells, with tumour volumes of 1774 ± 633 mm^3 and 646 ± 241 mm^3 after 20 days, respectively (Figure 4a). In contrast, mice injected with encapsulated sNRP-1 producing cells developed smaller tumours (27 ± 187 mm^3). A significant weight difference was also noted between the groups, with weights of 0.82 ± 0.22 g, 1.735 ± 0.48 g, and 0.43 ± 0.2 g for control, VEGF, and sNRP-1, respectively (Figure 4b). Staining of tumour sections with anti–CD31 antibodies revealed a dense capillary network in tumours retrieved from mice that were injected with VEGF-producing cells. A developed capillary network was also observed in tumours of mice that were injected with control cells. In contrast, the tumours in mice that were injected with sNRP-1–producing cells showed fewer and smaller capillaries, consistent with the necrosis of the tumours observed in these mice.

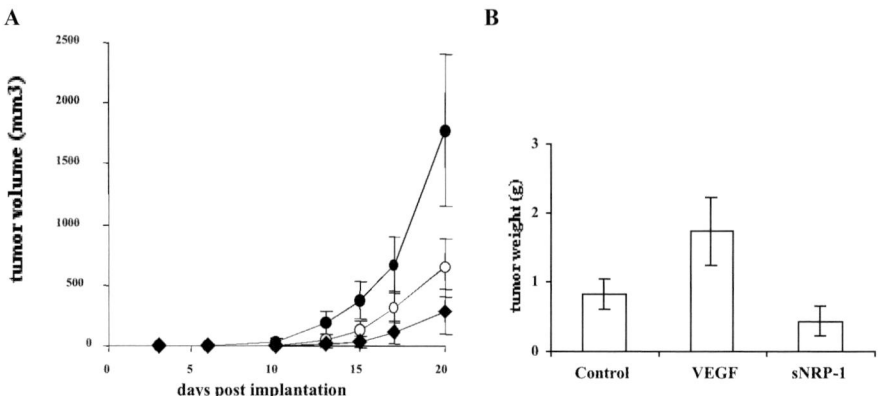

Figure 4. The effects of VEGF and sNRP-1 secreated from encapsulated NMuMG cells on AML progression in vivo: **A.** Tumour volume curves. M1 cells ($2x10^6$ per mouse) were injected subcutaneously to SCID mice. Capsules containing NMuMG (o), NMuMG/VEGF (●) and NMuMG/sNRP-1 (♦) cells ($5x10^5$ cells per mouse) were injected at the site of M1 cell injection, 72 hr. later. The mice (10 per group) were followed for tumour volumes at the indicated times. **B.** Final tumour weights. Tumours were harvested 21 days after M1 cell injection and weighted. The average weight and standard deviations were calculated for NMuMG (control). Errors are standard error of the mean (SEM). $p < 0.001$.

Recently, several other experimental studies using cell encapsulation for antiangiogenic therapy have been published [84-86]. Orive *et al.* immobilised VE-cadherin-secreting hybridoma cells in alginate-agarose microcapsules. The alginate and agarose solid beads were coated with poly-L-lysine and the 1B5 hybridoma cells were grown within the microcapsules for 9 days of culture, reaching a cumulative concentration of 1.7 g/ml [85]. According to the author, this antibody concentration inhibited microtubule formation (87%) in the *in vitro* angiogenesis Matrigel™ assay. In another study, myoblasts that had been genetically modified to secrete interleukin-2 linked to the Fv region of a humanised antibody with affinity to HER-2/neu were encapsulated in alginate-poly-L-lysine-alginate microcapsules [86]. The efficacy of this system was tested in a mouse model bearing HER-2/neu-positive tumours. The treatment led to a delay in tumour progression and prolonged survival of the animals. However, the long-term efficacy was limited by an inflammatory reaction against the implanted microcapsules probably because of the secreted cytokine and antigenic response against the xenogenic fusion protein itself. Over the short-term treatment (initial 2 weeks), efficacy was confirmed when a significant amount of biologically active interleukin-2 was detected systemically, and targeting of the fusion protein to the HER-2/neu-expressing tumour was shown by immunohistochemistry. The tumour suppression in the treated animals was associated with increased apoptosis and necrosis in the tumour tissue, thus demonstrating successful targeting of the antiproliferative effect to the tumours by this delivery paradigm.

7. Future perspective

Therapeutics delivery methods have continued to evolve over the past years. Extensive research is being conducted to improve non-invasive routes of protein delivery and system, which will deliver optimal quantities and bioactive protein to the site of interest. Commercially, potential administration of proteins using novel injection devices for daily protein administration such as insulin, continue to be the most used and least expensive delivery system. Depot delivery systems such as Nutropin Depot™ are also entering the pharmaceutical market. They may provide opportunities to decrease the number of injections thus improving patient compliance. Nevertheless, there are numerous issues, which need to be addressed before this system can be used for the delivery of therapeutics. In particular for anti-angiogenic inhibitors huge quantities, and high-grade purification protein are needed

Microencapsulation cell based delivery systems have the potential to overcome most of these issues. The progress made in the field of biomaterials together with the field of genetic engineering can contribute vastly to the development of this technology for local and systemic therapy. This technology could eliminate the need for immunosuppressives, which are a must in transplantation of cells and organ and allow the long-term delivery of therapeutic with no need of purification, stabilisation and large-scale production.

Yet, regardless of these enormous advantages that this technology can offer, the road to clinical studies using cell bioencapsulation is still facing problems, which need to be solved. Selection of cell type, cell for genetic engineering, cell support or polymer matrix and immune aspects need to be taken in account and studied carefully in order to bring these promising cell bioreactors to human clinical trials.

References

[1] Langer, R. (1998) Drug delivery and targeting. Nature 392: 5-10.
[2] Folkman, J. (1992) The role of angiogenesis in tumour growth. Semin. Cancer Biol. 3: 65-71.
[3] Folkman, J. and Shing, Y. (1992) Angiogenesis. J. Biol. Chem. 267: 10931-10934.
[4] Folkman, J. and D'Amore, P.A. (1996) Blood vessel formation: What is its molecular basis? Cell 87: 1153-1155.
[5] Folkman, J. (1971) Tumour angiogenesis: Therapeutic implications. N. Engle J. Med. 285: 1182-1186.
[6] Murray, G.I.; Duncan, M.E.; O'Neil, P.; Melvin, W.T. and Fothergill, J.E. (1996) Matrix metalloproteinase-1 is associated with poor prognosis in colorectal cancer. Nat. Med. 2: 461-462.
[7] Brooks, P.C.; Silletti S.; von Schalscha, T.L.; Friedlander, M. and Cheresh, D.A., (1998) Disruption of angiogenesis by PEX, a noncatalytic metalloproteinase fragment with integrin binding activity. Cell 92: 391-400.

[8] Bello, L.; Lucini, V.; Giussani, C.; Machluf, M.; Pluderi, M.; Nikas, D.; Zhang, J.; Tomei, G.; Villani, R.M.; Carroll, R.S.; Bikfalvi, A. and Black, PM. (2001) Simultaneous inhibition of glioma angiogenesis, cell proliferation, and invasion by a naturally occurring fragment of human metalloproteinase-2. Cancer Res. 61: 8730-8736.
[9] Arenberg, D.A.; Polverini, P.J.; Kunkel, S.L.; Shanafelt, A. and Strieter, R.M. (1997) *In vitro* and *in vivo* systems to assess role of CXC chemokines in regulation of angiogenesis. Methods Enzymol. 288: 190-220.
[10] Maione, T.E.; Gray, G.S.; Petro, J.; Hunt, A.J.; Donner, A.L.; Bauer, S.I.; Carson, H.F. and Sharpe, R.J. (1990) Inhibition of angiogenesis by recombinant human platelet-4 and related peptides. Science 247: 77-90.
[11] Jouan, V.; Canron, X.; Alemany, M.; Caen, J.P.; Quentin, G.; Plouet, J. and Bikfalvi, A. (1999) Inhibition of *in vitro* angiogenesis by platelet factor-4- derived peptides and mechanism of action. Blood 94: 984-993.
[12] Lu, D.; Jimenez, X.; Zhang, H.; Wu, Y.; Bohlen, P.; Witte, L. and Zhu, Z. (2001) Complete inhibition of vascular endothelial growth factor activities with a functional antibody directed against both VEGF kinas receptor, fumes-like tyrosine kinas receptor and kinas insert domain-containing receptor. Cancer Res. 61: 7002-7008.
[13] Dias, S.; Hattori, K.; Heissig, B.; Zhu, Z.; Wu, Y.; Witte, L.; Hicklin, D.J.; Tateno, M.; Bohlen, P.; Moore, M.A.S. and Rafii, S. (2001) Inhibition of both paracrine and autocrine VEGF/VEGFR-2 signaling pathways is essential to induce long-term remission of xenotransplanted human leukemias. Proc. Natl. Acad. Sci. USA 98: 10857-10862.
[14] Kim, K.J.; Li, B.; Winer, J.; Armanini, M.; Gillett, N.; Phillips, H.S. and Ferrara, N. (1993) Inhibition of vascular endothelial growth factor-induced angiogenesis suppresses tumour growth *in vivo*. Nature 362: 841-844.
[15] Strawn, L.M.; McMahon, G.; App, H.; Schreck, R.; Kuchler, W.R.; Longhi, M.P.; Hui, T.H.; Tang, C.; Levitzki, A.; Gazit, A.; Chen, I.; Keri, G.; Orfi, L.; Risau, W.; Flamme, I.; Ullrich, A.; Hirth, K.P. and Shawver, L.K. (1996) Flk-1 as a target for tumour growth inhibition. Cancer Res. 56: 3540-3545.
[16] Stearns, M.E.; Rhim, J. and Wang, M. (1999) Interleukin 10 inhibition of primary human prostate cell-induced angiogenesis: IL-10 stimulation of tissue inhibitor of metalloproteinase-1 and inhibition of matrix metalloproteinase(MMP)-2/MMP-9 secretion. Clin. Cancer Res. 5: 189-196.
[17] Luca, M.; Huang, S.; Gershenwald, J.E.; Singh, R.K.; Reich, R. and Bar-Eli, M. (1997) Expression of interleukin-8 by human melanoma cells up-regulates MMP-2 activity and increases tumour growth and metastasis. Am. J. Pathol. 151: 1105-1113.
[18] O'Reilly, M.S.; Holmgren, L.; Shing, Y.; Chen, C.; Rosenthal, R.A.; Moses, M.; Lane, W.S.; Cao, Y.; Sage, E.H. and Folkman, J. (1994) Angiostatin: A novel angiogenesis inhibitor that mediates the suppression of metastases by a Lewis lung carcinoma. Cell 79: 315-328.
[19] O'Reilly, M.S.; Boehm, T.; Shing, Y.; Fukai, N.; Vasios, G.; Lane, W.S.; Flynn, E.; Birkhead, J.R.; Olsen, B.R. and Folkman J. (1997) Endostatin: an endogenous inhibitor of angiogenesis and tumour growth. Cell 88: 277-285.
[20] Yamaguchi, N.; Anand-Apte, B.; Lee, M.; Sasaki, T.; Fukai, N.; Shapiro, R.; Que, I.; Lowik, C.; Timpl, R. and Olsen, B.R. (1999) Endostatin inhibits VEGF induced endothelial cell migration and tumour growth independently of zinc binding. EMBO J. 18: 4414-4423.
[21] Xiao, F.; Wei, Y. and Yang, L. (2002) A gene therapy for cancer based on the angiogenesis inhibitor, vasostatin. Gene Ther. 18: 1207-1213.
[22] Lange-Asschenfeldt, B.; Velasco, P.; Streit, M.; Hawighorst, T.; Pike, S.E.; Tosato, G. and Detmar, M., (2001) The angiogenesis inhibitor vasostatin does not impair wound healing at tumour inhibiting doses. J. Invest. Dermatol. 117: 1036-1041.
[23] Kamphaus, G.D.; Colorado, P.C.; Panka, D.J.; Hopfer, H.; Ramchandran, R.; Torre, A.; Maeshima, Y.; Mier, J.W.; Sukhatme, V.P. and Kalluri, R. (2000) Canstatin, a novel matrix-derived inhibitor of angiogenesis and tumour growth. J. Biol. Chem. 275: 1209-1215.
[24] Lee, B.M.; MacDonald, N.J. and Gubish, E.R. (2000) Angiostatin and endostatin:endogenous inhibitors of tumour growth. Cancer Metastasis Rev. 19: 181-190.
[25] Hagedorn, M. and Bikfalvi, A. (2000) Target molecules for anti-angiogenic therapy: from basic research to clinical trials. Crit. Rev. Onco./Hemato. 34: 89-110.
[26] Jain, R.K.; Schlenger, K.; Hockel, M. and Yuan, F. (1997) Quantitative angiogenesis assays: progress and problems. Nat. Med. 3: 1203-1208.

[27] Dawson, D.W.; Volpert, O.V.; Gillis, P.; Crawford, S.E.; Xu, H.; Benedict, W. and Bouck, N.P. (1999) Pigment epithelium-derived factor: a potent inhibitor of angiogenesis. Science 285: 245–248.
[28] Kisker, O.; Becker, C.M.; Prox, D.; Fannon, M.; D'Amato, R.; Flynn, E.; Fogler, W.E.; Sim, B.K.; Allred, E.N.; Pirie-Shepherd, S.R. and Folkman, J. (2001) Continuous administration of endostatin by intraperitonealy implanted osmotic pump improves the efficacy and potency of therapy in a mouse xenograft tumour model. Cancer Res. 61: 7669-7674.
[29] Peroulis, I.; Jonas, N. and Saleh, M. (2002) Antiangiogenic activity of endostatin inhibits C6 glioma growth. Int. J. Cancer 97: 839-845.
[30] Folkman, J. (1998) Antiangiogenic gene therapy. Proc. Natl. Acad. Sci. USA 95: 9064-9066.
[31] Kuo, C.J.; Farnebo, F.; Yu, E.Y.; Christofferson, R.; Swearingen, R.A.; Carter, R.; von Recum, H.A.; Yuan, J.; Kamihara, J.; Flynn, E.; D'Amato, R.; Folkman, J. and Mulligan, R.C. (2001) Comparative evaluation of the antitumour activity of antiangiogenicproteins delivered by gene transfer. Proc. Natl. Acad. Sci. USA 98: 4605-4610.
[32] Ding, I.; Sun, J.Z.; Fenton, B.; Liu, W.M.; Kimsely, P.; Okunieff, P. and Min, W. (2001) Intra-tumoural administration of endostatin plasmid inhibits vascular growth and perfusion in MCa-4 murine mammary carcinomas. Cancer Res. 61: 526-531.
[33] Regulier, E.; Paul, S.; Marigliano, M.; Kintz, J.; Poitevin, Y.; Ledoux, C.; Roecklin, D.; Cauet, G.; Calenda, V.; Homann, H.E. (2001) Adenovirus-mediated delivery of antiangiogenic genes as an antitumour approach. Cancer Gene Ther. 8: 45-54.
[34] Lin, P.; Buxton, J.A.; Acheson, A.; Radziejewski, C.; Maisonpierre, P.C.; Yancopoulos, G.D.; Channon, K.M.; Hale, L.P.; Dewhirst, M.W.; George, S.E. and Peters, K.G. (1998) Antiangiogenic gene therapy targeting the endothelium-specific receptor tyrosine kinase Tie2. Proc. Natl. Acad. Sci. USA 95: 8829-8834.
[35] Kuo, C.J.; Farnebo, F.; Yu, E.Y.; Christofferson, R.; Swearingen, R.A.; Carter, R.; von Recum, H.A.; Yuan, J.; Kamihara, J.; Flynn, E.; D'Amato, R.; Folkman, J. and Mulligan, R.C. (2001) Comparative evaluation of the antitumour activity of antiangiogenic proteins delivered by gene transfer. Proc. Natl. Acad. Sci. USA 98: 4605-4610.
[36] Liau, G.; Su, E.J. and Dixon, K.D. (2001) Clinical efforts to modulate new blood vessel formation in the adult; a comparison of gene therapy versus conventional approaches. Drug Discovery Today 6: 19-27.
[37] Harsh, G.R.; Deisboeck, T.S. and Louis, D.N. (2000) Thymidine kinase activation of ganciclovir in recurrent malignant gliomas: A gene-marking and neuropathological study. J. Neurosurg. 92: 804-811.
[38] Puumalainen, A.M.; Vapalahti, M.; Agrawal, R.S.; Kossila, M.; Laukkanen, J.; Lehtolainen, P.; Viita, H.; Paljarvi, L.; Vanninen, R. and Yla-Herttuala, S. (1998) Beta-galactosidase gene transfer to human malignant glioma in vivo using replication-deficient retroviruses and adenoviruses. Human Gene Therapy 9: 1769-1774.
[39] Temin, H.M. (1990) Safety considerations in somatic gene therapy of human disease with retrovirus vectors. Human Gene Therapy 1: 111-123.
[40] Ferber, D. (2001) Gene therapy: Safer and virus-free? Science 294: 1638-1642.
[41] Snyder, R.O.; Miao, C.H. and Patijn, G.A. (1997) Persistent and therapeutic concentrations of human factor IX in mice after hepatic gene transfer of recombinant AAV vectors. Nat. Genet. 16: 270-276.
[42] Hortelano, G.; Al-Hendy, A.; Ofosu, F.A. and Chang, P.L. (1996) Delivery of human factor IX in mice by encapsulated recombinant myoblasts: A novel approach towards allogeneic gene therapy of hemophilia B. Blood 87: 5095-5103.
[43] Dai, Y.; Roman, M.; Naviaux, R.K. and Verma, I.M. (1992) Gene therapy via primary myoblasts: Long-term expression of rat glial factor IX protein following transplantation in vivo. Proc. Natl. Acad. Sci. USA 89: 10892-10895.
[44] Peirone, M.A., Delaney, K., Kwiecin, J. and Fletch, P.L.A. (1998) Delivery of recombinant gene product to canines of neurons in the substantia nigra in vivo, with nonautologous microencapsulated cells. Human Gene Therapy 9: 195-206.
[45] Rinsch, C.; Regulier, E.; Deglon, N.; Dalle, B.; Beuzard, Y. and Aebischer P. (1997) A gene therapy approach to regulated delivery of erythropoietin as a function of oxygen tension. Human Gene Therapy 8: 1881-1889.
[46] Aebischer, P.; Pochon, N.A.; Heyd, B.; Deglon, N.; Joseph, J.M.; Zurn, A.D.; Baetge, E.E.; Hammang, J.P.; Goddard, M.; Lysaght, M.; Kaplan, F.; Kato, A.C.; Schluep, M.; Hirt, L.; Regli, F.; Porchet, F. and De Tribolet, N. (1996) Gene therapy for amyotrophic lateral sclerosis (ALS) using a polymer encapsulated xenogenic cell line engineered to secrete hCNTF. Human Gene Therapy 7: 851-860.

[47] Pallini, R.; Consales, A.; Lauretti, L. and Fernandez, L.E. (1997) Experiments in a Parkinson's rat model [letter; comment]. Science 277: 389-390.
[48] Joki, T.; Machluf, M.; Atala, A.; Zhu, J.; Seyfried, N.T.; Dunn, I.F.; Abe, T.; Carroll, R.S. and Black, P.M. (2001) Continuous release of endostatin from microencapsulated engineered cells for tumour therapy. Nat. Biotechnol. 19: 35-39.
[49] Cima, L.G.; Vacanti, J.P.; Vacanti, C.; Ingber, D.; Mooney, D. and Langer, R. (1991) Hepatocyte culture on biodegradable polymeric substrates. Biotechnol. Bioeng. 38: 145.
[50] Hirschi, K.K.; Skalak, T.C.; Peirce, S.M. and Little, C.D. (2002) Vascular assembly in natural and engineered tissues. Ann. NY Acad. Sci. 961: 223-242.
[51] Patrick, C.W. Jr.; Zheng, B.; Wu, X.; Gurtner, G.; Barlow, M.; Koutz, C.; Chang, D.; Schmidt, M. and Evans, G.R. (2001) Muristerone A-induced nerve growth factor release from genetically engineered human dermal fibroblasts for peripheral nerve tissue engineering. Tissue Eng. 7: 303-311.
[52] Kaushal, S.; Amiel, G.E.; Guleserian, K.J.; Shapira, O.M.; Perry, T.; Sutherland, F.W.; Rabkin, E.; Moran, A.M.; Schoen, F.J.; Atala, A.; Soker, S.; Bischoff, J.; Mayer, J.E. Jr. (2001) Functional small-diameter neovessels created using endothelial progenitor cells expanded *ex vivo*. Nat. Med. 7: 1035-1040.
[53] Langer, R. and Vacanti, J.P. (1993) Tissue engineering. Science 260: 920.
[54] Shofeng, Y.; Kah-Fai, L.; Jhaoui, D. and Chee-Kai, D. (2001) The design of scaffolds for use in tissue engineering. Part I. Traditional factors. Tissue Eng. 7: 679-689.
[55] Shapiro, L. and Cohen, S. (1997) Novel alginate sponges for cell culture and transplantation. Biomaterials 18: 583-590.
[56] Lim, F. and Sun, A.M. (1980) Microencapsulated islets bioartificial endocrine pancreas. Science 210: 908-910.
[57] Weber, C.J.; Zabinski, S.; Koschitzky, T.; Wicker, L.; Rajotte, R.; D'Agati, V.; Peterson, L.; Norton, J. and Reemtsma K., (1990) The role of CD34 helper T cells in destruction of microencapsulated cells. Transplantation 49: 396-404.
[58] Siebers, U.; Horcher, A.; Brandhorst, H.; Brandhorst, D.; Federlin, K.; Bretzel, R.G. and Zekorn, T. (1998) Time course of the cellular reaction toward microencapsulated xenogenic islets in the rat. Transplant Proc. 30: 494-495.
[59] Soon-Shiong, P.; Feldman, E.; Nelson, R.; Heintz, R.; Yao, Q.; Yao, Z.; Zheng, T.; Merideth, N.; Skjak-Braek, G.; Espevik, T.; *et al.* (1993) Long term reversal of diabetes by the injection of immunoprotected islets. Proc. Natl. Acad. Sci. U S A 90: 5843-5847.
[60] Parkash, S. and Chang, T.M.S. (1996) Microencapsulated genetically engineered live *E. coli* DH5 cells administered orally to maintain normal plasma urea level in uraemic rats. Nat. Med. 2: 883-887.
[61] Savelkoul, H.F.; van Ommen, R.; Vossen, A.C.; Breedland, E.G.; Coffman, R.L. and van Oudenaren, A. (1994) Modulation of systemic cytokine levels by implantation of alginate encapsulated cells. J. Imm. Met. 170: 185-196.
[62] Thorsen, F.; Read, T.A.; Lund-Johansen, M.; Tysnes, B.B. and Bjerkvig, R. (2000) Alginate-encapsulated producer cells: a potential new approach for the treatment of malignant brain tumors. Cell Transplant. 9: 773-783
[63] Emerich, D.F.; McDermott, P.E.; Krueger, P.M. and Winn, S.R. (1994) Intrastriatal implants of polymer-encapsulated PC12 cells:pro-effects on motor function in aged rats. Prog. Neuro-psychopharmacol. Biol. Psychiatry 18: 935–946.
[64] Emerich, D.F.; Plone, M.; Francis, J.; Frydel, B.R.; Winn, S.R. and Lindner, M.D. (1996) Alleviation of behavioural deficits in aged rodents following implantation of encapsulated GDNF- producing fibroblasts. Brain Res. 736: 99-110.
[65] Hoffman, D.; Breakefield, X.O.; Short, M.P. and Aebischer, P. (1993) Transplantation of a polymer-encapsulated cell line genetically engineered to release NGF. Exp. Neurol. 122: 100-106.
[66] Winn, S.R.; Hammang, J.P.; Emerich, D.F.; Lee, A.; Palmiter, R.D. and Baetge, E.E. (1994) Polymer-encapsulated cells genetically modified to secrete human nerve growth factor promote the survival of axotomized septal cholinergic neurons. Proc. Natl. Acad. Sci. USA 91: 2324-2328.
[67] Emerich, D.F.; Winn, S.R.; Hantraye, P.M.; Peschanski, M.; Chen, E.Y.; Chu, Y.; McDermott, P.; Baetge, E.E. and Kordower, J.H. (1997) Protective effect of encapsulated cells producing neurotrophic factor CNTF in a monkey model of Huntington's disease. Nature 386: 395-399.
[68] Aebischer, P.; Schluep, M.; Deglon, N.; Joseph, J.M.; Hirt, L.; Heyd, B.; Goddard, M.; Hammang, J.P.; Zurn, A.D.; Kato, A.C.; Regli, F. and Baetge, E.E. (1996) Intrathecal delivery of CNTF using encapsulated genetically modified xenogeneic cells in amyotrophic lateral sclerosis patients. Nat. Med. 2: 696-699.

[69] Deglon, N.; Heyd, B. and Tan, S.A. (1996) Central nervous system delivery of recombinant ciliary neurotrophic factor by polymer encapsulated differentiated C2C12 myoblasts. Human Gene Therapy 7: 2135-2146.
[70] Hammang, J.P.; Emerich, D.F.; Winn, S.R.; Lee, A.; Lindner, M.D.; Gentile, F.T.; Doherty, E.J.; Kordower, J.H. and Baetge, E.E. (1995) Delivery of neurotrophic factors to the CNS using encapsulated cells: developing treatments for neurodegenerative diseases. Cell Transplant. 4: S27-28.
[71] Emerich, D.F.; Cain, C.K.; Greco, C.; Saydoff, J.A.; Hu, Z.Y.; Liu, H. and Lindner, M.D. (1997) Cellular delivery of human CNTF prevents motor and cognitive dysfunction in a rodent model of Huntington's disease. Cell Transplant. 6: 249-266.
[72] Tresco, P.A. (1994) Encapsulated cells for sustained neurotransmitter delivery to the central nervous system. J. Control. Rel. 28: 253–258.
[73] Lindner, M.D.; Plone, M.A.; Mullins, T.D.; Winn, S.R.; Chandonait, S.E.; Stott, J.A.; Blaney, T.J.; Sherman, S.S. and Emerich, D.F. (1997) Somatic delivery of catecholamines in the striatum attenuate Parkinsonian symptoms and widen the therapeutic window of oral sinemet CR in rats. Exp. Neurol. 145: 130-140.
[74] Read, T.A.; Sorensen, D.R.; Mahesparan, R.; Enger, P.O.; Timpl, R.; Olsen, B.R.; Hjelstuen, M.H.; Haraldseth, O. and Bjerkvig, R. (2001) Local endostatin treatment of gliomas administered by microencapsulated producer cells. Nat. Biotechnol. 19: 29-34.
[75] Herbst, R.S.; Hess, K.R.; Tran, H.T.; Tseng, J.E.; Mullani, N.A.; Charnsangavej, C.; Madden, T.; Davis, D.W.; McConkey, D.J; O'Reilly, M.S.; Ellis, L.M.; Pluda, J.; Hong, W.K. and Abbruzzese, J.L. (2002) Phase I study of recombinant human endostatin in patients with advanced solid tumours. J. Clin. Oncol. 20: 3773-3784.
[76] Eder, J.P. Jr.; Supko, J.G.; Clark, J.W.; Puchalski, T.A.; Garcia-Carbonero, R.; Ryan, D.P.; Shulman, L.N.; Proper, J.; Kirvan, M.; Rattner, B.; Connors, S.; Keogan, M.T.; Janicek, M.J.; Fogler, W.E.; Schnipper, L.; Kinchla, N.; Sidor, C.; Phillips, E.; Folkman, J. and Kufe, D.W. (2002) Phase I clinical trial of recombinant human endostatin administered as a short intravenous infusion repeated daily. J. Clin. Oncol. 20: 3773-3784.
[77] Szary, J. and Szala, S. (2001) Intra-tumoural administration of naked plasmid DNA encoding mouse endostatin inhibits renal carcinoma growth. Int. J. Cancer 91: 835-839.
[78] Sauter, B.V.; Martinet, O.; Zhang, W.J.; Mandeli, J. and Woo, S.L. (2000) Adenovirus-mediated gene transfer of endostatin in vivo results in high level of transgene expression and inhibition of tumour growth and metastases. Proc. Natl. Acad. Sci. USA 97: 4802-4807.
[79] Schuch, G.; Machluf, M.; Bartsch ,G. Jr.; Nomi, M.; Richard, H.; Atala, A. and Soker S. (2002) In vivo administration of vascular endothelial growth factor (VEGF) and its antagonist, soluble neuropilin-1, predicts a role of VEGF in the progression of acute myeloid leukemia in vivo. Blood 100: 4622-4628.
[80] Soker, S.; Takashima, S.; Hua-Quan, M.; Neufeld, G. and Klagsbrun, M. (1998) Neuropilin 1 is expressed by endothelial and tumour cells as an isoform-specific receptor for vascular endothelial growth factor. Cell 92: 735-745.
[81] Gagnon, M.L.; Bielenberg, D.R.; Gechtman, Z.; Miao, H.Q.; Takashima, S.; Soker, S. and Klagsbrun, M. (2000) Identification of a natural soluble neuropilin-1 that binds vascular endothelial growth factor: in vivo expression and antitumour activity. Proc. Natl. Acad. Sci. USA 97: 2573-2578.
[82] Orive, G.; Hernandez, R.M.; Gascon, A.R.; Igartua, M.; Rojas, A. and Pedraz, J.L. (2001) Microencapsulation of an anti-VE-cadherin antibody secreting 1B5 hybridoma cells. Biotechnol. Bioeng. 76: 285-290.
[83] Ferrara, N. and Gerber, H.P. (2001) The role of vascular endothelial growth factor in angiogenesis. Acta Haematol. 106: 148-156.
[84] Orive, G.; Hernandez, R.M.; Gascon, A.R.; Calafiore, R.; Chang, T.M.; De Vos, P.; Hortelano, G.; Hunkeler, D.; Lacik, I.; Shapiro, A.M. and Pedraz, J.L. (2003) Cell encapsulation: Promise and progress. Nat. Med. 9: 104-108.
[85] Orive, G.; Hernandez, R.M.; Gascon, A.R.; Igartua, M.; Rojas, A. and Pedraz, J.L. (2001) Microencapsulation of an anti-VE-cadherin antibody secreting 1B5 hybridoma cells. Biotechnol. Bioeng. 76: 285-290.
[86] Cirone, P.; Bourgeois, J.M.; Austin, R.C. and Chang, P.L. (2002) A novel approach to tumour suppression with microencapsulated recombinant cells. Human Gene Therapy 13: 1157-1166.

CELL ENCAPSULATION THERAPY FOR MALIGNANT GLIOMAS

ANNE MARI ROKSTAD[1], ROLF BJERKVIG[2],
TERJE ESPEVIK[1] AND MORTEN LUND-JOHANSEN[3]
[1]Department of Clinical and Molecular Medicine,
Norwegian University of Science and Technology, Trondheim –
Fax: 47 73598801 – Email: anne.m.rokstad@medisin.ntnu.no
[2]Department of Anatomy and Cell Biology, University of Bergen
[3]Department of Neurosurgery, Haukeland University Hospital, Bergen

1. Introduction

Primary brain tumours consisting of atypical glial cells with areas of necrosis, vascular proliferation and mitoses are termed glioblastomas [1]. The conventional treatment for malignant gliomas is debulking surgery followed in a few weeks by fractionated radiotherapy. This treatment hardly ever leads to cure, and the median survival time is less than one year. Within 6-12 months after treatment most patients present with symptoms of recurrent disease (Figure 1) Relapsing tumours are occasionally treated with cytostatics or additional surgery. In recent years, an increasing number of phases I or II clinical trials have been carried out on patients with recurrent glioma. In many cases, the substances put into trials are biologically active molecules that differ from cytostatics, such as toxins, immunomodulators, and viruses (see http/www.cancer.gov.). Very few substances have made their way into phase III trials, and the results are in general disappointing [2]. One new concept which has not yet reached clinical trials is the grafting of alginate encapsulated cells, which produce molecules with anti-tumour activities, into the resection cavity in patients operated for gliomas. Tumour necrosis and prolonged survival has been achieved in rats bearing glioma cell-line tumours, when endostatin-secreting encapsulated cells were implanted into the tumour [3]. In the following sections, we discuss the potential use of cell immobilisation techniques to treat human glioma. First, a brief overview of the clinical and biological properties of glioblastomas is provided. Then, we present the current surgical and irradiation treatment strategies and discuss physiological parameters in the brain reacting to transplanted material, and to immunogenic agents. Furthermore, we present the current status for research devoted to cell encapsulation techniques and design cell lines producing therapeutic substances.

2. Glioma growth and invasiveness

Clinically and genetically, gliomas may be divided into primary and secondary neoplasms [4]. Primary glioblastomas, which arise *de novo*, constitute the majority of these tumours. They appear most often in elderly patients (-60 years) and have the worst prognosis. Secondary glioblastomas arise from already existing less aggressive lesion.

The spreading pattern of the glioblastomas is peculiar [5]. Usually there is no macroscopically clear boundary between the brain and the tumour. In this respect glioblastomas differ from metastatic tumours to the brain, where a defined cleavage plane separates the two tissues. Glioma cells show a high propensity for spread along white matter tracts. The main fibre connection between the two cerebral hemispheres, the callosal body, is a "highway" for the migrating cells that cross the midline and may give rise to a bilateral tumour. The mechanisms of this disseminated growth has been extensively studied, but are still poorly understood [6].

The infiltrative growth of the glioma cells within the brain is contradicted by the tumour's lacking capacity for extraneural spread. Tissue culture experiments indicate that the basement membrane, which is penetrable to metastatic tumour cells may be a non-permissive barrier to glioma cells [7].

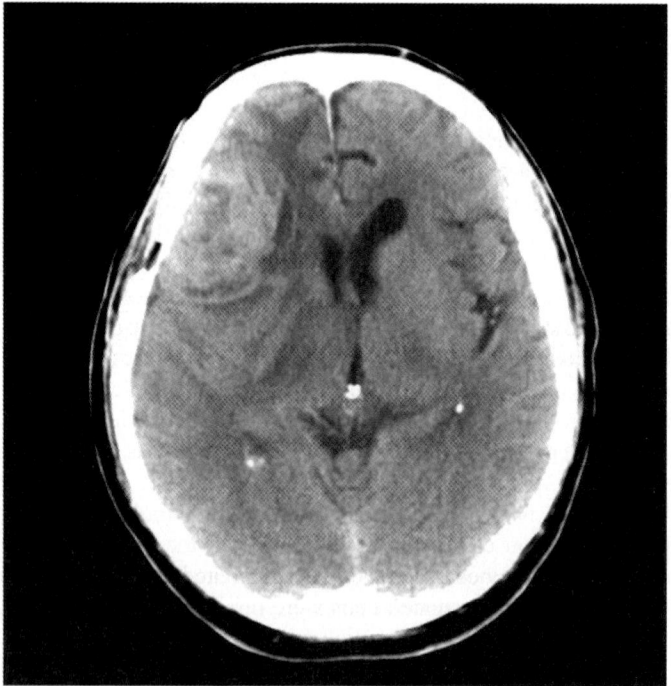

Figure 1. Recurrent glioma. Male, 48, operated and irradiated for frontal glioblastoma two years previously, presenting with local relapse. The CT scan shows a contrast-enhancing expansive tumour of the right hemisphere (left on CT scan), with a hypodense cyst deep to the lesion and peritumoural edema.

3. Glioma treatment

3.1. SURGERY

Debulking surgery for glioblastoma causes a reduction in raised intracranial pressure and neurological impairment caused by tumour mass effect [8]. In addition, removed tissue becomes available for diagnosis and scientific research. However, based on its local invasiveness glioblastoma can be regarded as a systemic brain disease that is impossible to remove surgically. The ineffectiveness of gross tumour removal in providing cure for glioblastoma was proven by the neurosurgeon Walter Dandy [9]. He demonstrated that even in patients operated with hemispherectomy, *i.e.* removal of the whole brain hemisphere carrying the tumour, relapsing tumour developed contralaterally. Noteworthy, this report written more than 70 years ago is of great relevance to any suggested loco-regional treatment modality for gliomas. It shows the challenge of pharmacologically distributing any anti-tumour therapeutic substance to the whole central nervous system. Patients with good performance, a relatively long progression-free period after the initial treatment, and a relapsing tumour in a favourable location, may be offered repeated surgery [10].

3.2. IRRADIATION

Postoperative irradiation leads to a prolonged survival in patients with malignant gliomas [11]. Postoperatively, fractionated irradiation accumulating to a total dose of 45-54 Gy is administered over a period of 3-4 weeks. Irradiation causes necrosis of tumour tissue and intratumoural vessels. The tumour bed becomes firm, and histology shows necrotic areas, fibrosis, gliosis, perivascular collagen and abundant inflammatory cells including microglia [12]. In the brain adjacent to the tumour bed, myelin damage is frequent, as is neuronal damage and focal necroses. The most serious complication of irradiation is delayed brain necrosis which occurs in a few per cent of the cases, usually many months or years after treatment [13]. The irradiation may kill any living material that is introduced into the tumour bed, including encapsulated cells. Therefore, it must be completed before any introduction of such.

3.3. CHEMOTHERAPY

The effect of traditional cytostatics in the treatment of gliomas is low, and recent prospective study indicates that some of the mostly used drugs have little or no effect [14]. The lacking effects of cytostatics have been attributed to distribution problems due to the blood-brain barrier, or to drug resistance. A local delivery system using drug-immersed wafers implanted into the tumour bed may prove effective, as may the new alkylating agent Temozolomide [15,16].

4. Glioma treatment with alginate bioreactors

Brain tumour cell growth and invasion imply complex interactions between malignant cells, neural cells, and endothelial cells, involving extracellular matrix components, proteases, growth factors and cell surface receptors [6]. The notion is that tumour growth requires the persistent formation of new blood vessels. Induction of angiogenesis appears during early stages of tumour development, suggesting that this event represents a potentially rate-limiting step [17]. Anti-angiogenic factors such as angiostatin [18], endostatin [19], or anti-angiogenic anti-thrombin III [20] can inhibit tumour growth in mice. Endostatin, a Mr 20,000 COOH-terminal fragment of collagen XVIII was initially isolated from the conditioned medium of haemangioendothelioma cells [19]. Several studies using endostatin of murine, rat or human origin have shown suppression of the growth of primary human and murine tumours as well as metastases [19,21-24]. Since glioblastoma is a highly vascularised, rapidly proliferating tumour, it is possible that it may respond to anti-angiogenic therapy. However, angiogenesis inhibitors are molecules that often have short half-lives *in vivo*. An advantage would be to have a continuous, intratumoural delivery system for the angiostatic molecules.

A new treatment concept is suggested were alginate encapsulated cells secreting proteins with anti-tumour properties can be implanted into the tumour resection cavity of malignant glioma [3,25,26]. An overview of the treatment concept is provided in figure 2. In recent studies, alginate beads containing engineered cells producing endostatin were implanted together with BT4C rat glioma cells into the rat brains of BD-IX rats. The animals in the treatment group survived significantly longer than the controls, and apoptotic cells and large necrotic areas were found within the treated tumours [3]. In another study human glioma cells were injected subcutaneously into mice ten days before insertion of alginate-poly-L-lysine (PLL)-alginate microcapsules with endostatin producing cells. After 21 days, the tumours in the treatment group were smaller and less vascularised than the controls [27]. The anti-angiogenic effect appeared to be directed specifically towards the neoplastic vessels [28]. A substantial body of data indicate that the encapsulated cells may be kept alive secreting recombinant proteins for months in rat brain parenchyma [3,25]. Recently several studies have demonstrated that the therapeutic effects of proteins are potentiated by the use of continuous local delivery systems. Anti-tumour effects of endostatin administered continuously exceeded those seen in animals receiving daily injections and the dose required was 8-10 fold lower [29]. Tumour growth has also been reduced by the use of an alginate microcapsule delivery system containing IL-2 secreting cells [30]. Furthermore, local treatment with antibodies against vascular endothelial growth factor (VEGF) was more effective in reducing the vascular permeability in tumour tissue than antibodies administered systemically [31]. In summary, biologically active molecules administered locally to animals may selectively inhibit growth of malignant tumours, including gliomas. Continuous release of such substances ensures a stable local effect. Further research should aim at refining the cellular delivery systems having future clinical use in mind.

4.1. OPTIMISING THE ALGINATE BIOREACTOR

The ideal alginate microcapsule and production cell-line has not yet been identified. Recently, we optimised the alginate microcapsule for endostatin producing 293 cells (293 endo cells) [32] used in the rat glioma study [3]. The 293 endo cells proliferated rapidly in the relatively strong alginate-PLL-alginate microcapsules. An example of the growth and distribution of live and dead 293 endo cells within alginate-PLL–alginate microcapsules is given in figure 3. The alginate bioreactors disintegrated approximately 30-40 days post-encapsulation leading to escape of transfected cells [32]. This may cause inflammatory reactions. It is shown that the mechanical stability of the alginate microcapsules is reduced after cell encapsulation [33], and that the cells interrupt the alginate gel-network further as they grow [34,35]. Improvement of the alginate bioreactor system may be achieved by strengthening the microcapsules and by finding cell-lines particularly suited for encapsulation. These strategies are discussed in the following paragraphs.

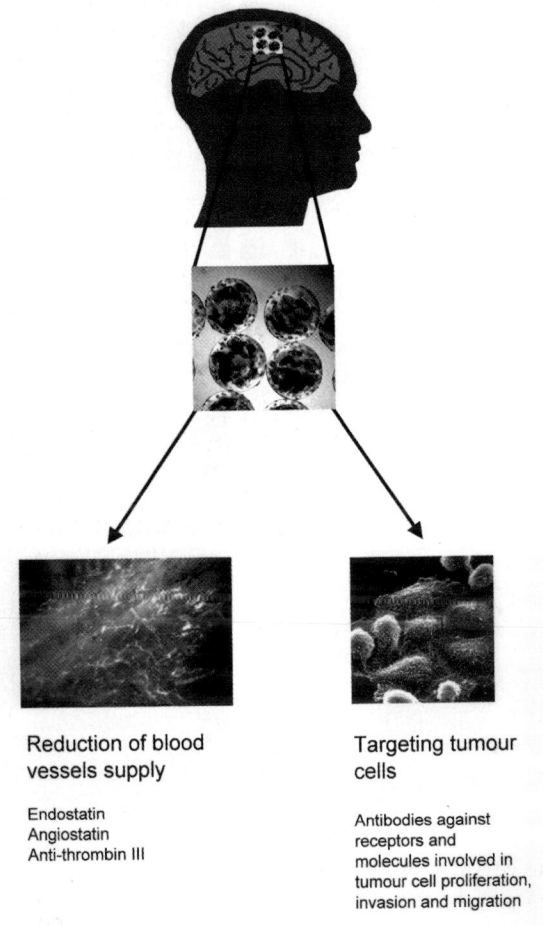

Figure 2. Proposed treatment of malignant brain tumours (gliomas). After surgery and irradiation removal, alginate microcapsules containing cells producing anti-angiogenic substances are inserted into the wall of the cavity. On the basis of tumour biopsies from the patient, a stage two treatment is individually tailored by inserting alginate microcapsules containing cells producing monoclonal antibodies directed against specific receptors and molecules involved in the tumour progression.

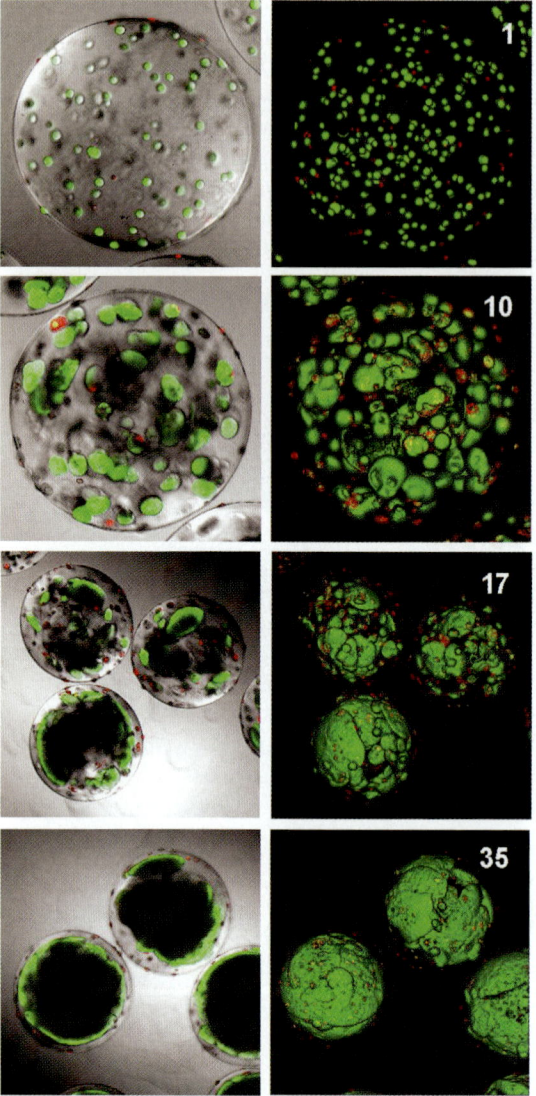

Figure 3. Confocal images showing the distribution of live and dead 293 endo cells within alginate-poly-L-lysine microcapsules gelled with a 50:1 mM $CaCl_2/BaCl_2$ solution. The growth of the 293 endo cells is shown between day 1 and 35 post-encapsulation. Green fluorescence is emitted from live cells whereas red fluorescence indicates dead cells. Optical sections through equator overlaid transmitted light are shown on the left side and three-dimensional re-constructions of optical cross sections are shown at the right side. The 293 endo cells producing endostatin proliferate well post-encapsulation. Ideally sized cell-clusters are seen around day 10 post-encapsulation. Mostly live cells are seen at all time points, with some dead cells in between the live cells. The dead cells are often associated with the outer layers of the microcapsules, indicating the involvement of PLL.

4.2. IMPROVEMENT OF THE ALGINATE MICROCAPSULES FOR PROLIFERATING CELLS

Alginates are algae-derived polysaccharides that gel in the presence of divalent cations. Through selection of alginate source, and by chemical and physical adjustments of the capsule synthesis, the properties of alginate microcapsules may be widely and selectively modified. Alginates constitute a family of unbranched binary copolymers of 1-4 linked β-D-mannuronic acid (M) and α-L-guluronic acid (G). The monomers are arranged in a pattern of blocks along the chain, with homopolymeric regions in the alginate chain (M and G blocks) interspersed with regions of alternating structures (MG blocks) [36]. The gel forming properties are a result of divalent cations (Ca^{2+}, Ba^{2+}, Sr^{2+}) that bind preferentially to G blocks in the alginate, and form bonds between these sites and equal sites in other alginate chains [36]. The composition of the alginate is therefore important for the gel forming properties. The highest mechanical strength is achieved by making beads of alginate with a G content higher than 70% ($F_G>0,7$) and an average length of G-blocks of about 15 [36]. Further increase in the gel strength can be achieved by forming inhomogeneous capsules [37,38], using barium or strontium ions in the gelling solution [36,39], and coating the capsule with a polycation like poly-L-lysine (PLL) [40]. PLL also reduces and controls the capsule permeability [41,42]. More information on alginate capsule properties and construction is found in the chapter "Microcapsule Formulation and Formation" in Fundamentals of Cell Immobilisation Biotechnology and in "Alginate as Immobilisation Matrix for Cells" [36].

Most functional studies on alginate microcapsules have been done using insulin producing islets and the most widely used capsule construction is the one introduced by Lim and Sun 22 years ago [43]. It consists of an alginate-PLL-alginate microcapsule with a dissolved core. Good results with islet transplantations in dogs and man were achieved by using a modified alginate-PLL-alginate microcapsule with a high content of G alginate ($F_G>0,64$) and a solid core [44,45].

Engineered cell-lines that divide continuously require more stabile alginate capsule types than those needed to keep low dividing islets encapsulated. Microcapsules constructed with a dissolved core [46] or a high M content [47] are not suited for *in vivo* transplantation of proliferating cells because of low stability. Only few attempts have been made to design alginate microcapsules suitable for dividing cells with the purpose of using them for delivery of therapeutic substances [32,48-51]. The information from these experiments can be summarised as follows. A high G content in the alginate, high molecular weight and high viscosity as well as increased concentration of alginate diminishes cell growth and protein release, while dissolving the capsule core lead to increased cellular growth. Therefore, if mechanical stability is needed, capsules with a high G content and a solid core should be the choice. PLL is crucial for the long-term *in vitro* mechanical stability, as proliferating cells tend to penetrate Ca-alginate beads [32,49]. Even by increasing the gel strength using high-G alginate and 10 mM Ba^{2+} as gelling solution we have seen that proliferating 293 endo cells are able to escape from the capsule beads. For a long-term culture in alginate microcapsules, a solid high G capsule coated with PLL or another polycation is necessary. However, it is desirable to reduce the PLL layer as much as possible since inflammatory reactions against empty alginate microcapsules derive from PLL [52]. Among the alternatives to PLL are poly ethylene glycol (PEG) [53,54] or chitosan [55].

Alginate modifying enzymes, which have recently become available, make it possible to tailor the alginate structure in a controlled manner. The functional properties of the alginate beads and alginate microcapsules are improved by using AlgE4 epimerases that converts M blocks to MG blocks [56,57]. Applying epimerised alginate for capsule formation results in a reduction of bead size, osmotic swelling, and permeability, thus the cell encapsulation capabilities of the beads are improved. A small increase in the G content is achieved by the epimerisation step, but the main improvement is probably caused by the flexible MG blocks that are introduced by the AlgE4 epimerases that contribute to more compact packing of the alginate chains. By using AlgE4 epimerised alginate and barium ions in the gelling solution, we have reduced the proliferation of 293 endo cells within alginate-PLL-alginate microcapsules [35]. The increased gel strength results also in a 3-fold reduction in secretion of endostatin. However, since these capsules are smaller in size, increasing the amount of microcapsules by 3-fold will only approximately double the total capsule volume. Although the microcapsules generated by epimerised alginate are strong, there is still a need for PLL as a mechanical barrier to keep the 293 endo cells within the alginate. AlgE4 epimerised alginate provide a good shielding of the PLL coat [57]. Therefore, host derived cellular overgrowth reactions of the alginate-PLL-alginate microcapsules become less prominent when AlgE4 epimerised alginate is used as an outer coat [47].

The growth pattern in the alginate microcapsules is also dependent of the cell-type [32,48]. The number of viable βTC3 cells decreases the first three weeks within alginate-PLL-alginate microcapsules containing 69%G and a concentration of 1.5% alginate [50]. In contrast, 293 endo cells encapsulated in alginate-PLL-alginate microcapsules with 68% G and 1.8% alginate start to grow immediately [32]. Therefore, the choice of alginate microcapsule is also dependent on the specific cell-line.

4.3. CHOOSING CELL-LINES SUITED FOR ENCAPSULATION

Ideally, encapsulated cells should be grown to small cell clusters that get access to nutrients and oxygen, while waste products and therapeutic substances diffuse out from the capsule. In addition, since dead cells may give rise to inflammatory reactions, their number should be as low as possible. An example of ideally sized cell clusters is given in figure 3 day 10.

One way to control growth and also the secretion rate is to use the Tet^{SWITCH} system described by Mazur et al. [58]. This system is based on the transfection with a multicistronic expression unit encoding both the product gene and a cytostatic cell cycle arresting gene under control of a tetracycline repressible (tet_{off}) promoter. The growth and production phase are decoupled such that a non-productive cell growth phase is induced by addition of tetracycline, while depletion of tetracycline leads to a non-growing state where the cells produce the protein of interest. The system was mainly developed for large-scale production of proteins. One obstacle with the system is a short function time as the growth-arrested production phase is reported to be seven days. A second problem is a tendency to increased genetic drift and outgrowth of proliferation-competent mutant cells. The cultivation period can however be extended by a selection technology based on surface antigen expression [59].

Normal cells have several mechanisms ensuring a tight contact inhibition of growth. This has been described in murine fibroblasts (NIH 3T3) [60], in a melanoma cell-line from canine (TLM1) [61], and for human embryonic fibroblasts (MRC-5) [62]. Cell-lines possessing contact inhibition may be good candidates for microencapsulation.

Skeletal muscle precursor cells, myoblasts possess unusual biological properties that make them well suited for the delivery of therapeutic products. Myoblasts have all the advantages of dividing cell-lines, like unlimited availability, possibility for screening of pathogens, suitability for stable gene transfer and, clonal selection. Furthermore, these cells can be terminally differentiated to a non-proliferating state by manipulating the growth conditions [63]. Human primary myoblasts can be isolated from biopsy or autopsy material, enriched and grown to large numbers and genetically engineered without loosing their potential to differentiate [64]. Myofibers implanted into muscle tissue in mice remain viable for at least six months showing stabile gene expression [64]. The established murine myoblast cell line C2C12 has been used in several experiments delivering therapeutic factors *via* alginate microcapsules or polyether-sulphone capsules [30,46,65-72]. C2C12 myoblasts transplanted in mice survive for at least 6 month [65] and 213 days [66], within alginate-PLL-alginate capsules with dissolved cores. The same alginate microcapsules containing myoblasts also function for at least 16 weeks in the brain of rodents [68], demonstrating the possibility of using microcapsules coated with PLL in the brain. The growth pattern of myoblasts within alginate microcapsules has been shown to be quite similar between three different capsule types with solid (gelled with Ba^{2+}) or dissolved cores [48]. However, no information about the M:G ratio or the differentiation state of the myoblasts was given. Recently, myoblast proliferation and differentiation in alginate gels has been demonstrated to vary according to the M:G ratio when the alginate was modified with a ligand (RGD) that promotes cell-adhesion [73]. Alginates with a high G ratio ($F_G=0.6$) were favourable for both myoblast growth and differentiation. This may be of interest for the future design of alginate bioreactors containing myoblasts.

4.4. THE BIOCOMPATIBILITY OF FOREIGN MATERIALS IN THE BRAIN

Since fibroblasts and immunocompetent cells are found in scarce amounts in the brain, it is likely that the implantation of foreign material may be better tolerated in the brain tissue than elsewhere in the body. However, experiments and clinical observations show that proliferation of glia (gliosis), and immunoreactivity both takes place in the brain as a response to damage of foreign material. In animal experiments, diverse implants such as aneurysm clips, fibrin glue, shunt material and various polymers for drug or cell delivery, are usually but not always well tolerated by native brain [74-77]. Gliosis is induced by mechanical damage of brain tissue, including surgery, irradiation, and by various neurodegenerative diseases [78]. Interleukin-6 producing encapsulated cells transplanted to the rat brains have been shown to induce massive gliosis [79]. Xenotransplants of foetal mouse brain cells into rat brains do not survive due to a MHC-mediated immune response from the host [80]. In humans, however, Parkinson patients receiving xenografts of foetal cells, may show long-term improvement of symptoms, which indicates graft survival [81,82]. It is therefore not unlikely that the brain is more permissive to foreign material than other organs.

When tumour tissue is removed from the brain, a cavity lined by damaged brain tissue with no pial covering is formed. In some cases, this cavity extends into cerebrospinal fluids (CSF) pathways such as the ventricles or the basal cisterns. It is likely that large particles such as alginate microcapsules may cause acute obstructive hydrocephalus if they escape into the CSF in large quantities. In some cases, the resection cavity may seal itself off and give rise to a cyst filled with a protein-rich liquid. Occasionally, the cyst becomes expansive due to continuous secretion and lack of drainage, and may cause mass effect. In other cases no such cyst is formed. The surgical trauma and the irradiation cause considerable changes in the area where the resection has taken place. Gradually, a hypergliotic scar is formed, consisting of reactive astrocytes, microglia, cells of the immune system, and, inevitably, an increasing population of recurring tumour cells [12]. Thus, the tissue into which the alginate microcapsules are to be inserted is different from a normal healthy brain and the response of that tissue to the insertion of the alginate microcapsules may differ similarly. Patients with brain tumours sometimes receive silicone catheters for CSF diversion. It is a well-known clinical fact that brain tumour patients with shunts have a high risk of developing shunt failures due to overgrowth of fibrous tissue around the tip of the ventricular catheter. In animal experiments, rat brains may harbour and maintain encapsulated cells over many months. When encapsulated cells were implanted into rats with brain tumours the implants were capable of causing an effect on the tumour, but the experiments were terminated after a few weeks [3,27].

In conclusion, it is difficult to predict the response of the postoperative, post-irradiated brain to the concerted trauma of the re-operation and the microcapsule implantation. Probably, graft survival is easiest to obtain if the alginate microcapsules may be implanted into a "clean" resection cavity without necrosis or fibrous tissue. However, gliosis, radiation necrosis and active immune mechanisms must be expected in the implantation area.

4.5. THE BIOCOMPATIBILITY OF THE ALGINATE BIOREACTORS

The ability of the host to accept the transplanted material depend not only on the material itself, but also on the host, its genetic dispositions and physical status [83]. As discussed earlier, gliosis, radiation and active immune mechanisms must be expected in the implantation area of the brain. In the following, we review the compatibility of the alginate itself, the alginate microcapsules and the encapsulated cells.

Purified alginate fulfils the criteria for use as food additives and for some pharmacological products [84]. Alginate used for transplantation must have a low impurity level. Therefore regulatory standards for alginate have been developed [85].

Soluble and gelled alginate rich in mannuronic acid ($F_M > 0,54$) induces an inflammatory response by stimulating monocytes to produce proinflammatory cytokines such as tumour necrosis factor (TNF), interleukin 1 (IL-1) and interleukin 6 (IL-6) [86,87]. In several studies it has been demonstrated that the TNF induction caused by high M alginate is not due to LPS contamination [86,88,89]. Soluble alginate rich in guluronic acid ($F_G = 0,64$) evokes a low IL-6 production and no TNF and IL-1 production while the gelled form provokes cytokine production to a lesser extent than the gelled high M alginate does [86]. For highly purified alginate rich in mannuronic acid (poly M, $F_M>0,92$) TNF stimulation occurs through the CD14 receptor [88,89], toll receptor (TLR) 2 and TLR4 on monocytes/macrophages [88,89]. The TNF inducing ability is reduced when poly M chains are degraded to a molecular weight of 5500 [90]. Finally, antibodies against alginate microcapsules containing high M but not high G are found [91]. On the basis of these results, we conclude that the leakage of mannuronic acid from the alginate microcapsules may provoke an inflammatory response, but this also depends on the molecular size and the concentration of the polymer.

Alginate beads provoke no or minor fibrotic reactions, but the PLL coating does trigger such reactions [47,52,53,92,93]. PLL leads to TNF production from human monocytes [52]. High concentrations of PLL induces necrosis of Jurkat cells [52] and also death of 293 endo cells and JJN3 cells [32]. Alginate-PLL-alginate microcapsules made with low G content give lower fibrotic reactions than capsules with a higher G content [92]. This is explainable because there is a lower interaction between PLL and guluronic acid than between PLL and mannuronic acid [40]. As already mentioned, a better shielding of the PLL coat has been achieved by using epimerised alginate [57].

The main reason for coating with PLL or another polycation is to increase capsule stability and reduce permeability. However, both alginate-barium beads with a high amount of G alginate [93-96] or with a high amount of M alginate [93] have given good results with islets transplantations. Duvivier-Kali postulates that lack of inflammatory cells at the capsule surface prevent secretion of cytokines and thus damage from locally produced cytokines [93]. Lanza et al. have suggested that prevention of cell-cell contact between immune cells and islets is important for survival, and also that immunologic effector molecules become trapped within the alginate network, thus preventing the cytotoxic effects [94]. The results from the above studies demonstrate that PLL may not be necessary for achieving protective effects of the alginate microcapsules at least for islet transplantations.

The origin of the encapsulated cells is an important determinant for their biocompatibility. Cells of allogeneic origin may trigger a cell-mediated immune response through histocompatibility (MHC) antigens. Xenogeneic cells provoke both cell- as well as antibody- and complement- mediated responses [97]. Different ways of modifying cells used in transplantation medicine are at present under development. Among these are the establishment of histocompatible cells [98] and self-immunomodulating cell lines [99].

Other factors influencing the biocompatibility of a device is the adsorption of proteins to the surface and the surface roughness. The surface show less tendency to promote cell attachment and growth if the adsorption of cell adhesion proteins is reduced [100]. It is shown that tubular diffusion chambers with smooth external surfaces gives less fibrotic overgrowth than rough surfaces [101]. Surface roughness and adsorption of protein may also be factors that contribute to the overgrowth reactions in high G *versus* high M capsules. As high M capsules swell more than high G capsules due to a lower binding of calcium and barium ions [37,47], the surface of a high M capsule may be softer.

5. Conclusion

Recent animal studies show that brain tumours in animals may respond to substances released from encapsulated cells. Although the brain is an immunologically privileged site, active immune mechanisms must be expected to be present in a brain tumour resection cavity. Finding an optimal bioreactor for the treatment of brain tumours is in its first phase of development. Alginate microcapsules have been optimised for 293 endo cells producing endostatin. An optimal alginate microcapsule must be functional for a long period, thus it is important to make a strong capsule. The most stable alginate microcapsule has a solid core with a high content of guluronic acid. Gelling with barium ions will further increase the strength. Coating with a polycation like PLL is necessary when highly proliferating cells are used. Enzymatically tailored alginate with an increased amount of MG blocks contributes to increased flexibility and more compact packing of the alginate chains that have further increased the capsule strength. Enzymatically tailored alginate also shield the PLL better, thus decreasing inflammatory reactions caused by PLL. Myoblasts that differentiate to a non-proliferate state or contact-inhibited cells may be better suited for encapsulation. The PLL coat may not be necessary when using cells with a limited growth-potential. Alginate beads should be made with an inhomogeneous distribution of the alginate. A higher concentration of the alginate in outer parts of the beads gives increased strength and probably makes a barrier for the proliferating cells. Further work along these strategies should improve the local delivery of anti-cancer proteins produced by encapsulated cells.

References

[1] Kleihues, P. and Cavenee, W.K. (2000) World Health Organization classification of tumours: Pathology and genetics of tumours of the central nervous system. IARC press.
[2] Rainov, N.G. (2000) A phase III clinical evaluation of herpes simplex virus type 1 thymidine kinase and ganciclovir gene therapy as an adjuvant to surgical resection and radiation in adults with previously untreated glioblastoma multiforme. Hum.Gene Ther. 11: 2389-2401.
[3] Read, T.A.; Sorensen, D.R.; Mahesparan, R.; Enger, P.O.; Timpl, R.; Olsen, B.R.; Hjelstuen, M.H.; Haraldseth, O., and Bjerkvig, R. (2001) Local endostatin treatment of gliomas administered by microencapsulated producer cells. Nat. Biotechnol. 19: 29-34.
[4] Watanabe, K.; Sato, K.; Biernat, W.; Tachibana, O.; von Ammon, K.; Yonekawa, Y.; Kleihues, P. and Ohgaki, H. (1996) Overexpression of the EGF receptor and p53 mutations are mutually exclusive in the evolution of primary and secondary glioblastomas. Brain Pathol. 6: 217-223.
[5] Burger, P.C.; Heinz, E.R.; Shibata, T. and Kleihues, P. (1988) Topographic anatomy and CT correlations in the untreated glioblastoma multiforme. J. Neurosurg. 68: 698-704.
[6] Bjerkvig, R.; Lund-Johansen, M. and Edvardsen, K. (1997) Tumor cell invasion and angiogenesis in the central nervous system. Curr. Opin. Oncol. 9: 223-229.
[7] Pedersen, P. H.; Rucklidge, G.J.; Mork, S.J.; Terzis, A.J.; Engebraaten, O.; Lund-Johansen, M.; Backlund, E.O.; Laerum, O.D. and Bjerkvig, R. (1994) Leptomeningeal tissue: a barrier against brain tumor cell invasion. J Natl. Cancer Inst. 86: 1593-1599.
[8] Harsh, G.R. and Wilson, C.B. (1990) Neuroepithelial tumors of the adult brain. Neurol. surg. 5: 3040-3136.
[9] Dandy, W.E. (1928) Removal of right cerebral hemisphere for certain tumors with hemiplegia. JAMA 90: 823-825.
[10] Azizi, A.; Black, P.; Miyamoto, C. and Croul, S.E. (2001) Treatment of malignant astrocytomas with repetitive resections: a longitudinal study. Isr. Med. Assoc. J 3: 254-257.

[11] Kristiansen, K.; Hagen, S.; Kollevold, T.; Torvik, A.; Holme, I.; Nesbakken, R.; Hatlevoll, R.; Lindgren, M.; Brun, A.; Lindgren, S.; Notter, G.; Andersen, A.P. and Elgen, K. (1981) Combined modality therapy of operated astrocytomas grade III and IV. Confirmation of the value of postoperative irradiation and lack of potentiation of bleomycin on survival time: a prospective multicenter trial of the Scandinavian Glioblastoma Study Group. Cancer 47: 649-652.
[12] Burger, P.C.; Mahley, M.S., Jr.; Dudka, L. and Vogel, F.S. (1979) The morphologic effects of radiation administered therapeutically for intracranial gliomas: a postmortem study of 25 cases. Cancer 44: 1256-1272.
[13] Kramer, S. (1968) The hazards of therapeutic iradiation of the central nervous system. Clin. Neurosurg. 15: 301-318.
[14] Anonymous. (2001) Randomized trial of procarbazine, lomustine, and vincristine in the adjuvant treatment of high-grade astrocytoma: a Medical Research Council trial. J. Clin. Oncol. 19: 509-518.
[15] Brem, H. and Gabikian, P. (2001) Biodegradable polymer implants to treat brain tumors. J. Control Release 74: 63-67.
[16] Stupp, R.; Dietrich, P.Y.; Ostermann; Kraljevic, S.; Pica, A.; Maillard, I.; Maeder, P.; Meuli, R.; Janzer, R.; Pizzolato, G.; Miralbell, R.; Porchet, F.; Regli, L.; de Tribolet, N.; Mirimanoff, R.O. and Leyvraz, S. (2002) Promising survival for patients with newly diagnosed glioblastoma multiforme treated with concomitant radiation plus temozolomide followed by adjuvant temozolomide. J. Clin. Oncol. 20: 1375-1382.
[17] Hanahan, D. and Folkman, J. (1996) Patterns and emerging mechanisms of the angiogenic switch during tumorigenesis. Cell 86: 353-364.
[18] O'Reilly, M.S.; Holmgren, L.; Shing, Y.; Chen, C.; Rosenthal, R.A.; Moses, M.; Lane, W.S.; Cao, Y.; Sage, E.H. and Folkman, J. (1994) Angiostatin: a novel angiogenesis inhibitor that mediates the suppression of metastases by a Lewis lung carcinoma. Cell 79: 315-328.
[19] O'Reilly, M.S.; Boehm, T.; Shing, Y.; Fukai, N.; Vasios, G.; Lane, W.S.; Flynn, E.; Birkhead, J.R.; Olsen, B.R. and Folkman, J. (1997) Endostatin: an endogenous inhibitor of angiogenesis and tumor growth. Cell 88: 277-285.
[20] O'Reilly, M.S.; Pirie-Shepherd, S.; Lane, W.S. and Folkman, J. (1999) Antiangiogenic activity of the cleaved conformation of the serpin antithrombin. Science 285: 1926-1928.
[21] Boehm, T.; Folkman, J.; Browder, T. and O'Reilly, M.S. (1997) Antiangiogenic therapy of experimental cancer does not induce acquired drug resistance. Nature 390: 404-407.
[22] Yamaguchi, N.; Anand-Apte, B.; Lee, M.; Sasaki, T.; Fukai, N.; Shapiro, R.; Que, I.; Lowik, C.; Timpl, R. and Olsen, B.R. (1999) Endostatin inhibits VEGF-induced endothelial cell migration and tumour growth independently of zinc binding. EMBO J. 18: 4414-4423.
[23] Perletti, G.; Concari, P.; Giardini, R.; Marras, E.; Piccinini, F.; Folkman, J. and Chen, L. (2000) Antitumour activity of endostatin against carcinogen-induced rat primary mammary tumours. Cancer Res. 60: 1793-1796.
[24] Sim, B.K.; MacDonald, N.J. and Gubish, E.R. (2000) Angiostatin and endostatin: endogenous inhibitors of tumour growth. Cancer Metastasis Rev. 19: 181-190.
[25] Thorsen, F.; Read, T.A.; Lund-Johansen, M.; Tysnes, B.B. and Bjerkvig, R. (2000) Alginate-encapsulated producer cells: a potential new approach for the treatment of malignant brain tumours. Cell Transplant. 9: 773-783.
[26] Visted, T.; Bjerkvig, R. and Enger, P.O. (2001) Cell encapsulation technology as a therapeutic strategy for CNS malignancies. Neuro.-oncol. 3: 201-210.
[27] Joki, T.; Machluf, M.; Atala, A.; Zhu, J.; Seyfried, N.T.; Dunn, I.F.; Abe, T.; Carroll, R.S. and Black, P.M. (2001) Continuous release of endostatin from microencapsulated engineered cells for tumour therapy. Nat. Biotechnol. 19: 35-39.
[28] Read, T.A.; Farhadi, M.; Bjerkvig, R.; Olsen, B.R.; Rokstad, A.M.; Huszthy, P.C. and Vajkoczy, P. (2001) Intravital microscopy reveals novel antivascular and antitumour effects of endostatin delivered locally by alginate-encapsulated cells. Cancer Res. 61: 6830-6837.
[29] Kisker, O.; Becker, C.M.; Prox, D.; Fannon, M.; D'Amato, R.; Flynn, E.; Fogler, W.E.; Sim, B.K.; Allred, E.N.; Pirie-Shepherd, S.R. and Folkman, J. (2001) Continuous administration of endostatin by intraperitoneally implanted osmotic pump improves the efficacy and potency of therapy in a mouse xenograft tumour model. Cancer Res. 61: 7669-7674.
[30] Cirone, P.; Bourgeois, J.M.; Austin, R.C. and Chang, P.L. (2002) A novel approach to tumour suppression with microencapsulated recombinant cells. Hum.Gene Ther. 13: 1157-1166.

[31] Lichtenbeld, H.C.; Ferarra, N.; Jain, R.K. and Munn, L.L. (1999) Effect of local anti-VEGF antibody treatment on tumour microvessel permeability. Microvasc. Res. 57: 357-362.
[32] Rokstad, A.M.; Holtan, S.; Strand, B.; Steinkjer, B.; Ryan, L.; Kulseng, B.; Skjak-Braek, G. and Espevik, T. (2002) Microencapsulation of cells producing therapeutic proteins: optimizing cell growth and secretion. Cell Transplant. 11: 313-324.
[33] Van Raamsdonk, J.M. and Chang, P.L. (2001) Osmotic pressure test: a simple, quantitative method to assess the mechanical stability of alginate microcapsules. J. Biomed. Mater. Res. 54: 264-271.
[34] Read, T.A.; Stensvaag, V.; Vindenes, H.; Ulvestad, E.; Bjerkvig, R. and Thorsen, F. (1999) Cells encapsulated in alginate: a potential system for delivery of recombinant proteins to malignant brain tumours. Int. J. Dev. Neurosci. 17: 653-663.
[35] Rokstad, A.M.; Strand, B.; Rian, K.; Steinkjer, B.; Kulseng, B.; Skjak-Braek, G. and Espevik, T. (2003) Evaluation of different types of alginate microcapsules as bioreactors for producing endostatin. Cell Transplant. 12: 351-364.
[36] Smidsrod, O. and Skjak-Braek, G. (1990) Alginate as immobilization matrix for cells. Trends Biotechnol. 8: 71-78.
[37] Thu, B.; Bruheim, P.; Espevik, T.; Smidsrod, O.; Soon-Shiong, P. and Skjak-Braek, G. (1996) Alginate polycation microcapsules. II. Some functional properties. Biomaterials 17: 1069-1079.
[38] Thu, B.; Gaserod, O.; Paus, D.; Mikkelsen, A.; Skjak-Braek, G.; Toffanin, R.; Vittur, F. and Rizzo, R. (2000) Inhomogeneous alginate gel spheres: an assessment of the polymer gradients by synchrotron radiation-induced X-ray emission, magnetic resonance microimaging, and mathematical modeling. Biopolymers 53: 60-71.
[39] Zekorn, T.; Horcher, A.; Sieber, U.; Schnettler, R.; Hering, B.; Zimmermann, U.; Bretzel, R.G. and Federlin, K. (1992) Barium-cross-linked alginate beads: A simple one-step method for successful immuno isolated transplantation of islets of langerhans. Acta Diabetol. 29: 99-106.
[40] Thu, B.; Bruheim, P.; Espevik, T.; Smidsrod, O.; Soon-Shiong, P. and Skjak-Braek, G. (1996) Alginate polycation microcapsules. I. Interaction between alginate and polycation. Biomaterials 17: 1031-1040.
[41] Kulseng, B.; Thu, B.; Espevik, T. and Skjak-Braek, G. (1997) Alginate polylysine microcapsules as immune barrier: permeability of cytokines and immunoglobulins over the capsule membrane. Cell Transplant. 6: 387-394.
[42] Goosen, M.F.A.; O'Shea, G.M.; Gharapetian, H.M. and Chou, S. (1985) Optimization of microencapsulation parameters: Semipermeable microcapsules as a bioartificial pancreas. Biotechnol. Bioeng. XXVII: 146-150.
[43] Lim, F. and Sun, A.M. (1980) Microencapsulated Islets as Bioartificial Endocrine Pancreas. Science 210: 908-910.
[44] Soon-Shiong, P.; Feldman, E.; Nelson, R.; Komtebedde, J.; Smidsrod, O.; Skjak-Braek, G.; Espevik, T.; Heintz, R. and Lee, M. (1992) Successful reversal of spontaneous diabetes in dogs by intraperitoneal microencapsulated islets. Transplantation 54: 769-774.
[45] Soon-Shiong, P.; Heintz, R.E.; Merideth, N.; Yao, Q.X.; Yao, Z.; Zheng, T.; Murphy, M.; Moloney, M.K.; Schmehl, M. and Harris, M. (1994) Insulin independence in a type 1 diabetic patient after encapsulated islet transplantation. Lancet 343: 950-951.
[46] Peirone, M.A.; Delaney, K.; Kwiecin, J.; Fletch, A. and Chang, P.L. (1998) Delivery of recombinant gene product to canines with nonautologous microencapsulated cells. Hum. Gene Ther. 9: 195-206.
[47] King, A.; Strand, B.; Rokstad, A.M.; Kulseng, B.; Andersson, A.; Skjak-Braek, G. and Sandler, S. (2003) Improvement of the biocompatibility of alginate/poly-L-lysine/alginate microcapsules by the use of epimerised alginate as a coating. J. Biomed. Mater. Res. 64A: 533-539.
[48] Peirone, M.; Ross, C.J.; Hortelano, G.; Brash, J.L. and Chang, P.L. (1998) Encapsulation of various recombinant mammalian cell types in different alginate microcapsules. J. Biomed. Mater. Res. 42: 587-596.
[49] Benson, J.P.; Papas, K.K.; Constantinidis, I. and Sambanis, A. (1997) Towards the development of a bioartificial pancreas: effects of poly-L-lysine on alginate beads with BTC3 cells. Cell Transplant. 6: 395-402.
[50] Constantinidis, I.; Rask, I.; Long, R.C., Jr. and Sambanis, A. (1999) Effects of alginate composition on the metabolic, secretory, and growth characteristics of entrapped beta TC3 mouse insulinoma cells. Biomaterials 20: 2019-2027.
[51] Stabler, C.; Wilks, K.; Sambanis, A. and Constantinidis, I. (2001) The effects of alginate composition on encapsulated betaTC3 cells. Biomaterials 22: 1301-1310.

[52] Strand, B.L.; Ryan, T.L.; In't Veld, P., Kulseng, B.; Rokstad, A.M.; Skjak-Brek, G. and Espevik, T. (2001) Poly-L-Lysine induces fibrosis on alginate microcapsules *via* the induction of cytokines. Cell Transplant. 10: 263-275.
[53] Vandenbossche, G.M.; Bracke, M.E.; Cuvelier, C.A.; Bortier, H.E.; Mareel, M.M. and Remon, J.P. (1993) Host reaction against empty alginate-polylysine microcapsules. Influence of preparation procedure. J. Pharm. Pharmacol. 45: 115-120.
[54] Sawhney, A.S. and Hubbel, J.A. (1992) Poly(ethylene oxide)-graft-poly(L-lysine) copolymers to enhance the biocompatibility of poly(L-lysine)-alginate microcapsule membranes. Biomaterials 13: 863-870.
[55] Gaserod, O.; Sannes, A. and Skjak-Braek, G. (1999) Microcapsules of alginate-chitosan. II. A study of capsule stability and permeability. Biomaterials 20: 773-783.
[56] Draget, K.I.; Strand, B.; Hartmann, M.; Valla, S.; Smidsrod, O. and Skjak-Braek, G. (2000) Ionic and acid gel formation of epimerised alginates; the effect of AlgE4. Int. J. Biol. Macromol. 27: 117-122.
[57] Strand, B.L.; Mørch, Y.; Syvertsen, K.; Espevik, T. and Skjak-Brek, G. (2003) Microcapsules made by enzymatically taillored alginate. J. Biomed. Mater. Res. 64A: 540-550.
[58] Mazur, X.; Eppenberger, H.M.; Bailey, J.E. and Fussenegger, M. (1999) A novel autoregulated proliferation-controlled production process using recombinant CHO cells. Biotechnol. Bioeng. 65: 144-150.
[59] Schlatter, S.; Bailey, J.E. and Fussenegger, M. (2001) Novel surface tagging technology for selection of complex proliferation-controlled mammalian cell phenotypes. Biotechnol. Bioeng. 75: 597-606.
[60] Meyyappan, M.; Wong, H.; Hull, C. and Riabowol, K.T. (1998) Increased expression of cyclin D2 during multiple states of growth arrest in primary and established cells. Mol. Cell Biol. 18: 3163-3172.
[61] Ritt, M.G.; Mayor, J.; Wojcieszyn, J.; Smith, R.; Barton, C.L. and Modiano, J.F. (2000) Sustained nuclear localization of p21/WAF-1 upon growth arrest induced by contact inhibition. Cancer Lett. 158: 73-84.
[62] Pani, G.; Colavitti, R.; Bedogni, B.; Anzevino, R.; Borrello, S. and Galeotti, T. (2000) A redox signaling mechanism for density-dependent inhibition of cell growth. J. Biol. Chem. 275: 38891-38899.
[63] Andres, V. and Walsh, K. (1996) Myogenin expression, cell cycle withdrawal, and phenotypic differentiation are temporally separable events that precede cell fusion upon myogenesis. J. Cell Biol. 132: 657-666.
[64] Blau, H.M.; Dhawan, J. and Pavlath, G.K. (1993) Myoblasts in pattern formation and gene therapy. Trends Genet. 9: 269-274.
[65] Al Hendy, A.; Hortelano, G.; Tannenbaum, G.S. and Chang, P.L. (1995) Correction of the growth defect in dwarf mice with nonautologous microencapsulated myoblasts--an alternate approach to somatic gene therapy. Hum. Gene Ther. 6: 165-175.
[66] Hortelano, G.; Al Hendy, A.; Ofosu, F.A. and Chang, P.L. (1996) Delivery of human factor IX in mice by encapsulated recombinant myoblasts: a novel approach towards allogeneic gene therapy of hemophilia B. Blood 87: 5095-5103.
[67] Chang, P.L.; Van Raamsdonk, J.M.; Hortelano, G.; Barsoum, S.C.; MacDonald, N.C. and Stockley, T.L. (1999) The *in vivo* delivery of heterologous proteins by microencapsulated recombinant cells. Trends Biotechnol. 17: 78-83.
[68] Ross, C.J.; Ralph, M. and Chang, P.L. (1999) Delivery of recombinant gene products to the central nervous system with nonautologous cells in alginate microcapsules. Hum. Gene Ther. 10: 49-59.
[69] Ross, C.J.; Ralph, M. and Chang, P. L. (2000) Somatic gene therapy for a neurodegenerative disease using microencapsulated recombinant cells. Exp. Neurol. 166: 276-286.
[70] Van Raamsdonk, J.M.; Ross, C.J.; Potter, M.A.; Kurachi, S.; Kurachi, K.; Stafford, D.W. and Chang, P.L. (2002) Treatment of hemophilia B in mice with nonautologous somatic gene therapeutics. J. Lab. Clin. Med. 139: 35-42.
[71] Regulier, E.; Schneider, B.L.; Deglon, N.; Beuzard, Y. and Aebischer, P. (1998) Continuous delivery of human and mouse erythropoietin in mice by genetically engineered polymer encapsulated myoblasts. Gene Ther. 5: 1014-1022.
[72] Dalle, B.; Payen, E.; Regulier, E.; Deglon, N.; Rouyer-Fessard, P.; Beuzard, Y. and Aebischer, P. (1999) Improvement of mouse beta-thalassemia upon erythropoietin delivery by encapsulated myoblasts. Gene Ther. 6: 157-161.
[73] Rowley, J.A. and Mooney, D.J. (2002) Alginate type and RGD density control myoblast phenotype. J. Biomed. Mater. Res. 60: 217-223.
[74] Cheng, H.; Almstrom, S. and Olson, L. (1995) Fibrin glue used as an adhesive agent in CNS tissues. J. Neural Transplant. Plast. 5: 233-243.

[75] Lawton, M.T.; Ho, J.C.; Bichard, W.D.; Coons, S.W.; Zabramski, J.M. and Spetzler, R.F. (1996) Titanium aneurysm clips: Part I - Mechanical, radiological, and biocompatibility testing. Neurosurgery 38: 1158-1163.
[76] Campioni, E.G.; Nobrega, J.N. and Sefton, M.V. (1998) HEMA/MMMA microcapsule implants in hemiparkinsonian rat brain: biocompatibility assessment using [3H]PK11195 as a marker for gliosis. Biomaterials 19: 829-837.
[77] Mofid, M.M.; Thompson, R.C.; Pardo, C.A.; Manson, P.N. and Vander Kolk, C. (1997) A.Biocompatibility of fixation materials in the brain. Plast. Reconstr. Surg. 100: 14-20.
[78] Graeber, M.B.; Blakemore, W.F. and Kreutzberg, G.W. (2002) Cellular pathology of the central nervous system. In: Graham, D.I. and Lantos, P.I. (Eds.) Greenfields neuropathology. Arnold, London; pp 123-191.
[79] Tilgner, J.; Volk, B. and Kaltschmidt, C. (2001) Continuous interleukin-6 application *in vivo* via macroencapsulation of interleukin-6-expressing COS-7 cells induces massive gliosis. Glia 35: 234-245.
[80] Mason, D.W.; Charlton, H.M.; Jones, A.J.; Lavy, C.B., Puklavec, M. and Simmonds, S. J. (1986) The fate of allogeneic and xenogeneic neuronal tissue transplanted into the third ventricle of rodents. Neuroscience 19: 685-694.
[81] Hagell, P.; Piccini, P.; Bjorklund, A.; Brundin, P.; Rehncrona, S.; Widner, H.; Crabb, L.; Pavese, N.; Oertel, W.H.; Quinn, N.; Brooks, D.J. and Lindvall, O. (2002) Dyskinesias following neural transplantation in Parkinson's disease. Nat. Neurosci. 5: 627-628.
[82] Piccini, P.; Lindvall, O.; Bjorklund, A.; Brundin, P.; Hagell, P.; Ceravolo, R.; Oertel, W.; Quinn, N.; Samuel, M.; Rehncrona, S.; Widner, H. and Brooks, D.J. (2000) Delayed recovery of movement-related cortical function in Parkinson's disease after striatal dopaminergic grafts. Ann. Neurol. 48: 689-695.
[83] Rihova, B. (2000) Immunocompatibility and biocompatibility of cell delivery systems. Adv. Drug Deliv. Rev. 42: 65-80.
[84] Skaugrud, O.; Hagen, A.; Borgersen, B. and Dornish, M. (1999) Biomedical and pharmaceutical applications of alginate and chitosan. Biotechnol. Genet. Eng. Rev. 16: 23-40.
[85] Dornish, M.; Kaplan, D. and Skaugrud, O. (2001) Standards and guidelines for biopolymers in tissue-engineered medical products: ASTM alginate and chitosan standard guides. American Society for Testing and Materials. Ann. N.Y. Acad. Sci. 944: 388-397.
[86] Otterlei, M.; Ostgaard, K.; Skjak-Braek, G.; Smidsrod, O.; Soon-Shiong, P. and Espevik, T. (1991) Induction of cytokine production from human monocytes stimulated with alginate. J. Immunother. 10: 286-291.
[87] Otterlei, M.; Sundan, A.; Skjak-Braek, G.; Ryan, L.; Smidsrod, O. and Espevik, T. (1993) Similar mechanisms of action of defined polysaccharides and lipopolysaccharides: characterization of binding and tumour necrosis factor alpha induction. Infect. Immun. 61: 1917-1925.
[88] Espevik, T.; Otterlei, M.; Skjak-Braek, G.; Ryan, L.; Wright, S.D. and Sundan, A. (1993) The involvement of CD14 in stimulation of cytokine production by uronic acid polymers. Eur. J. Immunol. 23: 255-261.
[89] Flo, T.H.; Ryan, L.; Latz, E.; Takeuchi, O.; Monks, B.G.; Lien, E.; Halaas, O.; Akira, S.; Skjak-Braek, G.; Golenbock, D.T. and Espevik, T. (2002) Involvement of toll-like receptor (TLR)2 and TLR4 in cell activation by mannuronic acid polymers. J. Biol. Chem. 277: 35489-35495.
[90] Berntzen, G.; Flo, T.H.; Medvedev, A.; Kilaas, L.; Skjak-Braek, G.; Sundan, A. and Espevik, T. (1998) The tumour necrosis factor-inducing potency of lipopolysaccharide and uronic acid polymers is increased when they are covalently linked to particles. Clin. Diagn. Lab. Immunol. 5: 355-361.
[91] Kulseng, B.; Skjak-Braek, G.; Ryan, L.; Andersson, A.; King, A.; Faxvaag, A. and Espevik, T. (1999) Transplantation of alginate microcapsules: generation of antibodies against alginates and encapsulated porcine islet-like cell clusters. Transplantation 67: 978-984.
[92] De Vos, P.; De Haan, B. and van Schilfgaarde, R. (1997) Effect of the alginate composition on the biocompatibility of alginate-polylysine microcapsules. Biomaterials 18: 273-278.
[93] Duvivier-Kali, V.F.; Omer, A.; Parent, R.J.; O'Neil, J.J. and Weir, G.C. (2001) Complete protection of islets against allorejection and autoimmunity by a simple barium-alginate membrane. Diabetes 50: 1698-1705.
[94] Lanza, R.P.; Kuhtreiber, W.M.; Ecker, D.; Staruk, J.E. and Chick, W.L. (1995) Xenotransplantation of porcine and bovine islets without immunosuppression using uncoated alginate microspheres. Transplantation 59: 1377-1384.
[95] Lanza, R.P.; Kuhtreiber, W.M.; Ecker, D.M.; Marsh, J.P.; Staruk, J.E. and Chick, W.L. (1996) A simple method for xenotransplanting cells and tissues into rats using uncoated alginate microreactors. Transplant. Proc. 28: 835.

[96] Lanza, R.P.; Ecker, D.M.; Kuhtreiber, W.M.; Marsh, J.P.; Ringeling, J. and Chick, W.L. (1999) Transplantation of islets using microencapsulation: studies in diabetic rodents and dogs. J. Mol. Med. 77: 206-210.
[97] Mikos, A.G.; McIntire, L.V.; Anderson, J.M. and Babensee, J.E. (1998) Host response to tissue engineered devices. Adv. Drug Deliv. Rev. 33: 111-139.
[98] Lanza, R.P.; Chung, H.Y.; Yoo, J.J.; Wettstein, P.J.; Blackwell, C.; Borson, N.; Hofmeister, E.; Schuch, G.; Soker, S.; Moraes, C.T.; West, M.D. and Atala, A. (2002) Generation of histocompatible tissues using nuclear transplantation. Nat. Biotechnol. 20: 689-696.
[99] Schneider, B.L.; Peduto, G. and Aebischer, P. (2001) A self-immunomodulating myoblast cell line for erythropoietin delivery. Gene. Ther. 8: 58-66.
[100] Horbett, T.A. and Schway, M.B. (1988) Correlations between mouse 3T3 cell spreading and serum fibronectin adsorption on glass and hydroxyethylmethacrylate-ethylmethacrylate copolymers. J. Biomed. Mater. Res. 22: 763-793.
[101] Lanza, R.P.; Butler, D.H.; Borland, K.M.; Staruk, J.E.; Faustman, D.L.; Solomon, B.A.; Muller, T.E.; Rupp, R.G.; Maki, T. and Monaco, A.P. (1991) .Xenotransplantation of canine, bovine, and porcine islets in diabetic rats without immunosuppression. Proc. Natl. Acad. Sci. 88: 11100-11104.

GENE THERAPY USING ENCAPSULATED CELLS

GONZALO HORTELANO
*Department of Pathology & Molecular Medicine, McMaster University, Hamilton, Ontario, L8N 3Z5 Canada – Fax 905 521-2613 –
Email: gonhort@mcmaster.ca*

1. Introduction

Since its development in the late 1970's, the powerful techniques of genetic engineering has had a major impact in numerous scientific disciplines, and are being exploited commercially in agriculture, food, and biotechnology industries [1,2]. Not surprisingly, genetic engineering techniques are gradually being implemented in medicine as well. Recombinant therapeutic products such as insulin [3] or blood coagulation factors (both VIII and IX) [4] have been commercially available for some time and are now rapidly substituting plasma-derived products in developed countries. Indeed, genetic engineering is making an increasing and noticeable impact in biomedicine.

The advent of effective genetic engineering techniques also made gene therapy possible. Gene therapy promises to revolutionise medicine in the XXI century. The concept of gene therapy is simple and has been around for a long time as the ultimate cure for disorders caused by known genetic mutations. Until now, medicine has attempted to treat the effects, but not the cause of genetic diseases. For example, the insufficient presence of functional factor VIII (FVIII) causes prolonged bleeding (haemophilia A). Traditionally, doctors have concentrated on treating the damage to the joints that successive bleeding episodes cause in haemophiliacs. Although physicians can now use FVIII preparations to stop bleeding episodes or as prophylaxis, the cause of the bleeding condition is still not addressed. Gene therapy proposes to introduce a functional FVIII gene in the patient, so the patient can produce the required amount of factor VIII to reach an effective long-term cure of the disease.

The first clinical gene therapy trial was conducted in 1990. It was designed to correct adenosine deaminase (ADA) deficiency in two young girls [5]. To date, more than 600 clinical trials have been approved in 5 continents all over the world. In spite of this effort, gene therapy has still to live up to its promises. The initial hype around this technology was followed by a period of disillusion and pessimism upon the death in 1999 of Jesse Gelsinger, a volunteer in a clinical gene therapy trial for ornithine transcarbamylase (OTC) deficiency [6]. Clearly, the difficulties associated with achieving effective long-term delivery of a functional gene were not fully recognised. Nevertheless, ten years after the first clinical trial, encouraging recent results [7] suggest that gene therapy potential is closer to than ever reality.

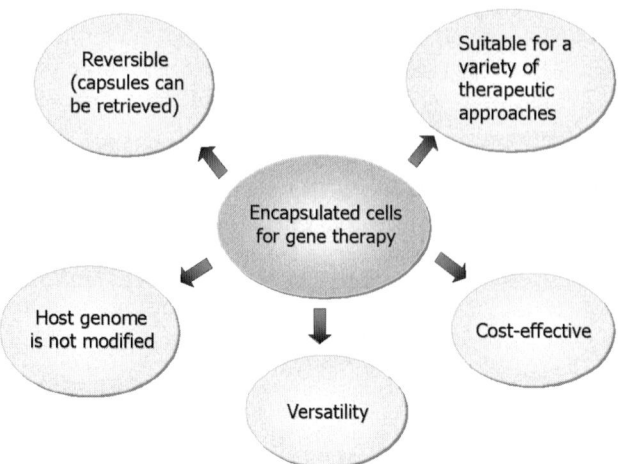

Figure 1. Advantages of encapsulated cells in gene therapy.

2. Gene therapy

There are two main approaches to gene therapy, depending on how the genetic modification of the cells is performed [8]. *In vivo* gene therapy strategies rely on the direct introduction of a functional gene into the patient. Viruses can infect mammalian cells, including human, very efficiently [9]. Therefore, viral vectors have become very popular for delivering a therapeutic gene in a host. However, virus-infected cells are efficiently targeted and cleared by the immune system. As a result, immune responses to the vector and/or the produced therapeutic product (transgene) can be expected, and re-administration of viral vectors is typically not possible. Lastly, a variety of safety concerns have been voiced against the direct injection of viral vectors for gene therapy [10,11]. Unfortunately, non-viral vectors have not yielded to date high levels of therapeutic product on a sustained basis [12].

In *ex vivo* gene therapy strategies, genetic modification of the cells occurs *in vitro*. In a typical *ex vivo* protocol, cells are obtained from the patient and genetically modified *in vitro*. The recombinant cells are then expanded using tissue culture techniques. Once the required number of cells is obtained, cells are transplanted back into the patient. The cells engraft in the host and produce the therapeutic product that is required by the patient. This gene therapy approach has been used in many clinical trials. However useful, a major disadvantage of *ex vivo* gene therapy is its dependency on the genetic modifications of cells for each and every patient, a costly and labour intensive proposition.

The implantation of encapsulated recombinant cells is a particular variation of *ex vivo* gene therapy. It is attractive because it eliminates the main disadvantage of *ex vivo* gene therapy. The encapsulation of non-autologous cells eliminates the need to perform genetic modification of cells for every patient. Instead, multiple patients can be treated with the same batch of capsules, therefore making this strategy cost-effective. As an additional safety feature, the microcapsules can be retrieved after implantation should it become necessary. Thus, this is arguably the only reversible gene therapy approach to date. It is worth noting that the genome of the host is never modified in this unique gene therapy approach (Figure 1).

Table 1. Examples of different encapsulated cells applied to gene therapy.

Cell Type	Application	Reference
Bacteria	Elimination of metabolites	[75]
Chromaffin	Neurological	[83]
Fibroblasts	Metabolic deficiencies	[37]
	Neurological	[53]
Hybridoma	Cancer	[77]
Kidney	Metabolic deficiencies,	[78]
	Cancer	[79]
	Neurological	[56]
Mesenchimal	Bone regeneration	[80]
Myeloma	Hepatic growth factor	[81]
Myoblasts	Metabolic deficiencies	[32]
	Cancer	[47]
	Neurological	[54]
Ovary	Metabolic deficiency	[82]
Virus producers	Cancer	[74]

3. Genetic engineering of cells

Strategies to use encapsulated cells for gene therapy applications entail some sort of genetic modification of the cells for the overproduction of a desired therapeutic product (Figure 2). Both viral and non-viral vectors have been successfully used in the genetic engineering of cells for encapsulation (Table 1). Because of the unique properties of cell encapsulation, it is possible to use expression vectors that would otherwise be ineffective if used in the context of an *in vivo* gene therapy strategy. For instance, viral vectors are typically antigenic in nature, and strong immune responses can occur due to the antigenic presentation of viral proteins encoded in vectors and produced in transduced cells. However, microcapsules can protect enclosed cells that have been transduced with viral vectors [13-14].

Additionally, the genetic modification of cells for encapsulation must be performed with expression vectors that incorporate appropriate genetic regulatory elements. The expression vector used to produce the therapeutic gene must be selected in accordance with the cell type to be employed. It is preferred that specific genetic regulatory elements, such as tissue-specific enhancers, are used so as to optimise therapeutic secretion. A number of tissue-specific enhancers have been described, such as muscle creatine kinase (MCK) or desmin for muscle cells [15]. In addition, the choice of promoter will dictate in a significant way the amount and duration of therapeutic delivery that can be expected. Thus, the use of strong viral promoters, such as CMV, is often associated with a vigorous, yet transient transgene delivery [16]. In contrast, the use of mammalian constitutive promoter, such as β-actin typically leads to sustained transgene delivery [15]. Again, the nature of the particular application will recommend a given strategy over the other.

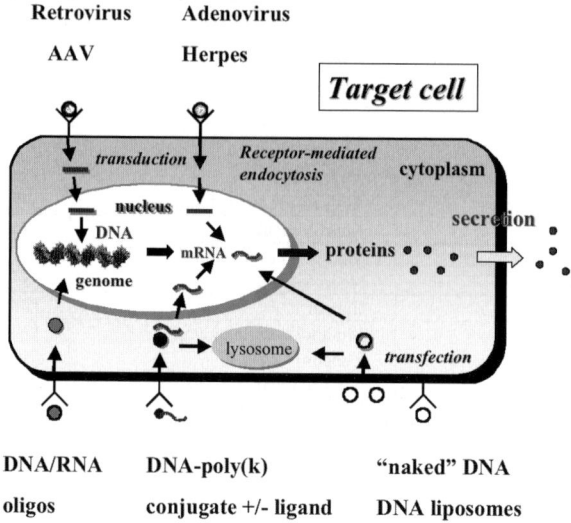

Figure 2. Strategies for genetically modifying cells.

3.1. VIRAL VECTORS

Viral vectors typically yield a higher transgene expression than non-viral vectors. A large variety of viral vectors have being considered for use in gene therapy strategies. However, the large majority of gene therapy protocols contemplate just one of the following viral vectors.

3.2. RETROVIRUS

Retroviral vectors were the first viral vectors to be used in gene therapy [17], and are currently still being used in many human trials. These vectors are mostly derived from the Moloney Murine Leukemia Virus (MoMLV), and have a loading capacity of ~7 kb of DNA. This capacity is enough to include the coding DNA sequence of most common therapeutic genes, plus a promoter and other necessary regulatory sequences. Retrovirus vectors contain two long terminal repeats (LTR) at both ends of the genome that facilitate its integration into the host genome [17]. Viral integration ensures long-term persistence of the transgene DNA in the targeted cell. LTR sequences also contain strong promoter elements that can be used to drive the expression of a therapeutic transgene in a host cell. Nonetheless, there is extensive evidence showing that transgene expression driven from the LTR promoter is often inactivated *in vivo* over time, resulting in transient therapeutic delivery in the host [18].

Since retroviral vectors require active cell division in order to enter the nucleus, only actively dividing cells can be effectively transduced with retroviruses. This property reduces the range of cell types and tissues that can be targeted with retroviral vectors, and is a significant limitation of this vector. However, although this limitation may affect the direct administration of retroviral vectors, it is not a major concern for most cell encapsulation applications. Nevertheless, the integration of retrovirus has the potential to interrupt genes, activate proto-oncogenes, and trigger unwanted genomic rearrangements in the host. Indeed, retroviral vectors have recently caused proliferative lymphoma in young patients enrolled in a gene therapy trial [19]. As opposed to *in vivo* gene therapy approaches, the genetic modification of encapsulated cells can be thoroughly characterised prior to their implantation, thus minimising this safety concern.

3.3. LENTIVIRUS

Lentivirus is part of the retroviridae family of viruses, and as such share many of the characteristics of retrovirus just outlined. Lentiviral vectors share with retrovirus a similar loading capacity and the ability to integrate into the host genome, as well as the same safety concerns described for retroviral vectors. The main difference between both viral vectors is that, unlike retroviral vectors, lentivirus does not require active cell division in order to achieve cell transduction. Thus, lentivirus can be used to introduce therapeutic genes in quiescent cells, such as stem cells. The best known lentivirus is arguably the Human Immunodeficiency Virus (HIV). Human lentiviral vectors are derived from the HIV genome [20]. In the process, the viral genes that code for proteins involved in HIV replication and pathogenesis have been removed from its genome. Recent successful attempts to generate a suitable packaging cell line for lentivirus made this vector a very attractive vehicle for gene therapy. Nonetheless, to date gene expression from lentiviral vectors has typically been lower than that from retrovirus.

3.4. ADENOVIRUS

Adenovirus is a rather benign virus that is associated in humans with mild flu-like symptoms. A large proportion of the human population has already been exposed to it, and thus it is considered to be quite safe. Adenovirus vectors can transduce a wide range of cell types, including quiescent cells. This characteristic opens the scope of applications for adenoviral vectors. In addition, adenoviral vectors can carry a very large load of foreign DNA of up to 35kb. Furthermore, transduction by adenovirus is very efficient, and transgene expression from this vector is very strong. Not surprisingly, the above properties have made adenoviral vectors extremely popular in gene therapy studies [21]. It is important to point out that adenoviral vectors are particularly immunogenic. Unlike retrovirus, adenoviral vectors do not integrate into the host genome, but rather stay as episomal autonomous DNA. As a result of its characteristics, adenovirus DNA is gradually lost with host cell division, and transgene expression does not typically persist long-term. Therefore, adenoviral vectors are still being use primarily in settings where strong, short-term transgene expression is required, or where the object is to elicit a strong immune response is even desirable, such as in the development of vaccines [22]. A notable exception is the so-called gutted [23] and gut-less [24] adenovirus, in which all viral genes have been removed. Essential genes for producing viral particles are instead supplied in trans by the packaging cell line. This new generation of adenovirus vectors is much less immunogenic. Indeed, there are reports of sustained expression of transgenes in immunocompetent mice using these vectors [25], opening the possibility to use them to treat chronic diseases such as haemophilia.

3.5. ADENO-ASSOCIATED VIRUS

Adeno-associated virus (AAV) is possibly the most promising viral vector available to date for *in vivo* gene therapy. AAV is a small parvovirus that requires the presence of a helper virus, such as adenovirus for efficient infection of cells. Similar to retrovirus, AAV genome is flanked by terminal repeats (ITR), allowing AAV to integrate in the host genome. Wild type AAV tends to integrate preferentially in chromosome 19, although this specificity appears to be somewhat lost upon vector modification. Perhaps the most interesting property of this vector for gene therapy is that, unlike adenovirus, AAV does not elicit a strong immune response from the host. Therefore, long-term transgene expression has been achieved using AAV vectors [26]. AAV genome is very small, just under 5 kb of DNA. This limited loading capacity restricts the use of AAV vectors to the expression of small to medium transgenes, and the inclusion of minimal regulatory elements. AAV is currently being tested in clinical trials for genetic diseases with promising results [27].

3.6. HERPES

Herpes virus has natural tropism for the central nervous system (CNS), and thus is particularly useful for the secretion of transgenes in CNS-derived cells. The best known member of this group of viruses is Herpes Simplex Virus [28]. Herpes has a very large genome, allowing the largest payload capacity of any of the vectors commonly used in gene therapy. Just like adenovirus, herpes does not integrate into the host genome, but rather stays as episomal DNA. Also like adenovirus, herpes vectors can transduce non-dividing quiescent cells. Nonetheless, the range of cell types that can be transduced with herpes is not as wide.

3.7. NON-VIRAL VECTORS

Non-viral vectors have become a serious alternative to currently used viral vectors [12]. They are generally less immunogenic than viral vectors, since they lack an antigenic capsid, and are less costly and cumbersome to manufacture. Non-viral vectors are generally believed to be safer than viral vectors. Nevertheless, current non-viral vectors are typically less effective in entering cells than viral particles, and as a result transgene expression is often lower. However, this limitation is particularly acute when *in vivo* gene therapy strategies are used. *Ex vivo* gene therapy approaches, including encapsulated cells, are not affected by this limitation nearly as much. The reduced transfection rate of cells with non-viral vectors can be overcome by screening transfected cells for high-secreting clones. Alternatively, it is also feasible to select for the entire pool of transfected cells. However, while established cell lines are relatively easily to transfect, a main concern is that primary cells are not as easily transfected by non-viral vectors. Alternative methods of genetic modification of cells, such as electroporation [29] or gene-gun [30] can be used to genetically modify cells suitable for encapsulation. Non-viral vectors are extremely attractive because of the ease of preparation and manipulation, the lack of size restrictions and their flexibility.

4. Selection of cells for encapsulation

A judicious choice of encapsulated cells is critical for the outcome of any gene therapy protocol that uses immobilised cells. Therefore, a serious effort should be directed toward this selection. Although not all cell types are suitable for encapsulation, a large number of different cell types have been successfully encapsulated in various polymers. Table 1 provides a non-exhaustive list of the applications of encapsulated recombinant cells to treat a variety of medical conditions.

The choice of cells will depend, to a large degree, on the intended application. Not every tissue, for instance, is capable of producing biologically active factor IX (FIX), the protein deficient in haemophilia B patients that is normally made by the liver. Not every cell is capable of performing the extensive post-translational modifications required to produce biologically active FIX. The quality of the therapeutic product, rather than its quantity is therefore a much more important consideration. At times, genetic modification of cells to overproduce a protein has met with negative results. In a previously published study that used fibroblasts secreting hFIX enclosed in implantable chambers, the use of particular expression vectors increased hFIX secretion several-fold. Unfortunately, a dramatic drop in the percentage of the secreted protein that was biologically functional accompanied this increase in secretion [31]. Consequently, there was no overall gain in the amount of functional hFIX that was available for therapeutic purpose. Therefore, the importance of thoroughly understanding the therapeutic product to be secreted cannot be overemphasised.

4.1. ESTABLISHED CELL LINES/PRIMARY CELLS

Most pre-clinical gene therapy studies of encapsulated cells have used established cell lines (Table 1). These are immortal cells of clonal origin, such as C2C12 mouse myoblasts, that can proliferate indefinitely without experiencing senescence while maintaining at least some of the original phenotype. Immortalised cells can be thoroughly characterised. These properties of established cell lines facilitate the consistency and reproducibility of the results, making such cell lines extremely useful in research. Most of the established cell lines used in pre-clinical gene therapy studies, such as C2C12 myoblasts are derived from primary tissue. Nonetheless, the important benefits of established cell lines have to be balanced against the safety concerns that they may pose. A well-known cell line such as C2C12 myoblasts has been widely used in gene therapy applications. However, while it has been reported to be safe in an immunocompetent host [32], it can also cause tumours in immunodeficient mice [33]. Thus, it is of paramount importance that the potential safety risks of the encapsulated cells be carefully evaluated, taking into consideration the intended recipient.

Primary cells can be considered safer than established cell lines, and are thus more appealing for use in clinical studies. Nonetheless, primary cells pose additional challenges of their own. The greatest challenge of primary cells is perhaps their availability. Since primary cells start to senesce after 30-40 cell divisions *in vitro*, their expansion potential is limited. Whereas established cell lines can be used as "universal" cell lines for an unlimited number of patients sharing a given disorder, each batch of primary cells can only be used to treat a defined number of patients, after which a new batch of cells must be obtained. Therefore, the use of primary cells requires access to a continuous source of cells. Other necessary steps for every batch preparation include cellular expansion, genetic modification, and a thorough characterisation of the recombinant cells. Genetic engineering of cells results in very diverse genetic modifications. Strong selection pressure to isolate high secreting clones may affect the safety of the resulting cells. The use of a pool of recombinant cells would reduce this risk, although the therapeutic output is reduced. Additionally, a pool of genetically diverse cells is not amenable to the thorough characterisation that is possible when all cells are genetically identical.

4.2. PROLIFERATIVE/QUIESCENT CELLS

Highly proliferative cell lines, such as fibroblasts, have been successfully used for the delivery of transgenes such as FIX [34], adenosine deaminase [35], human growth hormone [36], and β-glucuronidase [37] among others. Proliferative cells often continue to divide and multiply after encapsulation. The higher concentration of encapsulated cells is also reflected in an increase in the therapeutic output of the microcapsules. Proliferative cells are particularly appropriate for applications where a strong output of therapeutic product is required. However, it is important to consider that, over time, the proliferative cells can fill the entire capsular volume. When the concentration of encapsulated cells reaches certain level, deficiencies in nutrient diffusion can negatively affect cell viability [38]. This limitation is particularly relevant for the treatment of chronic diseases, like haemophilia, that require continuous long-term delivery of therapeutic products.

There are a number of mammalian cells that proliferate well under regular tissue culture conditions, but reduce their proliferation after encapsulation. Cells such as myoblasts have the capacity to differentiate into non-proliferative myotubes under certain circumstances. For instance, excessive concentration of cells in tissue culture and a low concentration of horse serum in the culture medium are known factors that trigger differentiation in myoblasts [39]. Encapsulation has a marked effect on myoblasts, restricting their proliferation after encapsulation [32]. Interestingly, cell viability remains high for long periods of time, for over 200 days [32]. This unique property of myoblasts is especially suitable for the treatment of chronic diseases. The therapeutic output from recombinant myoblasts can be sustained, particularly when a housekeeping constitutive promoter, is used [40]. Not surprisingly, encapsulated myoblasts have been used for the treatment of chronic diseases such as anaemia [41], haemophilia [32], growth retardation [36], and angiogenesis [42] among others.

4.3. ALLOGENEIC/XENOGENEIC CELLS

The main advantage of encapsulated cells is the immune protection offered to the enclosed cells. This property widens the range of cell types that can be considered for clinical application, to include allogeneic and xenogeneic cells. In addition to this advantage, it is cost-effective to use a well-characterised "universal" cell type to treat a large number of patients with a given condition. A number of polymers such as alginate and cellulose acetate offer significant immune protection to enclosed cells, and are commonly used in gene therapy applications (Table 1). Although current polymers do not offer as complete protection to xenogeneic cells as seen with allogeneic cells [43], a variety of pre-clinical and clinical studies have successfully used encapsulated xenogeneic cells. For instance, baby hamster kidney (BHK) cells have been used to deliver human ciliary neurotrophic factor (hCNTF) in humans with the neurological condition amyotrophic lateral sclerosis (ALS) [44]. Xenogeneic bovine chromaffin cells secreting natural endorphins have been used to mitigate chronic pain of patients with cancer [45]. Special consideration ought to be given to stem cells. The pluripotent potential of these cells can be exploited *in vitro* to obtain large quantities of cells with a given desired characteristic, and suitable for encapsulation [46]. Ultimately, the choice of cells to be encapsulated will be conditioned by the intended medical application.

5. Applications of encapsulated cells in gene therapy

A variety of implantable devices containing recombinant cells have been applied to the treatment of both genetic and acquired disease. Below there are examples of some of the most common applications given to date to encapsulated cells.

5.1. CANCER

A number of different strategies to fight cancer are being considered in pre-clinical and clinical studies. For instance, delivery of antibodies specific for cancer markers has therapeutic potential for targeting and eliminating tumour cells. Murine myoblasts genetically modified to secrete a humanised antibody with affinity to the HER-2/neu cancer antigen were encapsulated in alginate microcapsules. A murine model of cancer implanted with these capsules prolonged the life of the treated mice [47]. Another promising approach to treat cancer is to inhibit the growth of blood vessels (angiogenesis), a crucial step necessary for sustaining tumour growth. Recombinant cells secreting the antiangiogenic factor endostatin were enclosed in alginate microcapsules. A mouse model of brain cancer that received such microcapsules intracerebrally survived 84% longer than the control mice [48].

Microcapsules can also be used to deliver toxic metabolites that kill cancer cells. Embryonic kidney cells secreting CYP2B1 were encapsulated in cellulose sulphate microcapsules and implanted in mice with pancreatic cancer in direct contact with the tumour. Low-dose ifosfamide given to tumour bearing mice was converted to the toxic compound phosphoramide mustard by CYP2B1, and in so doing killed the adjacent cancer cells [49]. This strategy has been the basis of a human trial with promising results [50-51]. Finally, immunotherapy is considered an important strategy to fight cancer, based on priming the immune system to break the tolerance and target tumour cells. Geller *et al.*, [52] showed that tumour cells enclosed in immunoisolation devices (TheraCyte™, Baxter Corporation, IL) prevented tumour development in all treated mice in a cancer animal model. Further, 4 out of 5 treated mice survived a second challenge of tumour cells, while all control mice developed tumours. The authors' hypothesis is that soluble antigens diffused from the devices to prime the immune system against the enclosed tumour cells. Thus, the concept of cell encapsulation has shown to be versatile enough to target highly complex diseases such as cancer, using multiple different approaches.

5.2. NEUROLOGICAL CONDITIONS

In addition to cancer, encapsulated cells have also been used in many other biomedical settings. Most notably perhaps is the delivery of various neurotrophic factors to the CNS as a treatment for neurological disorders. Encapsulated fibroblasts secreting adenosine, an inhibitor of neuronal activity in the brain, offered a nearly complete protection against seizures in a rat model of partial epilepsy [53]. Similarly, recombinant C2C12 myoblasts secreting ciliary neurotrophic growth factor (CNTF) partially rescued motor neurons from axotomy-induced cell death, confirming the potential use of myoblasts as a source of neurotrophic factors for the treatment of neurodegenerative diseases [54]. Encapsulated cells secreting Glial cell line-derived neurotrophic factor (GDNF) have shown to be efficacious in protecting the nigral dopaminergic neurons against lesion-induced cell death in rodent as well as in primate models of Parkinson's disease [55].

The potential of cell encapsulation in gene therapy is reflected in the initiation of a phase I/II clinical trial of amyotrophic lateral sclerosis (or Lou Gehrig's disease). Patients received intrathecal implants of encapsulated recombinant baby hamster kidney (BHK) cells releasing human CNTF [56]. The xenogeneic cells survived for at least 20 weeks, and the secreted CNTF did not elicit any adverse side effect [56]. Other similar applications of implantable devices containing recombinant cells secreting therapeutic products for the CNS have shown promise in an animal model of Hungtington's disease [57]. Implantable polymeric devices containing bovine chromaffin cells that naturally secrete a mixture of analgesic compounds has been used in various pre-clinical studies of chronic pain [58]. These studies were continued in a human clinical trial of patients with chronic pain that failed to respond to standard treatment [45]. Therefore, encapsulated cells can be used to treat a variety of neurological conditions.

5.3. ERYTHROPOIETIN

Erythropoietin (epo) is responsible for the production of red blood cells. As such, it is widely used to treat anaemia. The transgenic mouse strain 134.3LC (Epo-TAg(H)) displays a severe chronic anaemia resembling that observed in human patients. Transgenic mice received subcutaneous implantation of encapsulated C2C12 myoblasts secreting murine epo. The hematocrit in the treated mice rose, indicating the delivery of functional epo. The clinical benefit of the treatment was sustained when the immunosuppressor anti-CD4+ monoclonal antibody was used to prevent an immune response in the host [41]. However, an excessive level of epo can also have undesirable side effects. Thus, Sommer *et al.* (2002) took these findings to a new exciting level by engineering the expression of the *EPO* gene under the control of a regulatable promoter [59]. In these animals, the hematocrit was regulated by the exogenous supply of the drug doxycycline. Patients suffering from β-Thalassemia also show signs of anaemia. The implantation of encapsulated cells secreting epo resulted in the clinical benefit of a murine model of β-Thalassemia [60].

5.4. ENCAPSULATED CELLS TO TREAT METABOLIC DISEASES

The implantation of encapsulated cells is a particularly suitable strategy to treat metabolic deficiencies. Promising results have been obtained in pre-clinical studies of growth hormone deficiency in dwarf mice [36], or β-glucuronidase deficiency in a mucopolysaccharidosis type VII mouse model [37], among others. The remainder of this chapter will be devoted to discuss the use of encapsulated cells to gene therapy of haemophilia.

Haemophilia is an X-linked recessive disorder caused by the deficiency of blood clotting factor VIII (FVIII, causing haemophilia A) and factor IX (FIX, causing haemophilia B), that combined affect about 1-10,000 live male births [61]. Severe haemophiliacs suffer life-long, spontaneous bleeding episodes. Common clinical presentations include haematomas, bleeding into the joints and intracranial haemorrhages, the latter being a common cause of death. Long-term complications include chronic haemophilic joint arthropathy, and progressive degeneration of the joints leading to severe crippling deformity. Haemophilia is a debilitating disease imposing a heavy burden on both patients and families. Many different mutations in the FVIII and FIX genes are known to cause haemophilia, these include sizeable deletions as well as point mutations causing amino acid substitution, premature termination, or mRNA splicing errors. In total, over 400 different mutations have been so far identified [62]. There appears to be no strict correlation between the type of mutation and the variability of clinical presentation [61].

Plasma coagulant activity (relative to pooled normal plasmas) varies from 0-1% in severely affected individuals, to 1-5% in moderate haemophiliacs, and 6-25% in mild haemophiliacs [63]. Patients with mild haemophilia require less intensive prophylaxis since they suffer fewer spontaneous bleeding episodes. Therefore, successful management of haemophilia requires that spontaneous bleeding episodes be prevented and controlled by periodic FVIII or FIX concentrate infusions. High-purity "virus-free" concentrates are now available through improved plasma fractionation and viral inactivation. However, the presence of additional and/or novel blood-borne pathogens can never be ruled out. Recombinant factors, the preferred therapeutic product for safety reasons have been approved since the 1990's [64]. However, patients on recombinant factor regimen still require frequent, life-long infusions. Hence, the development of a less invasive and more cost-effective therapy for haemophilia would enhance the long-term care for haemophiliacs.

Haemophilia is an excellent disease model for gene therapy. Firstly, the expression of FVIII and FIX is not very tightly regulated, and therefore not exposed to additional cellular regulations that may inhibit its secretion. Secondly, the delivery of even the smallest amount of coagulation factor can have a beneficial clinical effect. Indeed, Kay *et al.* observed phenotypic changes in haemophilic dogs treated with gene therapy that had levels of FIX as little as 0.1% of normal physiological levels [65]. Thirdly, the liver normally produces FVIII and FIX. Nevertheless, any gene therapy approach that can deliver biologically active proteins into the circulation, regardless of where they are produced, can be potentially considered. Finally, there are excellent animal models of haemophilia A and B, both murine [66,67] and canine [68,69] that accurately reproduce the human condition [70]. Not surprisingly then, haemophilia has been a very popular model for a myriad of gene therapy approaches, which have been used with varying degrees of success in pre-clinical and clinical studies [71,72].

5.5. GENE THERAPY OF HAEMOPHILIA B

Delivery of FIX is much more amenable than that of FVIII. As a result, development of gene therapy for haemophilia A has typically lagged behind that of haemophilia B. The main limitations for achieving effective gene therapy of haemophilia B have been the complexities of FIX biosynthesis, and the high circulating levels of FIX. The extensive and complex post-translational modifications required to produce biologically functional FIX limit the types of cells capable of efficiently performing such a task. In turn, the physiological concentration of FIX is rather high, between 3-5 $\mu g/ml$. The high concentration poses a significant challenge for the design of expression vectors able to induce the secretion of such high level of hFIX on a continuous basis [61].

Implantable alginate microcapsules have been used to deliver hFIX in mice. Encapsulated recombinant Ltk⁻ fibroblasts secreted hFIX that was 70% biologically functional. Furthermore, the cells remained viable for at least 2 weeks [34], therefore showing the feasibility of encapsulated cells to deliver biologically active FIX. C2C12 mouse myoblasts transfected with a hFIX plasmid and enclosed in alginate microcapsules were implanted intraperitoneally into mice, achieving detectable levels of hFIX for up to 14 days when antibodies to hFIX were detected [32]. However, this study also showed indirect evidence of continuous delivery of FIX for at least 200 days, indicating the potential of encapsulated myoblasts to deliver therapeutics for a sustained period of time.

The choice of expression vector is a key step in determining the level and extent of therapeutic delivery in the host. However, it is worth remembering that it is often difficult to predict the *in vivo* efficacy of a given plasmid based solely on their *in vitro* behaviour. In a comparative study, myoblasts were transfected with a variety of plasmids containing the hFIX cDNA under the control of different genetic regulators. The secretion of hFIX by transfected cells was evaluated *in vitro*, as well as in the circulation of mice implanted with encapsulated myoblasts. Interestingly, the cells secreting the highest amount of hFIX *in vitro* did not deliver the highest concentration of hFIX *in vivo* [15]. The highest concentration of hFIX achieved in this study (~60 ng/ml) is considered therapeutic, and if delivered to human patients would eliminate severe haemophiliacs. Furthermore, hFIX delivery persisted in nude immunodeficient mice for the entire length of the experiment (6 weeks).

The ultimate goal of any gene therapy strategy is obviously the treatment of patients. Toward this end, there are excellent mouse models of haemophilia B that closely resemble the clinical presentation of the human disease [70]. Encapsulated C2C12 myoblasts transfected with a plasmid containing hFIX cDNA were implanted into haemophilia mice, leading to transient levels of hFIX in treated mice [33]. A similar treatment of haemophilic immunodeficient mice yielded high levels of hFIX with a peak of ~600 ng/ml, or 12-20% of the normal physiological concentration (Figure 3). In addition, hFIX delivery was sustained for the entire experiment (11 weeks). Such a delivery in severe human patients would convert their phenotype to a mild form of the disease, with no spontaneous bleeding episodes and basically a normal life. Importantly, treated mice also showed correction of the disease as measured by hemostatic parameters. The blood clotting time (APTT) was reduced in the treated mice, showing that the implantation of encapsulated cells partially reversed the disease. The degree of correction seen in the treated mice was around 30%, in agreement with the concentration of circulating hFIX detected in these mice [33].

The encapsulated C2C12 myoblasts are transformed cells with known potential to develop malignancies in immunodeficient hosts. Indeed, 86% of the treated immunodeficient mice developed malignancy [33]. This finding highlights the fact that, in order to prevent these dangerous and undesirable side effects, it is critical to select the appropriate cell lines or cell types. Furthermore, the choice of cells must be conditioned to the intended host.

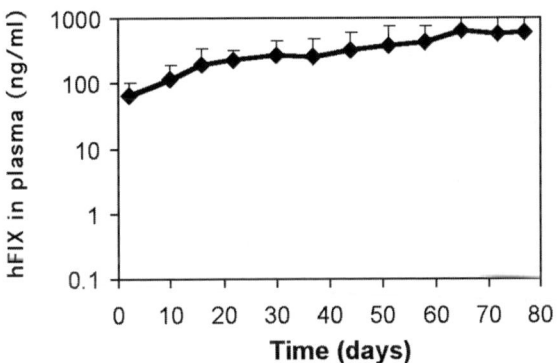

Figure 3. Delivery of human factor IX in haemophilic mice implanted with recombinant C2C12 myoblasts enclosed in alginate microcapsules [33].

Delivery of hFIX has also been achieved using polymeric implantable chambers. A human fibroblastic cell line (MSU 1.2) transduced with retroviral vector MFG-FIX secreted 0.9 µg hFIX/10^6 cells/day. The recombinant fibroblasts were then enclosed in an implantable device (TheraCyte™, Baxter Corporation, IL, USA). When athymic rats and mice were implanted with such devices, circulating levels of hFIX increased for 50 days, reaching peaks of 203 ng/ml (rats) and 597 ng/ml (mice), respectively [31]. Further, these animals had a sustained therapeutic level of hFIX, of >100 ng/ml, for at least 140 days. Thus, these findings agree with the studies using alginate microcapsules, supporting the notion that the use of encapsulated cells is suitable for the treatment of chronic diseases such as haemophilia B.

5.6. GENE THERAPY OF HAEMOPHILIA A

Gene therapy for haemophilia A (FVIII deficiency) is more challenging than that of haemophilia B, the main reason being the difficult expression of FVIII cDNA. The FVIII gene contains sequences that can inhibit its own expression, thus posing a significant challenge to gene therapists. In particular, expression vectors for FVIII able to sustain FVIII expression are difficult to design. It has been reported that expression vectors that result in a persistent secretion of other transgenes cannot sustain so with FVIII, instead leading to transient expression. The normal physiological concentration of FVIII in the circulation is low, between 100-200 ng/ml, up to 50-fold lower than FIX [61].

García-Martín *et al.* used C2C12 myoblasts transduced with the MFG-hFVIII vector [73]. Selected clones secreting high levels of hFVIII were then enclosed in alginate microcapsules and implanted intraperitoneally into mice. FVIII secretion was transient, being detected at therapeutic levels (>2 ng/ml) for a few days before becoming undetectable [73]. Concurrently, antibodies to hFVIII were detected in the circulation. These findings agree with the previous results of FIX delivery [15]. However, equally transient delivery was observed when encapsulated myoblasts secreting hFVIII were implanted in immunodeficient mice, incapable of eliciting an antibody response [73]. Since most implanted capsules could be retrieved and cell viability was high, these findings suggested an inhibition of hFVIII expression *in vivo*. Indeed, analysis of hFVIII expression by RT-PCR revealed a 3-5-fold reduction in hFVIII mRNA in encapsulated cells retrieved from implanted mice, as opposed to those encapsulated cells that were kept *in vitro*. This vector inhibition may explain the transient hFVIII delivery observed in mice [73]. Interestingly, the same MFG vector was shown to allow the sustained delivery of hFIX in immunodeficient mice for at least 6 weeks [15]. The difficulty in expressing hFVIII for sustained periods of time is not unique to cell encapsulation strategies, but a generalised significant challenge that gene therapists must overcome with novel expression vector designs.

6. Concluding remarks

With the aid of genetic engineering, cell encapsulation technology is poised to offer innovative solutions for the treatment of both genetic and acquired diseases. The versatility of this strategy is such that it can be used in a myriad of settings. Thus, encapsulated cells have been used to deliver a missing metabolic product [33], a toxic drug to kill malignant cells [49], to elicit an immune response to fight cancer [47], for the sustained delivery of viral particles [74], or the elimination of excessive amounts of urea [75].

The success of this strategy depends largely on a judicious choice of expression vector for the intended transgene, a suitable cell type for encapsulation, and the appropriate selection of immunoisolation polymer for the desired application. It is worth reinforcing the notion that each medical application presents unique characteristics and challenges to be solved. Therefore, it is not always possible to extrapolate a particular set of results to all medical conditions.

Advances in the understanding of diseases at the molecular level and the new molecular targets that are anticipated to come from the sequencing of the human genome [76] are expected to expand even further the applications of cell encapsulation in medicine.

References

[1] Chang, T.M.S. (1997) Artificial cells and bioencapsulation in bioartificial organs. Ann. NY Acad Sci. 831: 249-259.
[2] Huang, J.; Pray, C. and Rozelle, S. (2002) Enhancing the crops to feed the poor. Nature 418(6898): 678-684.
[3] Vajo, Z.; Fawcett, J. and Duckworth, W.C. (2001) Recombinant DNA technology in the treatment of diabetes: insulin analogs. Endocr. Rev. 22(5): 706-717.
[4] Mannucci, P.M. and Giangrande, P.L. (2000) Choice of replacement therapy for hemophilia: recombinant products only? Hematol. J. 1(2): 72-76.
[5] Anderson, W.F.; Blaese, R.M., and Culver, K. (1990) The ADA human gene therapy clinical protocol: Points to Consider response with clinical protocol, July 6, 1990. Hum Gene Ther. 1(3): 331-362.
[6] Teichler Zallen, D. (2000) US gene therapy in crisis. Trends Genet. 16(6): 272-275.
[7] Cavazzana-Calvo, M.; Hacein-Bey, S.; de Saint Basile, G.; Gross, F.; Yvon, E.; Nusbaum, P.; Selz, F.; Hue, C.; Certain, S.; Casanova, J.L.; Bousso, P.; Deist, F.L. and Fischer, A. (2000) Gene therapy of human severe combined immunodeficiency (SCID)-X1 disease. Science 288(5466): 669-672.
[8] Ledley, F.D. (1995) Hepatic gene therapy. In Chang, P.L. (Ed.) Somatic gene therapy, CRC Press, pp. 61-72.
[9] Lotze, M.T. and Kost, T.A. (2002) Viruses as gene delivery vectors: application to gene function, target validation, and assay development. Cancer Gene Ther. 9(8): 692-699.
[10] Marshall, E. (2001a) Gene therapy. Viral vectors still pack surprises. Science 294: 1640.
[11] Marshall, E. (2001b) Gene therapy. Panel reviews risks of germ line changes. Science 294: 2268-2269.
[12] Lechardeur, D. and Lukacs, G.L. (2002) Intracellular barriers to non-viral gene transfer. Curr Gene Ther 2(2): 183-194.
[13] Matthews, C.; Jenkins, G.; Hilfinger, J. and Davidson, B. (1999) Poly-L-lysine improves gene transfer with adenovirus formulated in PLGA microspheres. Gene Ther. 6(9): 1558-1564.
[14] Sailaja, G.; HogenEsch, H.; North, A.; Hays, J. and Mittal, S.K. (2002) Encapsulation of recombinant adenovirus into alginate microspheres circumvents vector-specific immune response. Gene Ther. 9(24): 1722-1729.
[15] Hortelano, G.; Xu, N.; Vandenberg, A.; Solera, J.; Chang, P.L. and Ofosu, F.A. (1999) Persistent delivery of factor IX in mice: Gene therapy for hemophilia B using implantable microcapsules. Human Gene Ther. 10(8): 1281-1288.
[16] Qin, L.; Ding, Y.; Pahud, D.R.; Chang, E.; Imperiale, M.J. and Bromberg, J.S. (1997) Promoter attenuation in gene therapy: interferon-gamma and tumor necrosis factor-alpha inhibit transgene expression. Hum Gene Ther. 8(17): 2019-2029.
[17] Buchschacher, G.L. Jr. (2001) Introduction to retroviruses and retroviral vectors. Somat. Cell Mol. Genet. 26(1-6): 1-11.
[18] Pannell, D. and Ellis, J. (2001) Silencing of gene expression: implications for design of retrovirus vectors. Rev. Med. Virol. 11(4): 205-217.
[19] Marshall, E. (2002) Gene Therapy a Suspect in Leukemia-like Disease. Science 298: 34-35.
[20] Yee, J.K. and Zaia, J.A. (2001) Prospects for gene therapy using HIV-based vectors. Somat. Cell Mol. Genet. 26(1-6): 159-174.
[21] Breyer, B.; Jiang, W.; Cheng, H.; Zhou, L.; Paul, R.; Feng, T. and He, T.C. (2001) Adenoviral vector-mediated gene transfer for human gene therapy. Curr. Gene Ther. 1(2): 149-162.
[22] Trudel, S.; Li, Z., Dodgson, C.; Nanji, S.; Wan, Y.; Voralia, M.; Hitt, M.; Gauldie, J.; Graham, F.L. and Stewart, A.K. (2001) Adenovector engineered interleukin-2 expressing autologous plasma cell vaccination after high-dose chemotherapy for multiple myeloma--a phase 1 study. Leukemia 15(5): 846-854.

[23] DelloRusso, C.; Scott, J.M.; Hartigan-O'Connor, D.; Salvatori, G.; Barjot, C.; Robinson, A.S.; Crawford, R.W.; Brooks, S.V. and Chamberlain J.S. (2002) Functional correction of adult mdx mouse muscle using gutted adenoviral vectors expressing full-length dystrophin. Proc. Natl. Acad. Sci. USA 99(20): 12979-12984.

[24] Kochanek, S.; Schiedner, G. and Volpers, C. (2001) High-capacity 'gutless' adenoviral vectors. Curr. Opin. Mol. Ther. 3(5): 454-463.

[25] Balague, C.; Zhou, J.; Dai, Y.; Alemany, R.; Josephs, S.F.; Andreason, G.; Hariharan, M.; Sethi, E. Prokopenko, E.; Jan, H.Y.; Lou, Y.C.; Hubert-Leslie, D.; Ruiz, L. and Zhang, W.W. (2000) Sustained high-level expression of full-length human factor VIII and restoration of clotting activity in hemophilic mice using a minimal adenovirus vector. Blood 95(3): 820-828.

[26] Owens, R.A. (2002) Second generation adeno-associated virus type 2-based gene therapy systems with the potential for preferential integration into AAVS1. Curr. Gene Ther. 2(2): 145-159.

[27] Kay, M.A.; Manno, C.S.; Ragni, M.V.; Larson, P.J.; Couto, L.B.; McClelland, A.; Glader, B.; Chew, A.J.; Tai, S.J.; Herzog, R.W.; Arruda, V.; Johnson, F.; Scallan, C.; Skarsgard, E.; Flake, A.W. and High, K.A. (2000) Evidence for gene transfer and expression of factor IX in haemophilia B patients treated with an AAV vector. Nat. Genet. 24(3): 257-261.

[28] Lilley, C.E.; Branston, R.H. and Coffin, R.S. (2001) Herpes simplex virus vectors for the nervous system. Curr. Gene Ther. 1(4): 339-358.

[29] Somiari, S.; Glasspool-Malone, J.; Drabick, J.J.; Gilbert, R.A.; Heller, R.; Jaroszeski, M.J. and Malone, R.W. (2000) Theory and in vivo application of electroporative gene delivery. Mol. Ther. 2(3): 178-187.

[30] Lin, M.T.; Pulkkinen, L.; Uitto, J. and Yoon, K. (2000) The gene gun: current applications in cutaneous gene therapy. Int. J. Dermatol. 39(3): 161-170.

[31] Brauker, J.; Frost, G.H.; Dwarki, V.; Nijjar, T.; Chin, R.; Carr-Brendel, V.; Jasunas, C.; Hodgett, D., Stone, W.; Cohen, L.K.; and Johnson, R.C. (1998) Sustained expression of high levels of human factor IX from human cells implanted within an immunoisolation device into athymic rodents. Hum. Gene Ther. 9(6): 879-888.

[32] Hortelano, G.; Al-Hendy, A.; Ofosu, F.A. and Chang, P.L. (1996) Delivery of human factor IX in mice by microencapsulated recombinant myoblasts: A novel approach towards allogeneic gene therapy of hemophilia B. Blood 87(12): 5095-5103.

[33] Hortelano, G.; Wang, L.; Xu, N. and Ofosu, F. (2001) Sustained and therapeutic delivery of human factor IX in nude hemophilia B mice implanted by encapsulated C2C12 myoblasts: concurrent tumorigenesis. Haemophilia 7(2): 207-213.

[34] Liu, H.W.; Ofosu, F.A. and Chang, P.L. (1993) Expression of human factor IX by microencapsulated recombinant fibroblasts. Hum Gene Ther. 4(3): 291-301.

[35] Hughes, M.; Vassilakos, A.; Andrews, D.W.; Hortelano, G.; Belmont, J.W. and Chang, P.L. (1994) Delivery of a secretable adenosine deaminase through microcapsules – a novel approach to somatic gene therapy. Hum Gene Ther. 5(12): 1445-1455.

[36] Al-Hendy, A.; Hortelano, G.; Tannenbaum, G.S. and Chang, P.L. (1995) Correction of the growth defect in dwarf mice with nonautologous microencapsulated myoblasts – an alternate approach to somatic gene therapy. Hum Gene Ther. 6(2): 165-175.

[37] Ross, C.J.; Bastedo, L.; Maier, S.A.; Sands, M.S. and Chang, P.L. (2000) Treatment of a lysosomal storage disease, mucopolysaccharidosis VII, with microencapsulated recombinant cells. Hum Gene Ther. 11(15): 2117-2127.

[38] Chang, P.L.; Van Raamsdonk, J.M.; Hortelano, G.; Barsoum, S.C.; MacDonald, N.C. and Stockley, T.L. (1999) The *in vivo* delivery of heterologous proteins by microencapsulated recombinant cells. Trends Biotechnol. 17(2): 78-83.

[39] Goto, S.; Miyazaki, K.; Funabiki, T. and Yasumitsu, H. (1999) Serum-free culture conditions for analysis of secretory proteinases during myogenic differentiation of mouse C2C12 myoblasts. Anal. Biochem. 272(2): 135-142.

[40] Gunning, P.; Leavitt, J.; Muscat, G.; Ng, S.Y. and Kedes L. (1987) A human beta-actin expression vector system directs high-level accumulation of antisense transcripts. Proc. Natl. Acad. Sci. USA 84(14): 4831-4835.

[41] Rinsch, C.; Dupraz, P.; Schneider, B.L.; Deglon, N.; Maxwell, P.H.; Ratcliffe, P.J.; Aebischer, P. (2002) Delivery of erythropoietin by encapsulated myoblasts in a genetic model of severe anemia. Kidney Int. 62(4): 1395-1401.

[42] Springer, M.L.; Hortelano, G.; Bouley, D.; Kraft, P.E.; Wong, J. and Blau, H.M. (2000) Induction of angiogenesis by implantation of encapsulated primary myoblasts expressing vascular endothelial growth factor. J. Gene Med. 2(4): 279-288.
[43] Rinsch, C.; Peduto, G.; Schneider, B.L. and Aebischer, P. (2001) Inducing host acceptance to encapsulated xenogeneic myoblasts. Transplantation 71(3): 345-351.
[44] Aebischer, P.; Schluep, M.; Deglon, N.; Joseph, J.M.; Hirt, L.; Heyd, B.; Goddard, M.; Hammang, J.P.; Zurn, A.D.; Kato, A.C.; Regli, F. and Baetge, E.E. (1996) Intrathecal delivery of CNTF using encapsulated genetically modified xenogeneic cells in amyotrophic lateral sclerosis patients. Nat. Med. 2(6): 696-699.
[45] Buchser, E.; Goddard, M.; Heyd, B.; Joseph, J.M.; Favre, J.; de Tribolet, N.; Lysaght, M. and Aebischer, P. (1996) Immunoisolated xenogenic chromaffin cell therapy for chronic pain. Initial clinical experience. Anesthesiology 85(5): 1005-1012.
[46] Piskin, E. (2002). Biodegradable polymeric matrices for bioartificial implants. Int. J. Artif. Organs 25(5): 434-440.
[47] Cirone, P.; Bourgeois, J.M.; Austin, R.C. and Chang, P.L. (2002). A novel approach to tumor suppression with microencapsulated recombinant cells. Hum. Gene Ther. 13(10): 1157-1166.
[48] Read, T.A.; Sorensen, D.R.; Mahesparan, R.; Enger, P.O.; Timpl, R.; Olsen, B.R.; Hjelstuen, M.H.; Haraldseth, O. and Bjerkvig, R. (2001) Local endostatin treatment of gliomas administered by microencapsulated producer cells. Nat. Biotechnol. 19(1): 29-34.
[49] Muller, P.; Jesnowski, R.; Karle, P.; Renz, R.; Saller, R.; Stein, H.; Puschel, K.; von Rombs, K.; Nizze, H.; Liebe, S.; Wagner, T.; Gunzburg, W.H.; Salmons, B. and Lohr, M. (1999 Injection of encapsulated cells producing an ifosfamide-activating cytochrome P450 for targeted chemotherapy to pancreatic tumors. Ann. NY Acad. Sci. 880: 337-351.
[50] Lohr, M.; Hummel, F.; Faulmann, G.; Ringel, J.; Saller, R.; Hain, J.; Gunzburg, W.H. and Salmons, B. (2002) Microencapsulated, CYP2B1-transfected cells activating ifosfamide at the site of the tumor: the magic bullets of the 21st century. Cancer Chemother. Pharmacol. 49 Suppl. 1: S21-24.
[51] Lohr, M.; Hoffmeyer, A.; Kroger, J.; Freund, M.; Hain, J.; Holle, A.; Karle, P.; Knofel, W.T.; Liebe, S.; Muller, P.; Nizze, H.; Renner, M.; Saller, R.M.; Wagner, T.; Hauenstein, K.; Gunzburg, W.H. and Salmons, B. (2001) Microencapsulated cell-mediated treatment of inoperable pancreatic carcinoma. Lancet 357(9268): 1591-1592.
[52] Geller, R.L.; Neuenfeldt, S.; Levon, S.A.; Maryanov, D.A.; Thomas, T.J. and Brauker, J.H. (1997) Immunoisolation of tumor cells: generation of antitumor immunity through indirect presentation of antigen. J. Immunother. 20(2): 131-137.
[53] Huber, A.; Padrun, V.; Deglon, N.; Aebischer, P.; Mohler, H. and Boison, D. (2001) Grafts of adenosine-releasing cells suppress seizures in kindling epilepsy. Proc. Natl. Acad. Sci. USA 98(13): 7611-7616.
[54] Deglon, N., Heyd, B.; Tan, S.A.; Joseph, J.M.; Zurn, A.D. and Aebischer, P. (1996) Central nervous system delivery of recombinant ciliary neurotrophic factor by polymer encapsulated differentiated C2C12 myoblasts. Hum. Gene Ther. 7(17): 2135-2146.
[55] Zurn, A.D.; Widmer, H.R. and Aebischer, P. (2001) Sustained delivery of GDNF: towards a treatment for Parkinson's disease. Brain Res. Rev. 36(2-3): 222-229.
[56] Zurn, A.D.; Henry, H.; Schluep, M.; Aubert, V.; Winkel, L.; Eilers, B.; Bachmann, C. and Aebischer, P. (2000) Evaluation of an intrathecal immune response in amyotrophic lateral sclerosis patients implanted with encapsulated genetically engineered xenogeneic cells. Cell Transplant. 9(4): 471-484.
[57] Emerich, D.F.; Cain, C.K.; Greco, C.; Saydoff, J.A.; Hu, Z.Y.; Liu, H. and Lindner, M.D. (1997) Cellular delivery of human CNTF prevents motor and cognitive dysfunction in a rodent model of Huntington's disease. Cell Transplant. 6(3): 249-266.
[58] Saitoh, Y.; Eguchi, Y.; Hagihara, Y.; Arita, N.; Watahiki, M.; Tsujimoto, Y. and Hayakawa, T. (1998) Dose-dependent doxycycline-mediated adrenocorticotropic hormone secretion from encapsulated Tet-on proopiomelanocortin Neuro2A cells in the subarachnoid space. Hum. Gene Ther. 9(7): 997-1002.
[59] Sommer, B.; Rinsch, C.; Payen, E.; Dalle, B.; Schneider, B.; Deglon, N.; Henri, A.; Beuzard, Y. and Aebischer, P. (2002). Long-term doxycycline-regulated secretion of erythropoietin by encapsulated myoblasts. Mol. Ther. 6(2): 155-161.
[60] Dalle, B.; Payen, E.; Regulier, E.; Deglon, N.; Rouyer-Fessard, P.; Beuzard, Y. and Aebischer, P. (1999) Improvement of mouse beta-thalassemia upon erythropoietin delivery by encapsulated myoblasts. Gene Ther. 6(2): 157-161.

[61] Hedner, U. and Davie, E.W. (1989) Introduction to hemostasis and the vitamin K-dependent coagulation factors. In Scriver, C.R.; Beaudet, A.L.; Sly, W.S. and Valle, D. (Eds.) The metabolic basis of inherited disease, vol. II, 6th edition, McGraw-Hill, New York; pp. 2107.
[62] Gianelli, F; Green, P.M.; High, K.A.; Sommer, S.; Poon, M.C.; Ludwig, M., Schwaab, R.; Reitsma, P.H.; Goossens, M. and Yoshioka, A. (1993) Haemophilia B: database of point mutations and short additions and deletions-third edition. Nucleic Acids Res. 21: 3075-3087.
[63] Roberts, H.R. and Eberst, M.E. (1993) Current management of hemophilia B. Hematology/Oncology clinics of North America 7(6): 1269-1279.
[64] Walker, I. (1997) Recombinant factor IX has been licensed for use. Hemophilia Today (Summer).
[65] Kay, M.A.; Rothenberg, S.; Landen, C.N.; Bellinger, D.A.; Leland, F.; Toman, C.; Finegold, M.; Thompson, A.R.; Read, M.S.; Brinkhous, K.M.; *et al.* (1993). *In vivo* gene therapy for hemophilia B: Sustained partial correction in factor IX deficient dogs. Science 262: 117-119.
[66] Bi, L.; Lawler, A.M.; Antonarakis, S.E.; High, KA; Gearhart, JD and Kazazian, HH Jr. (1995). Targeted disruption of the mouse factor VIII gene produces a model of haemophilia A. Nat Genet 10(1):119-121
[67] Lin, H.F.; Maeda, N.; Smithies, O.; Straight, D.L. and Stafford, D.W. (1997) A coagulation factor IX-deficient mouse model for human hemophilia B. Blood 90(10): 3962-3966.
[68] Giles, A.R.; Tinlin, S. and Greenwood, R. (1982) A canine model of hemophilic (factor VIII:C deficiency) bleeding. Blood 60(3): 727-730.
[69] Evans, J.P.; Brinkhous, K.M.; Brayer, G.D.; Reisner, H. and High, K.A. (1989) Canine hemophilia B resulting from a point mutation with unusual consequences. Proc. Natl. Acad. Sci. USA 86(24): 10095-10099.
[70] Wang, L.; Zoppe, M.; Hackeng, T.M.; Griffin, J.H.; Lee, K.F. and Verma, I.M. (1997) A factor IX-deficient mouse model for hemophilia B gene therapy. Proc. Natl. Acad. Sci. USA 94(21): 11563-11566.
[71] Kay, M. and High, K. (1999) Gene therapy for the hemophilias. Proc. Natl. Acad. Sci. USA 96: 9973-9975.
[72] Wang, L.; Takabe, K.; Bidlingmaier, S.M.; Ill, C.R. and Verma, I.M. (1999) Sustained correction of bleeding disorder in hemophilia B mice by gene therapy. Proc. Natl. Acad. Sci. USA 96(7): 3906-3910.
[73] Garcia-Martin, C.; Chuah, M.K.; Van Damme, A.; Robinson, K.E.; Vanzieleghem, B.; Saint-Remy, J.M.; Gallardo, D.; Ofosu, F.A.; Vandendriessche, T. and Hortelano, G. (2002) Therapeutic levels of human factor VIII in mice implanted with encapsulated cells: potential for gene therapy of haemophilia A. J. Gene Med. 4(2): 215-223.
[74] Saller, R.M.; Indraccolo, S.; Coppola, V.; Esposito, G.; Stange, J.; Mitzner, S.; Amadori, A.; Salmons, B. and Gunzburg, W.H. (2002) Encapsulated cells producing retroviral vectors for in vivo gene transfer. J. Gene Med. 4(2): 150-160.
[75] Prakash, S. and Chang, T.M. (1996). Microencapsulated genetically engineered live *E. coli* DH5 cells administered orally to maintain normal plasma urea level in uremic rats. Nat. Med. 2(8): 883-887.
[76] Birney, E.; Bateman, A.; Clamp, M.E. and Hubbard, T.J. (2001) Mining the draft human genome. Nature 409: 827-828.
[77] Dautzenberg, H.; Schuldt, U.; Grasnick, G.; Karle, P.; Muller, P.; Lohr, M.; Pelegrin, M.; Piechaczyk, M.; Rombs, K.V.; Gunzburg, W.H.; Salmons, B. and Saller, R.M. (1999) Development of cellulose sulfate-based polyelectrolyte complex microcapsules for medical applications. Ann. NY Acad. Sci. 875: 46-63.
[78] Lahooti, S. and Sefton, M.V. (2000) Effect of an immobilization matrix and capsule membrane permeability on the viability of encapsulated HEK cells. Biomaterials 21(10): 987-995.
[79] Joki, T.; Machluf, M.; Atala, A.; Zhu, J.; Seyfried, N.T.; Dunn, I.F.; Abe, T.; Carroll, R.S. and Black, P.M. (2001) Continuous release of endostatin from microencapsulated engineered cells for tumor therapy. Nat. Biotechnol. 19(1): 35-39.
[80] Weber, M.; Steinert, A.; Jork, A.; Dimmler, A.; Thurmer, F.; Schutze, N.; Hendrich, C. and Zimmerman, U. (2002) Formation of cartilage matrix proteins by BMP-transfected murine mesenchymal stem cells encapsulated in a novel class of alginates. Biomaterials 23(9): 2003-2013.
[81] Rokstad, A.M.; Holtan, S.; Strand, B.; Steinkjer, B.; Ryan, L.; Kulseng, B.; Skjak-Braek, G. and Espevik, T. (2002) Microencapsulation of cells producing therapeutic proteins: optimizing cell growth and secretion. Cell Transplant. 11(4): 313-324.
[82] Naganawa, Y.; Ohsugi, K.; Kase, R.; Date, I.; Sakuraba, H. and Sakuragawa, N. (2002) *In vitro* study of encapsulation therapy for Fabry disease using genetically engineered CHO cell line. Cell Transplant. 11(4): 325-329.

[83] Date, I.; Shingo, T.; Yoshida, H.; Fujiwara, K.; Kobayashi, K. and Ohmoto, T. (2000) Grafting of encapsulated dopamine-secreting cells in Parkinson's disease: long-term primate study. Cell Transplant. 9(5): 705-709.

ARTIFICIAL CELLS FOR BLOOD SUBSTITUTES, ENZYME THERAPY, CELL THERAPY AND DRUG DELIVERY

THOMAS MING SWI CHANG

Artificial Cells & Organs Research Centre, Departments of Physiology, Medicine and Biomedical Engineering, Faculty of Medicine, McGill University, 3655, Promenade Sir-William-Osler, Montreal, Quebec, Canada H3G 1H6 - Fax: 1-514-398-7452 - Email: artcel.med@mcgill.ca

Abstract

Artificial cells are being actively investigated for medical and biotechnological applications. The earliest routine clinical use of artificial cells is in the form of coated activated charcoal for haemoperfusion. Implantation of encapsulated cells are being studied for the treatment of diabetes, liver failure and the use of encapsulated genetically engineered cells for gene therapy. Blood substitutes based on modified haemoglobin are already in Phase III clinical trials in patients with as much as 20 units infused into each patient during trauma surgery. Artificial cells containing enzymes are being developed for clinical trial in hereditary enzyme deficiency diseases and other diseases. Artificial cell is also being investigated for drug delivery and for other uses in biotechnology, chemical engineering and medicine.

1. Introduction

Artificial cells evolves from Chang's initial studies to prepare artificial structures for bioencapsulation of enzymes, cells and other biologically active materials [1-4]. Once bioencapsulated, biologically active materials inside the artificial cells are prevented from coming into contact with external materials like leucocytes, antibodies or tryptic enzymes. Smaller molecules can equilibrate rapidly across the ultra thin membrane with large surface to volume relationship. A number of potential medical applications using artificial cells have been proposed [2-4]. The first of these developed successfully for routine clinical use is haemoperfusion [4]. After initial clinical trails for poisoning, kidney failure, and liver failure [5], it is now in routine clinical uses [6]. Some exciting recent developments include research and clinical trials on modified haemoglobin for blood substitutes; the use of artificial cells for enzyme therapy, cell therapy and gene therapy [7]. This review will highlight some examples of the increasing interests in the biotechnological approaches of artificial cells for clinical applications.

2. Artificial cells containing enzymes for inborn errors of metabolism and other conditions

Chang and Poznanksy have earlier implanted artificial cells containing catalase into acatalesemic mice, animals with a congenitic deficiency in catalase [8]. This replaces the deficient enzymes and prevented the animals from the damaging effects of oxidants. The artificial cells protect the enclosed enzyme from immunological reactions [9]. It was also showed that artificial cells containing asparaginase implanted into mice with lymphosarcoma delayed the onset and growth of lymphosarcoma [10]. The single problem preventing the clinical application of enzyme artificial cells is the need to repeatedly inject these enzyme artificial cells. To solve this problem, Bourget and Chang found that microencapsulated phenylalanine ammonia lyase given orally can lower the elevated phenylalanine levels in phenylketonuria [PKU] rats [11]. This is because of our more recent finding of an extensive recycling of amino acids between the body and the intestine [12]. This is now being developed for clinical trial in PKU [13,14]. In addition to PKU other examples our recent studies shows that oral artificial cells containing tyrosinase is effective in lowering systemic tyrosine levels in rats [15,16]. This has much potential for the treatment of the fatal skin cancer, melanoma. We are encouraged in this oral approach because of our preliminary clinical testing of oral microencapsulated xanthine oxidase as experimental therapy in Lesch-Nyhan Disease [17].

3. Artificial cells for cell therapy

Chang *et al.* reported the encapsulation of biological cells in 1966 based on a drop method and proposed that "protected from immunological process, encapsulated endocrine cells might survive and maintain an effective supply of hormone" [2,4]. Chang approached Conaught Laboratory to develop this for use in islet transplantation for diabetes. Sun from Conaught and his collaborators have developed this drop-method by using milder physical crosslinking [18]. This resulted in alginate-polylysine-alginate [APA] microcapsules containing cells. They show that after implantation, the islets inside artificial cells remain viable and continued to secrete insulin to control the glucose levels of diabetic rats [19]. Cell encapsulation for cell therapy has been extensively developed by many groups especially using artificial cells containing endocrine tissues, hepatocytes and other cells for cell therapy [7,18-26]. Microencapsulated genetically engineered cells has been carried out by many groups [19-26]. This has been studied for potential applications in amyotrophic lateral sclerosis, Dwarfism, pain treatment, IgG$_1$ plasmacytosis, Haemophilia B, Parkinsonism and axotomised septal cholinergic neurons. We have also studied the oral use of microencapsulated genetically engineered nonpathogenic *E. coli* DH5 cells containing *Klebsiella aerogenes* urease gene in renal failure rats [27-31]. We have been studying the use of implantation of encapsulated hepatocytes for liver support [32-39]. This included acute liver failure [32] and hyperbilirubinaemia in Gunn rats [34]. We developed a two-step cell encapsulation method to improve the APA method resulting in improved survival of implanted cells [37,38]. Using this two-step methods plus the use of co-encapsulation of stem cells and hepatocytes we have further increased the viability of encapsulated hepatocytes both in culture and also after implantation [39,40].

4. Red blood cell substitutes

4.1. POLYHEMOGLOBIN AS BLOOD SUBSTITUTES

Native haemoglobin [tetramer], breaks down into half molecules [dimers] after infusion causing renal toxicity and other adverse effects. Chang has extended his original approach of artificial cells containing haemoglobin and enzymes [1] to form polyhaemoglobin – a molecular version of artificial cells. This is based on his use of bifunctional agents like diacid [1,4] or later glutaraldehyde [41] to crosslink haemoglobin molecules into polyhaemoglobin. This gluataradehyde crosslinked polyhaemoglobin approach has been extensively developed more recently [40-42] Polyhaemoglobin consisting of 4 to 5 haemoglobin molecules stays longer in the circulation and they do not breakdown into dimers. One example is the recent report by Gould *et al.* on their ongoing clinical trials using pyridoxalated glutaraldehyde human polyhaemoglobin in trauma surgery. They show that this can successfully replace blood loss by maintaining the haemoglobin level with no reported side effects [46,47]. More recently, they have infused up to 20 units into individual trauma surgery patients. Another example is glutaraldehyde crosslinked bovine polyhaemoglobin that has been extensive tested in Phase III clinical trials [48]. This has been approved for veterinary medicine in the U.S. and for routine clinical use in South Africa. An o-raffinose polyhaemoglobin is also being developed and being tested for surgery that needs only low volume replacement [49]. All the above three polyhaemoglobins have been approved for compassionate uses in human and they are waiting for regulatory approval for routine clinical uses in human.

4.2. POLYHEMOGLOBIN CONTAINING CATALASE AND SUPEROXIDE DISUMUTASE

The present polyhaemoglobin shows promise especially for perioperative uses as in haemodilution, replacement of extensive surgical blood loss and other conditions with no potentials for ischemia-reperfusion injuries [42-50]. However, polyhaemoglobins do not contain red blood cell antioxidant enzymes like catalase and superoxide dismutase. Thus, for the resuscitation of sustained severe hemorrhagic shock or in reperfusion of ischemic organs as in stroke or in organ transplantation, the use of polyhaemoglobin may result in ischemia-reperfusion injuries [42]. We have therefore studied the crosslinking of superoxide dismutase and catalase with polyhaemoglobin (polyHb) to form PolyHb-SOD-CAT [51-55]. We found that when compared to polyHb, PolyHb-SOD-CAT, significantly decrease the release of haeme and iron from haemoglobin and also effectively removes oxygen radicals [52,53]. Reperfusion studies in a rat model of intestinal ischemia, shows that PolyHb-SOD-CAT resulted in negligible increase in oxygen radicals, unlike the high level that resulted from reperfusion using polyHb [54]. More recently [55], in a transient global cerebral ischemia rat model, we found that after

60 minutes of ischemia, reperfusion with polyHb resulted in significant increases in blood-brain barrier and the breakdown of blood-brain barrier. On the other hand, polyHb-SOD-CAT did no result in these adverse changes.

4.3. RECOMBINANT HUMAN HEMOGLOBIN

Although polyhaemoglobin is in the most advance stages of clinical trial, there are other modified haemoglobins [42-44,66-68]. Unlike polyhaemoglobin these are single tetrameric haemoglobin formed by intramolecular cross-linkage [59,60] or recombinant human haemoglobin [61,62]. Clinical trials on these show vasoactivity and other effects of nitrate oxide removal [61,62]. Doherty et al. [63] have therefore developed a new recombinant human tetrameric haemoglobin with markedly decrease affinity for nitric oxide. When infused into experimental animals, this did not cause vasoactivity.

4.4. OTHER NEW GENERATIONS OF MODIFIED HEMOGLOBIN BLOOD SUBSTITUTES

Polyhaemoglobin stays in the circulation with a half-time of only up to 27 hours. In order to increase this circulation time, Chang's original idea of a complete artificial red blood cell [1,2] is now being developed as third generation blood substitute. Thus submicron lipid membrane microencapsulated haemoglobin [64] is being explored especially more recently by the group of Tsuchida in Japan [58] and Rudolph in the U.S.A. [57]. The U.S. group has modified the surface properties to result in a circulation half-time of about 50 hours [65]. We are developing a new system based on biodegradable polymer and nanotechnology resulting in polylactide membrane haemoglobin nanocapsules of 80 to 150 nanometre diameter [66-69]. This is smaller than the lipid-vesicles and contains negligible amounts of lipids. We have included superoxide dismutase, catalase and also multi-enzyme systems to prevent the accumulation of methaemoglobin. Our recent studies show that surface modification using a polyethylene-glycol-polylactide copolymer, increased the circulation time of these Hb nanocapsules to double that of polyHb [69].

5. General

The above review contains a very brief overview of this rather large area. For more specific details, please refer to the references given. "Artificial Cells Biotechnology" is a rapidly evolving area and rapidly updating can be found at our McGill University website: www.artcell.mcgill.ca.

Acknowledgements

This author acknowledges the supports of the Canadian Institutes of Health Research, the "Virage" Centre of Excellence in Biotechnology from the Quebec Ministry, the MSSS-FRSQ Research Group award on Blood Substitutes in Transfusion Medicine from the Quebec Ministry of Health and the Bayer/Canadian Blood Agency/Hema Quebec/Canadian Institutes of Health Research Partnership Fund.

References

[1] Chang, T.M.S. (1964) Semipermeable microcapsules. Science 146: 524-525.
[2] Chang, T.M.S.; MacIntosh, F.C. and Mason, S.G. (1966) Semipermeable aqueous microcapsules: I. Preparation and properties. Can. J. Physiol. Pharmacol. 44: 115-128.
[3] Chang, T.M.S. (1966) Semipermeable aqueous microcapsules ["artificial cells"]: with emphasis on experiments in an extracorporeal shunt system. Trans. Am. Soc. Artif. Intern. Organs 12: 13-19.
[4] Chang, T.M.S. (1972) Artificial Cells. C.C. Thomas Publisher, Springfield.
[5] Chang, T.M.S. (1975) Microencapsulated adsorbent hemoperfusion for uremia, intoxication and hepatic failure. Kidney Int. 7: S387-S392.
[6] Winchester, J.F. (1989) Hemoperfusion. In: Maher, J.F. (Ed.) Replacement of Renal Function by Dialysis. Kluwer Academic Publisher, Boston; pp. 439-592.
[7] Chang, T.M.S. (1997) Artificial Cells. In: Dulbecco, R. (Editor-in-chief) Encyclopedia of Human Biology [2nd Edition]. Academic Press, Inc., San Diego (CA); pp. 457-463.
[8] Chang, T.M.S. and Poznansky, M.J. (1968) Semipermeable microcapsules containing catalase for enzyme replacement in acatalsaemic mice. Nature 218(5138): 242-245.
[9] Poznansky, M.J. and Chang, T.M.S. (1974) Comparison of the enzyme kinetics and immunological properties of catalase immobilized by microencapsulation and catalase in free solution for enzyme replacement. Biochim. Biophys. Acta 334: 103-115.
[10] Chang, T.M.S. (1971) The *in vivo* effects of semipermeable microcapsules containing L-asparaginase on 6C3HED lymphosarcoma. Nature 229(528): 117-118.
[11] Bourget, L. and Chang, T.M.S. (1986) Phenylalanine ammonia-lyase immobilized in microcapsules for the depleture of phenylalanine in plasma in phenylketonuric rat model. Biochim. Biophys. Acta. 883: 432-438.
[12] Chang, T.M.S.; Bourget, L. and Lister, C. (1995) New theory of enterorecirculation of amino acids and its use for depleting unwanted amino acids using oral enzyme-artificial cells, as in removing phenylalanine in phenylketonuria. Artif. Cells, Blood Substit. & Immobil. Biotechnol. 25: 1-23.
[13] Sarkissian, C.N.; Shao, Z.; Blain, F.; Peevers, R.; Su, H.; Heft, R.; Chang, T.M.S. and Scriver, C.R. (1999) A different approach to treatment of phenylketonuria: phenylalanine degradation with recombinant phenylalanine ammonia lyase. Proc. Natl. Acad. Sci. 96: 2339-2344.
[14] Liu, J.; Jia, X.; Zhang, J.; Xiang, G.; Hu, W. and Zhou, Y. (2002) Study on a novel strategy to treatment of Phenylketonuria. Artif. Cells, Blood Substit. & Immobil. Biotechnol. 30: 243-258
[15] Chang, T.M.S. and Yu, B. (2002) Composition for inhibiting tumour growth and methods thereof. US Provisional Patent Application 60/364,581 (March 18, 2002).
[16] Yu, B.L. and Chang, T.M.S. (2002) *In-vitro* kinetics of encapsulated tyrosinase. Artif. Cells, Blood Substit. & Immobil. Biotechnol. 30: 533-546.
[17] Palmour, R.M.; Goodyer, P.; Reade, T. and Chang, T.M.S. (1989) Microencapsulated xanthine oxidase as experimental therapy in Lesch-Nyhan Disease. Lancet 2(8664): 687-688.
[18] Lim, F. and Sun, A.M. (1980) Microencapsulated islets as bioartificial endocrine pancreas. Science 210: 908-909.
[19] Chang, T.M.S. (1995) Artificial cells with emphasis on bioencapsulation in biotechnology. Biotechnol. Ann. Rev. 1: 267-295.
[20] Orive, G.; Hernandez, R.M.; Gascon, A.R.; Calafiore, R.; Chang, T.M.S.; Vos, P. De; Hortelano, G.; Hunkeler, D.; Lacík, I.; Shapiro, A.M.J. and Pedraz, J.L. (2003) Cell encapsulation: promise and progress. Nat. Med. 9: 104-107.
[21] Kuhtreiber, W.M.; Lauza, P.P., Chick, W.L. (Eds.) (1999) Cell Encapsulation Technology and Therapy. Burkhauser, Boston.
[22] Hunkeler, D.; Prokop, A.; Cherrington, A.D.; Rajotte, R. and Sefton, M. (Eds.) (1999) Bioartificial Organs A: Technology, Medicine and Materials. Ann. N.Y. Acad. Sci., volume 875; pp.1-415.
[23] Chang, T.M.S. and Prakash, S. (2001) Procedure for microencapsulation of enzymes, cells and genetically engineered microorganisms. Mol. Biotechnol. 17: 249-260.

[24] Dionne, K.E.; Cain, B.M.; Li, R.H.; Bell, W.J.; Doherty, E.J.; Rein, D.H.; Lysaght, M.J. and Gentile, F.T. (1996) Transport characterization of membranes for immunoisolation. Biomaterials 17: 257-266.
[25] Chang, T.M.S. and Prakash, S. (1998) Therapeutic uses of microencapsulated genetically engineered cells. Mol. Med. Today 4: 221-227.
[26] Aebischer, P; Schluep, M.; Deglon, N.; Joseph, J.M.; Hirt, L.; Heyd, B.; Goddard, M.; Hammang, J.P.; Zurn, A.D.; Kato, A.C.; Regli, F. and Baetge, E.E. (1996) Intrathecal delivery of CNTF using encapsulated genetically modified xenogeneic cells in amyotrophic lateral sclerosis patients. Nat. Med. 2: 696-699.
[27] Prakash, S. and Chang, T.M.S. (1996) Microencapsulated genetically Engineered live *E. coli* DH5 cells administered orally to maintain normal plasma urea level in uremic rats. Nat. Med. 2(8): 883-887.
[28] Chang, T.M.S. (1997) Live *E. coli* cells to treatment uremia: replies to letters to the editor. Nat. Med. 3: 2-3.
[29] Prakash, S. and Chang, T.M.S. (1999) Growth kinetics of genetically engineered *E. coli* dh 5 cells in artificial cell APA membrane microcapsules: preliminary report. Artif. Cells, Blood Substit. & Immobil. Biotechnol. 27(3): 291-301.
[30] Chang, T.M.S. and Prakash, S. (2001) Microencapsulated genetically engineered microorganisms for clinical application. U.S. Patent 6,217,859, April 17 2001.
[31] Chang, T.M.S. and Prakash, S. (2001) Microencapsulated genetically engineered microorganisms for clinical application. Japanese Patent 3228941, September 7 2001.
[32] Chang, T.M.S. (2001) Bioencapsulated hepatocytes for experimental liver support. J. Hepatol. 34: 148-149.
[33] Wong, H. and Chang, T.M.S. (1986) Bioartificial liver: implanted artificial cells microencapsulated living hepatocytes increases survival of liver failure rats. Int. J. Artif. Organs 9: 335-336.
[34] Bruni, S. and Chang, T.M.S. (1989) Hepatocytes immobilized by microencapsulation in artificial cells: Effects on hyperbiliru-binemia in Gunn Rats. J. Biomat. Artif. Cells and Artif. Org. 17: 403-412.
[35] Wong, H. and Chang, T.M.S. (1988) The viability and regeneration of artificial cell microencapsulated rat hepatocyte xenograft transplants in mice. J. Biomat. Artif. Cells and Artif. Org. 16: 731-740.
[36] Wong, H. and Chang, T.M.S. (1991) Microencapsulation of cells within alginate poly-L-lysine microcapsules prepared with standard single step drop technique: histologically identified membrane imperfections and the associated graft rejection. Biomat. Artif. Cells and Immobil. Biotechnol. 19: 675-686.
[37] Wong, H. and Chang, T.M.S. (1991) A novel two-step procedure for immobilizing living cells in microcapsule for improving xenograft survival. Biomat., Artif. Cells and Immobil. Biotechnol. 19: 687- 698.
[38] Chang, T.M.S. and Wong, H. (1992) A novel method for cell encapsulation in artificial cells. USA Patent No. 5,084,350, Issued Jan. 28, 1992.
[39] Liu, Z. and Chang, T.M.S. (2000) Effects of bone marrow cells on hepatocytes: when co-cultured or co-encapsulated together. Artif. Cells, Blood Substit. & Immobil. Biotechnol. 28(4): 365-374.
[40] Liu, Z.C. and Chang, T.M.S. (2002) Transplantation of Co-encapsulated Hepatocytes and Marrow Stem Cells into rats. Artif. Cells, Blood Substit. & Immobil. Biotechnol. 30: 99-112.
[41] Chang, T.M.S. (1971) Stabilization of enzyme by microencapsulation with a concentrated protein solution or by crosslinking with glutaraldehyde. Biochem. Biophys. Res. Com. 44: 1531-1533.
[42] Chang, T.M.S. (1997) Blood Substitutes: Principles, Methods, Products and Clinical Trials. Vol.1. Karger, Basel.
[43] Chang, T.M.S. (1999) Artificial Blood: a prospective. TIBTECH 17: 61-67.
[44] Chang, T.M.S. (2002) Oxygen Carriers. Curr. Opin. Investig. Drugs 3(8): 1187-1190.
[45] Dudziak, R. and Bonhard, K. (1980) The development of hemoglobin preparations for various indications. Anesthesist 29: 181-187.
[46] Gould, S.A.; Moore, F.A.; Hoyt, D.B.; Burch, J.M.; Haenel, J.B.; Garcia, J.; DeWoskin, R. and Moss G.S. (1998) The first randomized tiral of Human polymerized hemoglobin as a Blood Substitute in Acute Trauma and Emergent Surgery. J. Am. College of Surgeons 187: 113-120.
[47] Gould, S.A.; Sehgal, L.R.; Sehgal, H.L.; DeWoskin, R. and Moss, G.S. (1998) The clinical development of human polymerized hemoglobin. In: Chang, T.M.S. (Ed.) Blood Substitutes: Principles, Methods, Products and Clinical Trials. Vol. 2. Karger, Basel; pp. 12-28.
[48] Pearce, L.B. and Gawryl, M.S. (1998) Overview of preclinical and clinical efficacy of Biopure's HBOCs. In: Chang, T.M.S. (Ed.) Blood Substitutes: Principles, Methods, Products and Clinical Trials. Vol. 2. Karger, Basel; pp. 82-98.
[49] Adamson, J.G. and Moore, C. (1998) Hemolink ™, an o-Raffinose crosslinked hemoglobin-based oxygen carrier. In: Chang, T.M.S. (Ed.) Blood Substitutes: Principles, Methods, Products and Clinical Trials. Vol. 2. Karger, Basel; pp. 62-79.

[50] Chang, T.M.S. (Ed.) (2000) Is there a need for blood substitutes in the new millennium and what can we expect in the way of safety and efficacy? Artif. Cells, Blood Substit. & Immobil. Biotechnol. 28(1): i-vii.
[51] D'Agnillo, F. and Chang, T.M.S. (1998) Polyhemoglobin-superoxide dismutase, catalase as a blood substitute with antioxidant properties. Nat. Biotechnol. 16(7): 667-671.
[52] D'Agnillo, F. and Chang, T.M.S. (1997) Modified hemoglobin blood substitute from Cross-linked hemoglobin-superoxide dismutase-catalase. US patent 5,606,025, Feb, 1997.
[53] D'Agnillo, F. and Chang, T.M.S. (1998) Absence of hemoprotein-associated free radical events following oxidant challenge of crosslinked hemoglobin-superoxide dismutase-catalase. Free Radic. Biol. Med. 24(6): 906-912.
[54] Razack, S.; D'Agnillo, F. and Chang, T.M.S. (1997) Effects of Polyhemoglobin-catalase-superoxide dismutase on oxygen radicals in an ischemia-reperfusion rat intestinal model. Artif. Cells, Blood Substit. & Immobil. Biotechnol. 25: 181-192.
[55] Powanda, D. and Chang, T.M.S. (2002) Cross-linked polyhemoglobin-superoxide dismutase-catalase supplies oxygen without causing blood brain barrier disruption or brain edema in a rat model of transient global brain ischemia-reperfusion. Artif. Cells, Blood Substit. & Immobil. Biotechnol. 30: 25-42.
[56] Winslow, R.M.; Vandegriff, K.D. and Intaglietta, M. (Eds.) (1997) Blood Substitutes: industrial opportunities and medical challenges. Birkhauser, Boston.
[57] Rudolph, A.S.; Rabinovici, R. and Feuerstein, G.Z. (Eds.) (1997) Red Blood Cell Substitutes. Marcel Dekker, Inc., N.Y.
[58] Tsuchida, E. (Ed.) (1998) Blood Substitutes: Present and Future Perspectives. Elservier, Amsterdam.
[59] Bunn, H.F. and Jandl, J.H. (1968) The renal handling of hemoglobin. Trans Assoc. Am. Physicians 81: 147-148.
[60] Nelson, D.J. (1998) Blood and HemAssist™ [DCLHb]: Potentially A complementary therapeutic team. In: Chang, T.M.S. (Ed.) Blood Substitutes: Principles, Methods, Products and Clinical Trials. Vol. 2. Karger, Basel; pp. 39-57.
[61] Hoffman, S.J.; Looker, D.L. and Roehrich, J.M. (1999) Expression of fully functional tetrameric human hemoglobin in *Escherichia coli*. Proc. Natl. Acad. Sci. USA 87: 8521-8525.
[62] Freytag, J.W. and Templeton, D. (1997) Optro™ [Recombinant Human Hemoglobin]: A Therapeutic for the Delivery of Oxygen and The Restoration of Blood Volume in the Treatment of Acute Blood Loss in Trauma and Surgery. In: Rudolph, A.S., Rabinovici, R. and Feuerstein, G.Z. (Eds.) Red Cell Substitutes; Basic Principles and Clinical Application. Marcel Dekker, Inc., New York; pp. 325-334.
[63] Doherty, D.H.; Doyle, M.P.; Curry, S.R.; Vali, R.J.; Fattor, T.J.; Olson, J.S. and Lemon, D.D. (1998) Rate of reaction with nitric oxide determines the hypertensive effect of cell-free hemoglobin. Nat. Biotechnol. 16: 672-676.
[64] Djordjevich, L. and Miller, I.F. (1980) Synthetic erythrocytes from lipid encapsulated hemoglobin. Exp. Hematol. 8: 584-586.
[65] Philips, W.T.; Klpper, R.W.; Awasthi, V.D.; Rudolph, A.S.; Cliff, R.; Kwasiborski, V.V. and Goins, B.A. (1999) Polyethylene glyco-modified liposome-encapsulated hemoglobin: a long circulating red cell substitute. J. Pharm. Exp. Therapeutics 288: 665-670.
[66] Yu, W.P. and Chang, T.M.S. (1996) Submicron polymer membrane hemoglobin nanocapsules as potential blood substitutes: preparation and characterization. Artif. Cells, Blood Substit. & Immobil. Biotechnol. 24:169-184.
[67] Chang, T.M.S. and Yu, W.P. (1998) Nanoencapsulation of hemoglobin and red blood cell enzymes based on nanotechnology and biodegradable polymer. In: Chang, T.M.S. (Ed.) Blood Substitutes: Principles, Methods, Products and Clinical Trials. Vol. 2. Karger, Basel; pp 216-231.
[68] Chang, T.M.S. and Yu, W.P. (1997) Biodegradable polymer membrane containing hemoglobin for blood substitutes. U.S.A. Patent 5,670,173, September, 23, 1997.
[69] Chang, T.M.S.; Powanda, D. and Yu, W.P. (2002) Biodegradable Polymeric Nanocapsules and uses thereof. PCT 2002.

PART 3

FOOD AND BEVERAGE APPLICATIONS

BEER PRODUCTION USING IMMOBILISED CELLS

VIKTOR NEDOVIĆ[1], RONNIE WILLAERT[2], IDA LESKOŠEK-ČUKALOVIĆ[1], BOJANA OBRADOVIĆ[3] AND BRANKO BUGARSKI[3]

[1]*Departement of Food Technology and Biochemistry, Faculty of Agriculture, University of Belgrade, Nemanjina 6, PO Box 127, 11081 Belgrade-Zemun, Serbia and Montenegro – Fax: +381 11 193659 – Email: vnedovic@EUnet.yu;*
[2]*Department of Ultrastructure, Flanders Interuniversity Institute for Biotechnology, Vrije Universiteit Brussel, Pleinlaan 2, B-1050 Brussel, Belgium – Fax: 32-2-6291963 – Email: Ronnie.Willaert@vub.ac.be;*
[3]*Department of Chemical Engineering, Faculty of Technology and Metallurgy, University of Belgrade, Karnegijeva 4, 11000 Belgrade, Serbia and Montenegro – Fax: +38111337038 – Email: bojana@elab.tmf.bg.ac.yu*

1. Introduction

The brewing of beer is one of the oldest applications in biotechnology; the oldest historical evidence of formal brewing dates back to about 6000 B.C. in ancient Babylonia. It was only at the end of the 19th and the beginning of the 20th century that brewing evolved to an efficient and well-controlled bioprocess. Increased understanding of brewing fermentation kinetics and mechanism led to design of new accelerated fermentation methods, which incorporate improved batch bioreactors ranging from open, relatively shallow tanks to large cylindroconical fermentors. Furthermore, advances were made in development of continuous beer fermentation processes attractive for many advantages, which continuous mode of operation offers as compared to batch operation such as greater efficiency in utilisation of carbohydrates and better use of equipment.

Since the beginning of the 20th century, many different systems using suspended yeast cells have been developed. The excitement for application of continuous beer fermentation led to development of various interesting systems especially during the 1950 and 1960's. These systems can be classified as: (i) stirred *versus* unstirred tank reactors, (ii) single-vessel systems *versus* systems consisting of a number of vessels connected in series, (iii) vessels which allow yeast to overflow freely with the beer ("open systems") *versus* vessels which have abnormally high yeast concentrations ("closed" or "semi-closed systems") [1-4]. However, these continuous beer fermentation processes were not commercially successful due to many practical problems, such as increased risk of contamination (not only during fermentation but also during storage of wort in supplementary holding tanks required for usually batch upstream and downstream brewing processes), variations in beer flavour [5] and poor understanding of the beer fermentation kinetics under continuous conditions. One of the well-known exceptions is successful implementation of a continuous beer production process in New Zealand by Morton Coutts (Dominion Breweries) still in use today [1,3].

Table 1. ICT requirements for beer production.

- High cell loading capacity
- Easy access to nutrient media
- Simple and gentle immobilisation procedure
- Immobilisation compounds approved for food applications
- High surface area-to-volume ratio
- Optimum mass transfer distance from flowing media to the support interior
- Mechanical stability (compression, abrasion)
- Chemical stability
- Highly flexible: rapid start-up after shut-down
- Sterilisable and reusable
- Suitable for conventional reactor systems
- Low shear experienced by cells
- Easy separation of carriers with immobilized cells from media
- Simple scale-up
- Economically feasible (low capital and operating costs)
- Desired flavour profile and consistent product quality
- Complete attenuation
- Controlled oxygenation
- Low risk of contamination
- Controlled yeast growth
- Wide choice of yeast

In the 1970's, there was a revival in development of continuous beer fermentation systems due to progress in research focused on immobilisation of living cells. Over the last 30 years, immobilised cell technology (ICT) for beer production has been extensively investigated and some systems have already reached commercial exploitation. Main advantages of using immobilised cells for production of beer are enhanced fermentation productivity due to higher biomass densities, improved cell stability, easier implementation of continuous operation, improved operational control and flexibility, facilitated cell recovery and reuse, and simplified downstream processing. Intensification of a particular fermentation process using ICT is generally industrialised if the acquired new characteristics result in a more economic system and the new technology can be readily scaled up [6,7]. Key parameters of this technology are selections of carrier material and method of immobilisation together with the bioreactor design. Determination of these parameters is directed by operational conditions such as temperature, pH, substrate composition, and fluid dynamics where special attention should be paid to mass transfer properties since limited nutrient supply can result in changes in yeast metabolism leading to inadequate flavour of the final product. Desirable features of immobilised cell systems applicable in the beverage industry are listed in Table 1 [8,9].

In this chapter, an overview of carrier materials, immobilisation methods and guidelines for bioreactor design aimed for applications in beer production is presented. In addition, examples of industrial applications of ICT in brewing are also described.

2. Carrier selection and design

Cell immobilisation can be classified into four categories based on the mechanism of cell localisation and the nature of support material: (i) attachment to the support surface, which can be spontaneous or induced by linking agents; (ii) entrapment within a porous matrix; (iii) containment behind or within a barrier; and (iv) self-aggregation, naturally or artificially induced [10-13]. It should be stressed that for beverage (and food) applications special attention must be paid to the selection of approved, food grade compounds or to prevention of any leakages of undesired compounds into the beverages. In Table 2, a list of the various carrier materials, which have been investigated for application in beer production, is presented.

Table 2. Cells immobilisation systems used for the production of beer.

Immobilisation system	Carrier material	Reference
Attachment to a surface	DEAE-cellulose	[14,16-29]
	Gluten pellets	[30,31]
	Spent grain	[54,55]
	Wood chips	[50-53]
Entrapment within porous support		
• Gel entrapment	Ca-alginate beads and microbeads	[32-41]
	Ca-pectate beads	[39,41]
	κ-Carrageenan beads	[42-44]
	Chitosan beads	[33]
	PVA beads	[45]
	PVA lens shaped particles	[46-49]
• Preformed support	Ceramic beads	[22,56]
	Diatomaceous earth	[57]
	Glass beads	[16,22-26,58-63]
	Silica beads	[16,23-26,64,65]
	Silicon carbide rods	[63,66-72]
	Sponge material	[73]
Self-aggregation	Continuous fermentation with cell recycle	[3,74]

Cell immobilisation by adsorption to a support material is a very popular method because it is simple, cheap and fast. Micro-organisms adsorb spontaneously on a wide variety of organic and inorganic supports. Binding of cells occurs through interactions such as Van der Waals forces, ionic bonds, hydrogen bridges or covalent interactions. Microbial cells exhibit a dipolar character and behave as cations or anions depending on the cell type and environmental conditions such as pH of the solution. Furthermore, cell physiology has a significant influence on the strength of adhesion. Various rigid support materials are available mainly aimed at applications in packed-bed reactors. DEAE-cellulose supports have been successfully used on industrial scale for the production of alcohol-free beer and maturation of green beer [14]. It is an inert, non-dissolving cellulose matrix, which has a non-uniform granular shape. Yeast cells are immobilised by ionic attraction. Compared to porous supports, the biomass loading capacity of DEAE-cellulose is considerable lower.

Cell immobilisation in porous matrices can be performed by two different basic methods. In the first, commonly regarded as gel entrapment, the porous matrix is synthesised *in situ* around the cells to be immobilised. In the second, cells are allowed to move into the preformed porous matrix. Generally, both methods provide cell protection from the fluid shear and higher cell densities as compared to surface immobilisation. A drawback in these systems can be mass transfer limitations. However, understanding of mass transfer phenomena within entrapment matrices may allow one to simultaneously provide different conditions at the carrier surface and in the interior, which could be attractive for co immobilisation of different cell types performing consecutive processes [15].

Over the last thirty years, most of the research concerning the immobilisation of living microbial cells was focused on gel entrapment. Polysaccharides (*e.g.* alginate, chitosan, pectate and carrageenan), synthetic polymers (*e.g.* polyvinylalcohol, PVA) and proteins (gelatine, collagen) can be gelled into hydrophilic matrices under mild conditions, thus allowing cell entrapment with minimal loss of viability. As a result, very high biomass loadings can be achieved. Gels are mostly used in form of spherical beads with diameters ranging from about 0.3 to 5 mm. However, a disadvantage of gels is a limited mechanical stability. It has been frequently observed that the gel structure is easily destroyed by growth of immobilised cells and carbon dioxide production. Moreover, Ca-alginate is weakened in the presence of phosphates. Long term use of Ca-alginate beads in continuous production (maturation) of beer resulted in loss of bead 3D-structure due to high phosphate contents in wort/beer. However, several methods were proposed for reinforcement of gel structures. For example, alginate gel can be strengthened by reaction with polyethyleneimine, glutaraldehyde crosslinking, addition of silica, genepin, and polyvinylalcohol or by partial drying of the gel [11].

Unlike gels, porous preformed supports can be inoculated directly from the bulk medium. In these systems, cells are not completely separated from the effluent, similarly as in the adsorption method. Cell immobilisation occurs by attachment to the internal surfaces, self-aggregation and retention in dead-end pockets within the material. Ideally, the colonised porous particles should retain some void spaces for flow so that mass transport of substrates and products could be achieved by both molecular diffusion and convection. Consequently, mass transport limitations are less stringent under optimal conditions as compared to gel entrapment method. However, fluid flow within the support can be realised only if cell adhesion is not very strong so that excessive biomass could be washed out from the matrix. When high cell densities are obtained, convection is no longer possible and the particles behave as dense cell agglomerates with high diffusion limitations. Yet the cell densities represented per unit of support volume are lower than those achievable by gel entrapment since the porous matrix material takes up significant volume fraction. As compared to gel particles, preformed carriers provide better mechanical properties and higher resistances to compression and disintegration.

Cell immobilisation behind or within a porous barrier includes systems with a barrier formed around cells such as microcapsules, and systems with cells contained within a compartment separated by a preformed membrane such as hollow fibre and flat membrane modules. However, microencapsulation is generally too expensive to be used in beer production. Entrapment behind preformed membranes represents a gentle immobilisation method since no chemical agents or harsh conditions are employed. Usually polymeric microfiltration or ultrafiltration membranes were used, although other types of membranes were also investigated, such as ceramic, silicone or ion exchange membranes. Mass transfer through the membrane is dependent on the pore size and structure as well as on the hydrophobicity/hydrophilicity and surface charge. A special design of a multichannel loop bioreactor has been used by the Belgian company Meura (Tournai) for production of lager and ale, and acidified wort [8]. Yeast cells are immobilised in porous sintered silicon carbide rods perforated with 19 or 37 channels for fluid flow. This immobilisation method can be regarded as containment behind a preformed barrier, or as entrapment in a porous preformed support.

Cell immobilisation by self-aggregation is based on formation of cell clumps or floccules, which can be naturally occurring as in the case of flocculent yeast strains, or induced by addition of flocculating agents. It is the simplest and the least expensive immobilisation method. Continuous beer fermentation technology using yeast flocculation and cell recycling, has been successfully exploited over almost 40 years by Dominion Breweries in New Zealand [3,13,75]. This fermentation system consists of a hold-up vessel followed by two stirred tank fermentors for the primary fermentation. Subsequently, the flocculent yeast cells are separated from the green beer in a conical settler by gravity. Yeast is then recycled back into the hold-up vessel to increase the cell density and to achieve better control of the fermentation rates.

3. Reactor design

Selection of the appropriate reactor type or configuration for an immobilised cell system must be based on critical issues such as supply and removal of gases and solutes in the liquid phase and removal of excess biomass formed. Cell aggregates can only be fully active if the external supply or removal rates match the internal transport, utilisation and production rates. High cell densities in the reactor put higher demands on nutrient supply and the transport rates. ICT bioreactors can be classified into three categories, depending on the location of immobilised cells: (i) mixed with suspended carriers, (ii) fixed carrier particles or large surfaces, and (iii) moving carrier surfaces [76,77].

For fermented food or beverage production, bioreactors of category (i) and (ii) are usually employed. Many reactor types can be modified to adapt to specific demands imposed by the substrate, organism or operational conditions. Beer production is based on the utilisation of yeast cells, while the acidification of wort according to the "Reinheitsgebot", employs lactic acid bacteria. Reactor configuration is related to the choice of cell carrier and various modifications and combinations of stirred tank, packed-bed, fluidised-bed, gas-lift and membrane reactors were proposed for different phases in beer production.

Operation mode of immobilised cell reactors can be batch and fed-batch running on a drain-and-fill basis. On the other hand, continuous operation eliminates the unproductive time in batch and fed-batch processes, associated with filling, emptying, cleaning and disinfection/sterilisation and start-up phase of the fermentation. Important aspects affecting the selection of reactor design and operation include reactor sterilisation and sterile transfer of immobilised biocatalysts. Regarding these features, it is desirable to use reactors that can be thoroughly sterilised and directly inoculated with cells or cell-aggregates such as membrane modules or reactors packed with preformed porous carriers. Gel entrapment and microencapsulation are more problematic in this respect as sterile transfer from the immobilisation equipment to the reactor is non-trivial, especially on a large scale [7]. Furthermore, primary beer fermentation systems are much more sensitive to contamination than the less nutritious secondary fermentation systems [78,79]. Enterobacteria and acetic acid bacteria have been detected in immobilised yeast bioreactors used for primary fermentation while lactic acid bacteria were found in reactors used for secondary fermentations. Additionally, wild yeasts may also contaminate both primary and secondary fermentation systems. Various methods are available to fight development of contaminants in an ICT bioreactor. Some methods, like sulphite addition (widely used in the wine industry) and heat treatment, are also used in suspended and immobilised cell systems. In addition, there are methods applicable only in immobilised cell systems, like the use of high dilution rates or harsher environmental conditions (*e.g.* pH, temperature, salt concentrations, ...) [78, 80].

4. ICT applications for the brewing industry

4.1. FLAVOUR MATURATION OF GREEN BEER

The main objective of flavour maturation (or secondary fermentation) is the removal of the vicinal diketones diacetyl and 2,3-pentanedione, and their precursors α-acetolactate and α-acetohydroxybutyrate. Vicinal diketones are formed by an oxidative decarboxylation from the excess α-acetohydroxy acids, which leak from the isoleucine-valine pathway. Diacetyl is reduced by yeast reductases to 2,3-butanediol via acetoin and 2,3-pentanedione to 2,3-pentanediol via acethylethylcarbinol. The conversion of the α-acetohydroxy acids to the vicinal diketones is the rate-limiting step. This reaction step is accelerated by heating the beer after yeast separation to 80 ~ 90°C for several minutes. The resulting vicinal diketones can be subsequently reduced by suspended as well as by immobilised yeast cells.

The traditional maturation process is characterised by a near-zero temperature, low pH and low yeast concentration, resulting in a maturation period of 3 to 4 weeks. ICT can reduce this period to 2 hours. An ICT maturation process using a packed-bed bioreactor with DEAE-cellulose granules has been successfully integrated in Synebrychoff Brewery (Finland) for the treatment of 1 million hl per year [20]. Later on this carrier was replaced by cheaper aspen wood chips for yeast cell immobilisation [79,81]. Alfa Laval and Schott Engineering developed a maturation system based on porous glass beads (Figure 1) [82]. This system has been implemented in several breweries in Finland, Belgium and Germany. The German company Brau & Brunnen has purchased and installed a 30000 hl/year pilot scale Alfa Laval maturation system in 1996 [83]. The same system has been implemented in a medium-sized German brewery (Schäff/Treuchtlingen) [84]. The obtained beers yielded overall good analytical and sensorial results.

A new concept in which heat treatment of green beer is replaced by enzymatic transformation of α-acetolactate to diacetyl was recently proposed by EasyProof Laborbedarf GmbH, CAVIS and TU Munchen-Weihenstephan, Germany [85,86]. This reaction catalysed by α-acetolactate decarboxylase immobilised in Multi-Layer-Capsules is carried out in a packed-bed reactor. The next phase in the process is the reduction of diacetyl in a second reactor packed with new open-porous swelling-glass Immopore carriers with immobilised yeast cells. This system enabled complete beer maturation in less then 4 hours. The obtained beer showed no significant differences as compared to conventionally matured beers.

Accelerated beer maturation can also be performed by increasing the maturation temperature and by pH reduction [87], and selecting a fast fermenting yeast strain. This can be realised by integrating the secondary fermentation in the primary fermentation using the cylindroconical vessel (CCV) as fermentor (which is used today in almost every brewery). In this way, the production of beer can be accomplished in less than 14 days in one CCV. This new fermentation technology can be used as cost effectively as ICT.

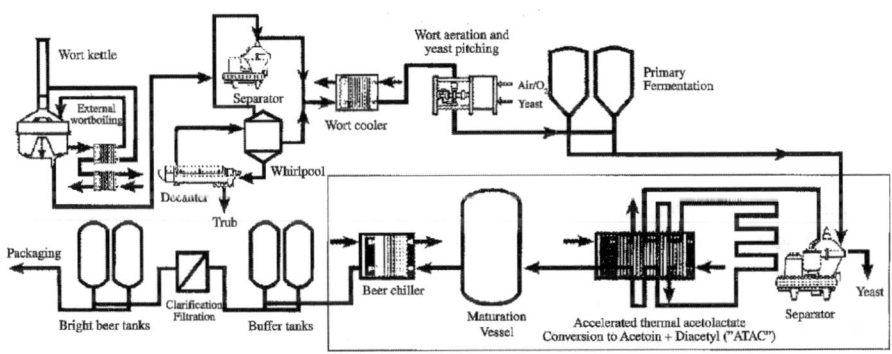

Figure 1. Process flowsheet for secondary fermentation of beer using the Alfa Laval system (from Willaert [91]).

4.2. PRODUCTION OF ALCOHOL-FREE OR LOW-ALCOHOL BEER

The traditional technology to produce alcohol-free or low-alcohol beer is based on the suppression of alcohol formation by arrested batch fermentation [88]. However, the resulting beers are characterised by an undesirable wort aroma since in this process, wort aldehydes have only been reduced to a limited degree [17,28,89]. Reduction of these wort aldehydes can be quickly achieved by short-contact with immobilised yeast cells at low temperatures without undesirable cell growth and ethanol production. A disadvantage of this short contact process is limited production of desirable esters, obtained only in small amounts.

Controlled ethanol production for low-alcohol and alcohol-free beers have been successfully achieved by partial fermentation using yeast immobilised on DEAE-cellulose, which was packed in a column reactor [17,21]. This technology was successfully implemented by Bavaria Brewery (The Netherlands) to produce malt beer on an industrial scale with capacity of 150000 hl/year [18]. Several other companies including Faxe (Denmark), Ottakringer (Austria) and a Spanish brewery, have also purchased this technology [83]. In Brewery Beck (Germany), a fluidised-bed pilot scale reactor (8 hl/day) filled with porous glass beads was used for the continuous production of non-alcohol beer [58,59,61]. Yeast cells immobilised in silicon carbide rods, arranged in a multichannel loop reactor (Meura, Belgium), have been used to produce alcohol-free beer at a pilot scale by Grolsch Brewery (The Netherlands) and Guinness Brewery (Ireland) [71].

4.3. PRODUCTION OF ACIDIFIED WORT USING IMMOBILISED LACTIC ACID BACTERIA

The objective of ICT application in the production of acidified wort is wort acidification according to the "Reinheidsgebot", before the start of the boiling process in the brewhouse. An increased productivity of acidified wort has been obtained using immobilised *Lactobacillus amylovorus* on DEAE-cellulose beads [18,19]. The pH of wort was reduced below 4.0 after contact times of 7 to 12 min using a packed-bed reactor in the downflow mode. The produced acidified wort was stored in a holding tank and used during wort production to adjust the pH.

4.4. CONTINUOUS MAIN FERMENTATION

The Japanese brewery Kirin developed a multistage continuous fermentation process for main beer fermentation [56,60,64]. The first stage is a stirred tank reactor (chemostat) for yeast growth, followed by packed-bed fermentors, and the final step consisting of a packed-bed maturation column (Figure 2). The first stage ensures adequate yeast cell growth with desirable free amino nitrogen consumption. Ca-alginate beads were initially selected as carrier material to immobilise the yeast cells; it was later replaced by ceramic beads ("Bioceramic®"). Beer could be produced in this process within three to five days.

Meura developed a system shown in Figure 3, consisting of an immobilised cell reactor where partial attenuation and yeast growth occurs, followed by a stirred tank for complete attenuation, ester formation and flavour maturation [67,90]. Silicon carbide rods are used in the first reactor as the immobilisation matrix. The stirred tank in this system is continuously inoculated by free cells, which escape from the immobilised cell reactor.

Another promising concept, a gas-lift bioreactor system (Figure 4), was successfully introduced in main beer fermentation studies by a research group at the University of Belgrade [34]. A gas-lift reactor retains the advantages of fluidised-beds, such as high loading of solids and good mass transfer properties and it is particularly suitable for applications with low density carriers. Laboratory and semi-pilot scale bioreactor systems were developed with alginate microbeads (0.8 mm in diameter) and lens shaped PVA particles as carriers for yeast cells [12,37,38,46-48]. Full beer attenuation in these studies was reached within 7.5 to 20 hours depending on solid loading (10 to 40%) [37,47]. Final beers had desired sensory and analytical profiles. A similar system that use κ-carrageenan immobilised yeast cells in an airlift reactor was developed by Labatt Breweries (Interbrew, Canada) in collaboration with the Dept. of Chemical and Biochemical Engineering at the University of Western Ontario [42-44,83]. Pilot scale research showed that in this system full attenuation was reached in 20-24 hours compared to 5-7 days required in the traditional batch fermentation. Flavour profile of the beer produced using ICT was similar to the batch fermented beer.

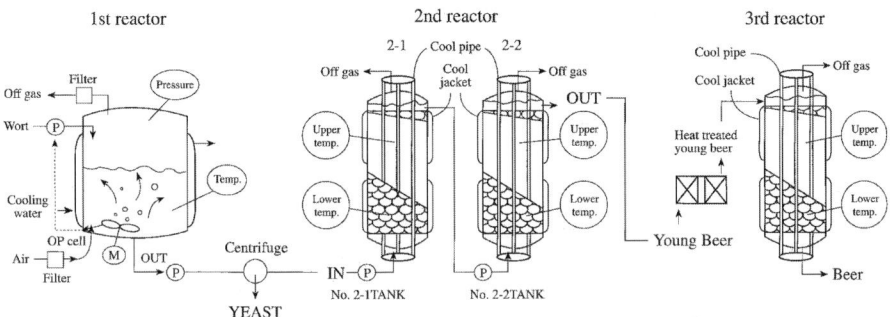

Figure 2: The Kirin three stage bioreactor system for continuous main fermentation and maturation (from Willaert [91]).

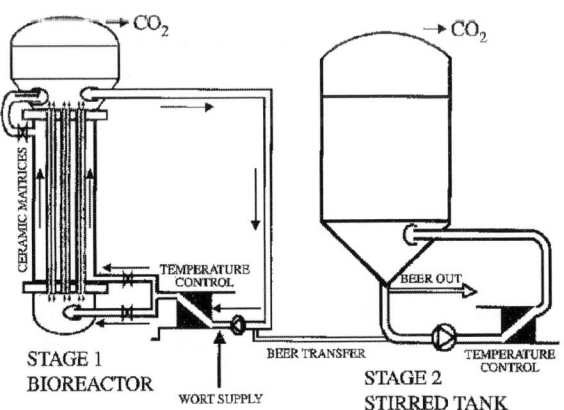

Figure 3: The Meura multichannel loop configuration (from Willaert [91]).

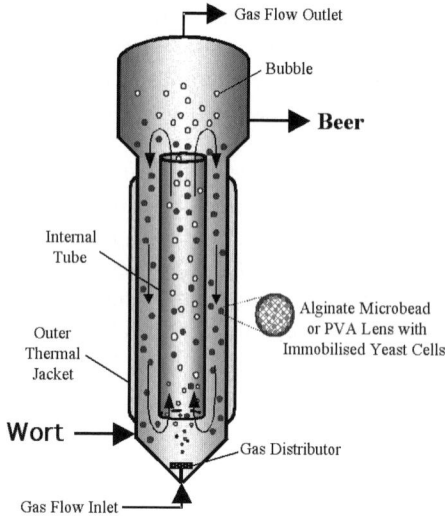

Figure 4. Gas-lift bioreactor system.

Hartwell Lahti and VTT Research Institute (Finland) developed a primary fermentation system using ICT on a pilot scale of 600 l/day [50]. This system is later on extended to include a continuous secondary fermentation unit with the same capacity [51]. Wood chips were used as the carrier material, which reduced the total investment cost by one third compared to other more expensive carriers. The results showed that in only 40 hours beer composition and flavour were very similar to beer produced by the traditional batch process.

Synebrychoff Brewery (Finland) in collaboration with Guinness, GEA Liquid Processing Scandinavia and Cultor Corporation of Finland developed a new ICT process in which the concentration of carbon dioxide is controlled in a fixed-bed reactor [29,52,53]. In this way, the CO_2 formed is kept dissolved and removed from the beer without foaming problems. DEAE-cellulose at the beginning, and wood chips later on, were used as carrier materials [52,53]. Good quality beer and constant flavour profile were achieved at a production time of 20 to 30 hours [53].

5. Summary

In this chapter, an overview of ICT applications in beer production is presented. ICT is well established for flavour maturation and alcohol-free and low-alcohol beer production on an industrial scale. In addition, several primary beer fermentation processes based on ICT are also developed on pilot and industrial scale. However, mass transfer limitations and process control are still issues that have to be resolved in order to obtain consistent quality of the beer. This chapter also outlines directions regarding the selection of carrier material, immobilisation method, and reactor design exploring possibilities for process optimisation.

6. References

[1] Hough, J.S.; Briggs, D.E.; Stevens, R. and Young, T.W. (1982) Malting and Brewing Science – Volume 2: Hopped wort and beer. Chapman and Hall, London.
[2] Wellhoener, H.J. (1954) Ein Kontinuierliches Gär- und Reifungsverfahren für Bier. Brauwelt 94(1): 624-626.
[3] Coutts, M.W. (1957) A continuous process for the production of beer. U.K. Patents 872,391-400.
[4] Bishop, L.R. (1970) A system of continuous fermentations. J. Inst. Brew. 76: 172-181.
[5] Thorne, R.S.W. (1968) Continuous fermentation in retrospect. Brew. Dig. 43(2): 50-55.
[6] Groboillot, A.; Boardi, D.K.; Poncelet, D. and Neufeld, R.J. (1994) Immobilization of cells for application in the food industry. Crit. Rev. Biotechnol. 14: 75-107.
[7] Keshavarz, T.; Bucke, C. and Lilly, M.D. (1996) Problems in scale-up of immobilized cell cultures. In: Wijffels, R.H.; Buitelaar, R.M.; Bucke, C. and Tramper, J. (Eds.) Immobilized cells: basics and applications. Elsevier, Amsterdam; pp. 505-510.
[8] Masschelein, C.A.; Ryder, D.S. and Simon, J.-P. (1994) Immobilized cell technology in beer production. Crit. Rev. Biotechnol. 14: 155-177.
[9] Willaert, R. (1996) Immobilised cell technology for the production of beverages and foods. Proc. of International Workshop on Bioencapsulation. Bioencapsulation Research Group, Postam, Germany; Talk 9, pp. 1-4.
[10] Karel, S.F.; Libicki, S.B. and Robertson, C.R. (1985) The immobilization of whole cells: engineering principles. Chem. Eng. Sci. 40: 1321-1353.
[11] Willaert, R. and Baron, G.V. (1996) Gel entrapment and micro-encapsulation: methods, applications and engineering principles. Rev. Chem. Eng. 12: 1-205.
[12] Nedović, V.A.; Obradović, B.; Leskošek-Čukalović, I. and Vunjak-Novaković, G. (2001) Immobilized yeast bioreactor systems for brewing - recent achievements. In: Thonart, Ph. and Hofman, M. (Eds.) Focus on Biotechnology Series, Vol. 4: Engineering and Manufacturing for Biotechnology. Kluwer Academic Publishers, Dordrecht; pp. 277-292.
[13] Pilkington, H.; Maragaritis, A.; Mensour, N. and Russell, I. (1998) Fundamentals of immobilised yeast cell for continuous beer fermentation: a review. J. Inst. Brew. 104: 19-31.
[14] Lommi, H. (1990) Immobilized yeast for maturation and alcohol-free beer, Brew. Dist. Int. 5: 22-23.
[15] Willaert, R.; Van De Winkel, L. and De Vuyst, L. (1999) Improvement of maltose and maltotriose brewery fermentation efficiency using immobilized cell technology. Proc. of the 27th European Brewery Convention Congress. EBC; pp. 663-670.
[16] Linko, M. and Kronlöf, J. (1991) Main fermentation with immobilized yeast. In: Proc. of the 23rd European Brewery Convention Congress. EBC; pp. 353-360.
[17] Collin, S.; Montesinos, M.; Meersman, E.; Swinkels, W. and Dufour, J.P. (1991) Yeast dehydrogenase activities in relation to carbonyl compounds removal from wort and beer. In: Proc. of the 23rd European Brewery Convention Congress. EBC; pp. 409-416.
[18] Pittner, H.; Back, W.; Swinkels, W.; Meersman, E.; Van Dieren, B. and Lomni, H. (1993) Continuous production of acidified wort for alcohol-free-beer with immobilized lactic acid bacteria. In: Proc. of the 24th European Brewery Convention Congress. EBC; pp. 323-329.
[19] Meersman, E. (1994) Biologische aanzuring met geïmmobiliseerde melkzuurbacteriën. Cerevisia Biotechnol 19(4): 42-46.
[20] Pajunen, E. (1995) Immobilized yeast lager beer maturation: DEAE-cellulose at Synebrychoff. In: EBC Monograph XXIV, EBC Symposium on immobilized yeast applications in the brewing industry; pp. 24-40.
[21] Van Dieren, D. (1995) Yeast metabolism and the production of alcohol-free beer. In: EBC Monograph XXIV, EBC Symposium on immobilized yeast applications in the brewing industry; pp. 66-76.
[22] Cashin, M.M. (1996) Comparative studies of five supports for yeast immobilisation by adsorption/attachment. J. Inst. Brew. 102: 5-10.
[23] Wackerbauer, K.; Fitzner, M. and Günther, J. (1996) Technisch-technologische Möglichkeiten mit immobilisierter Hefe. Brauwelt 45: 2140-2150.
[24] Wackerbauer, K.; Fitzner, M. and Lopsien, M. (1996) Untersuchungen mit dem neuen MPI-Bioreaktor-System. Brouwelt 46/47: 2250-2256.
[25] Wackerbauer, K.; Ludwig, A.; Mohle, J. and Legrand, J. (2003) Measures to improve long term stability of main fermentation with immobilized yeast. Proc. of the 29th European Brewery Convention Congress. EBC; pp. 1-11.

[26] Ludwig, A. and Wackerbauer, K. (2004) Quality improvement in continuous main fermentation with immobilized yeast. In: World Brewing Congress 2004 Proceedings, July 24-28, San Diego, California; ASBC and MBAA, ISBN: 0-9718255-5-6, CD Rom; O-81.
[27] Van Iersel, M.F.M.; Meersman, E.; Arntz, M.; Ronbouts, F.M. and Abee, T. (1998) Effect of environmental conditions on flocculation and immobilization of brewer's yeast during production of alcohol-free beer. J. Inst. Brew. 104: 131-136.
[28] Van Iersel, M.F.; Brouwer-Post, E.; Rombouts, F.M. and Abee, T. (2000) Influence of yeast immobilization on fermentation and aldehyde reduction during the production of alcohol-free beer. Enzyme Microb. Technol. 26: 602-607.
[29] Andersen, K.; Bergin, J.; Ranta, B. and Viljava, T. (1999) New process for the continuous fermentation of beer. In: Proceedings of the 27th European Brewery Convention Congress. EBC; pp. 771-778.
[30] Bardi, E.; Koutinas, A.A. and Kanellaki, M. (1997) Room and low temperature brewing with yeast immobilized on gluten pellets. Process Biochem. 32: 691-696.
[31] Bekatorou, A.; Koutinas, A.A.; Psarianos, K. and Kanellaki, M. (2001) Low-temperature brewing by freeze-dried immobilized cells on gluten pellets. J. Agric. Food Chem. 49: 373-377.
[32] Onaka, T.; Nakanishi, K.; Inoue, T. and Kubo, S. (1985) Beer brewing with immobilized yeast. Bio/Technol. 3: 467-470.
[33] Shindo, S.; Sahara, H. and Koshino, S. (1994) Suppression of α-acetolactate formation in brewing with immobilized yeast. J. Inst. Brew. 100: 69-72.
[34] Nedović, V.; Obradović, B.; Vunjak-Novaković, G. and Leskošek-Čukalović, I. (1993) Kinetics of beer fermentation with immobilized yeast cells in an internal loop air-lift bioreactor. Chem. Ind. 47: 168-172.
[35] Nedović, V.A.; Leskošek-Čukalović, I. and Vunjak-Novaković, G. (1996) Short-time fermentation of beer in an immobilized yeast air-lift bioreactor. In: Proc. 24th Conv. Inst. Brew., Singapore, Winetitles, Adelaide; p. 245.
[36] Nedović, V.A.; Leskošek-Čukalović, I.; Milošević, V. and Vunjak-Novaković, G. (1997) Flavour formation during beer fermentation with immobilized *Saccharomyces cerevisiae* in a gas-lift bioreactor. In: Godia, F. and Poncelet, D. (Eds.) Proc. Int. Workshop Bioencapsulation VI: "From fundamentals to industrial applications", Barcelona, Spain; T5.3, pp. 1-4.
[37] Nedović, V.; Obradović, B.; Leskošek-Čukalović, I.; Trifunović, O.; Pešić, R., and Bugarski, B. (2001) Electrostatic generation of alginate microbeads loaded with brewing yeast. Proc. Biochem. 37: 17-22.
[38] Nedović, V.A.; Obradović, B.; Leskošek-Čukalović, I.; Korać, A. and Bugarski, B. (2002) Alginate-immobilized yeast cells for continuous beer brewing in a gas-lift bioreactor. In: Proceedings of X International BRG Workshop on Bioencapsulation "Cell Physiology and Interactions of Biomaterials and Matrices", Prague, Czech Republic; pp. 152-155.
[39] Smogrovicova, D.; Domeny, Z.; Gemeiner, P.; Malovikova, A. and Sturdik, E. (1997) Reactors for continuous primary beer fermentation using immobilised yeast. Biotechnol. Techn. 11: 281-286.
[40] Smogrovicova, D.; Domeny, Z. and Svitel, J. (2001) Modeling of saccharide utilization in primary beer fermentation with yeasts immobilized in calcium alginate. Appl. Biochem. Biotechnol. 94: 147-158.
[41] Navratil, M.; Domeny, Z.; Sturdik, E.; Smogrovicova, D. and Gemeiner, P. (2002) Production of non-alcoholic beer using free and immobilized cells of *Saccharomyces cerevisiae* deficient in the tricarboxylic acid cycle. Biotechnol. Appl. Biochem. 35: 133-140.
[42] Mensour, N.; Margaritis, A.; Briens, C.L.; Pilkington, H. and Russell, I. (1995) Gas lift systems for immobilized cell systems. In: EBC Monograph XXIV, EBC Symposium on immobilized yeast applications in the brewing industry; pp. 125-133.
[43] Mensour, N.; Margaritis, A.; Briens, C.L.; Pilkington, H. and Russell, I. (1996) Applications of immobilized yeast cells in the brewing industry. In: Wijffels, R.H.; Buitelaar, R.M.; Bucke, C. and Tramper, J. (Eds.) Immobilized cells: basics and applications. Elsevier, Amsterdam; pp. 661-671.
[44] Pilkington, H.; Maragaritis, A.; Mensour, N.; Sobczak, J.; Hancock, I. and Russell, I. (1999) Kappa-carrageenan gel immobilization of lager brewing yeast. J. Inst. Brew. 105: 398-404.
[45] Shindo, S. and Kamimur, M.J. (1990) Immobilization of yeast with hollow PVA beads. J. Ferm. Bioeng. 70: 232-234.
[46] Nedović, V.A.; Obradović, B.; Bezbradica, D.; Leskošek-Čukalović, I. and Bugarski, B. (2003) Lentikats® as potential carriers for brewing yeast. In: Proceedings of XI International Workshop on Bioencapsulation "State of Art of Bio&Encapsulation Science and Technology", May 25-27, Strasbourg, France; P-1, pp. 1-4.

[47] Nedović, V.A.; Bezbradica, D.; Obradović, B.; Leskošek-Čukalović, I. and Bugarski, B. (2004) Primary beer fermentation by PVA-immobilized brewing yeast in a gas-lift bioreactor. In: World Brewing Congress 2004 Proceedings, July 24-28, San Diego, California; ASBC and MBAA, ISBN: 0-9718255-5-6, CD Rom; O-63.
[48] Bezbradica, D.; Stojanov, V.; Nedovic, V.; Obradovic, B.; Bugarski, B. and Leskosek-Cukalovic, I. (2003) Beer fermentation by PVA immobilized brewing yeasts in a gas-lift bioreactor. In: Sorvari, S. (Ed.) Proc. of the 1st International Congress on Bioreactor Technology in Cell, Tissue Culture and Biomedical Applications, Tampere, Finland; pp. 210-217.
[49] Smogrovicova, D.; Domeny, Z.; Navratil, M. and Dvorak, P. (2001) Continuous fermentation using polyvinyl alcohol entrapped yeast. In: Proc. of the 28th European Brewery Convention Congress. EBC; contribution number 50, pp. 1-9.
[50] Kronlöf, J. and Virkajärvi, I. (1999) Primary fermentation with immobilized yeast. In: Proc. of the 27th European Brewery Convention Congress. EBC; pp. 761-770.
[51] Kronlöf, J.; Virkajärvi, I.; Storgards, E.L.; Londesborough, J. and Dymond, G. (2000) Combined primary and secondary fermentation with immobilized yeast. In: World Brewing Congress 2000 Proceedings, July 27 - August 3, Orlando, Florida; P-56.
[52] Pajunen, E.; Tapani, K.; Berg, H.; Ranta, B.; Bergin, J.; Lommi, H. and Viljava, T. (2001) Controlled beer fermentation with continuous on-stage immobilized yeast reactor. In: Proceedings of the 28th European Brewery Convention Congress. EBC; contribution number 49, pp. 1-12.
[53] Tapani, K.; Soininen-Tengvall, P.; Berg, H.; Ranta, B. and Pajunen, E. (2003) Continuous primary fermentation of beer with immobilised yeast. In: Smart, K. (Ed.) Brewing yeast fermentation performance, Second Edition. Blackwell Publishing, Oxford; pp. 293-301.
[54] Branyik, T.; Vicente, A.; Cruz, J.M. and Teixeira, J. (2001) Spent grains – a new support for brewing yeast immobilisation. Biotechnol. Lett. 23: 1073-1078.
[55] Branyik, T.; Vicente, A.; Cruz, J.M. and Teixeira, J. (2002) Continuous primary beer fermentation with brewing yeast immobilized on spent grains. J. Inst. Brew. 108(4): 410-415.
[56] Inoue, T. (1995) Development of a two-stage immobilized yeast fermentation system for continuous beer brewing. In: Proc. of the 25th European Brewery Convention Congress. EBC; pp. 25-36.
[57] Narziss, L. and Hellich, P. (1971) Ein Beitrag zur wesentlichen Beschleunigung der Gärung und Reifung des Bieres. Brauwelt 111: 1491-1500.
[58] Aivasidis, A.; Wandrey, C.; Eils, H.-G. and Katzke, M. (1991) Continuous fermentation of alcohol-free beer with immobilized yeast cells in fluidized bed reactors. In: Proceedings of the 23rd European Brewery Convention Congress. EBC; pp. 569-576.
[59] Aivasidis, A. (1996) Another look at immobilized yeast systems. Cerevisia 21(1): 27-32.
[60] Yamauchi, Y.; Okamato, T.; Murayama, H.; Nagara, A.; Kashihara, T. and Nakanishi, K. (1994) Beer brewing using an immobilized yeast bioreactor design of an immobilized yeast bioreactor for rapid beer brewing system. J. Ferm. Bioeng. 78: 443-449.
[61] Breitenbücher, K. and Mistler, M. (1995) Fluidized-bed fermenters for the continuous production of non-alcoholic beer with open-pore sintered glass carriers. In: EBC Monograph XXIV, EBC Symposium on immobilized yeast applications in the brewing industry; pp. 77-89.
[62] Virkajärvi, I. and Kronlöf, J. (1998) Long-term stability of immobilized yeast columns in primary fermentation. J. Am. Soc. Brew. Chem. 56: 70-75.
[63] Tata, M.; Bower, P.; Bromberg, S.; Duncombe, D.; Fehring, J.; Lau, V.V.; Ryder, D. and Stassi, P (1999) Immobilized yeast bioreactor systems for continuous beer fermentation. Biotechnol. Prog. 15: 105-113.
[64] Yamauchi, Y. and Kashihara, T. (1995) Kirin immobilized system. In: EBC Monograph XXIV, EBC Symposium on immobilized yeast applications in the brewing industry; pp. 99-117.
[65] Yamauchi, Y.; Okamato, T.; Murayama, H.; Kajino, K.; Nagara, A. and Nogushi, K. (1995) Rapid maturation of beer using an immobilized yeast bioreactor. J. Biotechnol. 38: 109-116.
[66] Krikilion, P.; Andries, M.; Goffin, O.; Van Beveren, P.C. and Masschelein, C.A. (1995) Optimal matrix and reactor design for high gravity fermentation with immobilized yeast. In: Proc. of the 25th European Brewery Convention Congress. EBC; pp. 419-426.
[67] Andries, M.; Van Beveren, P.C.; Goffin, O. and Masschelein, C.A. (1996) Design and application of an immobilized loop bioreactor for continuous beer fermentation. In: Wijffels, R.H.; Buitelaar, R.M.; Bucke, C. and Tramper, J. (Eds.) Immobilized cells: basics and applications. Elsevier, Amsterdam; pp. 672-678.

[68] Andries, M.; Van Beveren, P.C.; Goffin, O.; Rajotte, P. and Masschelein, C.A. (2000) Results on semi-industrial continuous top fermentation with the Meura-Delta immobilized yeast fermenter. Brauwelt Int. II: 134-136.
[69] Van De Winkel, L.; Van Beveren, P.C. and Masschelein, C.A. (1991) The application of an immobilized yeast loop reactor to the continuous production of alcohol-free beer. Proc. of the 23rd European Brewery Convention Congress. EBC; pp. 577-584.
[70] Van De Winkel, L.; Van Beveren, P.C.; Borremans, E; Goossens, E. and Masschelein, C.A. (1993) High performance immobilized yeast reactor design for continuous beer fermentation. Proc. of the 24th European Brewery Convention Congress. EBC; pp. 307-314.
[71] Van De Winkel, L.; Mc Murrough, I.; Evers, G.; Van Beveren, P.C. and Masschelein, C.A. (1995) Pilot-scale evaluation of silicon carbide immobilized yeast systems for continuous alcohol-free beer production. In: EBC Monograph XXIV, EBC Symposium on immobilized yeast applications in the brewing industry; pp. 90-98.
[72] Van De Winkel, L. (1995) Design and optimization of a multipurpose immobilized yeast bioreactor system for brewery fermentations. Cerevisia 20(1): 77-80.
[73] Scott, J.A. and O'Reilly, A.M. (1995) Use of a flexible sponge matrix to immobilize yeast for beer fermentation. J. Am. Soc. Brew. Chem. 53: 67-71.
[74] Atkinson, B. and Taidi, B. (1995) Technical and technological requirements for immobilized systems. In: EBC Monograph XXIV, EBC Symposium on immobilized yeast applications in the brewing industry; pp. 17-22.
[75] Van De Winkel, L. and De Vuyst, L. (1997) Immobilized yeast cell systems in today's breweries and tomorrow's. Cerevisia 22(1): 27-31.
[76] Baron, G.V.; Willaert, R. and De Backer, L. (1996) Immobilised cell reactors. In: Willaert, R.; Baron, G.V. and De Backer, L. (Eds.) Immobilised living cell systems: modeling and experimental methods. John Wiley & Sons, Chichester; pp. 67-95.
[77] Chang, H.N. and Moo-Young, M. (1988) Analysis of oxygen transport in immobilized whole cells. In: Moo-Young, M. (Ed.) Bioreactor immobilized enzyme and cells. Elsevier Applied Science, London; pp. 33-51.
[78] Haikara, A.; Virkajärvi, I.; Kronlöf, J. and Pajunen, E. (1997) In: Proc. of the 26th European Brewery Convention Congress. EBC; pp. 439-446.
[79] Virkajärvi, I. (2002) Some developments in immobilized fermentation of beer during the last 30 years. Brauwelt Int. 20: 100-105.
[80] Champagne, C.P. (1996) Immobilized cell technology in food processing. In: Wijffels, R.H.; Buitelaar, R.M.; Bucke, C. and Tramper, J. (Eds.) Immobilized cells: basics and applications. Elsevier, Amsterdam; pp. 633-640.
[81] Anon. (2003) The Finnish flash. Brewer's Guardian 132(11): 30-32.
[82] Dillenhofer, W. and Ronn, D. (1996) Secondary fermentation of beer with immobilized yeast. Brauwelt Int.: 344-346.
[83] Mensour, N.; Margaritis, A.; Briens, C.L.; Pilkington, H. and Russell, I. (1997) New developments in the brewing industry using immobilized yeast cell bioreactor systems. J. Inst. Brew. 103: 363-370.
[84] Back, W.; Krottenthaler, M. and Braun, T. (1998) Investigations into continuous beer maturation. Brauwelt Int. III: 222-226.
[85] Nitzsche, F.; Hohn, G.; Meyer-Pittroff, R.; Berger, S. and Pommersheim, R. (2001) A new way for immobilized yeast systems: secondary fermentation without heat treatment. In: Proc. of the 28th European Brewery Convention Congress. EBC; contribution number 51, pp. 1-9.
[86] Blumelhuber, G.; Meyer-Pittroff, R. and Nitzsche, F. (2004) New results with an immobilized yeast system: secondary fermentation with Immopore. In: World Brewing Congress 2004 Proceedings, July 24-28, San Diego, California; ASBC and MBAA, ISBN: 0-9718255-5-6, CD Rom; O-61.
[87] McMurrough, I. (1995) Scope and limitations for immobilized cell systems in the brewing industry. In: EBC Monograph XXIV, EBC Symposium on immobilized yeast applications in the brewing industry; pp. 2-16.
[88] Narziss, L.; Miedaner, H.; Kern, E. and Leibhard, M. (1992) Technology and composition of non-alcoholic beers. Brauwelt Int. 4: 396.
[89] Debourg, A.; Laurent, M.; Goossens, E. and Van De Winkel, L. (1994) Wort aldehyde reduction potential in free and immobilized yeast systems. J. Am. Soc. Brew. Chem. 52: 100-106.

[90] Masschelein, C.A. and Andries, M. (1995) Future scenario of immobilized systems: promises and limitations. In: EBC Monograph XXIV, EBC Symposium on immobilized yeast applications in the brewing industry; pp. 223-241.
[91] Willaert, R. (2000) Beer production using immobilised cell technology. Minerva Biotecnol. 12: 319-330.

APPLICATION OF IMMOBILISATION TECHNOLOGY TO CIDER PRODUCTION: A REVIEW

ALAIN DURIEUX, XAVIER NICOLAY AND JEAN-PAUL SIMON
Unité de BioTechnologie, Institut Meurice, Campus Ceria, Av. E. Gryzon 1, 1070 Brussels, Belgium – Fax: 32-2-526.73.88 – Email:a.durieux@meurice.ubt.be

Abstract

Transformation of apple juice into cider is a complex process requiring activity of yeast and lactic acid bacteria to accomplish respectively alcoholic and malolactic fermentations. Despite the traditional aspect of cider, cidermaking industry researches improvement of the process to control the two fermentations and the production of flavouring components. During the two last decades interest was given to the application of the cell immobilisation technology to cider production. In this article, the advanced researches for yeast and *Oenococcus oeni* immobilisation are described.

1. Introduction

The fermentation of apple juice into cider is one of the oldest traditional beverage productions. Cider production is believed to have been practised for over 2000 years. Cider consumption is described during the Roman invasion of England. Beverage made from apple juice was a popular drink in the Celtic population. In the Middle Ages, several writings referred to cider in the temperate regions of Europe where apple tree cultivation is possible like France, Belgium, Germany, Sweden, Switzerland and North of Spain [1]. In some regions like Normandy and Brittany, cider is more popular than beer. The significant larger commercial production began in the 19th century. The apparition of larger cider production has required development of technologies to control and optimise the transformation of apple juice into cider.

Local producers today still keep the traditional process that consists in the natural fermentation by autochthon yeasts and bacteria associated to the fruit or to the cellar equipment [2]. This natural fermentation is a very unpredictable process in terms of desirable flavour compounds formation. The complex natural microflora described in the literature is represented by yeast strains like *Saccharomyces cerevisiae* (var. *uvarum*), *Kloeckera apicula*, *Pichia spp.*, *Hansenula spp.*, *Torulopsis spp.*, *Hanseniaspora valbyensis* and *Metschikowia pulcherrina* and by lactic acid bacteria like *O. oeni*, *Lactobacillus ssp.* or *Pediococcus ssp.* and some acetic acid bacteria

[3,4,5]. All these strains bring a contribution to different steps of the fermentation. The organoleptic profile of the cider after spontaneous fermentation is very variable and depends on factors like relative population of each microbial species associated to the fruit and the equipment, apple harvesting conditions, pre-treatment of the apple juice (sulphiting of apple juice, clarification with pectinolytic enzymes and fining with bentonite) and temperature of the apple juice at the initiation of the fermentation [6-9]. The growth sequence of each microbial population is very important. For example, an early development of lactic acid bacteria can cause alterations such as "piqûre lactique" due to heterolactic fermentation of sugars that leads to an increase in the volatile acidity and overproduction of acetic acid. Some lactic acid bacteria can also produce exocellular polysaccharides that thicken the consistency of the cider.

The use of selected pure cultures is well established in brewing industry and is adopted by winemaking and cidermaking industries [10-12]. The use of selected starter cultures allows a control of the whole fermentation process and uniform production of high quality cider through successive processes and seasons. The strains are selected in function of their influence on the organoleptic characteristics of the final product and of their ability to conduct the fermentation rapidly. Today several dozens of starters are available on the market and are intensively used in large cider producing companies all around the world.

The cider spoilage by bacteria is limited by sulphiting the apple juice during the pre-treatment. Use of heat-concentrated apple juice by several cidermaking companies appears as an attractive solution to avoid activity of the uncontrolled natural microflora; it requires dilution with water just before initiation of fermentation.

In the most industrial processes, transformation of apple juice into cider requires two successive fermentations. The first one is the classical alcoholic fermentation of sugars into ethanol proceeded by yeast strains like *Saccharomyces cerevisiae* species. The second one is the malolactic fermentation (MLF) that occurs after the decline phase of yeast or after a partial removing of the yeast by a filtration step. The onset of the MLF after the alcoholic fermentation permits to avoid the increase of acidity by the previous consumption of sugars.

The alcoholic fermentation is usually conducted at temperatures between 8 and 20°C. Beside the transformation of sugars (saccharose, fructose and glucose) into ethanol, the yeast is responsible for the synthesis of several compounds affecting the organoleptic profile [13,14]. The key products formed during alcoholic fermentation are fusel alcohols (isobutanol, isoamyl alcohol, propanol), esters (ethylacetate, isoamylacetate, isobutyl acetate and 2-phenyl acetate) and carbonyl compounds (diacetyl, 2,3-pentadione, acetaldehyde). The biosynthesis of higher alcohols by the yeast is linked to the amino acids metabolism [15]. Higher alcohols are generated as by-products of both anabolic (Genevois pathway) and catabolic metabolisms (Erlich pathway); it acts in restoring the redox balance involving NAD+/NADH cofactors. They can be produced during the amino acids biosynthesis pathway or by the deamination and decarboxylation of amino acids that are present in the medium. Both pathways may take place during the same fermentation process depending on the availability of amino acids in the medium. Yeast is able to switch from the degradative pathway to the biosynthetic pathway. Esters are produced by yeast during fermentation through a reaction between alcohols and acylCoA resulting from the activation of fatty acids by Co-enzyme A [16].

Carbonyl compounds in cider have very low flavour thresholds and are often considered as off-flavour at high concentration. Acetaldehyde is an intermediate in the formation of ethanol that can be secreted by the yeast cell before it undergoes the last reduction into ethanol by the alcohol dehydrogenase enzyme. Diacetyl and 2,3-pentadione are side-products of amino acids in yeast metabolism. They are produced by spontaneous oxidative decarboxylation of the corresponding acetohydroxyacids, α-acetolactate and α-acetohydroxybutyrate. These two last compounds are metabolite intermediates of the biosynthesis of valine and isoleucine. Acetohydroxyacids are first excreted from the yeast cells to the medium where they are chemically transformed into diacetyl and 2,3-pentadione. The yeast is able to reduce diacetyl into acetoin and 2,3-butanediol.

The malolactic fermentation is the bacterial conversion of L-malic into L-lactic acid and carbon dioxide. This fermentation is mainly accomplished by a lactic acid bacteria, *Oenococcus oeni,* which was previously classified in the general taxonomy as *Leuconostoc oenos* [17]. *O. oeni* proceeds the MLF in wine and in cider. This bacteria is well adapted to the low pH and high ethanol concentration that are encountered during apple juice fermentation [18]. This strain possesses the malolactic enzyme for the accomplishment of the malic acid conversion in a one-step reaction [19,20]. The malolactic fermentation becomes well known and the winemaking industry tends to inoculate wine with selected pure culture of *O. oeni* to promote the MLF [21]. By analogy, the cidermaking industry makes use of O. oeni starters to reduce the maturation step for which the initiation is problematic as it is depending on activity of the indigenous microflora. Reduction of the MLF duration is very important to limit exposure time of the fermented beverage to microbial spoilage. The main benefit of the malolactic fermentation is the partial deacidification by conversion of malic acid, a dicarboxylic acid present in apple juice into lactic acid, a mono carboxylic acid, resulting in a drop in titrable acidity and low increase of pH. Moreover, MLF improves the microbial stability and organoleptic quality of cider [22,23]. To ensure MLF, *O. oeni* population should reach at least 10^6 CFU/ml. Meanwhile the inoculation with selected *O.oeni* strains, the growth of this biomass is often limited by hostile conditions due to ethanol, fatty acids, acidic pH, inhibitors (high concentration of phenolic compounds and sulphiting of apple juice) and partial nutrient depletion of the medium by previous yeast propagation [24]. The *O. oeni* starters consist mainly in freeze-dried formulations. As applied to wine technology, a reactivation step in more adapted conditions (complemented apple juice medium under pH and temperature control) in some cases can ensure better viability of O. oeni and promote MLF. The pH seems to be the most critical parameter for the growth of *O. oeni*. It limits the use of energy (ATP) to produce biomass by favouring the use of energy to maintain the internal pH [25,26].

2. Cell immobilisation in cider production

Since the early 80's, the use immobilised cell technology in the production of fermented beverages had received great interest at laboratory and industrial pilot scale [21,27]. The opportunity to develop a continuous fermentation process by using immobilised cell reactor will offer several advantages comparatively to the traditional fermentation by free cells cultures such as higher biomass concentrations resulting in an increase of fermentation rates, reuse of the biomass for prolonged period without any propagation and sanitization step and reduction of bioreactor volumes decreasing the equipment cost [28]. The confinement of biomass in a porous matrix allows also to proceed at high dilution rate without any risk of biomass washing-out. Control of biomass population is more effective at high biomass concentration configuration and limits alteration of the beverage by spoilage. Automation of a continuous process could be easily carried out comparatively to batch fermentation. Nowadays application of immobilised cell technology is more studied in brewing and winemaking industry than in the cider making industry. The reasons lie in the larger production volumes of beer and wine in the world and in the fact that the cider making industry keeps the image of a traditional production process. Despite twenty years of research in the field of cell immobilisation, the majority of fermented beverage producers have not yet converted their batch processes to a continuous process due cost-effective of existing plants. In the following parts, a review of research on immobilisation technology applied to cider will be presented considering the performance of such configuration and the incidence in the flavouring components of cider.

2.1. IMMOBILISATION OF YEAST FOR APPLE JUICE FERMENTATION

Considering cell immobilisation, the first step consists in the selection of technology and matrices. Inclusion technology is well adapted for whole cell immobilisation because it allows mild conditions compatible with cells viability. Due to its contact with the beverage, the inclusion matrix has to be a non-toxic material, recognised as safe. Ca-alginate is the most widespread entrapment matrix and was tested for *S. cerevisae* inclusion by Dallmann *et al.* [29] to accomplish continuous fermentation of apple juice. This study aimed to produce an alcoholic beverage that can be distilled to produce brandy. For this purpose a cylindrical reactor filled with cell entrapping alginate beads was used. The cell density inside the beads was 10^9 cells ml^{-1} of gel. Only ethanol production was taken in consideration. Ethanol concentration reached 38.9 gl^{-1} and the volumetric productivity 6.3 $gl^{-1}h^{-1}$ at fermentation efficiency of 84.7%. To limit alginate dissolution, apple juice was completed with calcium chloride.

O'Reilly and Scott [30] have tested ion–exchange sponge to immobilise yeast for high gravity apple juice fermentation. Due to the negative charges of the *S. cerevisiae* outer surface, weakly basic sponge with diethyl amino (DE) functional groups showed the best cell loading capacity (8.3 10^8 cells/g sponge). Fermentation of apple juice was accomplished in a packed bed reactor with continuous circulation of the medium. The presence of DE sponge had a positive effect on the fermentation in terms of time to reach attenuation and final level of ethanol. This higher performance was attributed to an increased removal of carbon dioxide by the spongy material that provided an improved microenvironment acting as a nucleation site for carbon dioxide. Biomass leakage was noticed in this system. The DE material, was also used for co-immobilisation of yeast and a lactic acid bacteria [31], *L. plantarum*. Introduction of *L. plantarum* in the system at different steps of fermentation does not affect fermentation rate. Sensory evaluation of the product confirmed the quality of cider issued from immobilised systems and the beneficial impact of *L. plantarum*.

Several publications reported modifications of yeast metabolism in immobilised systems resulting in different volatile profiles for beer production [27,32]. The profile of volatile compounds was also modified by yeast immobilisation in case of apple juice fermentation. A lower production of fusel alcohols and isoamylacetate and higher concentration of diacetyl were described by Nedovic et al. [33] when S. bayanus was co-immobilised with O. oeni in alginate beads at 95% of sugars attenuation. These variations were not attributed to the presence of O. oeni but to a shift of yeast metabolism. Anabolic flux limitation of yeast cells in pseudo-stationary phase was proposed to justify the lower concentration of fusel alcohols (five times lower than for batch fermentation). The pseudo-stationary phase characterises the physiological state of the cells when the growth is limited by sterical hindrance inside the matrix resulting in a very slow growth [27]. Lower concentration in isoamylacetate was a result of lower isoamylalcohol availability. The larger concentration of diacetyl in the immobilised system was explained by diffusional mass transfer effect that prevents the transfer of diacetyl from the medium to the immobilised yeast after the chemical oxidative decarboxylation of α-acetolactate in the medium. The drawback of the use of alginate was the biomass leakage due to local overpressure in the beads generated by carbon dioxide production. The authors suggested that free yeast issued from the continuous reactor could be used for diacetyl uptake in a maturation tank. Similar variation in volatile compounds was reported (figure 1) when yeast was entrapped in Lentikats [34].

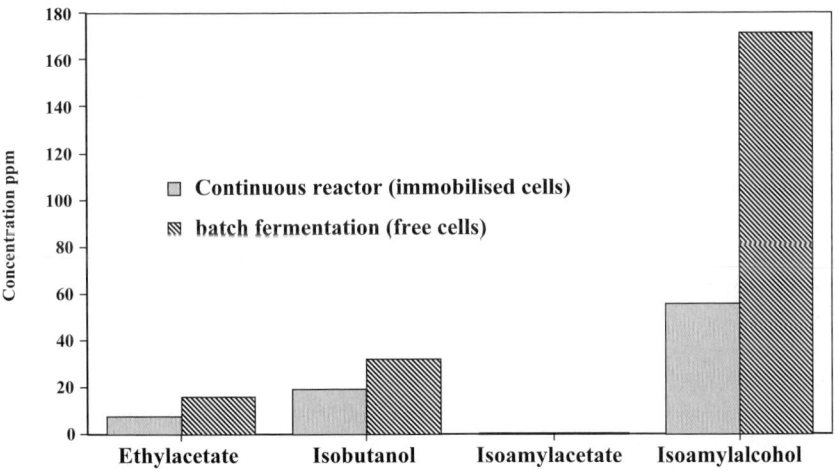

Figure 1. Comparison of the concentrations of higher alcohols and esters in two ciders: one produced in batch process and one produced in a continuous process (yeast immobilised in Lentikats), at sugar attenuation of 95% and temperature of 20°C [34].

Lentikat is a polyvinylalcohol matrix developed by Ding and Vorlop [35]; it is characterised by mild gelation conditions and nowadays used in several applications [36,37]. In this case, concentrations of vicinal diketones were around twenty times higher for the immobilised process at 95% of initial sugar consumption (figure 2). High concentrations of precursors α-acetolactate and α-acetohydroxybutyrate were detected. This feature suggested that limitation of anabolic pathway (biosynthesis of valine and isoleucine) and specially in acetohydroxyacid reducto-isomerase activity should be responsible for the precursors accumulation in the medium after their secretion. This matrix presents a better mechanical resistance than alginate. Residual sugar concentration can be easily adjusted by variation of the residence time in the continuous reactor and allows production of "soft" and "dry" cider.

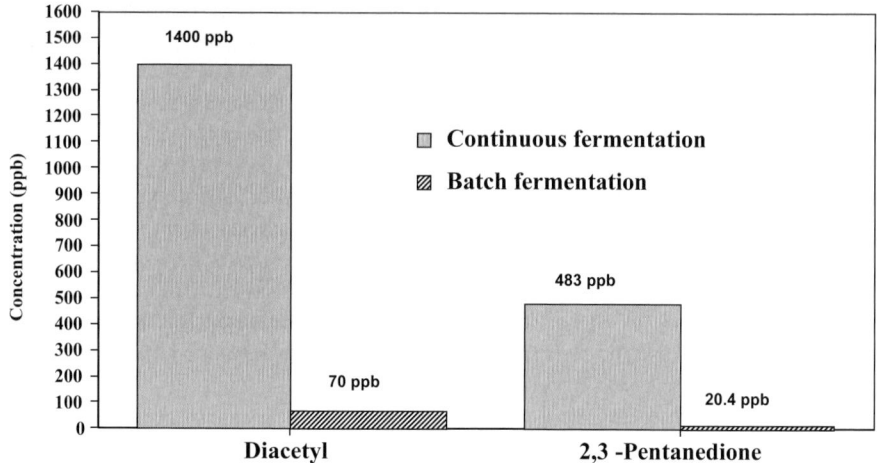

Figure 2. Comparison of the concentrations of carbonyl compounds in two ciders: one produced in batch process and one produced in a continuous process (yeast immobilised in Lentikats), at sugar attenuation of 95% and temperature of 20°C [34].

2.2. IMMOBILISATION OF *O. OENI* FOR MALOLACTIC FERMENTATION IN CIDER

As mentioned previously, the initiation of MLF appears as the main limiting factor in cider fermentation. Besides the use of a selected *O. oeni* starter, several alternative technologies are proposed in the literature to avoid any delay or failure in the accomplishment of the MLF in wine and cider [21]. Studies described the development of enzymatic reactor to carry out MLF [38]. However the bioconversion of malic acid by malolactic enzyme requires the addition of manganese and NAD^+ as cofactors that represent an additional cost for this technology. Progress in genetics allowed the cloning of malolactic enzyme and malate permease in yeast but in this case the legislation prevents the industrial transposition by forbidding the use of recombinant microorganism in beverage production [39]. Another alternative technology consists in the use of cell-recycle bioreactors favouring the obtaining of high free cell densities. These reactors are constituted by tangential flow or hollow fiber filters to separate the cells from the medium. Great attention is given to *O. oeni* immobilisation since the eighties. Most studies were applied to wine fermentations. The use of calcium alginate [40], kappa-carrageenan [41], cellulose sponge [42] and polyacrylamide [43] were investigated as immobilisation supports for MLF in wine.

Concerning the applications to cidermaking, feasibility to carry out MLF by *O. oeni* entrapped in alginate was firstly demonstrated [44]. Biomass concentrations between 10^9 and $8 \cdot 10^9$ CFU/ml were reached in this study depending on the inoculation level. Total uptake of malic acid (initial concentration of 6.5 g/l) in apple juice was measured in continuous reactor. Complete processing of apple juice into cider was performed by combining alginate beads loaded with *O. oeni* and others with *S. bayanus* in the same tubular reactor. Only sugars and malic acid consumption and ethanol production were taken in consideration in this study.

Cabranes *et al.* [45] accomplished the MLF with *O. oeni* immobilised in alginate simultaneously to the alcoholic fermentation and at the end of this fermentation. The alcoholic fermentation was carried out by free yeast cells. The fermentations were conducted at 12°C and 18°C. Yeast growth and ethanol production were not affected by the presence of immobilised *O. oeni*. Malic acid was fully metabolised at 18°C but residual amount of this acid was detected at 12°C. Acetic acid produced at the end of the fermentation was twice the concentration obtained with free *O. oeni* culture at 18°C but did not exceed 0.46 g/l. Fusel alcohol production was also not affected by the mode of *O. oeni* inoculation. Accumulation of methanol and ethylacetate was stimulated by simultaneous inoculation of yeast and *O. oeni*.

Table 1. Percentage of malic acid conversion and specific malic acid consumption for different apple juices at a residence time of 0.55 h [46].

pH of apple juices	% of initial malic acid converted by MLF	Specific malic acid consumption (g h^{-1})
4.46	98	11.4
3.95	98	11.4
3.36	90	11.0
2.90	84	10.2
2.30	33	4.0

Inclusion in Lentikats matrix was studied as an alternative to alginate [46]. This immobilisation technique was compatible with *O. oeni* viability and had thus allowed the conception of a continuous process. The experiments were conducted at different temperature and pH with apple juice or cider feeding of the reactor. Previous alcoholic fermentation did not affect *O. oeni* activity. Largest malic acid attenuation was noticed for temperature between 25 and 30°C but residual MLF was even detected at 5°C. The most surprising result was the malic acid conversion at very acidic pH (until pH = 2.30) as presented in table 1. The MLF does not occurred at pH below 3.9 with free cells. Cell physiology modification and immobilised cell microenvironment characterised by a pH gradient inside the matrix were proposed to explain the improved performance of *O. oeni* at acidic pH by respectively allowing generation of enough ATP to maintain cytoplasmic pH without any perturbation of the MLF and by restoring favourable pH in the direct environment of the cells. Deacidification levels were easily adjusted as function of the residence time in this continuous process. Herrero *et al.* [28] monitored organic acids and volatile compounds during MLF accomplished in batch process by *O. oeni* immobilised in alginate. For cell density of 2.9 10^8 CFU/ml of beads, biomass leakage was insignificant. Immobilised cells synthesised less acetic acid and consumed less quinic acid. Profile of other organic acids was similar for immobilised and free cells. Concerning volatile compounds, higher concentrations of fusel alcohols (propanol, 2-methylpropanol and butanol) were detected with immobilised cells. These features have beneficial incidence on the organoleptic profile of cider.

3. Conclusions

Immobilised cell technology offers a new alternative to control the transformation of apple juice into cider by ensuring a better control of the microbiology that defines the final product. The reputation of cider as traditional product could limit the transposition of this technology in the cidermaking industry but this aspect have to be considered in opposition with the several advantages of immobilisation cells technology comparatively to the present process. Economic studies have to be accomplished to verify the feasibility of continuous process with immobilised cells. Considering the scientific studies presented in this article, the most opportunity can be given to the malolactic fermentation with immobilised *O. oeni*. The difficulties to induce this last fermentation justify interest to immobilisation. The opportunity to separate the cell propagation phase and the accomplishment of the malolactic fermentation of cider with high cell density is the main advantage of immobilised system. Propagation phase could be conducted in adapted medium and optimal conditions of temperature and pH to obtain high concentration of biomass that can be immobilised after harvesting. Studies had demonstrated an improved performance of immobilised *O. oeni* comparatively to free cells cultures without affecting significantly the flavour profile. Concerning the use of immobilisation for the alcoholic fermentation, the modification of yeast metabolism affecting the flavours profile could limit its industrial transposition. Further studies are still necessary to elucidate this feature.

References

[1] Jarvis, B.; Forster, M.J. and Kinsella, W.P. (1995) Factors affecting the development of cider flavour. J. Appl. Bacteriol. Symp. Suppl. 79: 5S-18S.
[2] Laplace, J.M.; Apery, S.; Frère, J. and Auffray, Y. (1998) Incidence of indigenous microbial flora from utensils and surrounding air in traditional french cider making. J. Inst. Brew. 104: 71-74.
[3] Duenas, M.; Irastorza, A.; Fernandez, K.; Bilbao, A. and Huerta, A. (1994) Microbial populations and malolactic fermentation of apple cider using traditional and modified methods. J. Food Science 59: 1060-1064.
[4] Le Quère, J.M. and Drilleau, J.F. (1993) Microorganismes et typicité du cidre. Pomme 31: 16-19.
[5] Salih, A.G.; Le Quéré, J.M. and Drilleau, J.F. (1990) Lactic acid bacteria and malolactic fermentation in the manufacture of Spanish cider. J. Inst. Brew. 96: 369-372.
[6] Drilleau, J.F. (1979) L'utilisation des fruits à cidre et leurs transformations industrielles. Bios. 10: 22-25.
[7] Duenas, M.; Irastorza, A.; Fernandez, C.; Bilbao, A. and Del Campo, G. (1997) Influence of apple juice treatments on the cider making process. J. Inst. Brew. 103: 251-255.
[8] Herrero, M.; Cuesta, I.; Garcia, L.A. and Diaz, M. (1999) Changes in organic acids during malolactic fermentation at different temperatures in yeast-fermented apple juice. J. Inst. Brew. 105: 191-195.

[9] Herrero, M.; Cuesta, I.; Garcia, L.A. and Diaz, M. (1999) Organic acids in cider with simultaneous inoculation of yeast and malolactic bacteria : effect of fermentation temperature. J. Inst. Brew. 105: 229-232.
[10] Kosseva, M.R. (1999) Alternative biocatalysts for the malolactic fermentation: immobilised and/or lyophilised culture. Bulgarian Chemical Communications 31: 536-546.
[11] Cabranes, C.; Mangas, J.J. and Blanco, D. (1996) Controlled production of cider by induction of alcoholic fermentation and malolactic conversion. J. Inst. Brew. 102: 103-109.
[12] Lonvaud–Funel, A. (2001) Starters for wine industry. In: Durieux, A. and Simon, J.-P. (Eds.) Applied Microbiology. Kluwer Academic Publishers; pp. 31-47.
[13] Leguerinel., I; Cleret, J.J.; Bourgeois, C. and Mafart, P. (1988) Yeast strain and the formation of flavour components in cider. J. Inst. Brew. 96: 391-395.
[14] Leguerinel, I.; Mafart, P.; Cleret, J.J. and Bourgeois, C. (1989) Yeast strain and kinetic aspects of the formation of flavour components in cider. J. Inst. Brew. 95: 405-409.
[15] Hammond, J. (1986) The contribution of yeast to beer flavour. Brew. Guardian 115: 27-33.
[16] Peddie, H.A.B. (1990) Ester formation in brewery fermentation. J. Inst. Brew. 96: 327-331.
[17] Dicks, L.M.T.; Dellagio, F. and Collins, M.D. (1995) Proposal to reclassify *Leuconostoc oenos* as *Oenococcus oeni* (corrig.) gen. nov., comb. nov. Int. J. Syst. Bacteriol. 4: 395-397.
[18] Davis, C.R.; Wibowo, D.; Eschenbruch, R.E.; Lee, T.H. and Fleet, G.H. (1985) Practical implications of malolactic fermentation: a review. Am. J. Enol. Vitic. 36: 290-301.
[19] Lonvaud-Funel, A. and Strasser de Saad, A.M. (1982) Purification and properties of a malolactic enzyme from a strain of *Leuconostoc mesenteroides* isolated from grapes. Appl. Environ. Microbiol. 43: 357-361.
[20] Spettoli, P.; Nuti, M.P. and Zamaroni, A. (1984) Properties of malolactic activity purified from *Leuconostoc oenos* ML34 by affinity chromatography. Appl. Environ. Microbiol. 48: 900-901.
[21] Maicas, S. (2001) The use of alternative technologies to develop malolactic fermentation in wine. Appl. Microbiol. Biotechnol. 56: 35-39.
[22] Salou, P.; Loubière, P. and Pareilleux, A. (1994) Growth and energetics of *Leuconostoc oenos* during cometabolism of glucose with citrate or fructose. Appl. Environ. Microbiol. 60: 1459-1466.
[23] Martineau, B.; Henick-Kling, T. and Acree, T. (1995) Reassessment of the influence of malolactic fermentation on concentration of diacetyl in wines. Am. J. Enol. Vitic. 46: 385-388.
[24] Salih, A.G.; Drilleau, J.F.; Diviès, C. and Lenzi, P. (1987) Facteurs contribuant au contrôle de la transformation malolactique dans les cidres. Sciences des Aliments 7: 205-221.
[25] Champagne, C.P.; Gardner, N. and Doyon, G. (1989) Production of *Leuconostoc oenos* biomass under pH control. Appl. Environ. Microbiol. 55: 2488-2492.
[26] Maicas, S.; Gonzalez-Cabo, P.; Ferrer, S. and Pardo, I. (1999) Production of *Oenococcus oeni* biomass to induce malolactic fermentation in wine by control of pH and substrate addition. Biotechnol. Lett. 21: 349-353.
[27] Masschelein, C.A.; Ryder, D.S. and Simon, J.P. (1994) Immobilized cell technology in beer production. Crit. Rev. Biotechnol. 14: 155-177.
[28] Herrero, M.; Laca, A.; Garcia, L.A. and Diaz, M. (2001) Controlled malolactic fermentation in cider using O.oeni immobilized in alginate beads and comparison with free cell fermentation. Enzyme Microb. Technol. 28: 35-41.
[29] Dallmann, K.; Buzas, Z. and Szajani, B. (1988) Continuous fermentation of apple juice by immobilized yeast cells. Biotechnol. Lett. 9: 577-580.
[30] O'Reilly, A. and Scott, J.A. (1993) Use of an ion-exchange sponge to immobilise yeast in high gravity apple based (cider) alcoholic fermentations. Biotechnol. Lett.15: 1061-1066.
[31] Scott, J.A. and O'Reilly, A. (1996) Co-immobilization of selected yeast and bacteria for controlled flavour development in an alcoholic cider beverage. Process Biochem. 31: 111-117.
[32] van Iersel, M.F.M.; Brouwer-Post, E.; Rombouts, F.M. and Abee, T. (2000) Influence of yeast immobilization on fermentation and aldehyde reduction during the production of alcohol-free beer. Enzyme Microb. Technol. 26: 602-607.
[33] Nedovic, V.A.; Durieux, A.; Van Nedervelde, L.; Rosseels, P.; Vandegans, J.; Plaisant, A.M. and Simon, J.P. (2000) Continuous cider fermentation with co-immobilized yeast and *Leuconostoc oenos* cells. Enzyme Microb. Technol. 26: 834-839.

[34] Durieux, A.; Bodo, E.; Nedovic, V. and Simon, J.P. (2002) Effect of yeast and *Oenococcus oeni* immobilisation on the formation of flavour components for cider production. In: Proceedings of the international Workshop Bioencapsulation X: Cell physiology and interactions of biomaterials and matrices. Prague, Czech Republic, April 2002; pp. 54-57.
[35] Ding, W.A. and Vorlop, K.D. (1995) Gel aus Polyvinylalkohol und Verfahren zu seiner Herstellung. Patent DE 4327923.
[36] Jekel, M.; Buhr, A.; Wilke, T. and Vorlop, K.D. (1998) Immobilization of biocatalysts in Lentikats. Chem. Eng. Technol. 21: 275-278.
[37] Wittlich, P.; Themann, A. and Vorlop, K.D. (2001) Conversion of glycerol to 1,3-propanediol by a newly isolated thermophilic strain. Biotechnol. Lett. 23: 463-466.
[38] Vaillant, H.; Formisyn, P. and Gerbaux, V. (1995) Malolactic fermentation of wine: study of the influence of some physico-chemical factors by experimental design assays. J. Appl. Bact. 79: 640-650.
[39] Labarre, C.; Guzzo, J.; Cavin, J.F. and Diviès, C. (1996) Cloning and Characterisation of the genes encoding the malolactic enzyme and the malate permease of *Leuconostoc oenos*. Appl. Environ. Microbiol. 62: 1274-1282.
[40] Spettoli, P.; Bottacin, A.; Nuti, M.P. and Zamaroni, A. (1982) Immobilization of *Leuconostoc oenos* ML 34 in calcium alginate gels and its application to wine technology. Am. J. Enol. Vitic. 33: 1-5.
[41] Mc Cord, J.D. and Ryu, D.D.Y. (1985) Development of malolactic fermentation process using immobilized whole cells and enzymes. Am. J. Enol. Vitic. 36: 214-218.
[42] Maicas, S.; Pardo, I. and Ferrer, S. (2001) The potential of positively-charged cellulose sponge for malolactic fermentation of wine, using *Oenococcus oeni*. Enzyme Microb. Technol. 28: 415-419.
[43] Rossi, J. and Clementi, F. (1984) L-malic catabolism by polyacrylamide gel entrapped *Leuconostoc oenos*. Am. J. Enol. Vitic. 63: 100-102.
[44] Durieux, A.; Garre, V.; Mukamana, J.; Jourdain, J.M.; Silva, D.; Plaisant, A.M.; Defroyennes, J.P.; Foroni, G. and Simon, J.P. (1996) *Leuconostoc oenos* entrapment: applications to continuous malolactic fermentation. In: Wijffels, R.H.; Buitelaar, R.M.; Bucke, C. and Tramper, J. (Eds.) Immobilized Cells: Basics and Applications. Elsevier Sciences; pp. 679-686.
[45] Cabranes, C.; Moreno, J. and Mangas, J.J. (1998) Cider production with immobilized *Leuconostoc oenos*. J. Inst. Brew. 104: 127-130.
[46] Durieux, A.; Nicolay, X. and Simon, J.P. (2000) Continuous malolactic fermentation by *Oenococcus oeni* entrapped in Lentikats. Biotechnol. Lett. 22: 1679-1684.

WINE PRODUCTION BY IMMOBILISED CELL SYSTEMS

CHARLES DIVIES AND REMY CACHON
Laboratoire de Microbiologie UMR INRA 1082, ENSBANA, Université de Bourgogne, 1-esplanade Erasme, 21000 Dijon, France – Fax: 33 3 80 39 66 40 – Email: remy.cachon@ u-bourgogne.fr

1. Introduction

Winemaking is a very old endeavour because of the natural character of the product. Essentially, crushing of the grapes is all that is necessary; yeasts originating from the surface of the berries immediately start to grow, increasing from several thousand to several million cells per millilitre of juice as they conduct fermentation. The transformation of juice results in the production of wine. Winemaking involves two principal operations: first, preparation of the grape must to tailor its composition and to maintain the qualities of the grape at harvest, and, second, conducting microbial fermentation through rational exploitation of the biochemical activities of yeasts and lactic acid bacteria [1]. In winery, grape juices are inoculated with pure cultures of *Saccharomyces cerevisiae* at 1.10^6-5.10^6 cells/ml [2].

There is considerable diversity in bioreactor technology available to winemakers, leading to a wide variety of fermented wines [3]. Over the centuries, oenology has accumulated numerous pragmatic acts that in fact do not correspond to an optimisation of the winemaking process. This is why countries only recently engaged in this activity have been able to develop their wine industry with the aid of a more scientific approach and apply some of the newer advances in bioreactor technology. This article focuses on the application of immobilised cell technology in wine production. During the last 20 years, immobilised yeast cells have been explored in a view to reduce labour requirements, to simplify time-consuming procedures, and thereby to reduce costs.

2. Immobilised cell technology and heterogeneous bioreactors

The main techniques, which enable biomass confinement, are adsorption on a support, autoflocculation and entrapment in gels.

Adsorption is very simple. Lommi and Ahvenainen [4] proposed adsorption of yeasts on a non-compressible carrier having anion exchange properties (granulated DEAE cellulose). Such a reactor can be pressurised in order to maintain carbon dioxide in a dissolved state, and it can be regenerated. Bardi *et al.* [5] proposed wine making using gluten pellets with adsorbed yeasts. The biocatalyst was used for 28 repeated batch fermentations. Maicas *et al.* [6] used *Oenococcus oeni* adsorbed on positively charged cellulose sponge for malolactic fermentation of wine.

The flocculation of microbial cells to form a dense concentration of biomass is encountered naturally in a number of yeast strains [7,8]. It is a very attractive method of biomass retention, since it involves decanting, the most basic method of liquid-solid distribution [9,10]. In the case of yeasts, a biomass density of 60-110 g/l can be maintained with little or no physical constraints for the microorganisms except for the reactor vessel. The fermentation process requires a reactor configuration in the form of towers which house the biomass through which the liquid is passed. Industrial installations have been described in breweries [11].

Entrapment involves imprisoning living cells within a rigid network, which permits the diffusion of substrates and products, thereby making possible the growth and maintenance of active cells. Synthetic polymers (polyacrylamide, polyvinyl chloride, polyurethane) can be used as the entrapment matrix but natural polymers, such as alginate, carrageenan, chitosan and agar, enable polymerisation in very mild conditions and leave cell integrity intact [12-14]. Entrapment in alginate, for example, is a very simple process. A uniform suspension of the cells is prepared in a 2% sodium alginate solution. The suspension is then added drop wise to a solution of calcium chloride which catalyses the polymerisation of the alginate into a gel in the form of spherical beads. It is possible to obtain perfectly calibrated spheres between 0.2 and 2 mm in diameter. The beads can be housed as a fixed bed through which the liquid to be transformed is passed, or they may be agitated to give a fluidised bed system.

Heterogeneous reactors are subjected to limitations resulting from the transfer of solutes between the liquid medium and the surface of the immobilisation matrix (external transfer) and the interior of the immobilisation matrix containing the embedded cell mass (internal transfer). Published data have shown that these phenomena are primarily responsible for cellular activity in heterogeneous reactors [15,16]. These limitations to diffusion create substrate and product concentration gradients within the immobilisation matrix that will affect the kinetics and the efficiency of transformations by the immobilised cells. At low substrate concentrations, only a portion of the cells has access to the substrate and so diffusion will be the limiting factor. At high substrate concentrations, all the cells will have access to the substrate and the reaction can occur at a rate close to the maximal rate. If the product formed is inhibitory, diffusion limitations will limit its transfer to the exterior of the porous solid and create a concentration gradient. Thus, cells furthest from the surface will be subjected to a greater inhibition than those closer to the periphery of the support.

The effect of mass transfer on alcoholic fermentations with yeast cells immobilised in packed bed reactors has been illustrated by Radovitch [17]. Assuming that ethanol production was mass transfer limiting and on the basis of gel particle size, the author correlated the higher productivity for alginate beads than polyacrylamide cubes to the ratio of specific surface areas.

3. Potential of immobilised cell systems for applications in oenology

Many applications have been studied and published in the last 20 years. The publication of many patents on the use of immobilised microbial cells in the production of fermented beverages has shown the potential industrial interest for this technology.

The use of immobilised cells offers several advantages: improved productivity of fermentations (high cell density), adaptation to continuous processes that can be better optimised and controlled, simplified systems for removing microbial cells from batch processes, greater tolerance to inhibitory substances, smaller scale fermentation facilities (reduced capital and running costs), and possibilities of using a variety of microbial strains including genetically modified organisms. Some potential disadvantages must be also considered: cell overgrowth which increases turbidity of the fermented beverage, mechanical stability of the matrix used to immobilise microbial cells, loss of activity on prolonged operation.

To be attractive in wine production the method must be: cheap, easily performed in an industrial situation, not liable to cause oxidation of the wine, robust, not susceptible to contamination, able to impart correct flavour changes to the wine, must use commercially acceptable supports and organisms [18].

3.1. ALCOHOLIC FERMENTATION

The "A.P.V." system employing flocculating yeast cells has received some application in the brewing industry. The reactor containing the densely packed cells is a cylindrical tower with a conical base; geometry is 7:1-10:1 with a 0.9-2 m diameter. Cell concentrations of 50-80 g/l that produce 60 g/l per hour of ethanol are obtainable. These reactors have been used in breweries for the continuous production of beer [19], but it is very difficult to obtain a product with sensory qualities identical to those obtained with the conventional process. Other disadvantages of these reactors are complexity, lack of flexibility, risks of contamination, heterogeneity of the cell concentration, the lack of choice of usable strains and the excessive time required for start-up, i.e. two to three weeks to obtain a high cell density and stable operation.

In a general manner, the alcoholic fermentation by yeasts immobilised in alginate gel beads is accelerated, which has been related with changes in cell composition and function [20]. Using the same strain of S. cerevisiae and 2 mm diameter alginate spheres with a population of 2.10^9 cells/ml of gel, we measured specific ethanol production at 0.6 g/g of cells per hour in the immobilised form and 0.3 g/(g.h) for free cells. If smaller spheres are used, even better fermentation activities are possible. In addition, entrapment protects the yeasts against ethanol toxicity as well as against heavy metals, phenols, acidity, and extreme temperatures. For example, it has been suggested that low-temperature fermentations improve the quality of wine and reduce the toxicity [21,22]. Using immobilised cells at low temperatures, wine productivity was higher than for free cells, and the improvement increased as the temperature decreased [5].

While prospects for the entrapment technology appear encouraging, further research is needed to optimise reaction variables, improve the long-term stability of the reactors and to understand more about secondary metabolite production by yeasts under these conditions. Our own laboratory and pilot scale studies with entrapped cell of yeasts further demonstrate their potential value to oenology. We have been able to maintain the activity of 2.10^9 yeast cells *(S. cerevisiae)* per ml of gel if 1-3 mm diameter alginate spheres are used. This density is equivalent to 20-50 g/l of yeast taking the weight of one cell as 10^{-11} g. These beads of dense biomass can be conveniently handled and maintained active for repetitive and controlled use by relatively simple procedures. Such beads can be used in batch or continuous processes. Stepwise processes can also be used [23]. For example, using an obligate aerobic yeast in the first reactor lowers the sugar concentration of the fruit juice. In the second reactor, a fermentation yeast is used to carry out fermentation, so that a low alcohol product is obtained. Thus, the undesirable sweetness of low alcohol beverages can be overcome. Ogbonna et al. [24] *designed* a horizontal reactor for wine fermentations, which exploited the successive activities of cells of *Schizosaccharomyces* and *Saccharomyces,* which were immobilised on plates coated with alginate.

3.2. BOTTLE-FERMENTED SPARKLING WINES ("MÉTHODE CHAMPENOISE")

In the conventional process, a blend of dry stabilised wine is mixed with sugar (about 25 g) and the secondary fermentation is conducted in a bottle by the inoculation of yeast in liquid suspension. The secondary fermentation produces carbon dioxide up to 5 bar while yeast metabolism and yeast autolysis participate to the typical aroma and flavour. Subsequent sedimentation and removal of the yeast cells requires the lengthy and expensive procedure of "remuage". This operation occupies 35-40% of the space in cellars during 6 weeks. The sediment is removed by freezing the neck of the bottle (-25°C), which is then manually opened and the ice plug is squeezed out of the bottle by the internal pressure, it is the procedure of "dégorgement". This conventional method can be advantageously shortened using immobilised yeasts in place of free cells, and using this method, "remuage" requires only 20 sec/bottle.

Different methods have been proposed such as the use of entrapped yeasts [13,25-29] and bottle cap with membrane cartridge like "Millispark" of Millipore S.A. [30-32]) or other origin [33-35]. The kinetics of the "prise de mousse" (formation of champagne bubbles in the bottle) differentiates the processes. For the same initial population of yeast, the time to achieve the "prise de mousse" is doubled with the membrane method. Nevertheless, the membrane system proposed by Spooner [33] enables the opening of the bottle without cooling by slowly reducing the pressure prior to removing the closing of the bottle, which is performed by piercing the bottle cap. This method allows a slow pressure drop to obviate excess foaming and loss of material. The system of Quetsch [34] consists in immobilised biocatalyst stoppers, made of polyethylene or cork. A string passes through the top of the stopper to the cover of a housing with the immobilized yeast cells, enclosed in a micro-filter. The housing can be pulled into a stopper cavity where an elastic seal provides a tight seat.

In such systems the critical parameter is mass transfer, which modifies the reaction rate for immobilised cells only at the end of fermentation, because of the increase of cell density inside the matrix, which modifies the effective diffusion coefficients. Using cells immobilised in gel beads, the fermentation delay is close to the fermentation with free cells. In addition, it can be adjusted by the choice of the number of beads and of their specific area. When the specific area is doubled, the time required for finishing can be halved. To avoid cell leakage from beads, beads can be coated with a sterile calcium alginate layer [13,26,28]. Yokotsuka et al. [29] used also double-layer calcium alginate fibres. Godia et al. [27] have compared alginate and carrageenan gel beads. They observed that alginate showed a better structure to retain cells, 10-12 g/bottle was optimal to guarantee a clean wine free of cells. In the case of coated beads, the number of beads needed per bottle is reduced by half.

It has been shown that the organoleptic properties of the wine were equal or better than when the same wine was elaborated by the conventional technology [26-29]. The amino-acid composition (an important feature in champagne) of both wines do not differ appreciably, and no significant differences were detected in pH, ethanol, total acidity, volatile acidity and ethyl acetate.

Industrial utilisation of immobilised cells has been studied by Champagne Moët et Chandon which have proposed a process for large-scale production of immobilised cells [36]. For automation, the process using entrapped cells in gel beads requires the use of dried beads (dry matter 80-95%). Methods have been proposed by Diviès et al. [37] with the Champagne Moët et Chandon (for normal and coated beads) and by Hill [13], which also used a special system for transfer and dosing of beads in bottles [38]. This company has developed an industrial machine for the delivery of beads at the rhythm of 20,000 bottles per hour (and needs 1 m^3 of beads per day). An economic study undertaken for a plant producing 3,000,000 bottles yearly has demonstrated the competitiveness of this new process [39]. In 1992, about 500,000 bottles were produced using this technique. In 2001, the "Institut Oenologique des Vins de Champagne (IOC)" claimed three million bottles. The main problems for generalising this technology are now both conservation of freshly immobilised yeasts (10-15 days at 4°C) which might be resolved using bead's drying, and the investment cost for mixing beads and wine at the industrial scale.

The technology developed for champagne production can be transposed to other bottle-fermented sparkling beverages. Particularly interesting possibilities are the re-fermentation of wine supplemented before "prise de mousse" with an infusion of fruits obtained by hydro-alcoholic maceration [40], and second fermentation of fruit wine such as cider and pineapple wine [23]. These new products can thus be obtained under more thoroughly controlled conditions.

3.3. PRODUCTION OF SPARKLING WINES IN CLOSED REACTORS

Diviès and Deschamps [22] used a pressurised batch reactor to produce cider, sparkling wine or semi sparkling grape juice using yeasts immobilised in alginate gel beads. For example, 1,500,000 bottles of sparkling wine could be produced per year by a reactor (1 m^3) operating for 220 days. In a continuous system [41], a sparkling wine with composition and sensorial properties comparable to a product obtained conventionally was produced, but with a greatly enhanced productivity. In the fermentation process of Lommi and Ahvenainen [4], the reactor can be pressurised up to 14 bar to obtain a sparkling product. Similar continuous processes have been proposed by Sarishvili et al. [42] to produce "champagne"-like sparkling wines.

3.4. THE MALOLACTIC FERMENTATION

The malolactic fermentation of wine is the degradation of malic acid in lactic acid and carbon dioxide. It allows a reduction in acidity of wines and contributes to the development of subtle flavours that contribute to the sensorial quality of wines. It also stabilises wines and lowers the risk of fermentation in bottles. The main microbial strain involved in malolactic fermentation is *Oenococcus oeni* (its old name is *Leuconostoc oenos*). The malolactic fermentation can occur several weeks after alcoholic fermentation, and even in the case of wine inoculation by selected starters, there is no guarantee that the fermentation will occur. This is because wine is unfavourable for growth of microorganisms (pH < 3.0-3.5, ethanol > 10%, temperature < 15°C in northern countries, high SO_2 concentration, lack of nutrients...). In 1976, Diviès and Siess proposed an immobilised cell process using *Lactobacillus casei* entrapped in polyacrylamide gel lattice. This process operated for 12 months without loss of activity. Several reactor configurations have been tested [43-48]. Some problems have been related, associated with microbial contamination of the reactors, transfer of flavour taints to the wine, loss of activity on prolonged operation and leakage of cells from the solid support. This is in contradiction with Fleet and Costello [49] which published a patent on the malolactic fermentation of wine, the authors found that the bacteria can remain active for an indefinite time, thus allowing the process of the invention to operate continuously, or with interruptions. Kosseva *et al.* [50] obtained a good operational stability of calcium pectate gel beads and chemically modified chitosan beads, which led us to assume that such a technology is suitable for industrial application in wine making. For a working winery, the simpler method of adhesion (on oak chips) might be recommended to adapt the method to industrial practice [18].

4. Conclusion

Immobilised cell systems present the possibility of producing new styles of beverages (low alcohol content and very aromatic) and facilitate the conduct of fermentations where convenient removal of yeast cells is desired (*e.g.*, closed tanks, champagne method). Recent progress has enabled the industrial production of entrapped microorganisms and their availability to the industry. The natural course of wine fermentation is now known in a detailed manner, but fermentation processes are not yet fully controlled. Basic research is still needed to understand the biochemical behaviour of yeast and bacteria in the wine medium. During the last decades, there has been an important technology transfer from the sector of other industrial beverages toward oenology. The worldwide spread of wine production has led to new vineyards producing quality wine by the use of tested procedures: centrifugation, filtration, stainless steel tanks, temperature management, and inoculations. The seasonal nature of production is a hindrance to the development of innovations, especially in the technology of bioreactors adapted to all sectors of oenology. Research on the physiology of immobilised microorganisms should be continued, because data on their behaviour is insufficient to enable all possibilities to be optimally utilised. Immobilised cells have shown several possibilities to facilitate the conduct of fermentation, especially in the field of sparkling wines production. The use of this technology on an industrial scale needs further scale-up studies, and a good scientific knowledge of the effect of immobilisation on physiology of industrial strains (metabolic flux distribution, kinetic of autolysis). In the case of malolactic fermentation, opposite results have been published, and a better knowledge of the physiology of *Oenococcus oeni* is required, taking into account the physico-chemistry of the culture medium in which the malolactic fermentation is studied. In addition, there are recent advances on the physiological response of bacteria to environmental stress, and it can be expected that new strains, well adapted to wine medium, will be proposed in the next years.

References

[1] Barre, P.; Blondin, B.; Dequin, S.; Feuillat, M.; Sablayrolles, J.M. and Salmon, J.M. (1998) La levure de fermentation alcoolique. In: Oenologie, Technique et Documentation, Lavoisier, Paris (France); pp. 414-495.
[2] ITV (1994) Choix et emploi des micro-organismes en oenologie. ITV - Communication, Paris (France).
[3] Flanzy C. (1998) Oenologie. Techniques et Documentation, Lavoisier, Paris (France).
[4] Lommi, H. and Ahvenaimen, J. (1990) Method using immobilized yeast to produce ethanol and alcoholic beverages. European Patent # 0361165.
[5] Bardi, E.P.; Bakoyianis, A.; Kounitas, A. and Kanellaki, M. (1996) Room temperature and low temperature wine making using yeast immobilized on gluten pellets. Process Biochem. 31: 425-430.
[6] Maicas, S.; Pardo, I. and Ferrer, S. (2001) The potential of positively-charged cellulose sponge for malolactic fermentation of wine, using *Oenococcus oeni*. Enzyme Microb. Technol. 28: 415-419.
[7] Stewart, G.G. and Russell, I. (1986) One hundred years of yeast research and development in the brewing industry. J. Inst. Brew. 92: 537-558.
[8] Calleja, G.B. (1987) Cell aggregation. In:The yeasts. Vol. 2. Academic Press, London; pp. 164-298,
[9] Ghommidh, C. and Navarro, J.M. (1986) Floculation-fermentation. In: Bioreacteurs, Société Française de Microbiologie, Paris (France); pp. 89-112.
[10] Salou, P.; Sablayrolles, J.M. and Barre, J.M. (1988) Production de levures d'intérêt oenologique. Revue Française d'Oenologie 114: 29-34.
[11] Maule, D.R. (1986) A century of fermenter design. J. Inst. Brew. 92: 137-145.
[12] Diviès, C. (1975) Procédé enzymatique utilisant des microorganismes inclus. French Patent #2320349.
[13] Hill, F. (1990) Immobilization of yeast in alginate beads for production of alcoholic beverages. U.S. Patent #5070019.
[14] Groboillot, A.; Boadi, D.K.; Poncelet, D. and Neufeld, R.J. (1994) Immobilization of cells for application in the food industry. Crit. Rev. Biotechnol. 14: 75-107.
[15] Cachon, R.; Molin, P. and Diviès, C. (1995) Modeling of continuous pH-stat stirred tank reactor with *Lactococcus lactis* ssp. *lactis* bv. *diacetylactis* immobilized in calcium alginate gel beads. Biotechnol. Bioeng. 47: 567-574.
[16] Cachon, R.; Lacroix, C. and Diviès, C. (1997) Mass transfer analysis for immobilized cells of *Lactococcus lactis* sp. Using both simulations and *in-situ* pH measurements. Biotechnol. Tech. 11: 251-255.
[17] Radovith, J.M. (1985) Mass transfer effects in fermentations using immobilized whole cells. Enzyme Microb. Technol. 7: 2-10.
[18] Janssen, D. (1993) Immobilized-cell malolactic fermentation. Vitic. Enol. Sci. 48: 219.
[19] Portno, A.D. (1978) Fermentation and storage. In: Symposium European Brewing Convention in Zoeterwoude, Elsevier Scientific Publishers, London (GB); pp. 145-154.
[20] Galazzo, J.L. and Bailey, J.E. (1990) Growing *Saccharomyces cerevisiae* in calcium-alginate beads induces cell alterations which accelerate glucose conversion to ethanol. Biotechnol. Bioeng. 36: 417-426.
[21] Bardi, E.P; Koutinas, A.A.; Psarianos, K. and Kanellaki, M. (1997) Volatile by-products formed in low temperature wine making by immobilized yeast cells. Process Biochem. 32: 579-584.

[22] Iconomopoulou, M.; Psarianos, K.; Kanellaki, M. and A.A. Koutinas (2002) Low temperature and ambient temperature wine making using freeze dried immobilized cells on gluten pellets. Process Biochem. 37: 707-717.
[23] Diviès, C. and Deschamps, P. (1986) Procédé et appareillage pour la mise en œuvre de réactions enzymatiques et application à la préparation de boissons fermentées. French Petent #2601687.
[24] Ogbonna, J.M.; Amano, Y.; Nakamura, K.; Yokotsuka, K.; Shimaza, Y.; Wtanabe, M. and Hara, S. (1989) A multistage bioreactor with replaceable bioplates for continuous wine fermentation. Am. J. Enol. Vitic. 40: 292-298.
[25] Bidan, P.; Diviès, C. and Dupuy, P. (1978) Procédé perfectionné de préparation de vins mousseux. French petent FR2432045A1
[26] Fumi, M.D.; Trioli, G.; Colombi, M.G. and Colagrande, O. (1988) Immobilization of *Saccharomyces cerevisiae* in calcium alginate gel and its application to bottle-fermented sparkling wine production. Am. J. Enol. Vitic. 39: 267-272.
[27] Godia, F.; Casas, C. and Sola, C. (1991) Application of immobilized yeast cells to sparkling wine fermentation. Biotechnol. Prog. 7: 468-470.
[28] Busova, K.; Magyar, I. and Janky, F. (1994) Effect of immobilized yeasts on the quality of bottle-fermented sparkling wine. Acta Alimantaria 23: 9-23.
[29] Yokotsuka, K.; Yajima, M. and Matsudo, T. (1997) Production of bottle-fermented sparkling wine using yeast immobilized in double-layer gel beads or strands. Am. J. Enol. Vitic. 48: 471-481.
[30] Lemonnier, J. (1986) Cartouche tubulaire filtrante et son application à la fabrication de vin mousseux en bouteille. European Patent #0210915.
[31] Lemonnier, J. (1990) Insert for second fermentation of champagne or wines produced by the traditional method. European Patent #0428428.
[32] Lemmonier, J. (1992) Cartouche de fibres creuses microporeuses pour la fermentation de boissons sucrées. European Patent #0555603.
[33] Spooner, J.E. (1973) Method for producing champagne. U.S. Patent #4009285.
[34] Quetsch, K-H. (1990) Immobilised bio:catalyst bottle stopper - for sparkling wine with string to retract the filter housing inside the stopper. German Patent #3931906.
[35] Poirat, D. (1996) Procédé pour faire fermenter un liquide en bouteille, et cage amovible pour sa mise en œuvre. French Patent #2745586.
[36] Duteurtre, B.H.J.; Ors, P.; Charpentier, M.S. and Midoux, N. (1984) Procédé et installation de fabrication de goutelettes essentiellement sphériques de dimension approximativement uniforme et de diamètre supérieur à au moins environ deux millimètres à partir d'un matériau susceptible de s'écouler, solidifiable, coagulable ou gélifiable contenant de préférence des particules en particulier formées par les levures ou microorganismes de fermentation. French Patent #2570959.
[37] Diviès, C.; Lenzi, P.; Beaujeu, J. and Herault, F. (1989) Process for preparing micro-organisms incorporated within substantially dehydrated gels, gels obtained and their use in the preparation of fermented beverages. World Patent #9000602.
[38] Duteurtre, B.H.J.; Coulon, P.A. and Goutard, R.M. (1984) Procédé et appareil de transfert et/ou de dosage de particules fragiles, à cadence élevée, en vue, par exemple, de l'insertion dans un récipient. French Patent #2571135.
[39] Valade, M. and Rinville, X. (1990) Approches techniques et économiques des différents procédés de prise de mousse en bouteille. In: Les nouvelles techniques de tirage, Association des techniciens supérieurs en viticulture et oenologie, Epernay (France); pp. 34-39,
[40] Lenzi, P. and Cavin, J.F. (1985) Procédé de fabrication de boissons fermentées aromatisées. French Patent #2580665.
[41] Fumi, M.D.; Bufo, M.; Trioli, G. and Colagrande, O. (1989) Bulk sparkling wine production by external encapsulated yeast bioreactor. Biotechnol. Letters 11: 821-824.
[42] Sarishvili, N.G.; Storchevoi, N.; Vaganov, V.M. and Reiblat, B.B. (1987) Process for producing sparkling wines. U.S. Patent #4981700.
[43] Crapisi, A.; Nuti, M.P.; Zamorani, A. and Spettoli, P. (1987) Improved stability of immobilized *Lactobacillus* sp. Cells for the control of malolactic fermentation in wine. Am. J. Enol. Vitic. 38: 310-312.
[44] Crapisi, A.; Spettoli, P.; Nuti, M.P. and Zamorani, A. (1987) Comparative traits of *Lactobacillus brevis, Lactobacillus fructivorans* and *Leuconostoc oenos* immobilized cells for the control of malolactic fermentation in wine. J. Appl. Bacteriol. 63: 513-521.

[45] Spettoli, P.; Bottacin, A.; Nuti, M.P. and Zamorani, A. (1982) Immobilization of *Leuconostoc oenos* ML34 in calcium alginate gels and its application to wine technology. Am. J. Enol. Vitic. 33: 1-5.
[46] Cuenat, P. and Villetaz, J.C. (1984) Essais de fermentation malolactique des vins par des bactéries lactiques immobilisées du genre *Leuconostoc*. Revue suisse viticulture arboriculture horticulture 16: 145-151.
[47] Rossi, J. and Clementi, F. (1984) L-malic acid catabolism by polyacrylamide gel entrapped *Leuconostoc oenos*. Am. J. Enol. Vitic. 35 : 100-102.
[48] Naouri, P.; Chagnaud, P.; Arnaud, A.; Galzy, P. and Mathieu, J. (1991) A new technology for malolactic bioconversion in wine. J. Wine Res. 2: 5-20.
[49] Fleet, G.H. and Costello, P.J. (1991) Malolactic fermentation of wine. U.S. Patent #5104665.
[50] Kosseva, M.; Beschkov, V.; Kennedy, J.F. and Lloyd, L.L. (1998) Malolactic fermentation in Chardonnay wine by immobilised *Lactobacillus casei* cells. Process Biochem. 33: 793-797.

IMMOBILISED CELL TECHNOLOGIES FOR THE DAIRY INDUSTRY

CHRISTOPHE LACROIX[1], FRANCK GRATTEPANCHE[2], YANN DOLEYRES[1] AND DIRK BERGMAIER[2]
[1]*Laboratory of Food Biotechnology, Institute of Food Science and Nutrition, Swiss Federal Institute of Technology, ETH Zentrum, 8092 Zurich, Switzerland – Fax: +41 1 632 1403 –*
Email: christophe.lacroix@ilw.agrl.ethz.ch; [2]*Dairy Research Centre STELA, Pavillon Paul Comtois, Université Laval, Québec, PQ, Canada, G1K 7P4*

1. Introduction

Traditionally, milk fermentations are conducted in batch bioreactors using freely suspended microbial cells. Lactic acid bacteria (LAB) are widely used in the production of fermented dairy products such as cheeses or fermented milks and creams because of their technological, nutritional and eventual health properties. The production of organic (mainly lactic and acetic) acids and the resulting acidification is essential for the production, development of typical flavour and preservation of these products. Other inhibitory compounds such as bacteriocins can also increase shelf-life and safety of the products. The transformation of lactose by lactic cultures improves the digestibility and various metabolic and enzymatic activities of LAB lead to the production of volatile substances, which contribute to flavour, aroma and texture developments in fermented dairy products. Probiotics are defined as microbial cells which transit the gastrointestinal tract and which, in doing so, benefit the health of the consumer [1]. Among these micro-organisms, lactobacilli and bifidobacteria are already used in many probiotic dairy products including milk, yogurt, ice cream, and cheese.

There are numerous publications on immobilisation of LAB, emphasising the importance and interest in this new technology. Cell immobilisation has been shown to offer many advantages for biomass and metabolite productions compared with free-cell (FC) systems such as: high cell density, reuse of biocatalysts, retention of plasmid bearing cells, improved resistance to contamination, stimulation of production and secretion of secondary metabolites and physical and chemical protection of the cells. The purpose of this review is to describe recent developments in the application of immobilised cells (IC) in the dairy sector. The readers are also invited to refer to a previous review on this subject by Champagne *et al.* [2].

2. Immobilisation techniques

Different methods have been used for immobilising LAB (Table 1): physical entrapment in polymeric networks, attachment or adsorption to a preformed carrier, membrane entrapment and microencapsulation. All these techniques have similar purposes: either to retain high cell concentrations within the bioreactor or to protect cells from a hostile environment. For industrial applications in the dairy industry, the carrier material must be non-toxic, readily available and affordable. It should lead to high-cell loading and the cells should have a prolonged viability in the support.

2.1. ENTRAPMENT WITHIN POLYMERIC NETWORKS

For food applications, the most widely used immobilisation technique is the entrapment of cells within a food-grade porous polymeric matrix (Table 1). In many applications, controlled-size polymer droplets are produced using extrusion or emulsification, under mild conditions. Thermal (κ-carrageenan, gellan, agarose, gelatin) or ionotropic (alginate, chitosan) gelation of the droplets are used to produce spherical gel biocatalysts. These polymers are readily available and widely accepted for use as additives in the dairy industry. Gel entrapment is a relatively simple method resulting in usually spherical beads with diameters ranging from 0.3 to 3.0 mm with high biomass concentration. However, polysaccharide or protein gels generally exhibit limited mechanical stability over extended fermentation periods [2]. A careful selection of polymer composition is necessary, according to the conditions of the fermentation [3]. For dairy applications with LAB, we have shown that a mixed gel of deionized kappa-carrageenan and locust bean gum (LBG) exhibited good mechanical stability during long-term continuous fermentation in milk and whey for more than 7 weeks [4-6]. However, the large scale production of beads under aseptic conditions still remains an important issue for the industrialisation of immobilised cells in the food sector.

2.2. ADSORPTION TO A PREFORMED CARRIER

Immobilisation of cells by adhesion or adsorption to a preformed matrix is based on their affinity for solid surfaces during growth and is achieved by keeping the carrier and the actively growing cells in contact for a defined period. This technique imitates what takes place in a natural environment, where cells are almost always associated with surfaces and grow in a biofilm [7]. Since the carrier is inert and usually no additional chemicals are involved, the immobilisation is carried out under very mild conditions resulting in high IC viability. Solid particles are more resistant to compression and disintegration but usually IC densities are lower than with gel particles. However, there are only few solid supports that can be used in food processing and the compatibility of a material for application in food fermentation needs to be carefully assessed. Bergmaier et al. [8] recently reported the efficient immobilisation of *Lactobacillus rhamnosus* RW 9595M in porous silicon rubber, with a very high viable biomass concentration in the colonised supports estimated by DNA analysis at $8.5 \; 10^{11}$ cfu.ml^{-1}. However, this support is expensive and its utilisation may be only justified for high value-added products and long utilisation of the supports. Other preformed carriers tested with LAB are listed in Table 1.

Table 1. Immobilised lactic acid bacteria by entrapment or adsorption techniques.

Support	Species	Maximum cell concentration[a]	Reference
Polymeric networks			
κ-carrageenan - LBG[c]	Lactobacillus casei	$5.1\ 10^{11}$ cfu.ml^{-1}	[17]
κ-carrageenan - LBG[c]	Lactococcus lactis ssp. (three strains)	$1.3\ 10^{11}$ cfu.g^{-1}	[34]
Gellan gum	Bifidobacterium longum	$6.8\ 10^{10}$ cfu.g^{-1}	[35]
Ca-alginate	Lactococcus lactis ssp.	$2.0\ 10^{11}$ cfu.ml^{-1}	[109]
Ca-alginate	Lactococcus lactis ssp.	$3.8\ 10^{11}$ cfu.g^{-1}	[109]
Preformed carriers			
Sintered porous glass	Lactobacillus casei	0.105 g.g^{-1}	[110]
Alumina beads	mixed lactic acid bacteria	n.p.[b]	[111]
Sintered porous glass, ceramic particles	Lactobacillus rhamnosus	34.4 g.l^{-1}	[11]
Foam glass (polyethyleneimine-coated)	Lactobacillus rhamnosus	0.490 g.g^{-1} or 97 g.l^{-1}	[112]
Plastic composite	Lactobacillus casei	$2.3\ 10^{9}$ cfu.g^{-1}	[113,114]
Polypropylene (chitosan treated)	Lactobacillus plantarum	n.p.[b]	[42]
Porous silicone rubber (ImmobaSil®)	Lactobacillus rhamnosus	$8.5\ 10^{11}$ cfu.ml^{-1}	[8,88]

[a] cfu.ml^{-1} or g of support; g biomass.l^{-1} or g^{-1} of support
[b] n.p. = not presented
[c] LBG = locust bean gum

2.3. MEMBRANE ENTRAPMENT

In a membrane bioreactor, the cells are not held in a porous matrix, but are retained by a semipermeable membrane barrier. Hollow-fiber bioreactors, in which the cells are confined on one side of the porous fiber and the soluble medium components on the other, have been successively used for lactic acid production, and in some cases, yielded high volumetric productivity [9]. However, the application of membrane reactors is limited by high capital investment, complex maintenance of the sophisticated equipment and membrane fouling, particularly with whey permeate, which results in a rapid decrease of process performance [10].

2.4. MICROENCAPSULATION

A microcapsule consists of a semipermeable, spherical, thin, and strong membrane surrounding a liquid core, with a diameter varying from a few microns to 1 mm [11]. Different materials have been used for microcapsulation of LAB and probiotic cultures (Table 2), although many of these technologies do not fall within the strict definition of microencapsulation, but rely on gel entrapment techniques. One important challenge for cell encapsulation is the large size of microbial cells (typically 1-4 μm) or particles of freeze-dried culture (more than 100 µm). This characteristic limits cell loading for small capsules or, when large size capsules are produced, can negatively affect the textural and sensorial properties of food products in which they are added. In almost all cases, gel entrapment using natural biopolymers such as calcium alginate, κ-carrageenan and gellan gum has been favoured by researchers. However, although promising on a laboratory scale, many developed technologies for producing gel beads still present serious difficulties for large-scale production of food-grade microencapsulated microorganisms[12]: low production capacity and large bead diameters for the droplet extrusion methods; transfer from organic solvents and large size dispersion for the emulsion techniques; difficulty to maintain asepsis in the process.

Table 2. Microcapsules with potential use in the dairy industry.

Matrix	Species	Reference
Alginate	*Lactobacillus plantarum*	[102]
Alginate	*Lactobacillus delbrueckii* ssp. *bulgaricus*	[100]
κ-carrageenan LBG	*Bifidobacterium longum*	[103]
Alginate	*Lactobacillus acidophilus* *Bifidobacterium longum*	[99]
κ-carrageenan	*Bifidobacterium longum*	[96]
Alginate	*Bifidobacterium bifidum* *Bifidobacterium infantis*	[105]
κ-carrageenan	*Bifidobacterium bifidum*	[107]
Phtalate cellulose acetate	*Bifidobacterium pseudolongum*	[115]
Alginate	*Bifidobacterium longum*	[108]
Gellan/xanthan	*Bifidobacterium infantis*	[116]
Alginate	*Lactobacillus acidophilus* *Bifidobacterium lactis*	[117]
Milk fat	*Bifidobacterium longum*	[13]
Milk fat/whey proteins	*Bifidobacterium breve*	[15, 16]
Whey proteins	*Bifidobacterium longum*	[16]

Spray-coating of a lyophilised culture with milk fat has been used to protect bacteria against oxygen. Champagne *et al.* [13] showed that milk-fat coating of *Bifidobacterium longum* improved cell viability and activity during storage compared with FC and cells encapsulated in gelatine or xanthan gum. The technology patented by Sunhoara *et al.* [14] uses the same principle of a protecting hydrophobic phase. Freeze-dried bacteria are suspended in a fat phase (coconut oil) and co-extruded using a three channel concentric needle to produce capsules with an external enteric layer made of gelatine and pectin, which are subsequently freeze-dried. Although this method is apparently efficient for protecting sensitive bacteria such as *B. longum*, during storage, its application is difficult due to the complexity of the extrusion technique that limits the scale of production, and the large size of the capsules (0.3 to 8 mm).

Water-insoluble dry microcapsule preparations with small (< 100 µm) and controlled particle size are desirable for incorporating immobilised bacteria in food products for various reasons, including higher stability, easier handling and storage of the cultures, and limited effects on sensorial properties of foods. The preparation of multiphase water-insoluble food-grade microcapsules designed for stabilisation and controlled release of sensitive probiotic cultures, using emulsification and spray-drying was recently described [15,16]. Soluble whey-protein polymers are used as coating material and form upon rehydration a stabilizing film around milk fat globules containing micronised powder of freeze-dried bacteria. The encapsulation of *B. breve* and *B. longum* as freeze-dried or fresh cultures was studied in water-insoluble food-grade microcapsules produced by emulsion and/or spray-drying, using milk fat and/or denatured whey proteins as immobilisation material [16]. Dispersion of fresh cells in heat-treated whey-protein suspension followed by spray-drying was shown to be the least destructive immobilisation technique tested, with a survival rate of 25.7 ± 0.1% and 1.4 ± 0.2% for *B. breve* and *B. longum*, respectively. The micronisation treatment required to reduce the powder particle size of freeze-dried bacteria before microencapsulation decreased both cell viability and possibly cell resistance to other stresses during subsequent processing, such as heat during spray-drying.

3. Microbial growth in gel beads

3.1. BIOMASS DISTRIBUTION IN GEL BEADS

The performance of IC bioreactors, in which viable cells are entrapped in polysaccharide gel beads, depends on cellular reaction kinetics, internal and external mass transfer of substrates and inhibition products, and also on cell release from the bead surface to the bulk medium. Consequently conditions are more favourable for cell growth close to the bead surface due to diffusion limitations for both substrates and inhibitory products, mainly lactic acid in the case of LAB [17], which results in a sharp pH gradient in beads [18,19].

Mathematical models taking into account this competitive diffusion-reaction phenomenon described the nonuniform cell growth in the colonised gel beads [19-23]. The progressive formation of the peripheral dense cell layer (with a thickness varying from 100 to 400 µm) and a core with low cell growth and viability during successive batches or continuous cultures with LAB immobilised in gel beads has been demonstrated by transmission electron [4,24] and confocal microscopy using fluorescent probes [23] or fluorescent-labelled polyclonal antibodies [25-27]. The thickness of this cell layer and the subsequent biomass loading of the gel depend on the bacteria, the obstruction effect, the diffusion rates in the cell layer, as well as other parameters of fermentation such as pH, concentration of substrates and products, and temperature. Consequently, bead diameter is an important parameter for the activity of the biocatalysts. Reducing this diameter increases the volumetric productivity of IC bioreactors because a larger fraction of beads is colonised, resulting in a higher biomass concentration in the bioreactor containing a defined volume of beads. However, reducing the size of gel beads below 1 mm may result in mechanical instability of the gel during long-term continuous cultures; as reported in Audet *et al.* [28] cell growth at the bead centre destroyed gel structure. Therefore, the choice of bead diameter should be a compromise between productivity, mechanical and separation characteristics.

3.2. CELL RELEASE FROM BEADS

Cell release from gel beads in the liquid medium occurs spontaneously due to the formation of the high biomass-density peripheral layer at the bead surface with a high cell-growth activity. During continuous culture of immobilised LAB, microscopic observations showed that peripheral gel cavities containing the microcolonies are disrupted by forces resulting from cell growth and shear forces due to mechanical agitation and multiple bead contacts in the bioreactor [4,29]. Consequently, at steady state, a biofilm of fixed biomass is formed, in which cell growth and cell release in the culture medium are balanced. This phenomenon was used for the continuous milk inoculation and prefermentation processes [30-33], and for the continuous fermentation of whey or other media for lactic biomass production [26,34,35].

Different parameters may influence the cell growth and release from the bead surface to the bulk medium during fermentation, including the type of microorganism and growth characteristics, the thickness and the density of the biofilm formed, the composition and size of the gel beads, and the hydrodynamic conditions in the bioreactor. High-stirring speed negatively affects biomass volumetric productivity. Indeed, despite an improvement in the internal and external mass transfers, high-stirring reduced the overall activity of the bioreactor by decreasing the peripheral cell layer thickness [33,36]. Other fermentation parameters such as temperature, pH, medium composition and dilution rates greatly influence IC growth and consequently cell release from the peripheral layer of beads to the fermentation medium [5,6,34,37-39].

The specific activity of entrapped cells is decreased compared with FC due to diffusion limitations in gel beads. Norton et al. [38] estimated the specific growth rate for immobilised *Lb. helveticus* during continuous culture at 12 to 18% of that for FC, whereas Cachon [40] reported a low value of 9% for entrapped *Lactococcus lactis* ssp. *lactis* biovar. *diacetylactis* when compared with the maximum specific growth rate of FC. Arnaud et al. [17] showed that the average specific growth rate of *Lb. casei* in colonised gel beads used for continuous culture was largely dependent on the mixing rate in a continuous stirred tank reactor (CSTR), and corresponded to approximately 17 and 62% of FC when agitation in the bioreactor was set at 50 and 150 rpm, respectively. The low effectiveness factor of the gel beads, defined as the ratio of the reaction rate in the gel bead to the reaction rate in the absence of diffusion limitations clearly emphasises the importance of size reduction in order to increase the surface volume ratio of gel beads as well as the volumetric productivity.

3.3. CROSS-CONTAMINATION PHENOMENON

The dynamics of the complex microbiological equilibrium in a continuous reactor with LAB mixed strains immobilised in κ-carrageenan/LBG gel beads have been recently characterised. Our recent data for fermentation in milk [4] or supplemented whey permeate (SWP) [5,6,34] demonstrated that extensive cross contamination between beads, initially containing a pure culture, occurred during continuous fermentations performed over extended time periods (5 to 9 weeks). Nevertheless, the process showed high biological stability and high reproducibility, and no strain became dominant or was eliminated. This was clearly demonstrated in continuous IC fermentations using three strains of *L. lactis* in whey permeate by changes in individual bead composition, stability of cell populations in the reactor effluent and in bulk bead samples, and by constant acidifying activity of the effluent biomass [5,6,34].

We proposed a theoretical model of cell release that could explain cross-contamination [41]. This model suggests that after the disruption of gel microcolonies near the bead surface as a result of pressure due to cell colony expansion, collisions and shearing forces in the reactor, the partially emptied cavities composed of a viscoelastic polysaccharide gel material would close again, entrapping samples from the bulk medium, thus incorporating contaminating flora into the active layer of the bead. Then, a new colonisation cycle of the cavity would begin and strain competition inside the microcavities would largely depend on local conditions, such as the concentration of undissociated lactic acid and pH. Controlled abrasion of gel beads in the stirred tank reactor might be an important factor for culture stability as it may bring to the bead surface cells of the pure culture that were located deeper in the gel, therefore acting as a reservoir for pure cultures that were separately entrapped.

To experimentally validate this model and identify factors of cross-contamination, Prioult *et al.* [25] developed an immunofluorescent method involving double-colour labelling and confocal microscopy to specifically detect cells of *L. diacetylactis* and *B. longum* co-immobilised in gel beads. A quantitative analysis was then developed to locate and quantify microcolonies of the two strains in beads [27]. This method was used to demonstrate the microbial dynamics and cross-contamination in gel beads during continuous production of a model mixed-lactic starter composed of a probiotic non competitive culture, *B. longum,* and a competitive LAB culture, *L. diacetylactis*, immobilised separately in gel beads [26]. After 14 days, the strain ratio (*i.e.* 7:1 for *L. diacetylactis/B. longum*) in bulk bead samples was similar to that measured in individual beads. The determination of the spatial distribution of the two strains in gel beads by immunofluorescence and confocal laser-scanning microscopy showed that bead cross-contamination was limited to a 100 µm peripheral layer.

4. Changes of culture characteristics during immobilised-cell fermentations

Several studies have shown that cells produced using IC technology exhibit changes in growth and physiology as well as morphology compared with cells produced during conventional FC cultures. Cachon *et al.* [19] observed differences in cell physiology during batch cultures with and without pH control and continuous cultures with free and immobilised *L. lactis*, depending on the culture mode. The redox states, enzymatic pool and intracellular pH differed for IC and FC cultures. A shift in the metabolic pathway from homofermentative to heterofermentative has also been observed during continuous cultures with immobilised *Lactobacillus* [42]. Moreover, different authors have observed an increased tolerance of IC to product inhibition [43], alcohols [44,45], phenols [46-48], antibiotics [49,50], or quaternary ammonium sanitizers [51]. Not only immobilisation but also prolonged culture of IC resulted in an increased cell tolerance to antibiotics and sanitizers. Recently, Doleyres *et al.* [26,50] studied the effect of immobilisation and long-term continuous-flow culture on probiotic and technological characteristics of cultures produced in the effluent medium. Tolerance of free cells produced in the effluent medium to various stresses including freeze-drying, oxygen peroxide, simulated gastro-intestinal conditions, nisin, and antibiotics markedly increased with culture time and were generally significantly higher after 6 days than those of cells produced by conventional FC-batch fermentations. A reversibility of the acquired tolerance of *B. longum*, but not for *L. diacetylactis*, to antibiotics was shown during successive FC batch cultures initially inoculated with cells taken from the effluent of continuous IC fermentation.

Trauth *et al.* [51] tentatively explained this phenomenon of increased cell tolerance by a possible modification of the cell membrane and physiology, and by cell proximity in a saturated matrix which gives protection to the cell membrane. In contrast, the increased tolerance to various environmental stresses reported by Doleyres *et al.* [50] when cells were immobilised was not associated with cell or strain specific mechanisms or physical protection by cell contact and high density in the gel matrix, since in this study tolerance was measured with released cells and for different stresses and strains. This result suggests that the non-specific increase in cell resistance during long-term continuous IC fermentations might be partly due to a reversible increase in thickness of the cell membrane, as proposed by Trauth *et al.* [51] to account for an increased tolerance to quaternary ammonium sanitizers of IC.

Cell growth in gel beads is limited by diffusional limitations of both substrates and inhibitory products, in this case lactic and acetic acids, as presented above. This leads to the development of steep inhibitory product, pH and biomass gradients in colonised beads [18,19,23,26] which can induce a non-specific stress adaptation of IC. Presently, the only technique used to increase LAB and probiotic-cell resistance to stresses occurring during production, storage, use in dairy products or digestion relies on incubation of FC under starving or other stressful conditions, such as heat, high concentrations of salt, bile salts, or hydrogen peroxide, or low pH [52,53]. Recently, enhanced cell-to-cell signalling and cell-matrix interactions leading to coordinated behaviour of immobilised microorganisms have been reported [54]. Additional studies are needed to better understand the effects of the local microenvironment of IC combined with long-term cultures leading to a progressive increase of resistance characteristics. Data from Doleyres *et al.* [50] showed that IC technology combined with long-term continuous culture can be used to efficiently produce, in a one step process, cells with enhanced tolerance to environmental stresses. In addition, cells produced by continuous IC cultures, which are in exponential or early stationary growth phase, exhibited both a high viability and metabolic activity compared with starving cells produced by conventional batch cultures.

A dramatic change in culture physiology and morphology was also observed in a 32-day chemostat culture with *Lb. rhamnosus* RW-9595M immobilised by adsorption on solid porous silicone rubber supports, used for exopolysaccharide (EPS) production in SWP [55]. Despite the very high IC concentrations obtained in this study ($3.7 \cdot 10^{11}$ cfu.ml^{-1} of support), the mean soluble EPS production in the fermented broth remained very low (138 mg.l^{-1}) compared with that obtained by conventional FC batch (1718 mg.l^{-1}) or continuous (1808 mg.l^{-1} for a dilution rate of 0.3 h^{-1}) cultures in the same medium. However, a large production of macroscopic aggregates (1-2 mm) containing high viable-cell ($4.3 \cdot 10^{12}$ cfu.g^{-1}, estimated by DNA analysis) and non-soluble exopolysaccharide (14.2%, w/w) concentrations was observed. The loss of soluble-EPS production capacity of cells in the effluent of the IC system was reversible and EPS production (1742 mg.l^{-1}) was fully recovered after four successive pH-controlled batch cultures. Low soluble-EPS concentration resulted in low viscosity in fermented broths, which may facilitate cell propagation and separation. In addition, aggregates produced by continuous IC cultures in SWP can be easily separated and then used as a symbiotic product with very high active cell and EPS concentrations.

5. Biological stability

5.1. PSYCHROTROPH CONTAMINANTS

Extended refrigeration of milk at the farm and in plants promotes the selective development of psychrotrophic bacteria, which is the most important group of microorganisms in dairy products from a food spoilage perspective. These bacteria secrete lipolytic and proteolytic enzymes leading to bitterness, rancidity and various off-flavors, as well as reduced yields and heat stability [56]. Lactic acid bacteria can be added to raw milk to prevent the growth of psychrotrophic bacteria by producing H_2O_2, which activates the lactoperoxidase system [57]. Champagne *et al.* [2] reviewed the application of IC technology for seeding raw milk and decreasing contamination by psychrotrophs. Although IC technology has a good potential in this application, the engineering questions raised by these authors, particularly those relating to mechanical stability of gel beads, have still not been resolved. To improve the technique, it is necessary to select food grade bacteria exhibiting a high H_2O_2 production while not acidifying milk.

The development of psychrotrophs during a 7-week continuous prefermentation with immobilised LAB (four strains) in pasteurised milk was studied [4]. The prefermentation was performed at 26°C and the pH was maintained at 6.0 by the addition of fresh milk. The psychrotrophic population in pasteurised and prefermented milks were highly parallel. The high stability of the continuous IC fermentation can be explained by very high dilution rates and short milk residence times of milk in the bioreactor (less than 6 minutes under the conditions of fermentation). Except for when the milk was highly contaminated, the mean psychrotrophic counts in prefermented milk were always low compared with the released lactic acid flora, averaging 4.1 10^4 and 4.1 10^8 cfu.ml^{-1}, respectively [4]. Even though in this study the microbiological quality of pasteurised milk was low, with psychrotroph counts exceeding 6 10^6 cfu.ml^{-1} for three days, the presence of psychrotrophs on gel beads was only detected after 5 weeks and reached 10^5 cfu.ml^{-1} at the end of the 7 week experiment, *i.e.* a very low level compared with the entrapped LAB population (1.5 10^{11} cfu.ml^{-1}) [4]. The contamination of gel beads by psychrotrophs was explained by the cross-contamination model discussed in a previous section.

5.2. BACTERIOPHAGE CONTAMINANTS

Bacteriophages can infect and destroy LAB and are a serious threat in dairy fermentations. They lead to a rapid decrease or stoppage of milk acidification.

Bacteriophages occur in raw milk, generally in very low numbers (e.g., 1 or less per liter), but the main source of contamination in large plants is whey. They proliferate when bacteria are grown in large quantities, for example in starters or during milk fermentations. The burst size, i.e., the number of phages that are released on average from an infected cell can be considerable, up to about 100 phages per cell for LAB [58]. Several phages of LAB can survive low pasteurisation of milk, i.e., 15 s at 72°C and cause serious problems in dairy plants which can result in economic losses.

Different studies have shown that LAB acidification rate is much less affected by phages in IC than in FC cultures [58-62]. The diffusion of bacteriophages in polysaccharidic gels used for cell immobilisation is prevented by size exclusion effects [61]. An alginate gel of 40 g.l^{-1} has a smaller pore size than the head of commonly isolated phages in cheese plants (i.e. 49 nm). However, although IC in gel beads are protected from phages, this protection does not extend to FC released from the beads. Therefore phage contamination during continuous IC fermentations over extended periods still represents a serious threat, particularly for the applications in the dairy industry, and is therefore an important question to address.

Champagne et al. [59] showed that the population of bacteriophages in an IC system increased to reach 10^9 pfu.ml^{-1} when beads were used for repetitive batch incubation. With L. lactis immobilised in a calcium alginate gel coating a spiral mesh, Passos et al. [61] found that a dilution rate of 2.6 h^{-1} prevented phage built-up but was insufficient to eliminate the phage completely from the bioreactor after 15 to 28 h continuous culture in milk. During continuous fermentations of milk with immobilised L. lactis at a dilution rate of 0.5 h^{-1}, the decrease in FC population (1 log) after phage infection was small compared with that for FC continuous culture [58]. Nevertheless, at this dilution rate, phages with a latent period of only 30 min were not washed from the continuous reactor and could multiply and infect sentitive cells. However, even when higher dilution rates ranging from 10 to 30 h^{-1} were used, a complete washout of phages inoculated at low (10^2 pfu.ml^{-1}) or high levels (10^5 pfu.ml^{-1}) in the continuous IC reactor was not observed [58,62]. The level of phage in milk effluent within only 6 h after infection reached from 5 10^6 to 10^8 pfu.ml^{-1} for dilution rates decreasing from 30 to 10 h^{-1} [62]. After 48 h, all treatments had effluent populations around 5 10^7 pfu.ml^{-1}, indicating that dilution rate in the tested range did not have a significant effect on phage population. High phage counts in gel beads (approx. 5 10^9 pfu.g^{-1}) after 48 h culture in milk, which were approximately 2 log higher than free-phage counts in the medium, suggest phage implantation at the bead surface with constant phage release into the medium. According to the cross-contamination model presented earlier in this review, phages are entrapped in surface cavities of the gel, multiply in the presence of sensitive cells and are continuously released into the culture medium. Phage entrapment only resulted in a slight decrease of IC population, and had therefore only a limited effect on the FC and total lactococci population in the system, but subsequent acidification of the phage-contaminated culture was markedly delayed [62]. Phage entrapment also induced the development of phage-resistant cells, representing up to 1.9% of the total bead population, but estimated at 10-23% for the surface population 48 h after phage contamination [58, 62]. The inoculated milk showed an acidifying activity related to its phage-resistant bacterial population.

With respect to using IC for the production of lactic starters, phage contamination could be controlled by the use of phage-resistant media and/or conditions that prevent phage contamination (disinfection of equipment and high heat treatment of milk). For continuous inoculation/acidification of milk, another alternative might be to immobilise phage-resistant or multiple phage-unrelated strains to alleviate the effect of phage contamination of processing milk that is only pasteurised. The population dynamics of mixed cultures in this perspective need to be examined.

5.3. PLASMID STABILITY

Important LAB characteristics are plasmid-encoded, such as: phage resistance, antibiotic resistance, lactose or citrate utilisation, and bacteriocins and EPS production [63-65]. Plasmids can also be transferred by genetic manipulations to produce recombinant bacteria with improved characteristics. However, plasmid instability often limits the use of natural and recombinant cultures. It has been shown that cell immobilisation can lead to improved plasmid stability of recombinant bacteria [66], but very little work has been done with natural LAB.

The effect of cell immobilisation on plasmid stability was shown in two recombinant strains of L. lactis [67]: after 370 and 540 generations, 10% of IC still contained pIL252 or pIL253, respectively, whereas only 50 and 210 generations were needed to reach this level in FC cultures. Huang et al. [68] studied the production of pediocin and the stability of the related plasmid in continuous cultures with Pediococcus acidilactici UL5. From 10 to 28% of low-bacteriocin-producing variant cells appeared after 144 h of FC cultures carried out at various dilution rates and pH, while almost no such variants were detected after 192 h of IC cultures. The variant cells produced 8 to 64-fold less pediocin than the parent cells. A decreased quantity of plasmid DNA suggested that the decreased pediocin activity of variant cells resulted from a decrease in plasmid copy number. This improved plasmid stability in the IC system could be attributed to the limitation of cell division resulting from the environmental conditions within the gel and to low gene expression as observed in the case of recombinant cells [67,69]. The absence of competition within beads between recombinant and plasmid-free cells [70] or the discrete location of cells in the immobilisation matrix [71,72] might account for the high plasmid stability of IC. When compared with other methods for improving plasmid stability during continuous fermentations, cell immobilisation presents several advantages: (a) easy to perform when compared to time-consuming genetic methods; (b) appropriate for long-term continuous cultures; (c) directly applicable to food processing; (d) apparently not influenced by culture conditions, such as pH and dilution rate.

6. Applications of immobilised cell technology

6.1. BIOMASS PRODUCTION

Cell release from gel beads occurs spontaneously as microcolonies form near the surface of the biocatalysts. In some applications, such as entrapped biomass production, cream fermentations and metabolite productions, cell release is not desirable. Steps for reducing FC levels have been presented by Champagne et al. [2]. However, repeated use of beads tends to increase the level of FC, and although different treatments have been tried, the rate of cell release is only reduced during the early stages of the fermentation [2,73]. In this section, methods used to promote cell release for continuous starter production or inoculation of milk are discussed, with emphasis on recent data.

6.1.1. Starter production

6.1.1.1. Lactic acid bacteria starter. The LAB are largely used in single and mixed cultures for the production of cheeses, fermented milks, yoghurts, and cultured butter. The use of concentrated LAB starter cultures for bulk tank or direct vat inoculation has eliminated the multiple subculture steps that were traditionally needed to build an inoculum from a mother culture. The main objectives of producing concentrated starters are to obtain a high number of living cells containing the necessary enzymes to function effectively during manufacturing of cultured milk products; and also to maintain a given strain balance in mixed starter cultures. Although lactic starters are traditionally produced by batch fermentation, the accumulation of metabolic end products limits cell growth. The use of continuous fermentations may overcome this limitation, but they are more susceptible to contamination and loss of plasmid-mediated characteristics and they rapidly lead to the disappearance or domination of strains in a mixed-strain continuous fermentation [67,74]. One promising alternative is the use of IC technology to produce mixed lactic starters in continuous fermentation [34]. The high IC concentration results in a very high productivity and decreases contamination risks due to the high dilution and inoculation rates provided by cell release from beads. Immobilisation also improves plasmid stability [67,68].

The production of mixed-strain mesophilic lactic starters was studied during continuous fermentations of SWP, with three strains of lactococci separately immobilised in κ-carrageenan/LBG gel beads in a stirred tank reactor [34]. The effect of pH, temperature and dilution rate and their interactions on the starter composition and activity in the effluent, bead population and lactic acid production were studied by periodically changing the set points during continuous fermentation. The process showed a high biological stability and cell productivity (maximum cell productivity of $5.3 \ 10^{12}$ cfu.l^{-1}.h^{-1}) over the tested period exceeding 50 days. By varying pH, dilution rate (D) and temperature (T), a large range of strain ratios could be obtained, while starter activity remained constant. However, the enumeration of individual beads, initially immobilising a single strain, did not reveal a pure bacterial culture after one week of fermentation. Nevertheless, the biomass redistribution led to a strain ratio in beads that was close to the initial bead ratio in the reactor.

The microbiological and mechanical stabilities of gel beads during prolonged fermentations are critical properties of an IC process, and industrial applications are largely dependent on these properties. During a 7-week continuous fermentation operated with a single set of conditions and three strains of Lactococcus sp., the stability of the IC lactic starter production process was demonstrated [5,6]. Nonetheless, a high proportion of one of the strains, *L. diacetylactis* MD, was found in the total population when the continuous fermentation was operated with a single set or varying conditions [5,34]. The MD free and immobilised populations accounted for 50% or more of the total free and immobilised populations at the end of the 8-week continuous fermentation although MD beads represented only 1/3 of the bead inoculum. The MD population on individual beads originally immobilising a pure culture of the other strains of the mixed culture accounted for 60 to 80% of bead population after 6-8 weeks. The control of strain ratios is critical for starter quality. Lamboley et al. [6] showed that the ratios of different strains in the effluent from the continuous IC reactor were relatively unaffected by the bead ratios used to inoculate the reactor. This data might be explained by the cross-contamination of the gel beads, previously presented in this review, and led to a redistribution of cultures among the different beads initially immobilising a pure culture. Therefore, the control of culture composition in the effluent might be better achieved by selecting fermentation set points (pH, T, D), as shown by Lamboley et al. [34], rather than by changing the ratio of the different beads. The dominance of *L. diacetylactis* in mixed cultures containing *L. cremoris* and/or *L. lactis* was observed both during continuous whey fermentation and milk prefermentation with IC with various strain combinations [4,6,75]. This dominance might be associated with the ability of *L. diacetylactis* to co-metabolise glucose or lactose and citrate, thereby creating a more favorable micro-environment for IC. In a mixed culture of non-isogenic proteinase positive and proteinase negative strains of *Lactococcus lactis* ssp., the technique of cell immobilisation in κ-carrageenan/LBG gel beads was used to produce active mixed starters with the desired and controlled proportions of strains [76].

6.1.1.2. Probiotic cultures. Probiotics are now used to prepare a large variety of pharmaceuticals and fermented milk foods, such as fresh cheeses, fermented milks and health supplements. However, many probiotic cultures, such as bifidobacteria, are fastidious and non-competitive bacteria that are very sensitive to environmental parameters, such as oxygen and acidity, and require complex and expensive media for propagation, with the addition of growth-promoting factors, due to their stringent growth requirements [77].

Doleyres et al. [35] studied the production of *B. longum* ATCC 15707 cells in MRS medium supplemented with whey permeate (MRS-WP). Continuous fermentation of MRS-WP with *B. longum* immobilised in gellan gum gel beads produced the highest cell concentrations (4.9 10^9 cfu.ml^{-1}) in the effluent at a dilution rate of 0.5 h^{-1}. A very high maximal volumetric productivity (6.9 10^9 cfu.ml^{-1}.h^{-1}) was obtained for a dilution rate of 2.0 h^{-1}, and was approximately 9.5-fold higher than in FC batch cultures at optimal pH of 5.5 (7.2 10^8 cfu.ml^{-1}.h^{-1}). In the only other study on continuous production of bifidobacteria with IC, *B. infantis* immobilised in κ-carrageenan/LBG gel beads was used to continuously ferment skim milk (10% solids) supplemented with 1% yeast extract [78]. Cell counts in the cultured milk increased from 1.0 to 2.2 10^9 cfu.ml$^-$ for dilution rates decreasing in the range from 1.0 to 0.5 h^{-1}. Compared with data from Doleyres et al. [35], the lower cell production might be due to different growth rates for the two strains and to the fact that MRS-WP medium is better suited for growth of bifidobacteria than milk-based media.

Doleyres et al. [26] recently demonstrated that cell immobilisation in polysaccharide gel beads can be used to continuously and stably produce a mixed lactic culture containing a non-competitive strain of bifidobacteria. The production of a mixed lactic culture containing *L. diacetylactis* MD and *B. longum* ATCC 15707 was studied during a 17-

day continuous IC culture at different temperatures between 32 and 37°C. The two-stage fermentation system was composed of a first reactor (R1) containing cells of the two strains separately immobilised in κ-carrageenan/LBG gel beads and a second reactor (R2) operated with free cells released from the first reactor. The system allowed to continuously produce a concentrated mixed culture with a strain ratio whose composition depended on temperature and fermentation time. A stable mixed culture (with a 22:1 ratio of *L. diacetylactis* and *B. longum*) was produced at 35°C in the effluent of R2, whereas the mixed culture was rapidly disbalanced in favour of *B. longum* at a higher temperature (37°C) or L. diacetylactis at a lower temperature (32°C). Strain redistribution in beads originally immobilising pure cultures of L. diacetylactis or *B. longum* was observed as presented before in this review.

The production of concentrated suspensions of *B. bifidum* in a whey-based medium using a continuous stirred tank reactor attached to an ultrafiltration device was studied by Corre *et al.* [79]. *B. bifidum* maximal productivity ($2 \cdot 10^8$ cfu.ml^{-1}.h^{-1}) was approximately 35-fold lower than that obtained by Doleyres *et al.* [35] with continuous IC fermentation of *B. longum*, for similar cell concentrations in the ultrafiltration retentate ($5 \cdot 10^9$ cfu.ml^{-1}) and in the effluent of the IC reactor. In addition, long-term stability of cultures with membrane reactors is limited by membrane fouling, particularly with whey-based media.

Recent data reported on IC technology clearly shows that it can be used to continuously and stably produce mixed-strain starters, eventually containing fastidious micro-organisms, such as bifidobacteria, with a high volumetric productivity and high biomass concentrations in the outflow of the continuous fermentation, even at high dilution rates exceeding the specific growth rate.

6.1.2. Prefermentation of milk

Starter culture preparation is of paramount importance in the manufacture of fermented dairy products. Any failure in the starter preparation will, in most cases, lead to detrimental effects on quality, affecting the appearance, texture and flavour of the end product. Traditionally, batchwise processing has been used, but increasing demand for these products has resulted in a large increase in size of fermentation tanks, leading to variations in product quality.

The continuous inoculation-prefermentation of milk for yoghurt production in a stirred tank reactor by separately entrapped cells of *Lb. bulgaricus* and *Streptococcus thermophilus* in Ca-alginate gel beads was the first dairy application of IC [30-32], and was previously presented by Champagne *et al.* [2]. In this system, incoming milk was used to control pH in the IC reactor. The milk feed rate was therefore a direct function of the acidification activity in the fermenter. With a pH set at 5.7, the residence time of milk in the reactor was very short, 8-9 minutes, and lactic acid production was increased 3.3-fold compared with free-cell mixed cultures [31]. The ratio cocci:rod remained stable in the prefermented milk during the 10 day culture, equal to 1 which is the optimum value for yoghurt manufacture. The high cell confinement due to entrapment might explain the high cell and lactic acid production in the bioreactor [32]. This technology allowed a reduction in fermentation time by approximately 50 and 20%

compared with freeze-dried strains and a liquid yoghurt culture, and the resulting yoghurt qualities were evaluated as satisfactory [32].

Passos and Swaisgood [80] studied the continuous inoculation of milk to pH 5.7 by a strain of *L. lactis* immobilised in a Ca-alginate film on a stainless steel spiral mesh. The low lactate and cell volumetric productivity of this system compared with continuous FC culture was explained by mass transfer limitation due to the absence of stirring. This condition resulted in a high lactate concentration and low pH on the gel surface, and leed to the formation of a clotted milk film on the gel surface after only three days of continuous operation.

Fresh cheese starters are complex cultures composed of a mixture of several mesophilic LAB (typically four strains), including acidifying, aromatic and gas-producing strains. The inoculation-prefermentation of milk for fresh cheese production with a mixed culture of four strains separately entrapped in κ-carrageenan/LBG gel beads was extensively studied in a well-mixed stirred tank reactor. Very high lactic acid and biomass productions of up to 28 $g.h^{-1}$ and $7.7 \; 10^{12}$ $cfu.l^{-1}.h^{-1}$, respectively, when the IC bioreactor was operated at 30°C, pH 6.2 and a bead load of 40% (v/v) [33]. The corresponding dilution rate was approximately 44 h^{-1} and could be further increased by approximately 40% by adjusting the pH to 6.4. In these conditions, a high and efficient inoculation of milk with the four strains was observed from $5 \; 10^7$ to $1.5 \; 10^8$ $cfu.ml^{-1}$. The total fermentation time to produce the fresh cheese curd (pH 4.8) was considerably reduced by up to more than 50% compared with the traditional industrial process [4,33], or by 10-15% of that of batch fermentation under optimum laboratory conditions [81]. This fermentation time reduction was explained by the high inoculation level and lower pH of prefermented milk compared with the classical batch fermentation. Moreover, cells released from gel beads are exponentially growing cells with no lag time compared with freeze-dried or bulk starter cultures used in cheese manufacture. The effects of major operating parameters for the fermentation were tested on microbial composition of prefermented milk harvested from the reactor and on process performance during long term experiments (up to 9 weeks), including pH of fermented milk, temperature, bead load, milk (pasteurised and UHT) and composition, and mixing rate [4,33,75,81]. The very high production rate (a reactor of 100 l can produce daily over 100,000 l of prefermented milk inoculated with $1-2 \; 10^8$ $cfu.ml^{-1}$, or 1 million liters of milk inoculated at the usual rate used for fresh cheese production, *i.e.* approx. 10^7 $cfu.ml^{-1}$), the microbial stability and the extensive reduction of fermentation times are among the major advantages of this new process. Another interesting feature of continuous IC fermentation is the high flexibility of the process which can be easily and repeatedly (daily or weekly) interrupted [75]. For example, after a three-day interruption, both the start-up time of the system and the time to reach a pseudo-steady state were very short, less than 30 min. Inoculation of milk by IC bioreactor exhibited more reproducible initial ratios between strains [81]. No sensory significant difference was found between two drained curds obtained with inoculated fresh milk or prefermented milk for sensory properties, susceptibility to syneresis, firmness and modulus of elasticity [81].

6.2. METABOLITE PRODUCTION

Industrial cheese manufacturing yields large volumes of whey as a by-product. In many plants, the valuable whey proteins are concentrated by ultrafiltration and incorporated into cheese or dried and sold as food ingredients. The ultrafiltration step yields large volumes of low-value whey permeate, which has limited uses. Due to its high lactose (about 50 g.l^{-1}) and mineral contents (9 g.l^{-1}), whey permeate may be used as a culture medium for the production of lactic starter cultures or metabolites, such as lactic acid.

6.2.1. Lactic acid production

Lactic acid is widely used as an acidulant and preservation agent in foods and as a precursor for production of emulsifiers, such as stearoyl-2-lactylates, for the baking industries [82]. New applications such as its use as a monomer for production of biodegradable plastics and as an environment-friendly chemical and solvent will increase future lactic acid demand [83]. Whey permeate must be supplemented with a source of metabolizable nitrogen and vitamins, such as yeast extract (YE), in order to satisfy nutritional requirements of LAB. However, the cost of YE contributes largely to lactic acid production costs [10] and, therefore, minimising YE addition is an important goal for process optimization.

Table 3. Summary of lactic acid production in whey permeate with IC systems.

Type[a]	Carrier[b]	Organism	Supplement[c] concentration (g l^{-1})	Productivity (g l^{-1} h^{-1})	Sugar utilisation (%)	Reference
Batch						
	agar	Lb. casei	MC, 60	0.64	94	[118]
	alginate	Lb. casei	YE, 4	0.86	94	[119]
Continuous						
PB	alginate	Lb. helveticus	YE, 15	8	82	[120]
CFBR	SGB	Lb. casei	YE, 5 HRWP, 50	10	93	[110]
CSTR	κ-carr./ LBG	Lb. helveticus	YE, 10	41.3[e] 28.5[f]	60[e] 46[f]	[38]
			YE, 1	15	33	
			YE, 0	10	22	
IC+FC CSTR	κ-carr./ LBG	Lb. helveticus	YE, 1	13.5	98	[84]
two IC CSTR	κ-carr./ LBG	Lb. helveticus	YE, 10	22	98	[23]
CPBR	alginate	Lb. helveticus	YE, 15	4	68	[121]
PBR-R	PEI	Lb. casei	YE, 2.5 + HWP, 5.3 [d]	4.6	100	[112]

[a] CFBR: continuous fluidized bed reactor, CPBR: continuous packed bed reactor, CSTR: continuous stirred tank reactor, PB = packed bed reactor, PBR-R = packed bed reactor with recycle loop for pH control (batch mode). [b] PEI = polyethyleneimine, SGB = sintered glass beads. [c] HRWP= hydrolyzed retentate from whey ultrafiltration, HWP = hydrolyzed whey proteins, MC = mustard oil cake, YE= yeast extract. [d] 70 g l^{-1} glucose was added for 100 g l^{-1} total sugar. [e] 54% (v/v) gel beads is the IC reactor. [f] 33% (v/v) gel beads is the IC reactor.

Studies on lactic acid production in SWP using IC are presented in Table 3. High lactic acid productivities and long-term stability have been obtained during continuous IC fermentation of YE-supplemented whey permeate by *Lb. helveticus* immobilised in κ-carrageenan/LBG gel beads, but with limited conversion of lactose [84]. Complete sugar conversion was only obtained when an additional FC reactor was placed in series with the IC reactor [38]. In this two-stage IC/FC process, an overall lactic acid productivity of 13.5 $g.l^{-1}.h^{-1}$ was reached during continuous culture in whey permeate containing 10 $g.l^{-1}$ YE, with 1 $g.l^{-1}$ residual sugar at an overall dilution rate of 0.27 h^{-1}. The authors also concluded that operation at low yeast extract levels (1-3 $g.l^{-1}$), or using periodic addition of higher YE concentrations, could lead to a cost-effective operation. The productivity of a two-stage IC reactor (19-22 $g.l^{-1}.h^{-1}$ with low residual sugar) largely exceeded that measured for the two-stage IC/FC process [23] and for FC batch cultures (approximately 2.0 $g.l^{-1}$) with the same strain and experimental conditions [85]. Higher productivities in SWP, with 10 $g.l^{-1}$ YE, and high sugar utilisation close to 100%, were only measured in cell-recycle processes, with 35 $g.l^{-1}.h^{-1}$ for *Lb. helveticus* [9] and and 85 $g.l^{-1}.h^{-1}$ for *Lb. bulgaricus* [86]. However, the application of membrane reactors is limited by high capital investment, complex maintenance of the sophisticated equipment and membrane fouling, particularly with whey permeate, which results in a rapid decrease of process performance [10].

6.2.2. Exopolysaccharide production

The EPS produced by LAB have received increasing attention in recent years. They play an important role in the manufacturing of fermented dairy products such as yogurts, drinking yogurts, cheeses, fermented creams and milk-based desserts. The EPS are very attractive for use as food additives because of their contribution to the texture, mouthfeel, taste perception and stability of the final product. They may act as texturizers and stabilisers and so decrease syneresis and improve product stability. Furthermore, EPS may contribute to human health as prebiotics or due to their antitumor, antiulcer, immunomodulating or cholesterol-lowering activities [87]. However, compared with dextran-producing or Gram-negative EPS producers, the low production of EPS by most LAB is a constraint for their commercial use as food additives [64].

Very little research has been done on cell immobilisation for EPS production. Bergmaier et al. [8] used the mucoid properties of *Lb. rhamnosus* RW 9595M for cell immobilisation by adsorption on solid porous supports (ImmobaSil®). The production of EPS was investigated during pH-controlled IC repeated-batch cultures in SWP. A high immobilised biomass of 8.5 10^{11} $cfu.ml^{-1}$ support was tested after colonisation by DNA analysis. During repeated IC cultures, a high EPS concentration (1808 $mg.l^{-1}$) was measured after four cycles for a short incubation period of 7 h. The high biomass in the IC system increased maximum EPS volumetric productivity (258 $mg.l^{-1}.h^{-1}$ after 7 h culture) compared with FC batch cultures (110 $mg.l^{-1}.h^{-1}$ after 18 h culture).

A continuous 38-day fermentation was carried out in a two-stage bioreactor under the same conditions, with a first fermentation stage containing immobilised cells and a second fermentation stage in series that was continuously inoculated by cells released from the first one [88]. Despite very high biomass concentrations in both reactors, the average total soluble EPS production of the fermentation system was very low

(138 mg.l^{-1}), with no effect of temperature, dilution rate and carbon to nitrogen ratio, compared with that measured in batch cultures with FC (2350 mg.l^{-1} after 18 h incubation) or IC (1808 mg.l^{-1} after 7 h) and continuous culture with FC (1808 mg.l^{-1} at a dilution rate of 0.3 h^{-1}). This result was explained by important changes in cell morphology and physiology, and the formation of very large aggregates containing very high cell and insoluble EPS concentrations, as presented previously in this review. These data show the high potential of the strain, *Lb. rhamnosus* RW9595M, and of IC technology for the production of EPS as a functional food ingredient.

6.2.3. Bacteriocin production

Bacteriocins are ribosomally-synthesised, extracellularly released proteins or protein complexes with a bactericidal activity against closely related bacteria, and for some bacteriocins, against a wide range of Gram-positive bacteria [89]. In recent years, there has been considerable interest in bacteriocins produced by LAB, and for diverse applications such as biopreservatives, to improve quality and innocuity of food products. However, the major limiting factor in using bacteriocins as food preservatives is their low yield during production.

Cell immobilisation has been used to increase cell density for continuous bacteriocin fermentation. However, nisin production tested during continuous IC fermentations was always low compared to traditional batch cultures with FC. Nisin production using Ca-alginate immobilised *L. lactis* in a continuous packed-bed reactor was only 150 IU.ml^{-1} [90]. Nisin production measured during continuous cultures of *L. lactis* IO-1 adsorbed on photo-crosslinked resin (ENTG) [91] or for *L. diacetylactis* UL719 entrapped in κ-carrageenan/LBG gel beads [92] was 2.5 to 10-fold lower than that obtained with pH controlled FC batch cultures. Even though a very high cell concentration (exceeding 10^{11} cfu.ml^{-1} beads) was obtained in the IC reactor (approximately 10-fold higher than that measured in continuous FC cultures), maximum nisin-Z production for both continuous FC and IC cultures in SWP were very similar [92]. Nisin-Z production with ENTG-adsorbed cells was also approximately 25% lower than that measured for continuous FC cultures for the same 0.1 h-1 dilution rate [91]. Similar data were reported for continuous pediocin production using *Pediococcus acidilactici* UL5 immobilised in κ-carrageenan/LBG gel beads compared with batch and continuous FC cultures [68,93].

To explain the low bacteriocin production during continuous FC and IC cultures compared with FC batch cultures, Desjardins *et al.* [92] and Huang *et al.* [68] assumed that some processes involved in the biosynthesis of the mature peptides, such as post-translational reactions, transport and externalisation, and cleavage of leader peptides may be limiting under steady-state conditions. Bertrand *et al.* [94] clearly showed that the evolutionary conditions produced during batch cultures are required for high nisin production by IC. A very high nisin Z production (8200 IU.ml^{-1}) was measured in the broth after 1-h cycles during repeated-cycle pH-controlled batch (RCB) cultures with *L. diacetylactis* UL719 immobilised in κ-carrageenan/LBG gel beads in SWP, with a corresponding volumetric productivity of 5730 IU.ml^{-1}.h^{-1}. This productivity is much higher than maximum nisin productivities reported in literature or maximum productivities obtained for FC batch cultures (850 IU.ml^{-1}.h^{-1}), and FC (460 IU.ml^{-1}.h^{-1}) or IC (1760 IU.ml^{-1}.h^{-1}) continuous cultures, using the same strain and fermentation conditions. In addition, the cost for inoculum and bioreactor preparation could be decreased compared with batch cultures, and RCB cultures would allow better use of fermentation and downstream processing equipment. The stability of RCB cultures was demonstrated for twenty-four and thirty-six 1-h cycles carried out over 3 and 6-day periods, respectively. A similar IC-RCB fermentation was successfully used to produce a high concentration of pediocin by *P. acidilactici* UL5 (unpublished data).

6.3. CELL PROTECTION

Probiotic bacteria are "living microorganisms, which upon ingestion in certain numbers, exert health benefits beyond inherent basic nutrition" [95]. To produce therapeutic benefits, the minimum suggested level of viable of probiotic bacteria in a food product should be $\geq 10^7$ cfu.ml^{-1} or g^{-1} of a product at the time of consumption [96]. Despite the importance of viability of these beneficial organisms, several surveys have shown large fluctuations and poor viability of probiotic bacteria, especially bifidobacteria, in yogurt preparations, which is the main probiotic carrier [97]. Acidity, pH, concentration of lactic and acetic acids, hydrogen peroxide, and dissolved oxygen content have been identified to have an effect on viability during manufacture and storage of yoghurt [98]. Moreover, because viable and biologically active microorganisms are usually required at the target site in the host, it is essential that the probiotic be able to withstand the host's natural barriers against ingested bacteria.

Encapsulation of probiotic cells has been suggested by several authors to enhance cell resistance to freezing and freeze-drying. A higher stability was reported for *Lb. acidophilus* and *B. longum* immobilised in alginate beads compared with FC during storage of frozen dairy desserts [99] and for *B. longum* in ice cream [100]. In this latter study, the addition of cryoprotective agents (glycerol and mannitol) in the alginate solution increased the protective effects of immobilisation and gave survival rates as high as 90% compared with only 40% for FC. In addition, an optimum capsule size between 30 and 100 μm was suggested for food applications, since a smaller diameter resulted in less cell protection and a larger diameter gave texture and sensory defects in the products. However, ambiguous data were reported for immobilisation and freeze-drying, with a positive effect on cell viability for *L. lactis* [101] and *Lb. plantarum* [102] immobilised in Ca-alginate, but no effect on *B. longum* in κ-carrageenan/LBG [103]. Champagne *et al.* [13] used spray-coating in milk fat to encapsulate freeze-dried cells of *B. longum*. Although a higher viability was noted for milk-fat-encapsulated cells compared with gelatin and xanthan gum, *B. longum* encapsulated in butter oil did not improved survival in frozen yoghurt [104].

Cell encapsulation also improved probiotic viability in fermented dairy products. Immobilisation of *B. longum* in κ-carrageenan [96], *B. bifidum* and *B. infantis* in Ca-alginate [105], and *B. infantis* in gellan-xanthan beads [106] allowed to maintain high-cell concentrations during 5-week storage of yoghurt, with no change in sensorial properties [96]. Picot and Lacroix [16] studied the encapsulation of *B. breve* and *B. longum* as freeze-dried or fresh cultures in water-insoluble food-grade microcapsules produced by emulsion and/or spray-drying, using milk fat and/or denatured whey proteins as immobilisation material. Viable counts of *B. breve* entrapped in whey protein microcapsules using this method were significantly higher than those of FC after

28 days in yoghurt stored at 4°C (+ 2.6 log cycles), but no effect was observed for *B. longum*. The encapsulation of *B. bifidum* in κ-carrageenan beads maintained cell viability for as long as 24-weeks of cheddar cheese ripening, with no negative effects on texture, appearance and flavour [107].

Different encapsulation methods exhibited protective effects on cell viability during *in vitro* tests simulating gastric or intestinal digestion. Exposure to a simulated gastric juice at pH 2.5 caused the FC viable count to drop from 1.2 10^9 cfu.ml^{-1} to undetectable levels in 30 min, while the IC viable count decreased by only 0.67 log cycle within the same period [106]. Lee and Heo [108] reported that survival of *B. longum* ATCC 15707 immobilised in calcium alginate beads in simulated gastric juices and bile salt solution was better with higher gel concentrations, with an effect of bead size. Very large beads (*i.e.* > 2 mm) provided more protection for *B. longum* ATCC 15707 in simulated gastric juices and bile salt solution. Spray drying encapsulation in whey-protein suspension protected *B. breve*, but not *B. longum*, during and after sequential exposure to simulated gastric and intestinal juices (+ 2.7 log cycles) compared with FC [16].

Different data obtained for the protective effects of immobilisation might be due to strain-specific effects of immobilisation, but also to differences in methodology of the different studies since numerous experimental factors may affect cell survival. In particular, the technological properties of cells should be taken into account in the selection of encapsulation method, such as heat resistance for spray-dray encapsulation.

7. Conclusions

Many advantages have been demonstrated for IC systems that may be applied to LAB and probiotic bacteria in the dairy and starter industries. Since the last review on this theme by Champagne *et al.* [2], important advances have been made on characterizing and demonstrating the biological stability of the IC bioreactor during long operation periods. Specific effects of immobilisation on the physiology and technological characteristics of both entrapped and released cells have been recently shown. Application of this research could be particularly important for the production of probiotic bacteria, functional dairy products containing high concentrations of viable bacteria and bioingredients from LAB with important functional properties for use in foods and health. Immobilisation can efficiently protect cells, making this approach potentially useful for delivery of viable bacteria to the gastrointestinal tract of humans *via* dairy fermented products. It may be anticipated that application of IC technology in the dairy sector will begin with these special cultures which are difficult to propagate and use with the traditional culture techniques, and which are used to produce high-value dairy products with positive effects on consumers' health.

References

[1] Stanton, C.; Gardiner, G.; Meehan, H.; Collins, K.; Fitzgerald, G.; Lynch, P.B. and Ross, R.P. (2001) Market potential for probiotics. Am. J. Clin. Nutr. 73: 476S-483S.

[2] Champagne, C.P.; Lacroix, C. and Sodini-Gallot, I. (1994) Immobilized cell technologies for the dairy industry. CRC Crit. Rev. Biotechnol. 14: 109-134.

[3] Artignan, J.M.; Corrieu, G. and Lacroix, C. (1997) Rheology of pure and mixed kappa-carrageenan gels in lactic-acid fermentation conditions. J. Texture Stud. 28: 47-70.
[4] Sodini, I.; Boquien, C.Y.; Corrieu, G. and Lacroix, C. (1997) Microbial dynamics of co- and separately entrapped mixed cultures of mesophilic lactic acid bacteria during the continuous prefermentation of milk. Enzyme Microb. Technol. 20: 381-388.
[5] Lamboley, L.; Lacroix, C.; Artignan, J.M.; Champagne, C.P. and Vuillemard, J.C. (1999) Long-term mechanical and biological stability of an immobilized cell reactor for continuous mixed-strain mesophilic lactic starter production in whey permeate. Biotechnol. Prog. 15: 646-654.
[6] Lamboley, L.; Lacroix, C. and Champagne, J.C. (2001) Effect of inoculum composition and low KCl supplementation on the biological and rheological stability of an immobilized-cell system for mixed mesophilic lactic starter production. Biotechnol. Prog. 17: 1071-1078.
[7] Dunne, W.M., Jr. (2002) Bacterial adhesion: seen any good biofilms lately? Clin. Microbiol. Rev. 15: 155-166.
[8] Bergmaier, D.; Lacroix, C. and Champagne, C.P. (2002) Exopolysaccharide production during batch cultures with free and immobilized *Lactobacillus rhamnosus* RW-9595M. J. Appl. Microbiol., 95: 1049-1057.
[9] Kulozik, U. and Wilde, J. (1999) Rapid lactic acid production at high cell concentrations in whey ultrafiltrate by *Lactobacillus helveticus*. Enzyme Microb. Technol. 24: 297-302.
[10] Tejayadi, S. and Cheryan, M. (1995) Lactic acid from cheese whey permeate. Productivity and economics of a continuous membrane bioreactor. Appl. Microbiol. Biotechnol. 43: 242-248.
[11] Gonçalves, L.M.D.; Barreto, M.T.O.; Xavier, A.M.B.R.; Carrondo, M.J.T. and Klein, J. (1992) Inert supports for lactic acid fermentation - a technological assessment. Appl. Microbiol. Biotechnol. 38: 305-311.
[12] Poncelet, D. and Neufeld, R.J. (1996) Fundamentals of dispersion in encapsulation technology. In: Wijffels, R.H.; Buitelaar, R.M.; Bucke, C. and Tramper, J. (Eds.) Immobilized cells: Basics and applications. Elsevier Science, Amsterdam, The Netherlands; pp. 47-54.
[13] Champagne, C.P.; Raymond, Y.; Mondou, F. and Julien, J.P. (1995) Studies on the encapsulation of *Bifidobacterium longum* cultures by spray-coating or cocrystallization. Bif. Microflora. 14: 7-14.
[14] Sunohara, H.; Ohno, T.; Shibata, N. and Seki, K., inventors; Morishita Jintan Co. Ltd, assignee (1995) Dec. 26. Process for producing capsule and capsule obtained thereby. U.S. patent 5,478,570.
[15] Picot, A. and Lacroix, C. (2003) Effect of dynamic loop mixer operating conditions on o/w emulsion used for cell encapsulation. Lait. 83 : 237-250.
[16] Picot, A. and Lacroix, C. (2003) Encapsulation of bifidobacteria in whey protein-based microcapsules and survival in simulated gastrointestinal conditions and in yoghurt. Intern. Dairy J. (in press).
[17] Arnaud, J.-P.; Lacroix, C. and Castaigne, F. (1992) Counterdiffusion of lactose and lactic acid in κ-carrageenan/locust bean gum gel beads with or without entrapped lactic acid bacteria. Enzyme Microb. Technol. 14: 715-724.
[18] Masson, F.; Lacroix, C. and Paquin, C. (1994) Direct measurement of pH profiles in gel beads immobilizing *Lactobacillus helveticus* using a pH sensitive microelectrode. Biotechnol. Tech. 8: 551-556.
[19] Cachon, R.; Antérieux, P. and Diviès, C. (1998) The comparative behaviour of *Lactococcus lactis* in free and immobilized culture processes. J. Biotechnol. 63: 211-218.
[20] Monbouquette, H.G.; Sayles, G.D. and Ollis, D.F. (1990) Immobilized cell biocatalyst activation and pseudo-steady state behavior: Model and experiment. Biotechnol. Bioeng. 35: 609-629.
[21] Wijffels, R.H.; De Gooijer, C.D.; Kortekass, S. and Tamper, J.H. (1991) Growth and substrate consumption of *Nitrobacter agilis* cells immobilized in carrageenan : Part 2. Model evaluation. Biotechnol. Bioeng. 38: 544-550.
[22] Yabannavar, V.M. and Wang, D.I.C. (1991) Analysis of mass transfer for immobilized cells in an extractive lactic acid fermentation. Biotechnol. Bioeng. 37: 544-550.
[23] Schepers, A.W. (2003) Modelling of growth and lactic acid production of *Lactobacillus helveticus* during continuous free and immobilized cell cultures. Ph.D. Dissertation, 20428, Université Laval, Quebec, Canada.
[24] Arnaud, J.P. and Lacroix, C. (1991) Diffusion of lactose in κ-carrageenan / locust bean gum gel beads with or without entrapped growing bacteria. Biotechnol. Bioeng. 38: 1041-1049.
[25] Prioult, G.; Lacroix, C.; Turcotte, C. and Fliss, I. (2000) Simultaneous immunofluorescent detection of coentrapped cells in gel beads. Appl. Environ. Microbiol. 66: 2216-2219.
[26] Doleyres, Y.; Fliss, I. and Lacroix, C. (2003) Continuous production of lactic starters containing probiotics using immobilized cell technology. Biotechnol. Prog. (in press).

[27] Doleyres, Y.; Fliss, I. and Lacroix, C. (2002) Quantitative determination of the spatial distribution of pure and mixed strain immobilized cells in gel beads by immunofluorescence. Appl. Microbiol. Biotechnol. 59: 297-302.
[28] Audet, P.; Lacroix, C. and Paquin, C. (1991) Continuous fermentation of a whey supplemented whey permeate medium with immobilized *Streptococcus salivarius* subsp. *thermophilus*. Int. Dairy J. 1: 1-15.
[29] Arnaud, J.P.; Lacroix, C. and Choplin, L. (1992) Effect of agitation rate on cell release rate and metabolism during continuous fermentation with entrapped growing *Lactobacillus casei* subsp. *casei*. Biotechnol. Techn. 6: 265-270.
[30] Prévost, H. and Diviès, C. (1987) Fresh fermented cheese production with continuous prefermented milk by a mixed culture of mesophilic lactic streptococci entrapped in Ca-alginate. Biotechnol. Lett. 9: 789-791.
[31] Prévost, H. and Diviès, C. (1988) Continuous prefermentation of milk by entrapped yogurt bacteria. I. Development of the process. Milchwissenschaft 43: 621-625.
[32] Prévost, H. and Diviès, C. (1988) Continuous prefermentation of milk by entrapped yogurt bacteria. II. Data for optimization of the process. Milchwissenschaft 43: 716-719.
[33] Sodini-Gallot, I.; Corrieu, G.; Boquien, C.Y.; Latrille, E. and Lacroix, C. (1995) Process performance of continuous inoculation and acidification of milk with immobilized lactic acid bacteria. J. Dairy Sci. 78: 1407-1420.
[34] Lamboley, L.; Lacroix, C. and Champagne, J.C. (1997) Continuous mixed strain mesophilic lactic starter production in supplemented whey permeate medium using immobilized cell technology. Biotech. Bioeng. 56: 502-516.
[35] Doleyres, Y.; Paquin, C.; LeRoy, M. and Lacroix, C. (2002) *Bifidobacterium longum* ATCC 15707 cell production during free- and immobilized-cell cultures in MRS-whey permeate medium. Appl. Microbiol. Biotechnol. 60: 168-173.
[36] Arnaud, J.P.; Lacroix, C.; Foussereau, C. and Chopin, L. (1993) Shear stress effects on growth and activity of *Lactobacillus delbrueckii* subsp. *bulgaricus*. J. Biotechnol. 29: 157-175.
[37] Champagne, C.P.; Girard, F. and Rodriguez, N. (1993) Production of concentrated suspensions of thermophilic lactic acid bacteria in calcium-alginate beads. Int. Dairy J. 3: 257-275.
[38] Norton, S.; Lacroix, C. and Vuillemard, J.C. (1994) Reduction of yeast extract supplementation in lactic acid fermentation of whey permeate by immobilized cell technology. J. Dairy Sci. 77: 2494-2508.
[39] Gobbetti, M. and Rossi, J. (1993) Continuous fermentation with free-growing and immobilized multistaters to get a kefir production pattern. Microbiologie-Aliments-Nutrition 11: 119-127.
[40] Cachon, R. (1993) Etude du comportement cinétique d'une bactérie lactique modèle en culture libre ou immobilisée dans des billes de gel. Thèse de doctorat, Université de Bourgogne, Dijon, France.
[41] Lacroix, C.; Sodini, I. and Corrieu, G. (1996) Microbiological stability of an immobilized cell bioreactor with mixed lactic acid bacteria during continuous fermentation of milk. In: Wijffels, R.H.; Buitelar, R.M.; Bucke, C. and Tramper, J. (Eds.) Immobilized cells: Basics and applications. Elsevier Science, Amsterdam, The Netherlands; pp. 600-607.
[42] Krishnan, S.; Gowda, M.C.; Misra, M.C. and Karanth, N.G. (2001) Physiological and morphological changes in immobilized *L. plantarum* NCIM 2084 cells during repeated bacth fermentation for production of lactic acid. Food Biotechnol. 15: 193-202.
[43] Krisch, J. and Szajani, B. (1997) Ethanol and acetic tolerance in free and immobilized cells of *Saccharomyces cerevisiae* and *Acetobacter aceti*. Biotechnol. Lett. 19: 525-528.
[44] Curtain, C. (1986) Understanding and avoiding ethanol inhibition. Trends in Biotechnol. 4: 110.
[45] Holcberg, I. and Margalith, P. (1981) Alcoholic fermentation by immobilized yeast at high sugar concentration. Eur. J. Appl. Microbiol. Biotechnol. 13: 133-140.
[46] Diefenbach, R.; Keweloh, H. and Rehm, H.J. (1992) Fatty acid impurities in alginate influence the phenol tolerance of immobilized *Escherichia coli*. Appl. Microbiol. Biotechnol. 36: 530-534.
[47] Heipieper, H.J.; Keweloh, H. and Rehm, H.J. (1991) Influence of phenols on growth and membrane permeability of free and immobilized *Escherichia coli*. Appl. Environ. Microbiol. 57: 1213-1217.
[48] Keweloh, H.; Heipieper, H.J. and Rehm, H.J. (1989) Protection of bacteria against toxicity of phenol by immobilization in calcium alginate. Appl. Microbiol. Biotechnol. 31: 383-389.
[49] Jouenne, T.; Tresse, O. and Junter, G.A. (1994) Agar-entrapped bacteria as an *in vitro* model of biofilms and their susceptibility to antibiotics. FEMS Microbiol. Lett. 119: 237-242.
[50] Doleyres, Y.; Fliss, I. and Lacroix, C. (2003) Changes of lactic and probiotic culture characteristics during continuous immobilized-cell fermentation with mixed strains. submitted.

[51] Trauth, E.; Lemaitre, J.P.; Rojas, C.; Diviès, C. and Cachon, R. (2001) Resistance of immobilized lactic acid bacteria to the inhibitory effect of quaternary ammonium sanitizers. Lebensm.-Wiss. U.-Technol. 34: 239-243.
[52] Reilly, S.S. and Gilliland, S.E. (1999) *Bifidobacterium longum* survival during frozen and refrigerated storage as related to pH during growth. J. Food Sci. 64: 714-718.
[53] Desmond, C.; Stanton, C.; Fitzgerald, G.F.; Collins, K. and Ross, R.P. (2002) Environmental adaptation of probiotic lactobacilli towards improvement of performance during spray drying. Int. Dairy J. 12: 183-190.
[54] Shapiro, J.A. and Dworkin, M. (Eds.) (1997) Bacteria as multicellular organism. Oxford Univ. Press, New York.
[55] Bergmaier, D. (2002) Production d'exopolysaccharides par fermentation avec des cellules immobilisées de *Lb. rhamnosus* RW-9595M d'un milieu à base de perméat de lactosérum. Ph.D. Dissertation, 20383, Université Laval, Quebec, PQ, Canada.
[56] Dieckelmann, M.; Johnson, L.A. and Beacham, I.R. (1998) The diversity of lipases from psychrotrophic strains of Pseudomonas: a novel lipase from a highly lipolytic strain of *Pseudomonas fluorescens*. J. Appl. Microbiol. 85: 527-536.
[57] Barrett, N.E.; Grandison, A.S. and Lewis, M.J. (1999) Contribution of the lactoperoxidase system to the keeping quality of pasteurized milk. J. Dairy Res. 66: 73-80.
[58] Lapointe, M.; Champagne, C.P.; Vuillemard, J.C. and Lacroix, C. (1996) Effect of dilution rate on bacteriophage development in an immobilized cell system used for continuous inoculation of Lactococci in milk. J. Dairy Sci. 79: 767-774.
[59] Champagne, C.P.; Girard, F. and Morin, N. (1988) Bacteriophage development in an immobilized lactic acid bacteria system. Biotechnol. Lett. 10: 463-468.
[60] Steenson, L.R.; Klaenhammer, T.R. and Swaisgood, H.E. (1987) Calcium alginate-immobilized cultures of lactic Streptococci are protected from bacteriophages. J. Dairy Sci. 70: 1121-1127.
[61] Passos, F.M.L.; Klaenhammer, T.R. and Swaisgood, H.E. (1994) Response to phage infection of immobilized lactococci during continuous acidification of skim milk. J. Dairy Res. 61: 537-544.
[62] Macedo, M.G.; Champagne, C.P.; Vuillemard, J.C. and Lacroix, C. (1999) Establishment of bacteriophages in an immobilized cells system used for continuous inoculation of Lactococci. Int. Dairy J. 9: 437-445.
[63] Fitzgerald, G.F. and Hill, C. (1996) Genetics of starter cultures. In: Cogan, T.M. and Accolas, J.P. (Eds.) Dairy starter cultures. VCH Publishers, New-York, NY; pp. 25-46.
[64] De Vuyst, L. and Degeest, B. (1999) Heteropolysaccharides from lactic acid bacteria. FEMS Microbiol. Rev. 23: 153-177.
[65] Teuber, M.; Meile, L. and Schwarz, F. (1999) Acquired antibiotic resistance in lactic acid bacteria from food. Antonie Van Leeuwenhock 76: 115-137.
[66] Barbotin, J.N. (1994) Immobilization of recombinant bacteria. A strategy to improve plasmid stability. Ann. N.Y. Acad. Sci. 721: 303-309.
[67] D'Angio, C.; Beal, C.; Boquien, C.Y. and Corrieu, G. (1994) Influence of dilution rate and cell immobilization on plasmid stability during continuous cultures of recombinant strains of *Lactococcus lactis* subsp. *lactis*. J. Biotechnol. 34: 87-95.
[68] Huang, J.; Lacroix, C.; Daba, H. and Simard, R.E. (1996) Pediocin 5 production and plasmid stability during continuous free and immobilized cell cultures of *Pediococcus acidilactici* UL5. J. Appl. Bacteriol. 80: 635-644.
[69] Kumar, P.K. and Schugerl, K. (1990) Immobilization of genetically engineered cells: a new strategy for higher stability. J. Biotechnol. 14: 255-272.
[70] Nasri, M.; Sayadi, S.; Barbotin, J.N.; Dhulster, P. and Thomas, D. (1987) Influence of immobilization on the stability of pTG201 recombinant plasmid in some strains of *Escherichia coli*. Appl. Environ. Microbiol. 53: 740-744.
[71] Nasri, M.; Sayadi, S.; Barbotin, J.N. and Thomas, D. (1987) The use of the immobilization of whole living cells to increase stability of recombinant plasmid in *Escherichia coli*. J. Biotechnol. 6: 147-157.
[72] Dincbas, V.; Hortacsu, A. and Camurdan, A. (1993) Plasmid stability in immobilized mixed cultures of recombinant *Escherichia coli*. Biotechnol. Prog. 9: 218-220.
[73] Klinkenberg, G.; Lystad, K.Q.; Levine, D.W. and Dyrset, N. (2001) Cell release from alginate immobilized *Lactococcus lactis* ssp. *lactis* in chitosan and alginate coated beads. J. Dairy Sci. 84: 1118-1127.

[74] Gilliland, S. E. (1985) Concentrated starter cultures. In: Gilliland, S.E. (Ed.) Bacterial starter cultures for food. CRC Press Inc.: Boca Raton, FL; pp. 145-157.
[75] Sodini, I.; Boquien, C.Y.; Corrieu, G. and Lacroix, C. (1997) Use of an immobilized cell bioreactor for the continuous inoculation of milk in fresh cheese manufacturing. J. Ind. Microbiol. Biotechnol. 18: 56-61.
[76] Audet, P.; St-Gelais, D. and Roy, D. (1995) Production of mixed cultures of non-isogenic *Lactococcus lactis* ssp. *cremoris* using immobilized cells. Milchwissenschaft 50: 18-22.
[77] Ibrahim, S.A. and Bezkorovainy, A. (1994) Growth-promoting factors for *Bifidobacterium longum*. J. Food Sci. 59: 189-191.
[78] Ouellette, V.; Chevalier, P. and Lacroix, C. (1994) Continuous fermentation of a supplemented milk with immobilized *Bifidobacterium infantis*. Biotechnol. Tech. 8: 45-50.
[79] Corre, C.; Madec, M.N. and Boyaval, P. (1992) Production of concentrated *Bifidobacterium bifidum*. J. Chem. Technol. Biotechnol. 53: 189-194.
[80] Passos, F.M.L. and Swaisgood, H.E. (1993) Development of a spiral mesh bioreactor with immobilized Lactococci for continuous inoculation and acidification of milk. J. Dairy Sci. 76: 2856-2867.
[81] Sodini, I.; Lagace, L.; Lacroix, C. and Corrieu, G. (1998) Effect of continuous prefermentation of milk with an immobilized cell bioreactor on fermentation kinetics and curd properties. J. Dairy Sci. 81: 631-638.
[82] Vickroy, T.B. (1985) Lactic acid. In: Moo-Young M. (Ed.) Comprehensive Biotechnology, vol. 3. Pergamon Press, Oxford, UK; pp. 761-776.
[83] Datta, R.; Tsai, S.P.; Bonsignore, P.; Moon, S.H. and Frank, J.R. (1995) Technological and economic potential of poly(lactic) acid and lactic acid derivatives. FEMS Microbiol. Rev. 16: 221-231.
[84] Norton, S.; Lacroix, C. and Vuillemard, J.C. (1994) Kinetic study of a continuous fermentation of whey permeate by immobilized *Lactobacillus helveticus*. Enzyme Microb. Technol. 16: 457-466.
[85] Schepers, A.W.; Thibault, J. and Lacroix, C. (2002) Multiple factor kinectic analysis and modeling of *Lactobacillus helveticus* growth and lactic acid production during pH-controlled batch cultures in whey permeate/yeast extract medium. Part I: Multiple factor kinetic analysis. Enzyme Microb. Technol. 30: 176-186.
[86] Mehaia, M.A. and Cheryan, M. (1986) Lactic acid from whey permeate in a membrane recycle bioreactor. Enzyme Microb. Technol. 8: 289-292.
[87] Ruas-Madiedo, P.; Hugenholtz, J. and Zoon, P. (2002) An overview of the functionnality of exopolysaccharides produced by lactic acid bacteria. Int. Dairy J. 12: 163-171.
[88] Bergmaier, D.; Lacroix, C. and Champagne, C.P. (2003) Exopolysaccharide production during chemostat cultures with free and immobilized *Lactobacillus rhamnosus* RW-9595M. J. Appl. Microbiol. submitted.
[89] Jack, R.W.; Tagg, J.R. and Ray, B. (1995) Bacteriocins of gram-positive bacteria. Microbiol. Rev. 59: 171-200.
[90] Wan, J.; Hickey, M.W. and Mawson, R.F. (1995) Continuous production of bacteriocins, brevicin, nisin and pediocin using calcium alginate immobilized bacteria. J. Appl. Bacteriol. 79: 671-676.
[91] Sonomoto, K.; Chinachoti, N.; Endo, N. and Ishizaki, A. (2000) Biosynthetic production of nisin Z by immobilized *Lactococcus lactis* IO-1. J. Mol. Catal. B. Enzym. 10: 325-334.
[92] Desjardins, P.; Meghrous, J. and Lacroix, C. (2001) Effect of aeration and dilution rate on nisin Z production during continuous fermentation with free and immobilized *Lactococcus lactis* UL719 in supplemented whey permeate. Int. Dairy J. 11: 943-951.
[93] Goulhen, F.; Meghrous, J. and Lacroix, C. (1999) Production of a nisin/pediocin mixture by pH-controlled mixed-strain batch cultures in supplemented whey permeate. J. Appl. Microbiol. 86: 399-406.
[94] Bertrand, N.; Fliss, I. and Lacroix, C. (2001) High nisin-Z production during repeated-cycle batch cultures in supplemented whey permeate using immobilized *Lactococcus lactis* UL719. Int. Dairy J. 11: 953-960.
[95] Guarner, F. and Schaafsma, G.J. (1998) Probiotics. Int. J. Food Microbiol. 39: 237-238.
[96] Adhikari, K.; Mustapha, A.; Grün, I.U. and Fernando, L. (2000) Viability of microencapsulated bifidobacteria in set yogurt during refrigerated storage. J. Dairy Sci. 83: 1946-1951.
[97] Schillinger, U. (1999) Isolation and identification of lactobacilli from novel-type probiotic and mild yoghurts and their stability during refrigerated storage. Int. J. Food Microbiol. 47: 79-87.
[98] Dave, R.I. and Shah, N.P. (1997) Viability of yoghurt and probiotic bacteria in yoghurts made from commercial starter cultures. Int. Dairy J. 7: 31-41.
[99] Shah, N.P. and Ravula, R.R. (2000) Microencapsulation of probiotic bacteria and their survival in frozen fermented dairy desserts. Aust. J. Dairy Technol. 55: 139-144.

[100] Sheu, T.Y.; Marshall, R.T. and Heymann, H. (1993) Improving survival of culture bacteria in frozen desserts by microentrapment. J. Dairy Sci. 76: 1902-1907.
[101] Champagne, C.P.; Gaudy, C.; Poncelet, D. and Neufeld, R.J. (1992) *Lactococcus lactis* release from calcium alginate beads. Appl. Environ. Microbiol. 58: 1429-1434.
[102] Kearney, L.; Upton, M. and Mc Laughlin, A. (1990) Enhancing the viability of *Lactobacillus plantarum* inoculum by immobilizing the cells in calcium-alginate beads incorporating cryoprotectants. Appl. Environ. Microbiol. 56: 3112-3116.
[103] Maitrot, H.; Paquin, C.; Lacroix, C. and Champagne, C.P. (1997) Production of concentrated freeze-dried cultures of *Bifidobacterium longum* in κ-carrageenan-locust bean gum gel. Biotechnol. Techn. 11: 527-531.
[104] Modler, H.W. and Villa-Garcia, L. (1993) The growth of *Bifidobacterium longum* in a whey-based medium and viability of this organism in frozen yogurt with low and high levels of developed acidity. Cult. Dairy Prod. J. 28: 4-8.
[105] Hussein, S.A. and Kebary, K.M.K. (1999) Improving viability of bifidobacteria by microentrapment and their effect on some pathogenic bacteria in stirred yogurt. Acta Aliment. 28: 113-131.
[106] Sun, W. and Griffiths, M.W. (2000) Survival of bifidobacteria in yogurt and simulated gastric juice following immobilization in gellan-xanthan beads. Int. J. Food Microbiol. 61: 17-25.
[107] Dinakar, P. and Mistry, V.V. (1994) Growth and viability of *Bifidobacterium longum* in cheddar cheese. J. Dairy Sci. 77: 2854-2864.
[108] Lee, K.Y. and Heo, T.R. (2000) Survival of *Bifidobacterium longum* immobilized in calcium alginate beads in simulated gastric juices and bile salt solution. Appl. Environ. Microbiol. 66: 869-873.
[109] Prévost, H. and Diviès, C. (1992) Cream fermentations by a mixed culture of Lactococci entrapped in two- layer calcium alginate gel beads. Biotechnol. Lett. 14: 583-588.
[110] Krischke, W.; Schröder, M. and Trösch, W. (1991) Continuous production of L-lactic acid from whey permeate by immobilized *Lactobacillus casei* subsp. *casei*. Appl. Microbiol. Biotechnol. 34: 573-578.
[111] Iwasaki, K.-I.; Nakajima, M. and Sasahara, H. (1992) Porous alumina beads for immobilization of lactic acid bacteria and its application for repeated-batch fermentation in soy sauce production. J. Ferment. Bioeng. 73: 375-379.
[112] Senthuran, A.; Senthuran, V.; Mattiasson, B. and Kaul, R. (1997) Lactic acid fermentation in a recycle batch reactor using immobilized *Lactobacillus casei*. Biotechnol. Bioeng. 55: 841-853.
[113] Velazquez, A.C.; Pometto, A.L., 3rd; Ho, K.L. and Demirci, A. (2001) Evaluation of plastic-composite supports in repeated fed-batch biofilm lactic acid fermentation by *Lactobacillus casei*. Appl. Microbiol. Biotechnol. 55: 434-441.
[114] Ho, K.-L.G.; Anthony, L.; Pometto, I.; Hinz, P.N.; Dickson, J.S. and Demirci, A. (1997) Ingredient selection for plastic composite supports for L-(+)-lactic acid biofilm fermentation by *Lactobacillus casei* subsp. *rhamnosus*. Appl. Environ. Microbiol. 63: 2516-2523.
[115] Rao, A.V.; Shiwnarain, N. and Maharaj, I. (1989) Survival of microencapsulated *Bifidobacterium pseudolongum* in simulated gastric and intestinal juices. Can. Inst. Food Sci. Technol. J. 22: 345-349.
[116] Wenrong, S. and Griffiths, M.W. (2000) Survival of bifidobacteria in yoghurt and stimulated gastric juice following immobilization in gellan-xanthan beads. Int. J. Food Microbiol. 61: 17-25.
[117] Fàvaro Trindade, C.S. and Grosso, C.R.F. (2000) The effect of the immobilization of *Lactobacillus acidophilus* and *Bifidobacterium lactis* in alginate on their tolerance to gastro-intestinal secretions. Milchwissenschaft 55: 496-499.
[118] Tuli, A.; Khanna, P.K.; Marwaha, S.S. and Kennedy, J.F. (1985) Lactic acid production from whey permeate by immobilized *Lactobacillus casei*. Enzyme Microb. Technol. 7: 164-168.
[119] Roukas, T. and Kotzekidou, P. (1991) Production of lactic acid from deproteinized whey by coimmobilized *Lactobacillus casei* and *Lactococcus lactis* cells. Enzyme Microb. Technol. 13: 33-38.
[120] Boyaval, P. and Goulet, J. (1988) Optimal conditions for production of lactic acid from cheese whey permeate by Ca-alginate-entrapped *Lactobacillus helveticus*. Enzyme Microb. Technol. 10: 725-728.
[121] Roy, D.; Goulet, J. and LeDuy, A. (1987) Continuous production of lactic acid from whey permeate by free and calcium alginate entrapped *Lactobacillus helveticus*. J. Dairy Sci. 70: 506-513.

FOOD BIOCONVERSIONS AND METABOLITE PRODUCTION

P. HEATHER PILKINGTON
Technology Development Americas, Labatt – Interbrew, 197 Richmond Street, London, Ontario, Canada, N6A 4M3 – Fax: (519) 667-7115 – Email: heather.pilkington@labatt.com

1. Introduction

This chapter discusses the application of immobilised cell technology to food bioconversions and metabolite production. It is divided into two parts:
- Part 1. Food bioconversions are defined as the change of one chemical into another by a living organism, resulting in material that can be consumed as food.
- Part 2. Food metabolites refer to substances produced in the metabolism of food materials or food waste that are intended for use as additives or ingredients in the food industry.

In 1994, Norton and Vuillemand [1] published a comprehensive review on food bioconversions and metabolite production using immobilised cell technology. This chapter will summarise past work and highlight recent developments.

Food bioconversions such as the production of beer, cider, wine, dairy, meat, bioflavourings and probiotics are covered in other sections of this book and will not be covered here. As well, other non-food fermentations *e.g.* fuel ethanol and industrial antibiotics are discussed elsewhere in this book.

A distinguishing feature of food applications as compared to other industrial or technical applications of immobilised cell bioconversions and metabolite production, is the need to meet the requirements of government food safety legislation. For use in the food industry, the type of immobilisation matrix, chemicals, microorganisms and processing steps used must be evaluated in terms of food safety legislation. The general categories of immobilisation methods include surface attachment, entrapment in porous matrices, and containment behind a barrier/encapsulation. Several good reviews of food-specific encapsulation/entrapment techniques have been published recently [2-6]. Natural polymers such as starch derivatives, maltodextrins, gum arabic, agar, cellulosics, chitosan, gelatin, carrageenan, alginates, low methoxyl (LM) pectin and gellan gum, are examples of entrapment/encapsulation materials suitable for food applications. Synthetic polymers such as polyacrylamide, used to prepare capsules for medical applications, are not food approved [3]. In addition to the materials employed for immobilisation, Fu [7] highlights the importance of the batch to batch consistency between separately prepared immobilised cell cultures. The physiological condition and corresponding metabolite production of immobilised cells is impacted by the microenvironment created within the immobilisation matrix. Thus the same microenvironment must be created during each immobilisation preparation in order to ensure batch to batch product consistency.

Whole cell immobilisation, in addition to allowing for an increase in cell density within the bioreactor, enhances cell-to-cell contact and is thought to affect cellular physiology in a way that causes the cells to produce levels of certain metabolites higher than those produced by freely suspended cells [7]. Consequently volumetric productivity of immobilised cell bioreactors can be higher than freely suspended cells. Immobilisation also can provide the advantage of easy separation of cells in cases where the product or metabolite of interest is contained within the liquid phase. Immobilised cells can be easily reused in subsequent operations by draining and refilling the bioreactor. In cases where non-growth associated metabolites or enzymes are being produced, immobilised cells that are not actively growing are advantageous. Because immobilised cells are retained by a matrix, continuous bioreactors can be operated above the nominal washout rate for freely suspended cells without loss of cell mass. It has also been reported that cell immobilisation provides protection from toxic substances, increased plasmid stability and increased catalytic activity Cassidy [8].

In some cases the bioconversions and metabolites produced in the sections to follow could be performed using free or immobilised enzyme preparations as an alternative to immobilised cells. Immobilised enzymes are not within the scope of this work. Immobilisation of whole cells offers the following advantages over enzymes:
- Immobilised enzymes generally demonstrate poor stability relative to whole cells.
- The high costs of enzyme separation for immobilisation within bioreactors can make this strategy commercially non-viable.
- Enzymes often require cofactors that need to be regenerated.
- Immobilised whole cells offer the ability to perform bioconversions that involve complex biosynthetic pathways and large amounts of energy.
Enzymes are generally limited to mono- or di-enzymatic bioconversion steps.

Recent publications [9-11] refer to the increasing demand by consumers for foods and food ingredients that are perceived to be healthy and natural *versus* artificial in origin. Food bioconversions and metabolites produced by microorganisms, such as fermented beverages, natural flavourings and pigments, address this trend. Cell immobilisation technology offers more efficient methods for producing these foods and food ingredients to meet this consumer demand.

2. Food bioconversions

2.1. SAKE PRODUCTION

Sake is a traditional alcoholic beverage in Japan and is produced using rice as its main raw material. The sake production process is traditionally referred to as a "parallel" batch fermentation due to the progressive hydrolysis of starch and slow fermentation at low temperature occurring simultaneously.

Because of the density of the main fermentation or "moromi", the traditional brewing process is not easy to apply to a bioreactor system. Therefore, Nunokawa and Hirotsune [12] of Ozeki Corporation in Japan performed the saccharification step in advance and the resulting clarified sugar solutions were fed into a bioreactor packed with immobilised yeast cells to produce continuously fermented sake. Some differences from the traditional sake were noted including lowered alcohol (about 10% v/v for immobilised vs 15-20% v/v for traditional batch) and the presence of a pleasant fruity flavour. Yeast cells were immobilised in alginate gel beads. The pilot scale immobilised yeast reactor system consisted of a preincubation tank and a column containing yeast cells immobilised in alginate gel beads. The preincubation tank was used in order to prevent contamination during long-term operation of the system. The preincubation tank was maintained at 2×10^8 suspended yeast cells/ml and > 4% v/v alcohol, making long-term operation possible without contamination. The optimisation of dissolved oxygen concentration during fermentation was found to be key for the quality of the resulting sake. Nunokawa and Hirotsune [12] were able to produce sake continuously after an initial 3-day preincubation period, compared to 20 days for batch production of sake.

2.2. SOYA SAUCE PRODUCTION

Immobilised cell technology has been investigated for shortening the lengthy soy sauce production process. In a previous review [1], the challenges noted by researchers included achieving a taste profile match to conventionally produced soy sauce and finding a carrier suitable for food applications that was stable in the soy sauce fermentation medium during long term continuous fermentation periods. The standard soy sauce production process begins with the mixing of cooked soybean and roasted wheat, which is inoculated with spores of *Aspergillus sojae* and/or *Aspergillus oryzae* and is cultured for 2 days to make "koji". The koji is mixed with brine to make "moromi". The enzymes in the "koji", such as proteinases, petidases and amylases, hydrolyse the components of the "moromi" mash. Including the full fermentation and aging of the "moromi", the entire process for making soy sauce is more than 6 months in duration [13].

Osaki et al. [14] were first to investigate the use of immobilised cells for continuous production of soy sauce. Alginate gel immobilised *Pediococcus helophilus*, *Saccharomyces rouxii*, *Torulopsis versatilis* were used for fermentation and processing time was successfully reduced to two weeks, including enzymatic hydrolysis and the refining process. The continuously produced liquid had a similar composition to commercial coy sauce, however the aroma profile was not a match. Hamada et al. [15] were able to produce an aroma profile comparable to commercial products using a gas lift bioreactor system and *Zygosaccharomyces rouxii* immobilised in alginate beads through optimization of aeration, pH and temperature during fermentation. Unfortunately, the system was not suitable for long-term continuous operation due to instability of the alginate beads.

To overcome the stability issues with alginate gels, ceramic beads were utilized by Horitsu et al. [16] for continuous fermentation in two series bioreactors, with *Z. rouxii* cells immobilised in the first bioreactor and *C. versatilis* cells in the second. Total fermentation time was shorted to six days with no difference in soy sauce properties sensory compared to commercial. High ethanol productivity for soy sauce production has been realized by Iwasaki et al. [17] using a hollow fiber membrane bioreactor where the fermented product was removed from the bioreactor through the microfiltration membrane. Ethanol productivities of 0.78-1.23 g/l/h were achieved in this system compared to 0.02–0.05 g/l/h and 0.2-0.22 g/l/h for freely suspended cell batch and other immobilised cell bioreactors, respectively.

2.3. MEAD PRODUCTION

Mead is traditionally produced by batch fermentation of honey (fructose, glucose, and sucrose). Fructose is generally the most abundant simple sugar found in honey and the mead fermentation process is longer than most beer fermentations, where other sugars are higher in concentration. In 1985, Qureshi and Tamhane [18] used immobilised *Saccharomyces cerevisiae* cells immobilised in alginate gel beads to continuously produce mead, achieving more than 3 months of operation.

2.4. REMOVAL OF MALIC ACID FROM COFFEE BEANS

There has been interest by a large coffee company in removing malic acid from unroasted (green) coffee beans. Malic acid was thought to increase the secretion of stomach acid following coffee consumption [19]. Bertkau et al. [19], used a pilot scale immobilised cell bioreactor and a countercurrent pulse extraction column to remove malic acid from green coffee beans before roasting. A high-soluble-solids water extract was recycled from the green coffee beans through a spiral-wound immobilised cell bioreactor. Food-grade bacteria (*Lactobacillus spp.* or *Leuconostoc spp.*), contained within the immobilised cell bioreactor, metabolized the malic acid in the green coffee bean extract. The extract that was demalated in the immobilised cell bioreactor was then recirculated to extract the malic acid from the green coffee beans. This occurred during the countercurrent flow of the "malic acid-lean" extract over the beans. Using this process, the malic acid concentration in the beans was reduced by over 80%. The immobilised cell bioreactor consisted of a spiral-wound ribbed NC120 Microporous Plastic Sheet (MPS®) manufactured by FMC Corporation Biosupport Materials. MPS® is a proprietary PVC matrix with embedded porous silica particles suitable for immobilizing cells for food-related applications. The study states that this extract recycle process may have additional uses for removing undesirable organics or for the addition of water-soluble components, and is especially applicable to processing grains and beans [19].

2.5. DE-BITTERING OF CITRUS JUICE

Undesirable bitterness can be developed in citrus juice from limonin formed during storage. Because of the acid pH in citrus juice and because the enzymes which degrade limonin have highest activity as alkaline pH, there has been research activity on microorganisms the produce enzymes that degrade limonin at low pH. Some researchers have been successful using immobilised cells of *Rhodococcus fascians* [20] and *Corynebacterium fascians* [21] to degrade limonin for application to citrus juice.

2.6. "UGBA" FOOD SNACK PRODUCTION

In Africa, "Ugba" is a popular snack food made through a solid-state fermentation of the seed of the African oil bean tree (*Pentaclethra macrophylla* Bentham). Ugba is a nutritious food, rich in protein, carbohydrates, lipids and vitamins, with appealing taste characteristics. In a paper by Isu and Ofuya [22], a study was undertaken to improve on the traditional production methods and quality of the Ugba production process. Pure cultures of *Bacillus subtillus* were immobilised on edible cowpea granules in order to ferment the African oil beans. As a result of immobilisation, fermentation time was significantly reduced from approximately 72 hours freshly grown cells in broth culture down to 48 hours using immobilised cells. This was thought to be due to higher cell density and the positive effects of immobilisation on cell wall permeability and metabolism. The Ugba produced using the immobilised cell process was reported to have favourable organoleptic properties. The immobilised cell cultures were maintained in an active physiological state for up to 6 months [22].

2.7. REMOVAL OF SIMPLE SUGARS

Fermentation using immobilised cells is one of several methods used to remove simple sugars such as glucose, fructose and sucrose, from food-grade processes, offering the advantage of high cell densities and the opportunity for continuous processing. The removal of glucose, fructose and sucrose from food-grade oligosaccharides and removal of glucose from egg are described in the next sections.

2.7.1. Purification of food-grade oligosaccharides

Food-grade oligosaccharides are used as food ingredients, providing helpful modifications to the flavour and physicochemical characteristics of foods. They also have some properties that provide health benefits to consumers. Non-digestible purified oligosaccharides (NDOs) display low cariogenicity and may be used as low-calorie substitutes. There are a few NDOs that function as prebiotics by stimulating the growth of probiotic bacteria in the intestines. Commercially produced food-grade oligosaccharides are mixtures including oligosaccharides, monosaccharides, some disaccharides, and glycosidic linkages. The advantages of removing simple sugars from the oligosaccharide products have the following advantages: increased viscosity, reduced sweetness and hygroscopicity, and results in fewer Maillard reactions during heat processing. The lack of simple sugars in NDOs lowers the cariogenicity and caloric content, permitting the oligosaccharides to be used in diabetic foods [23].

The most common method for removal of simple sugars from oligosaccharide mixtures is chromatographic processing. Following chromatographic purification, the product retains some simple sugars, typically comprising 5-10% of the total carbohydrates, with oligosaccharide yields of 70-80%. *Zymomonas mobilis* bacterium has a narrow carbon substrate range, fermenting only sucrose, glucose and fructose and does not synthesise the carbohydrases needed to hydrolyse most oligosaccharides. This bacterium rapidly ferments glucose and fructose to ethanol and carbon dioxide with negligible by-product formation (minor amounts of sorbitol and fructo-oligosaccharides formed) [23].

Crittenden and Playne [23] examined the potential of using *Z. mobilis* immobilised in alginate bead for the removal of glucose, fructose and sucrose from inulin-, fructo-, malto-, isomalto-, and gentio-oligosaccharide solutions. The total carbohydrate concentration of each test solution was 300 g/l. An isomalto-oligosaccharide solution containing 500 g/l total carbohydrate was also tested. At 300 g/l, complete fermentation of the simple sugars in the oligosaccharide solutions occurred. For the 500 g/l isomalto-oligosaccharide solution, 91% removal of glucose occurred. Fermentation stopped before sugar consumption was complete in this case. This was though to be due to ethanol toxicity at 14% v/v. Immobilisation allows for future continuous purification and the potential of co-immobilizing *Z. mobilis* with the enzymes for oligosaccharide synthesis to produce a single stage process with simultaneous oligosaccharide synthesis and purification [23].

2.7.2. Glucose removal from egg

During the manufacture of dehydrated egg, the glucose present in the raw egg causes undesirable browning. The glucose can be removed enzymatically with glucose oxidase and catalase, but due to the high cost, the most common approach is to use yeast to ferment the glucose. D'Souza and Godbole [24] immobilised yeast in alginate gel beads to remove glucose from egg through fermentation. The immobilised cells were used in 30 repeated batch treatments over 9 days without loss in glucose removal activity [24].

2.8. SUGAR CONVERSIONS

The various sugar conversions utilising immobilised whole cells is reviewed by Norton and Vuillemard [1]. Table 1 summarizes the various studies using immobilised cells for sugar conversions.

The production of High Fructose Syrup is one of the most important enzymatic sugar conversions, being used for the sweetening of soft drinks. Fructose has a high sweetening index relative to glucose, making it an important sugar in the sweeteners market. Commercially, fructose is chemically or enzymatically produced by isomerisation of glucose or sucrose hydrolysis (or inversion). The hydrolysis of sucrose to fructose and glucose makes sweet syrups that are less likely to crystallise and have more stability.

Levan is a fructose polymer that can be used as a fructose source and texture-forming compound. In a study by Bekers *et al.* [25], levan and ethanol were simultaneously produced in a sucrose medium using immobilised cells of *Zymomonas mobilis*. Chien *et al.* [9] utilised mycelia of *Aspergillus japonicus* immobilised in gluten for batch and continuous production of fructooligosaccharides from sucrose. These fructose oligomers can be used as a non-digestible sweetener in food and also are physiologically beneficial because they improve the population of Bifidusbacteria in the intestines.

During continuous production in a packed-bed reactor, the highest productivity of frucotooligosaccharides of 173 g/l/h was achieved at a flow rate of 0.8 ml/min with the mass fraction of oligosaccharides at 0.31. The study found that in order to obtain a more concentrated product a reactor residence time of more than 4 hours was required. Gluten was found to be an effective immobilisation matrix for *A. japonicus* mycelia, with the gluten-immobilised cell preparation displaying stability during long term operation [9].

Table 1. Sugar conversions studied using immobilised cells.

Sugar conversion	Cell type	Immobilisation matrix	Reference
High fructose syrup production			
Glucose isomerisation	*Arthrobacter*	κ-Carrageenan	[59]
Sucrose hydrolysis/inversion	*Saccharomyces sp.*	Polyacrylamide	[60]
Sucrose hydrolysis/inversion	*Saccharomyces sp.*	Granules	[61]
Sucrose hydrolysis/inversion	Yeast fragments	Gelatin / glutaraldehyde	[62]
Inulin hydrolysis	*Kluyveromyces marxianus*	Alginate	[63]
Levan production	*Zymomonas mobilis*	Alginate, alumina, wire spheres	[25]
Fructooligosaccharides production	*Aspergillus japonicus*	Gluten	[9]
Lactose hydrolysis	*Kluyveromyces fragilis*	Alginate	[64]
Lactose hydrolysis	*Kluyveromyces bulgaricus*	Alginate	[65]
Lactose hydrolysis	*Kluyveromyces fragilis*	Gelatin fibres	[66]
Cellobiose hydrolysis	*Pichia eschellsii*	Alginate	[67]
Cellobiose hydrolysis	*Trichosporon pullulans*	Alginate	[68]

The hydrolysis of lactose is catalysed by β-galactosidase (lactase) enzyme, yielding galactose and glucose. The removal of lactose from milk and milk products makes these products acceptable for lactose-intolerant individuals who do not secrete enough β-galactosidase enzyme in their intestinal tract. Examples of lactose hydrolysis processes using immobilised cells are given in Table 1.

2.9. HYDROLYSIS OF TRIGLYCERIDES AND PROTEINS IN MILK

Lipolysed milk fat, which has a butter- or cheese-like flavour that can be used in different foods, is produced by lipase-catalyzed hydrolysis of triglycerides in milk fat. Commercial processes typically utilise free lipase enzyme for the production of lipolysed milk fat. Permeabilised *Rhizopus delemar* cells immobilised on polyurethane foam were used for the hydrolysis of milk fat and olive oil in a continuous stirred tank reactor [26]. Strong adherence of fungal mycelia to the polyurethane foam matrix was observed and the immobilised cell biocatalysts had lipase activity that was 3.3 times greater than that of freely suspended cells. The lipase within the immobilised cells had similar properties to purified extracellular lipase produced by *Rhizopus delemar*, but was more thermostable than the free enzyme. The immobilised cell reactor system provided good control of the degree of hydrolysis of the triglycerides in the olive oil and milk fat [26].

Hydrolysed milk protein has nutritional value and other beneficial physiological effects. Free and immobilised cells of *Serratia marcescens* were used as an enzyme source for continuous production of small peptides from buttermilk protein [27]. It was found that in both the free and alginate-immobilised cell systems, the large protein residues were significantly hydrolyzed, but more fee amino acids were released in the immobilised cell reactor than the free cell reactor [27].

2.10. VITAMINS

The compound L-sorbose has been used as an intermediate for the synthesis of vitamin C. Because D-sorbitol is converted to L-sorbose by the enzyme D-sorbitol dehydrogenase, bound to the cell membrane of *Gluconobacter suboxydans,* oxygen demand is comparatively large. Kim *et al.* [28] immobilised *Gluconobacter suboxydans* cells in alginate gel beads for the production of L-sorbose from D-sorbitol with the aim of improving the diffusion of oxygen through the beads. It was demonstrated that reducing bead size and making holes in the beads could enhance oxygen diffusion rates. However, diffusion limitations within the beads still existed. Co-immobilisation of the oxygen carrier n-dodecane and cells in the alginate beads helped to further surmount the oxygen limitations within the beads. Kim *et al.* [28] also showed that enhanced diffusion of oxygen into the beads produced an increase in L-sorbose yield.

Buzzini and Rossi [29] used alginate immobilised *Candida tropicalis* for the semi-continuous and continuous production of riboflavin from concentrated rectified grape must as the sole carbohydrate source. During continuous production, the productivity of riboflavin was in the range of 120 mg/l/day, corresponding to vitamin concentrations in the effluent of 400-600 mg/l. The immobilised cell system operated for more than one month without significant loss of flavinogenic activity. The greatest value of sugar bioconversion and riboflavin production occurred at the lowest dilution rate ($0.008 \, h^{-1}$) tested [29].

3. Metabolite production

3.1. AMINO ACIDS

There has been a significant amount of research on the use of immobilised cell technology for the production of amino acids, as summarised in Table 2. Amino acids are used as food additives and comprise a multibillion-dollar market [30].

The taste enhancer monosodium glutamate (MSG) is made from L-glutamic acid. Because of its sweet taste, L-alanine is also used as a food additive [31]. Aspartame, the low calorie synthetic sweetener, is a dipeptide of L-aspartic acid and L-phenylalanine methyl ester, causing the production of L-aspartic acid and L-phenylalanine to receive significant research attention [32]. The amino acids L-tryptophan, L-phenylalanine and L-lysine can be used as dietary supplements.

The physiologically active natural form of amino acids is the L-isomer and thus this is the necessary form for food. A recent study [31] examined the production of optically pure L-alanine using κ-carrageenan immobilised cells of *Pseudomonas sp.* BA2 with L-aminoacylase activity. The immobilised cells were used to produce L-alanine from N-acetyl-DL-alanine in a batch reactor. The *Pseudomonas sp.* BA2 did not express D-aminoacylase activity.

Table 2. *Studies on the production of amino acids using immobilised cells.*

Amino acid	Cell type	Immobilisation matrix	Reference
L-Alanine	*Pseudomonas sp.*	κ-Carrageenan	[31]
	E. coli/Pseudomonas dacunhae	κ-Carrageenan	[69-70]
L-Arginine	*Serratia marcescens*	κ-Carrageenan	[71]
L-Aspartic Acid	*Escherichia coli*	Polyacrylamide gel	[32]
	Escherichia coli	κ-Carrageenan	[72-73]
	Escherichia coli	Vermiculite	[30]
	Escherichia coli	Latex	[74]
	Escherichia coli	Polyurethane foam	[75]
L-Glutamic Acid	*Corynebacterium glutamicum*	Polyacrylamide gel	[76]
	Brevibacterium ammoniagenes	Alginate	[77]
	Corynebacterium glutamicum	Euchema gel	[78]
	Corynebacterium glutamicum	Porous sintered glass	[79]
L-Isoleucine	*Serratia marcescens*	κ-Carrageenan	[80]
L-Lysine	*Corynebacterium glutamicum*	PVA-Cryogel	[81]
L-Phenylalanine	*Escherichia coli, Rhodotorula rubra*	Vermiculite	[30]
	Paracoccus denitrificans	κ-Carrageenan	[82]
	Corynebacterium sp./Paracoccus denitrificans	κ-Carrageenan	[83]
L-Serine	*Protomonas extorquens*	Alginate	[84]
L-Tryptophan	*Escherichia coli*	Polyacrylamide gel	[85-86]
	E. coli/Pseudomonas putida	Alginate	[87]

During the production processes for amino acids, some amino acids can be used as a precursor for others. For example, L-aspartic acid is a substrate for the production of L-alanine. Takamatsu and Tosa [33] used two immobilised microorganisms in a single closed column reactor at high pressure to produce L-alanine efficiently from ammonium fumarate. *E. Coli* and *P. dacunhae* cells were immobilised in κ-carrageenan gel beads. The reaction scheme is as follows [33]:

$$\text{Fumaric acid} + NH_3 \underset{aspartase}{\overset{E.\ coli}{\rightleftarrows}} \text{L-aspartic acid} \xrightarrow[L\text{-aspartate }\beta\text{-decarboxylase}]{P.\ dacunhae} \text{L-alanine} + CO_2$$

3.2. ORGANIC ACIDS

The use of immobilised cells has been proposed for a variety of organic acid metabolites relevant to the food industry, including acetic, citric, fumaric, lactic, malic, gluconic, kojic and propionic acid. In 1994, Norton and Vuillemand [1] reviewed recent progress in organic acid production using immobilised cells. Mori [34] has published an in-depth review of acetic acid production using immobilised cells and, more recently, provided a summary table of bioreactors and cell carriers used for acetic acid production [35].

The production of citric, malic, tartaric and gluconic acid by fermentation for food applications was reviewed in detail by Milsom [36]. Citric acid production was studied by Sakurai *et al.* [37] using *Aspergillus niger* cells immobilised on cellulose beads. The effects of pre-culture period, initial sugar concentration, and the time interval for repeated batch fermentation cycles on citric acid production were studied. It was found that citric acid production rates and yields were strongly affected by the pre-culture time and conditions. An initial sucrose concentration of 100 g dm^{-3} was best from the perspective of citric acid production rate and yield. A slight effect of repeated batch cycle time interval, however, an 8-day interval was deemed to be most cost-effective because it had the lowest amount of residual sugar [37].

Lactic acid is mainly used in the food industry as an acidulant and preservative because of its mild acidic taste that does not dominate other flavours in foods. Both single stage and multi-stage reactor configurations have been used for the continuous production of lactic acid with varying degrees of success in terms of lactic acid productivity [38-39]. As previously mentioned, cell immobilisation increases resistance to physicochemical stress, an important advantage for the scale-up of fermentations. In a full-scale lactic acid production plant, a potential stress factor is the presence of sanitiser residue in tanks and piping that taint the bacteria culture medium and reduce activity. Trauth *et al.* [40] investigated the inhibitory effect of quaternary ammonium sanitizers (QAS) on the fermentation activity of lactic acid bacteria *Lactococcus lactis*. It was found that by immobilizing the lactic acid bacteria in calcium alginate gel beads the inhibitory effects of QAS on cell growth and acidification rate was reduced. As the degree of cell colonisation of the beads and the number of successive acidifications increased, the acidification rate and resistance to the QAS also increased [40].

3.3. ALCOHOLS

Ethanol production has been extensively studied using immobilised cell technology, however, most has been focused on beer, wine or cider production in the beverage sector or fuel ethanol production. All of these categories are covered in other sections of this book. Details on the production of glycerol, xylitol, sorbitol and mannitol using immobilised cells can be found in a previously published review [1].

There has been a relatively small amount of literature on the production of food-grade ethanol using immobilised cells [41-44]. In order to improve on the food safety aspects of immobilisation techniques, de Alteriis *et al.* [41] replaced the standard cross-linking agents such as glutaraldehyde or formaldehyde with oxidised starch. High ethanol productivity and good mechanical stability were observed using the oxidised starch cross-linked gelatin immobilised yeast cells [41]. In a recent study [44], Saccharomyces cerevisiae and Zymomonas mobilis were immobilised in agar gel for the production of food-grade ethanol. Comparisons of higher alcohol and by-product production were made between the two cell types using rye mash and glucose fermentation media. It was found that *Zymomonas mobilis* produced 5 times less higher alcohols than the yeast, and over 4 times fewer by-products and a higher ethanol efficiency [44].

3.4. ENZYMES

A range of enzymes has been produced using immobilised whole cells including glucoamylase, α-amylase, β-amylase, pullulanase, which convert starch to glucose and maltose [1]. These enzymes are of special use in the brewing industry where starches can be broken down into sugars that yeast can assimilate during fermentation in the production of beer. Also in the food industry, immobilised cells have been reported for production of catalase [45], cellulases [46], invertase [47], lactase [48], lipases [49], proteases [50-54], and chitinolytic enzymes [55]. Chitinolytic enzymes can be used in the bioconversion of chitin-rich materials such as shellfish waste for the production of chito-oligosaccharides and N-acetyl-β-D-glucosamine for the food and feed industry and the pharmaceutical or chemical purposes. Fenice et al. [55], used immobilised cells of the fungus *Penicillium janthinellum* P9 for the production of chitinolytic enzymes. Cells immobilised in macroporous cellulose carriers were tested in repeated-batch and continuous production of chitinolytic enzymes in a fluidized bed reactor. They found that the time necessary to obtain maximum enzyme production in repeated-batch processes using immobilised cells was reduced by 50% compared to free cells. In continuous culture using immobilised cells the maximum enzyme activity achieved (450 U l^{-1}) was much higher than those obtained in repeated-batch processes with free cells in shaken culture (305 U l^{-1}) and immobilised cells in either shaken culture (342 U l^{-1}) or bioreactor (338 U l^{-1}). An increase in volumetric productivity was also observed [55].

3.5. BACTERIOCINS

Bacteriocins are proteins or protein complexes that display bacteriocidal activity against closely related bacteria. They are used as biopreservatives to improve the quality and innocuity of products in the food industry. Nisin is a bacteriocin synthesised by *Lactococcus lactis* strains and is used in food to prevent the growth of pathogens, spores and contaminant bacteria. The literature shows that there has been interest in immobilizing *Lactococcus lactis* bacteria in order to increase the cell density in the reactor and to enable continuous production of Nisin at high dilution rates [56-58].

References

[1] Norton, S. and Vuillemard, J.-C. (1994) Food bioconversions and metabolite production using Immobilised Cell Technology. Crit. Rev. Biotechnol. 14(2): 193-224.
[2] Gibbs, B.F.; Kermasha, S.; Alli, I. and Mulligan, C.N. (1999) Encapsulation in the food industry: A Review. Int. J. Food Sci. Nutr. 50: 213-224.
[3] King, A.H. (1995) Encapsulation of food ingredients: A review of available technology, focusing on hydrocolloids. In: Risch, S.J. and Reineccius, G.A. (Eds.) Encapsulation and Controlled Release of Food Ingredients. ACS Symposium Series 590; pp. 26-39.
[4] Navratil, M.; Gemeiner, P.; Klein, J.; Sturdik, E.; Malovikova, A.; Nahalka, J.; Vikartovska, A.; Domeny, Z. and Smogrovicova, D. (2002) Properties of hydrogel materials used for entrapment of microbial cells in production of fermented beverages. Art. Cell, Blood Subs. Immob. Biotech. 30(3): 199-218.

[5] Risch, S.J. (1995) Encapsulation: overview of uses and techniques. In: Risch, S.J. and Reineccius, G.A. (Eds.) Encapsulation and Controlled Release of Food Ingredients. ACS Symposium Series 590; pp. 2-8.
[6] Thies, C. (2001) In: Vilstrup, P. (Ed.) Microencapsulation of Food Ingredients. Letterhead Publishing, Surrey (UK); pp. 1-54.
[7] Fu, T.-J. (1998) Safety considerations for food ingredients produced by plant cell and tissue culture. Chemtech 28(1): 40-46.
[8] Cassidy, M.B.; Lee, H. and Trevors, J.T. (1996) Environmental applications of immobilised cells: a review. J. Ind. Microbiol. 16: 79-101.
[9] Chien, C.-S.; Lee, W.-C. and Lin, T.-J. (2001) Immobilisation of *Aspergillus japonicus* by entrapping cells in gluten for production of fructooligosaccharides. Enzyme Microb. Technol. 29: 252-257.
[10] Gardner, N.; Champagne, C.P. and Gelinas, P. (2002) Effect of yeast extracts containing propionic acid on bread dough fermentation and bread properties. Food Microbiol. Safety 67(5): 1855-1858.
[11] Lee, S.-L.; Cheng, H.-Y.; Chen, W.-C. and Chou, C.-C. (1999) Effect of physical factors on the production of γ-decalactone by immobilised cells of *Sporidiobolus salmonicolor*. Proc. Biochem. 34: 845-850.
[12] Nunokawa, Y. and Hirotsune, M. (1993) Production of soft sake by an immobilised yeast reactor system. In: Tanaka, A.; Tosa, T. and Kobayashi, T. (Eds.) Industrial Application of Immobilised Biocatalysts. Marcel Dekker, Inc., New York (USA); pp. 235-253.
[13] Motai, H.; Hamada, T.; Fukushima, Y. (1993) Application of a bioreactor system to soy sauce production. In: Tanaka, A.; Tosa, T. and Kobayashi, T. (Eds.) Industrial Application of Immobilised Biocatalysts. Marcel Dekker, Inc., New York (USA); pp. 315-335.
[14] Osaki, K.; Okamoto, Y.; Akao, T.; Nagata, S. and Takamatsu, H. (1985) Fermentation of soy sauce with immobilised whole cells. J. Food Sci. 50: 1289-1292.
[15] Hamada, T.; Ishiyama, T. and Motai, H. (1989) Continuous fermentation of soy sauce by immobilised cells of *Zygosaccharomyces rouxii* in an airlift reactor. Appl. Microbiol. Biotechnol. 31: 346-350.
[16] Horitsu, H.; Wang, M.G. and Kawai, K. (1991) A modified process for soy sauce fermentation by immobilised yeasts. Agric. Biol. Chem. 55: 269-271.
[17] Iwasaki, K.-I.; Nakajima, M. and Sasahara, H. (1991) Rapid ethanol fermentation for soy sauce production using a microfiltration membrane reactor. J. Ferment. Bioeng. 72(5): 373-378.
[18] Qureshi, N. and Tamhane, D.V. (1985) Production of mead by immobilised whole cells of *Saccharomyces cerevisiae*. Appl. Microbiol. Biotechnol. 21: 280-281.
[19] Bertkau, G.H.; Murphy, S.M. and Sabella, F.J. (1999) combined immobilised cell bioreactor and pulse column technology as a novel approach to food modification. Process Biochem. 34: 221-229.
[20] Martinez-Madrid, C.; Manjon, A. and Iborra, J.L. (1989) Degradation of limonin by entrapped *Rhodococcus fascians* cells. Biotechnol. Lett. 11: 653-658.
[21] Hasegawa, S.; Verdercook, C.E.; Choi, G.Y.; Hermann, Z. and Ou, P. (1985) Limonoid debittering of citrus joice sera by immobilised cells of *Corynebacterium fascians*. J. Food Sci. 50: 330-332.
[22] Isu, N.R. and Ofuya, C.O. (2000) Improvement of the traditional processing and fermentation of african oil bean (*Pentaclethra macrophylla* Bentham) into a food snack – 'Ugba'. Int. J. Food Microbiol. 59: 235-239.
[23] Crittenden, R.G. and Playne, M.J. (2002) Purification of food-grade oligosaccharides using immobilised cells of *Zymomonas mobilis*. Appl. Microbiol. Biotechnol. 58: 297-302.
[24] D'Souza, S.F. and Godbole, S.S. (1989) Removal of glucose from egg prior to spray drying by fermentation with immobilised yeast cells. Biotechnol. Lett. 11(3): 211-212.
[25] Bekers, M.; Laukevics, J.; Karsakevich, A.; Ventina, E.; Kaminska, E.; Upite, D.; Vina, I.; Linde, R. and Scherbaka, R. (2001) Levan-ethanol biosynthesis using *Zymomonas mobilis* Cells immobilised by attachment and entrapment. Process Biochem. 36: 970-986.
[26] Chen, J.-P. and McGill, S. D. (1992) Enzymatic hydrolysis of triglycerides by *Rhizopus delemar* immobilised on Biomass Support Particles. Food Biotechnol. 6(1): 1-78.
[27] Vuillemard, J.C.; Goulet, J. and Amiot, J. (1988) Continuous production of small peptides from milk proteins by extracellular proteases of free and immobilised *Serratia marcescens* cells. Enzyme Microb. Technol. 10: 2-8.
[28] Kim, H.J.; Kim, J.H. and Shin, C.S. (1999) Conversion of D-sorbitol to L-sorbose by *Gluconobacter suboxydans* cells co-immobilised with oxygen carriers in alginate beads. Process Biochem. 35: 243-248.
[29] Buzzini, P. and Rossi, J. (1989) Semi-continuous and continuous riboflavin production by calcium-alginate-immobilised *Candida tropicalis* in concentrated rectified grape must. World J. Microbiol. Biotechnol. 14: 377-381.

[30] Hamilton, B.K.; Hsiao, H.; Swann, W.E.; Anderson, D.M. and Delent, J.J. (1985) Manufacture of L-amino acids with bioreactors. Trends Biotechnol. 3: 64-68.
[31] Santoyo, A.B.; Rodriguez, J.B.; Carrasco, J.L.G.; Gomez, E.G.; Rojo, I.A. and Teruel, M.L.A. (1998) Production of optically pure L-alanine by immobilised *Pseudomonas sp.* BA2 Cells. J. Chem. Technol. Biotechnol. 73: 197-202.
[32] Sato, T.; Mori, T.; Tosa, T.; Chibata, I.; Furui, M.; Yamashita, K. and Sumi, A. (1975) Engineering analysis of continuous production of L-aspartic acid by immobilised *Escherichia coli* Cells in fixed bed reactor. Biotechol. Bioeng. 17: 1797-1804.
[33] Takamatsu, S. and Tosa, T. (1993) Production of L-alanine and D-aspartic acid. In: Tanaka, A.; Tosa, T. and Kobayashi, T. (Eds.) Industrial Application of Immobilised Biocatalysts. Marcel Dekker, Inc., New York (USA); pp. 25-35.
[34] Mori, A. (1985) Production of vinegar by immobilised cells. Proc. Biochem. 6: 67-74.
[35] Mori, A. (1993) Vinegar production in a fluidized-bed reactor with immobilised bacteria. In: Tanaka, A.; Tosa, T. and Kobayashi, T. (Eds.) Industrial Application of Immobilised Biocatalysts. Marcel Dekker, Inc., New York (USA); pp. 291-313.
[36] Milsom, P.E. (1987) Organic acids by fermentation, especially citric acid. In: King, R.D. and Cheetham, P.S.J. (Eds.) Food Biotechnology I. Elsevier Applied Sciences Publishers, Ltd., Reading (UK); pp. 273-307.
[37] Sakurai, A.; Itoh, M.; Sakakibara, M.; Saito, H. and Fujita, M. (1997) Citric acid production by *Aspergillus niger* immobilised on porous cellulose beads. J. Chem. Technol. Biotechnol. 70: 157-162.
[38] Bruno-Barcena, J.M.; Ragout, A.L.; Cordoba, P.R. and Sineriz, F. (1999) Continuous production of L(+)-lactic acid by *Lactobacillus casei* in two-stage Systems. Appl. Microbiol. Biotechnol. 51(3): 316-324.
[39] Goksungur, Y. and Guvenc, U. (1999) Production of lactic acid from beet molassus by calcium alginate immobilised *Lactobacillus delbrueckii* IFO 3202. J. Chem Technol. Biotechnol. 74: 131-136.
[40] Trauth, E.; Lemaitre, J.-P.; Rojas, C.; Divies, C. and Cachon, R. (2001) Resistance of immobilised lactic acid bacteria to the inhibitory effect of quaternary ammonium sanitizers. Lebensm.-Wissensch.Technol. 34: 239-243.
[41] De Alteriis, E.; Parascandola, P. and Scardi, V. (1990) Oxidized starch as a hardening agent in the gelatin-immobilisation of living yeast cells. Starch 42: 57-60.
[42] Gilson, C.D. and Thomas, A. (1995) Ethanol production by alginate immobilised yeast in a fluidized bed reactor. J. Chem. Tech. Biotechnol. 62: 38-45.
[43] Lee, S.W.; Yajima, M. and Tanaka, H. (1993) Use of food additives to prevent contamination during fermentation using a co-immobilised mixed culture system. J. Ferment. Bioeng. 75: 389-391.
[44] Nowak, J.; Czarnecki, Z. and Kaminski, E. (2000) Bacterial and yeast by-products fermentation in ethanol fermentation of glucose medium and rye mashes. Pol. J. Food Nutr. Sci. 9/50(4): 49-51.
[45] Seip, J.E. and Di Cosimo, R. (1992) Optimization of accessible catalase activity in polyacrylamide gel-immobilised *Saccharomyces cerevisiae*. Biotechnol. Bioeng. 40: 638-642.
[46] Cahill, G.; Walsh, P.K. and Ryan, T.P. (1990) Studies on the production of β-Glucanase by free and immobilised recombinant yeast cells. In: de Bont, J.A.M.; Visser, J.; Mattiasson, B. and Tramper, J. (Eds.) Physiology of Immobilised Cells. Elsevier Science Publishers, Amsterdam (The Netherlands); pp. 405-410.
[47] Parascandola, P.; de Alteriis, E. and Scardi, V. (1993) Invertase and acid phosphatase in free and gel-immobilised cells of *Saccharomyces cerevisiae* grown under different cultural conditions. Enzyme Microbiol. Technol. 15: 42-49.
[48] Zhang, X.; Bury, S.; DiBiasio, D. and Miller, J. (1989) Effects of immobilisation on growth, substrate consumption, β-galactosidase induction, and byproduct formation in *Escherichia coli*. J. Ind. Microbiol. 4: 239-246.
[49] Nakashima, T.; Fukuda, H.; Nojima, Y. and Nagai, S. (1989) Intracellular lipase production by *Rhizopus chienensis* using biomass support particles in a circulating bed fermentor. J. Ferment. Bioeng.; 68: 19-24.
[50] Aleksieva, P.; Petricheva, E.; Konstantinov, E.; Robeva, C. and Mutafov, S. (1991) Acid proteinases production by *Humicola lutea* cells immobilised in polyhydroxyethylmethacrylate gel. Acta Biotechnol. 11: 255-261.
[51] El Aassar, S.A.; Ed Badry, H.M. and Abdel Fattah, A.F. (1990) The biosynthesis of proteases with firbinolytic activity in immobilised cultrues of *Penicillium chrysogenum* H9. Appl. Microbiol. Biotechol. 33: 26-30.

[52] Okita, W.B. and Kirwan, D.J. (1987) Protease production by immobilised *Bacillus licheniformis*. Ann. NY Acad. Sci. 506: 256-259.
[53] Vuillemard, J.C.; Terre, S.; Benoit, S. and Amiot, J. (1988) Protease production by immobilised growing cells of *Serratia marcescens* and *Myxococcus xanthus* in calcium alginate gel beads. Appl. Microbio. Biotechnol. 27: 423-431.
[54] Zahran, A.S. and Zayed, G. (1992) Production of extracellular protease by immobilised and free cells of *Flavobacterium* sp. R23. Milchwiss. 48: 18-21.
[55] Fenice, M.; Di Giambattista, R.; Raetz, E.; Leuba, J.-L. and Federici, F. (1998) Repeated-batch and continuous production of chitinolytic enzymes by *Penicillium janthinellum* immobilised on chemically-modified macroporous cellulose. J. Biotechnol. 62: 119-131.
[56] Bertrand, N.; Fliss, I. and Lacroix, C. (2001) High nisin-Z production during repeated-cycle batch cultures in supplemented whey permeate using immobilised *Lactococcus lactis* UL719. Int. Dairy J. 11: 953-960.
[57] Desjardins, P.; Meghrous, J. and Lacroix, C. (2001) Effect of aeration and dilution rate on Nisin Z production during continuous fermentation with free and immobilised *Lactococcus lactis* UL719 in supplemented whey Permeate. Int. Dairy J. 11: 943-951.
[58] Sonomoto, K.; Chinachoti, N.; Endo, N. and Ishisaki, A. (2000) Biosynthetic production of Nisin Z by Immobilised *Lactococcus lactis* IO-1. J. Mol. Catal. B. Enzymatic 10: 325-334.
[59] Bazaraa, W.A. and Hamdy, M.K. (1989) Fructose production by immobilised *Arthrobacter* Cells. J. Ind. Microbiol. 4: 267-274.
[60] Ghosh, S. and D'Souza, S.F. (1989) Crushing strength as a tool for reactor height determination for invertase-containing yeast cells immobilised in polyacrylamide. Enzyme Microb. Technol.; 11: 376-378.
[61] Hasal, P.; Vojtisek, V.; Cejkova, A.; Kleczek, P. and Kofronova, O. (1992) An immobilised whole yeast cell biocatalyst for enzymatic sucrose hydrolysis. Enzyme Microb. Technol. 14: 221-229.
[62] Horbach, U. and Hartmeier, W. (1989) Immobilised invertase on the basis of matrix-embedded yeast fragments. Gordian 89: 134-136.
[63] Bajpai, P. and Margaritis, A. (1986) Optimization studies for production of high fructose syrup from jerusalem artichoke using calcium alginate immobilised cells of *Kluyveromyces marxianus*. Process Biochem. 21: 16-18.
[64] Rao, B.Y.K.; Godbole, S.S. and D'Souza, S.F. (1988) Preparation of lactose free milk by fermentation using immobilised *Saccharomyces fragilis*. Biotechnol. Lett. 10: 427-430.
[65] Decleire, M.; Van Huynh, N.; Motte, J.C. and De Cat, W. (1985) Hydrolysis of whey by whole cells of *Kluyveromyces bulgaricus* immobilised in calcium alginate gels in hen white. Appl. Microbiol. Biotechnol. 22: 438-441.
[66] Castillo, E. and Casas, L.T. (1990) Reutilization of free and immobilised *Kluyveromyces fragilis* yeast cells with a controlled permeabilization treatment. In: de Bont, J.A.M.; Visser, J.; Mattiasson, B. and Tramper, J. (Eds.) Physiology of Immobilised Cells. Elsevier Science Publishers, Amsterdam (The Netherlands); pp. 213-218.
[67] Jain, D. and Ghose, T.K. (1984) Cellobiose hydrolysis using *Pichia etchellsii* cells immobilised in calcium alginate. Biotechnol. Bioeng. 26: 340-346.
[68] Adami, A.; Cavazzoni, V.; Trezzi, M. and Craveri, R. (1988) Cellobiose hydrolysis by *Trichosporon pullulans* cells immobilised in calcium alginate. Biotechnol. Bioeng. 32: 391-395.
[69] Takamatsu, S.; Umemura, I.; Yamamoto, K.; Sato, T.; Tosa, T. and Chibata, I. (1982) Productin of L-alanine from ammonium fummarate using two immobilised microorganisms. Eur. J. Appl. Microbiol. Biotechnol. 15: 147-152.
[70] Yamamoto, K.; Tosa, T. and Chibata, I. (1980) Continuous production of L-alanine using *Pseudomonas dacunhae* immobilised with carrageenan. Biotechnol. Bioeng. 22: 2045-2054.
[71] Fujimura, M.; Kato, J.; Tosa, T. and Chibata, I. (1984) Continuous production of L-arginine using immobilised growing *Serratia marcescens* cells: effectiveness of supply of oxygen gas. Appl. Microbiol. Biotechnol. 19: 79-84.
[72] Sato, T.; Nishida, Y.; Tosa, T. and Chibata, I. (1979) Immobilisation of *Escherichia coli* cells containing aspartase activity with κ-carrageenan. Biochim. Biophys. Acta. 570: 179-186.
[73] Tosa, T.; Sato, T.; Mori, T.; Yamamoto, K.; Takata, I.; Nishida, Y. and Chibata, I. (1979) Immobilisation of enzymes and microbial cells using carrageenan as matrix. Biotechnol. Bioeng. 21: 1697-1709.
[74] Bunning, T.J.; Lawton, C.W.; Klei, H.E. and Sundstrom, D.W. (1991) Physical property improvement of a pellicular biocatalyst. Bioprocess Eng. 7: 71-75.

[75] Fusee, M.C. (1987) Industrial production of L-aspartic acid using polyurethane-immobilised cells containing aspartase. In: Mosbach, K. (Ed.) Methods in Enzymology. Academic Press, Orlando (USA); pp. 463-471.
[76] Slowinski, W. and Charm, S.E. (1973) Glutamic acid productoin with gel-entrapped *Corynebacterium glutmicum*. Biotechnol. Bioeng. 15: 973-979.
[77] Lu, W.M. and Chen, W.C. (1988) Production of L-glutamate using entrapped living cells of *Brevibacterium ammoniagenes* with calcium alginate gels. Proc. Natl. Sci. Counc. ROC(A) 12: 400-406.
[78] Li, Y.F.; Huang, Y.; Lee, L.F.; Sui, P. and Wen, Q.Q. (1990) Production of glutamic acid by immobilised cells. Ann. NY. Acad. Sci. 613: 883-886.
[79] Henkel, H.J.; Johl, H.J.; Trosch, W. and Chmiel, H. (1990) Continuous production of glutamic acid in a three phase fluidized bed with immobilised *Corynebacterium glutamicum*. Food Biotechnol. 4: 149-154.
[80] Wada, M.; Uchida, T.; Kato, J. and Chibata, I. (1980) Continuous production of isoleucine using immobilised growing *Serratia marcescens* Cells. Biotechnol. Bioeng. 22: 1175-1188.
[81] Velizarov, S.G.; Rainina, E.I.; Sinitsin, A.P. and Varfolomeyev, S.D. (1992) Production of L-lysine by free and PVA-cryogel immobilised *Corynebacterium glutamicum* cells. Biotechnol. Lett. 14: 291-296.
[82] Nakamichi, K.; Nabe, K.; Nishida, Y. and Tosa, T. (1989) Production of L-phenylalanine from phenylpyruvate by *Paracoccus denitrificans* containing aminotransferase activity. Appl. Microbiol. Biotechnol. 30: 243-246.
[83] Nishida, Y.; Nakamishi, K.; Nabe, K. and Tosa, T. (1987) Continuous production of L-phenylalanine from acetomidocinnamic acid using co-immobilised cells of *Corynebacterium sp.* and *Paracoccus denitrificans*. Enzyme Microbiol. Technol. 9:479-483.
[84] Sirirote, P.; Yamane, T. and Shimizu, S. (1988) L-serine production from methanol and glycine with an immobilised Methyltroph. J. Ferment. Technol. 66: 291-297.
[85] Bang, W.G.; Behrendt, U.; Lang, S. and Wagner, F. (1983) Continuous production of L-tryptophan from indole and L-serine by immobilised *Escherichia coli* Cells. Biotechnol. Bioeng. 25: 1013-1025.
[86] Decottignies-Le, M.P.; Calderon-Seguin, R.; Vandecasteele, J.P. and Azeard, R. (1979) Synthesis of L-tryptophan by immobilised *Escherichia coli* Cells. Eur. J. Appl. Microbiol. Biotechnol. 7: 33-44.
[87] Ishiwata, K.I.; Fukuhara, N.; Shimada, M. Makiguchi, N. and Soda, K. (1990) Enzymatic production of L-tryptophan from DL-serine and indole by a coupled reaction of tryptophan synthase and amino acid racemase. Biotechnol. Appl. Biochem. 12: 141-149.

IMMOBILISED-CELL TECHNOLOGY AND MEAT PROCESSING

LINDA SAUCIER AND CLAUDE P. CHAMPAGNE
Food Research and Development Centre,
Agriculture and Agri Food Canada, 3600 Casavant Blvd West,
Saint-Hyacinthe (Québec) Canada J2S 8E3 – Fax: 450-778-8461 –
Email: saucierl@agr.gc.ca; champagnec@agr.gc.ca

Introduction

Meat is a highly perishable food commodity because it provides a suitable substrate for a wide variety of micro-organisms (bacteria, yeasts, and mould). Moreover, since nutrients and moisture are present in sufficient quantity, pH is close to neutral, and the redox potential at the surface is higher than at the core, meat is a good culture medium. Compounds such as glucose, glycogen and amino acids are readily available for use and can sustain the growth of indigenous flora up to 10^9 cells per cm^2 [1-2]. In whole muscle meat, microbial growth occurs on the surface because edible tissues of healthy animals are either sterile or contain very low microbial populations [3]. In ground meat, micro-organisms are introduced into the mass and grow, but are confined and do not spread as easily as in a liquid matrix.

Meat carcasses get contaminated by organisms from the hide/skin of the animal, gut content, workers' hands, and the slaughter environment at a level of up to 10^4 bacteria/cm^2 after dressing [2]. In meat, the initial microflora is mainly mesophilic and the prevailing micro-organisms will vary with storage conditions (Table 1). Under refrigeration temperature, a psychrotrophic microflora develops and micro-organisms grow in number. Under aerobic conditions, *Pseudomonas* spp. prevail and constitute 50 to 90% of the overall microbial population because of a better growth rate than the other micro-organisms. *Enterobacteriaceae* prevail under poor refrigeration (10°C) conditions and spoil the meat [2, 4-5]. When the pH is higher than 5.5, the growth of *Enterobacteriaceae*, *Brochothrix thermosphacta* and the psychrotrophic pathogen *Yersinia enterocolitica* is favoured, such as in fat tissue and in meat from animals that were stressed before they were slaughtered. In the latter case, post-mortem production of lactic acid is reduced, because the glycogen reserve is depleted [2, 6].

Many pathogenic micro-organisms (i.e., essentially bacteria) can be found and can grow on meat and meat products. In North America, roughly one-third of the food-borne outbreaks has been linked to the consumption of contaminated meats. The pathogenic strains that present the greatest risk with respect to meat- and poultry-borne diseases are *Salmonella* spp., *Campylobacter* spp., verotoxigenic *Escherichia coli*, *Listeria monocytogenes*, and *Toxoplasma gondii* [7].

Intrinsic factors (occurring naturally in food; *e.g.*, nature of constituents, pH and buffering capacity, redox potential, etc.) and extrinsic factors (applied to food systems; *e.g.*, temperature, preservatives, modified atmosphere, etc.) influence microbial growth to various extents. The use of these extrinsic factors in meat processing, especially in combination, has increased, in some instances, meat shelf life from a few days to a few months. The use of multiple antimicrobial systems is referred to as the "hurdle technology." Heat was soon recognized as an effective means of improving shelf life. Meat and fish preserved better when they are cooked, because the microbial population is reduced through the killing effect of heat. Other means used in meat preservation include cold temperature, modified atmosphere packaging, and chemical preservatives (*e.g.*, nitrite).

1. Historical use of meat fermentation

Fermentation has been used as a form of food preservation since biblical times. It is now known that the antimicrobial activity of the fermenting microflora is responsible for this effect, which includes the production of inhibitory substances, such as organic acids, CO_2 gas, and antimicrobial peptides (*i.e.*, bacteriocins; [8-9]). Hence, undesirable micro-organisms are controlled by creating an environment that is unfavourable to their survival. The ultimate efficacy of microbial control and quality in fermented meat hinges on a combination of antimicrobial hurdles (*e.g.*, low pH, low a_w, salt and nitrite concentration, etc).

Before the development of specialised equipments to control the fermentation environment, climatic conditions were important for proper fermentation and drying in order to secure the production of safe and palatable products. Although fermented meat has been produced in many countries around the world, the mountainous regions of Spain and Italy were particularly well suited because of their low relative humidity. In Hungary, winter provided the proper conditions for meat fermentation and drying [10].

Proper fermentation conditions are critical for lactic acid bacteria (LAB) to prevail over the pseudomonads and the *Enterobacteriacea* population, and to reduce the pH by the formation of lactic acid from glucose in the meat or added to it. The first attempts to improve the development of the desirable LAB population consisted in the inoculation, up to 5%, of the fresh batter with a previous one that had led to a product of desirable quality, a process known as "back-slopping". However, a sound food-fermentation industry cannot rely on such an opportunistic process. The studies on meat microbiology and the development of commercial starter cultures, in the beginning of the 1900s, secured a sustainable industry by assuring product quality and process reliability [10].

After fermentation, the product can be dried to different degrees, or not, as it is the case in spreadable German products such as Teewurst, Braunschweig and Mettwurst. Smoking and, to a lesser extent, cooking (*e.g.*, summer sausage) are used as adjunct treatments to enhance organoleptic properties, shelf life and safety [10]. However, North American regulations stipulate that for a meat product to be shelf stable, it must comply with one of the following conditions: a) a pH equal or smaller than 4.6, or b) an aw equal or smaller than 0.85, or c) a combination of pH equal or smaller than 5.3, a_w equal or smaller than 0.90 and equal or greater than 100 ppm of nitrite, or d) a scientific demonstration of product stability [11-12]. Products that fail to meet these requirements must be refrigerated.

Food-borne outbreaks associated with fermented meat are mostly linked to salmonellosis, *Staphylococcus aureus* toxicosis and, more recently, to enterohaemorrhagic *Escherichia coli* strains (*i.e.,* O157:H7 serotype). Although the prevalence and risk of *Trichinella spirallis* infection are low, if not inexistent, in many countries (*e.g.,* Canada), precaution and surveillance are still in effect due to their presence in rodents and in other wild animals, which may lead to sporadic infections. Before fermentation, conditions such as high water activity, high pH, and low level of LAB promote the development of *Enterobacteriacea*, including *Salmonella*. However, the presence of salt, a proper fermentation by the starter cultures, and the drying process limit their number. A traditional fermentation/drying process can provide a reduction of *E. coli* O157:H7 of 1 to 2 log unit [13-14]. Following the recent outbreaks of *E. coli* O157:H7 in the US (1994), Australia (1995), and Canada (1998, 1999), the US and Canadian government inspection bodies now require that manufacturers demonstrate that their process assures the control of *E. coli* O157:H7, and *Salmonella*, since it may also be found under conditions where *E. coli* O157:H7 is not controlled [12, 14].

Low levels of *S. aureus* can be found in meat, but it is at high concentrations (*i.e.,* >10^7 CFU/g) that the toxin is present in sufficient quantity to pose a health risk [15]. The organism is not affected by the presence of salt and nitrite, but it does not compete well with other organisms, including the starter cultures, especially under anaerobic conditions. Several conditions found in fermented meat products provide an effective barrier against *S. aureus*. At an a_w below 0.92, toxin production is not observed. Once a pH of 5.3 is reached, growth and toxin production are stopped and a temperature equal to or higher than 15.6°C is necessary for bacterial growth and toxin production. The relation between the fermentation temperature and the time spent to reach a pH of 5.3 is successfully used to control *S. aureus* and to monitor the safety of the fermentation process [12, 15].

2. New approach to meat preservation

Even now, new means of meat preservation are investigated. Improvements of product shelf life have an important economic impact on the agri-food sector by reducing losses attributed to spoilage. Because of market globalisation, food commodities may travel great distances for a long period in order to reach new markets. This means that when problems occur, they can affect many people in many areas, which makes them more difficult to confine. Consumer demand plays a major role in the way our food supply evolves and it is currently driven towards foods that are minimally processed, "natural" and free of additives, but still safe and convenient to use [16].

Micro-organisms are omnipresent. The ones found in food are part of an ecological system that is modulated by storage conditions. The type and number of micro-organisms prevailing will vary depending on weather the conditions are favourable or not. The judicious use of these conditions ensures a better control of spoilage and pathogenic organisms, a process referred to as "microflora management" which means to have the appropriate type and quantity of micro-organisms at a specific time during storage [17]. Antimicrobial systems that reduce the total number of micro-organisms (*e.g.,* heating, irradiation, high pressure, etc.) will also reduce pathogen-related risks. However, they also create an environment devoid of competition for any micro-organisms contaminating the meat post-treatment. When these contaminants are pathogenic organisms, the consequences on product safety are serious. This concept of microflora management is a novel approach used to improve the quality and safety of food. The success of this type of management in extending the shelf life of meats is evinced by the effectiveness of the modified atmosphere packaging (MAP) technology.

A lactic microflora develops in meat placed in an anaerobic environment because of a greater tolerance to carbon dioxide (CO_2) than the pseudomonads and the Enterobacteriacea. CO_2 affects microbial growth by extending the lag phase and increasing the generation time. In vacuum packaged meat, the residual O_2 is used up through muscle and microbial respiration and a maximum concentration of 30% CO_2 can be reached. A minimum of 20% CO_2 is required to inhibit microbial growth [18]. The reason meat exhibits an extended storage life under anaerobic conditions is because the LAB that grow on it cause spoilage only when maximum populations are reached, whereas the aerobic spoilage bacteria cause putrefactive odours earlier in the growth cycle. The defect caused by LAB is described as "souring", which is less offensive than the putrefaction that develops aerobically [2]. The prevalence of LAB exerts an inhibitory pressure on several food-borne pathogens and improves the safety of the product [19], but this ecological equilibrium is very precarious (Nattress, personal communication).

Bacterial preparations used as protective cultures have shown promise in extending the shelf life of meat by inhibiting undesirable micro-organisms without changing its sensory properties. Table 2 presents the current protective cultures for meat products that are available commercially.

Table 2. Protective cultures commercially available for meat products[a].

Commercial name	Strains	Utilisation
Bactoferm™ B-2	*Lactobacillus sakei*	Vacuum and modified atmosphere packaged meat products
Bactoferm™ B-FM	*Staphylococcus xylosus* *Lactobacillus sakei*	Vacuum and modified atmosphere packaged fresh meat products (*e.g.,* fresh sausages)
Bactoferm™ B-SF-43	*Leuconostoc carnosum*	*Listeria* controlling culture for vacuum and modified atmosphere packaged meat products

[a] Information kindly provided by Dr Lone Andersen at Chr. Hansen, Denmark.

Commercial single-strain culture of *Lactobacillus sakei* (Bactoferm-B2), formerly *Lactobacillus alimentarius* BJ-33 (FloraCarn L-2), has been successfully used for meat preservation. *L. sakei* was tested in a wide variety of meat products, including ham, ground beef, fresh sausages and frankfurters, and the culture was able to suppress various undesirable bacteria [20-24]. However, the use of *L. sakei* as a protective culture is limited by its sensitivity to heat [20]. According to the manufacturer, the culture can be added directly to fresh uncooked products, such as fresh sausages or bacon, or by dipping or spraying a solution of *L. sakei* after heat treatment on cooked meat products. However, post-processing addition of the culture is not possible for a product cooked and sold in its original casing.

3. Probiotic cultures and health

The use of probiotic cultures in meat fermentation has been proposed by several investigators [25-27]. LAB fermentation was originally carried out to preserve foods, by acidification, or to develop specific flavours or textures. However, there is increasing evidence that the consumption of LAB may affect the composition of indigenous gut microflora and have several beneficial effects on human health. In this respect, many studies have shown the potential role of LAB in improving lactose assimilation, food digestibility, and immune response, or in the prevention of intestinal infections, vaginal infections, hypercholesterolaemia, cancer, food allergies, and constipation [28-31]. However, it is important to point out that any postulated benefits from consumption of "probiotics" (*i.e.,* food cultures that have these types of beneficial effects on human health) can only be substantiated after extensive human clinical studies. To be more specific, probiotic cultures were originally defined as "a live microbial supplement which beneficially affects the host (animal) by improving its intestinal microbial balance" [32]. However, it has recently been suggested to use the following definition: "defined, live micro-organisms which, when administered in adequate amounts, confer a beneficial physiological effect on the host." In summary, probiotics are cultures that are added to foods for their health-promoting properties not for technological purposes (acidification, flavour, colour, preservation). Foods that contain probiotics could thus become "functional foods."

Although this review focuses on the usefulness of adding probiotics in the processing of meats, it is worth mentioning that adding cultures to feeds may help reduce the faecal counts of unwanted micro-organisms in farm animals. One example is the 3.4 to 5.3 reduction in faecal content of *Salmonella* in broiler chicks following the administration of alginate-entrapped cultures [33].

4. Immobilisation/encapsulation and starter production

Encapsulation technologies are widely applied to ingredients used for the production of meat products [34]. Examples include encapsulation of the meat-tenderising agent bromelain in liposomes [35], encapsulated organic acids for meat sticks [11, 36] encapsulated salts [37], encapsulated bacteriocins such as nisin and pediocin [38-42] and encapsulated cured-meat pigments [43-44]. Therefore, encapsulation technologies are applied to various ingredients in meat processing and this section will examine its potential use in starter cultures.

Cultures added to meat batters are primarily selected for their acidification properties, their effects on flavour and texture as well as their ability to improve product shelf life and safety. Such cultures will be referred to as "technological cultures."

Another group of cultures, the probiotics, could be used to confer certain health-promoting properties. The two types of cultures will be discussed separately.

4.1. TECHNOLOGICAL CULTURES FOR MEAT FERMENTATION

The main technological cultures used in meat fermentation are presented in Table 3. To our knowledge, all commercial starters are prepared as free-cell cultures and are not currently available in an encapsulated form. Consequently, the starter industry has not yet taken up the advantages of ICT in cultures. However, many studies highlight the benefits of immobilising and encapsulating LAB [45-46].

Table 3. Main technological cultures used in meat fermentations [47].

Group	Genera	Species
Bacteria	Lactobacillus	alimentarius, bavaricus, carnis, curvatus, delbrueckii subsp. lactis, pentosus, sakei
	Pediococcus	acidilactici, cerevisiae, pentosaceus
	Micrococcus	caseolyticus, varians
	Staphylococcus	carnosus, simulans, xylosus
	Streptomyces	griseus
Yeast	Debaryomyces	hansenii
Molds	Penicillium	candidum, nalgiovense

The commercialisation of LAB starter cultures requires some preservation treatment in order for cells to retain their viability and fermentative activity during transport and storage. Two forms of meat starters are thus commercially available: frozen and freeze-dried.

Bacterial survival during freezing or frozen storage is dependent on interacting variables such as species/strain, growth conditions, culture age, and the medium surrounding the cells during freezing and thawing [48]. The freezing/thawing process exposes cells to dynamic environmental conditions. There are numerous studies on the effect of freezing on LAB designed for meat fermentations [49-51], but there are none on the effect of ICT and on the survival of these particular cultures. Data from dairy cultures [52] show the potential benefits of ICT on survival to freezing, but this remains to be confirmed for meat-designed starters.

Immobilising the bacteria in a polymer gel may provide protection during the freeze-drying process and a means of controlling the environment to which cells are exposed during their rehydration after direct inoculation into the fermentation substrate. However, the beneficial effect of immobilisation depends on the protective medium. Immobilised cells were less resistant to freeze-drying than free cells when milk was the only protective ingredient, but the opposite was found when glycerol or adonitol was added to milk (Table 4). Higher survival to freeze-drying in immobilised cells seems partially related to lower damage to cell wall components (Table 4). Possible explanations could be a benefit arising from the mechanical support provided by the gel matrix and enhanced repair to stressed or damaged cells by preventing loss or dissipation of essential components required for cell wall repair [53]. In addition to medium composition, the moment of immobilisation strongly influences the survival of immobilised cells. Cultures that are freeze-dried immediately after centrifugation and immobilisation sometimes show higher mortality levels than free cells [54]. These data show the potential of ICT to increase LAB survival upon freeze-drying, but also points out the limits of the technique. One study reports similar survival rates (38%) of free and alginate-immobilised *L. sakei* [55]. Unfortunately, the data on *L. pantarum* and *L. sakei* are the only ones found in the literature on the effect of immobilisation on the survival of meat starters during freeze-drying, and more studies are needed to extend the technology to other cultures.

Table 4. Effect of immobilisation in alginate gel on cellular damages and survival to freeze-drying of *Lactobacillus plantarum* [45, 56].

Medium	Survival (%)		% cells with wall damage	
	Free	Immobilised	Free	Immobilised
Milk	8.8	1.1	-	-
Milk + glycerol	36.1	44.7	38	14
Milk + adonitol	64.6	85.9	41	30

Cultures need to be stable while in transit as well as during storage. At 30°C, a standard freeze-dried *Pediococcus acidilactici* probiotic culture loses 85% of its viability over 4 months of storage, while the micro-encapsulated products only show a 31% drop in viability under the same conditions [57]. The encapsulation procedure is carried out by coating the particles of a typical freeze-dried culture with a lipid-based compound. There are unfortunately no reports on the acidifying activity of encapsulated cultures during dry sausage manufacture. At first glance, lipid-encapsulated starters would seem better adapted for the additions of probiotics rather than technological strains, because the lipid coating could delay the release of the cells in the meat batter.

In view of the potential of alginate-based ICT for meat starters, why has the starter-producing industry not adopted ICT yet? One reason is the presence of bead particles in the fermented products. This would go unnoticed in most dry sausages, but sensory aspects must not be underestimated. Furthermore, alginate can affect the texture of meats, albeit sometimes beneficially [58]. The use of ICT in cultures could thus potentially modify both the appearance and texture of a product. A second reason is the need to add a technological step of immobilisation or encapsulation to the manufacturing process, which complicates the process and increases costs. Evidently, the potential benefits of ICT on starter survival to freezing or drying as well as for acidifying activity do not yet sufficiently offset the increased production costs. In order to avoid the necessity of a separate immobilisation step in the starter production process, an attempt was made to combine the inoculation of a 1500-L fermentor and the immobilisation process [59]. Other systems that could also be considered in such a fashion would be the "multiple syringe" units of Innova Nisco (Switzerland) as well as the Jet-cutting method [60]. In such processes, the beads could be formed inside the fermentor and the cultures could be grown inside the beads; this would eliminate the need for a concentration step, such as centrifugation or filtration, and reduce costs. Consequently, the added step of immobilisation in the starter production process can be compensated by the removal of the concentration step. A diagram of such a process is presented in Figure 1. However, overall biomass yields are lower in bead-grown cultures and a fraction of the cells is released from the beads into the medium [61]. These problems will need to be resolved if ICT is to be made more attractive to meat starter producers. This being said, the cell-containing beads could be air-dried instead of freeze-dried (Figure 1). Air-drying is much less expensive than freeze-drying. *Lactobacillus plantarum* cultures can be successfully air-dried and post-drying residual acidifying activities between 70 to 85% are reported [62]. Immobilisation on potato starch enables the use of fluid bed drying [63], which is a rapid and low-cost method used industrially for yeast. Therefore, ICT could offer novel and ultimately less expensive means of producing meat LAB starters. Some work has been done on air-drying of LAB that can be used in meat fermentations and this warrants further examination. Some additional aspects of the potential advantages of ICT in cultures will be discussed further in the section on "Meat fermentation using immobilised cells."

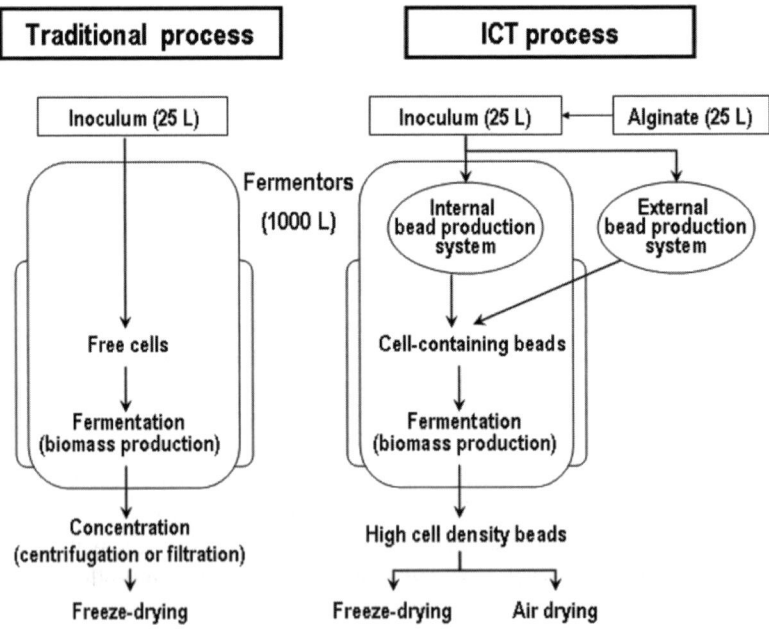

Figure 1. Production of dried meat starters by the traditional process and an immobilised cell technology process.

4.2. PROBIOTIC CULTURES FOR MEATS

One way of developing probiotic-containing meats is to carry out the fermentation with the probiotic strains. *L. gasseri,* of the *L. acidophilus* group, achieves the best fermentation performance in sausages. In such strategies, high fermentation temperatures are required since the probiotics grow well at 37°C. Another approach is to add the desired population of probiotic cultures directly to the product without expecting the probiotic strain to contribute to acidification. In such cases, the fermentation can be carried out below 30°C.

When added to fermented foods, many probiotic cultures die during storage, particularly if the products are acidic [64], as in fermented sausages. LAB can be grown to be more resistant to acid environments [65], but micro-encapsulation seems to be the best method. Probiotic cultures can be micro-encapsulated with lipid compounds to provide protection against oxygen, moisture, and acidity [57]. Unfortunately, no data is available on the stability of such probiotics in fermented sausages.

It should be mentioned that some probiotic cultures have shown inhibitory activities against *S. aureus* [66]. Therefore, immobilised probiotics could potentially be used for their health-promoting and preservation properties.

Although no data on the stability of ICT probiotics in sausages is available, there is data with respect to storage of probiotics at the plant prior to inoculation in foods. As mentioned previously, micro-encapsulated *Pediococcus acidilactici* probiotic cultures exhibit greater stability during storage at 30°C [57] than the traditional freeze-dried product. Such products are available in the market and should be considered in attempts to develop novel probiotic-containing meats.

5. Applications of ICT cultures in meat

5.1. MEAT FERMENTATION USING IMMOBILISED CELLS

Contrary to the dairy sector, there is little research on ICT in meat fermentation. Nonetheless, an earlier review outlining the major factors influencing meat fermentation using ICT has been published [45]. Traditionally, the meat batter is composed of lean meat and fat, salt, spices, nitrate and/or nitrite, ascorbic acid or other inhibitors of nitrosamine formation, phosphate and a carbon source such as sugar [67]. The composition of the batter, the fermentation conditions, and the use of adjunct processes (*e.g.,* smoking, cooking) will vary according to the final shelf life and organoleptic properties desired. Because the meat is not sterilised before inoculation, its microbial quality and the manufacturing conditions must be controlled in order to ensure the prevalence of LAB during the fermentation process and the storage period before consumption and to assure safety of the final product. The solid/semisolid nature of meat restricts the movement, or mass transfer, of both cells and nutrients creating a "natural immobilisation" phenomenon [45]. This is also the main reason the reduction of starter culture efficacy from bacteriophage attack is not a major problem in meat fermentation, compared to fermented liquid [68-69], although bacteriophages of meat starter cultures, such as *L. plantarum* and *Staphylococcus carnosus,* have been isolated

[70-71]. When nutrients are entrapped in the immobilisation matrix (*e.g.*, alginate beads) with the cells, improvements in the growth of the starter culture can be achieved, which leads to a higher fermentation rate and provides the encapsulated cells with a competitive advantage [45,72]. Furthermore, meat is not particularly rich in fermentable sugar and different sources (*e.g.*, white or brown sugar, maple syrup, honey) are often added to increase its content in order to improve the fermentation rate and acid production [73].

Kearney *et al.* [74] compared the fermentation of meat with free and immobilised cultures of *L. plantarum* or *Pediococcus pentosaceus*. A batter constituted of beef and pork meat (1.5:1.0 w/w ratio), sodium nitrite (0.0156%), salt (3%) and dextrose (1%) was fermented with either the commercial free-cell preparation or the immobilised cells. The two starter cultures showed similar fermentation patterns. On the one hand, cell-count differences between free and immobilised cells remained well below a log unit difference during the whole fermentation process, and the final level of lactic acid was equivalent in both cases. On the other hand, fermentation rate was much faster with the encapsulated cells, where pH 5.0 was reached sooner. The authors suggest that the greater fermentation performance of the immobilised cells was caused by available nutrients (*i.e.*, skim milk) and more protective rehydration conditions with the beads, although cell counts were equivalent in both groups. A faster fermentation rate may indeed suggest a greater efficacy in controlling the growth of pathogens, especially *S. aureus*, but the demonstration that meat fermentation using ICT provides an equivalent level of safety, compared to a conventional free-cells system, remains to be determined. The same is true for organoleptic properties. It was suggested that the heterogeneous nature of fermentation meats would not be affected by the presence of ground lyophilised alginate beads, but, to our knowledge, no sensory-evaluation studies have been published thus far [45].

Contrary to liquid fermentation, starter cultures encapsulated in alginate cannot be recuperated for continuous cell utilization in fermented sausage production. Other advantages, however, still make the technology attractive and justify the extra cost. These advantages include stability of the starter culture (*e.g.*, reduction of cell damage after lyophilisation, improvement of plasmid retention, reduction of bacteriophage attack), protection against inhibitory compounds and conditions found in the medium, improvement of acid production, protection against shear forces in the cutter/mixer, increased competitiveness (ecological competence) of the starter cultures by a better control of their environment, *etc.* [45-46].

Given that fermentation is a time-consuming process, direct acidification, or chemical acidulation, using organic acids can be used to reduce the pH of the meat batter [11]. However, the acid should be encapsulated in a lipid matrix to control its release, otherwise protein denaturation will lead to crumbly preparations as a result of the water being released from the meat. Each encapsulation matrix is designed to melt at a specific temperature so as to avoid rapid pH drop before a proper gelification of meat proteins occurs [3]. Several encapsulated acids, including lactic, citric and glucono-δ-lactone, are available commercially for that purpose. The application of organic acids immobilised in a calcium alginate gel has also been studied by Siragusa and Dickson [75-76] for the control of various pathogens on lean beef tissue.

5.2. IMPROVING THE USE OF PROTECTIVE CULTURES *VIA* ICT

The protective cultures available commercially are non-spore formers such as lactic acid bacteria and *Staphylococcus* strains (Table 2), which are therefore not particularly resistant to heat. To overcome the lack of heat resistance in protective cultures, the product was initially inoculated at high levels prior to heat treatment. In ham and frankfurters cooked at a core temperature of 72°C, single-strain protective cultures have been used at an inoculation level of 10 and 7 $\log_{10}$ cfu/g, respectively [21,77]. In an acidified chicken meat model cooked at a core temperature of 55°C, an initial inoculum of 5 log cfu/g was used to obtain a final cell concentration of 4 log cfu/g after heat treatment [78]. This procedure is not very practical as it is costly. Furthermore, if lysis of the inactivated portion of the inoculum occurs, the endogenous enzymes released in the meat medium could have an impact on the sensory properties of the product by reacting with meat constituents (*i.e.,* lipids and proteins). Hence, for cooked meat applications, means of shielding the protective culture need to be developed in order for them to survive the heat treatment when added directly into the raw batter.

At the Food Research and Development Centre, we investigated the possibility of immobilising cells in lyophilised alginate beads supplemented with glycerol as a means to protect *L. sakei* against heat and other antimicrobial barriers encountered in processed meat. The protective effect of immobilised freeze-dried cells (IFDC) technology was studied both in a MRS broth culture and in an acidified chicken meat model (pH 5.0) cooked at a core temperature of 55°C [55]. To our knowledge, this is the first time ICT was investigated as a means to improve the resistance of protective culture against heat treatment and other antimicrobial barriers in meat.

The protective culture used in our experiments was *L. sakei* (Bactoferm-2; Chr. Hansen, Horshom, Denmark) supplied by the manufacturer in freeze-dried form with glucose as a carrier. The lyophilised culture was added directly into the alginate solution containing 1.8 M glycerol. The alginate beads were lyophilised and then grinded into small particles. This IFDC was added directly into the liquid medium or into the meat batter. Results indicate that IFDC were more resistant to a heat treatment of 55°C for 15 min than the free-cells, in a MRS broth acidified to pH 5.0 and 5.7. However, no significant differences were observed when the medium was not acidified (pH 6.7). Hence, ICT could potentially improve the resistance of microbial cells against inhibitory conditions found in processed meat. The amount of available water in food during heating has been shown to have a major influence in heat resistance of micro-organisms as it increases with decreasing water activity, moisture, or humidity [79-81]. Furthermore, the environment in which lyophilised cells are rehydrated influences the level of survival and activity [82-83]. When the lyophilised cultures (free-cells) are directly inoculated in liquid media, the rehydration is usually completed in a suboptimal environment (*e.g.,* low pH) within a few seconds, so that the cells are quickly subjected to changes (*e.g.,* osmotic or acidic shock) occurring in the liquid medium. Furthermore, other investigators have suggested that when cell rehydration is performed in the controlled environment of the alginate beads, with a protective agent incorporated into the beads (i.e., glycerol or adonitol with skim milk), the survival and activity retention is increased [56]. Although the alginate beads are expected to rehydrate slowly, and therefore maintain a lower water content for a longer time, assays with rehydration times of up to 120 min prior to heating did not affect the survival of the encapsulated culture in broth culture. This suggests that chemical conditions (*e.g.,* pH, glycerol, etc), rather than the variation in moisture content of the beads microenvironment, are responsible for the observed beneficial effect. In fact, glycerol has been reported to play a role in overcoming the harmful effects of low pH levels in freeze-dried cultures [84].

Unexpectedly, when a similar experiment was carried out in the acidified chicken meat model, the improved resistance conferred by IFDC was not observed. The dynamics in a meat system is very different from that of a liquid culture. In meat, the culture is expected to rehydrate slowly and, in this case, the diffusion of the nutrients into the beads, especially those of large molecular weights, needed to repair damage and resume growth might be a limiting factor. In fact, Kearney and his collaborators [74] found that the incorporation of nutrients, such as skim milk, into the alginate bead could increase cell viability. Moreover, Bashan [72] reported a higher bacterial numbers when skim milk is incorporated into the beads. Migration of these nutrients is probably more difficult in beads than in free-cells in a meat matrix and inclusion of those nutrients into the beads should be investigated further with respect to the use of IFDC in meat systems. Furthermore, in cooked meat products, the medium in which micro-organisms are heated has a marked effect on their resistance to heat treatment. Certain meat constituents and conditions such as salt percentage, pH, polysaccharide concentration, curing agent use, and water and fat content, have been reported to potentially influence heat resistance [85]. For instance, many authors suggest that spores and vegetative cells surrounded by fatty materials are more heat resistant due to the lower water activity in fat [86-88]. Investigations with respect to the effect of the chicken meat model constituents indicate that the wheat flour and the tapioca starch binders have some protective effect, but the meat itself also substantially improve heat resistance in free cells. Indeed, increasing fat, glucose, or starch content in a medium is known to increase heat resistance or recovery of bacterial cells [88-90].

A pure culture of *L. sakei* was isolated from the commercial culture source and was stored at -80°C in a MRS broth supplemented with 20% glycerol. When this culture was subcultured twice and then lyophilised in our facilities (hence a process and sources of cryoprotectant most probably different from the ones used by the manufacturer), cells were more resistant to heat treatment than those of commercial preparations. These results indicate that production parameters for both cells and beads production should be optimised for the application of protective cultures in meat. Indeed, the amount of water remaining after drying as well as the various protective substances (*e.g.,* skim milk, sucrose, trehalose, glycerol, adonitol) used in the suspending medium have been shown to influence the viability of bacterial cells, and warrants further investigation [83,91-94].

5.3. OTHER APPLICATIONS OF ICT IN MEAT

Another area of work to consider, with respect to muscle food, is protein solubilisation for different purposes including waste management, supplement for animal feed, production of functional ingredients with properties such as water-holding capacity, emulsion, foaming, etc. [95]. Protein hydrolysates can be produced by enzymes isolated from plants (*e.g.,* papain, fucin), animals (*e.g.,* trypsin, pancreatin) or micro-organisms (*e.g.*, pronase, alcalase; [95]). Microbial enzymes have several advantages compared to plant or animal enzymes. They are cheaper to produce and mass production of microbes and enzyme isolation are easier. On the one hand, the proteolytic enzyme can be used directly or immobilised to different substrate such as polyethylene terephthalate, glass, and chitosan [95]. On the other hand, the use of whole microbial cells does not require enzyme extraction and purification. These savings are particularly important when the end product has little commercial value (*e.g.*, waste treatment). Furthermore, enhanced enzymatic activity by entrapped cells compared to free cells has been reported [96].

Few studies have been carried out on the solubilisation of muscle food using ICT, but some work has been done with fish mince meat by Venugopal and his collaborators [97]. In this study, mechanically deboned fish meat was used after partial deodorisation by boiling in water for 15 min at low pH. Odour-bearing compounds, extracted in water, were removed using a screw press resulting in mince meat with a content of 50% moisture and 1% lipid. Fish meat was hydrolysed in a 8% (w/v) water based meat slurry (50°C, pH 7-9) by the secreted protease of *Bacillus magaterium*, *Aeromonas hydrophila*, or *Pseudomonas marinoglutinosa* cells immobilised in calcium alginate beads containing 10^{11} cells/g. A reactor especially designed to recover the beads from the meat slurry was used. Their results indicate that proteins were best hydrolysed by *B. magaterium* followed by *A. hydrophila*, *P. marinoglutinosa* being the least effective. A ratio of 4 g of meat for 3 g of wet beads was optimum for protein solubilisation. Beads retained their proteolytic activity when stored at 4°C for up to 30 days and could be used for two cycles. Beads disintegrated when the pH exceeded 8.5. A continuous system where the supernatant was withdrawn at a rate of 10 ml/h gave greater solubilisation compared to a stationary system without withdrawal, suggesting that end-product inhibition occurred. To our knowledge, protein solubilisation by ICT cells is not currently used industrially.

6. Conclusion

So far, three main applications of ICT have been identified for meat or muscle food systems: fermentation, shielding of protective and probiotic cultures, and solubilisation of constituents, especially proteins. The main advantages of immobilised cultures are increased fermentation rates, when rehydrated in meat batters, as well as preventing death of the cultures during processing steps or during storage. However, much remains to be done to optimise parameters for both cells and beads preparation in order for the technology to be used to its full potential in meat. With respect to probiotics, immobilisation and encapsulation are increasingly being used to improve the health-promoting properties of selected cultures in the development of functional foods. Their use in cholesterol-lowering activity for example, will eventually find application. In this event, ICT will answer some of the technological problems associated with the use of probiotics in foods. Understanding bacterial cell physiology under conditions such as freezing, drying, thawing, rehydration is mandatory since they have a considerable impact on the final viability and functionality of the cells. Benefits such as better product quality, process efficacy, improved safety and additional health benefits will eventually outweigh the cost associated with the use of ICT.

References

[1] Greer, G.G. (1989) Red meats, poultry and fish. In: McKellar, R.C. (Ed.) Enzymes of psychrotrophs in raw food. CRC Press, Inc., Boca Raton, FL; pp 267-292.
[2] Dainty, R.H. and Mackey, B.M. (1992) The relationship between the phenotypic properties of bacteria from chill-stored meat and spoilage processes. J. Appl. Bacteriol. 73: 103S-114S.
[3] Urbain, W.M. and Campbell, J.F. (1987) Meat preservation. In: Price, J.F. and Schweigert, B.S. (Eds.) The science of meat and meat products, 3rd edition. Food & Nutrition Press, inc. Westport, CT, USA; pp. 371-412.
[4] Gill, C.O. and Newton, K.G. (1978) The ecology of bacterial spoilage of fresh meat at chill temperatures. Meat Sci. 2: 207-217.
[5] Stiles, M.E. (1991) Modified atmosphere packaging of meat, poultry and their products. In: Ooraikul, B. and Stiles, M.E. (Eds.) Modified atmosphere packaging of food. Ellis Horwood Ltd., Chichester, England; pp. 118-147.
[6] Lambert, A.D; Smith, J.P. and Dodds, K.L. (1991) Shelf life extension and microbiological safety of fresh meat – a review. Food Microbiol. 8: 267-297.
[7] Todd, E.C.D. and Harwig, J. (1996) Microbial risk analysis of food in Canada. J. Food Prot. 59 suppl., 10-18.
[8] Ray, B. and Daeschel, M. (1992) Food biopreservatives of microbiological origin. CRC Press Inc., Boca Raton, FL.
[9] Roller, S. (1995) The quest for natural antimicrobials as novel means of food preservation: status report on a European research project. Int. Biodeterior. Biodegrad. 36: 333-345.
[10] Zeuthen, P. (1995) Historical aspects of meat fermentations. In: Campbell-Platt, G. and Cook, P.E. (Eds.) Fermented meats, chap. 3. Blackie Academic & Professional, London, UK; pp.53-68.
[11] Quinton, R.D.; Cornforth, D.P.; Hendricks, D.G.; Brennand, C.P. and Su, Y.K. (1997) Acceptability and composition of some acidified meat and vegetable stick products. J. Food Sci. 62: 1250-1254.
[12] Canadian Food Inspection Agency (1999) Fermented meat products. In: Meat hygiene manual, 4, Meat and Processed Animal Products Division, Government of Canada, Ottawa, Canada, pp. 111-114.
[13] Hinkens, J.C.; Faith, N.G.; Lorang, T.D.; Bailey, P.; Buege, D.; Kaspar, C.W. and Luchansky, J.B. (1996) Validations of pepperoni processes for control of Escherichia coli 0157:H7. J. Food Prot. 59: 1260-1266.
[14] Incze, K. (1998) Dry fermented sausages. Meat Sci. 49: S169-S177.
[15] Lücke, F.-K. (1998) Fermented sausages. In: Wood, B.J.B. (Ed.) Microbiology of fermented foods, Vol. 2, Chap. 14. Blackie Academic & Professional, London, UK.
[16] Rhodehamel, E.J. (1992) FDA's concerns with sous vide processing. Food Technol. 46(12): 73-76.
[17] Saucier, L. (1999) Meat safety: Challenges for the future. Outlook Agric. 28: 77-82.
[18] Lambert, A.D.; Smith, J.P. and Dodds, K.L. (1991) Effect of headspace CO2 concentration on toxin production by Clostridium botulinum in MAP, irradiated fresh pork. J. Food Prot. 54, 588-592.
[19] Stiles, M.E. (1994) Potential for biological control of agents of foodborne disease. Food Res. Int. 27: 245-250.
[20] Andersen, L. (1995) Biopreservation with FloraCarn L-2. Fleischwirtschaft 75: 1327-1329.
[21] Kotzekidou, P. and Bloukas, J.G. (1996) Effect of protective cultures and packaging film permeability on shelf-life of sliced vacuum-packed cooked ham. Meat Sci. 42: 333-345.
[22] Andersen, L. (1997) Bioprotective culture for fresh sausage. Fleischwirtschaft 77: 635-637.
[23] Juven, B.J.; Baredoot, S.F.; Pierson, M.D.; McCaskill, L.H. and Smith, B. (1998) Growth and Survival of Listeria monocytogenes in vacuum-package ground beef inoculated with Lactobacillus alimentarius FloraCarn L-2. J. Food Prot. 61: 551-556.
[24] Kotzekidou, P. and Bloukas, J.G. (1998) Microbial and sensory changes in vacuum-packed frankfurter type sausage by Lactobacillus alimentarius and fate of inoculated Salmonella enteritidis. Food Microbiol. 15: 101-111.
[25] Hugas, M. and Monfort, J.M. (1997) Bacterial starter cultures for meat fermentation. Food Chem. 59 (4): 547-554.
[26] Arihara, K.; Ota, H.; Itoh, M.; Kondo, Y.; Sameshima, T.; Yamanaka, H.; Akimoto, M.; Kanai, S. and Miki, T. (1998) Lactobacillus acidophilus group lactic acid bacteria applied to meat fermentation. J. Food Sci. 63 (3): 544-547.

27] Hammes, W.P. and Hertel, C. (1998) New developments in meat starter cultures. Meat Sci. 49 (Suppl. 1): S125-S138.
[28] Kailasapathy, K. and Rybka, S. (1997) L. acidophilus and Bifidobacterium ssp. - their therapeutic potential and survival in yogurt. Aust. J. Dairy Technol. 52: 28-35.
[29] Mattila-Sandholm; Blum, S.; Collins, J.K.; Crittenden, R.; de Vos, W.; Dunne, C.; Fondén, R.; Grenov, G.; Isolauri, E.; Kiely, B.; Marteau, P.; Morelli, L.; Ouwehand, A.; Reniero, R.; Saarela, M.; Salminen, S.; Saxelin, M.; Schiffrin, E.; Shanahan, F.; Vaughan, E. and von Wright, A. (1999) Probiotics: towards demonstrating the efficacy. Trends in Foods Science and Technology 10: 393-399.
[30] De Roos, N.M. and Katan, M.B. (2000) Effects of probiotic bacteria on diarrhea, lipid metabolism, and carcinogenosis: a review of papers published between 1988 and 1998. Am. J. Clin. Nutr. 71: 405-411.
[31] Shah, N.P. (2000) Some beneficial effects of probiotic bacteria. Bioscience Microflora 19: 99-106.
[32] Fuller, R. (1989) Probiotics in man and animals. J. Appl. Bacteriol. 66: 365-378.
[33] Hollister, A.G.; Corrier, D.E.; Nisbet, D.J. and DeLoach, J.R. (1994) Effect of cecal cultures encapsulated in alginate beads or lyophilized in skim milk and dietary lactose on Salmonella colonization in broiler chicks. Poultry Sci. (USA) 73(1): 99-105.
[34] Cahill, S.M.; Uption, M.E. and McLoughlin, A.J. (2001) Bioencapsulation technology in meat preservation. In: Durieux, A. and Simon, J.-P. (Eds.) Focus on Biotechnology, Vol. 2: Applied Microbiology. Kluwer Academic Publishers, Dordrecht, The Netherlands; pp 239-266.
[35] Lee, D.H.; Jin, B.H.; Hwang, Y.I. and Seung-Cheol-Lee, S.C. (2000) Encapsulation of bromelain in liposome. Journal of Food Science and Nutrition 5: 81-85.
[36] Stier, R. (1997) Case study: Slim Jim beef jerky. Food Process. 58(3): 56.
[37] Pszczola, D.E. (1997) Encapsulated salt. Food Technol. 51:80.
[38] Degnan, A.J. and Luchansky, J.B. (1992) Influence of beef tallow and muscle on the antilisterial activity of pediocin AcH and liposome-encapsulated pediocin AcH. J. Food Prot. 55: 552-554.
[39] Degnan, A.J.; Buyong, N. and Luchansky, J.B. (1993) Antilisterial activity of pediocin AcH in model food systems in the presence of an emulsifier or encapsulated within liposomes. Int. J. Food Microbiol. 18: 127-138.
[40] Fang, T.J. and Cyi-CL (1995) Inhibition of Listeria monocytogenes on pork tissue by immobilized nisin. J. Food Drug Anal. 3: 269-274.
[41] Cutter, C.N. and Siragusa, G.R. (1996) Reduction of Brochothrix thermosphacta on beef surfaces following immobilization of nisin in calcium alginate gels. Lett. Appl. Microbiol. 23: 9-12.
[42] Cutter, C.N. and Siragusa, G.R. (1997) Growth of Brochothrix thermosphacta in ground beef following treatments with nisin in calcium alginate gels. Food Microbiol. 14: 425-430.
[43] Shahidi, F. and Pegg, R.B. (1991) Encapsulation of the pre-formed cooked cured-meat pigment. J. Food Sci. 56: 1500-1504.
[44] O'Boyle, A.R.; Aladin Kassam, N.; Rubin, L.J. and Diosady, L.L. (1992) Encapsulated cured-meat pigment and its application in nitrite-free ham. J. Food Sci. 57: 807-812.
[45] McLoughlin, A. and Champagne, C.P. (1994) Immobilized cells in meat fermentations. CRC Crit. Rev. Biotechnol. 14(2): 179-192.
[46] Anonymous (2000) Immobilize cells to facilitate meat fermentation. Emerging Food R&D Report 11: 7-8.
[47] Gélinas, P. and Houde, A. (1998) Catalog of food microbial starters in North-America. Fondation des Gouverneurs CRDA, St-Hyacinthe, Canada, p. 119.
[48] Ray, B. and Speck, M.L. (1973) Freeze injury in bacteria. Crit. Rev. Clin. Lab. Sci. 4: 161-166.
[49] Gehrke, H.H.; Pralle, K. and Deckwer, W.D. (1992) Freeze drying of microorganisms - influence of cooling rate on survival. Food Biotechnol. 6 (1): 35-49.
[50] Coppola, R.; Iorrizzo, M.; Sorrentino, A.; Sorrentino, E. and Grazia, L. (1996) Survival after freezing of mesophilic lactobacilli isolated from fermented meat and sourdough. Industrie Alimentari 35 (347): 349-351, 356.
[51] Peter, G. and Reichart, O. (2001) The effect of growth phase, cryoprotectants and freezing rates on the survival of selected micro-organisms during freezing and thawing. Acta Aliment. 30 (1) 89-97.
[52] Sheu, T.Y.; Marshall, R.T. and Heyman, H. (1993) Improving survival of culture bacteria in frozen desserts by microentrapment. J. Dairy Sci. 76: 1902-1907.
[53] Necas, O. and Svoboda, A. (1985) Cell wall regeneration and protoplast reversion. In: Peberdy, J. F. and Ferenczy, L. (Eds.) Fungal Protoplasts. Marcel Dekker, Basel; p. 115.
[54] Champagne, C.P.; Mondou, F.; Raymond, Y. and Brochu, E. (1996) Effect of immobilization in alginate on the stability of freeze-dried Bifidobacterium longum. Bioscience Microflora 15(1): 9-15.

[55] Lemay, M.-J.; Champagne, C.P.; Gariépy, C. and Saucier, L. (In press) A comparison of the effect of meat formulation on the heat resistance of free or encapsulated culture of Lactobacillus sakei. J. Food Sci.
[56] Kearney, L.; Upton, M. and McLoughlin, A. (1990) Enhancing the viability of Lactobacillus plantarum inoculum by immobilizing the cells in calcium-alginate beads incorporating cryoprotectants. Appl. Environ. Microbiol. 56: 3112-3116.
[57] Siuta-Cruce, P. and Goulet, J. (2001) Improving probiotic survival rates. Food Technol. 55: 36-42.
[58] Ensor, S.A.; Sofos, J.N. and Schmidt, G.R. (1990) Optimization of algin/calcium binder in restructured beef. Journal of Muscle Foods 1: 197-206.
[59] Champagne, C.P.; Blahuta, N.; Brion, F. and Gagnon, C. (2000) A vortex-bowl disk atomizer system for the production of alginate beads in a 1500 L fermenter. Biotechnol. Bioeng. 68: 681-688.
[60] Pruesse, U.; Fox, B.; Kirchhoff, M.; Bruske, F.; Breford, J. and Vorlop, K. D. Biotechnol. Tech. (1998) The Jet Cutting Method as a new immobilization technique. 12: 105-108.
[61] Champagne, C.P.; Girard, F. and Rodrigue, N. (1993) Production of concentrated suspensions of thermophilic lactic acid bacteria in calcium alginate beads. Int. Dairy J. 3 (3): 257-275.
[62] Linders, L.J.M.; Meerdink, G. and Riet, K. van't. (1996) Influence of temperature and drying rate on the dehydration inactivation of Lactobacillus plantarum. Food and Bioproducts Bioprocessing 74: 110-114.
[63] Lievense, L.C.; Verbeek, M.A.M.; Meerdink, G. and Riet, K. van't. (1990) Inactivation of Lactobacillus plantarum during drying. II. Measurement and modelling of the thermal inactivation. Bioseparation 1: 161-170.
[64] Shah, N.P.; Lankaputhra, W.E.V.; Britz, M.L. and Kyle, W.S.A. (1995) Survival of Lactobacillus acidophilus and Bifidobacterium bifidum in commercial yoghurt during refrigerated storage. Int. Dairy J. 5: 515-521.
[65] Hartke, A.; Bouché, S.; Giard, J.-C.; Benachour, A.; Boutibonnes, P. and Auffray, Y. (1996) The lactic acid stress response of Lactococcus lactis subsp. lactis. Curr. Microbiol. 33: 194-199.
[66] Sameshima, T.; Magome, C.; Takeshita, K.; Arihara, K.; Itoh, M. and Kondo, Y. (1998) Effect of intestinal Lactobacillus starter cultures on the behaviour of Staphylococcus aureus in fermented sausages. Int. J. Food Microbiol. 41: 1-7.
[67] Dabin, E. and Jussiaux, R. (1994) Le saucisson sec. Erti, Paris, France.
[68] Nes, I.F. and Sorheim, O. (1984) Effect of infection of a bacteriophage in a starter culture during the production of salami dry sausage: a model study. J. Food Sci. 49: 337-340.
[69] Trevors, K.E.; Holley, R.A. and Kempton, A. G. (1984) Effect of bacteriophage on the activity of lactic starter cultures used in the production of fermented sausages. J. Food Sci. 49: 650-651, 653.
[70] Trevors, K.E.; Holley, R.A. and Kempton, A.G. (1983) Isolation and characterization of a lactobacillus plantarum bacteriophage isolated from a meat starter culture. J. Appl. Bacteriol. 54: 281-288.
[71] Goetz, F.; Popp, F. and Schliefer, K.H. (1984) Isolation and characterization of a virulent bacteriophage from Staphylococcus carnosus. FEMS (Fed. Eur .Microbiol. Soc.) Microbiol. Lett. 23: 303-307.
[72] Bashan, Y. (1986) Alginate beads as synthetic inoculant carriers for slow release of bacteria that affect plant growth. Appl. Environ. Microbiol. 51: 1089-1098.
[73] Townsend, W.E. and Olson, D.G. (1987) Cured meats and cured meat products processing. In: Price, J.F. and Schweigert, B.S. (Eds.) The science of meat and meat products, 3rd edition. Food & Nutrition Press, inc. Westport, CT, USA; pp. 431-456.
[74] Kearney, L.; Upton, M. and McLoughlin, A. (1990) Meat fermentation with immobilized lactic acid bacteria. Appl. Microbiol. Biotechnol. 33: 648-651.
[75] Siragusa, G.R. and Dickson, J.S. (1992) Inhibition of Listeria monocytogenes on beef tissue by application of organic acids immobilized in calcium alginate gel. J. Food Sci. 57: 293-296.
[76] Siragusa, G.R. and Dickson, J.S. (1993) Inhibition of Listeria monocytogenes, Salmonella thyphimurium and Escherichia coli O157:H7 on beef muscle tissue by lactic or acetic acid contained in alginate gels. J. Food Saf. 13:147-158.
[77] Bloukas, J.G.; Paneras, E.D. and Fournitzis, G.C. (1997) Sodium lactate and protective culture effects on quality characteristics and shelf-life of low-fat frankfurters produced with olive oil. Meat Sci. 45: 223-238.
[78] Lemay, M.-J.; Choquette, J.; Delaquis, P.J.; Gariépy, C.; Rodrigue, N. and Saucier, L. (2002) Antimicrobial effect of natural preservatives in a cooked and acidified chicken meat model. Int. J. Food Microbiol. 78: 217-226.
[79] Fox, K. and Eder, B.D. (1969) Comparison of survivor curves of Bacillus subtilis spores subjected to wet and dry heat. J. Food Sci. 34: 518-520.

[80] Archer, J.; Jervis, E.T.; Bird, J. and Gaze, J. E. (1998) Heat resistance of Salmonella weltevreden in low moisture environment. J. Food Prot. 61: 969-973.
[81] Jay, M.J. (2000) High temperature food preservation and characteristics of thermophilic microorganisms. In: Jay, M.J. (Ed) Modern food microbiology. 6th ed., Gaithersburg: Aspen publishers Inc.; pp 341-362.
[82] Sinha, R.N.; Shukla, A.K.; Madan, L.A.L. and Rangaanathan, B. (1982) Rehydratation of freeze-dried cultures of lactic streptococci. J. Food Sci. 47: 668-679.
[83] de Valdéz, G.F.; Giori, G.S.; Ruiz Holgado, A.P. and Oliver, G. (1985) Effect of the rehydration medium on the recovery of freeze-dried lactic acid bacteria. Appl. Environ. Microbiol. 50: 1339-1341.
[84] Lamprech, E.D. and Foster, E.M. (1963) The survival of starters organisms in concentrated suspensions. J. Appl. Bacteriol. 26: 359-369.
[85] Stumbo, C.R. (1973) Thermal resistance of bacteria. In: Stumbo, C.R. (Ed.) Thermobacteriology in food processing. 2nd ed. Academic press, London; pp 93-120.
[86] Senhaji, A.F and Loncin, M. (1977) The protective effect of fat on heat resistance of bacteria (I). J. Food Technol. 12: 203-216.
[87] Gaze, J.E. (1985) The effect of oil on the heat resistance of Staphylococcus aureus. Food Microbiol. 2: 277-283.
[88] Ahmed, N.M.; Conner, D.E. and Huffman, D.L. (1995) Heat resistance of Escherichia coli O157: H7 in meat and poultry as affected by product composition. J. Food Sci. 60: 606-610.
[89] Gonzalez, I.; Lopez, M.; Mazas, M.; Gonzalez, J. and Bernardo, A. (1997) Thermal resistance of Bacillus cereus spores as affected by additives in the recovery medium. J. Food Saf. 17: 1-12.
[90] Annous, B.A. and Kozempel, M.F. (1998) Influence of growth medium on thermal resistance of Pediococcus sp. NRRL B-2354 (Formerly Micrococcus freudenreichii) in liquid foods. J. Food Prot. 61: 578-581.
[91] de Valdéz, G.F.; Giori, G.S.; Pesce, A.; Ruiz Holgado, A.P. and Oliver, G. (1983) Comparative study of the efficiency of some additives in protecting lactic acid bacteria against freeze-drying. Cryobiology 20: 560-566.
[92] de Valdéz, G.F.; Giori, G.S.; Ruiz Holgado, A.P. and Oliver, G. (1983) Protective effect of adonitol on lactic acid bacteria subjected to freeze-drying. Appl. Environ. Microbiol. 45: 302-304.
[93] de Valdéz, G.F.; Giori, G.S.; Ruiz Holgado, A.P. and Oliver, G. (1985) Effect of drying medium on residual moisture content and viability of freeze-dried lactic acid bacteria. Appl. Environ. Microbiol. 49: 413-415.
[94] Leslie, S.B.; Israeli, E.; Lighthart, B.; Crowe, J.H. and Crowe, L.M. (1995) Trehalose and sucrose protect both membranes and proteins in intact bacteria during drying. Appl. Environ. Microbiol. 61: 3592-3597.
[95] Venugopal, V. (1994) Production of fish protein hydrolyzates by microorganisms. In: Fisheries processing: biotechnological applications. Chapman and Hall, London, UK; pp. 223-243.
[96] Younes, G.; Nicaud, J.-M. and Guespin-Michel, J. (1984) Enhancement of extracellular enzymatic activities produced by immobilized growing cells of Myxococcus xanthus. Appl. Microbiol. Biotechnol. 19: 67-69.
[97] Venugopal, V.; Alur, M.D. and Nerkar, D.P. (1989) Solubilization of fish proteins using immobilized microbial cells. Biotechnol. Bioeng. 33: 1098-1103.

BIOFLAVOURING OF FOODS AND BEVERAGES

RONNIE WILLAERT[1], HUBERT VERACHTERT[2],
KAREN VAN DEN BREMT[2], FREDDY DELVAUX[2] AND
GUY DERDELINCKX[2]
[1]*Department of Ultrastructure, Flanders Interuniversity Institute for Biotechnology, Vrije Universiteit Brussel, Pleinlaan 2, B-1050 Brussel, Belgium – Fax: 32-2-6291963 – Email: Ronnie.Willaert@vub.ac.be;*
[2]*Centre for Malting and Brewing Science, Katholieke Universiteit Leuven, Kasteelpark Arenberg 22, B-3001 Heverlee, Belgium – Fax: 32-16-321997 – Email: Guy.Derdelinckx@agr.kuleuven.ac.be*

1. Introduction

Natural flavours and fragrances play nowadays an important role in the quality of food and beverages. Due to food-processing operations such as premature harvesting, extended storage and physical treatment, aromas may be lost and the addition of flavour supplements to foodstuff is often required [1]. Additionally, consumers are more concerned about food quality and prefer natural food additives to chemically synthesised compounds. As a result, the "natural" label allocated by the European and US Food legislation represents a strong marketing advantage.

Because the isolation of natural flavours from plants is limited and the flavour market is increasing rapidly, alternative sources for natural flavours are searched for. Therefore, the biotechnological generation of aroma compounds, by means of microorganisms, isolated enzymes or plant cell cultures, receives much attention. Currently, about 100 flavour molecules of biological origin are available to the food industry [2]. Among them, natural lactones represent one of the major large-volume flavour products. For, example, the annual potential market for natural γ-decalactone is estimated to be 10 tons, which corresponds to a fermentation volume of about 2500 m^3. This volume represents a substantial market for such a high-value, fine chemical product.

In this chapter, the biological production of natural flavours will be reviewed and it will be demonstrated that this can be a good alternative for the use of plant extracts and that immobilised cell technology can be introduced successfully in this field to guarantee an economical exploitation.

2. Definition of bioflavouring

When discussing the recently developed concept of bioproduction of natural flavours, the term "natural" has to be clearly defined at first. In the USA, a distinction between natural and artificial flavour compounds is made and according to the "Code of Federal Regulations" a natural flavour is ... "the essential oil, oleoresin, essence or extractive, protein hydrolysate, distillate of any product of roasting, heating or enzymolysis, which contains the flavouring of constituents derived from a spice, fruit juice, vegetable or vegetable juice, edible yeast, herb, bud, bark, root, leaf or similar plant material, meat, seafood, poultry, eggs, dairy products or fermented products thereof, whose significant function in food is rather flavouring than nutrition."

With respect to the bioproduction of flavour compounds, it can be concluded from this definition that products obtained by reactions of enzymes or microorganisms can be considered natural as long as natural raw materials are used and the resulting molecule itself occurs in nature. Products that occur in nature but that are produced in a non-natural process are called "nature-identical" [3].

3. Processes for flavour production

For the production of flavour compounds, three distinct methods can be used. Flavours can be produced chemically, extracted from plants or produced by means of microorganisms, plant cell cultures or isolated enzymes. The chemical production of synthetic, nature-identical compounds is no longer accepted as consumer friendly. As an alternative, natural flavours can be obtained from plant material. However, some major disadvantages can be associated with the use of plant extracts. The important constituents of plants are often present in very small amounts or/and in bound form like glycosides. Therefore, their isolation and concentration is rather difficult. Moreover, the quality and supply of the plant material can be variable, due to uncontrollable environmental factors such as drought, heat and cold, plant diseases and sometimes the unstable political climate of the country from which the plants have to be imported. Taken into account these problems together with the increasing demand for natural flavours, which causes a shortage in several plant resources, the price of several plant oils or aromas is very high [4, 5].

Bioproduction can offer a good alternative for flavour production. First, biotechnical processes require less drastic process conditions and are consequently less damaging to the environment than the chemical processes. Secondly, they offer a good alternative for the use of plant extracts from an economical point of view, provided that the biochemistry and the metabolic regulation of the biocatalyst are well understood and profitable process conditions can be obtained.

The advantages and disadvantages of the different methods of flavour production can be illustrated with some examples. Raspberry ketone (threshold is $1-10$ ppb [6]), for instance, is one of the characteristic flavour compounds of raspberries. It can be produced chemically by condensation of p-hydroxy-benzaldehyde with acetone. However, natural raspberry ketone is preferred to the nature-identical compound. Natural raspberry ketone can be obtained from plant extracts, mainly from raspberries.

Only 3.7 mg of raspberry ketone is gained from 1 kg of fresh berries, which makes isolation and purification of this flavour compound unprofitable. Bioproduction can offer a solution for this problem (Figure 1). Raspberry ketone can be produced from betuligenol (rhododendrine) as raw material. This can be found in rhododendron, birch and maple. The first step in the bioproduction of raspberry ketone is the hydrolysis of betuloside using an intrinsic or extrinsic β-glucosidase. A commercial β-glucosidase from e.g. *Aspergillus niger* can be used to catalyse this reaction. Bioconversion of the intermediate betuligenol into raspberry ketone consists of a dehydrogenation of the secondary alcohol into the corresponding ketone. This reaction can be carried out by the secondary alcohol dehydrogenase of *Candida boidinii* [7].

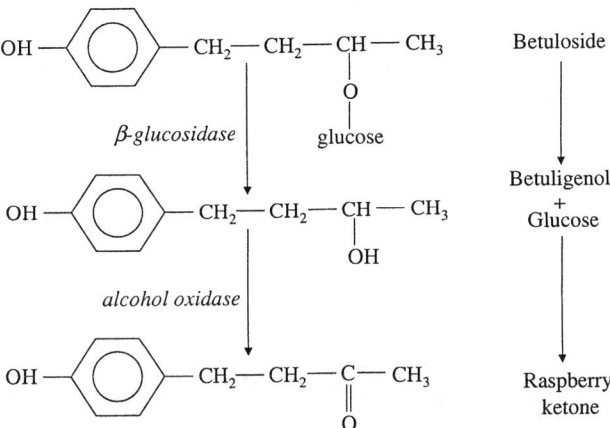

Figure 1. Bioproduction of raspberry ketone [7].

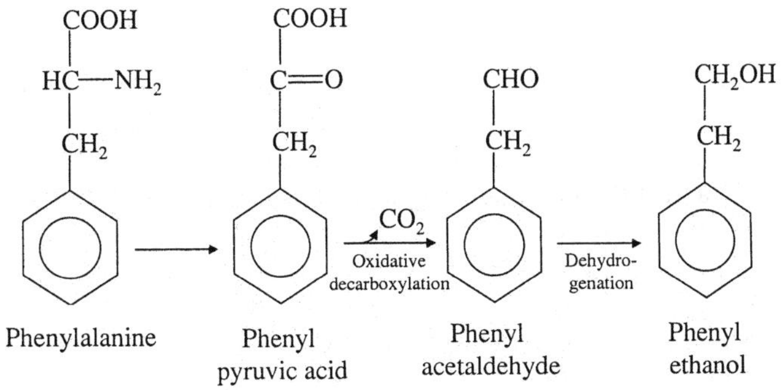

Figure 2. Bioproduction of phenethyl alcohol from phenylalanine by mutants of Saccharomyces cerevisiae *(based on [12])*.

A second example illustrating the fact that bioproduction is a promising alternative for flavour production is the synthesis of phenethyl alcohol (threshold = 125 ppm). As for raspberry ketone, phenethyl alcohol can be produced chemically but the naturally obtained product is about 50 to 100 times more valuable. For the bioproduction of phenethyl alcohol several possibilities exist. During fermentation by *Saccharomyces cerevisiae*, phenethyl alcohol can be recuperated from the fermentation gases by using resins [8]. *Kluyveromyces marxianus* is also reported to produce phenethyl alcohol from phenylalanine. About 170 mg/l of phenethyl alcohol can be recuperated from the fermentation medium by means of XAD2 adsorber techniques [9]. More recently, Fabre *et al.* [10] showed the extraction of phenethyl alcohol from the culture medium of *K. marxianus* with supercritical CO_2. Another possibility for the bioproduction of phenethyl alcohol is the use of mutants of *S. cerevisiae*, selected for their production of high amounts of phenethyl alcohol on a specific culture medium containing 1.17% of yeast carbon base, 0.5% of casamino acids and 15% of glucose [11]. About 2 g/l of phenethyl alcohol can be obtained without any specific optimisation of the process. Phenylalanine is converted through different steps into the desired alcohol (see Figure 2).

4. Bioproduction of natural flavours

It has become clear from the previous examples that the bioproduction of natural flavours is possible and can have some important advantages compared with plant extracts. Many microbial processes have been described that are able to produce interesting aroma compounds. However, the number of industrial applications is limited. Reasons for this are in most cases low final product yield, low biotransformation rates, substrates and/or end products inhibition, toxicity towards the microorganisms themselves and difficulties of recovery from the bioreaction mixture. Therefore, process intensification using high cell density immobilised systems in combination with the appropriate process design, is a prerequisite for an economic successful bioflavour bioproduction process.

When discussing the bioproduction of natural flavours a distinction can be made between different approaches, based on the biocatalyst that is used, i.e. plant cell cultures, microorganisms and isolated enzymes.

4.1. MICROORGANISMS

It is well known that microorganisms can produce off-flavours in food and beverages. However, also the production of compounds introducing pleasant, often fruity, flavours in food products is already known for a long time. For more than thousand years, humans have used microorganisms in the production of flavours in food and beverages. Production of wine and beer dates already from early times (probably more than 5000 years B.C.). However, people were unaware of the role of microorganisms as biocatalytic agents in flavour production. Thanks to the increased knowledge in microbiology and biochemistry, the relationship between microorganisms and the typical flavour of fermented food and beverages was only recognised in the beginning of the 20th century [2]. One of the first works on aroma producing microorganisms was published by Omelianski in 1923 [13]. Due to new developments in biotechnology, new techniques – i.e. genetic engineering and metabolic engineering [14] – could be used to improve the production of specific flavour compounds.

The production of a complex flavour during fermentation was the first step in the bioproduction of flavours by microorganisms. In a wide range of traditional fermented foods and beverages, a large variety of microorganisms is responsible for the typical flavour of the product. Gueuze beer [15], kefir [16], sourdough [17], soy sauce [18] and dry sausages [19] are only a few examples of products in which a typical mixed microbial culture is responsible for the production of flavour compounds [20]. Whereas the microflora in these fermented food and beverages have a positive effect on the flavour of the product, other microorganisms can cause off-flavours. Contamination of dairy products with *Pseudomonas fragi*, for instance, will result in a fruity aroma defect [21].

The production of flavours by microorganisms is not only limited to the synthesis of complex flavours during fermentation. Along with the evolution of microbiology and biotechnology in general, microorganisms were used in the production of specific single flavour compounds. These can be found by *de novo* synthesis or by conversion of a well-chosen substrate into the desired compound.

4.1.1. De novo biosynthesis

A wide range of microorganisms is able to produce flavours by *de novo* synthesis starting from simple ingredients. Several examples will illustrate this production process (some examples of processes using immobilised cells are tabulated in Table 1).

4.1.1.1. Biosynthesis of lactones. Lactones derived from hydroxy fatty acids (see also further) belong to the most important class of flavour components in many fruits, milk products, and fermented foods [2]. The olfactory properties of such lactones depend on chain length, degree of saturation, size of the lactone ring and chirality. Organoleptic descriptions vary from fruity, peach, and coconut to fatty or even flowery. Such lactones are used extensively in the food industry. Particular yeast strains can produce lactones via de novo synthesis. Tahara et al. [22,23] report the presence of the peach-like smelling compounds γ-decalactone (4-decanolide) and 4-hydroxy-cis-6-dodecenoic acid γ-lactone (cis-6-dodecen-4-olide) in the culture medium of Sporodiobolus salmonicolor (formerly Sporobolomyces odorus).

4.1.1.2. In situ bioflavouring. An alternative approach to the use of immobilised cells for bioflavour production, has been presented by Kogan and Freeman [24]. They co-immobilised microbial cells already employed for food/beverage treatment (e.g. baker's yeast) with natural precursor(s) for bioflavour production, in macro (millimetre size) beads made of a food-grade matrix (e.g. alginate, carrageenan). Following an appropriate incubation period with or without external liquid medium, the bioflavour, or mixture of bioflavours, is generated in situ. This formulation is stored, either cooled or frozen, until used as a food additive bearing natural flavours. A feasibility study demonstrated this approach with ethanol production by baker's yeast co-immobilised with glucose medium. Complex bioflavour generation was also demonstrated by baker's yeast co-immobilised with apple juice, generating cider flavours. Beads providing beer taste were also readily made via co-immobilisation of commercial brewing yeast with malt. Furthermore, the potential inherent in bioflavour generation by co-immobilisation of filamentous fungi with an emulsion of oily precursor was demonstrated by γ-decalactone production from castor oil.

Table 1. Some examples of bioflavouring by de novo biosynthesis.

Application/Production	Microorganism	Immobilisation method	Reference
Citric acid	Aspergillus niger	Hollow fibre	[25]
		Agarose beads	[26]
	Candida lipolytica	Membrane cell recycle	[27]
	Saccharomycopsis lipolytica	Membrane cell recycle	[28]
	Yarrowia lipolytica	Ca-alginate beads	[29]
		Poly(dimethyldiallyl-ammonium chloride)	[30]
γ-Decalactone	Tyromyces sambuceus	Alginate macrocapsule	[24]
Diacetyl	Lactococcus lactis	Ca-alginate fibres	[31]
	L. lactis + catalase	Ca-alginate fibres	[32]
4-Ethylguaiacol	Candida verstalis	Alginate beads	[33,34]
Gluconic acid	Aspergillus niger	Ca-alginate beads	[35]
Glutamic acid	Brevibacterium ammoniagenes	Ca-alginate beads	[36]
	Corynebacterium glutamicum	κ-Carrageenan	[37]
		Euchema gel	[38]
		Porous glass	[39]
Kojic acid	Aspergillus oryzae	Ca-alginate beads	[40,41]
		Membrane reactor	[42]
Lactic acid	Lactobacillus cremoris	Membrane cell recycle	[43]
	Lactobacillus delbrueckii	Hollow fibre	[44]
	Lactobacillus helveticus	Membrane cell recycle	[45]
		Ca-alginate beads	[46]
	Lactobacillus rhamnosus	Membrane cell recycle	[47]
	Lactococcus lactis	Cross-linked chitosan microcapsules	[48]
		Cross-linked gelatin microcapsules	[49]
		Alginate/poly-L-lysine microcapsules	[50]
Mead fermentation	Saccharomyces cerevisiae	Alginate beads	[51]
Meat fermentation	Lactobacillus plantarum + Pediococcus pentosaceus	Ca-alginate beads	[52]
Soy sauce fermentation	Pediococcus halophilus, Saccharomyces rouxii, Torulopsis versatilis	Alginate beads	[53]
	Zygosaccharomyces rouxii	Ca-alginate beads	[18,54]
		Membrane cell recycle	[55]

4.1.1.3. Production of flavour metabolite organic acids. Organic acids serve as acidulants in the food industry and are employed to enhance or modify flavour. Two flavour metabolites of filamentous fungal origin, citric acid and gluconic acid, are produced by large-scale fermentation [56]. One fungal organic acid, kojic acid, serves as a chemical precursor of the potent flavour enhancers maltol and ethylmaltol.

Citric acid occurs naturally in almost all unprocessed food materials and has unrestricted generally recognized as safe (GRAS) status for use in foods. This organic acid is widely used as an additive in foods and beverages owing to its versatility and multifunctional nature. It serves not only as a flavour enhancer, but also as an acidulant, pH regulator, preservative, chelating agent, stabilizer and antioxidant. Citric acid, as well as other organic acids, is used extensively by the food industry to adjust the acid flavour of a wide variety of soft drinks, fruit and vegetable juices, canned fruits, wines and wine coolers, ciders, and powdered beverage mixes [57, 58]. The main benefit of added citric acid in beverages is its ability to provide a pleasant tartness and acid taste that complements fruit, berry and other flavours.

Most citric acid is produced by *Aspergillus niger* (see Table 1). However, some processes use yeast cells, especially *Yarrowia lipolytica* has been used. Citric acid production by fermentation has to be performed under very demanding conditions (for reviews see Milson [59] and Mattey [60]). As a result the use of immobilised cell fermentation improved the productivity considerable. Immobilised cell technology allowed a continuous production under non-growth conditions, which improved the oxygen transfer (due to a lower viscosity), and consequently reduced the agitation costs. Long-term utilisation of the starved biocatalysts requires periodic cell regeneration to counteract the decrease in activity [61].

Table 2. Some examples of bioflavouring by bioconversion with microorganisms.

Application/production	Microorganism	Immobilisation method	Reference
2,3-Butanediol	*Klebsiella oxytoca*	Glass wool	[64]
Citrus juice debittering/ limonin degradation	*Rhodococcus fascians*	κ-Carrageenan beads	[65,66]
Gluconic acid	*Zymomonas mobilis*	Hollow fibre	[67]
Glucose isomerisation	*Arthrobacter* sp.	κ-Carrageenan beads	[68]
	Saccharomyces cerevisiae	Ca-alginate beads	[69]
Isomaltulose	*Erwinia rhapontici*	Ca-alginate beads	[70]
Malic acid	*Brevibacterium ammoniagenes*	Polyacrylamide beads	[63]
	B. flavum	κ-Carrageenan beads	[71,72]
	Candida rugosa	Polyvinyl alcohol	[73]
	Pichia membranaefaciens	Polyacrylamide beads	[74]
	Saccharomyces cerevisiae	Agarose beads	[75]
Olive oil hydrolysis	*Rhizopus delemar*	Polyurethane foam	[76]
Vanillic acid	*Pseudomonas fluorescens*	Ca/Ba-alginate	[77]

Other organic acids can also serve as food acidulants and flavour compounds. Lactic acid was once produced by fermentation with the filamentous fungi *Rhizopus oryzae* at high yield, but is currently produced with homofermentative lactic acid bacteria or by large-scale chemical synthesis [60]. Some examples of the production of lactic acid using immobilised cells can be found in Table 1. As can be noted, a wide variety of methods have been applied. Although fumaric acid was produced by fermentation in the 1940s and 1950s in the USA [99], it is now produced exclusively by chemical synthesis. Kautola and Linko proposed in 1989 the production of fumaric acid from xylose by immobilised *Rhizopus arrhizus* cells [62]. In the best conditions, immobilised cells produced 3.5 times more fumaric acid than free cells. The bioconversion of fumaric acid into malic acid was the first successful industrial process for the production of organic acids by immobilised cells [62] (see Table 2). Malic acid is also made with bacterial fumarase, or it can be obtained by chemical synthesis. Tartaric acid is usually extracted from wine or chemically synthesised. It can also be produced by fermentation or bioconversion from *cis*-epoxysuccinate [100,101].

4.1.2. Bioconversion by microorganisms

Specific substrates that are added to the culture medium can be converted into specific flavour compounds. The list of microorganisms that are able to convert a well-chosen substrate into a specific flavour is very extended [5]. However, from the large number of compounds that can be produced microbiologically, only a few are produced on an industrial scale. Up to now, only compounds with a high added value like γ- and δ-lactones, vanilla extracts and various esters are taken into production. This is due to the high costs of the processes since these are characterised by a low production yield (caused by a still insufficient knowledge of the biosynthetic pathway that plays a key role in the conversion of substrates into flavour compounds and an inadequate process design). Fundamental research on the microorganisms in order to get a better control of their metabolic pathways, and on the development of alternative production processes, like immobilized cell systems (Table 2) and solid-state fermentation, can help to improve the process yield [12]. Additionally, the use of inexpensive raw materials as substrates in the bioconversion processes is another requirement to obtain a high added value.

4.1.2.1. Biosynthesis of vanillin. The biosynthesis of vanillin can illustrate this problem. Possible routes for vanillin synthesis are shown in Figure 3. The first possible route is the oxidation of the natural stilbene isorhapontin. This phenolic stilbene is commonly found in spruce bark, which could be used as raw material. The oxidation of isorhapontin to vanillin is catalysed by the stilbene dioxygenase which is present in some Pseudomonas sp. This process has already resulted into numerous patents.

Another interesting route for the production of vanillin could be the transformation of eugenol by strains of *Corynebacterium* sp. and *Pseudomonas* sp. This is the principal constituent of clove oil, which makes eugenol an inexpensive commercially available raw material. However, production yield is still very poor since the metabolic pathway is not yet well understood [12,78,79]. Ramachandra Rao and Ravishankar [79] discussed in detail economic and safety aspects of the biotechnological production of vanilla.

The reactor design and process optimisation for the bioconversion of vanillin into vanillic acid by Ca/Ba-alginate immobilised *Pseudomonas fluorescens* is described by Baré *et al.* [77]. By using a two-phase reactor (water/dodecanol), the vanillic acid production was increased compared to the one-phase reactor, with a conversion rate near 80%.

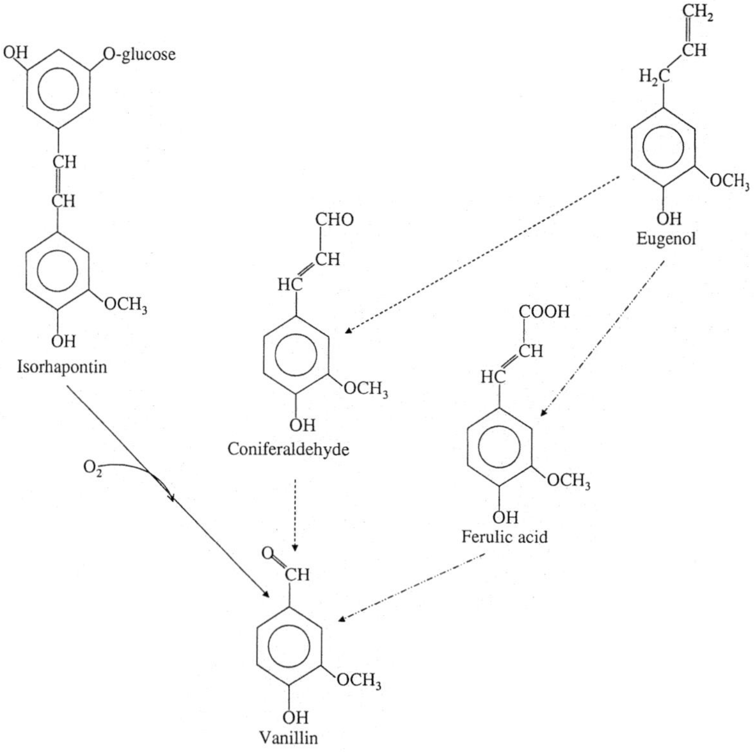

Figure 3. Proposed pathway for the bioproduction of vanillin [12].

4.1.2.2. Biosynthesis of flavours starting from fatty acids. A considerable number of valuable flavours and fragrances are produced by microorganisms starting from fatty acids. These are derived from plant oils or animal oils as raw material (Table 3). They are cheap and available in large amounts which will contribute to the high added value of the resulting flavour compound.

As an example, the filamentous fungus *Botryodiplodia theobromae* is able to synthesise jasmonic acid, a molecule with a sweet-floral, jasmine-like odour, starting from linolenic acid (Figure 4). It was found that jasmonic acid, an endogenous growth regulator in plants, was synthesised in *Botryodiplodia theobromae* in a similar way as in plants. Generally, unsaturated fatty acids are degraded by the lipoxygenase pathway and here linolenic acid is converted by lipoxygenase into linolenic acid hydroperoxide. This compound is converted into allene oxide, which cyclises. β-Oxidation and reduction of the double bond results in the formation of jasmonic acid [2].

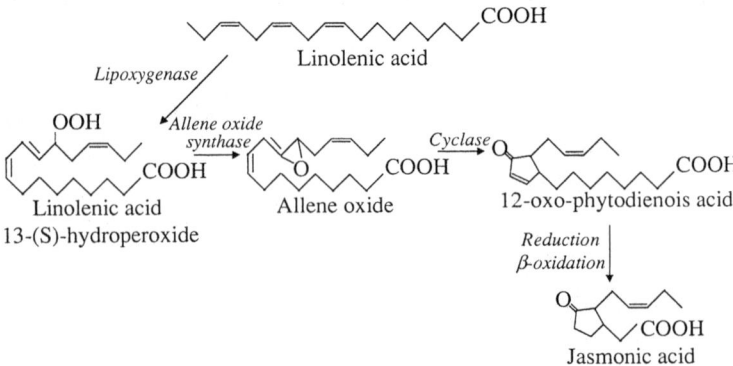

Figure 4. Bioconversion of linolenic acid to jasmonic acid by Botryodiplodia theobromae *(Häusler & Münch, 1997).*

Table 3. Commercial sources of precursor fatty acids [2].

Precursor fatty acids (chain length : unsaturation)	Commercial source
Saturated fatty acids	
C4 to C12	Milk fat
C8 to C12	Coconut oil fractions
C10 to C18	Butter fat
Unsaturated fatty acids	
C18:1 oleic acid	Plant oils, animal fats
C18:2 linoleic acid	Most plant oils
C18:3 linolenic acid	Linseed oil
C20:4 arachidonic acid	Animal fats
C20:5 eicosapentaenoic acid (EPA)	Fish oil
C22:6 docosahexaenoic acid (DHA)	Fish oil
Hydroxylated acids	
C18:1-OD ricinoleic acid	Castor oil
C16:0-OH 11-hydroxypalmitic acid	Jalap root (sweet potato)

A very important group of flavour compounds produced from hydroxy fatty acids are lactones. *Sporobolomyces salmonicolor* (formerly *S. odorus*) can convert ricinoleic acid into γ-decalactone (Figure 5). The bioproduction of γ-decalactone is an important process. Not only because the natural product is preferred by consumers over the chemical product, but also because of the stereochemical purity of the bioproduct (γ-decalactone is a chiral molecule). Whereas γ-decalactone has a defined stereospecificity in fruit [(R)-isomer in peaches and most other fruit, (S)-isomer in mango], the chemical synthesis of this molecule will result in a racemic mixture. Microorganisms capable of forming optically active γ-decalactone have been selected [80]. The largest problem for the industrial production of γ-decalactone is the limited availability of hydroxy fatty acids. The only source of considerable amounts of hydroxy fatty acids is castor oil in which ricinoleic acid is the main constituent [2].

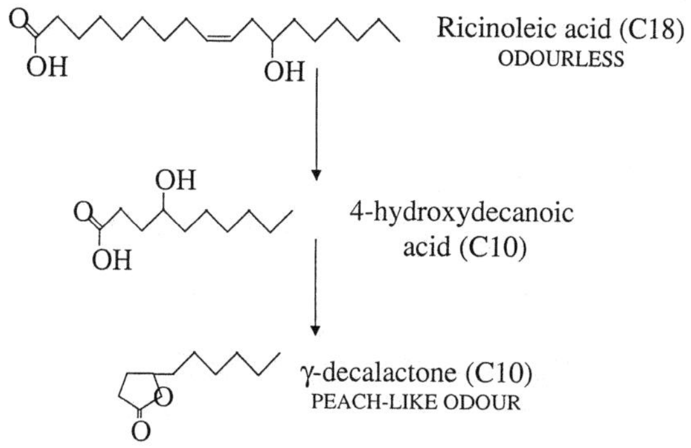

Figure 5. Formation of the peach-flavoured lactone γ-decalactone by Sporobolomyces odorus *[80].*

4.1.2.3. Biosynthesis of aldehydes. The oxidation of primary alcohols can result in the formation of aldehydes that can sometimes have pleasant flavour characteristics; even some simple unbranched aldehydes have a fresh, fruity flavour. They can be grouped according to their chain length. Aldehydes with a chain length of 2 to 5 carbon atoms have generally a green flavour, while aldehydes with a chain length of 6 to 11 carbon atoms are often characterised by a bitter, orange peel flavour. The threshold concentration of these compounds decreases with increasing chain length from, for example, 1 ppm for n-butanal (flavour: melon, green leaves, varnish) to 6 ppb for n-decanal (flavour: bitter, aldehyde, orange peel). Additionally, when these unbranched aldehydes have a mono-unsaturated function, an extra pleasant flavour is obtained [81].

Biological oxidation of primary alcohols into aldehydes can be accomplished by methylotrophic yeasts. Their alcohol oxidase (necessary for growth on methanol) is also able to convert other alcohols into aldehydes. Nozaki *et al.* [82,83] described the high chemo selective production of aldehydes from alcohols by the methylotrophic yeast *Candida boidinii*.

4.1.2.4. Bioflavouring of beer. The flavour maturation of green beer is an example where immobilised cell technology (ICT) has been used successfully on an industrial scale for the process intensification of the beer fermentation process (see also the chapter about "Beer production using immobilised cells" in this volume). During the classical lager beer maturation process, vicinal diketones (i.e. diacetyl and 2,3-pentanedione) are reduced by active yeast cells into less-flavour-active compounds. Diacetyl is the most flavour-active diketone which is responsible for an undesirable buttery flavour. ICT allowed to reduce the traditional maturation period of 3 to 4 weeks to less than 24 hours.

The use of ICT for the bioflavouring of beer is rather an old technique. In earlier times, however, the term "immobilised cells" was not used since the fermentation process was poorly understood. Immobilised cells play an important role in the bioflavouring of old style beers, like the Belgian beers "lambic" and the acidic ales, where the secondary fermentation is performed in wooden barrels. Different yeast and bacterial strains are immobilised in the wood and results in a rather complex maturation process. Today, wood shavings are used in some Belgian breweries to increase the cell density in the fermentation tanks and to speed-up the maturation process in a more controlled way.

Corran and Mulvey [84] describe a patented continuous maturation process for the production of the traditional acid beers "porter" and "stout" using immobilised *Lactobacillus pastorianus* and the yeast *Brettanomyces*. This yeast plays also an important role in the fermentation of traditional lambic, Belgian acidic beers and the trappist beer "Orval". Some wines of great reputation are also infected with *Brettanomyces*, which contribute to the complexity and delicacy of the flavour. *Brettanomyces* is also involved in the natural fermentation of tequila [85]. This yeast has a unique esterase activity which is responsible for the production of the esters ethyl acetate, ethyl lactate and ethyl caprate in high amounts. Additionally, high amounts of higher fatty acids (C_8-C_{12}) are formed [86,87]. *Brettanomyces* is also characterised by a dextrinase and β-glucosidase activity, and has a high ethanol tolerance. The idea of using *Brettanomyces* during a secondary fermentation as a controlled process step dates from the beginning of the twentieth century. N. Hjelte Clausen (a brewing microbiogist at Carlsberg Brewery in Copenhagen) has deposited a patent in 1903 for using this yeast to perform the secondary fermentation of English beers.

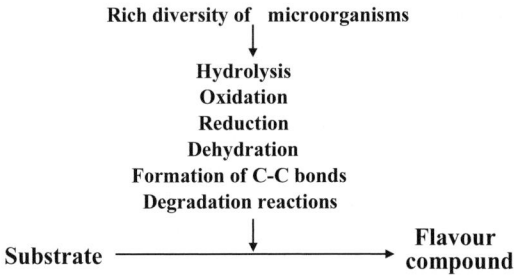

Figure 6. Flavouring potential of non-Saccharomyces yeasts [91].

Recently, new methodologies have been investigated to increase the flavour of beer using well selected yeasts [88,89,90,91]. The new flavours are introduced by a secondary fermentation using non-*Saccharomyces* yeasts or by linking a secondary fermentation with substrates obtained from plants. The use of non-*Saccharomyces* yeasts can complement the metabolic activity of the *Saccharomyces* yeast, which has been used to perform the primary fermentation by being more active and specific or by catalyzing new reactions (Figure 6). A practical example is the hydrolysis of monoterpene alcohols present in hops and coriander (added to flavour some beers, like

Belgian White Beer) by a β-glucosidase. These molecules can be hydrolysed to a sugar part and powerful flavour molecules as linalool (threshold about 100 ppb), geraniol (threshold between 50 and 500 ppb), nerol and α-terpineol.

GRAS ("Generally Recognised As Safe")-yeasts, which are known to produce a typical flavour in other beverages than beer or in food, can be selected and tried out in the secondary fermentation of beer. Several non-*Saccharomyces* yeasts have been proposed as potential interesting microorganisms to give beer a unique flavour (some examples are given in Table 4). The flavouring potential of glycosides has been reported in wine making and fruit processing [92]. For the bioconversion of glycosides, different β-galactosidases from plants and microorganisms were tested. The β-galactosidases from *Candida molischiana* and *C. wickerhamii* were the most efficient for flavour enrichment of Muscat wine and apricot juice.

Table 4. Potential beer bioflavouring yeasts.

Yeast	Activity	Reference
Brettanomyces/Dekkera	Esterase	[93]
	β-D-galactosidase	[94]
	Dextrinase	[95]
Candida boidinii	Alcohol oxidase	[82]
Candida molischiana	β-D-galactosidase	[96]
Candida methanolovescens	β-D-galactosidase and alcohol oxidase	[89,97]
Candida wickerhamii	β-D-galactosidase	[92]
Hanseniaspora vineae	β-D-galactosidase	[91]
Hansenula wickerhamii	β-D-galactosidase	[91]
Torulopsis molischiana	β-D-galactosidase	[91]

Aldehydes (and ketones), which may have flavour characteristics, can also be produced by the oxidation of alcohols. The oxidation of alcohols by microorganisms growing on such carbon substrates generally proceeds through their oxidation to aldehydes (or ketones when growing on secondary alcohols) and acids. Especially methanol is a good substrate for many microorganisms, the so-called methylotrophs. These microorganisms use an alcohol oxidase to produce an aldehyde. This aldehyde, being formaldehyde when methanol is used, is then further degraded to carbon dioxide to generate energy for growth, but a large part must be used for the synthesis of sugars and other anabolic intermediates. When other alcohols are oxidised by this alcohol oxidase, interesting flavour aldehydes may be produced. The microbial oxidation of alcohols by *Candida boidinii* has been recently described [82]. It has been shown that the synthesis of alcohol oxidase in methylotrophs is repressed by glucose [83].

Candida methanolovescens, like *C. molischiana* and *C. wickerhamii*, is a methylotrophic yeast, which not only hydrolyses glycosides into alcohols but also converts alcohols into flavour aldehydes. It produces a β-galactosidase and an alcohol oxidase. A UV-mutant strain, derepressed for alcohol oxidase synthesis, could be isolated [88]. This strain has been recently evaluated for its ability to use glycosides as starting material for flavour enrichment of non-alcoholic beer [89]. It was demonstrated that this strain could hydrolyse the glucoside salicin into salicyl alcohol and could produce the almond-flavoured salicyl aldehyde in the beer (Figure 7).

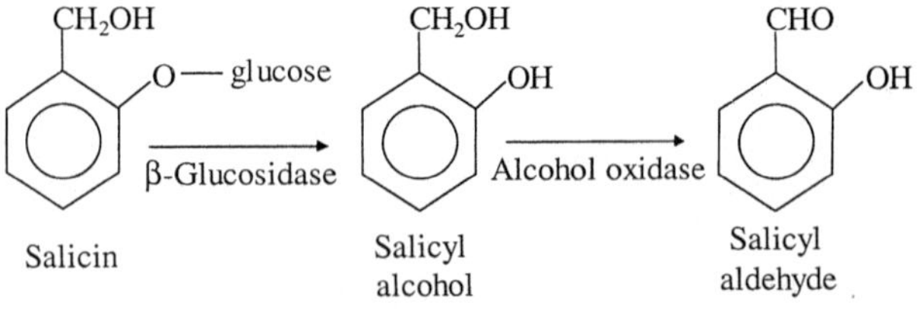

Figure 7. Bioconversion of salicin into salicin alcohol and salicyl aldehyde.

The use of non-*Saccharomyces* yeasts to perform the secondary beer fermentation necessitates the development of an adapted fermentation/bioconversion process [98]. A fast, well-controlled bioconversion step is wanted. This can be accomplished by using a high cell density fermentation/bioconversion, since a high volumetric productivity can be obtained with these systems. The use of an immobilised cell system is here very appropriate too.

References

[1] King, U. and Berger, R.G. (1998) Biotechnological production of flavours and fragrances. Appl. Microbiol. Biotechnol. 49: 1-8.
[2] Häusler, A. and Münich, (1997) Microbial production of natural flavours. ASM News 63: 551-559.
[3] Cheetham, P.S.J. (1997) Combining the technical push and the business pull for natural flavours. Trends Biotechnol. Adv. Biochem. Eng./Biotechnol. 55: 1-49.
[4] Whitaker, R.J. and Evans, D.A. (1987) Plant biotechnology and the production of flavour compounds. Food Technol. 41: 86-101.
[5] Janssens, L.; De Poorter, H.L.; Schamp, N.M. and Vandamme, E.J. (1992) Production of flavours by microorganisms. Proc. Biochem. 27: 195-215.
[6] Larsen, M. and Poll, L. (1990) Odour threshold of some important aroma compounds in raspberries. Z. Lebensm.-Unters. Fors. 191: 129-131.
[7] Hugueny, P.; Dumont, B.; Ropert, F. and Belin, J.M. (1995) The raspberry ketone, a biotechnological way for its production. Proceedings Bioflavour 95, Dijon (France), pp. 269-273.
[8] Arnould, M.; Bes, M.; Harmegnies, F.; Delvaux, F.; Escudier, J.-L; Derdelinckx, G. (2001) Recovery of volatiles from gasses of fermenting wort. Proceedings of the 28th European Brewery Convention congress, Budapest; pp. 764-772.
[9] Janssens, L. (1991) Microbiële produktie van fruitaroma's. PhD thesis, University of Gent, Belgium.
[10] Fabre, C.E.; Condoret, J.S. and Marty, A. (1999) Extractive fermentation of aroma with supercritical CO_2. Biotechnol. Bioeng. 64: 392-400.
[11] Akita, O.; Ida, T.; Obata, T. and Hara, S. (1990) Mutants of *Saccharomyces cerevisiae* producing a large quantity of β-phenethyl alcohol and β-phenethyl acetate. J. Ferment. Bioeng. 69: 125-128.
[12] Feron, G.; Bonnarme, P. and Durand, A. (1996) Prospects in the microbial production of food flavours. Trends Food Sci Technol. 7: 285-293.
[13] Omelianski, V.L. (1923) Aroma-producing microorganisms. J. Bacteriol. 8: 393-419.
[14] Stephanopoulos, G.N.; Aristidou, A.A. and Nielsen, J. (1998) Metabolic engineering: principles and methodologies. Academic Press, San Diego, USA.
[15] Verachtert, H. (1992) General properties of mixed microbial cultures. In: Korhola, M. and Backström, V. (Eds.) Microbial contaminants. Foundation for Biotechnical and Industrial Fermentation Research, Vol. 7, Helsinki, Finland; pp. 429-478.

[16] Hallé, C.; Leroi, F.; Dousset, X. and Pidoux, M. (1994) Les Kéfirs – Des associations bactéries lactiques – levures. In: de Roissart, H. and Luquet, F.M. (Eds.) Bactéries lactiques – Aspects fondamentaux et techniques. Vol. 2, Lorica Uriage, France; pp. 169-182.
[17] Halm, M.; Lillie, A.; Sorensen, A.K. and Jakobsen, M. (1993) Microbiological and aromatic characteristics of fermented maize dough for kenkey production in Ghana. Int. J. Food Microbiol. 19: 135-143.
[18] Hamada, T.; Sugishita, M.; Fukushima, Y.; Fukase, T. and Motai, H. (1991) Continuous production of soy sauce by a biorector system. Process Biochem. 26: 39-45.
[19] Montel, M.C.; Reitz, J.; Talon, R.; Berdagué, J.-L. and Rousset-Akrim, S. (1996) Biochemical activities of *Micrococcaceae* and their effects on the aromatic profiles and odours of a dry sausage model. Food Microbiol. 13: 489-499.
[20] Verachtert, H.; Kumara, H.M.C.S. and Dawoud, E. (1989) Yeast in mixed cultures with emphasis on lambic beer brewing. In: Verachtert, H. and De Mot, R. (Eds.) Yeast – Biotechnology and Biocatalysis. Marcel Dekker, New York, USA; pp. 429-478.
[21] Pereira, J.N. and Morgan, M.E. (1958) Identity of esters produced in milk cultures of *Pseudomonas fragi*. J. Dairy Sci. 41: 1201-1205.
[22] Tahara, S.; Fujiwara, K.; Ishizaka, H.; Mizutani, J. and Obata, Y. (1972) γ-Decalactone – One of constituents of volatiles in cultures broth of *Sporobolomyces odorus*. Agric. Biol. Chem. 36: 2585-2587.
[23] Tahara, S.; Fujiwara, K. and Mizutani, J. (1973) Neutral constituents of volatiles in cultured broth of *Sporobolomyces odorus*. Agric. Biol. Chem. 37: 2855-2861.
[24] Kogan, N. and Freeman, A. (1994) Development of macrocapsules containing bioflavors generated *in situ* by immobilized cells. Process Biochem. 29: 671-677.
[25] Chung, B.H. and Chang, H.N. (1988) Aerobic fungal cell immobilization in a dual hollow-fiber bioreactor: continuous production of citric acid. Biotechnol. Bioeng. 32: 205-212.
[26] Khare, S.K.; Jha, K. and Gandhi, A.P. (1994) Use of agarose-entrapped Aspergillus niger cells for the production of citric acid from soy whey. App. Microbiol. Biotechnol. 41: 571-573.
[27] Rane, K.D. and Sims, K.A. (1995) Citric acid production by *Candida lipolytica* Y 1095 in cell recycle and fed-batch fermentors. Biotechnol. Bioeng. 46: 325-332.
[28] Enzminger, J.D. and Asenjo, J.A. (1986) Use of cell recycle in the aerobic fermentative production of citric acid by yeast. Biotechnol. Lett. 8: 7-12.
[29] Rymowicz, W.; Kautola, H.; Wojtatowicz, M.; Linko, Y.-Y. and Linko, P. (1993) Studies on citric acid production with immobilized *Yarrowia lipolytica* in repeated batch and continuous air-lift bioreactors. Appl. Microbiol. Biotechnol. 39: 1-4.
[30] Förster, M.; Mansfeld, J.; Schellenberger, A. and Dautzenberg, H. (1994) Immobilization of citrate-producing *Yarrowia lipolytica* cells in polyelectrolyte complex capsules. Enzyme Microb. Technol. 16: 777-784.
[31] Takahashi, M.; Ochi, H.; Kanko, T.; Suzuki, H. and Tanaka, H. (1990) Diacetyl production by immobilized citrate-positive *Lactococcus lactis* subsp. *lactis* 3022 in the fibrous Ca-alginate gel. Biotechnol. Lett. 12: 569-574.
[32] Ochi, H.; Takahashi, M.; Kaneto, T.; Suzuki, H. and Tanaka, H. (1991) Diacetyl production by co-immobilized citrate-positive *Lactococcus lactis* subsp. *lactis* 3022 and homogenized bovine liver in alginate fibers with double gel layers. Biotechnol. Lett. 13: 505.
[33] Hamada, T.; Ishiyama, T. and Motai, H. (1990) Continuous production of 4-ethylguaiacol by immobilized cells of salt-tolerant *Candida versatilis* in an air-lift reactor. J. Ferment. Bioeng. 69: 166.
[34] Hamada, T.; Ishiyama, T. and Motai, H. (1990) Contribution of immobilized and free cells of salt-tolerant *Zygosaccharomyces rouxii* and *Candida versatilis* to the production of ethanol and 4-ethylguaiacol. Appl. Microbiol. Biotechnol. 33: 624-628.
[35] Moresi, M.; Perente, E.; Ricciardi, A. and Lanorte, M. (1996) Effect of dissolved oxygen concentration on pH-controlled fed-batch gluconate production by immobilised *Aspergillus niger*. In: Wijffels, R.H.; Buitelaar, R.M.; Bucke, C. and Tramper, J. (Eds.) Immobilized cells: basics and applications. Elsevier Sciences B.V., Amsterdam, The Netherlands; pp. 370-378.
[36] Lu, W.M. and Chen, W.C. (1988) Production of L-glutamate using entrapped living cells of *Brevibacterium ammoniagenes* with calcium alginate gels. Proc. Natl. Sci Counc. ROC(A). 12: 400.
[37] Kim, H.S. and Ryu, D.D.Y. (1982) Continuous glutamate production using an immobilized whole-cell system. Biotechnol. Bioeng. 24: 2167-2174.
[38] Li, Y.F.; Huang, Y.; Lee, L.F.; Sui, P. and Wen, Q.Q. (1990) Production of glutamic acid by immobilized cells. Ann. N.Y. Acad. Sci 613: 883-886.

[39] Henkel, H.J.; Johl, H.J.; Trosch, W. and Chmiel, H. (1990) Continuous production of glutamic acid in a three phase fluidized bed with immobilized *Corynebacterium glutamicum*. Food Biotechnol. 4: 149.
[40] Kwak, M.Y. and Rhee, J.S. (1992) Cultivation characteristics of immobilized *Aspergillus oryzae* for kojic acid production. Biotechnol. Bioeng. 39: 903-906.
[41] Kwak, M.Y. and Rhee, J.S. (1992) Controlled mycelial growth for kojic acid production using Ca-alginate-immobilized fungal cells. Appl. Microbiol. Biotechnol. 26: 578-583.
[42] Nakanishi, K.; Wakisaka, Y.; Tanaka, T. and Sakiyama, T. (1996) Production of kojic acid using mold by newly developed membrane)surface liquid culture. Proceedings 5th World Congress Chemical Engineering, vol. II; pp. 271-276.
[43] Bibal, B.; Vayssier, Y.; Goma, G. and Pareilleux, A. (1991) High-concentration cultivation of *Lactococcus cremoris* in a cell-recycle reactor. Biotechnol. Bioeng. 37: 746-754.
[44] Vick Roy, T.B.; Blanch, H.W. and Wilke, C.R. (1982) Lactic acid production by *Lactobacillus delbrueckii* in a hollow fiber fermenter. Biotechnol. Lett. 4: 483-488.
[45] Kulozik, U.; Hammelehle, B.; Pfeifer, J. and Kessler, H.G. (1992) High reaction rate continuous bioconversion process in a tubular reactor with narrow residence time distributions for the production of lactic acid. J. Biotechnol. 107-116.
[46] Øyaas, J.; Storro, I. and Levine, D.W. (1996) Uptake of lactose and continuous lactic acid fermentation by entrapped non-growing *Lactobacillus helveticus* in whey permeate. Appl. Microbiol. Biotechnol. 46: 240-249.
[47] Xavier, A.M.R.B.; Gonçalves, L.M.D.; Moreira, J.L. and Carrondo, M.J.T. (1995) Operational patterns affecting lactic acid production in ultrafiltration cell recycle bioreactor. Biotechnol. Bioeng. 45: 320-327.
[48] Groboillot, A.F.; Champagne, C.P.; Darling, G.D.; Poncelet, D. and Neufeld, R.J. (1993) Membrane formation by interfacial cross-linking of chitosan for microencapsulation of *Lactococcus lactis*. Biotechnol. Bioeng. 42: 1157-1163.
[49] Hyndman, C.L.; Groboillot, A.F.; Poncelet, D.; Champagne, C.P. and Neufeld, R.J. (1993) Microencapsulation of *Lactococcus lactis* within cross-linked gelatin membranes. J. Chem. Tech. Biotechnol. 56: 259-263.
[50] Larisch, B.C.; Poncelet, D.; Champagne, C.P. and Neufeld, R.J. (1994) Microencapsulation of *Lactococcus lactis* subsp. *cremoris*. J. Microencapsul. 11: 189-195.
[51] Qureshi, N. and Tamhane, D.V. (1985) Production of mead by immobilized whole cells of *Saccharomyces cerevisiae*. Appl. Microbiol. Biotechnol. 21: 280.
[52] Kearney, L.; Upton, M. and McLoughlin, A. (1990) Meat fermentations with immobilized lactic acid bacteria. Appl. Microbiol. Biotechnol. 33: 648-651.
[53] Osaka, K.; Okamoto, Y.; Akao, T.; Nagata, S. and Takamatsu, H. (1985) Fermentation of soy sauce with immobilized whole cells. J. Food. Sci. 50: 1289.
[54] Hamada, T.; Ishiyama, T. and Motai, H. (1989) Continuous fermentation of soy sauce by immobilized cells of *Zygosaccharomyces rouxii* in an airlift reactor. Appl. Microbiol. Biotechnol. 31: 346-350.
[55] Iwasaki, K.-I.; Nakajima, M. and Sasahara, H. (1991) Rapid ethanol fermentation for soy sauce production using a microfiltration membrane reactor. J. Ferment. Bioeng. 72: 373-378.
[56] Bigilis, R. (1992) Flavor metabolites and enzymes from filamentous fungi. Food Technol. November: 151-161.
[57] Goldberg, I.; Peleg, Y. and Rokem, J.S. (1991) Citric, fumaric, and malic acids. Goldberg, I. and Williams, R. (Eds.) Biotechnology and food ingredients, Van Nostrand Reinhold, New York, USA; pp. 349-374.
[58] Bigilis, R. (1991) Fungal metabolites in food processing. Arora, D.K.; Mukerji, K.G. and Marth, E.H. (Eds.) Handbook of applied mycology. Marcel Dekker Inc., New York, USA; pp. 415-443.
[59] Milson, P.E. (1987) Organic acids by fermentation, especially citric acid. In: King and Cheetham, P.S.J. (Eds.) Food Biotechnology I. Elsevier Applied Science Publishers, Reading, U.K.; pp. 273.
[60] Mattey, M. (1992) The production of organic acids. Crit. Rev. Biotechnol. 12: 87-132.
[61] Norton, S. and Vuillemard, J.-C. (1994) Food bioconversions and metabolite production using immobilized cell technology. Crit. Rev. Biotechnol. 14: 193-224.
[62] Kautola, H. and Linko, Y.Y. (1991) Fumaric acid production from xylose by immobilized *Rhizopus arrhizus* cellc. Appl. Microbiol. Biotechnol. 31: 448.
[63] Yamamoto, K.; Tosa, T.; Yamashita, K. and Chibata, I. (1976) Continuous production of L-malic acid by immobilized *Brevibacterium ammoniagenes* cells. Eur. J. Appl. Microbiol. 3: 169.

[64] Champluvier, B.; Francart, B. and Rouxhet, P.G. (1989) Co-immobilization by adhesion of β-galactosidase in nonviable cells of *Kluyveromyces lactis* with *Klebsiella oxytoca*; conversion of lactose into 2,3-butanediol. Biotechnol. Bioeng. 34: 845.
[65] Martinez-Madrid, C.; Manjon, A. and Iborra, J.L. (1989) Degradation of limonin by entrapped *Rhodococcus fascians* cells. Biotechnol. Lett. 11: 653.
[66] Manjon, A.; Iborra, J.L. and Martinez-Madrid, C. (1991) pH control of limonin debittering with entrapped *Rhodococcus fascians* cells. Appl. Microbiol. Biotechnol. 35:176-179.
[67] Paterson, S.L.; Fane, A.G. and Fell, C.J.D. (1988) Sorbitol and gluconate production in a hollow fibre membrane reactor by immobilized *Zymomonas mobilis*. Biocatalysis 1: 217-229.
[68] Bazaraa, W.A. and Hamdy, M.K. (1989) Fructose production by immobilized *Arthrobacter* cells. J. Ind. Microbiol. 4: 267.
[69] Koren, D.W. (1992) Production of fructose and ethanol by selective fermentation of glucose-fructose mixtures. PhD thesis, University of Ottawa, Canada.
[70] Cheetham, P.S.J.; Imber, C.E. and Isherwood, J. (1982) The formation of isomaltulose by immobilized *Erwinia rhapontici*. Nature 299: 628-630.
[71] Takata, L.; Yamamoto, K.; Tosa, T. and Chibata, I. (1980) Immobilization of *Brevibacteriem flavum* with carrageenan and its application for continuous production of L-malic acid. Enzyme Microb. Technol. 2: 30.
[72] Chibata, I.; Tosa, T. and Shibatani, T. (1992) The industrial production of optically active compounds by immobilized biocatalysts. In: Collins, A.N.; Sheldrake, G.N. and Crosby, J. (Eds.) Chirality in industry. John Wiley & Sons, Chichester, UK; pp. 251-370.
[73] Yang, L.W.; Wang, X.Y. and Wei, S. (1992) Immobilization of *Candida rugosa* cells having high fumarase activity with polyvinyl alcohol. Ann. N.Y. Acad. Sci. 672: 563-633.
[74] Rossi, J. and Clementi, F. (1985) L-Malic acid production by polyacrylamide gel entrapped *Pichia membranaefaciens*. Biotechnol. Lett. 7: 329
[75] Neufeld, R.J.; Peleg, Y.; Rokem, J.S.; Pines, O. and Goldberg,I. (1991) L-Malic acid formation by immobilized *Saccharomyces cerevisiae* amplified for fumarase. Enzyme Microb. Technol. 13: 991-996.
[76] Chen, J.P. and McGill, S.D. (1992) Enzymatic hydrolysis of triglycerides by *Rhizopus delemar* immobilized on biomass support particles. Food Biotechnol. 6: 1.
[77] Baré, G.; Delaunois, V.; Rikir, R. and Thonart, P. (1994) Bioconversion of vanillin into vanillic acid by *Pseudomonas fluorescens* strain BTP9. Appl. Biochem. Biotechnol. 45/46: 599-610.
[78] Hagedorn, S. and Kaphammer, B. (1994) Microbial biocatalysis in the generation of flavour and fragrance chemicals. Ann. Rev. Microbiol. 48: 773-800.
[79] Ramachandran Rao, S. and Ravishankar, G.A. (2000) Vanilla flavour: production by conventional and biotechnological routes. J. Sci. Food Agric. 80: 289-304.
[80] Cheetham, P.S.J. (1993) The use of biotransformations for the production of flavours and fragrances. Trends Biotechnol.11: 478-488.
[81] Schauenstein, E.; Esterbauer, H.; Zollner, H. (1977) Aldehydes in biological systems – Their natural occurrence and biological activities. In: Lagnado, J.R. (Ed.), Pion Ltd., London, UK; pp. 205.
[82] Nozaki, M.; Washizu, Y.; Suzuki, N. and Kanisawa, T. (1995) Microbial oxidation of alcohols by *Candida boidinii*. In: INRA (Ed.) Bioflavour 95, Paris, France; pp. 255-260.
[83] Egli, T.; Haltmeier, T. and Fiechter, A. (1982) Regulation of methanol oxidizing enzymes in *Kloeckera* sp. 2201 and *Hansenula polymorpha*, a comparison. Arch. Microbiol. 131: 174-175.
[83] Nozaki, M.; Suzuki, N. and Washizu, Y. (1996) Microbial oxidation of alcohols by *Candida boidinii*: selective oxidation. In: Takeoka, G.R.; Reranishi, R.; Williams, P.J. and Kobayashi, A. (Eds.) Biotechnology for improved foods and flavors, ASC Symposium Series 637, American Chemical Society, Washington, USA; pp. 188-195.
[84] Corran, H.S. and Mulvey, J.E. (1963) Improvements relating to the brewing of beer. English patent 984473.
[85] Lachance, M.A. (1995) Yeast communities in a natural tequila fermentation. Antonie Van Leeuwenhoek 68: 151-160.
[86] Spaepen, M.; Van Oevelen, D. and Verachtert, H. (1978) Fatty acids and esters produced during the spontaneous fermentation of lambic and gueuze. J. Inst. Brew. 84: 278-282.
[87] Spaepen, M.; Van Oevelen, D. and Verachtert, H. (1979) Higher fatty acids (HFA) and HFA-ester content of spontaneously fermented Belgian beers and evaluation of their analytical determination. Brauwissenschaft 32: 1-6.

[88] Van den Bremt, K.; Gasarasi, G.; Delvaux, F. and Derdelinckx, G. (1999) Bioflavouring by refermentation. Cerevisia 24(4): 31-39.
[89] Van den bremt, K.; Delvaux, F.R.; Verachtert, H. and Derdelinckx, G. (2001) Biogeneration of flavors: performance of *Candida methanolovescens* strains in nonalcoholic beer. J. Am. Soc. Brew. Chem. 59: 80-83.
[90] Derdelinckx, G.; Van Den Bremt, K.; Delvaux, F. and Verachtert, H. (2000) Bioflavouring by refermentation (part II) – Refermentation of beer: past, present and future. Cerevisia 25(1): 41-49.
[91] Derdelinckx, G.; Van Den Bremt, K.; Masschelein, C.A.; Verachtert, H. and Delvaux, F. (2000) Biogeneration of flavour (part III) – Novel methods for beer maturation. Cerevisia 25(2): 57-66.
[92] Gueguen, Y.; Chemardin, P.; Janbon, G.; Arnaud, A. and Galzy, P.A. (1996) J. Agric. Food Chem. 44: 2336-2340.
[93] Spaepen M. and Verachtert, H. (1982) Esterase activity in the genus *Brettanomyces*. J. Inst. Brew. 88: 11-17.
[94] Spindler, D.; Grohmann, K. and Wyman, C.E. (1992) Simultaneous saccharification and fermentation (SSF) using cellobiose fermenting yeast *Brettanomyces custersii*. US Patent 5100791.
[95] Shantha Kumara, H.M.C. and Verachtert, H. (1991) Identification of lambic superattenuating micro-organisms by the use of selective antibiotics. J. Inst. Brew. 97: 181-185.
[96] Freer, S.N. and Skory, C.D. (1996) Production of beta-glucosidase and diauxic usage of sugar mixtures by *Candida molischiana*. Can. J. Microbiol. 42: 431-436.
[97] Van den Bremt, K. (2001) Production of flavour aldehydes from glucosides by *Candida methanolovescens*. PhD Dissertation, Katholieke Universiteit Leuven, Leuven, Belgium.
[98] Willaert, R. (2001) New trends in brewing technology: wort boiling, bioflavouring and beer mixed drinks. Proceedings conference "Malting and Brewing Technology", Hogeschool Gent, Gent, Belgium.
[99] Prescot, S.C. and Dunn, D.G. (1959) The acetic acid bacteria and some of their biochemical activities. In: Industrial microbiology. McGraw-Hill, New York, USA; pp. 428.
[100] Yoshio, K.; Hisayoshi, Y.; Ko, I.; Noriaki, F. and Yoshio, Y. (1987) Verfahren zur Herstellung von Weinsäure. BRD Patentschrift 2600589.
[101] Yuichi, M.; Kiyohiko, Y.; Hitoshi, T.; Kenji, F. and Yusuke, I. (1980) Verfahren zur enzymatischen Herstellung von *L*-weinsäure aus *cis*-Epoxybernsteinsäure. German Patent 2605921.

PART 4

INDUSTRIAL BIOCHEMICAL PRODUCTION

PRODUCTION OF ETHANOL USING IMMOBILISED CELL BIOREACTOR SYSTEMS

ARGYRIOS MARGARITIS AND PETER M. KILONZO
Department of Chemical and Biochemical Engineering, University of Western Ontario, London, Ontario N6A 5B9, Canada –
Fax: 519-661-3498 – Email: amarg@uwo.ca

1. Introduction

Crude oil and its derivatives are the sources of the vast proportion of fuels used today. However, the 1973 oil embargo and accompanying increases in oil prices, along with the prospective of fossil fuels exhaustion in the near future [1-3] triggered an extensive interest in search for alternative sources of liquid fuels. Amongst these alternatives is the production of ethanol by fermentation process using renewable plant products. The production of industrial ethanol by fermentation started in the 19th century and used as an automotive fuel in the late 1930s. The production, however, declined in the late 1940s as a result of competition with ethanol produced chemically by direct hydration of cheap oil-based ethylene.

Because of the increasing environmental concerns in recent years regarding greenhouse gas emission (1996, Habitat II Agenda), interest is again turning to fermentation ethanol as a motor fuel. The use of biomass derived ethanol as a gasoline substitute or supplement has been implemented with a great zeal in Brazil and in the U.S. In 2002, production in Brazil reached about 7.0 billion gallons (BG) and 2.2 BG in the U.S [4]. According to the U.S. gasohol program, fuel ethanol production is expected to reach 2.3 BG in 2004 increasing to 5.0 BG in 2012. To boost this program, the U.S administration is providing market, loan and tax credits for ethanol made from renewable sources.

The cost of fermentation ethanol is, however, higher than gasoline. Ethanol's wholesale price in the U.S. is about US$1.10/gal. This compared with an average price of about 70¢/gal for gasoline at the refinery [4]. However, as a gasoline additive, ethanol in the U.S. obtains an excise tax reduction that amounts to 5.3¢/gallon for winter gasoline that contains 10% ethanol. This is equivalent to a subsidy of 53¢/gal of ethanol. The selection of a suitable and cheap raw material is therefore central to the economics of any new ethanol plant. Furthermore, since ethanol is a low cost, high volume product, the economic success of such a product depends very much on the capital and operating costs.

Currently, most of the ethanol produced employs conventional batch fermentation technology. In Brazil, the Melle Boinet process in which yeast cells are separated and reused after acid rinsing is industrialised to almost 70% of large-scale ethanol production [5]. These processes are associated with complex operation requiring the traditional "fill and shut" batch process, the dangers of culture contamination, a general "instability" of the system from an operations standpoint, unrealised cost and productivity advantages in the process plant, and the fear of genetic drift of the yeast strain [6]. Moreover, these devices have low volumetric productivities and require long fermentation times [7]. In addition the high capital and operating costs, the continual start up and shut down nature of such processes makes them difficult to automate and therefore high labour costs ensue.

Consequently, the use of continuous processes, which are simple to operate, with low energy requirements, and allowing almost complete utilisation of the expensive substrates, will significantly lower the operating costs. Capital cost on the other hand, may be reduced by using mechanically simple, small bioreactors with high rates of ethanol production. Superior ethanol productivities may be achieved by employing a high concentration of yeast or bacterial cells (the catalysts for the ethanol production reaction) within the bioreactor. However, the production of ethanol using conventional continuous processes has limitations of maintaining high cell concentrations in the bioreactor. Such limitations are associated with low volumetric productivities and long fermentation times in these systems [7]. Present trends in immobilised yeast or bacterial cell bioreactor systems nowadays provide the ethanol industry with a method not only for maintaining high cell concentrations in the bioreactors, but also reducing processing time without sacrificing product quality.

The purpose of this chapter is to provide a state-of-the art review of published literature, and describe the use of immobilised cell bioreactor systems in the production of fuel ethanol and compare the efficiency of such systems vis-à-vis freely suspended cell systems used in traditional ethanol fermentation systems.

2. Immobilised cell systems

Most microorganisms tend to attach themselves to solid-liquid interfaces [8]. Cell immobilisation is encountered in many different bioprocesses, such as wastewater treatment, vinegar production, bioleaching of mineral ores [9,10]. Some of the biotechnological processes carried out using immobilised cells include the continuous production of L-aspartic acid with immobilised *E. coli* [11,12], production of L-malic acid with immobilised *Brevibacterium ammoniagenes,* isomerization of glucose to fructose by immobilised cells of *Streptomyces* spp., and production of L-alanine by immobilised *Pseudomonas dacunhae*. Moreover, the industrial production of high fructose corn syrup, L-aspartic acid and L-malic-acid has been achieved using immobilised microbial systems [13,14].

The production of ethanol from glucose requires the sequential action of different enzymes in addition to the two coenzyme pairs, ATP/ADP and NADP/NAD. In order to achieve such a conversion, the constant regeneration of the coenzymes is essential. This may only be achieved if the immobilised cell is maintained in a viable state.

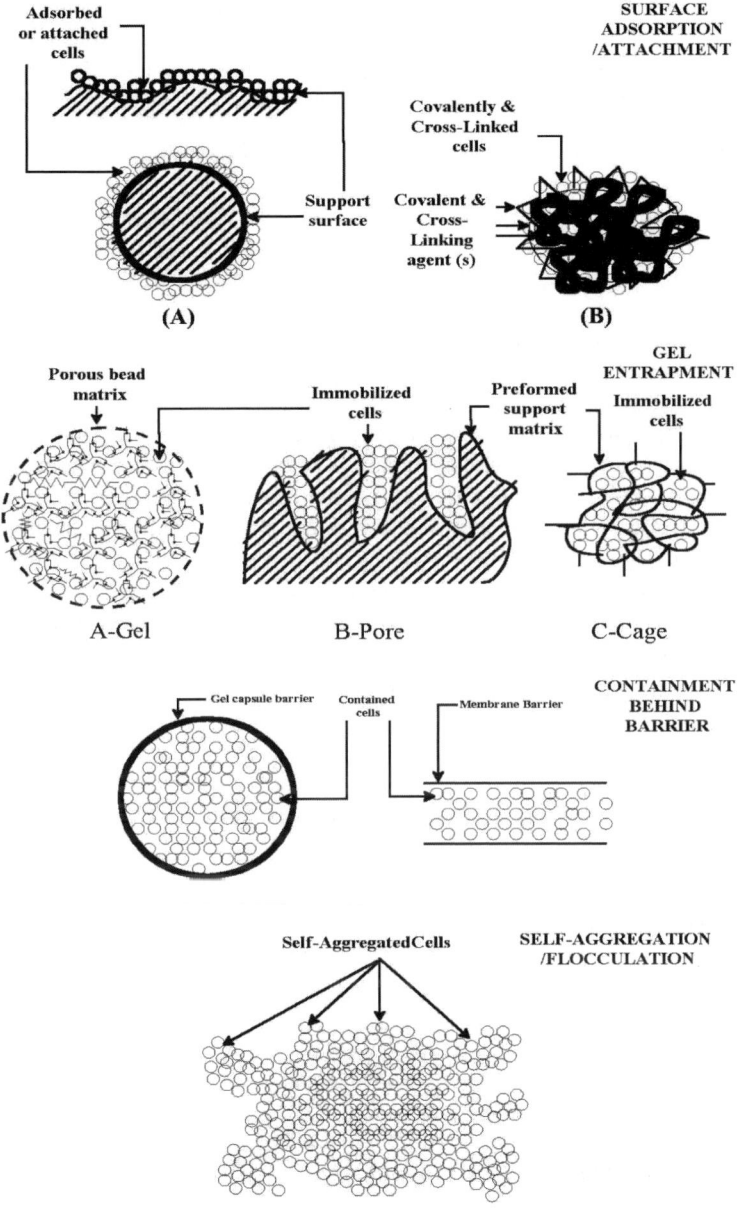

Figure 1. Basic immobilised cell systems (adapted from Dervakos & Webb [17], Masschelein et al. [23]).

Thus, although several techniques have been developed for the immobilisation of cells (see reviews [15-17]) not all of these can be readily applied to fermentation systems requiring viable cells. It is therefore essential that any technique developed for the immobilisation of live cells, must be mild enough to retain the viability of cells.

2.1. BASIC PRINCIPLES OF CELL IMMOBILISATION

Cell immobilisation may be defined as the physical confinement or localization of viable microbial cells to a certain defined region of space in such a way as to limit their free migration, while retaining their desired catalytic activities for repeated and continuous use [16-19].

This is most usually achieved by significantly increasing the effective size or density of the cells by aggregation or by attachment of the cells to some support surface. A number of different techniques and immobilisation support materials have been proposed for cell immobilisation *in vitro*. Figure 1 illustrates basic methods for cell immobilisation. Four main groups of immobilised cells systems can be distinguished in terms of the immobilisation technique and the nature of the support material used [16,19]: Adsorption or attachment to a pre-formed carrier surface by adhesion or cross-linking agents, physical entrapment within porous matrix, cell containment behind a barrier and self-aggregation by flocculation (natural) or cross-linking agents [16,20-22].

Several investigators [23,24,25] have indicated that the success of large scale industrial systems is dependent on the cell immobilisation technique. Table 1 summarises the criteria used to select most optimum immobilisation technique.

Table 1. Selection criteria for cell immobilisation.

- Have minimal adverse effects on the desired biocatalytic properties;
- Safe and simple and relatively inexpensive;
- Require few process steps and ingredients;
- Generate no by-products requiring special disposal procedures;
- Amenable to scale-up;
- Adaptable to automation;
- Stable at the operating pH and temperature;
- Economical
- Capable of regeneration following deactivation which may occur after a long time of operation;
- Mild enough to ensure cofactor regeneration capability;
- High biomass concentrations that can be achieved within the bioreactor and retained at the level for an extended period of time.

During the last 30 years, many different types of matrices for immobilisation have been developed, which includes porous and non-porous pre-formed, materials such as, wood chips [26], diatomaceous earth, volcanic rocks, ceramics, stainless steel, porous brick, porous sintered or sheet glass, porous silica, DEAE cellulose [26], PVC chips, polyurethane cubes, cotton cloth, glass fibre, and plant cell matrices. Several polymeric matrices have been used for cell immobilisation, such as, Calcium alginate, kappa-carrageenan, [27], polyacrylamide, gelatin [28] and epoxy resin, agar, chitosan, and silica sols. The choice of cell-supporting material for any specific application ideally should meet the several important criteria summarised in Table 2.

Table 2. Criteria for selection of cell immobilisation matrices.

- Retain the desired biocatalytic activity of the cells;
- No reaction with substrates, nutrients or products;
- Retain their physical integrity and be insoluble under the bioprocess reaction conditions;
- Permeable to reactants and products;
- Have large specific area per unit volume;
- Have high diffusion coefficients for substrates, nutrients and products;
- Provide appropriate hydrophilic-hydrophobic balance for nutrients, reactants and products;
- Resistant to microbial degradation, and excellent mechanical strength;
- Retain chemical and thermal stability under bioprocess and storage conditions;
- Elastic enough to accommodate the growth of cells;
- Have functional groups for cross-linking;
- Generally recognized as safe for food and pharmaceutical bioprocess applications;
- Generally non-toxic and available in adequate quantities with consistent quality and acceptable price;
- Easy and simple to handle in the immobilisation procedure;
- Environmentally safe to dispose of, and/or be capable of recycling;
- The manufacturing system is efficient, easy to operate, and give good yields.

2.2. CELL IMMOBILISATION BY ADSORPTION

The attachment of cells to non-porous or porous solid supports may be accomplished by two different methods:
- adsorption of cells to support by virtue of electrostatic interactions
- attachment of cells to supports by covalent bonding or by using cross-linking agents,

whereas in the special case of pelleting or aggregation the biomass itself can serve as immobilisation matrix.

Any surface in contact with nutrient medium and suspended microorganisms will be colonized due to cell adsorption, and it is the most natural method of cell immobilisation and the formation of biofilm. One of the limitations of this immobilisation method is that cells may be detached from the surface due to high shear forces during fluid flow.

The affinity by microbial cells to surfaces depends on the chemical nature of the cell wall surface [29]. Various physical and chemical factors affect cell adsorption to solid surfaces, which include, the type of support (particle size, porosity, charge, composition), microorganisms (strain, age, surface charge) and environmental conditions (pH, temperature, liquid flow rate, and ionic strength). The cationic or anionic character of microbial cells also depends on the pH of the solutions and environmental conditions [16]. Van der Waals forces, electrostatic interactions, covalent bonding, and hydrophobic interactions, play a key role in adsorption. In some cases, the initial attraction between the cell and the support surface, is followed by a cellular secretion of the adhesion biopolymers that stabilise the attachment [23].

There are several applications of cell immobilisation employing glass as an adsorption support. Sintered glass, Raschig rings can be used for spontaneous adsorption. Sintered glass with 60% porosity, 60 to 100 nm pores, and specific surface area of 0.4 m^2/g has produced good results. Lommi *et al.* [30] reported the use of granular DEAE cellulose particles in the immobilisation of *Saccharomyces cerevisiae* cells by adsorption. Table 3 summarises ethanol production results by different investigators using immobilised cells by adsorption.

Table 3. Ethanol production by cells immobilised by adsorption to a surface.

Type of support	Micro-organism	Type of bioreactor	Substrate conc. (g/l)	Sugar utilisation (%)	Ethanol conc. (g/l)	Productivity (g/l/h)	Operation time (days)	Ref.
Wood chips	S. cerevisiae NRRL-Y-132	Vertical packed bed	Glucose 164	98	76	21.2	–	[31]
Porous ceramic particles	S. cerevisiae UG-5	Vertical packed bed + solvent extraction	Glucose 409	100	—	1.03 $(V_T)^a$	–	[32]
Silica gel particles	Z. mobilis ATCC 10988	Tapered fluidised bed	Glucose 50	–	–	21.2	–	[33]
Anion exchange resin XE-352	S. cerevisiae	Vertical packed bed	Glucose 120	94	–	53.1 $(V_L)^b$	–	[34]
Wood chips	S. cerevisiae	CSTR in series	Glucose 100	70-90	–	60 $(V_L)^b$	29	[35]
Boro-silicate glass fibre	Z. mobilis ATCC 10988	Vertical packed bed	Glucose 160	–	64	150 $(V_T)^a$	28	[36]
Glass beads	S. cerevisiae ATCC 24860	Vertical packed bed	Glucose 150	99	–	43 $(V_L)^b$	–	[37]

^{a} V_T: total bioreactor volume; ^{b} V_L: liquid bioreactor volume.

2.3. CELL IMMOBILISATION BY COVALENT BONDING

In order to circumvent the problems associated with adsorbed cell system, the use of cross-linking agents and covalent bonding of cells to inert supports has been investigated.

Modified supports used have included cross-linked gelatin [38], cross-linked chitosan beads [39], and ceramics pretreated with polyethyleneimine [40]. Porous materials enhance the surface area available for cell binding [41]. However, due to toxicity of most of the chemical reagents used, covalent binding of cells to a carrier is not frequently employed. This technique does have the advantage that cells are linked to a uniform surface by bonds, which are stable for long periods so that cell leakage from the bioreactor is minimised.

2.4. CELL IMMOBILISATION BY PHYSICAL ENTRAPMENT WITHIN POROUS MATRICES

Cell immobilisation by entrapment is based on the inclusion of cells within a rigid polymeric matrix. Mosbach & Mosbach [32] were the first to report the successful entrapment of whole cells within a polymeric gel matrix with retention of enzyme activity. Updike et al. [42] and Mosbach & Larson [43] subsequently entrapped live cells and demonstrated that the cells can be made to grow while present within the support.

Physical entrapment has been a well-accepted method for live cell immobilisation and has been used extensively for the continuous production of ethanol. In considering a suitable matrix for live cell immobilisation, the following characteristics are desirable:
- mixing of the cell suspension and polymer solution;
- gel formation should occur under mild conditions without the use of toxic chemicals in order to maintain the viability of cells;
- the gel should have good mechanical stability under the operating conditions;
- the pore size within the gel matrix must be sufficiently large enough to allow the diffusion of the substrates and products, but should be small enough to retain the cells within the matrix;
- the overall cost of the entrapment procedure should be low.

In the production of ethanol, entrapment of cells has been achieved within five different porous matrices, namely, Ca-alginate [20,44-51], κ-carrageenan, [27,52], polyacrylamide, gelatin [53] and epoxy resin. In achieving successful entrapment, three different techniques were used to induce gelation depending on the properties of the polymer used:
- entrapment in ionic polymer networks by ionic cross-linking of linear polyanions (alginate) via multivalent cations;
- entrapment by precipitation caused by changes in pH and temperature or solvent changes (gelatin and carrageenan);
- entrapment within covalent polymeric matrices by cross-linking-copolymerisation (polyacrylamide gels) or by polycondensation of prepolymers (epoxide).

2.4.1. Calcium alginate matrix

Alginic acid is a heteropolysaccharide of L-guluronic acid and D-mannuronic acid extracted from various species of brown algae. Depending on the source, the composition and the sequence in L-guluronic acid and D-mannuronic acid vary widely. The monomers are arranged in a pattern of blocks along the chain, with homopolymeric regions interspersed with regions of alternating structure. After processing, alginates are available as water-soluble sodium alginates. When the water-soluble sodium alginate is mixed with multivalent counterions (Ca^{2+}, Al^{3+}, Zn^{2+}, Co^{2+}, Ba^{2+}, Fe^{2+}, or Fe^{3+}), gelation occurs by displacement of sodium ion.

Alginate beads are formed by droplet or emulsification methods using 2 to 4% Na-alginate sterile solutions prior to introducing the cell suspension. Gel beads are produced by dropping the cell/polymer suspension through a syringe needle (0.22 to 2 mm I.D.), to free-fall approximately 20 cm into a $CaCl_2$ (20 to 100 mM) hardening solution. It is possible to scale-up the extrusion technique by using a resonating vibrating nozzle, which causes a breakup of the jet into drops at a frequency interval of 200-700 Hz [54-57].

The emulsification process involves the dispersion of the cell/polymer suspension into a vegetable oil. The cell/alginate droplets are gelled by adding either $CaCl_2$ solution [53] or by internal gelation [58]. In the latter case, an insoluble calcium vector such as the carbonate salt is added prior to emulsion formation. The liberation of Ca is initiated by the addition of an oil-soluble acid such as acetic acid, reducing the alginate pH from 7.5 to approximately 6.5. Following gelation, the beads are partitioned into water and washed to remove residual oil. Very high biomass loadings can be achieved and polymer-cell mixture can be formed in different shapes and sizes. The most common forms are small beads about 0.3 to 5 mm in diameter [16]. Small diameter beads are generally preferred because of the more favourable mass transfer characteristics for the entrapped cells. Furthermore, the immobilisation reagents are of low costs, making the procedure attractive for large scale application. The porosity and diffusion properties of the alginate pellets are influenced by the concentration of sodium alginate and or calcium chloride solutions employed [59,60]. Thus, at relatively high concentrations of the immobilizing reagents, the pellets formed are readily diffusible to low molecular weight (MW) compounds (MW < 5,000), but not to the higher MW enzymes (MW > 100,000), whereas by lowering the concentration of the gelating agents, pellets of high porosity can be formed to allow the diffusion of high MW compounds. In addition, the type of alginate used is also known to affect the diffusivity of various molecules into and out of the gels.

This technique characteristically allows a considerably higher biomass loading than immobilisation in or preformed supports. The significance of this immobilisation method is that the matrix is porous enough for substrates and products to traverse where a level of cell retention is maintained within the immobilisation matrix.

This retention of cells should obviously be as complete as possible. In ionotropic gels, such as alginate and chitosan, the maximum pore size is 15 nm [23], which is well below the size of any microbial cell and that, therefore, helps to prevent cell leakage. Nevertheless, cellular outgrowth is one of the central problems of most gel-entrapped systems and considerable work has focused on overcoming this problem. To this end, coated beads with double layer polymers have been proposed by several investigators [61,62].

The mechanical properties of Ca-alginate pellets have been studied [59]. The mechanical strength of the gel matrix is important for minimising gel splitting or stripping of cells from the matrix due to the evolution of CO_2 by the immobilised cells during fermentation. These studies have shown that the pellets so formed are resilient and versatile. Due to the even size and spherical shape of the alginate pellets and good mechanical properties, packed bed columns have been used with low intrinsic pressure drops. Operational stability of immobilised cells in algal polysaccharides can be improved by treatment with hardening agents such as periodate, glutaraldehyde, or hexamethylenediamine [11] or polyethyleneimine, silica, and or polyvinylalcohol [16].

Abrasion caused by particle-to-particle contact, particularly in fluidised bed or stirred tank reactors, has been a further problem. Particle compression, as in packed bed bioreactors, may also lead to immobilised cell aggregate breakdown and has been a further reason for the optimisation of mechanical strength of the gel particle. Thus, due to the several proven good properties of the alginate gels, yeast and bacterial cells have been entrapped for the continuous production of ethanol.

The performance characteristics of various systems employing alginate entrapped yeast and bacterial cells for continuous ethanol production have been studied using alginate entrapped yeast, ethanol productivities of up to 46 g/l/h has been achieved [63]. One of the highest bioreactor volumetric ethanol productivities of 102 g/l/h has been achieved by Margaritis et al. [44] using Ca-alginate immobilised cells of Z. mobilis in a packed bed bioreactor system. Additionally, long-term operational stability of such systems has been shown to be possible provided that cell regeneration is periodically carried out. The use of alginate as a suitable cell entrapment matrix has gained widespread popularity in recent years due to the simplicity of this method, it is non-toxic, less expensive, reversible, and has good mechanical properties [64]. Alginate is also resistance to acidic and basic environments and also stable even at thermophilic fermentation temperature [65].

Due to these proven features of alginate entrapped yeast cell system, it has been considered for large-scale production of ethanol in Japan. Pilot scale studies using alginate-entrapped yeasts were carried out with two fluidised bioreactors, 200 m^3 each in volume, which yielded 2000 l per day of ethanol for several months. The alginate gel beads (2 to 4 mm diameter), occupying 50 to 60% of the bioreactor volume, were kept in a fluidised state by using a mixture of air and recycled CO_2. Conversion rates of 95% of the theoretical were achieved with an effluent ethanol content of 10% (w/v) when the bioreactor was fed with dilute cane molasses at a residence time of 3.3 h.

A major disadvantage of the use of calcium alginate as an immobilisation matrix is that moderate concentrations of calcium chelating agents and certain cations such as phosphates, EDTA, Mg^{2+} and K^+ disrupt the gel [59] by solubilising the Ca^{2+}. Phosphate, which is assimilated by yeast cells via an active transport mechanism is essential for the control of the biosynthesis of lipids and carbohydrates, maintenance of cell wall integrity and yeast viability [66]. The incorporation of phosphate as a nutrient for the regeneration of fermentation in an immobilised yeast cell system is therefore essential. Unfortunately, the presence of phosphate in calcium alginate entrapped cell system causes the swelling of beads. This occurs due to the breaking of the bonds between Ca and alginate, resulting in a loss of the mechanical stability of the gel and finally to its complete disruption.

Several techniques have been suggested to stabilise and to maintain the integrity of calcium alginate gel beads. Veliky and Williams [67] observed that the treatment of calcium alginate beads with certain high molecular weight cationic polymers (e.g., polyethyleneimine), led to the formation of surface coated material possessing improved stability in the presence of phosphate. The treatment procedure did not reduce the ethanol production rate of the calcium alginate entrapped cells, but did, however, inhibit the respiration of entrapped cells.

Birnbaum et al. [68] applied the stabilised system to the conversion of glucose to ethanol in 10 mM phosphate, pH 7.0, using S. cerevisiae. The experiment showed that contrary to untreated preparation, which completely dissolved in 72 h, the stability of the treated preparation was unaffected releasing practically no cells into the effluent. Several other materials for improving the physical and chemical stability of the calcium alginate gel have also been described. The use of divalent cation other than Ca^{2+} as gel inducing agents has been suggested. Haug and Smidstrod [64] found that Ba was a much more efficient gel inducing cation than Ca. Paul and Vignais [70] demonstrated that Sr^{2+} and especially Ba^{2+} produced gels of greater mechanical and chemical stability than Ca^{2+} and that barium alginate bead could maintain their structural stability for relatively long period of time in the presence of a 10 mM phosphate buffer solution. Thus, with stabilised alginate beads, phosphate may be readily incorporated into the reaction medium either as an essential nutrient or as cheap buffer without fear of gel disruption. Table 4 summarises ethanol production results by different investigators using immobilised cells by entrapment in alginate gels.

Table 4. Ethanol production by cells immobilised entrapment in alginate gels.

Type of support	Micro-organism	Type of bioreactor	Substrate conc. (g/l)	Sugar utilisation (%)	Ethanol conc. (g/l)	Productivity (g/l/h)	Operation time (days)	Ref.
Ca-alginate	Z. mobilis	Column, perforated plates	Glucose 100	96	50	15 $(V_T)^a$	33	[71]
Ca-alginate	S. cerevisiae	Vertical packed bed	Molasses 175RS[c]	78	70	17.5 $(V_T)^a$	60	[72]
Ca-alginate	S. cerevisiae	Fluidised bed	Glucose 100	94	47	21.2 $(V_L)^b$	–	[73]
Ca-alginate	Z. mobilis ATCC 10988	Horizontal packed bed	Glucose 100	97	44.2	71 $(V_L)^b$	16	[48]
Ca-alginate	K. marxianus UCD (FST) 55-82	Vertical packed bed	Jerusalem artichoke extract 101RS	92	42.4	80 $(V_L)^b$	12	[49]
Al-alginate	S. formosensis M-111	Multi-stage fluidised bed	Molasses 162RS	–	78	28	–	[74]

[a] V_T: total bioreactor volume; [b] V_L: liquid bioreactor volume; [c] RS: reducing sugars.

2.4.2. Carrageenan matrix

Carrageenan is a polysaccharide isolated from seaweed composed of unit structures of β-D-galactose sulfate and 3,6,anhydro-α-D-galactose. It has a molecular weight of 100,000 to 800,000 and due to its non-toxicity it has been widely used as a food additive [75]. In an attempt to develop a suitable technique for enzyme and cell immobilisation, Takata et al. [76] screened a wide variety of natural and synthetic polymers. Carrageenan was found to have amongst the best overall characteristics for cell immobilisation and was subsequently used for the immobilisation of a wide variety of enzymes and cells.

Carrageenan, which is available commercially at a low cost, can be easily induced to gel by contact with a solution containing a number of gel inducing agents, such as, metal ions, amines and water-miscible organic solvents. However, due to its mildness and simplicity, cooling and/or contacting with an aqueous solution containing K^+ ions may preferably carry out gelation. Furthermore, the pore size of the gel matrix is small enough for the passage of low molecular weight substrates and products. The structure is mechanically stable and if necessary, preparations of greater stability have been obtained by further treatment with hardening reagents such as glutaraldehyde, polyacrylamide [77], and hexamethylenediamine.

Entrapped cell systems are also subjected to mass transfer limitations imposed by the additional diffusion barrier created by the support matrix. This is especially true when low substrate concentrations or high molecular weight substrates are used. Thus the productivity of a given bead may depend largely on the activity of cells existing on the surface of the particle [78]. Additionally, high levels of the toxic product ethanol may accumulate within the gel and therefore reduce the efficiency of the system. However, the use of small diameter beads, [44] and also, open porous matrices [79] may substantially improve the mass transfer of substrate and product, into and out of the matrix, respectively.

Table 5. Ethanol production by cells immobilised by entrapment in carrageenan matrices.

Type of support	Micro-organism	Type of bioreactor	Substrate conc. (g/l)	Sugar utilisation (%)	Ethanol conc. (g/l)	Productivity (g/l/h)	Operation time (days)	Ref.
κ-carrageenan	Z. mobilis Z-M4	Vertical packed bed	Glucose 150	97	54.4	16 $(V_T)^a$	33	[84]
κ-carrageenan	S. cerevisiae IFO 2363	Vertical packed bed	Glucose 250	95	114	43.8 $(V_L)^b$	60	[78]
κ-carrageenan	Saccharomyces bayanus 60	Vertical packed bed	Glucose 175	82.3	69.3	7 $(V_L)^b$	30	[85]
κ-carrageenan	Z. mobilis ATCC 10988	Horizontal packed bed	Glucose 108	86	41.7	101 $(V_L)^b$	9	[20]

^{a} V_T: total bioreactor volume; ^{b} V_L: liquid bioreactor volume.

In ethanol fermentation systems, CO_2 gas is also a major product. This gaseous product is characterized by its relatively low solubility in aqueous media and therefore the diffusion of the gaseous product out of the matrix can be rate limiting [80]. Thus, if the diffusion of CO_2 out of the matrix of slower compared to its production, CO_2 will accumulate and be saturated within the matrix. Eventually, gas bubbles may be formed within the matrix, which results in the disruption of the entrapment matrix, and therefore, sets some severe restrictions to the applicability of such systems.

Using alginate and agar-agar entrapped yeast cell systems; expressions of the theoretical maximum productivity of this system have been derived [80]. The productivity has been shown to be a strong function of the critical gas concentration within the matrix (CGC), which in turn depends on the mature of the matrix used. Thus the mechanically weak and brittle matrices (*e.g.*, agar-agar) are disrupted easily, whereas stronger and elastic matrices, such as calcium alginate, can withstand high-pressure differences without gel disruption.

CO_2 has also been shown to inhibit the aerobic growth of yeast aerobically [81] and also during anaerobic fermentation [82]. These effects have been extensively evaluated in a recent review by Jones & Greenfield [83]. The CO_2 gas accumulation within a matrix, may not only cause gel disruption, but also adversely affect the metabolism of entrapped cells. Thus, by employing small, porous beads, facilitating the rapid removal of CO_2 gas, combined with the use of mechanically strong and flexible matrices, the undesirable effects due to CO_2 may be minimised in entrapped cell systems. The application of cells immobilised in carrageenan for continuous ethanol fermentations are shown in Table 5

2.4.3. Polyacrylamide gel matrix

Polyacrylamide was the first support used for the entrapment of cells and enzymes and it was shown that cells so immobilised could retain their enzymatic activity and cell viability [43]. Since then, entrapment in polyacrylamide has been the most frequently employed technique for enzyme and cell immobilisation. In general, entrapment of cells in polyacrylamide involves the polymerisation of an aqueous solution of acrylamide monomers in which microorganisms are suspended. The polymerisation of acrylamide takes place by a free radical process in which linear polyacrylamide chains build up. The inclusion of a bi-functional agent, which has two saturated double bonds, results in cross-links between the polymer chains [86]. The porosity of the gel is a function of the degree of cross-linking which in turn depends on the relative amounts of the acrylamide monomer and the bi-functional cross-linking agent used. The cell containing polymeric gel can be granulated for use as a column packing.

Due to the toxicity of the acrylamide monomer, the temperature used for the polymerisation process, the duration of the contact of cells with the monomers, and the time required for gel formation, dictate the cell viability. Thus, An increase in temperature to 20°C or higher, and an increase in the duration of immobilisation to 20 minutes or more, leads to a sharp decline in the number of viable yeast cells [87]. Siess & Divies [88] observed that polymerisation process may destroy 40 to 80% of the cells depending on their physiological state. Thus, *S. cerevisiae* cells in the stationary phase of growth were much more resistant to the polymerisation reaction than those in the exponential phase of growth.

2.4.4. Epoxy resin matrix

Epoxy polymeric networks have been obtained by well-defined polycondensation reactions. The characteristics of these materials are their high mechanical strength and chemical stability under variable conditions [89]. The covalent polymeric network is formed by polycondensation of epoxy beads. This was achieved by preparing calcium alginate coated epoxy beads containing the epoxy precursor, polyfunctional amines, and microbial cells. Polycondensation was allowed to take place within the beads and the particles dried using mild ventilation. The calcium alginate component of the beads was subsequently leached out with a phosphate buffer, which led to the formation of capillary channels throughout the matrix.

Table 6. Ethanol production by cells immobilised by entrapment in polyacrylamide, gelatin and epoxy resin matrices.

Type of support	Micro-organism	Type of bioreactor	Substrate conc. (g/l)	Sugar utilisation (%)	Ethanol conc. (g/l)	Productivity (g/l/h)	Operation time (days)	Ref.
Pectin	S. cerevisiae CMI-120 26602	Horizontal packed bed	Molasses 160RS^c	90	70	40 $(V_L)^b$	20	[53]
Pre-polymerised poly-acrylamide	S. cerevisiae	Vertical packed bed	Glucose 200	90	92	51.1 $(V_L)^b$	14	[95]
Pre-polymerised poly-acrylamide polymerised on agar or carrageenan	S. cerevisiae ATCC 24553	Vertical packed bed	Glucose 100	95	42	50.5 $(V_L)^b$	60	[77]
Gelatin	S. uvarum	Vertical packed bed	Molasses 160RS^c	95	45.2	144 $(V_L)^b$	30	[96]

^{a} V_T: total bioreactor volume; ^{b} V_L: liquid bioreactor volume; ^{c} RS: reducing sugars

2.4.5. Gelatin polymer matrix

Unlike epoxy and polyacrylamide gels, gelatin is a natural polymer. It was first used by Gianfreda et al. [90] for immobilisation of S. cerevisiae cells having invertase activity. Composite beads of calcium alginate-gelatin containing entrapped yeast cells were treated with a phosphate buffer to leach out the calcium alginate component. The residual gelatin matrix was then cross-linked with 0.01 M glutaraldehyde for a period of 1.5 h at about 10°C.

Viable yeast cells have also been entrapped in glass beads, glass sheets coated with SiO_2-gel, sponges, natural zeolite and pectin for ethanol production [91-94]. The main application of cells immobilised in matrices other than alginate or carrageenan for continuous ethanol fermentations can be found in Table 6.

2.5. CELL IMMOBILISATION BY CONTAINMENT BEHIND A MEMBRANE BARRIER

Margaritis and co-workers [97-100] reported the first application of immobilised yeast cells for the production of ethanol at high cell densities using a rotating sintered stainless steel microporous membrane inside the bioreactor vessel. This novel bioreactor called the "rotorfermenter" has several advantages compared to ordinary bioreactor systems used to produce ethanol with freely suspended yeast cells.

Other membrane based bioreactors have been reported to be ten times the maximum obtainable by calcium alginate. Diffusion limitations involving the build-up of inhibitory products are one of the limitations of membrane supports. The membranes are typically made of polymers (e.g., polyvinylchloride, polypropylene, or polysulfon). The membrane should be hydrophilic for easy exchange of nutrients and products and mechanically strong to withstand the pressure differentials. Two basic configurations are used: sheets and hollow fibre cartridges. The sheet-type usually has low strength and requires additional support. Hollow fibre cartridges provide the membrane with strength and a higher surface-to-volume ratio, which are the characteristics of hollow fibre bioreactors.

Problems that arise with hollow fibre systems are: the supply and removal of gases, as CO_2 gas bubbles can block the flow in the tubes. Unrestricted cell growth can restrict the nutrient flow, and unchecked adhesion can cause fouling of the membranes. To control growth, antibiotics, mutants, and nutrient starvation have been employed.

Table 7. Ethanol production using immobilised cells by retention behind membrane barriers.

Type of bioreactor	Micro-organism	Substrate conc. (g/l)	Sugar utilisation (%)	Ethanol conc. (g/l)	Productivity (g/l/h)	Operation time (days)	Ref.
Dialysis membrane reactor	S. cerevisiae ATCC 4126	Glucose 200	78	82.7	2.1	–	[102]
Rotor fermentor	S. cerevisiae ATCC 4126	Glucose 104	100	–	27.3	–	[100]
Hollow fibres	S. cerevisiae NRRL-Y-132	Glucose 100	85	40	10	4	[103]

An interesting variation of standard containment methodology is called encapsulation. Cells are usually first entrapped in a spherical gel, but then the sphere is coated with a polymer such as polyethyleneimine (PEI). The gel is then dissolved, and the cells are left in suspension but are contained behind the polymer barrier that reportedly allows only molecules of less than 100,000 Da to pass through. The porosity of this polymer shell can be modified by changing the polymer composition [101]. The mechanical containment of cells within a bioreactor may be achieved by incorporating porous material, microfilters or dialysis membranes in the bioreactor. Some of the application of cells immobilised by retention behind membrane barrier for continuous ethanol fermentations are shown in Table 7.

2.6. CELL IMMOBILISATION BY SELF-AGGREGATION

The flocculent nature of certain yeasts and *Z. mobilis* cells has provided a means of maintaining high cell populations within a bioreactor without the need for added materials to support film growth and retain cells. It is therefore a mode of immobilisation in which the addition of the inert materials is not necessary [26,104]. Nevertheless, the nature of flocculation and the influence of factors such as cell age, ionic strength, and pH have not been completely established. The characteristics of various flocculent cell systems [105] and their application in brewing [106] have been adequately reviewed. These properties have been exploited for the production of fuel ethanol.

Various bioreactor configurations have been employed for ethanol using flocculent cells. These include, stirred tank bioreactors, slant tubes, vertical packed columns, gas-lift fluidised bioreactors, and tower bioreactors. The most popular reactor is the tower bioreactor. The tower bioreactor consists, essentially, of a vertical cylindrical column with a conical base through which the medium is fed. The fermentation proceeds as the medium rises through the dense floc of yeast cells. A gas separation device is incorporated at the top of the column, which allows a volume, free from turbulence and therefore permitting yeast settling to occur [107].

Table 8. Ethanol production by self-aggregated cells.

Micro-organism	Type of bioreactor	Substrate conc. (g/l)	Sugar utilisation (%)	Ethanol conc. (g/l)	Productivity (g/l/h)	Operation time (days)	Ref.
Z. mobilis ZM401	CSTR + settler + cell recycle	Glucose 100	95	49	39	–	[109]
Z. mobilis NRR B-12526	Vertical tapered column	Glucose 110	96	46	39	–	[27]
S. cerevisiae NCYC 1257	Tower bioreactor	Sucrose	95	58	16.1	–	[110]
Z. mobilis WR6	CSTR	Glucose 100	97	47.6	80	–	[111]
S. uvarum	Tower bioreactor	Glucose 170	95	72	15.1	48	[66]

The major drawbacks of the tower fermentation system are complexity, lack of flexibility, contamination risks, heterogeneity with respect to cellular distribution in the reactor, restriction in the choice of yeasts, and a starting time of 2 to 3 weeks to build up the desired high cell density and to achieve a stable operation [107]. Flocs break-up and floatation of cells occurs due to CO_2 gas release, and poor mixing characteristics. Table 8 summarises ethanol production results by different investigators using self-aggregated cells.

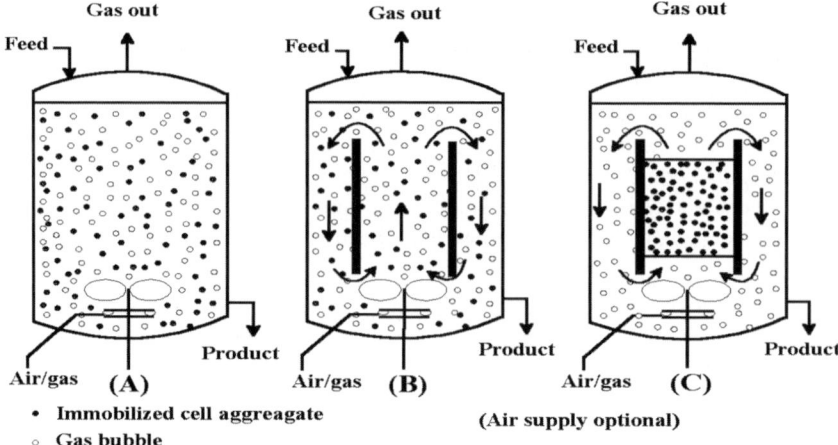

Figure 2. Stirred tank reactors: (A) simple tank reactor, (B) draft tube reactor, (C) packed draft tube tank reactor.

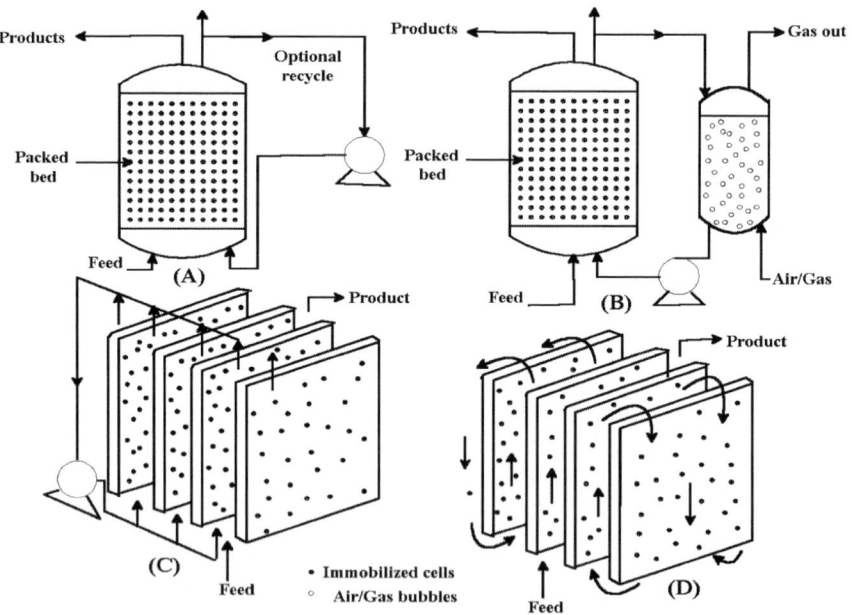

Figure 3. Packed and sheet reactors: (A) packed bed, (B) packed bed with external aeration, (C) sheet reactor with external circulation, (D) sheet reactor with internal circulation.

Table 9. Classification of immobilised cell bioreactors.

Bioreactor type	Advantages	Disadvantages
SUSPENDED PARTICLES		
Stirred tank reactor	flexible, variable mixing intensity, suitable for high viscosity	high power consumption, shear damage to particles, high costs
Air/gas-lift/ bubble column	no moving parts, simple, high solid fraction, high mass transfer, good heat transfer	low local mixing intensity, only for low viscosities
Fluidised bed	no moving parts, simple, low cost, very high solids contents, good heat transfer, variable mixing for solids and liquid	difficult matching of feed & fluidisation rates, requirements on particle density (dense support), good local mixing intensity, only for low viscosities
FIXED PARTICLES		
Fixed bed/monolith	simple, low cost, plug flow characteristics possible, large surface to volume ratio	plugging by solids at low flow rates, high pressure drop, channelling problems
Membranes	very high cell densities, very high productivities, perfusion operation, possible, simultaneous product separation possible, separate feed of gas & liquid, low shear	sterilisation problems, microbial damage, membrane perforation, low capacities only, high cost
MOVING SURFACES		
Rotating surfaces (disc, cylinder or packing)	Low shear on biofilm, batch or continuous, excellent aeration, high productivity,	Power consumption

3. Immobilised cell bioreactor types

The main objectives of immobilisation are to increase the bioreactor productivity with improved cell stability, better substrate utilisation and easier downstream processing. Immobilisation also allows for continuous operation and alternative reactor configurations. Unless the volumetric productivity of the bioreactor (amount of product formed per unit volume of bioreactor and time) is strongly increased, the added cost of some immobilisation techniques and in some cases higher complexity versus free culture will not be justified, especially for large-scale, low added value products. Thus, associated with cell immobilisation is the bioreactor configuration and a good choice of the reactor system is essential for a successful fermentation.

The choice of the bioreactor is related to the type of immobilisation, to the metabolism of cells, and the mass and heat transfer requirements. For example, the resistance of the matrix to shear stress, the size of the beads that affects the mass transfers properties, or the oxygen transfer requirements of the cells may determine the type of reactor. The common immobilised cell bioreactor types used to date for ethanol production are given in Figures 2, 3, 4, 5, 6 and 7.

Although for each type of immobilised cell system a variety of reactor types can be selected, optimal performance requires a careful matching of immobilisation method and bioreactor configuration. Design of the cell aggregate and selection of conditions in the reactor should also go hand in hand.

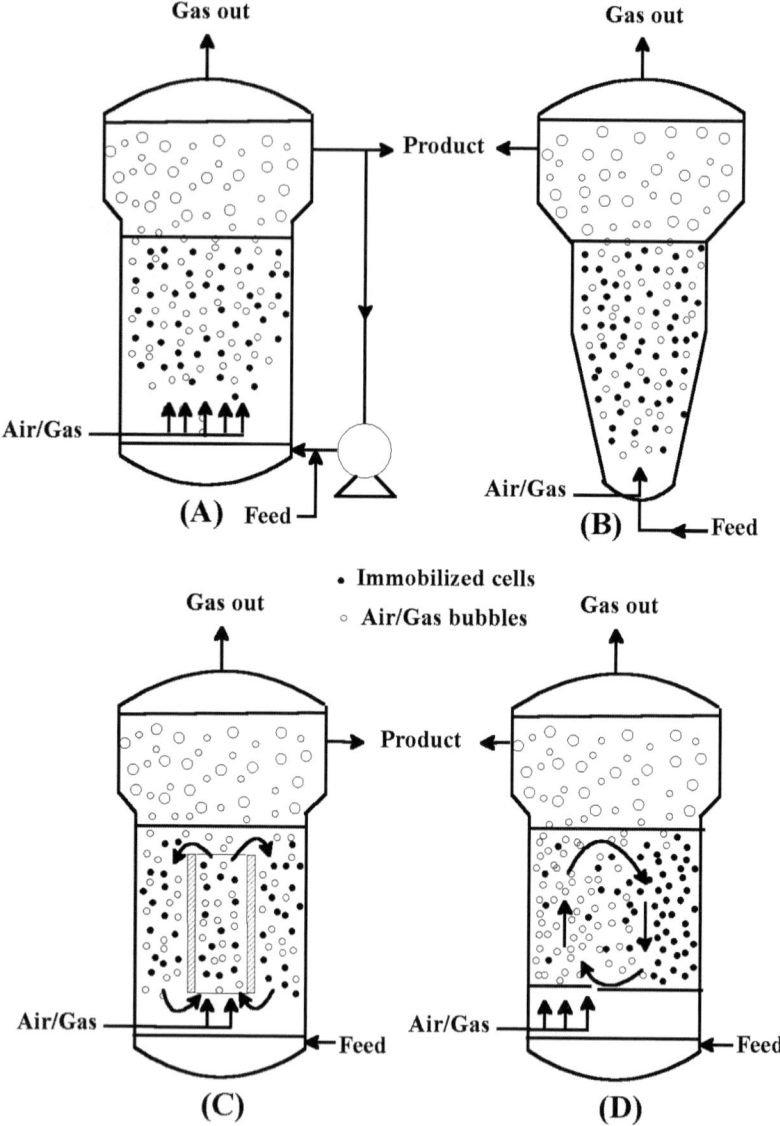

Figure 4. Fluidised bed reactors: (A) without draft tube, (B) tapered, (C) with draft tube, (D) circulating bed.

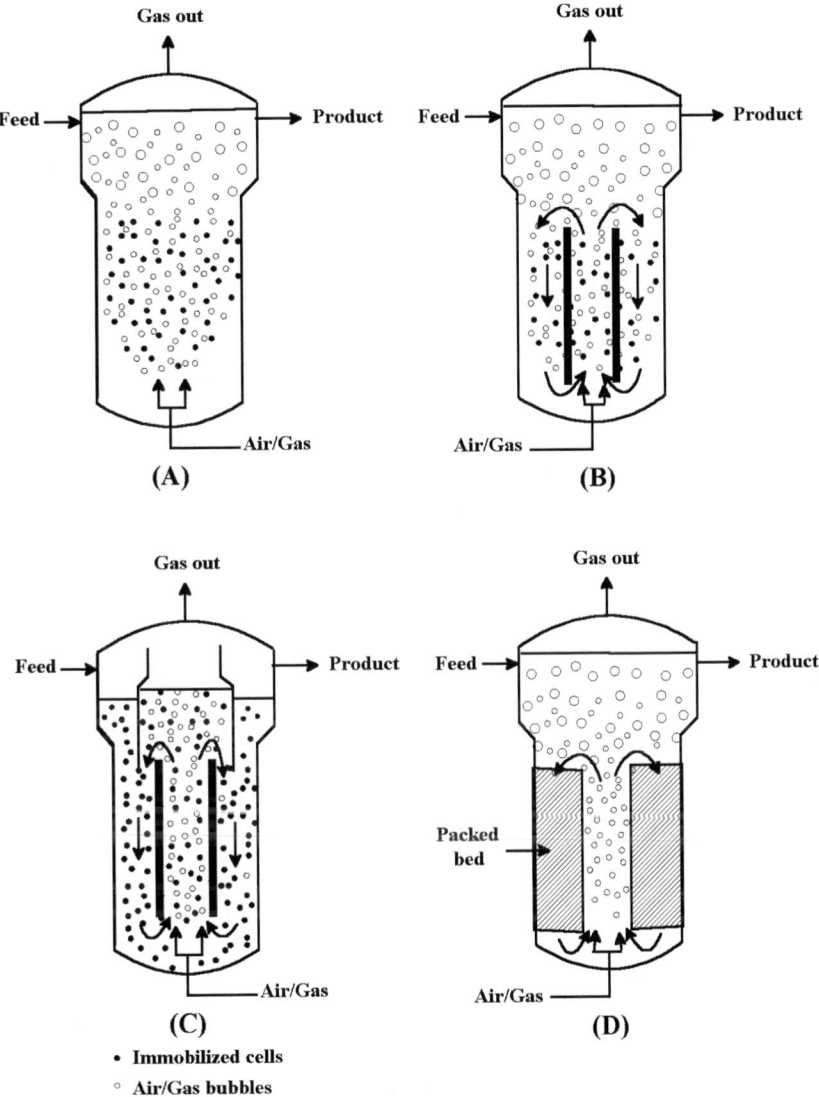

Figure 5. Bubble column and air/gas lift loop reactors: (A) bubble column, (B) air/gas lift loop reactor, (C) air/gas lift loop with particle separator, (D) packed bed air/gas lift loop reactor.

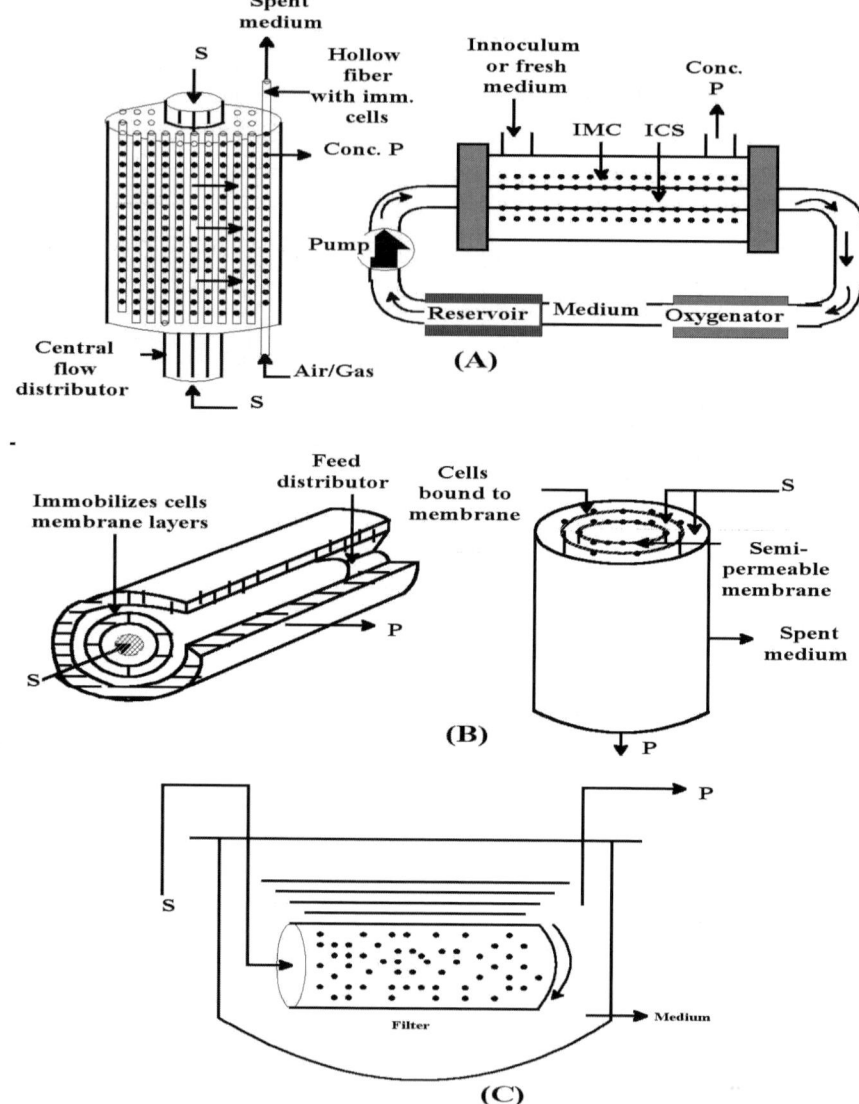

Figure 6. Membrane reactors and rotorfermentor: (A) hollow fibre reactor, (B) spiral-wound flat sheet reactor, (C) rotorfermentor.

Table 9 gives a classification of common and alternative bioreactor types, used with immobilised cells [16]. Some of the major advantages and disadvantages are given as well. Three categories can be distinguished, depending on the location of the cell aggregates [112]: suspended particles, fixed particles and moving surfaces.

An added advantage to these immobilised cell bioreactors is that they give much faster fermentation times compared to the existing free cell fermentation. The proven good features of immobilised cell bioreactor systems have been applied to many applications in food and beverages production.

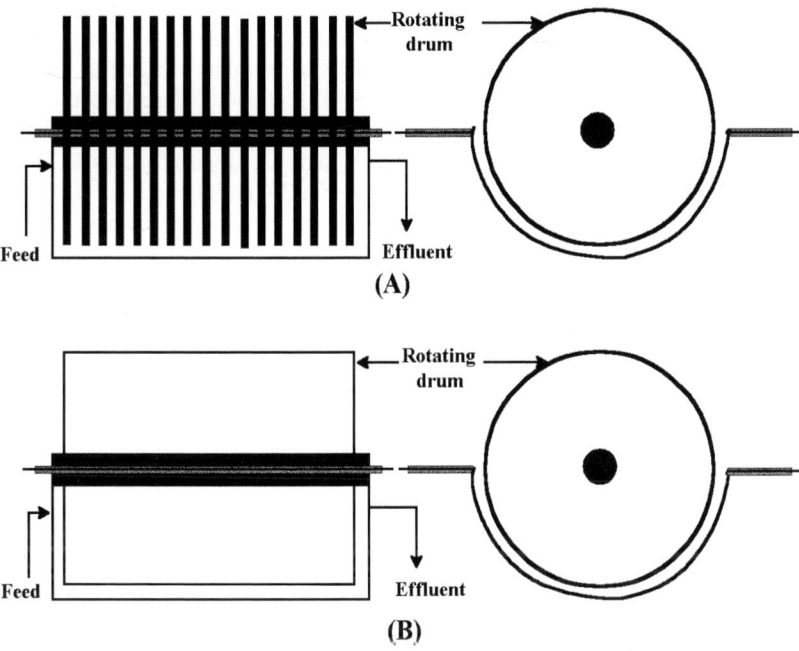

Figure 7. Rotating biological contactors: (A) rotating disc reactor, (B) rotating cylinder reactor.

4. Ethanol production from non-conventional feedstock using immobilised cell systems

Sugars needed for the production of ethanol (sucrose, glucose or fructose) may be derived from three major classes of raw materials, *i.e.* sugar containing feedstock, starchy feed materials and cellulosics. Most commonly, sucrose, derived from sugar cane, and starch, derived from corn are two major feedstock currently employed for the industrial production of ethanol in Brazil and the U.S., respectively. These raw materials are, however, expensive and therefore dictate the overall high price of ethanol. Thus, the use of other cheaper carbohydrate sources, including lactose-derived from cheese whey, inulin-type polyfructans derived from Jerusalem artichokes, and mono-and disaccharides (glucose, xylose, and cellobiose) derived from lignocelluloses, are being extensively evaluated for the fermentative production of ethanol.

4.1. THE PRODUCTION OF ETHANOL FROM CHEESE WHEY

Cheese whey is a major product of cheese production. Approximately, half of the annual world whey production is currently wasted, corresponding to about 2.5 million tons of lactose, which is a potential source for about 1 million tons of ethanol [72]. Furthermore, cheese whey is highly polluting and constitutes a major waste disposal problem for most manufacturers [113,114]. Thus, the disposal of whey by way of fermentation to produce ethanol is receiving increasing interest. Several authors have devised procedures to this effect that begin with β-D-galactosidase (lactase) treatment of whey to hydrolyse the lactose before fermentation [115,116]. A second approach involves the direct fermentation of whey with lactose-utilising yeasts [117]. Liquid whey contains about 5% of lactose resulting in about 2% (w/v) ethanol concentration in conventional batch fermentation. Unfortunately, relatively few yeasts are able to ferment lactose. *Kluyveromyces fragilis* and *Saccharomyces fragi* are known to directly ferment lactose to ethanol, and have been used in an immobilised state for the continuous processing of whey. Production of ethanol from fermentation of whey using immobilised cells is given in Table 10.

Table 10. Ethanol production from cheese whey using immobilised cells.

Micro-organism	Type of bioreactor	Support material	Feed lactose (g/l)	Lactose utilisation (%)	Ethanol conc. (g/l)	Productivity (g/l/h)	Ref.
K. fragilis	Vertical packed bed	Ca-alginate beads	50	80-90	–	5.2	[72]
S. fragi	Tapered fluidised bed	Cellulose acetate particles	100	78	–	2.3	[118]
S. cerevisiae + β-galactosidase	Vertical packed bed	Ca-alginate beads	45	60	–	4.6	[119]
S. cerevisiae + β-galactosidase	Vertical packed bed	Ca-alginate beads	50	–	–	5.0	[120]

4.2. THE PRODUCTION OF ETHANOL FROM JERUSALEM ARTICHOKES

The possibility of growing energy crops on marginal land, as an additional feedstock for ethanol production appears to be promising in view of the fact that such processes will not threaten food supplies [121]. One such carbohydrate rich plant is the Jerusalem artichoke (*Helianthus tuberosus*). The Jerusalem artichoke contains an inulin-type polyfructan, which consist of linear chains of about 30 fructose units terminated by a glucose molecule [122,123]. In addition to its high carbohydrate yields [124], the Jerusalem artichoke, unlike most other traditional energy crops, has low fertiliser requirements, grows well on poor secondary land and is resistant to frost and plant diseases [125].

Yeasts with inulase activity and high fermentation capacity have been utilised for the direct production of ethanol from Jerusalem artichokes by simultaneous enzyme hydrolysis of inulin and fermentation to ethanol. This process has been extensively studied in Canada [44,47,126] and by other workers in France [127] and the U.S., and also reviewed by Margaritis & Merchant [20]. Results of ethanol production from Jerusalem artichoke using immobilised cells are illustrated in Table 11.

Table 11. Ethanol production from Jerusalem artichoke using immobilised cells.

Micro-organism	Type of bioreactor	Support material	Feed sugars (g/l)	Sugars utilisation (%)	Ethanol conc. (g/l)	Productivity (g/l/h)	Ref.
K. fragilis	STR fed-batch	Ca-alginate beads	100	98	–	6.9	[50]
D. polymorphus	STR fed-batch	Ca-alginate beads	100	99	–	5.9	[126]
K. marxianus	Vertical packed bed	Ca-alginate beads	101	80	–	104.0	[141]
K. marxianus	Vertical packed bed	Ca-alginate beads	200	83	–	56.8	[44]

4.3. THE PRODUCTION OF ETHANOL FROM CELLULOSE AND CELLOBIOSE

Cellulose is a major constituent of plant material, which is being constantly replenished by photosynthesis. Due to its abundance and low cost, it has a great potential as a source of fuel if the cellulose can be economically converted to its constituent glucose residues. The enzymatic saccharification and alcohol production from cellulose has been the subject of numerous investigations and has been adequately reviewed by several researchers [127,128]. The enzymatic hydrolysis of cellulose is affected by the synergistic action of the exo-cellulase, endo-cellulase and β-glucosidase. Cellobiose, the intermediate, and glucose, the end product of cellulose saccharification does, however, inhibit the enzyme reaction. Attempts have been made to overcome the end-product inhibition of the cellulolytic enzymes by designing one step process for the simultaneous saccharification and fermentation [129]. One such process that is being actively perused is the use of cellulase enzymes co-immobilised with yeast for the production of ethanol from cellulose or cellobiose.

Hagerdahl & Mosbach [130] devised a procedure for the continuous production of ethanol from cellobiose by using β-glucosidase co-immobilised with baker's yeast cells. The alginate- β-glucosidase complex was mixed with cells of S. cerevisiae and the mixture precipitated by calcium ions in the form of 2 mm calcium alginate beads, which were the packed in a small vertical column. Cellobiose (5% w/v) was fed continuously into the bioreactor, which was maintained at either 35°C or 22°C. At 35°C, the theoretical yield of ethanol was reached after two days, but, thereafter the activity of the system declined rapidly. Thus, after 2 weeks of operation, only 10% of the maximum activity was retained. When the bioreactor was maintained at 22°C, a steady state effluent ethanol concentration of 1.5% (w/v), corresponding to about 60% of the theoretical yield, was maintained at least for three weeks. When the enzyme loading in the bioreactor was increased by a factor of three, a 2% (w/v) effluent ethanol concentration could be maintained for at least four weeks.

Hagerdahl et al. [131] covalently coupled β-glucosidase (from *Aspergillus niger*) to alginate and co-entrapped *S. cerevisiae* cells within calcium alginate beads by the method described above. A mixture of cellobiose (5% w/v) and glucose (5% w/v) was pumped through a tapered fluidised bed bioreactor. Almost complete conversion of the feed to ethanol was achieved when operated at a dilution rate of 0.25 h^{-1}.

Sun et al. [132] evaluated the performance of co-immobilised *S. cerevisiae* and amyloglucosidase (AG) in a fluidized bed bioreactor. The immobilised *S. cerevisiae* cells were immobilised in covalently cross-linked gelatin (6 wt%) and chitosan (0.25 wt%) with AG. Cross-linking was accomplished by glutaraldehyde. The bioreactor performed well, and demonstrated no significant loss of activity or physical integrity during 10 weeks of continuous operation. The bioreactor was easily operated and required no pH control. The productivities ranged between 25 and 44 g ethanol/l/h.

Hartmeir [133] developed a new method to co-immobilize baker's yeast with cellulolytic enzymes. In this method, an mixture of pressed yeasts and sorbitol was suspended in a 10% (w/v) cellulase enzyme solution for 15 min during which time, the enzymes were transferred on to the cell surface. Tannin precipitation and glutaraldehyde cross-linking followed this. The resultant yeast-cellulase immobilisate was mixed with Perlite B and packed in a vertical column between two filter sheets. When a 1% (w/v) cellobiose medium was fed continuously into the bioreactor, almost theoretical yields of ethanol were achieved. Filter paper was also used as the raw material for this continuous fermentation by a proceeding saccharification step carried out separately in an ultrafiltration device containing soluble cellulase. The ultrafiltration obtained was then fed into the packed bed enzyme-microbe bioreactor. A theoretical yield of ethanol from 1 g of cellulose was achieved when the system was operated at low flow rates.

Kierstan et al. [134] immobilised β-glucosidase (derived from thermophilic fungus *Talaromyces emersonii*) on to Sepharose ®. The Sepharose-immobilised β-glucosidase was replaced by an equal aliquot of water. The Ca-alginate beads were then tested for their ethanol producing capacity in a 200 mM cellobiose incubation medium. Preliminary batch data illustrated, that unlike the control system in which no ethanol was detected, efficient production of ethanol took place in the experimental gel system containing both yeast and immobilised β-glucosidase. Although most of the studies discussed above have been carried out on a modest scale and the results obtained are only preliminary, the systems do however, appear to offer some advantages. Thus, in these systems, the glucose formed by the hydrolysis of cellobiose by glucosidase is immediately removed from the system by the yeast cells, which convert the glucose to ethanol. Such proximal arrangement reduces product inhibition of the β-glucosidase enzyme normally exerted by glucose. However, ethanol has also been known to inhibit the enzymatic degradation of cellulose by the denaturation of the cellulase enzymes.

This is due to the hydrophobicity of ethanol and this effect becomes more pronounced at elevated temperatures.

Furthermore, the conversion of the insoluble macromolecular substrate cellulose, by the immobilised cell-enzyme system bound to solid matrices may not be feasible. This problem emanates from the fact that such systems are operated under diffusion restrictions and that steric-hindrance only permits the smaller molecules (*e.g.*, glucose and cellobiose) to come into contact with enzyme or cell, and not the water insoluble macromolecular particles of cellulose. Another characteristic of such systems is the possible accumulation of the products in the vicinity of the bound biocatalyst. Thus, such local enrichment may further reduce the catalytic efficiency of the immobilised cell-enzyme system [131].

In order to reduce the diffusion restrictions apparent in the solid-liquid two-phase systems and product inhibition effects, several workers have employed aqueous two phase systems and their potential applications have been reviewed by Lilly [135]. The phase systems are formed when aqueous solutions of two different water-soluble phases, both of which have a high water content and are therefore biocompatible and non-toxic to the cells or enzymes. The products are then either evenly distributed between the two phases, or preferentially partitioned to one of the phases.

Hahn-Hagerdal *et al.* [131] have studied the feasibility of such "soluble" immobilised biocatalytic systems for the production of ethanol from cellulose. The aqueous two-phase system consisted of dextran T-40 and Carbowax PEG 800. The surface tension between these two phases is very low so that even slight agitation results in a fine disperse emulsion. When left to stand, such a mixture forms a top and a bottom phase. In this system, the yeast cells and the cellulose substrate particles contained adsorbed cellulolytic enzymes, partitioned to the bottom phase whereas the end- and intermediary-products of the bioconversion of cellulose were evenly distributed between the two phases. The enzymatic degradation of cellulose (5% w/v) was carried out at 40°C and the enzymes utilized were cellulase (from *Trichoderma reesei*) and β-glucosidase (from *Aspergillus niger*).

4.4. PRODUCTION OF ETHANOL FROM XYLOSE

The hydrolytic products of cellulosic materials include hexoses and pentoses. The efforts described in the preceding section have been directed to the utilization of the six-carbon component. The pentose content of some cellulose biomass may reach as high as 65% of the total carbohydrate [136] and therefore, the utilisation of both, the pentoses (xylose) and hexoses (glucose) is crucial for the economic production of ethanol from such feedstock. Unfortunately, the pentose fraction is unfermentable by ordinary brewers' yeast and efforts have therefore been directed towards isolating suitable microorganisms capable of directly fermenting xylose to ethanol. Some microorganisms that have been found to posses such characteristics include *Pachysolen tannophilus* [137], *Candida* sp. [138] and several strains of *Kluyveromyces marxianus* [52]. In most of these cases, the rate of fermentation is slow and may require more than 3 days for the complete utilization of xylose.

Maleszka *et al.* [139] observed that enhanced rates of ethanol production were achieved when cells of *Pachysolen tannophilus* were immobilised within Ca-alginate beads, for the continuous production of ethanol from xylose. Under favourable operating conditions, the investigators observed that the immobilised cells retained at least 50% of its initial activity after 26 days of operation. Chiang et al. [140], employed two different immobilisation system, for the production of ethanol from xylose, and they either used a single immobilised microorganism or alternatively co-immobilised the enzyme glucose isomerase with the microorganism. The purpose of glucose isomerase in the latter system was to convert xylose to xylulose, a more suitable fermentation substrate for yeasts. The capabilities of alginate entrapped cells of *Fusarium oxysporum, Mucor sp.* and *S. cerevisiae* in fermenting the pentose to ethanol was compared [140]. *S. cerevisiae* was found to have the best fermentation rate on D-xylulose. Thus, by using a separate glucose isomerase column, an ethanol concentration of 32 g/l was obtained from 100 g/l xylose. To date, most of the studies are only preliminary and substantial improvements in the fermentation systems are needed to achieve higher ethanol yields.

Acknowledgement

This work was supported by an individual Natural Science and Engineering Research Council of Canada (NSERC) operating research grant awarded to Dr. A. Margaritis.

References

[1] Yu, B.; Zhang, F.; Zheng, Y. and Wang, Pu. (1996) Alcohol fermentation from the mash of dried sweet potato with its dregs using immobilised yeast. Process Biochem. 31: 1-6.
[2] Debnath, S.; Bannergiee, M. and Majumdar, S.K. (1990) Production of alcohol from starch by immobilised cells of *Saccharomyces diastaticus* in batch and continuous process. Process Biochem. 25: 43-46.
[3] Norton, S. and Vuillemard, J.-C. (1994) Food bioconversion and metabolite production using immobilised cells technology. Crit. Rev. Biotechnol. 14(2): 193-224.
[4] Hairston, D.W. (2002) Ethanol: On easy street, for now. Chem. Eng. August: 36-38.
[5] Maiorella, B.L.; Blanch, H.W. and Wilke, C.R. (1985) Economic evaluation of alternative ethanol fermentation processes. Biotechnol. Bioeng. 26: 1003-1025.
[6] Tata, M.; Bower, P.; Bromberg, S.; Duncombe, D.; Fehring, F.; Lau, V.; Ryder, D. and Stassi, P. Immobilised yeast bioreactor systems for continuous beer fermentation. Biotechnol. Progress 15: 105-113.
[7] Krishnan, M.S.; Taylor, F.; Davison, B.H. and Nghiem, N.P. (2000) Economic analysis of fuel ethanol production from corn starch using fluidized bed bioreactor. Biores. Technol. 75: 99-105.
[8] Messing, R. A. (1981) Support-bound microbial cells. Appl. Biotechnol. 6: 167-177.
[9] Abbott, B.J. (1977) Immobilised cells. In: Perlman, D. (Ed.) Annual Reports on Fermentation Processes, Vol. 1; Academic Press, New York; pp. 205.
[10] Venkatsubramanian, K. and Veith W. R. (1980) Immobilised microbial cells. In: Bull, M. J. (ed.) Progress in Industrial Microbiology, Vol. 15, Elsevier Scientific Publishing, Amsterdam.
[11] Chibata, I.; Tosa, T., and Tokata, I. (1988) Methods Enzymology; pp.135-146.
[12] Sato, T.; Mori, T.; Tosa, T.; Chibata, I.; Kurui, M.; Yamashita, K. and Sumi, A. (1975) Engineering analysis of continuous production of L-Aspartic acid by immobilised *Escherichia coli* cells in fixed beds. Biotechnol. Bioeng. 17: 1797-1806.
[13] Chibata, I. (1979) Development of enzyme engineering – application of immobilised cell system. In: Linko, P. and Larinkari, J. (Eds.) Food Process Engineering, Vol. 2, Applied Science Publishers, London; pp. 1-25.
[14] Linko, P. (1979) Immobilised biological systems for continuous fermentation. In: Linko, P. and Larinkari, J. (eds.) Food Process Engineering. Vol. 2, Applied Science Publishers, London; pp. 27-54.

[15] Klein, J. (1988) Matrix design for microbial cell immobilisation in bioreactor immobilised enzymes and cells. In: Moo-Young, M. (ed.) Fundamental and Applications. Elsevier Applied Sciences, pp. 1-19.
[16] Willaert, R.G.; Baron, G.V. and De Backer, L. (1996) Immobilised living cell systems. John Wiley & Sons, Chichester.
[17] Dervakos, G.A. and Webb, C. (1991) On the merit of viable-cell immobilisation. Biotechnol. Adv. 9: 560-612.
[18] Chibata, I. and Tosa, T. (1983) Immobilised cells: historical background. Applied Biochemistry and Bioengineering 4: 1-9.
[19] Pilkington, P.H.; Margaritis, A. and Mensour, N.A. (1998) Mass transfer characteristics of immobilised cells used in fermentation processes. Crit. Rev. Biotechnol. 18(2-3): 237-255.
[20] Margaritis, A. and Merchant, F.J.A. (1984) Advances in ethanol production using immobilised cell systems. CRC Crit. Rev. Biotechnol. 1: 339-393.
[21] Godia, F.; Casas, C.; Castellano, B. and Sola, C. (1987) Immobilised cells behavior of carrageenan entrapped yeast during continuous ethanol fermentation. Appl. Microbiol. Biotechnol. 26: 342-351.
[22] Akin, C. (1987) Biocatalysis with immobilised cells. Biotechnol. Genet. Eng. Rev. 5: 319-367.
[23] Masschelein, C.A., Ryder, D.S. and Simon, J.-P. (1994) Immobilised cell technology in beer production. Crit. Rev. Biotechnol. 14(2): 155-177.
[24] Mensour, N.A.; Margaritis, A.; Briens, C.L.; Pilkington, H. and Russel, I. (1996) Application of immobilised yeast cells in the brewing industry. In: Wijffels, R.H.; Buitelaar, R.M.; Bucke, C. and Tramper, J. (eds.) Immobilised cells: basics and applications. Elsevier Scuence, The Netherlands. pp. 661-671.
[25] Hayes, S.A.; Power, J. and Ryder, D.S. (1991) Immobilised cell technology for brewing: a progress report. Brewers Dig. 66(9): 14(11): 28-33.
[26] Durand, G. and Navarro, J.M. (1978) Immobilised microbial cells. Process Biochem. 14: 21-28.
[27] Lee, K. J.; Skotnicki, M. L. and Rogers, P. L. (1982) Kinetic studies on a flocculent strain of *Zymomonas mobilis*, Biotechnol. Lett. 4: 615-624.
[28] Ohlson, S.; Larsson, P.O. and Mosbach, K. (1980) Eur. J. Appl. Microbiol. Biotechnol. 19: 261-272.
[29] Mustranta, A.; Pere, J. and Poutanen, K. (1987) Comparison of different carriers for adsorption of *Saccharomyces cerevisiae* and *Zymomonas mobilis*. Enzyme Microb. Technol. 9: 272-276.
[30] Lommi, H.; Gronqvist, A. and Pajunen, E. (1990) Immobilised yeast cell reactor speeds beer production. Food Technol. May: 127-135.
[31] Geankoplis, C.J. (1993) Transport Processes and unit operations. Third Edition. Predice Hall P.T.R., New Jersey.
[32] Mosbach, K. and Moshbach, R. (1966) Entrapment of enzymes and microorganisms in synthetic cross-linked polymers and their applications in column techniques. Acta. Chem. Scand. 20: 2807-2816.
[33] Scott, C.D.; Hancher, C. W. and Arcuri E. J. (1981) Tapered fluidized bed bioreactors for environmental control and fuel production. In: MooYoung, M. (ed.) Advances in Biotechnology, Vol. 1, Pergamon Press, Toronto; pp. 651-662.
[34] Dale, B.E. (1991) Ethanol production from cereal grains. In: Lorence, K.J. and Kulp, K. (Eds.) Food Sci. Technol. Handbook of cereal Sci. and Technol. Marcel Dekker, Inc., New York; pp. 863-870.
[35] Minier, M. and Goma, G. (1982) Ethanol production by extractive fermentation. Biotechnol. Bioeng. 24: 1565-1575.
[36] Arcuri, E.J.; Worden, R.M. and Shumate, S.E. (1980) Ethanol productivity by immobilised cells of *Zymmomonas mobilis*. Biotechnol. Lett. 2: 499-504.
[37] Sitton, O. C. and Gaddy, J. L. (1980) Ethanol production in an immobilised cell reactor. Biotechnol. Bioeng. 22: 1735-1744.
[38] Doran, P.M. and Bailey, J.E. (1986) Effects of immobilisation on growth fermentation properties, and macromolecular composition of *Saccharomyces cerevisiae* attached to gelatin. Biotechnol. Bioeng. 28: 73-81.
[39] Shinonaga, M.A.; Kawamura, Y. and Yamane, T. (1992) Immobilisation of yeast cells with cross-linked chitosan beads. J. Ferment. Bioeng. 74: 90-98.
[40] Guoqiang, D.; Kaul, R. and Mattiason, B. (1992) Immobilisation of *Lactobacillus casei* cells to ceramic material pretreated with polyethyleneimine. Appl. Microbiol. Biotechnol. 37: 305-314.
[41] Ahmad, S. and Johri, B.N. (1992) Immobilisation of *Rhodococcus equi* DSM 89-133 onto porous celite beads for cholesterol side-chain cleavage. Appl. Microbiol. Biotechnol. 37: 468-476.
[42] Updike, S. J.; Harris, D. R. and Shrago, E. (1969) Microorganisms, alive and imprisoned in a polymer cage. Nature: 1122-1130.

[43] Mosbach, K. and Larson P. O. (1970) Preparation and Application of polymer entrapped enzymes and microorganisms in microbial transformation processes, with special reference to steroid 11-β-Hydroxylation and Δ-Dehydrogenation. Biotechnol, Bioeng. 12: 19-29.
[44] Margaritis, A.; Bajpai, P. K. and Cannnell, E. (1981) Optimization studies for the bioconversion of Jerusalem Artichoke tubers to ethanol and microbial biomass. Biotechnol. Lett. 3: 595-604.
[45] Margaritis, A. and Wallace, J.B. (1982) The use of immobilised cells of *Zymomonas mobilis* in a novel fluidized bioreactor to produce ethanol. Biotechnol. Bioeng. Symp. No. 12: 147-159.
[46] Margaritis, A. and Wallace, J.B. (1984) Novel bioreactor systems and their applications. Biotechnol. 2: 447-453.
[47] Margaritis, A. and Bajpai, P. (1981) Repeating batch production of ethanol from Jerusalem Artichoke tubers using recycled immobilised cells of *Kluyveromyces fragilis*. Biotechnol. Lett. 3: 679-685.
[48] Margaritis, A. and Wallace, J.B. (1982) The use of immobilised yeast systems in the brewing industry. Brewers Guardian 121: 18-35.
[49] Margaritis, A. and Bajpai, P. (1982) Direct fermentation of D-xylose to ethanol by *Kluyveromyces marxianus strains,* Appl. Environ. Microbiol. 44: 1039-1048.
[50] Margaritis, A. and Bajpai, P. (1982) Continuous ethanol production from Jerusalem Artichoke tubers II. Use of immobilised cells of *Kluyveromyces marxianus*. Biotechnol. Bioeng. 24: 1483-1493.
[51] Margaritis, A. and Bajpai, P. (1982) Continuous ethanol production from Jerusalem Artichoke tubers. I. Use of free cells of *Kluyveromyces marxianus,* Biotechnol. Bioeng., 24: 1473-1483.
[52] Margaritis, A. and Rowe, G. E. (1983) Ethanol production using *Zymomonas mobilis* immobilised in different carrageenan gels. Dev. Ind. Microbiol. 24: 329-337.
[53] Nilsson, K.; Birnbaum, S.; Flygare, S.; Linse, L.; Shroder, U.; Jeppson, U.; Larsson, P.O.; Mosbach, K. and Brodelius, P. (1983) A general method for the immobilisation of cells with preserved viability. Eur. J. Appl. Microbiol. Biotechnol. 17: 319-327.
[54] Hulst, A.G.; Tramper, J. ; Van't Riet, K. and Westerbeek, J.M.M. (1985) A new technique for production of immobilised biocatalyst in large quantities. Biotechnol. Bioeng., 27: 870-876.
[55] Wijffels, R.H. and Tramper, J. (1995) Nitrification by immobilised cells. Enzyme Microb. Technol., 17: 482-492.
[56] Serp, D.; Cantana, E.; Heinzen, C.; von Stockar, U. and Marioson, I.W. (2000) Characterization of an encapsulation device for the production of monodisperse alginate beads for cell immobilisation. Biotechnol. Bioeng. 70: 41-53.
[57] Senuma, Y.; Lowe, C.; Zweifel, Y.; Hilborn, J.G., and Marison, I. (2000) Alginate hydrogel microspheres and microcapsules prepared by spinning disk atomization. Biotechnol. Bioeng. 67: 616-622.
[58] Poncelet de Smet, B.; Poncelet, D. and Neufeld, R.J. (1993) Emerging techniques, materialism, and applications in cell immobilisation. In: Goosen, M.F.A. (ed.) Fundamentals of animal cell encapsulation and immobilisation. CRC Press Inc., Boca Raton, FL; pp. 297-313.
[59] Cheetham B.J. (1979) Physical studies on the mechanical stability of columns of calcium alginate gel pellets containing entrapped microbial cells. Enzyme Microb. Technol. 1: 183-191.
[60] Kierstan, M. and Bucke, C. (1977) The immobilisation of microbial cells, subcellular organelles, and enzymes in calcium alginate gels. Biotechnol. Bioeng. 19: 387-396.
[61] Kondo, A.L. (1979) In-Liquid Curing Coating Process. In: van Valkenburg, W. (ed.) Microcapsules. Proceedings and Technology; pp. 59.
[62] Wang, H.Y. and Hettwer, D.J. (1982) Cell immobilisation in κ-carrageenan with triphosphate. Biotechnol. Bioeng. 24: 1827-1838.
[63] Shiotani, T. and Yamane, T. (1982) A horizontal packet bed bioreactor to reduce CO_2 holdup in the continuous production of ethanol by immobilised yeast cells, Eur. J. Appl. Microbiol. 13: 96-105.
[64] Vogelsang, C.; Wijffels, R.H. and Østgaard, K. (2000) Rheological properties and mechanical stability of new gel-entrapment system applied in bioreactors. Biotechnol. Bioeng. 70: 247-253.
[65] Windholz, M. (1976) Alginate alginic acid. *The Merck Index*, Merck and Co., New Jersey; pp. 34-45.
[66] Jones, R. P.; Pamment, N. and Greenfield, P. F. (1981) Alcohol fermentation by yeasts-the effect of environmental and other variables. Process Biochem. April/May: 42-53.
[67] Veliky, I. A. and Williams, R. E. (1981) The production of ethanol by *Saccharomyces cerevisiae* immobilised in polycation-stabilized calcium alginate gels, Biotechnol. Lett. 3: 275-284.
[68] Birnbaum, S.; Larson, P.O. and Mosbach, K. (1981) Stabilization of calcium alginate gel. In: Abstr. of Commun, 2nd European Congr. for Biotechnol., 5-10 April, Eastbourne, England.

[69] Haug, A. and Smidsrod, O. (1965) The effect of divalent metals on the properties of alginate solutions. II. Comparison of different metal ions. Acta. Chem. Scand. 19: 341-350.
[70] Paul, F. and Vignais, P. M. (1980) Photophosphorylation in bacterial chromatophores entrapped in alginate gel improvement of the physical and biochemical properties of gel beads with barium as the gel-inducing agent. Enzyme Microb. Technol. 2: 281-291.
[71] Groboillot, A.; Boadi, D.K.; Poncelet, D. and Neufeld, R.J. (1994) Immobilisation of cells for application in the food industry. Crit. Rev. Biotechnol. 14(2): 75-107.
[72] Linko, Y. Y. and Linko, P. (1981) Continuous ethanol production by immobilised yeast reactor. Biotechnol. Lett. 3: 21-28.
[73] Cho, G.H.; Choi, C.V.; Choi, Y.D. and Han, M.H. (1981) Continuous ethanol production by immobilised yeast in a fluidized reactor. Biotechnol. Lett. 3: 667-675.
[74] Freitas, C. and Teixeira, J.A., Hydrodynamic studies in an airlift reactor with an enlarged degassing zone. *Bioprocess Engineering* 18: 267-279 (1998).
[75] Chibata, I. and Tosa, T. (1980) Immobilised microbial cells and their applications. T.I.B.S. 88: 23-45.
[76] Takata, I.; Tosa, T. and Chibata, I. (1977) Screening of matrix suitable for immobilisation of microbial cells. J. Solid phase Biochem. 2: 225-235.
[77] Kuu, W. Y., and Pollack, J. A. (1982) Strengthening Immobilised Biocatalysts by Diffusion and Gel Phase Polymerisation. Presented at the 75th Annual AIChE meeting, 14-18 Nov., L.A., California.
[78] Wada, M.; Kato, J. and Chibata, I. (1981) Continuous production of ethanol in high concentration using immobilised growing yeast cells. Eur. J. Appl. Microbiol. Biotechnol. 11: 67-78.
[79] Wang, H.Y.; Lee, S.S.; Tabach, Y. and Cawthon, L. (1982) Maximizing microbial cell loading in immobilised cell systems. Biotechnol. Bioeng. Symp., John Wiley & Sons, New York, pp. 139-150.
[80] Krouwel, P. G. and Kossen, N. W. F. (1980) Gas production by immobilised micro-organisms: theoretical approach. Biotechnol. Bioeng. 22: 681-692.
[81] Chen, S.L. and Gutmanis, F. (1976) CO_2 inhibition of yeast growth in biomass production. Biotechnol. Bioeng. 18: 1455-1464.
[82] Kunkee, R.E. and Ough, C.S. (1966) Multiplication and fermentation of *Saccharomyces cerevisiae* under carbon dioxide pressure in wine. Appl. Microbiol. 14: 643-651.
[83] Jones, R. P. and Greenfield, P. F. (1982) Effect of carbon dioxide on yeast growth and fermentation, Enzyme Microbiol. Technol. 4: 210-228.
[84] Birnbaum, S. and Larson, P.O. (1982) Application of magnetic immobilised microorganisms: ethanol production by *Saccharomyces cerevisiae*. Appl. Biochem. Biotechnol. 7: 55-64.
[85] Amin, G. and Verachtert, H. (1982) Comparative study of ethanol production by immobilised cell systems using *Zymmomonas mobilis* and *Saccharomyces cerevisiae*. Eur. J. Appl. Microbiol. Biotechnol. 11: 67-75.
[86] Öztop, H.N.; Öztop, A.Y.; Isikver, Y. and Saraydin, D. (2002) Immobilisation of *Saccharomyces cerevisiae* on to radiation crosslinked HEMA/Aam hydrogels for production of ethyl alcohol. Proc. Biochem. 37: 651-657.
[87] Koshcheenko, K. A. (1981) Living immobilised cells as biocatalysts of transformation and biosynthesis of organic compounds. Appl. Biochem. Microbiol. 17: 351-360..
[88] Siess, M. H. and Divies, C. (1981) Behavior of *Saccharomyces cerevisiae* cells entrapped in a polyacrylamide gel performing alcoholic fermentation, Eur. J. Appl. Microbiol. Biotechnol. 12: 10-19.
[89] Klein, J and Eng, H. (1979) Immobilisation of microbial cells in epoxy carrier systems. Biotechnol., Lett. 1: 171-180.
[90] Gianfreda, L., Parascandola, P., and Scardi, V., (1980) A new method of whole microbial cell immobilisation. Eur. J. Appl. Microbiol. Biotechnol. 11: 6-14.
[91] Giordano, R.L.C.; Hirano, P.C.; Goncalves, L.R.B. and Netto, W.S. (2000) Study of biocatalyst to produce ethanol from starch. Appl. Biochem. Biotechnol. 84-86: 643-654.
[92] Shindo, S.; Takata, S.; Taguchi, H. and Yoshimura, N. (2001) Development of novel carriers using zeolite and continuous ethanol fermentation with immobilised *Saccharomyces cerevisiae* in a bioreactor. Biotechnol. Lett. 23: 2001-2004.
[93] Melin, E. and Shieh, W.K. (1992) Continuous ethanol production from glucose using *Saccharomyces cerevisiae* immobilised on fluidized microcarriers. Chem. Eng. J. 50: B17-B22.
[94] Inama, L.; Dire, S. and Carturan, G. (1993) Entrapment of viable microorganisms by SiO_2 sol-gel layers on glass surfaces: Trapping, catalytic performance and immobilisation durability of *Saccharomyces cerevisiae*. J. Biotechnol., 30: 197-210.

[95] Pines, G. and Freeman, A. (1982) Immobilisation and characterization of *Saccharomyces cerevisiae* in cross-linked, prepolymerized polyacrylamide-hydrazide, Eur. J. Appl. Microbiol. Bitechnol. 16: 75-886.
[96] Siva Raman, H.; Seetarama Rao, B.; Pundle, A. V. and Siva Raman, C. (1982) Continuous ethanol production by yeast cells immobilised in open pore gelatin matrix. Biotechnol. Lett. 4: 359-366.
[97] Margaritis, A. and Wilke, C. R. (1972) Engineering analysis of the rotorfermenter. Dev. Ind. Microbiol. 13: 159-165.
[98] Margaritis, A. (1975) A study of the rotorfermentor and the kinetics of ethanol fermentation. Ph.D. thesis, University of California, Berkeley; also as The Lawrence Berkeley Lab Report, LBL-3278.
[99] Margaritis, A. and Wilke, C. R. (1978) The rotor fermenter. I. Description of the apparatus power requirements, and mass transfer characteristics, Biotechnol. Bioeng. 20: 709-718.
[100] Margaritis, A. and Wilke, C. R. (1978) The rotor fermenter. II. Application to ethanol fermentation. Biotechnol. Bioeng. 20: 727-7735.
[101] Guisely, K.B. (1989) Chemicals and physical properties of agal polysaccharides used for cell immobilisation. Enzyme Microbiol. Technol. 11: 706-714.
[102] Sitton, O. C.; Magruder, G. C.; Book, N. L. and Gaddy, J. L. (1980) Comparison of immobilised cell reactor and CSTR for ethanol production. Biotechnol. Bioeng., Symp. No. 10: 213-222.
[103] Inloes, D.S.; Michael, A.S.; Robertson, C.R. and Martin, A. (1985) Appl. Microbiol. Biotechnol. 23: 85-91.
[104] Kolot, F. B. (1980) Immobilised microbial systems: present state of development. Dev. Ind. Micribol., 21: 295-303.
[105] Atkinson, B. and Daoud, I.S. (1976) Microbial flocs and flocculation in fermentation process engineering. Adv. Biochem. Eng. 4: 41-124.
[106] Greenshields, R.N. and Smith, E.L. (1974) The tubular reactor in fermentation. Process Biochem. 8: 11-23.
[107] Royston, M.G. (1966) Tower fermentation of beer. Proc. Biochem. 1(4): 12-21.
[108] Portno, A.D. (1978) Continuous fermentation in the brewing industry. European Brewing Convention Symposium on fermentation and storage, Monograph V; pp. 145-152.
[109] Prince, I. G. and Barford, J. P. (1982) Continuous tower fermentation for power ethanol production, Biotechnol. Lett. 4: 263-273.
[110] Strandberg, G. W.; Donaldson T. L. and Arcuri, E. J. (1982) Continuous ethanol production by a flocculant strain of *Zymomonas mobilis*, Biotechnol. Lett. 4: 347-356.
[111] Fan, L.-S. (1989) Gas-liquid-solid fluidization engineering. Butterworths Series In Chemical Engineering. Butterworths, Boston.
[112] Karel, S.F.; Libicki, S.B. and Robertson, C.R. (1985) The immobilisation of whole cells: Engineering principles. Chem. Eng. Sci. 40: 1321-1354.
[113] Dohan, L.A.; Baret, J.L.; Pain, S. and Delalande, P. (1980) Lactose hydrolysis by immobilised lactase: semi-industrial experience. In: Linko, P. and Larinkari, J. (Eds.) Food Process Engineering, Vol. 2. Applied Science Publishers, London; pp. 137-155.
[114] Compare, A.L. and Griffith, W.L. (1976) Fermentation of waste materials to produce industrial intermediates. Dev. Ind. Microbiol. 17: 247-255.
[115] Gawel, J. and Kosikowski, F.V. (1978) Improving alcohol fermentation in concentrated Ultrafiltration permeates of cottage cheese whey. J. Food. Sci. 43: 1717-1724.
[116] O'Leary, V. S.; Sutton, C.; Bencivengo, M.; Sullivan, B. and Holsinger, V. H. (1977) Influence of lactose hydrolysis and solids concentration in alcohol production by yeast in acid whey ultrafiltrate, Biotechnol. Bioeng. 19: 1689-1697.
[117] Izaguirre, M. E. and castello, F. J. (1982) Selection of lactose-fermenting yeast for ethanol production from whey. Biotechnol. Lett. 4(4): 257-262.
[118] Chen, H.C. and Zall, R.R. (1982) Continuous fermentation of whey into alcohol using an attached film expanded bed reactor. Process Biochem. 18: 20-27.
[119] Hahn-Hagerdal, B. and Mattiason, B. (1982) Shift in metabolism towards ethanol production in *Saccharomyces cerevisiae* by addition of metabolic inhibitors. Biotechnol. Bioeng. Symp. No. 12: 193-212.
[120] Linko, P. (1980) Immobilised live cells. In: Moo-Young, M. (ed.) Advances in Biotechnology. Vol. 1, Pergamon Press, Toronto. pp. 711-738.
[121] Hayes, R.D. (1981) Energy crops-what little we know. Proc. 3rd Bioenergy R&D Semin., 24-25 March, Ottawa, Canada.

[122] Bacon, J.S.D. and Edelmanc, J. (1951) The carbohydrates of the Jerusalem Artichoke and other composite. Biochem. J. 48: 114-126.
[123] Kierstan, M. (1980) Production of fructose syrups from inulin. Process Biochem. 15(4): 24-32.
[124] Dorrel, D.G. and Chubey, B.B. (1977) Irrigation, fertilizer, harvest dates and storage effects on the reducing sugar and fructose concentration of Jerusalem Artichoke tubers. Can. J. Plant. Sci. 57: 591-601.
[125] Fleming, S.A. and Grootwassink, J.W.D. (1979) Preparation for high-fructose syrup from the tubers of the Jerusalem artichoke *(Helianthis tuberosus L.)*. CRC Critical Reviews Food Sci. Nutr. 12: 1-13.
[126] Margaritis, A.; Bajpai, P., and Lachance, A. (1983) The use of free and immobilised cells of *Debaromyces polymorphus* to produce ethanol from Jerusalem Artichoke tubers. J. Ferment. Technol., 61: 533-542.
[127] Bisaria, V.S. and Ghose, T.K. (1981) Biodegradation of cellulosic materials: substrates, microorganisms, enzymes and products. Enzyme Microb. Technol. 3: 90-104.
[127] Guiraud, J.P. and Galzy, P. (1981) Alcohol production by fermentation of Jerusalem artichoke extract. Paper Presented at the 28th IUPAC Congress, August, Vancouver, Canada; pp. 16-22.
[128] Tangnu, S.K. (1982) Process development for ethanol production based on enzymatic hydrolysis of cellulosic biomass. Process Biochem. 17: 36-45.
[129] Savarese, J. J., and Young, S. D. (1978) Combined enzyme hydrolysis of cellulose and yeast fermentation. Biotechnol. Bioeng. 20: 1291-1311.
[130] Hahn-Hagerdal, B. and Mosbach, K. (1979) The production of ethanol from cellobiose using baker's yeast coimmobilised with β-glucosidase, In: Linko, P. and Larinkari, J. (eds.) Food Process Engineering, Vol. 2, Applied Science Publishers, London; pp. 129-154.
[131] Hahn-Hagerdal, B.; Mattiasson, B. and Albertsson, P. A. (1981) Extractive bioconversion in aqueous two-phase systems. A model study on the conversion of cellulose to ethanol. Biotechnol. Lett. 3: 53-58.
[132] Sun, M.Y.; Bienkowski, P.R., Davison, B.H.; Spurrier, M.A. and Webb, O.F. (1997) Performance of co-immobilised yeast and amyloglucosedase in a fluidized bed reactor for ethanol production. Appl. Biochem. Biotechnol. 63-65: 483-493.
[133] Hartmeir, W. (1981) Basic trials on the conversion of cellulosic material to ethanol using yeast co-immobilised with cellulolytic enzymes, In: Moo-Young, M. (ed.) Advances in Biotechnology, Vol. 3, Pergamon Press, Toronto; pp. 377-396.
[134] Kierstan, M.; McHale, A. and Coughlan, M. P. (1982) The production of ethanol from cellobiose using immobilised β-Glucosidase co-entrapped with yeast in alginate gels. Biotechnol. Bioeng. 24: 1461-1463.
[135] Lilly, M. D. (1982) Two-liquid-phase biocatalytic reactions. J. Chem. Tech. Biotechnol. 32: 162-172.
[136] Rosenberg, S. L. (1980) Fermentation of pentose sugars to ethanol and other neutral products by microorganisms, Enzyme Microb. Technol. 2: 185-194.
[137] Schneider, H.; Wang, P. Y.; Chan, Y. K. and Maleszka, R. (1981) Conversion of D-xylose into ethanol by the yeast *Pachysolen tannophilus*. Biotechnol. Lett. 3: 89-98.
[138] Gong, C.S.; McCraken, L.D. and Tsao, G.T. (1981). Direct fermentation of D-xylose into ethanol by the yeast *Pachysolen tannophilus*. Biotechnol. Lett. 3: 89-95.
[139] Maleszka, R.; Veliky, I. A. and Schneider, H. (1981) Enhanced rate of ethanol production from D-xylose using recycled or immobilised cells of *Pachysolen tannophilus*. Biotechnol. Lett. 3: 415-421.
[140] Chiang, L.C.; Hsiao, H.Y.; Flickinger, M.C. and Tsao, G.T. (1982) Ethanol production from pentoses by immobilised microorganisms. Enzyme Microb. Technol. 4: 93-103.
[141] Margaritis, A. and Bajpai, P. (1982) Continuous ethanol production from Jerusalem Artichoke tubers. II. Use of free cells of *Kluyveromyces marxianus*. Biotechnol. Bioeng. 24: 1483-1493.

PRODUCTION OF BIOPHARMACEUTICALS THROUGH MICROBIAL CELL IMMOBILISATION

TAJALLI KESHAVARZ
Fungal Biotechnology Group, Biotechnology Department, University of Westminster, 115 New Cavendish Street, London W1W 6UW London UK – Fax: 44 20 79115087 – Email: T.Keshavarz@westminster.ac.uk

1. Introduction

This article aims to provide an overview of cell immobilisation for production of biopharmaceuticals. Pharmaceutical industry has grown immensely in size, product type and market over the last fifty years. Over this period, it has benefited from a better understanding of cell physiology and metabolism, and technological developments covering strain selection all the way to downstream processing. The advances in molecular biology and biochemical and process engineering have ensured higher productivity of commercially demanding biopharmaceuticals. Traditionally pharmaceuticals have been synthesised chemically or obtained through chemical processes. However, after the discovery of penicillin and its rapid development to industrial-scale production, biopharmaceutical industry's efforts have been directed to the exploitation of other microbes, as well as plants and animals. Research has been focused on the discovery of natural products beneficial for human and animal health.

Cell immobilisation, since the earliest reports [1-7] has made extensive contribution, as a process intensification method, to overproduction and extension of production of a range of metabolites. However, in majority of cases, achievements have been confined to research at laboratory-scale and there are only a few examples of successful process-scale operations utilising cell immobilisation technology at the industrial-scale [8,9]. The success in this field is even more limited when "growing cells" are used.

This article is an attempt to explore examples of cell immobilisation for production of biopharmaceuticals. In this context, microbial cells are chosen deliberately and exclusively as another chapter is devoted to immobilisation of other cell types. Examples of biopharmaceuticals have been taken essentially from articles on the production of antibiotics. Some examples of bioconversion of steroids through immobilised cells are also included. The coverage of nutraceuticals is outside the scope of this article.

2. The use of immobilised microorganisms in production of antibiotics

A range of antibiotics has been produced from fungi (mainly *Penicillium* sp.) and bacteria (mainly *Streptomyces* sp.) over the last thirty years. Varieties of matrices including hydrogels for entrapment and inert particles for adsorption have been used for this purpose. Different types of bioreactors and modes of fermentation has been adopted and investigated. But in all cases, the studies have remained within the realm of research laboratories and no industrial-scale production has been reported. In this part of the article, production of different antibiotics by immobilised cell systems will be discussed. The reasons for lack of industrial-scale production will be discussed later in this chapter.

2.1. PENICILLIN PRODUCTION

One of the earliest reports on the production of penicillin came from Tokyo Institute of Technology in 1979 [10]. The authors reported the immobilisation of *Penicillium chrysogenum* ATCC 12690 mycelium in three matrices: polyacrylamide gel, collagen membrane and calcium alginate. It was found that penicillin G production using reciprocal shakers for 5 hours was highest using calcium alginate beads but due to the fragility of the beads in the presence of phosphate ion, the reuse of the immobilised cells was not possible. Collagen system was not satisfactory due to the adverse effect of glutaraldehyde which was used as a tanning agent. The polyacrylamide immobilisation was the most favourable choice but there was a need for further studies to optimise the concentrations of the polymerisation reagents to reduce their inactivation of the immobilised mycelium. Comparative repeated shaken flask fermentations carried out on the washed and immobilised mycelium showed that on repeated incubation, immobilised system had superior performance in terms of penicillin G relative activity. The importance of the matrix pore size and aeration was reported in the production of penicillin G by the immobilised system.

Mild entrapment using calcium alginate was adopted to investigate the possibility to improve penicillin G production by immobilised vesicles isolated from the protoplasts of *Penicillium chrysogenum* PQ-96 [11]. It was found that penicillin G activity by the immobilised vesicles was only 44% of that of the native ones after incubation for one hour but on storage at 4°C, the immobilised preparation could retain the activity for up to 240 hours while the native preparations lost completely their penicillin G activity after 60 hours.

Morphological changes due to immobilisation were studied on two strains of *Penicillium chrysogenum* ATCC 12690 and S1 (an industrial strain). These strains were immobilised in calcium alginate beads and their morphology was compared with that of free-cell culture [12]. Starting with spores, shaken flask cultures were grown in rotary shakers at 100 rpm and 5 cm stroke. Samples were observed under a light microscope as well as scanning electron microscope. The authors reported differences between the two strains in terms of the development of morphology inside the beads and improvement in penicillin production activity compared to the free cell culture. They suggested that some of the factors for the improved activity could be the formation of micropellets inside the beads, increase in branching frequency in immobilised system and reduced entanglement of mycelia due to alginate gel separating them. They also mentioned, as two of the favourable factors, the improvement of mass transfer rate and slow growth-rate under immobilisation.

The two *Penicillium* strains (ATCC 12690 and S1) were investigated under semi-continuous [13] and continuous [14] immobilised culture conditions for production of penicillin. The immobilisation matrix was calcium alginate in both cases. Under the semi-continuous conditions, the immobilised conidia were grown in 450 ml flasks and a 400 ml bubble column. Chemically defined medium as well as a complex medium were used in these studies. In all cases, production yield based on biomass was higher in the complex medium and with the industrial strain as would be expected. However, penicillin production reduced in all cases after two production cycles. This was suggested to be the consequence of oxygen diffusion limitation in the immobilised-cell system and autolysis of the free-cell system. In any case this diminished production was a serious drawback.

The high penicillin-producing strain S1 was used for further studies under continuous culture conditions. Here, two bioreactor types, the previously used bubble column (400 ml) and a 2.5 l conical bubble fermentor were adopted for the studies with a complex medium. The cultures were switched from batch to continuous after 72 hours of growth using either glucose or lactose as the carbon source. However, the bubble column operation of the immobilised-cell cultures was terminated after 96 and 120 hours respectively for the glucose-fed and lactose-fed cultures respectively. This was due to the fast reduction in concentration of penicillin. The production in the conical bubble fermentor, however, was more successful. Lower dilution-rates were achievable in this reactor as the operational problems were reduced. Under a dilution-rate of 0.0156 h^{-1} the penicillin concentration stabilised around 45% of the starting concentration and there was an increase for a period of 10 days in the total amount of penicillin. However, the system was uneconomical due to operational problems, the high risk of contamination and high concentration of unutilised medium constituents.

Deo and Gaucher [15] also investigated entrapped cell cultures of *P. chrysogenum* (ATCC 12690) for production of penicillin G under semi-continuous and continuous modes of fermentation. The semi-continuous fermentation was carried out in a 500 ml flask containing a 50 ml glucose-based growth medium and 400 beads of κ-carrageenan carrying *P. chrysogenum* conidia. The immobilised culture was washed after 72 hours and shaken in starvation medium (no glucose) for 16-18 hours. The starved immobilised cells were washed again and resuspended in replacement media (two formulations with lower glucose concentration). For the continuous culture, a 200 ml aerated fluidised-bed reactor was adopted to be used for three runs with a variety of culture conditions including media, air flow-rate, pH and product removal rate. The authors suggested that the fluidised-bed reactor with immobilised-cells offered a clear alternative for penicillin production as the concentration, and yield from glucose, of penicillin was higher than the free-cell culture. However, they acknowledged that the replacement of the well-established free-cell fermentation facilities of the antibiotics industries with immobilised-cell systems was unlikely.

A study was carried out on the morphology of *Penicillium* using viable spores [16]. In this study, spores of Penicillium chrysogenum P2 were immobilised in κ-carrageenan and the effect of viable spore loading per bead ($10-10^4$) was investigated on the rate of spore germination and the development of morphology and biomass within the beads when grown in 2 l shaken flask in an orbital shaker (250 rpm, 2.5 cm throw). It was found that by using a lower spore loading, the initial concentration of biomass at the internal periphery of the beads was reduced. This resulted in improved mass transfer (and particularly oxygen transfer into the beads) with subsequent improvement in the final biomass throughout each bead. An optimal number of 50 viable spores per bead resulted in the highest biomass concentration per bead. The authors referred also to the development of free-cells in the immobilised-cell system and reported the effect of the spore loading on the emergence and development of the free-cells.

Immobilisation of *Penicillium chrysogenum* P2 in κ-carrageenan proved useful as a means to show physiological differences between the free- and immobilised-cell cultures [17]. A significant extension in the duration of expression of the isopenicillin N synthase gene (pcbC), a reduction in the penicillin G maximum specific productivity, and an increase in the average specific productivity were reported for the immobilised-cell system in comparison to the free-cell culture. While the free-cell culture was carried out in a 2 l stirred-tank reactor and a 4 l air-lift reactor was used for the immobilised-cell culture, the differences in the production seem to be mainly a result of the immobilisation rather than the bioreactor types.

The feasibility of the attachment of spores and growth of *Penicillium chrysogenum* strains, including ATCC 12690, was investigated using 2 l rotating disk fermentor fitted with different disc materials in different sets of experiments [18]. The heat sterilisable materials included polypropylene, polycarbonate, nylon, polysulfon and stainless steel. All the disc materials except stainless steel supported attachment and growth of all the *Penicillium chrysogenum* strains; polycarbonate and polypropylene being the best. Subsequent batch fermentation studies were carried out with the ATCC 12690 strain immobilised on polycarbonate disks due to their superior characteristics during heat sterilisation. The authors attributed relatively high yield of penicillin G to the mild and "natural" method of immobilisation although diffusional limitations due to excessive film growth was a serious drawback.

The first large-scale immobilised-cell fermentation of penicillin was reported in 1984 [19]. Contrary to most of the studies based on entrapment, the authors used Celite particles as the suitable support for the immobilisation of *Penicillium chrysogenum* P2 spores through adsorption. Pre-sterilised Celite (20 kg) particles (180-500 µl) were added to a pre-sterilised 200 l working volume tower fermentor and were dried by air. Penicillin spores were prepared in solution and a 60 l spore suspension was added to the dried Celite in the fermentor. Batch fermentation commenced after addition of a complex medium and continued for a period of 300 hours. A free-cell fermentation (no addition of Celite) was also carried out in the same fermentor for comparison. An increased oxygen transfer per power input and higher penicillin productivity was achieved with the immobilised cell reactor as a result of lower levels of gas-liquid oxygen transfer limitation.

A comparison was made between the entrapped and adsorbed immobilised *Penicillium chrysogenum* (ATCC 26818) cultures under semi-continuous and continuous modes of fermentation using a defined medium [20]. κ-Carrageenan and Celite R-630 were used for entrapment and adsorption of the spores respectively. The average κ-carrageenan bead size and Celite particles average diameter were 2.7-3.2 mm and 410 µm respectively. In all cases where Celite R-630 was used, the number of spores per gram Celite was 24 10^8. The spores were adsorbed onto the wet particles and after their germination and colonisation; the immobilised cells were transferred into the shaken flasks or the bioreactor as necessary. The semi-continuous fermentations were carried out in 500 ml shaken flasks containing 40 ml of a production medium and 5 grams of colonised Celite. Repeated batches were carried out for 15 passages (semi-continuous). For the continuous culture a 1.2 l fluidised-bed reactor was used with the same production medium and two different flow-rates but the inoculum was scaled-up to 50 grams colonised Celite. The fermentations were carried out for a period of 300 hours. It was found that although semi-continuous runs with κ-carrageenan immobilised cells initially produced higher concentrations of penicillin G, the concentrations decreased in the subsequent passages reaching similar concentrations obtained with Celite immobilised cells. As the volume of the carrageenan beads was almost four times bigger than the Celite particles, the fermentations carried out with the latter were much more productive on a volume basis. Under the continuous culture using Celite, there was a sharp decrease in the penicillin production-rate after an early increase for all medium flow-rates. The authors mentioned that the production compared favourably with an industrial process using the same generation of microbes. However, the system was not working under optimised conditions and further improvements in medium development and feeding strategy would help increasing further the penicillin production.

Penicillium chrysogenum ATCC 26128 immobilised on an unspecified Celite grade was grown in a small (75 mm diameter and 200 mm height) reactor to investigate the potential of the immobilised system for penicillin fermentation [21]. The spores of the fungus were first immobilised in a shaken flask and transferred aseptically, to the reactor after 72 hours when the hyphae had developed on the Celite particles. A complex production medium was used for semi-continuous and repeated fed-batch fermentations which lasted for 130 hours. The working volume for the semi-continuous fermentation was 300 ml with the addition of 100 ml fresh medium after removal of the same volume of culture every 8 hours. For the fed-batch operation, the starting volume was 250 ml with a feed of 12.5 ml/h again; 100 ml of the culture was removed every 8 hours. The authors found that biomass concentration and maximum specific growth-rate in the shaken flasks increased with increasing the concentration of the Celite particles. When 10% Celite was used, the biomass and specific growth-rate were 1.5 and 1.6 time, respectively, that of free cell culture. As for the fluidised-bed fermentations, repeated fed-batch proved to be superior to the semi-continuous fermentation but the bio-particle size and liberated free cell concentration was higher in the latter. In order to extend the duration of the production phase, nutrient limitation conditions for the fed-batch fermentation were considered in two separate sets of fermentation: nitrogen limitation and phosphate limitation. Morphological differences to the immobilised hyphae were observed. Fluffy loose bio-particles with increased buoyancy developed under nitrogen limited conditions. This resulted in poor mixing and forced termination of fermentation. Phosphate limited cultures, however, resulted in compact smooth pellets which helped in improved mixing. The penicillin concentration was kept at 80% of the maximum level for over two weeks and the overall penicillin productivity was improved by 3-4 times compared to the conventional batch fermentation. The fermentor could be run under these conditions for up to 30 days without serious operational problems.

In an effort to provide a systematic approach to the large-scale Celite immobilised- cell penicillin fermentation, preliminary studies were carried out on Celite R-633 to investigate the feasibility and efficiency of the use of Celite in a wet or dry state. The optimum time was also investigated for the maximum adsorption of spores onto the Celite [22]. The studies were carried out both in shaken flasks (250 ml) and a fluidised-bed reactor (750 ml). It was found that within 2 hours almost 90% of the spores were adsorbed onto the Celite particles in both types of vessels and there was little difference between the % spore uptake by dry or pre-wetted Celite.

The information obtained by the previous study, was used for the large-scale immobilised cell fermentation of *Penicillium chrysogenum* P2 and its temperature sensitive cell division cycle (Cdc) mutant P2-95 under batch and continuous modes respectively [23]. Parallel batch fermentations were carried out in 6 l and 250 l working volume airlift reactors. The *Penicillium* spores were added to the R-633 Celite pre-sterilised in water filled fermentor. After two washes to discard the unbound spores, the water was drained and a semi-defined medium was added to the fermentor. The batch fermentations were run for 450 hours in both reactors and biomass, penicillin G and δ-(L-α-aminoadipyl)-L-cysteinyl-D-valine (ACV) profiles were prepared for this period. The development of the free cells in the immobilised system was reported for the first time. The immobilised and free biomass concentrations were almost the same in both reactors. However, there were differences between the levels of penicillin G and ACV in the small and large reactors but the total concentrations of the two metabolites were nearly the same after 300 hours. The variable design of the large-scale airlift reactor used for this study was described giving the configurations for batch and continuous modes of operation. The paper covered two new features for the first time: firstly, an immobilised-cell continuous fermentation was carried out at a large-scale (320 l) and secondly, a temperature-sensitive mutant, *Penicillium chrysogenum* P2-95, was used. This helped in controlling the biomass growth and reduced the developed of free cells. After 500 hours both ACV and penicillin G were still being produced [24].

In a comparison between immobilisation by adsorption on Celite and entrapment in polysaccharide gels for both small-scale and large-scale (250 l) fermentation [25], the fermentor designs, equipment and procedures for immobilisation were discussed based on two microbial model systems: vegetative cells of a lysine-overproducing *Bacillus subtilis* strain and spores of *Penicillium chrysogenum* P2. The authors emphasised the importance of considerations regarding problems of scale-up prior to small-scale experiments if eventual industrial exploitation is envisaged. They stressed also the importance of "specifically designed strains" suitable for immobilisation.

A recent publication [26] has investigated the feasibility of continuous production of penicillin from a strain of *Penicillium chrysogenum* in a 2.4 l external loop airlift reactor utilising a transverse magnetic field. The sterilisable spherical support particles consisted of a ferromagnetic core of magnetite covered with a stable layer of activated carbon or zeolite layers. After pre-treatment and sterilisation, 270 grams of magnetic beads were mixed with 800 ml of *Penicillium chrysogenum* spore suspension with a concentration of $1.1 \cdot 10^8$ spores per ml. The immobilised cells were washed and left in a shaken flask containing growth medium. After 48 hours, the beads were colonised with the cells and were transferred to the fermentor containing penicillin production medium. A flow-rate of 15 ml per hour was adopted. It was claimed that penicillin production was improved compared to the conventional systems due to the changes in the hydrodynamic behaviour of magnetised bed.

2.2. CEPHALOSPORIN-C PRODUCTION

Yong-Ho Khang et al. [27,28] investigated another β-lactam antibiotic, cephalosporin-C (CPC) production by calcium alginate immobilised cells of the fungus *Cephalosporium acreminium* (ATCC 36225). In the first study [27], the growth of the fungus and production of the antibiotic was investigated in 2 l stirred fermentor under free- and immobilised- cell culture conditions where the cells produced cephalosporin-C in defined "rest" media under a range of oxygen concentrations. It was found that the cell growth-rate and antibiotics production-rate reduced under the immobilised cell conditions but the specific antibiotic production-rate increased under oxygen saturation conditions. Based on this finding, the authors suggested that the whole cell immobilisation of *Cephalosporium acreminium* was desirable as the potential reduction in the reactor volume would be beneficial for the industry.

Co-immobilisation of the oxygen demanding fungus *Cephalosporium acreminium* with oxygen producing alga *Chlorella pyrenoidosa* was investigated [28] to exploit the symbiotic relation between the two organisms for potential increased levels of CPC. The symbiotic relation between *Cephalosporium acreminium* and *Chlorella pyrenoidosa* was established through preliminary studies. The optimal ratio for the mixed culture fermentations was investigated by changing the biomass concentration of the organisms in the mixed cultures. The two organisms were grown separately to produce a biomass of 6 g/l each. This biomass was used as the inoculum for of free mixed culture and co-immobilised culture studies. It was found that a cell dry weight ratio of 1:8 (fungus:alga) was the most suitable for highest production of CPC both under free and immobilised cell culture conditions. The increase in the production-rate was higher for the immobilised cell culture. However, the duration of the reported fermentations was quite short (45 hours).

Three different supports, calcium alginate, bagasse and silk sachets, were investigated using a small (70.3 cm by 3.8 cm) packed-bed reactor to find an optimal immobilisation matrix and a suitable continuous culture process for production of CPC by *Cephalosporium acreminium* NICM 1069 [29]. Silk sachets proved to be superior to bagasse and calcium alginate in terms of reduction in cell growth-rate and increased specific β-lactam antibiotic production-rate when compared to the free cell culture for a period of 100 hours. Over this period, the leakage of cells from the carriers was less in the case of calcium alginate and highest from bagasse. The optimal cell-to-carrier ratio and medium flow-rate for the highest production of the antibiotic was reported at 3:2 and 30 ml/h respectively. When an output recycle was adopted, the production of CPC increased linearly up to a 100% recycle ratio.

The same strain (*Cephalosporium acreminium* NICM 1069) was used in a 1.4 l internal draft tube air-lift reactor to investigate cephalosporin production by free and immobilised cells. In addition to sachets (used in the previous study), inorganic SIRAN™ (Siran) particles (600-1000 μm) were prepared for adsorption of spores [30]. Results of batch fermentation revealed that the growth-rate of the cells immobilised in Siran was higher than the cells grown in silk sachets which in turn were higher than the free cells. The specific β-lactam antibiotic production followed the same pattern as the growth-rate. In the continuous culture runs, CPC production increased with decrease in dilution-rate and there was an optimal dilution-rate for which the specific CPC production-rate was at its highest. This suggested a critical residence time for the production.

In yet another comparative study [31] *Cephalosporium acreminium* C10 (ATCC 48272) was used in a 5 l stirred tank reactor and a 1.7 l tower reactor containing a defined medium for free- and –immobilised cell fermentation respectively to produce CPC. A vegetative culture grown from spores in shaken flasks was used to inoculate the 5 l fermentor. The immobilisation was carried out also on the pre-grown vegetative cells using alginate gel (20 g/l) and alumina (10 g/l, particles less than 325 mesh) as the support matrix. There was a global reduction in the process-rate of immobilised cells compared to the free cells and this was attributed to the oxygen diffusion limitation at the particle level. The authors pointed out that despite reduction in the productivity of CPC in the immobilised cell culture, when the carbon consumption was taken into account; the productivity was in fact 1.4 times higher in the shaken flask studies. The results from the immobilised batch fermentations in the tower reactor showed also a higher productivity based on the consumption of the carbon source compared to the free-cell culture carried out in the stirred tank reactor. It was emphasised that although the respiration rate is a limiting step in the immobilised *Cepalosporium acreminium* C10 fermentation process [32], further analysis of the results of the respiration-rate shows that the limitation does not affect the final CPC yield. This is an important claim indicating that industrial production of the antibiotic is feasible.

At a theoretical level, different models [32,33] have been proposed on the effect of oxygen mass transfer for production of CPC but the industrial relevance of the findings seems preliminary at this stage as the calculations are based on experiments carried out at very small-scale (shaken flasks and 2 l bioreactors).

Streptomyces clavuligerus NRRL 3585 cells were immobilised in a synthetic polymer for production of antibiotics (cephalosporins). This early work [34] reported that the resting cells could retain their β-lactam antibiotics activity when they were entrapped in pre-polymerised acrylamide partially substituted with acylhydrazide and cross-linked with various dialdehydes. Under this immobilisation method significant improvement in the specific antibiotic production was observed compared to the direct polymerisation of acrylamide monomers in the presence of cells.

2.3. OTHER ANTIBIOTICS

Streptomyces T 59-235 (a tylosin, a macrolide antibiotic, producer) and *Streptomyces tendae* Tu 901 (a nikkomycin, a nucleoside peptide antibiotic, producer) were employed for immobilisation studies using different matrices including calcium alginate, agar, carrageenan and polyacrylamide [35]. The experiment design and methodology was the same in both cases. The immobilisation was carried out on the mycelia of the strains grown in liquid culture. Free and immobilised batch fermentations were carried out in 100 ml baffled flasks while continuous culture was implemented in a 700-ml fluidised-bed reactor. The best results were obtained with calcium alginate where fermentations continued for 300 hours without liberation of free cells from the matrix. There was no production in immobilised-cell fermentations using polyacrylamide or gelatine due to the usage of glutaraldehyde and toxicity of polyacrylamide monomers. The study showed the importance of initial matrix loading (concentration of cells per bead) in relation to antibiotics production. It was also shown that production of the antibiotics was dependent on the flow-rate of the medium in continuous immobilised culture.

Porous glass particles (2-2.24 mm diameter, 125-315 µm pore size) were used as a matrix to investigate the effect of oxygen on nikkomycin production in immobilised cell continuous culture of *Streptomyces tendae* Tü901/8c [36]. It was found that the extent of the development of the biomass layers on the particles was critical in the production of the antibiotic by the oxygen-demanding organism. In batch fermentation, the cells were developed on the glass particles in a fluidised bioreactor. The specific nikkomycin productivity decreased as the thickness of the biomass layer increased. The use of a phosphate-free medium helped in controlling the fungal growth. When oxygen supply to the cells was optimal, the nikkomycin production continued for more than 200 hours.

Rifamycin is an example of the ansamycins, the complex macrolactam antibiotics. Rifamycin production through repeated batch fermentation was tried by immobilisation of *Amycolatopsis mediterranei* CBS 42575 cells on glass wool (8 µm diameter) [37] reduction of batch time from 96 to 48 hours did not decrease rifamycin production.

In addition to hydrogels and adsorption particles, other means have been tried for cell immobilisation. Hollow fibre membrane modules could be used simultaneously as matrices for cell immobilisation and bioreactors for fermentation of the immobilised cells. *Nocardia mediterranei* ATCC 21789 was immobilised in a hollow fibre bioreactor which was run continuously and successfully for more than 50 days [38]. The bioreactor was set up by using polypropylene hollow fibres as carriers of the medium. The fibres were placed inside silicone tubing and the cells occupied the area outside the tubing. A bundle of silicone tubing was placed inside a glass tube cover. The air or oxygen as necessary passed through the glass tube. The antibiotic production increased by 22-30 folds that of a comparable batch system. The specific productivity of rifamycin, however, was 30-40% of the free cell culture carried out in the shaken flasks. It was suggested that medium design and pH control play important roles in improving the productivity of rifamycin in hollow fibre system.

A 2-litre bubble column was used for production of the β–lactam antibiotic, thienamycin, under continuous immobilised cell culture conditions [39]. The design of the reactor made it possible for the liberated free cells to leave the reactor. This helped in reducing the viscosity. *Streptomyces cattleya* immobilised on Celite particles was used in batch and continuous fermentation using both synthetic and complex media. The study involved changing the media for growth and production and changing the dilution-rate. The study showed that thienamycin could be produced continuously under immobilised cell culture conditions for more than 1300 hours. The importance of the medium composition and the dilution-rate on the rate of the antibiotic production was emphasised.

Streptomyces parvullus ATCC 12434 was immobilised in calcium alginate to investigate the possibility to improve actinomycin D productivity in continuous culture using a 600 ml working volume airlift fermentor [40]. This antibiotic is highly toxic to humans and is mainly used in research. Two types of defined media (nitrogen starvation, carbon and nitrogen starvation) were used to eliminate catabolite repression. Through a series of batch, fed-batch and continuous fermentations, the authors concluded that immobilised continuous culture at low dilution-rate was superior to the other two types of fermentation. Long-term production was possible with a carefully designed periodic feeding of the production media.

Erythromycin is produced by *Streptomyces erythraeus* and belongs to the macrolide group of antibiotics. In a comparative study carried out in 250 ml shaken flasks, the antibiotic production was compared between conventional free cell fermentation, washed cell fermentation (similar to immobilised cell fermentation but with free cells) and immobilised cell fermentation [41]. Cell entrapment was carried out through immobilisation of wet mycelia in calcium alginate. It was found that total antibiotic production and average specific (and volumetric) productivity per cycle were higher with the immobilised cell system. In addition, the production process was significantly extended (total fermentation time: 720, 192, and 96 hours respectively for the immobilised cell, washed cell, and free cell fermentations).

Tetracyclines are polyketides; wide-spectrum antibiotics used both for human and animal health. Their production through cell immobilisation has been used by different groups employing adsorption and entrapment techniques. A hollow-fibre bioreactor was used to grow *Streptomyces aureofaciens* ATCC 12416c cells for production of tetracycline [42]. While long-term production of the antibiotic was not achieved, it was demonstrated that it was possible to provide much needed oxygen to the *Streptomyces* cells in a hollow fibre system. The reactor was designed such that silicone tubules were assembled inside polypropylene fibres where *Streptomyces* cells were immobilised. The fibres were packaged inside a tubular glass shell. Gas was pumped through the silicone tubules and nutrient medium ran along the bioreactor outside the polypropylene fibres. Free-cell batch fermentations were carried out in 500 ml shaken flasks to provide base-line information. The immobilised cell fermentation continued for 130 hours (diffusion mode) during which complex growth medium was changed to a synthetic defined medium. The decline in the concentration of tetracycline after 70 hours was attributed to solute transport limitation or an inhibitory feedback mechanism due to the accumulation of tetracycline in the system. It was suggested that further studies were needed on the metabolic state of the cells to obtain stable steady-state operation behaviour.

In two separate studies, *Streptomyces rimosus* Pfizer 18234-2 and *Streptomyces rimosus* TM 55 (CCRC 960061) were immobilised in calcium alginate to investigate oxytetracycline (OTC) production [43,44]. The combined results of the two studies show that under repeated batch fermentations in shaken flasks it was possible to maintain the antibiotic production for 28 days, the relative yield of the antibiotic produced by the immobilised cells was twice that of the free cells, the growth-rate of the cells decreased but the specific productivity oxytetracycline increased. The OTC production depended on the inoculum density and bead diameter.

The same strain, *Streptomyces rimosus* Pfizer 18234-2, was also immobilised on glass wool (8 μm) by adsorption and the effect of inoculum on production of biomass and OTC was investigated in batch cultures carried out in 250 ml shaken flasks containing 50 ml of a complex medium [45]. Spore inoculum and vegetative inoculum with different ages were used for both free and immobilised cell fermentations and 500 mg glass wool was used for the immobilised runs in each flask. Immobilisation brought forward the starting point for growth and OTC production as well. The duration of the immobilised cell fermentation was extended through adoption of repeated batch; during which there was no discernible decrease in OTC production for 40 days. Similar strategy with free cell culture could extend the culture for 20 days with decreasing production between the batches.

Response surface methodology was applied to the immobilised cell fermentations of *Streptomyces aureofaciens* strain 5711 when the mycelia were immobilised in κ-carrageenan in 500 ml shaken flasks [46]. A defined medium was used for production of both chlortetracycline (CTC) and oxytetracycline (OTC). The concentration of the two antibiotics and the pH of the culture were considered as the dependent variables while the two independent variables were considered to be the concentration of κ-carrageenan and the potassium chloride (acting as gelling agent). The aim of the experiment was to investigate, systematically, the combined effect of the matrix concentration and the concentration of the gelling agent on the production of the two antibiotics. It was found that both the matrix and the gelling agent concentrations had pronounced effect on the yield of the antibiotics.

Production of cephamycin C, yet another β–lactam antibiotic, was investigated by applying two matrices for immobilisation of *Streptomyces clavuligerus* PS2: a) calcium alginate and b) sponge [47]. The seed used for both immobilisation matrices was a 48 hours grown culture. This culture was mixed with 10 X 10 mm sponge pieces for immobilisation by adsorption while entrapment in calcium alginate beads was carried out applying the classical method. The fermentations were carried out as repeated batches. Every 120 hours, fresh medium replaced the old one and there were a total of 5 repeats in both cases. It was found that compared to calcium alginate, sponge was a better matrix supporting higher antibiotic production. Moreover, the homogeneity of the immobilised preparations was better with the sponge.

Streptomyces kasugaensis, a kasugamycin (an aminoglycoside antibiotic) producer was adsorbed onto Celite particles and fermented under continuous culture [48]. The productivity of the antibiotic was between 14-23 folds higher than the traditional free cell batch fermentations. The stable operation was maintained for more than 820 hours.

3. The use of immobilised microorganisms in biotransformation in two liquid phase systems

In addition to the generally agreed benefits of cell immobilisation in aqueous cultures, the immobilised cell systems offer advantages when used in water insoluble medium. Industrial use of microorganisms for biotransformations related to biopharmaceutical products includes catalytic activity of microorganisms on steroids. In this section, some examples of such usage are given.

Nocardia rhodochrous immobilisation resulted in notable improvement in the stability of steroid Δ^1-dehydrogenation [49]. *Arthrobacter simplex* NCIB 8929 cells were used in free cell culture and as immobilised cells in calcium alginate beads for Δ^1-dehydrogenation of hydrocortisone in a 70 ml working volume aqueous two-liquid-phase stirred tank reactor [50]. Extended reactor operation with high purity steroid product was achieved with the immobilised cell system.

Several types of matrices (agar, polyurethane foam, κ-carrageenan and Celite) and solvent systems were investigated for immobilisation of *Arthrobacter simplex* ATCC 6946 to find out the optimal matrix and the related technique to be used in Δ^1-dehydrogenation of the water insoluble corticosteroid 6-α-methyl-hydrocortisone-21-acetate [51]. The most suitable system was found to be polyurethane foam with *n*-decane-1-ol as a conversion medium. The problem with immobilisation through entrapment was the unfavourable partition and diffusion which affected productivities negatively. On the other hand, solvent toxicity encountered in the surface adsorption process affects the activity.

Penicillium raistrickii i 477 was used as free cell or as immobilised cell in calcium alginate for 15α-hydroxylation of 13-ethyl-gon-4-en-3,17-dione (GD) [52]. When β-cyclodextrin was used either as a sole carbon source or as a supplement in a variety of glucose containing media an increase was found in product formation compared with methanol containing media. 11α-hydroxylation of progesterone to form 11α-hydroxy progesterone was carried out as a model system using a novel cell immobilisation method [53]. The method was based on immobilisation of the spores of *Aspergillus ochraceus* NRRL 405 in polyurea-coated alginate. Fifty ml of a reaction medium which contained saccharose and a range of different organic solvents containing progesterone were added to a 500 ml shaken flask and free or immobilised cells were added to the flask and reaction-rate was estimated. The use of polyurea coated immobilised cells in ethyl acetate-water biphasic medium resulted in faster reaction-rate. Substrate concentration could be increased without significant inhibition and majority of hydroxyl progesterone produced was in the organic phase facilitating separation. The damage inflicted by the solvents to the cells limited the steroid transformation in the two-phase system. The authors suggested that if the toxicity problem could be eliminated, the method used in their study would have good potential for industrial application.

In a follow-up study, the stability of alginate gels was investigated for 11α-hydroxylation of progesterone by *Aspergillus ochraceus* NRRL 405 [54]. Four different hardening methods employing polyetheleneimine-glutaraldehyde, periodate-polyetheleneimine, hexamethylenediamine-glutaraldehyde and polyacrylamide were used to improve alginate gel stability. In addition, as Fe^{+3} enhances the activity of *A. ochraceus for* 11α-hydroxylation of progesterone, Ca^{+2} was replaced with Fe^{+3} as the ion in the gelling agent. The use of polyacrylamide hardening procedure gave the best result and the beads stayed stable for 36 hours in presence of 0.1 M phosphate buffer. The productivity and the maximum yield were also improved for the 11α-hydroxylation of progesterone.

β-sitosterol is considered to be a cheap and available material for production of pharmaceutical steroids. Three matrices (Celite, polyurethane foam, and κ-carrageenan) were investigated to immobilise *Mycobacterium* sp. NRRL B-3805 for side-chain cleavage of β-sitosterol using several organic solvents (octanol, dodecane, decanol, *bis*(2-ethylhexyl) phthalate, didecyl-phthalate, 2,6-dimethyl-4-heptanone, methyl nonyl ketone, oleyl alcohol and isopropyl palpitate) as media for bioconversion [55]. After an extensive study, it was found that using Celite 80-120 mesh as the matrix and *bis*(2-ethylhexyl) phthalate as solvent resulted in the best specific productivity for 4-androstene-3, 17dione. It was also found that the activity and stability of the system was temperature dependent and the biocatalyst would deactivate faster at higher temperatures. The optimum temperature for the activity and stability was shown to be 25°C.

4. Concluding remarks

It is now more than forty years since the first report on cell immobilisation was published. Considering the advantages that have been attributed repeatedly to this technique through numerous publications, it is perhaps surprising that cell immobilisation has been scaled-up only to a few industrial bioprocesses at production scale. In the biopharmaceutical industry using microbial cells, there are no examples. Some of the reasons for this are listed below:

- Pharmaceutical industry benefits from established and robust systems within which free cell fermentations have been carried out repeatedly with continuous and gradual improvements to the original process. Any new process which requires radical changes to the existing system faces resistance.
- There is a high degree of complexity (and risk) associated with the immobilised cell process. This is not received well within the industry. Complexity invites potential operational errors.
- There is a need for extra pieces of equipment and detailed procedures with extra steps to be added (or to replace) to a process which is currently working smoothly.
- This, together with the complexity factor, means that labour with new expertise is required to work in an untested territory. This is an extra cost.
- Most of the background studies, as is shown in this article, are at small-scale and only couple of large-scale work has been reported so far. This is not sufficient for the industry to contemplate new large-scale processes based on cell immobilisation.
- There is no comprehensive economic assessment of the new process to convince the traditional industry. Only a big margin of benefit would make the industry to start thinking about a potential move into the new process.
- The continuous search for new products (*e.g.* bioactive compounds) by the industry makes rather less attractive the use of a new approach for the production of an old compound.
- Although different research groups have worked on a range immobilisation techniques, even those who have worked with the same matrix, have produced results. This is due to the use of different species, different makes of matrices with different characteristics, different designs and sizes of reactors and, at times, different strains of the same species. Lack of a "universal" portrait of the immobilised cell systems does not help in formulising opinion for large-scale production.
- Regulatory and validation issues are very important factors that might prove too costly and inhibitory to the development of an industrial-scale microbial cell immobilisation process for the production of biopharmaceuticals.

Considering the above points, the prospect of immobilised cell technology for direct use at large-scale in biopharmaceutical industry does not seem to be promising. However, should new discoveries in this field demand the design and construction of new production plants, the potential of the cell-immobilisation might be realised and exploited. As it stands, cell immobilisation can have an *indirect* role in providing valuable information regarding the physiology of the cells under particular environments (*e.g.* oxygen/nutrients limited conditions). The technology will continue to develop at research level for the foreseeable future with regards usage of matrices, cells as biocatalysts and the interaction between the cells and matrices.

References

[1] Hattori, T. and Furusaka, C. (1960) Chemical activities of *E. coli* adsorbed on a resin. J. Biochem. 48: 831-837.
[2] Franks, N.E (1971) Catabolism of L-arginine by entrapped cells of *Streptococcus faecalis* ATCC 8043. Biochim. Biophys. Acta 252: 246-254.
[3] Mosbach, K. and Larsson, P. (1970) Preparation and application of polymer entrapped enzymes and microorganisms in microbial transformation processes with special reference to steroid 11-b-hydroxylation and d1-dehydrogenation. Biotechnol. Bioeng. 12: 19-27.
[4] Tosa, T.; Sato, T.; Mori, T. and Chibata, I. (1974) Immobilised aspartase containing microbial cells: preperation and enzymic properties. Appl. Microbiol. 27: 878-885.
[5] Chibata, I. and Tosa, T. (1977)Transformations of organic compounds by immobilized microbial cells. Adv. Appl. Microbiol. 22: 1-27.
[6] Kierstan, M and Bucke, C. (1977) The immobilization of microbial cells, subcellular organelles, and enzymes in calcium alginate gels. Biotechnol. Bioeng. 19: 387-397.
[7] Wada, M.; Kato, J. and Chibata, I. (1979) A new immobilization of microbial cells. European J. Appl. Microbiol. 8: 241-247.
[8] Shibatani, T. (1996) Industrial applications of immobilized biocatalysts in Japan. In: Wijffels, R.H.; Buitelaar, R.M.; Bucke, C. and Tramper, J. (Eds.) Immobilized Cells: Basics and Applications. Elsevier, Amsterdam (The Netherlands); pp. 585-591.
[9] Champagne, C.P. (1996) Immobilized cell technology in food processing. In: Wijffels, R.H.; Buitelaar, R.M.; Bucke, C. and Tramper, J. (Eds.) Immobilized Cells: Basics and Applications. Elsevier, Amsterdam (The Netherlands); pp. 633- 640.
[10] Morikawa, Y.; Karubi, I. and Suzuki. S. (1979) Penicillin G production by immobilized whole cells of *Penicillium chrysogenum*. Biotechnol. Bioeng. 21: 261-270.
[11] Kurzatkowski, W.; Kurylowicz, W. and Paszkiewicz, A. (1982) Penicillin G production by immobilized fungal vesicles. European J. Appl. Microbiol. Biotechnol. 15: 211-213.
[12] El-Saed, A.M.M. and Rehm, H. J. (1986) Morphology of *Penicillium chrysogenum* strains immobilized in calcium alginate beads and used in penicillin fermentation. Appl. Microbiol. Biotechnol. 24: 89-94.
[13] El-Saed, A.M.M. and Rehm, H.J. (1987) Semicontinuous penicillin production by two *Penicillium chrysogenum* strains immobilized in calcium alginate beads. Appl. Microbiol. Biotechnol. 26: 211-214.
[14] El-Saed, A.M.M. and Rehm, H.J. (1987) Continuous penicillin production by *Penicillium chrysogenum* strains immobilized in calcium alginate beads. Appl. Microbiol. Biotechnol. 26: 215-218.
[15] Deo, Y.M. and Gaucher G.M. (1984) Semicontinuous and continuous production of penicillin G by *Penicillium chrysogenum* cells. Biotechnol. Bioeng. 26: 285-295.
[16] Mussenden, P.J.; Keshavarz, T. and Bucke, C. (1991) The effects of spore loading on the growth of *Penicillium chrysogenum* immobilized in κ-carrageenan. J. Chem. Tech. Biotechnol. 52: 275-282.
[17] Mussenden, P.J.; Keshavarz, T.; Saunders, G. and Bucke, C. (1993) Physiological studies related to the immobilization of *Penicillium chrysogenum* and penicillin production. Enzyme Microb. Technol. 15: 2-7.
[18] Karhoot, J.M.; Anderson, J.G. and Blain, J.A. (1987) Production of penicillin by immobilized films of *Penicillium chrysogenum*. Biotechnol. Lett. 9: 471-474.
[19] Wang, D.I.C.; Meier, J. and Yokoyama, K. (1984) Penicillin fermentation in a 200-litre tower fermentor using cels cofined to microbeads. Appl. Biochem. Biotechnol. 9: 105-116.
[20] Jones, A.; Wood, D.N.; Razniewska, T. Gaucher, M. and Behie, L.A. (1986); Continuous production of penicillin-G by *Penicillium chrysogenum* cells immobilized on Celite biocatalyst support particles. Can. J. Chem. Eng. 64: 547-552.

[21] Kim, J.H.; Oh, D.K.; Park, S.K. and Park, Y.H. (1986) Production of penicillin in a fluidised-bed bioreactor using a carrier-supported mycelial growth. Biotechnol. Bioeng. 28: 1838-1844.
[22] Keshavarz, T.; Walker, E.; Eglin, R.; Lilley, G.; Holt, G.; Bull, A.T. and Lilly, M.D. (1989) Immobilization of *Penicillium chrysogenum*: spore growth on Celite. Appl. Microbiol. Biotechnol. 30: 487-491.
[23] Keshavarz, T.; Eglin, R.; Walker, E.; Bucke, C.; Holt, G.; Bull, A.T. and Lilly, M.D. (1990) The large-scale immobilization of *Penicillium chrysogenum*: batch and continuous operation in an air-lift reactor. Biotechnol. Bioeng. 36: 763-770.
[24] Lilly, M.D.; Keshavarz, T.; Bucke, C.; Bull, A.T. and Holt, G. (1990) Pilot-scale studies of immobilized *Penicillium chrysogenum*. In: de Bont, J.A.M.; Visser, B.; Mattiasson, B. and Tramper, J. (Eds.) Physiology of immobilized cells. Elsevier, Amsterdam (Netherlands); pp. 369-375.
[25] Al-Qodah, Z. (2000) Continuous production of antibiotics in an airlift fermentor utilising a transverse magnetic field. Appl. Biochem. Biotechnol. 87: 37-55.
[26] Keshavarz, T.; Bucke, C. and Lilly, T. (1996) Problems in scale-up of immobilized cell cultures. In: Wiffels, R.H.; Buitelaar, R.M.; Bucke, C. and Tramper, J. (Eds.) Immobilized Cells: Basics and Applications. Elsevier, Amsterdam (Netherlands); pp. 505-510.
[27] Khang, Y.H.; Shanker, H. and Senatore, F. (1988) Comparison of free and immobilized *Cephalosporium acreminium* for β-lactam antibiotic production. Biotechnol. Lett. 10: 719-724.
[28] Khang, Y.H.; Shanker, H. and Senatore, F. (1988) Enhanced β-lactam antibiotic production by coimmobilization of fungus and alga. Biotechnol. Lett. 10: 867-872.
[29] Kundu, S.; Mahapatra, A.C.; Srivastava, P. and Kundu K. (1992) Studies on cephalosporin-C production using immobilized cells of *Cephalosporium acreminium* in a packed bed reactor. Process Biochem. 27: 347-350.
[30] Srivastava, P. and Kundu, S. (1998) A comparative evaluation of cephalosporin C production using various immobilization modes. J. Gen. Appl. Microbiol. 44: 113-117.
[31] Araujo, M.L.G.C.; Giordano, R.C. and Hokka, C.O. (1999) Studies on the respiration rate of free and immobilized cells of *Cephalosporium acreminium* in cephalosporin C production. Biotechnol. Bioeng. 63: 593-600.
[32] Araujo, M.L.G.C.; Oliviera, R.C. and Hokka, C.O. (1996) Comparative studies on cephalosporin C production process with free and immobilized cells of *Cephalosporium acreminium* ATCC 48272. Chem. Eng. Science. 51: 2835-2840.
[33] Khang, Y.H.; Shanker, H. and Senatore, F. (1988) Modelling the effect of oxygen mass transfer on β-lactam antibiotic production by immobilized *Cephalosporium acreminium*. Biotechnol. Lett. 10: 861-866.
[34] Freeman, A. and Aharonowitz, Y. (1981) Immobilization of microbial cells in crosslinked, prepolymerized, linear polyacrylamide gels: antibiotic production by immobilized *Streptomyces clavuligerus* cells. Biotechnol. Bioeng. 23: 2747-2759.
[35] Veelken, M. and Pape, H. (1982) Production of tylosin and nikkomycin by immobilized *Streptomyces* cells. Appl. Microbiol. Biotechnol. 15: 206-210.
[36] Trück, H.U.; Chmiel, H.; Hammes, W.P. and Trösch, W. (1990) Effects of oxygen supply on the production of nikkomycin with immobilized cells of *Streptomyces tendae*. Appl. Microbiol. Biotechnol. 34: 1-4.
[37] Abu-Shady, M.R.; el-Diwany, A.I.; Farid, M.A. and el-Enshasy, H.A. (1995) Studies of rifamycin production by *Amycolatopsis mediterranei* cells immobilized on glass wool. J. Basic Microbiol. 35: 279-284.
[38] Chung, B.H.; Chang, H.N. and Kim, I.H. (1987) Rifamycin B production by *Nocardia mediterranei* immobilized in a dual hollow fibre bioreactor. Enzyme Microb. Technol. 9: 345-349.
[39] Arcuri, E.J.; Slaff, G. and Greasham, R. (1986) Continuous production of thienamycin in immobilized cell systems. Biotechnol. Bioeng. 28: 842-849.
[40] Dalili, M. and Chau, P.C. (1988) Production of actinomycin D ith immobilized *Streptomyces parvullus* under nitrogen and carbon starvation conditions. Biotechnol. Lett. 10: 331-336.
[41] Bandyopadhyay, A.; Das, A.K and Mandal, S.K. (1993) Erythromycin production by *Streptomyces erythraeus* entrapped in calcium alginate beads. Biotechnol. Lett. 15: 1003-1006.
[42] Robertson, C. and Kim, I.H. (1985) Dual aerobic hollow-fibre bioreactor of *Streptomyces aureofaciens*. Biotechnol. Bioeng. 27: 1012-1020.
[43] Farid, M.A.; el-Diwany, E.I. and el-Enshasy, H.A. (1994) Production of oxytetracycline by immobilized *Streptomyces rimosus* cells in calcium alginate gels. Acta Biotechnol. 14: 303-309.

[44] Yang, S. and Yueh, C.Y. (2001) Oxytetracycline production by immobilized *Streptomyces rimosus*. J. Microbial. Immunol. Infect. 34: 235-242.
[45] el-Enshasy, H.A.; Farid, M.A. and el-Diwany, A.I. (1996) Oxytetracycline production by free and immobilized cells of *Streptomyces rimosus* in batch and repeated batch cultures. In: Wijffels, R.H.; Buitelaar, R.M.; Bucke, C. and Tramper, J. (Eds.) Immobilized Cells: Basics and Applications. Elsevier, Amsterdam (The Netherlands); pp. 437-443.
[46] Terual, M.L.A.; Gontier, E.; Bienaime, C.; Saucedo, J.E.N. and Barbotin, J.N. (1997) Response surface analysis of chlortetracycline and tetracycline production with κ-carrageenan immobilized *Streptomyces aureofaciens*. Enzyme Microb. Technol. 21: 314-320.
[47] Devi, S. and Sridhar, P. (2000) Production of Cephamycin C in repeated batch operations from immobilized *Streptomyces clavuligerus*. Process Biochem. 36: 225-231.
[48] Kim, C.J.; Chang, Y,K.; Chun, G.T.; Jeong, Y.H. and Lee, S.J. (2001) Continuous culture of immobilized *Streptomyces* cells for kasugamycin production. Biotechnol. Progress 17: 453-461.
[49] Yamani, T.; Nakatani, H.; Sada, E.; Omata, T.; Tanaka, A. and Fukui, S. (1979) Steroid bioconversions in water insoluble organic solvents: Δ^1-dehydrogenation by microbial cells and by cells entrapped in hydrophilic and lipophilic gels. Biotechnol. Bioeng. 21: 1887-1903.
[50] Hocknull, M.D. and Lilly, M.D. (1990) The use of free and immobilized *Arthrobacter simplex* in organic solvent/aqueous two-liquid-phase reactors. Appl. Microbiol. Biotechnol. 33: 148-153.
[51] Pinheiro, H.M. and Cabral, J.M.S. (1992) Screening of whole-cell immobilization procedures for the Δ^1-dehydrogenation of steroids in organic medium. Enzyme Microb. Technol. 14: 619-624.
[52] Schlosser, D.; Irrang, S and Schmauder, H.-P. (1993) Steroid hydroxylation with free and immobilized cells of *Penicillium raistrickii* in the presence of β-cyclodextrrin. Appl. Microbiol. Biotechnol. 39: 16-20.
[53] Houng, J.-Y.; Chiang, W.-P. and Chen K.-C. (1994) 11α hydroxylation of progestrone in biphasic media using alginate-entrapped *Aspergillus ochraceus* gel beads coated with polyurea. Enzyme Microb. Technol. 16: 485-491.
[54] Chen K.-C.; Yin, W.-S. and Houng, J.Y. (1994) 11α hydroxylation of progesterone using modified alginate-immobilized cells. Enzyme Microb. Technol. 16: 551-555.
[55] Dias, A.C.P.; Cabral, J.M.S. and Pinheiro, H.M. (1994) Sterol side-chain cleavage with immobilized *Mycobacterium* cells in water-immiscible organic solvents. Enzyme Microb. Technol. 16: 708-714.

PRODUCTION OF BIOLOGICS FROM ANIMAL CELL CULTURES

JAMES WARNOCK AND MOHAMED AL-RUBEAI

Animal Cell Technology Group, School of Engineering, Chemical Engineering, University of Birmingham, Birmingham, B15 2TT United Kingdom – Fax: +44 [0] 121 414 3888 – Email: m.al-rubeai@bham.ac.uk

1. Introduction

The production of biologics is the central core of animal cell biotechnology. It was estimated that the direct and indirect activities of the biotechnology industry contributed $47 billion in revenues to the US economy in 1999 [1]. Immobilisation technologies have allowed for the production of large biomasses, which forms the substrate from which a wide range of products can be produced, as listed in Table 1. Animal cells are used for the production of these products to allow for the correct post-translational processing, including glycosylation, the formation of disulfide-bond, γ-carboxylation and other modifications to occur, which cannot be achieved using other eukaryotic or bacterial cells. However, the relatively low biomass productivity compared to bacterial cultures and the low product expression rate has meant it has become necessary to scale-up their production in order to meet the high demand for these products. Many biologics, such as viral vaccines, are produced from anchorage dependent cell lines. Although developmental work can be done in monolayer cultures, using either tissue culture flasks or roller bottles, it is not a desirable method for large-scale manufacture. This is because scale-up is achieved by increasing the unit number of either flasks or bottles, which is extremely labour intensive and susceptible to contamination [2]. Modern facilities have overcome these problems by the use of expensive robots; however, it is still not possible to control environmental conditions such as pH or dissolved oxygen levels [3]. The most successful method for the scale-up of anchorage-dependent cells has been the use of microcarriers, first reported by van Wezel [4], which provide a high surface area to volume ratio and can be used in traditional stirred tank bioreactors.

The use of suspension cell lines, such as hybridomas, for the production of monoclonal antibodies has been widely used. Scale-up of these systems is relatively simple and the maximum cell densities achieved in fed-batch cultures is approximately 2 x 10^7 cells/ml. However, in order to produce the required amount of product high volume bioreactors are necessary. Aside from this, the simple design and operation, along with flexibility and ease of licensing, these systems are the most popular for manufacturing. The immobilisation of cells allows for far greater cell densities to be achieved, whilst retaining these features. It also allows for long-term perfusion cultures which are economically advantageous as time is saved in turnaround and seed build-up, and because of the higher densities it can reduce the capital equipment and plant costs as smaller vessels can be used (50-100 litres are equivalent to 4000 litres for suspension cultures). Immobilised cell cultures can also enhance cell specific productivity, thus giving even greater savings [5]. When developing a production process it is desirable to avoid nutrient limitation and accumulation of toxic metabolites such as lactate and ammonia. This problem can be overcome by the use of medium perfusion. While spin filters have offered limited success, immobilisation has been adopted as the best method [5].

The immobilisation of cells for the production of biologics can be divided into three main areas. The first is cell retention within a compartment, allowing the free passage of medium. Examples of these systems include hollow-fibre and ceramic matrix bioreactors. While it is possible to achieve a 50-fold to 100-fold increase in unit cell biomass they are difficult to scale-up and are limited to 1-2 litres. The second method is cell encapsulation. This technique uses alginate or agarose beads to entrap the cells and allows the product to diffuse into the culture medium. Finally, the most common system is cell entrapment, which uses macroporous microcarriers as a cell immobilisation substrate. The advantage of encapsulation and entrapment is that these processes allow cells to be grown in conventional bioreactors and the potential scale-up of these systems is far greater than with cell retention devices [6].

Table 1. Biological products from animal cells [7-11].

Viral Vaccines	RotaShield™, RabAvert™, VAQTA™
Antibodies	Monoclonal antibodies e.g. Capromab pendetide/ProstaScint™
Interferons	Human IFN-γ
Enzymes	Fibrinolytic
Whole cells	Autologous cultured chondrocytes
Insecticides	Insect viruses
Immunoregulators	Human Interleukin-2
Hormones	Insulin
Growth Factors	Epidermal growth factor (EGF), nerve growth factor (NGF)

2. Cell retention

As the potential of animal cells for the production of biologics became apparent, many investigators looked to develop new bioreactors in order to overcome the problems of hydrodynamic forces caused by mechanical agitation and bubble burst that were believed to exist in the traditional stirred microcarrier or suspension cultures. One system that was developed by Knazek et al. [12], was the use of ultrafiltration capillary fibres. In this bioreactor cells attach to the outer surface of semi-permeable fibres, growing in the extra-capillary space (ECS) while medium is circulated through the inter-capillary space (ICS) or lumen. Nutrients are able to diffuse through the fibres, usually made of cellulose acetate, while toxic metabolites diffuse into the ECS and are carried away from the cells. A unit consists of thousands of fibres housed in a cylinder. A similar device is the ceramic matrix bioreactor. This is a cylinder of porous ceramic with square channels passing through the cylinder. Cells are inoculated into the channels and either attaches to the surface or are immobilised in the ceramic pores. Medium is passed through the channels to provide nutrients and remove metabolites. In a ceramic system cells are directly bathed in the recirculating medium, whereas in a hollow-fibre bioreactor cells are only exposed to a slow stream of permeate [3]. The product accumulates within the ECS and is harvested intermittently.

There are numerous advantages to using hollow-fibre bioreactors (HFBR) and they have received much attention for the commercial production of monoclonal antibodies and recombinant proteins [13-15]. The main advantage of HFBRs is the high cell densities that can be achieved. Numerous studies have reported near tissue densities of $>10^8$ cells/ml [13,16]. These high densities in turn lead to increased production and concentration of products and enhance the specific productivity in a relatively small bioreactor [17-19]. Tzianbos and Smith [20] reported that a flat HFBR was used to produce monoclonal antibodies from problematic cell lines and yielded a 200-300 fold increase in the concentration of antibody harvested when compared to levels obtained in static culture. Furthermore, it was possible to maintain antibody production over a two-month period, demonstrating the long-term potential for HFBRs. This has been verified in studies conducted by Kessler *et al.* [13] who consistently produced a MAb for 85 days and Fassnacht *et al.* [21] that produced antibody for a 6 week period from two different murine hybridoma cell lines. Marx [22] tested 31 hybridoma cell lines of murine, human, and rat origin and found that the average culture times were 56 days, 84 days and 67 days, respectively, demonstrating that long term cultivation and production are possible in HFBR.

A further advantage of HFBRs is the total amount of serum used in production can be lower than that used in other types of bioreactor. The high molecular weight of serum makes it difficult to diffuse into the ECS and therefore only low concentrations are required in the culture medium, and in some cases it can be totally removed. Serum is the most expensive component in the production of biologics and may cause complications in the downstream processing of products. Therefore, removal of serum from any process is highly desirable not only for economical reasons but also removal of components of animal origin in the manufacture of therapeutic and diagnostic products are favoured by regulatory authorities. It has been shown that a reduction in serum concentration to 2.5% can also lead to an increase in antibody production in hybridoma cell lines, although productivity decreases at concentrations ≤ 1% [13,20].

Whilst HFBRs offer a number of advantages, these systems are associated with several disadvantages including poor cell viability, large diffusional gradients which limit scale-up and effect product quality, and difficulties in culturing anchorage dependent cell lines. As medium enters the lumen, it encounters a positive transmembrane pressure and is able to permeate into the ECS. As this pressure decreases along the length of the fibre, it becomes negative towards the outlet, causing medium to flow back into the lumen ICS. This phenomenon is celled Starling flow [23]. This results in a higher cell density at the inlet end compared to the outlet or, with suspension cells, a packed cell mass accumulating at the outlet end [6]. The resulting presence of large numbers of dead cells limits culture performance in two ways. Firstly, there is a need for high rates of cell proliferation to replace lost cells, which will reduce the specific productivity [24]. Secondly, the release of proteases, DNA and other cellular components will lead to product degradation and will complicate downstream processing [21]. Aside from the adverse effect on cell growth, these gradients also affect the quality of secreted products. Glucose concentration and culture pH have both been shown to affect the glycosylation patterns of secreted proteins [25,26]. The delivery of oxygen is a particular problem and is considered to be the main limiting factor in cell growth and protein production [15]. In many HFBR systems high molecular weight proteins, either introduced with the cell inoculum (*e.g.* serum) or produced by the cells (*e.g.* protein products) are unable to permeate through the fibres and accumulate within the ECS, where product proteins are periodically harvested. The secondary flow within the ECS transports these proteins to the downstream end of the bioreactor where they accumulate and increase the local osmotic pressure. Downstream polarization of medium proteins can skew cell growth in this direction and cause a decrease in bioreactor productivity [27]. However, this polarisation can concentrate the product proteins and harvesting from the downstream ECS port can facilitate product recovery and purification [16]. One way of overcoming these gradients is the use of flatbed hollow-fibre bioreactors. An example of this system is the Tecnomouse (Integra Biosciences, St Albans, UK). This bioreactor incorporates traditional hollow fibre technology with a direct oxygenation technique across a silicone membrane and thereby separates the aeration from the medium supply, and was designed for the production of antibodies from hybridomas. Typical antibody yields in the Tecnomouse are ~1.0-2.0 mg/day [21]. The Tecnomouse has also been used to culture transformed B- and T-cell lines and CHO cells to produce cytokines [20] and primary rat hepatocytes for potential use as a component of a bioartificial liver system [28]. However, as with conventional HFBRs scale-up can only be achieved by increasing the number of units.

Hollow-fibre culture has mainly been used for the culture of suspension cells. This is due to the fact that anchorage-dependent cells do not readily attach to the cellulose acetate fibres. The cultivation of such cell lines can be achieved by coating the fibre walls with cell attachment material, such as poly-D-lysine [3] or by using fibres made of polypropylene [6]. It may also be possible to use microcarrier beads as a cell substratum and load these into the ECS. Williams *et al.* [15] successfully cultured CHO K1 cells on Cytodex™ 1 microcarriers in the ECS of a Setec Tricentric™ HFBR cartridge (Livermore, CA), and reached a final cell number of 1.4×10^9.

3. Cell encapsulation

The principle of cell encapsulation was originally developed by Nilsson and Mosbach [29] and Lim and Sun [30] and can be used for the large-scale production of monoclonal antibodies [31]. Although suspension cells can theoretically be immobilised in macroporous microcarriers, in practice the cell loading capacity is usually relatively low due to the non-adherent nature of the cells [3]. Thus, for suspension cells, such as hybridomas, it is more manageable to entrap cells rather than allow them to attach to solid particles. Suspended cells are mixed with alginate or agarose before droplets are formed to produce spherical particles. Alginates are naturally occurring polysaccharides extracted from seaweed, which have been commonly used in the biotechnology industry for numerous applications. In the case of sodium alginate beads, the particles are coated with polylysine, which provides a semi-permeable layer, allowing the transport of nutrients into the bead and toxic metabolites out of the bead. The sodium alginate is then solubilized with sodium citrate, releasing the cells into suspension. During the course of the culture, cells will proliferate within the capsule and produce antibodies, which are also retained within the structure, and very high concentrations of both can be achieved. A similar method is the use of calcium alginate only. This allows antibodies to diffuse out of the bead into the culture medium while cells are maintained within the capsule. Agarose beads can also be used where cells are encapsulated within a honeycomb matrix within the gel. However, these tend to have a wide size distribution and a lower mechanical strength compared to alginate gels [6].

The main advantages of using alginate beads are:
- Encapsulated cells can be cultivated in conventional fermentors. Consequently, volumetric scale-up has greater potential than in hollow-fibre systems.
- The technology permits the cost effective production of monoclonal antibodies that are functional and highly purified.
- The process uses materials that are well established for use in the production of therapeutic agents.

Cell densities of $>10^8$ cells/ml have been reported using sodium alginate beads for microcapsule culture, which is significantly higher than the densities achieved in suspension culture [31]. Calcium alginate beads, however, were only shown to reach densities of ~7 x 10^6 cells/ml with a low viability of 56% through a 46 day run [32].

One problem of major importance in all immobilised systems is the limited mass transfer rates [33]. Significant gradients of oxygen, nutrients and metabolites exist, resulting in uneven distribution of cell viability, metabolic activity and growth rate in the beads. Another disadvantage of using sodium alginate gels is that it is limited to batch operation with a culture time of 10-15 days. After this the capsules are lysed to allow the release of the product. The other methods of cell encapsulation do offer longer culture periods, as the product is able to diffuse out of the bead and into the culture medium. However, it is difficult to maintain cell viability above 50%. These beads tend to be relatively large (0.5-1 mm) which causes severe nutrient limitation in the centre, resulting in a necrotic region within the particle. It has been recommended that the bead diameter should be < 0.5 mm to maintain uniform distribution of metabolically active cells [34].

4. Cell entrapment

Whilst cell immobilisation techniques such as hollow fibre bioreactors and gel encapsulation allow for high cell densities and increased product formation, the problems associated with their scale-up and the complexity of these systems has limited their suitability for industrial-scale processes [35,36]. Originally, the alternative to these high density/low volume cultures was low density/high volume cultures using either suspension cells or solid microcarriers grown in stirred tank bioreactors. The disadvantage of these systems is that they are not cost effective in terms of capital expenses, time required for inoculum preparation and low cell densities and product titres, and cells were exposed to hydrodynamic forces and mechanical stresses during production, which limited cell specific product formation. The development of macroporous microcarriers has given manufacturers the option of having medium volume/high densities cultures with high specific productivities during long-term cultivation. This immobilisation technique is passive in nature; cells can be seeded into a culture and will attach to the external surface of the carriers and grow in to the internal pores, thus being protected from mechanical damage caused by bead-bead collisions, bead-impeller collisions or bead-eddy interactions. Suspension cells can also be immobilised on macroporous carriers by forming aggregates within the porous structure. Hence, macroporous carriers can be used for both suspension and anchorage-dependent cell lines. The main advantages of macroporous microcarriers are summarised in Table 2.

Table 2. Advantages of macroporous microcarriers.

Unit cell density of 20- to 100-fold higher than solid microcarriers
Support both attached and suspension cells
In situ 100-250 seed expansion reducing scale-up steps
Suitable for stirred, fluidised or packed bed bioreactors
Short diffusion paths into a sphere
Good scale-up potential
Cells protected from hydrodynamic forces and mechanical stress
Capable of long-term continuous culture
Immobilisation in 3-D configuration, which is easily derivatised

Despite these numerous advantages some possible limitations exist. Following the growth of cells to tissue-like densities, the glucose consumption rate has been seen to decrease as a result of necrotic cores forming [37]. Cells initially penetrate relatively deeply into the porous matrix. However, over time an outer shell of highly dense cells forms, limiting transfer of oxygen and nutrients to the centre of the particle [34,38]. In addition to this, toxic metabolites such as lactate and ammonia are unable to diffuse out of the particle and CO_2 production will cause a drop in the pH. Consequently, this microenvironment induces cell death and a necrotic core develops, reducing the overall cell density and viability. This decreasing viability may therefore lead to unsatisfactory product yields. In a study by Preissmann et al. [39] an assessment of limiting parameters for cells growing on macroporous microcarriers in a fluidised bed bioreactor was carried out. As no significant concentration gradients could be detected oxygen transfer to and into the carriers was examined. They found a significant transfer resistance within the laminar boundary film at the surface of the carrier and that 40% air saturation in the bulk liquid could not provide efficient oxygenation during the exponential growth phase.

A further limitation in the use of macroporous microcarriers is in the production of viral vaccines. Viral production relies on the ability of virus particles to infect cells, replicate, and then release their progeny, usually by cell lysis. However, the accessibility of cells to viral infection may be a limiting factor when using macroporous microcarriers. Berry *et al.*, [40] compared the production of reovirus type-1 and type-3 from Vero cells cultivated on solid and macroporous microcarriers. Overall, the productivity in macroporous carriers was significantly lower than that from solid microcarriers and was likely to be due to the inaccessibility of entrapped cells. They also concluded that the maximum cell yields were significantly lower when compared to values determined for solid microcarrier cultures.

There are currently a large number of macroporous microcarriers available (Table 3), which have been developed for stirred tank, packed-bed and fluidised bed bioreactors. The production of biologics within these bioreactor systems is described below.

Table 3. Characteristics of commercial porous microcarriers and their mode of use.

Microcarrier	Manufacturer	Matrix	Cell lines used
Cultisphere G	Percell Biolytica	Gelatine	CHO
Cytoline	Amersham Pharmacia Biotech	Polyethylene and silica	CHO
			Hybridoma
Cytopore	Amersham Pharmacia Biotech	Cross-linked cotton cellulose	CHO
			BHK
Fibra-cel	Bibby Sterilin	Polyester non-woven fibre and polypropylene	CHO
			Hybridoma
			BHK
			Vero
			FLYRD
			TEFLYRD
Immobasil	Ashby Scientific	Silicone rubber	CHO
			HEK 293
			BHK
			FLYRD
			Hybridoma
Microsphere	Verax Corp.	Collagen	CHO Hybridoma
Siran	Schott Glasswerke	Glass	CHO Hybridoma

4.1. STIRRED TANK BIOREACTORS

Although many alternatives have been developed, the traditional stirred tank bioreactor has emerged as industry's technology of choice [7]. The attractiveness of stirred tank bioreactors stems from their simplicity and ease of monitoring and controlling scale-up. Although they are commonly used for suspension cell culture, the use of macroporous microcarriers means that the same bioreactors can be used for immobilised anchorage-dependent cell lines. Suspension cells can also be entrapped within macroporous beads to increase the cell density. These cells are less dependent on attachment factors and Xiao *et al.* [41] reported that the serum concentration could be reduced to as little as 0.1% to grow hybridomas, which are normally grown in 5-10% serum. This is an important consideration with respect to downstream processing and culture economics.

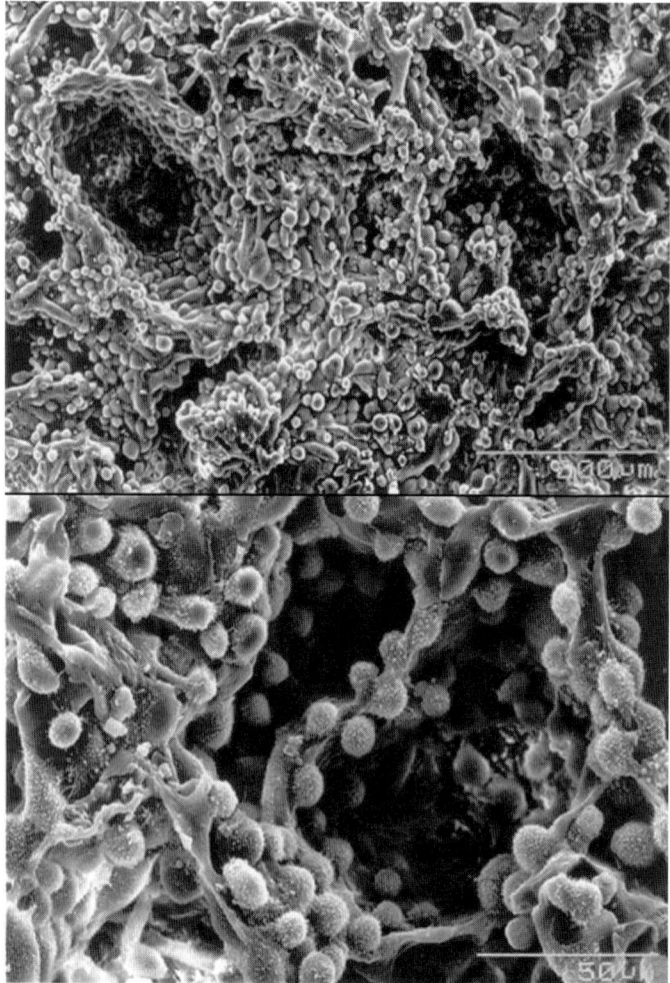

Figure 1. Scanning electron micrograph of CHO cells immobilised on Immobasil FS macroporous microcarriers

A typical characteristic of stirred tanks, and a major concern in animal cell technology, are the high mechanical forces that exist within the vessel. Not only are cells subjected to hydrodynamic forces from agitation when attached to solid microcarriers, but they are also vulnerable to high levels of energy released by bubble burst. Macroporous microcarriers offer protection from these elements, provided that cells are able to migrate towards the centre of the particle. Many fibroblasts can move as fast as one centimetre a day on a smooth surface [3] but the internal surface of these matrices is rarely flat or smooth and is quite tortuous for cells. However, once cells are protected from external forces the mixing of the stirred tank can be increased. This creates a completely homogeneous environment and allows for a greater oxygen and nutrient transfer rate to the immobilised cells [41], resulting in high cell densities that range between 1 and 6 x 10^7 cells/ml [9,37,42], depending on cell line and microcarrier used. Figure 1 shows CHO cells that have been grown to a high density on Immobasil FS microcarriers.

It has been observed that a decrease in specific growth rate occurs when immobilised suspension and anchorage dependent cells are grown in stirred tank perfusion cultures [42,43]. Growth suppression has been proved to enhance antibody production in hybridoma cells. Therefore, immobilisation of cells may improve protein synthesis in stirred tank bioreactors. Certainly, the high cell densities can produce high yields of product. Stirred tanks have been used in the production of membrane anchored recombinant proteins, urokinase-type plasminogen activator, prourokinase, IgG and IgA monoclonal antibodies and vesicular stomatitis virus [9,41,44-47].

The most appealing aspect of stirred tank bioreactors is the ease at which they can be scaled up in excess of 1000 litres, and this is undoubtedly why they have been so extensively used in industry. However, the inoculum density required is relatively high (when compared to microbial cultures). This problem is further increased when anchorage dependent cells are used if they have to be trypsinised at each step of the scale-up process. It has been shown, though, that bead to bead cell transfer is possible with cytopore microcarriers [41]. A scale-up ratio of more than 20 was achieved meaning that only three scale-up steps are necessary for a 1000 l culture. This discovery has made the use of macroporous microcarriers a viable option for the commercial production of recombinant products from genetically engineered cells.

4.2. FLUIDISED-BED BIOREACTORS

The most widely used culture system for porous microcarriers is the fluidised-bed bioreactor [10]. The principle of their operation is that microcarriers of higher density than the culture medium are suspended by the upward flow of the medium, which is circulated through the bed. The height of the bed will increase as fluid flow increases. Medium circulation and fluid flow through the bed are achieved by using either a peristaltic pump or agitation. The advantage of this system is that carriers are separated from direct aeration and agitation, which avoids local high shear forces. In addition, fluidised-bed bioreactors provide a very high mass transfer interface since both cells and fluid are moved [39] making it a suitable production system for larger scales.

Earlier studies into the potential of fluidised bed reactors for the production of biologics focussed on the use of porous glass beads. These studies conducted by Kratje and co-workers [10,48] examined the potential of fluidised bed reactors as production systems using hybridomas suspension cells and anchorage dependent baby hamster kidney cells for the production of interleukin-2. In these studies, cell densities of 3.8 x 10^8 cells/ml intrasphere volume were reached, whilst the IL-2 production rate was 0.75 mg/l/day. In spite of the cell density being over 18-fold higher, the productivity showed a 1.9 fold decrease when compared to a homogeneous stirred culture. In contrast, the production rate of hybridoma cells cultured in the fluidised bed remained appreciably higher than in stirred tank culture. The difference was assumed to be caused by BHK cells changing their production characteristics during long-term cultivation. This phenomenon had previously been reported in hybridomas cells [49]. An alternative reason is that, unlike hybridoma cells, specific cell productivity is related to specific growth rate. It has been shown by Leelavatcharamas *et al.* [50] that there is a linear relationship between the production of interferon-γ and the specific growth rate of CHO cells. Therefore, in long-term cultures where the cells remain in the stationary phase over an extended period, productivity will be low as the specific growth rate is essentially zero.

Another substrate for the immobilisation of cells grown in fluidised bed bioreactors is polyethylene and silica microcarriers, such as Cytoline (Amersham Pharmacia Biotech, Sweden). These have been successfully used in the production recombinant human proteins, especially human interferon-γ, and erythropoietin [11,51]. These microcarriers have been specifically designed for use in the Cytopilot fluidised bed bioreactor and have proved to offer great potential for the production of biologics. Kong et al. [52] compared CHO cell cultures in this system with solid microcarrier culture in a stirred tank bioreactor and reported that the product output rate was approximately three times higher in the fluidised bed, with the specific production rate being 5.5 times higher. The volumetric productivity of the stirred tank reached 5.6 mg/l/day during the growth period. However, after the maximum cell density was reached the volumetric productivity dropped and was maintained around 1.65 mg/l/day. In contrast, the fluidised bed bioreactor maintained a consistent volumetric productivity of 22.8 mg/l/day. It was suggested that the isolated aeration and agitation in the fluidised bed bioreactor reduced the amount of hydrodynamic and mechanical stress imposed on the cells. Thus, more cellular energy was utilised in product formation, resulting in the higher and more consistent productivity. The total product yield was reported as 34.2 mg/day and 13.2 mg/day for the 2 litre fluidised bed bioreactor and the 10 litre stirred tank bioreactor, respectively.

When comparing the number of cells in each square meter of microcarrier surface area, the fluidised bed bioreactor was five times lower than the stirred tank reactor. This shows that there is further potential to increase the cell number and improve the productivity of the system. In order to unlock this potential, it is necessary to ascertain the limiting factor for growth and how to overcome this limitation. Investigations carried out by Preissmann *et al.* [39] determined that the transport of oxygen to cells within the macroporous matrix cultivated in a fluidised bed bioreactor was severely limited, leading to poor process performance. Although no limitations in nutrient concentration were detected, the beads were not believed to be an ideal substrate to grow cells to tissue like densities.

Efficient glycosylation is an important factor in the development of a bioprocess production method. The existence of gradients that cause decreases in dissolved oxygen, nutrients or pH may have implications for product quality, as well as cell densities. In a recent study, the relative proportion of glycosylation site occupancy for interferon-γ was not seen to change over time during a 500-hour culture. Hence, it was concluded that long-term perfusion cultures of CHO cells in a fluidised bed bioreactor could produce IFN-γ with a consistent and highly comparable degree of glycosylation [51]. Further studies, carried out by Wang et al. [11], have also examined the productivity and quality of erythropoietin (EPO) from a recombinant CHO cell line, cultured in a Cytopilot fluidised bed bioreactor. EPO is particularly vulnerable to a decline in sialylation due to the extensive level of glycosylation and heterogeneity. It was therefore important to show that the glycosylation profile of EPO in the Cytopilot did not differ from other culture systems. Results have shown that no difference was observed in electrophoretic product profiles between fluidised bed, stirred tank and stationary cultures. In addition, the significantly higher productivity rate in the fluidised bed bioreactor, when compared to stirred tank reactors, makes it an attractive option for the large-scale production of recombinant proteins.

A potential problem that has been observed during the long-term cultivation of cells in medium containing 10% foetal calf serum is the clumping of carriers. This has been observed in recent work carried out by A. Slade and co-workers (Oxford BioMedica, UK). This clumping causes a reduction in bed height, which requires increased agitation and fluid flow to maintain sufficient fluidisation. The reason for carriers clumping is not fully understood but it may be due to cell bridging. This is an incident usually associated with solid microcarriers, where cells become simultaneously attached to two beads, thus creating a bridge. A possible means of overcoming this problem is to reduce the concentration of serum used in the process. This could be done step wise during the course of the experiment, as serum is necessary for cell attachment and growth during the initial stages. The clumping of carriers has not been observed to have an adverse effect on productivity. It does, however, limit the duration of the process as excessive clumping has an adverse effect on reactor performance.

4.3. PACKED-BED BIOREACTORS

Packed-bed bioreactors have been used for the cultivation and production of a wide range of cell lines and biologics. Some recent examples include monoclonal antibody, anti-leukaemic factor from stromal cells, recombinant proteins (*e.g.* recombinant Ca^{2+} binding receptor and HIV-1 gp120), and retrovirus vectors for use in gene therapy [9,35,53-56]. Since the initial conception of packed-bed bioreactors for the production of biologics in the 1970's, they have grown in acceptance and their potential as a viable system for commercial manufacture of animal cell products. Stirred tank and fluidised bed bioreactor systems have been scaled up in excess of 1000 litres and have been widely used in industrial processes [57]. However, at high cell densities, cells detach from the carriers and along with cell debris cause complications in product purification during downstream processing. Packed bed bioreactors have the advantage that they are capable of generating high cell densities with a low concentration of free cells in suspension. This is possible due to the low shear forces present in the system. Cells are immobilised within porous carriers that may be porous ceramic beads [58], porous glass beads [59], or polyester disks [9,35,54], which are packed and retained in a cylindrical vessel through which culture medium is re-circulated. As mentioned earlier, a characteristic problem in intensive production systems is the reduced transport rate of limiting nutrients, such as oxygen to immobilised cells, which restricts their final density. Packed-bed technology has overcome this problem by the use of intraparticle convective flow. Through the development of a simple hydrodynamic model based on the Blake-Kozeny equation, Park and Stephanopoulos [56] significantly improved the transport of oxygen in a packed bed, allowing the maintenance of cell viability and productivity whilst sustaining a low shear environment. In their evaluation of the model they achieved cell densities of 5.1 x 10^8 cells/ml. They also calculated the specific insulin productivity to be 0.88 x 10-5 µU/cell/hour. The model developed allows packed bed bioreactors to be scaled-up several-fold before oxygen becomes limiting. Any oxygen limitations that do occur primarily exist close to the bioreactor exit. Another recent development to overcome this problem has been to provide oxygenation to the bed through silicone tubing. This simple but effective method has been used for the successful culture of hepatocytes and the production of retrovirus particles from the human packaging cell lines, FLYRD18 and TEFLYRD [60,61].

An alternative system is the Celligen® packed-bed bioreactor, developed by New Brunswick Scientific Co. (New Jersey, USA). Conventional packed beds require a conditioning vessel to control culture parameters such as temperature, dissolved oxygen and pH. The culture medium is circulated through the packed bed and the product may be concentrated and continuously or periodically harvested from the conditioning vessel for prolonged periods. An important parameter in the optimisation of these arrangements is the circulation rate; if the rate is too low then gradients in pH, nutrients and metabolites will appear along the bed affecting cell viability, productivity and product quality, as can be seen in hollow fibre bioreactors [14]. If the circulation rate is too high, it may have an adverse affect on cell attachment or product formation. This problem is eliminated in the Celligen® packed bed bioreactor by the fact that the bed is contained in a basket within a stirred tank bioreactor. Mixing in the bioreactor is achieved with either a cell-lift impeller, which allows bubble-free medium to flow through the packed bed [35] or a double screen concentric cylindrical cage impeller. Shi et al. [62] have shown that the latter impeller is able to increase convective mass transfer, cell concentration and MAb product concentration. The significant improvements were attributed to the increased surface area allowing for convective oxygen transfer and protection of cells from hydrodynamic stresses. Both methods create a homogeneous environment for cells to grow and product can be directly removed from the culture vessel, which allows for easier downstream processing and product purification.

One drawback of large-scale packed bed bioreactors using anchorage-dependant cells is the generation of seed stock. Cultures are commonly inoculated at a density of 2 x 10^5 cells/ml of bed volume, which could be as high as 20 litres in a 100-litre vessel at commercial scale. Scale-up is typically done from monolayer cultures using either T-flasks or roller bottles due to difficulties in harvesting cells from macroporous microcarriers and low cell densities achieved from solid microcarrier culture. Thus, while the production culture may be simple and cost effective the scale-up steps may be less economical and susceptible to contamination. In a recent study, a method has been devised whereby cells are scaled-up on Cytopore macroporous microcarriers and the carriers are directly inoculated into the packed bed bioreactor, alleviating the need to detach cells from the carriers [57]. Using this method, it was possible to achieve a cell density of 2 x 10^7 cells/ml bed volume and thrombopoietin production was 1.3-1.8 mg/l/day, compared to 0.76-1.1 mg/l/day for macroporous microcarrier perfusion culture. Cultures were maintained in excess of 30 days.

A further disadvantage of packed bed bioreactors is the inaccessibility of carriers, making it difficult to monitor viable cell number. Cellular metabolism can be followed by online measurements of pH and dissolved oxygen and offline determinations of glucose, lactate, ammonia and glutamine. The substrate uptake rates and production rates are conventionally used as cell growth markers [63]. In general, the cell concentration has a linear correlation with glucose uptake rate at high cell densities. Thus, it is possible to calculate the former from the latter [64]. However, this yield coefficient is not a true constant and will vary for each cell line and for each bioreactor system. Therefore, it should be determined before being used to estimate cell number. Alternatively, cell number can be determined through measuring the oxygen uptake rate (OUR). By placing oxygen probes upstream and downstream of the packed bed and measuring the percentages of dissolved oxygen, it is possible to determine the OUR, which can be directly correlated with cell number. The density of cells within an immobilised culture is an important parameter for many purposes including the calculation of the specific productivity of a system.

Table 4. Advantages and disadvantages of cell immobilisations systems for the production of biologics.

Production system	Advantages	Disadvantages
Hollow fibre bioreactor	>10^8 cell /ml	Poor cell viability
	200-300 fold increase in product	Diffusion gradients exist
	Reduced serum concentration	Limited scale-up
		Difficult to culture anchorage-dependent cells
Alginate beads	Uses conventional fermentors	Limited mass transfer rates
	Cost effective	Sodium alginate beads limited to batch culture
	Uses materials well established in biotech industry	Low cell viability
Stirred tank bioreactor	Simple to operate	Lower cell concentration compared to other systems
	Cells protected from mechanical stresses	Unable to support high microcarrier density
	Growth suppression leads to enhanced productivity	
	Scale-up >4000 litres	
Fluidised bed bioreactor	Carriers separated from agitation and aeration	Specific productivity reduced in some cell lines
	Consistent volumetric productivity	Oxygen limitations may occur
	Consistent product quality	Carriers clump at high serum concentrations
	Good scale-up potential	
Packed bed bioreactor	Generate and retain high cell densities	Difficult to sample carriers to measure cell density
	Intraparticle flow restricts O_2 limitations	Difficult to generate anchorage-dependent cells for inoculation

5. Conclusions

Traditional stirred tank reactors using suspension cells are still the preferred choice of biologic production in the biotech industry. However, due to the relatively low cell densities, immobilised cell cultures are growing in interest for the production of high value biologics such as viral vaccines and gene therapy products. Immobilised cultures offer many advantages such as near tissue cell densities leading to higher productivity, increased specific production rates, separation of products from cells, improving the efficiency of the downstream processing, serum concentration can be reduced making the process more economical and immobilisation techniques can be used for both suspension and anchorage-dependent cells.

A number of systems can be used in the production of biologics from immobilised cell cultures. The advantages and disadvantages of these systems are summarised in table 4.

References

[1] Ernst & Young. (2000) The economic contributions of the biotechnology industry to the US economy. http://www.bio.org/news/ernstyoung.pdf. (accessed 29th October 2002). Biotechnology Industry Organizer.
[2] Butler, M. (1987) Growth limitations in microcarrier cultures. Adv. Biochem. Eng. Biotechnol. 34: 57-84.
[3] Hu, W.S. and Peshwa, M.V. (1991) Animal Cell Bioreactors - Recent advances and challenges to scale-up. Can. J. Chem. Eng. 69: 409-420.
[4] van Wezel, A.L. (1967) Growth of cell-strains and primary cells on micro-carriers in homogeneous culture. Nature 216 (110): 64-65.
[5] Griffiths, B. (1992) Alternative strategies to the scale-up of animal cells. Ann. NY Acad. Sci. 665: 364-370.
[6] Griffiths, J.B. (1988) Overview of cell culture systems and their Scale-up. In: Spier, R.E. and Griffiths, J.B. (Eds.), Animal Cell biotechnology. Academic Press Limited, London, UK; pp. 179-220.
[7] Chu, L. and Robinson, D.K. (2001) Industrial choices for protein production by large-scale cell culture. Curr. Opin. Biotechnol. 12 (2): 180-187.
[8] Griffiths, J.B. (1985) Cell products: An overview. In: Spier, R.E. and Griffiths, J.B. (Eds.), Animal Cell biotechnology. Academic Press Limited, London, UK; pp. 3-12.
[9] Hu, Y.C.; Kaufman, J.; Cho, M.W.; Golding, H. and Shiloach, J. (2000) Production of HIV-1 gp120 in packed-bed bioreactor using the vaccinia virus/T7 expression system. Biotechnol. Prog. 16 (5): 744-750.
[10] Kratje, R.B. and Wagner, R. (1992) Evaluation of production of recombinant Human Interleukin-2 in fluidized bed bioreactor. Biotechnol. Bioeng. 39: 233-242.
[11] Wang, M.D.; Yang, M.; Huzel, N. and Butler, M. (2002) Erythropoietin production from CHO cells grown by continuous culture in a fluidized-bed bioreactor. Biotechnol. Bioeng. 77 (2): 194-203.
[12] Knazek, R.A.; Gullino, P.M.; Kohler, P.O. and Dedrick, R.L. (1972) Cell culture on artificial capillaries: An approach to tissue growth in Vitro. Science 178: 65-67.
[13] Kessler, N.; Thomas, G.; Gerentes, L.; Delfosse, G. and Aymard, M. (1997) Hybridoma growth in a new generation hollow fibre bioreactor: antibody productivity and consistency. Cytotechnol. 24: 109-119.
[14] Thelwell, P.E. and Brindle, K.M. (1999) Analysis of CHO-K1 cell growth in a fixed bed bioreactor using magnetic resonance spectroscopy and imaging. Cytotechnol. 30: 121-132.
[15] Williams, S.N.O.; Callies, R.M. and Brindle, K.M. (1997) Mapping of oxygen tension and cell distribution in a hollow-fiber bioreactor using magnetic resonance imaging. Biotechnol. Bioeng. 56 (1): 56-61.
[16] Koska, J.; Bowen, B.D. and Piret, J.M. (1997) Protein transport in packed-bed ultrafiltration hollow-fibre bioreactors. Chem. Eng. Sci. 52 (14): 2251-2263.
[17] Fassnacht, D. and Portner, R. (1999) Experimental and theoretical considerations on oxygen supply for animal cell growth in fixed-bed reactors. J. Biotechnol. 72 (3): 169-184.
[18] Glacken, M.W.; Fleischaker, R.J. and Sinskey, A.J. (1983) Large-scale production of mammalian cells and their products: Engineering principles and barriers to scale-up. Ann. NY Acad. Sci. 413: 355-373.
[19] Knight, P. (1989) Hollow fiber bioreactors for mammalian cell culture. Bio/Technol. 7: 459-461.
[20] Tzianbos, A.O. and Smith, R. (1995) Use of hollow fibre bioreactor for production in problematic cell lines. UK Product Review, 32.
[21] Fassnacht, D.; Rössing, S.; Singh, R.P.; Al Rubeai, M. and Pörtner, R. (1999) Influence of bcl-2 on antibody productivity in high cell density perfusion cultures of hybridoma. Cytotechnol. 30: 95-105.
[22] Marx, U. (1998) Capitalising on Capillaries. Laboratory Technology International.

[23] Starling, E.H. (1896) On the absorption of fluids from the convective tissue space. J. Physiol. 19: 312-326.
[24] Al Rubeai, M.; Emery, A.N.; Chalder, S. and Jan, D.C. (1992) Specific monoclonal antibody productivity and the cell cycle-comparisons of batch, continuous and perfusion cultures. Cytotechnol. 9 (1-3): 85-97.
[25] Borys, M.C.; Linzer, D.I. and Papoutsakis, E.T. (1993) Culture pH affects expression rates and glycosylation of recombinant mouse placental lactogen proteins by Chinese hamster ovary (CHO) cells. Biotechnol. (NY) 11(6): 720-724.
[26] Hayter, P.M.; Kirkby, N.F. and Spier, R.E. (1992) Relationship between hybridoma growth and monoclonal antibody production. Enzyme Microb. Technol. 14 (6): 454-461.
[27] Piret, J.M. and Cooney, C.L. (1990) Mammalian cell and protein distributions in ulltrafiltration hollow-fiber bioreactors. Biotechnol. Bioeng. 36: 902-910.
[28] Bratch, K. and Al Rubeai, M. (2001) Culture of primary rat hepatocytes within a flat hollow fibre cassette for potential use as a component of a bioartificial liver support system. Biotechnol. Lett. 23: 137-141.
[29] Nilsson, K. and Mosbach, K. (1980) Preparation of immobilized animal cells. FEBS Lett. 118 (1): 145-150.
[30] Lim, F. and Sun, A.M. (1980) Microencapsulated islets as bioartificial endocrine pancreas. Science 210 (4472): 908-910.
[31] Rupp, R.G. (1985) Use of cellular microencapsulation in large-scale production of monoclonal antibodies. In: Feder, J. and Tolbert, W.R. (Eds.), Large-scale mammalian cell culture. Academic Press Inc., London, UK; pp. 19-38.
[32] Al Rubeai, M.; Musgrave, S.C.; Lambe, C.A.; Walker, A.G.; Evans, N.H. and Spier, R.E. (1990) Methods for the estimation of the number and quality of animal cells immobilized in carbohydrate gels. Enzyme Microb. Technol. 12 (6): 459-463.
[33] Al Rubeai, M. and Spier, R.E. (1989) Quantitative cytochemical analysis of immobilised hybridoma cells. Appl. Microbiol. Biotechnol. 31: 430-433.
[34] Al Rubeai, M.; Rookes, S. and Emery, A.N. (1990) Studies of cell proliferation and monoclonal antiody synthesis and secretion in alginate-entrapped hybridoma cells. In: de Bont, J.A.M.; Visser, J.; Mattiasson, B. and Tramper, J. (Eds.), Physiology of Immobilised Cells. Elsevier Science Publishers, Amsterdam, The Netherlands; pp. 181-188.
[35] Wang, G.; Zhang, W.; Jacklin, C.; Freedman, D.; Eppstein, L. and Kadouri, A. (1992) Modified CelliGen-packed bed bioreactors for hybridoma cell cultures. Cytotechnol. 99 (1-3): 41-49.
[36] Yamaji, H.; Fukuda, H.; Nojima, Y. and Webb, C. (1989) Immobilisation of anchorage-independent animal cells using reticulated polyvinyl formal resin biomass support particles. Appl. Microbiol. Biotechnol. 30: 609-613.
[37] Kennard, M.L. and Piret, J.M. (1994) Glycolipid membrane anchored recombinant protein production from CHO cells cultured on porous microcarriers. Biotechnol. Bioeng. 44: 45-54.
[38] Yamaji, H. and Fukuda, H. (1992) Growth and death behaviour of anchorage-independent animal cells immobilized within porous support matrices. Appl. Microbiol. Biotechnol. 37: 244-251.
[39] Preissmann, A.; Wiesmann, R.; Buchholz, R.; Werner, R.G. and Noe, W. (1997) Investigations on oxygen limitations of adherant cells growing on macroporous microcarriers. Cytotechnol. 24: 121-134.
[40] Berry, J.M.; Barnabe, N.; Coombs, K.M. and Butler, M. (1999) Production of reovirus type-1 and type-3 from Vero cells grown on solid and macroporous microcarriers. Biotechnol. Bioeng. 62 (1): 12-19.
[41] Xiao, C.; Huang, Z.; Li, W.; Hu, X.; Qu, W.; Gao, L. and Liu, G. (1999) High density and scale-up cultivation of recombinant CHO cell line and hybridomas with porous microcarrier Cytopore. Cytotechnol. 30: 143-147.
[42] Yamaji, H. and Fukuda, H. (1994) Growth kinetics of animal cells immobilized within porous support particles in a perfusion culture. Appl. Microbiol. Biotechnol. 42: 531-535.
[43] Wagner, R.; Marc, A.; Engasser, J.M. and Einsele, A. (1992) The use of lactate dehydrogenase (LDH) release kinetics for the evaluation of death and grwoth of mammalian cells in perfusion reactors. Biotechnol. Bioeng. 39: 320-326.
[44] Kennard, M.L.; Piret, J.M. (1995) Membrane anchored protein production from spheroid, porous, and solid microcarrier Chinese Hamster Ovary cell cultures. Biotechnol. Bioeng. 47: 550-556.
[45] Nikolai, T.J. and Hu, W.S. (1992) Cultivation of mammalian cells on macroporous microcarriers. Enzyme Microb. Technol. 14: 203-208.

[46] Schweikart, F.; Jones, R.; Jaton, J-C. and Hughes, G.J. (1999) Rapid structural characterisation of a murine monoclonal IgA α chain: heterogeneity in the oligosaccharide structures at a specific site in samples produced in different bioreactor systems. J. Biotechnol. 69: 191-201.
[47] Yamaji, H. and Fukuda, H. (1997) Continuous IgG production by myeloma cells immobilized within porous support particles. J. Ferment. Bioeng. 83 (5): 489-491.
[48] Kratje, R.B.; Reimann, A.; Hammer, J. and Wagner, R. (1994) Cultivation of recombinant baby hamster kidney cells in a fluidized bed bioreactor system with porous borosilicate glass. Biotechnol. Prog. 10 (4): 410-420.
[49] Frame, K.K. and Hu, W.S. (1990) The loss of antibody productivity in continuous culture of hybridoma cells. Biotechnol. Bioeng. 35: 469-476.
[50] Leelavatcharamas V.; Emery, A.N. and Al Rubeai, M. (1994) Growth and interferon-gamma production in batch culture of CHO cells. Cytotechnol. 15(1-3): 65-71.
[51] Goldman, M.H.; James, D.C.; Rendall, M.; Ison, A.P.; Hoare, M. and Bull, A.T. (1998) Monitoring recombinant human interferon-gamma N-glycosylation during perfused fluidized-bed and stirred-tank batch culture of CHO cells. Biotechnol. Bioeng. 60 (5): 596-607.
[52] Kong, D.; Cardak, S.; Chen, M.; Gentz, R. and Zhang, J. (1999) High cell density and productivity culture of Chinese Hamster Ovary cells in a fluidized bed bioreactor. Cytotechnol. 29: 215-220.
[53] Kadouri, A. and Zipori, D. (1989) Production of anti-leukemic factor from Stroma cells in a stationary bed reactor on a new cell support. In: Spier, R.E.; Griffiths, J.B.; Stephenne, J. and Rooy, P.J. (Eds.), Advances in animal cell biology and technology for bioprocesses. Courier International Ltd., Tiptree, Essex, UK; pp. 327-330.
[54] Kang, S-H.; Lee, G.M. and Kim, B-G. (2000) Justification of continuous packed-bed reactor for retroviral vector production from amphotopic ψCRIP murine producer cell. Cytotechnol. 34: 151-158.
[55] Kaufman, J.; Wang, G;. Zhang, W.; Valle, M.A. and Shiloach, J. (2000) Continuous production and recovery of recombinant Ca^{2+} binding receptor from HEK 293 cells using perfusion through a packed bed bioreactor. Cytotechnol. 33: 3-11.
[56] Merten, O.W.; Cruz, P.E; Rochette, C.; Geny-Fiamma, C.; Bouquet, C.; Goncalves, D.; Danos, O. and Carrondon, J. (2001) Comparison of different bioreactor systems for the production of high titer retroviral vectors. Biotechnol. Prog. 17 (2): 326-335.
[57] Cong, C.; Chang, Y.; Deng, J.; Xiao, C.; Su, Z. (2001) A novel scale-up method for mammalian cell culture in packed-bed bioreactor. Biotechnology Lett. 23: 881-885.
[58] Park, S. and Stephanopoulos, G. (1993) Packed bed bioreactor with porous ceramic beads for animal cell culture. Biotechnol. Bioeng. 41: 25-34.
[59] Chiou, T.W.; Murakami, S. and Wang, D.I.C. (1991) A fiber-bed bioreactor for anchorage-dependent animal cell culture: Part 1. Bioreactor design and operations. Biotechnol. Bioeng. 37: 755-761.
[60] McTaggart, S. and Al Rubeai, M. (2000) Effects of culture parameters on the production of retroviral vectors by a human packaging cell line. Biotechnol. Prog. 16 (5): 859-865.
[61] Warnock, J.N. (2002) Optimisation of retrovirus production systems for gene therapy applications. Thesis, The University of Birmingham, UK.
[62] Shi, Y.; Ryu, D.D.Y. and Park, S. (1992) Performance of mammalian cell culture bioreactor with a new impeller design. Biotechnol. Bioeng. 40 (2): 260-270.
[63] Rodrigues, M.T.A.; Vilaça, P.R.; Garbuio, A.; Takagi, M.; Barbosa, Jr S.; Léo P.; Laignier, N.S.; Silva, A.A.P. and Moro, A.M. (1999) Glucose uptake rate as a tool to estimate hybridoma growth in a packed bed bioreactor. Bioprocess Eng. 21: 543-546.
[64] Portner, R.; Bohmann, A.; Ludemann, I. and Markl, H. (1994) Estimation of specific glucose uptake rates in cultures of hybridoma cells. J. Biotechnol. 34 (3): 237-246.

STABILISATION OF PROBIOTIC MICROORGANISMS

An overview of the techniques and some commercially available products

HELMUT VIERNSTEIN[1], JOSEF RAFFALT[2]
AND DIETHER POLHEIM[2]
[1]*Institute of Pharmaceutical Technology and Biopharmaceutics, University of Vienna, Althanstrasse 14, 1090 Vienna, Austria - Fax: 43 1 4277 9554 - Email: helmut.viernstein@univie.ac.at;*
[2]*F. Joh. Kwizda GmbH. R&D, Dr. Karl Lueger-Ring 6, 1010 Vienna, Austria – Fax: 43 1 534 68 280 – Email: j.raffalt@Kwizda.co.at, d.polheim@Kwizda.co.at*

1. Introduction

Beneficial microorganisms and their effects have been of growing interest during the last two decades [1,2]. The term probiotic was actually coined to distinguish beneficial microorganisms originating from intestinal bacteria, certain fungi and yeast strains from other common bacteria [3-5]. There is a clear tendency in research and food technology to separate probiotic bacteria from the trivial and age old food preserving homo- and heterofermentative microbes [6], for example *Lactobacillus bulgaricus* or *Lactococcus lactis* for milk fermentation.

Research in probiotics concentrates on modes of action, host-microbe interactions and such specific questions as the influences on the human or animal immune system [7-10].

It is commonly accepted within the scientific community that probiotic effects are beneficial interactions of living microbial organisms with a hosts' existing intestinal flora and physiology. Therefore, the problem of a functional probiotic formulation is reduced to questions of protective technical measures to keep probiotic organisms alive.

When speaking of stability in connection with probiotic microorganisms, the term viability is more appropriate. The so-called stabilisation of microorganisms can be demonstrated only if they still show metabolic activity after storage and intake by a new host. Colony counting techniques make use of this relation and provide us with the log numbers counted in colony forming units per gram or millilitre to compare the living organisms before and after storage. An average rate of loss found for sophisticated formulations under excellent storage conditions is one log unit of cell reduction per year [11]. In other words even well-formulated products may lose 90% of active microorganisms within a year.

Stability is a decisive value for production and distribution systems. For the food and pharmaceutical industries, periods of one year are a minimum requirement to supply a marketable dry product in the probiotic segment. For fresh and therefore refrigerated probiotics, storage time is limited to four up to six weeks.

Only culture producers can successfully handle long time storage of cultures. This techniques are based on deep-freezing technologies as well as the necessary know how for thawing after storage. Such a technique is generally beyond the capabilities of supermarkets or consumers.

The parameters for a successful one-year storage of a dry probiotic product are summarized in table 1.

Table 1. Parameters and their limits for one-year storage of dry probiotic products.

Parameter	Limits
Most reliable storage temperature	5 - 10°C
Water activity	0.1 – 0.25
Oxygen	Has to be removed from storage package
Carbon dioxide	Has to be removed from storage package

Research teams in chemistry, microbiology, food technology, pharmaceutical technology, biochemistry, some veterinarian institutes and many more researchers from commercial companies have investigated approaches to the best method of stabilisation. In some cases subtle technical systems have been invented to meet a products target. Most formulations currently available on the market rely on pharmaceutical standard techniques for drug delivery [12]. Capsule fillings, sachets and tablets are well known, accepted by the consumer, and, more importantly inexpensive to produce. The more sophisticated, newer stabilisation techniques are so far only feasible on the workbench and in relatively small-scale production [13].

It is essential to accept biological limitations for production as well as formulation of microorganisms originating from their genetically determined metabolic peculiarities and their phenotypes [14]. Because of these strain dependent limits, production of probiotic microorganisms relies on the two classic ways of bacterial culture distribution. One way is the storage and delivery of fresh concentrated, chilled or frozen culture for direct use. This has the advantage of very limited loss of viability, but the limit of a short storage time. Stability of such a product is directly comparable to that of milk products.

The other way is the manufacturing with of state of the art drying techniques to give microorganisms more stability and flexibility for the intended use in probiotic products. Spray drying and freeze-drying are well-accepted techniques with minor technology variations depending on equipment details. End result of such a drying process, and of subsequent downstream processes like milling and screening are powder particles of microorganisms and additives such as the ones shown in figure 1.

Dried microorganisms lose their classic shapes during drying and appear as irregular and crystalline, sometimes amorphous glass-like solids. Milling and other physical preparation steps are applied to disintegrate dried aggregates into smaller particles. These technical powders of freeze dried or spray dried microorganisms have a large surface and a certain porosity. They are very hygroscopic. This production type of dry and stabilised organisms has limitations for the necessary storage conditions. Careful separation from sources of water as well as volatile organic or inorganic compounds is essential.

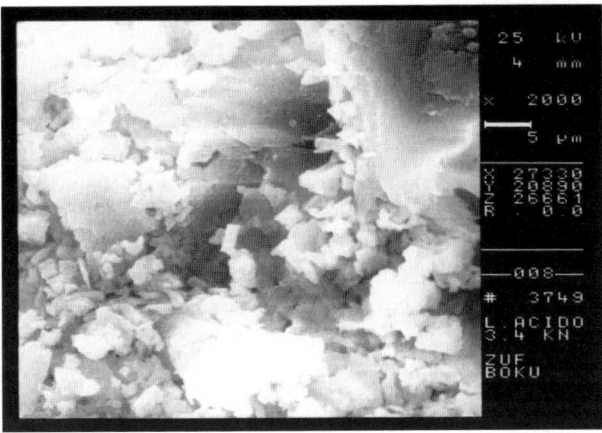

Figure 1. Solid particles of a dried Lactobacillus acidophilus (Reflection electron microscopy, 2000 x magnification).

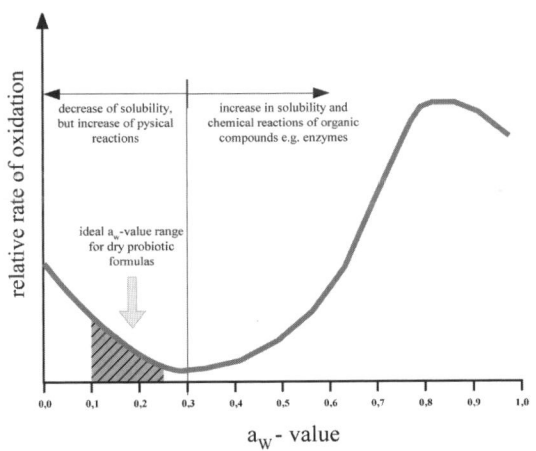

Figure 2. Relative fatty acid oxidation rate correlated to water activity (a_W).

In view of the physical needs of such stabilised microorganisms, some formulations and distribution channels are inappropriate, because any dried organism will always take up water from whatever source available to it [15]. This caveat is valid for all possible technical uses in probiotic foods, feeds and in the pharmaceutical area. Food technology provides exact information on the reactions of proteins, fatty acids and other compounds in dry systems and following uptake of water. A dry but viable microorganism consists of proteins and lipids similar to those found in certain foods for example dried meat or vegetables.

Figure 2 provides a graphical explanation on the relative oxidation rate of fatty acids in relation to the activity of water [16]. The activity of water is dimensionless, but provides a measure for the availability of water in a system.

A given system containing organic components like proteins of viable cells will always follow the relation depicted above. The desiccation process lowers oxidation of fats as well as other processes like enzyme activity, but should never reach zero level. Otherwise physical damage or death of the organism would result [17]. Enzymes and proteins need at least monolayers of water to maintain their shapes and functionality. If there is no water uptake possible for the organism, the protein functions remain intact until physical damage happens to the proteins. The degradation process for such a system is therefore irreversible. This is what we encounter for dried microorganisms even under the best storage conditions.

A formulator of probiotic products can only choose between more water resulting in higher metabolic activity and a shorter time of storage stability of his active ingredient, or a lower water content, a relatively longer storage time, but also higher risk of physical damage for the microbial cells.

2. Standard pathways for production of microorganism and related stabilisation techniques

A demand for pure and reliable single strain starter cultures was developed in the dairy industry about a century ago. Fermentation technology is part of the core business for this industry and it is the cultivation technology of choice for mass production of microorganisms. The pathway in figure 3 depicts the most common way of production and indicates where technically feasible stabilisation measures may take place.

For the broad range of available probiotic formulations and products, the freeze-dried microorganisms are the appropriate choice.

Freeze drying, also called lyophilisation, is state of the art technology for production of dry but viable cells of microorganisms [18,19].
Rehydration, the act of water uptake, converts the dried microbial cells back into functional and proliferating organisms. Some crucial factors for the success of this process are summarised in table 2.

Pure microbial cell matter does not have sufficient resistance against the physical impact of freezing and drying processes. However, a variety of chemical and biological materials are used as additives to protect cells against damage. Best known for their positive effects on microbial cells are sugars and skim milk powder. Table 3 summarises a selection of substances currently used in varying combinations as cryoprotectants for probiotics.

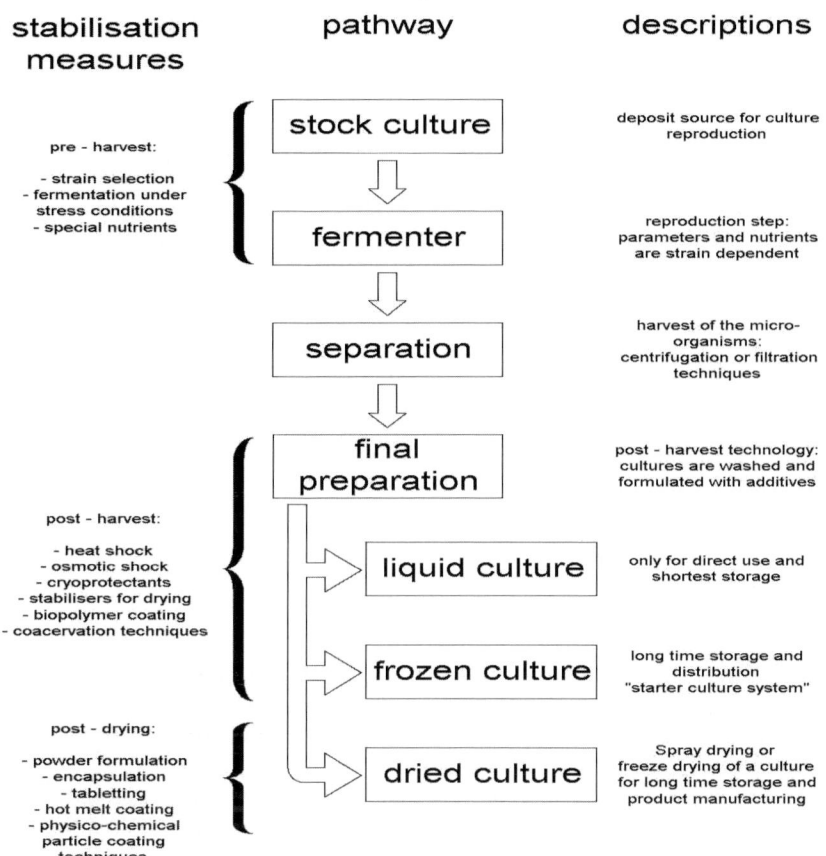

Figure 3. Pathways of standard microorganism production, brief description and stabilisation measures.

Table 2. Factors affecting rehydration of dried microorganisms [20-23].

- Growth media contents, for example salts, proteins and vitamins
- Growth parameters, for example pH, temperature, duration, buffer system
- Harvesting technology
- Cryoprotectants and other additives
- Freeze drying / spray drying parameters
- Osmotolerance of a strain
- Ability to repair cell wall damage
- Oxygen tolerance of a strain
- Storage technology
- Speed of rehydration after storage
- Water temperature during rehydration

Table 3. A selection of cryoprotectants for probiotic microorganisms [24-30]

Adonitol	Skim milk powder
Alpha-tocopherol	Maltodextrin
Beta-glycerophosphate	Malt extract
Bovine albumin	Mannitol
Calcium alginate	Milk fat
Calcium carbonate	m-Inositol
Casein hydrolysate	Na-alginate
Carnithine	Na-glutamate
Dextran	Pectin
Dimethylsulphoxide	Polyethyleneglycol
Gelatine	Sucrose
Glucose	Sorbitol
Glycerol	Trehalose
Glycogen	Tween
Lactose	Xanthan gum
L-ascorbic acid	Yeast extract
L-asparagine	Whey powder
L-cystein	

3. Stabilisation methods

Stabilisation processes should be considered with respect to the time of their application within the production chain of a probiotic and the resulting product. Every stabilising action within the process of fermentation and up to the freezing or drying of a microorganism is a primary stabilisation method.

All measures taking place during formulation and production of a probiotic product are secondary methods. In terms of positive effects, a primary stabilisation for example the application of cryoprotectants has a clear advantage over secondary methods for example the addition of antioxidants, oligosaccharides or desiccants to a powder formulation later in a process. Risk of unintended water uptake increases with every additional process step and manipulation of the dry microorganism powders.

Most manufacturing processes for dry probiotic foods and feeds use dried microorganism.

3.1. CAPSULE FORMING PROCESSES

Technically, a thin solid wall material that encloses a solid or liquid core forms a capsule. A range of mechanical and physico-chemical capsule forming processes is defined for the pharmaceutical area. The core material is used as a carrier for the active ingredient. On the other hand the wall material protects and controls the release of the active ingredient.

In the case of viable microorganisms as the active content, the capsule forming processes are limited to a limited range of methods. Organic solvents and metal catalysts, reactive monomers, bactericidal substances and in general all other materials reacting with viable microorganisms are dangerous for the viability. The processes must be carried out without ultrasound devices, microwave or heating. Only processes injecting a microorganism-loaded core material into the capsule-forming bath are so far in use. Capsules must be harvested and undergo several steps of washing and rinsing before they are ready for drying and storage.

Water must be removed from the capsules core and wall material as well. Water activity in a capsule system has to fall to a minimum of 0,1. Since the drying physics of encapsulated materials are more complicated compared to a freeze drying process, most microorganism products produced by these processes are short lived and unstable.

Systems utilising a water resistant core, for example fat and a capsule forming material, do slightly better, because they use dried microorganisms embedded in fat. Thus, the microorganisms do not come into direct contact with moisture.

An important aspect concerns the question of residual moisture inside a capsule. During drying, a capsule normally becomes more and more protective. But even with ideal wall materials there will always be water entrapped inside the capsule. A capsule forms therefore a very small container with its own internal water activity. The entrapped water interacts easily with the core components, because of the short distances to move.

LBC ME10

This is a Swiss veterinary product also known as UFA-Antifex. The Strain *Enterococcus faecium* SF68 is the active organism. It is distributed as a probiotic feed additive for pigs, poultry and cattle. Veterinary prescriptions are not required. The product is shipped in One-liter plastic packages. According to the producers' information, there is a multi layer coating around the bacterial core to protect the active organism. Refrigeration below 8°C is recommended for long time storage. Microbial analyses give correct numbers for the relevant strain in the fresh and appropriately stored product. The manufacturers claim that LBC ME 10 can be used for the production of cereal feed pellets. There was no scientific proof available for this claim. The product is currently available on the Swiss and European markets.

Bifina®

Bifina® consists of edible hardened oil as a core material and carrier for dried *Bifidobacteria*. This core is coated with two additional layers of fat and gelatine plus pectin. The Japanese JINTAN's seamless capsule technology is used for the encapsulation process [31,32]. The encapsulated *Bifidobacteria* are blended with oligosaccharide granules. Bifina® is currently available as a pharmaceutical product.

BiActon®

The French veterinary product BiActon® is a probiotic feed additive for pig fattening farms. Veterinary prescription is not required. The strain *Lactobacillus farciminis* MA 67 – 4R is the active organism. According to the producers' information, a microencapsulation process and a further aggregation of the dried bacteria protect the strain.

3.2. SPECIFIC SOLUTIONS

LifeTop™ Cap

This concept of a special bottle cap is named LifeTop™ Cap and can be used for beverages and mineral water. It was presented by the Swedish company BioGaia in 2002 [33]. It works by the separation of the bacteria from a liquid bottle filling until the moment of consumption and has nothing to do with stabilisation processes of bacteria. Each LifeTop™ Cap provides *Lactobacillus reuteri* powder, which is placed in a protective blister inside the cap. The bacteria are mixed with the liquid when the top of the cap is pushed through the blister.

LifeTop™ Straw

A LifeTop™ Straw from the same company, which can be attached to bottles in the same way as a regular drinking straw, works similar. A prepared straw contains roughly 100 million *Lactobacillus reuteri* in an oil droplet. The bacteria or any other ingredients are automatically released and mixed with the liquid during consumption.

3.3. EXTRUSION

From a current standpoint extrusion does not offer a practical solution for the stabilisation of dried but viable bacteria. The feed industry makes use of extrusion techniques to form feed particles for easy and dust-free handling. These particles – feed pellets – can have probiotic microorganisms as active ingredient for example as a growth promoter.

Since physical impact on the biological materials exceeds limits for their survival, only spore forming bacteria of the genus *Bacillus* have been shown to be suitable in feed extrusion up to now. The technique of extrusion is not practicable for *Lactobacilli*, *Bifidobacteria* and *Enterococcus* strains.

The technical term extrusion describes a variety of machine-dependent techniques. In general, materials are forcibly pressed through nozzles or steel dyes to form irregular and oblong particles. Depending on its ability to stick or plastify, physical impact on the material has to be high or low. In some cases the materials chosen have to be wetted or blended with a binder prior to the process. Drying, screening and the removal of dust are technical steps regularly applied to extruded materials.

The food industry and feed producers use extrusion to form uniform matrices from heterogeneous raw materials. Their technical systems use screw extruders as standard. The system works with high pressure, heat and moisture only.

In the feed industry, a different type of semi-continuous extrusion, sometimes called pelletisation, is the standard process to form larger particles from milled and blended grain or corn. Binders may be added as required. For example, molasses from sugar production are commonly used in this industry.

3.4. HARD (GELATINE) CAPSULES

Hard (gelatine) capsules are an ideal dosage system for powders and granules in the pharmaceutical and food industries. Filling equipment is in wide use and contract producers are available. Film coating may be applied to the capsules for protection against gastric juice as it is used for Mutaflor®, Ardeypharm GmbH, 58313 Herdecke, Germany, or for controlled release. The standard packaging systems for hard (gelatine) capsules are blisters or plastic bottles.

The advantage of capsules for preparations with dried probiotic bacteria comes mainly from the powder formulation (see 3.8.). The filling material can be adjusted to the specific needs for bacterial shelf life. The capsule filling process is a mechanical process with low physical impact on the bacterial material.

Bioflorin® capsules

The Swedish pharmaceutical product Bioflorin® contains the *Enterococcus faecium* SF68 strain as the active ingredient. This product is used for therapy of intestinal disorders after antibiotic treatment, during change of diet, against diarrhoea or as a dietary supplement. Cool storage is recommended for this product.

Probio-Tec®QUATRO-cap-4

This is a size 3 Capsule filled with a probiotic powder blend produced by Chr. Hansen. The blend contains four cultures in a base formulation of dextrose and magnesium stearate. The Capsules are shipped in aluminium tubes with desiccant. A shelf life period of two years at room temperature is given for this product and a minimum of 4×10^9 viable bacterial cells are guaranteed at the time of manufacture [34].

Omniflora® N

Omniflora® N is a German pharmaceutical product. Strains of *Lactobacillus gasseri* and *Bifidobacterium longum* are the active ingredients. This product is used for therapy of intestinal disorders after antibiotic treatment, during change of diet, against diarrhoea or as a dietary supplement. The blistered Omniflora® N capsules are shipped in a box. Cool storage is recommended for this product.

BioTura®

BioTura® is a Danish pharmaceutical product. Strains of *Lactobacillus acidophilus* and *Bifidobacterium* species are the active ingredients. BioTura® is used for the therapy of intestinal disorders after antibiotic treatment, during change of diet, against diarrhoea or as a dietary supplement. The blistered BioTura® capsules are shipped in a box. Cool storage is recommended for this product.

3.5. LIQUID FORMULATIONS

Liquid formulations are mentioned in this article for reasons of a complete overview only. For obvious biological and technical reasons, they are limited by the need for refrigeration and by short storage periods. This is the classic area for fresh food products, for example yoghurt.

Formulations comprising bacteria in water are very short lived. Since water is available to the bacteria, it takes low temperatures to keep metabolism at low rate. This technique is applicable to commercial dairy products.

Bacteria-in-oil–formulations are a coming technique in the area of biological pesticide production for sprayer applications. Oil has the advantage of being a natural barrier to moisture and oxygen. This allows the preparation of bacteria-in-oil-suspensions with dried bacteria or with fresh fermented and harvested bacteria. Combined with refrigeration, at least a modest shelf life can be achieved.

3.6. TECHNIQUES FOR SOLID PARTICLE COATING AND AGGREGATION

A technology capable of forming a more or less seamless surface on a solid particle can be used to create a protective coating around dried microorganisms for stabilisation. Provided the viability limits are not exceeded during processing, such techniques can be employed.

In most cases a fluid bed equipment, fitted with top or bottom spray nozzles, is used to coat solid particles. Within a containment, particles are in constant, but directed move. Solutions of coating material are sprayed with or against the stream of particles. The particle surface is covered by liquid coating material. Air is used as both the carrier and drying medium at the same time. It removes the water used as solvent.

Depending on particle size and coating, the system makes a perfect surface or a mix of coating and aggregation. Ideally, a perfect seamless coating protects every single particle.

A major disadvantage of particle coating techniques in combination with dry microorganisms is the fact that only water can be used as a solvent for coating materials. The hygroscopic nature of these dry particles results in the uptake of moisture from this source. Therefore, careful coating application and quick water removal are essential. Otherwise entrapped water initiates degradation processes in the solid particles.

One modified process derived from conventional spray coating is the use of meltable materials as a coating. The process is regularly named (hot) melt coating. Typical materials for such a coating are waxes, fats and some polymer types. For particles of dried microorganisms, it is essential to use materials with low melting points to keep the influence of heat within limits. The (hot) melt coating has a broad range of applications in prebiotics, mainly as a coating for oligosaccharides and other hygroscopic sugars.

3.7. PASTES AND OINTMENTS

Pastes and ointments are a technically very effective way to stabilise dried viable microorganisms. Especially the oil–based formulations protect organic contents against moisture. Combined with refrigeration, these formulations show excellent shelf life data. A key point for such formulations remains the oil. It has to be saturated and free of any volatile impurities or metal ions. Regularly used formulation materials are peanut oil in combination with starch, sugars, types of celluloses, antioxidants, silica, and alginates.

Wide use of oil pastes is realised in the feed industry for products combining microorganisms, vitamins and trace elements for pigs and calves. For human nutrition, such oil pastes are unpopular and the food industry does not use this technique.

Lactiferm® Ironpaste

This is a probiotic paste formulation combined with iron from Sweden. A need for veterinary prescription in Europe is dependent on national regulations. It is a white paste with the strain *Enterococcus faecium* M74 as the active organism. Plastic syringes are used for storage and application. Refrigeration is recommended for the long time storage. The product is currently available in several European countries.

3.8. POWDER FORMULATIONS

Powders are basic types of pharmaceutical and food formulations for direct use or as starting point for technical processes. For dried probiotic bacteria or other types of protected bacterial particles, for example microcapsules, a powder formulation offers certain advantages for stabilisation. Protection in gastric juice can be achieved when bacteria are mixed with gel-forming polysaccharides that are insoluble at low pH but are dissolved in the upper part of the small intestine; there, the probiotic bacteria are rehydrated and released in their fully biologically active state.

Formulation components can be selected for their low water activity. Particle size is adjustable to the requirements of formulations in most cases. Low physical impact allows to keep damage to the bacteria powder to a minimum. The production can be performed in closed systems as a protection against moisture.

A variety of systems allows the use of readily prepared dry powder mixtures for tablets, hard gelatine capsule fillings, and sachet fillings.

Antibiophilus sachets and capsules

Antibiophilus is an Austrian pharmaceutical powder formulation. There is no prescription for this product required. Two types of dosage forms, the sachet and the hard gelatine capsule are in use. The active ingredient is an antibiotic resistant *Lactobacillus acidophilus* strain. Antibiophilus is used for the therapy of intestinal disorders after antibiotic treatment, during change of diet, against diarrhoea or as a dietary supplement. The powdery product has to be mixed with a liquid for proper consumption.

Effidigest® sachets

This is a powder formulation from France and it belongs to the category of nutraceuticals. It is used to support the intestinal flora after diarrhoea or as a dietary supplement. The Powder has to be mixed with a liquid for proper consumption. Cool storage is recommended for this product.

Lactiferm®

Lactiferm® is a powder formulation from Sweden. It is mainly in use as a probiotic feed additive. Veterinary prescription for this product is not required. The strain

Enterococcus faecium M74 is the active organism. Lactiferm® is available in a variety of concentrations and packaging systems. Cool storage below 8°C is recommended.

Effervescent formulation

A PCT patent [35] application was made for effervescent formulations containing dried bacteria as the active organism. The Effervescent powder or granulate formulation functions as a protective system for the viable microorganisms during storage, rehydration and during stomach transit. The technology is adaptable for sachets, tablets and capsule filling as well.

3.9. TABLET PROCESSING

Tablets are still the most common pharmaceutical dosage forms. Performed on multi station rotary presses and with only few limits for shapes and volumes of the final solid tablet, the process of powder compression has revolutionised pharmaceutical dosage. Relative to other dosage types, tabletting is inexpensive and is therefore of interest for the production of probiotic bacteria loaded tablets. Only dried bacteria can be used in the process.

In general, it has to be accepted that the preparation of tablets is not the ideal process for products comprising viable bacteria, because the excess forces used to compact a powdery or granulated pre-mixture into the solid dosage form cause physical damage in the dry cell walls [36]. This was a reason for using tabletting in pharmaceutics for microbiologically delicate materials as a hygienic measure – contrary to the idea of a probiotic product [37].

Nonetheless, various formulation components help to reduce risk for cell damage and wall disruption. Skim milk powder, micro-crystalline cellulose, denatured dry milk proteins and other newer pharmaceutical tablet components help to overcome limitations for dry bacteria in tablets. In addition, alginate or cellulose derivatives like hydroxypropyl-methylcellulose acetate succinate (HPMC-AS) and HPMC-phthalate can be used as matrix-forming excipients providing resistance to gastric juice [38,39].

The Nutraceutix-exclusive Cryotabletting^{TM} process was developed to counteract the harmful effect of heat on the bacteria during tabletting, resulting in a higher survival rate of the microorganisms [40].

The packaging concepts – especially gas filled blister systems and plastic bottles with desiccating stopper filling – are a further reason for use of tablets to deliver bacteria. These systems partly reduce the risk of damage during storage, transport and sales.

Not mentioned in the following examples are those tablet products containing cell wall fragments, cell contents or dried dead cells.

Bion® 3 and Multibionta®

This pharmaceutical dietary supplement is shipped under two different brand names: Bion® 3 in Germany and Continental Europe, and Multibionta® in UK and other countries. It is an oval three layer tablet with film coating. The tablet layers can be distinguished by colour. According to the label the tablets contain a combination of probiotic cultures, vitamins, minerals and trace elements. The Package is a combination of a plastic bottle and a box for labelling and consumer information. There is no bacteria specific information about necessary storage conditions given.

Gynoflor®

Gynoflor® is a Swiss pharmaceutical product, available only by prescription. It is a vaginal tablet used to re-establish the physiological flora. The active organism is a *Lactobacillus acidophilus* strain. The package is a combination of aluminium blister and a box. The storage is recommended within a range of 2 to 8°C.

Paidoflor®

This German pharmaceutical product is only available by prescription. It is formulated as tablet for intestinal therapy after antibiotic treatment, during a change of infant diets or as a dietary supplement. The active organism is a *Lactobacillus acidophilus* strain. The package is a combination of aluminium blister and a box. According to the recommended storage conditions 8°C should not be exceeded.

Lacto

Lacto is the name for a variety of Italian pharmaceutical products and probiotic dietary supplements. Their main use is for intestinal therapy after antibiotic treatment. The active organisms are a mixture of strains of *Lactobacillus acidophilus*, *Lactobacillus bulgaricus*, a *Bifidobacterium* species and a spore forming *Lactobacillus*. The packaging system for the Lacto B tablets is a combination of blister and box.

4. Summary

Careful tests are required to assure the stability of microorganisms and to develop stable products for commercial distribution.

A producer of intermediate formulations or products for the end user can only rely on the physical stability and viability of the immobilised microorganisms. After production, such products are regularly distributed through channels such as pharmacies or supermarkets. There, a product is needed with organisms remaining viable on the shelves for months and if possible for years. The customer in turn will rely on the imprinted shelf life date and on the activity of the organism in the product.

The quest for systems directly applicable for the production with lyophilised microorganisms stimulated the use and adaptation of mainly pharmaceutical technologies, the search for new coating materials, and some basic changes in fermentation and freeze-drying technologies. Pharmaceutical powder formulations with the additional use of enteric coating techniques are resulting in end products such as tablets, sachet fillings or capsules.

New methods such as the encapsulation into biological matrices like fats and gelatine, or hot melt coating processes, have so far not gained a sufficient level of confidence for their use in mass production. Their main disadvantages are the complicated techniques and high costs. This of course hampers upscaling, wider use and the availability of less expensive probiotic consumer products.

So far products comprising viable microorganisms with reasonable shelf live can be best formulated with basic pharmaceutical technologies, but the organisms remain sensitive to moisture effects, daylight, temperature, free radicals, antibiotics and some other influences. Depending on the strains used, it is important to manipulate lyophilised microorganisms at low relative humidity, low temperature, and in the absence of the any harmful influences to maintain viable cells in the product for a long time. But, independent from most stabilisation techniques in regular use, the majority of products cannot be stored without proper refrigeration.

The authors apologize to those companies of which products are not mentioned or outdated products are cited in this book article. Following the production and marketing of probiotic and prebiotic formulas in food and pharmaceuticals over years, we found that the products enter and leave the market in quick succession. Especially the probiotic dietary supplements from internet shops show this characteristic. Further on we failed in several cases to get fresh product samples for test use. With that limitations we concentrated on those products that we considered to be the most relevant examples for the basic techniques in stabilisation of microorganisms.

References

[1] Mitsuoka, T. (1992) The Human Gastrointestinal Tract. In: Wood, J.B. (Ed.) The Lactic Acid Bacteria in Health and Disease. Chapman & Hall; ISBN 0751403080; pp. 69–114.

[2] Fuller, R. (1997) Probiotics 2, Applications and practical aspects. 1st Edition, Kluwer Academic Publishers; ISBN 0412736101; pp. 1–8.

[3] Fuller, R. (1992) History and development of probiotics. In: Fuller, R. (Ed.) Probiotics – The scientific basis. Chapman&Hall; ISBN 0412408503; pp. 1–8.

[4] Fuller, R. (1989) Probiotics in man and animals. J. Appl. Bacteriol. 66: 365–378.

[5] Gibson, G.R. (2000) Introduction. In: Gibson, G.R. and Angus, F. (Eds.) LFRA Ingredients Handbook Prebiotics and Probiotics. Leatherhead Publishing, Surrey, UK; ISBN 0905748824; pp. 1–12.

[6] Lee, J.-K. (1999) Introduction, 1.2. Probiotic Microorganisms. In: Lee, J.-K.; Nomoto, K.; Salminen, S. and Gorbach, S.L. (Eds.) Handbook of probiotics. John Wiley & Sons, Inc., USA; ISBN 047119025X; pp. 1-22.

[7] Gill, H.S.(1998) Stimulation of the immune system by lactic cultures. Int. Dairy J. 8: 535–544.

[8] Isolauri, E. (1994) Lactic acid bacteria and the immune response. IDF Nutrition Newsletter 141(3): 10.

[9] De Simone, C.; Vesely, R.; Bianchi Salvadori, B. and Jirillo E. (1993) The role of probiotics in modulation of the immune system in man and in animals. Int. J. Immunother. 1: 23–28.

[10] Marteau, P. (1995) Impact of ingested lactic acid bacteria on the immune response in man. In: Lactic acid bacteria, Actes du colloque Lactic 94, Caen, 7–9 Septembre 1994; Presses Universitaires de Caen, Caen; pp. 31–42.

[11] Ritzén, L.G. (2002) A need for improved stability of freeze-dried Lactobacilli. In: Cell Physiology and Interactions of Biomaterials and Matrices, Cost 840 & X International BRG Workshop on Bioencapsulation, Prague, Czech Republic, 26–28 April 2002; pp. 99–101.

[12] Geiss, H.K. (2002) Probiotika. Medizinische Monatsschrift für Pharmazeuten 25 (6): 186–192.

[13] Thies, C. (1996) A survey of microencapsulation processes. In: Benita, S. (Ed.) Microencapsulation, Methods and Industrial Applications. Marcel Dekker, Inc.; ISBN 0824797035; pp. 1–19.

[14] Mattila-Sandholm, T.; Myllarinen, P.; Crittenden, R.; Mogensen, G.; Fonden, R. and Saarela, M. (2002) Technological challenges for future probiotic foods. Int. Dairy J. 12: 173–182.

[15] Laulund, S. (1994) Commercial aspects of formulation, production and marketing of probiotic products. In: Gibson, S. (Ed.) Human Health: The Contribution of Microorganisms. Springer Series in Applied Biology; ISBN 0387198717; pp. 159–173.

[16] Taoukis, P. and Labuza, T.P. (1996) Summary: Integrative concepts, Chapter 17. In: Fennema, O.R. (Ed.) Food Chemistry. 3rd Edition; Marcel Dekker, Inc; pp. 1013–1042.
[17] de Valdez, G.F.; de Giori, S.G.; de Ruiz Holgado, A.P. and Oliver, P. (1985) Effect of drying medium on residual moisture content and viability of freeze-dried lactic acid bacteria. Appl. Environ. Microb. 51: 413–415.
[18] Champagne, C.P.; Gardner, N.; Brochu, E. and Beaulieu, Y. (1991) The freeze–drying of lactic acid bacteria. A review. Can. Inst. Sci. Technol. J. 24(3/4): 118-128.
[19] Jennings, T.A. (1999) Introduction. In: Jennings, T.A. (Ed.) Lyophilization, Introduction and Basic Principles. Interpharm Press, Denver, USA; ISBN 1574910817; pp. 1–13.
[20] Linders, L.J.M.; Meerdink, G. and Van't Riet, K. (1997) Effect of growth parameters on the residual activity of *Lactobacillus plantarum* after drying. J. Appl. Microb. 82: 683–688.
[21] Gilliland, S.E. (1977) Preparation and storage of concentrated cultures of lactic streptococci. J. Dairy Sci. 60(5): 805–809.
[22] Wright, C.T. and Klaenhammer, T.R. (1981) Calcium-induced alteration of cellular morphology affecting the resistance of *Lactobacillus acidophilus* to freezing. Appl. Environ. Microb. 47: 807–815.
[23] Rhyänen, E.L. (1991) Über den einfluss der gefriergeschwindigkeit auf lebensfähigkeit und stoffwechselaktivität gefrorener und gefriergetrockneter *Lactobacillus acidophilus* – Kulturen. Finish J. Dairy Sci. 49(2): 14–36.
[24] Bozoglu, T.F. and Candan Gurakan, G. (1998) Freeze drying injury of *Lactobacillus acidophilus*. J. Food Protect. 52(4): 259–260.
[25] de Valdez, G.F.; de Giori, S.G.; de Ruiz Holgado, A.P. and Oliver, P. (1983) Comparative study of the efficiency of some additives in protecting lactic acid bacteria against freeze–drying. Cryobiology 20: 560–566.
[26] Kearney, L.; Upton, M. and McLoughlin, A. (1990) Enhancing the viability of *Lactobacillus plantarum* inoculum by immobilizing the cells in calcium alginate beads incorporating cryoprotectants. App. Environ. Microb. 10: 3112–3116.
[27] Leslie, S.B.; Israeli, E.; Lighthart, B.; Crowe, J.H. and Crowe, L.M. (1995) Trehalose and sucrose protect both membranes and proteins in intact bacteria during drying. Appl. Environ. Microb. 61: 3592–3597.
[28] Morin, N.; Bernier-Cardou, M. and Champagne, C.P. (1992) Production of concentrated *Lactococcus lactis subsp. Cremoris* suspensions in calcium alginate beads. Appl. Environ. Microb. 58: 545–550.
[29] Castro, H.P.; Teixeira, P.M. and Kirby, R. (1997) Evidence of membrane damage in *Lactobacillus bulgaricus* following freeze drying. J. Appl. Microbiol. 82: 87–94.
[30] Gandhi, D.N. and Shahani, K.M. (1994) Survival of *Lactobacillus acidophilus* in freeze dried acidophilus powder using two different media. Indian J. Microb. 34: 45–47.
[31] Morishita, T.; Sunohara, H. and Sonoi, S. (1982) Method and apparatus for encapsulation of a liquid or meltable solid material. US – Patent 4,422,985.
[32] Morishita, J. (2002) Jintan capsule technology. Capsule Product Devision. http://www.jintanworld.com; Japan.
[33] BioGaia (2002) New approach to good health. http://www.biogaia.se; Sweden.
[34] Hansen, Chr. (2002) Chr. Hansen BioSystems. http://chbiosystems.com; Denmark.
[35] Viernstein, H. (2000) Formulations having probiotically active microorganisms. PCT – Patent WO 00/07571.
[36] Hoover, D.G. (1993) Pressure effects on biological systems. Food Technol. 47: 150–155.
[37] Blair, T.C. (1991) On the mechanisms of kill of microbial contaminants during tablet compression. Int. J. Pharm. 72: 111–115.
[38] Stadler, M. and Viernstein, H. (2001) Tablet formulations of viable lactic acid bacteria. Sci. Pharm. 69: 249-255.
[39] Viernstein, H., Polheim, D. and Laulund, S. (1999) Probiotically acting formulation. Patent AT 405 235.
[40] Nutraceutix (2001) Nutraceutics Probiotics. http://www.nutraceutix.com/probiotics. Redmond, USA.

GROWTH OF INSECT AND PLANT CELLS IMMOBILISED USING ELECTRIFIED LIQUID JETS

MATTHEUS F. A. GOOSEN
School of Science and Technology, University of Turabo, PO Box 3030, Gurabo, Puerto Rico, USA, 00778 – Fax: (787) 744-5427 –Tel: (787) 743-7979 Ext. 4167 – Email: mgoosen@suagm.edu

1. Introduction

In many bio-processing laboratories, major focus has been placed on attempting to find cell culture methods, which can increase the concentration of cells and cell products and permit cost-effective large-scale production. Methods of animal cell culture have been developed (mainly for mammalian cells) and involve the use of hollow fibres, gel entrapment, ceramic cartridges and microcarriers [1]. It is generally recognised that compared to microbial systems, large-scale mammalian cell suspension culture has been limited, to some extent, by relatively low cell densities. The concentration of extracellular proteins such as monoclonal antibodies and growth factors produced by this method is low and purification from growth media is difficult. The same problems are also encountered in insect culture. Inlow and co-workers [2], for example, reported maximum insect cell densities of 5.5×10^6 cells/ml in a spinner flask. Hink [3], working with *Trichoplusia ni* and *Spodoptera frugiperda* cells, obtained maximum polyhedra (i.e. recombinant viruses AcNPV) concentrations of *ca.* 2.2×10^8 polyhedra/ml medium at a cell density of 3.8×10^6 cells/ml medium (*i.e. ca.* 60 polyhedra/cell). Knudson and Tinsley [4], using *S. frugiperda* cells, reported 12-40 polyhedra (150-500 IFU)/cell.

Microencapsulation, an alternative cell immobilisation technique originally developed for use as an artificial pancreas for the treatment of diabetes, has been employed industrially for the enhanced production and recovery of high-value biologicals from animal cells. The encapsulation technique entraps viable cells within semipermeable polysaccharide-polycation microcapsules (for example, alginate/poly-L-lysine (PLL)). The capsule membrane selectively allows small molecules such as nutrients and oxygen to freely diffuse through, but prevents the passage of large molecules and cells. Posillico [5] reported the use of microencapsulation for the commercial production of monoclonal antibodies. However, while cell densities of ca. 1×10^8 cells/ml capsules were obtained after three weeks of culture, they reported that the cells appeared to grow preferentially near the interior surface of microcapsule membrane and speculated that this could have been due to mass transfer limitations during culture and/or to the presence of a viscous intracapsular alginate solution.

The first step in the making of a microcapsule is droplet formation. Let us consider one technique, electrostatic droplet generation (*i.e.* electrified liquid jet), which has become the primary method used in our laboratory.

2. Droplet generation using an electrified liquid jet

Electrified liquid jets (*i.e.* electrostatic atomisation) and electrostatically assisted atomisation have been employed in a variety of areas, including paint spraying [6], electrostatic printing [7], and cell immobilisation [1,8]. Recently, micro and nano capsules as small as 0.15 micrometers, for example, have been produced using 0.7 mm ID needles [9]. The effect of electrostatic forces on mechanically atomised liquid droplets was first studied in detail by Lord Rayleigh [10,11] who investigated the hydrodynamic stability of a jet of liquid with and without an applied electric field.

When a liquid is subjected to an electric field, a charge is induced on the surface of the liquid. Mutual charge repulsion results in an outwardly directed force. Under suitable conditions, for example, extrusion of a liquid through a needle, the electrostatic pressure at the surface forces the liquid drop into a cone shape. Surplus charge is ejected by the emission of charged droplets from the tip of the liquid. The emission process depends on such factors as the needle diameter, distance from the collecting solution, and applied voltage (strength of electrostatic field) [12]. Under most circumstances, the electrical spraying process is random and irregular, resulting in drops of varying size and charge that are emitted from the capillary tip over a wide range of angles. However, when the electrostatic generator configuration has been adjusted for liquid pressure, applied voltage, electrode spacing and charge polarity, the spraying process can become quite regular and periodic.

2.1. PRODUCTION OF ALGINATE BEADS USING ELECTRIFIED LIQUID JETS

The section starts with a detailed experimental description of electrostatic droplet generation for those not familiar with the technique [12] (Figure 1). Attach a syringe pump to a vertical stand. Use a 10 ml plastic syringe and 22- or 26- gauge stainless steel needles. A variable high voltage power supply (0-30 kV) with low current (less than 0.4 mA) is required. We have used a commercial power supply model 230-30R from Bertan (Hicksville, NY). Prepare 1.5% (w/v) $CaCl_2$ in saline (0.85 g NaCl in 100 ml distilled water). Saline can be replaced with distilled water if an alginate solution without cells is being extruded. Place the $CaCl_2$ solution in a petri dish on top of an adjustable stand. The stand allows for fine-tuning of the distance between the needle tip and collecting solution. Prepare 1 to 4% (w/v) low viscosity sodium alginate by dissolving alginate powder with stirring in a warm water bath. Slowly add the 1 to 4 g sodium alginate to 100 ml warm saline solution (or distilled water), stirring continuously. It may take several hours to dissolve all of the alginate. Add about 8 ml of the alginate solution to a 10-ml plastic syringe, put back the plunger, and attach the syringe to the upright syringe pump. Make sure that the stainless steel needle, 22 gauge, is firmly attached and the syringe plunger is in firm contact with the moveable bar on the pump. Position the petri dish (or beaker) containing CaCl₂ so that the needle tip is about 3 cm from the top of the CaCl₂ hardening solution. This is the primary reason for using an adjustable stand.

Attach the positive electrode wire to the stainless steel needle and the ground wire to the collecting solution. The wires may need some additional support to prevent them from bending the needle. Switch on the syringe pump and wait for the first few drops to come out of the end of the needle. This could take a minute or two. Doing it this way also ensures that the needle is not plugged. After the first drop or two has been produced, switch on the voltage supply. Make sure that the voltage is set low, less than 5 kV. If this is the first time that you have tried electrostatic droplet generation, raise the voltage slowly and observe what happens to the droplets. The rate at which they are removed from the needle tip increases until only a fine stream of droplets can be seen. The changeover from individual droplets to a fine stream can be quite dramatic. The most effective electrode and charge arrangement for producing small droplets is a positively charged needle and a grounded plate. Two other arrangements are also possible; positively charged plate attached to needle, and a positively charged collecting solution. Make sure that the positive charge is always on the needle. This ensures that the smallest microbead size is produced at the lowest applied potential. With a 22-gauge needle and an electrode spacing of 2.5-4.8 cm there will be a sharp drop in microbead size at about 6 kV. This can be noticed visually by observing the droplets coming from the needle tip. Standard commercially available stainless steel needles can be employed. However, when going from a 22- to 26-gauge (or higher) needle, needle oscillation may be observed. This needle vibration will produce a bimodal bead size distribution with one peak around 50 µm diameter beads and another around 200 µm.

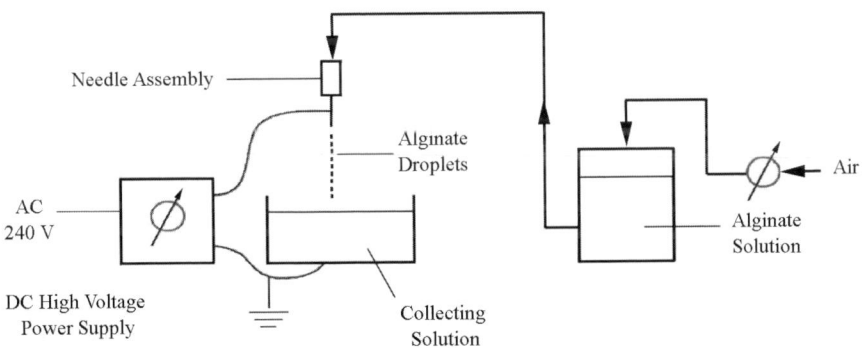

Figure 1. Electrostatic droplet generator system. Reprinted from Goosen et al. 1997 [21].

If a syringe pump is not available, remove the syringe plunger and attach an air line with a regulator to the end of the syringe. Varying the air pressure on the regulator can control the alginate extrusion rate.

Lumps of sodium alginate often form if the powder is added all at once to the warm saline. Sprinkle the alginate powder into the saline a small amount at a time with gentle mixing. Once it has dissolved (up to 1-2 h), allow the viscous solution to cool and then transfer it to several plastic test tubes, cap and store in the refrigerator until needed. This prevents bacterial growth. If the alginate solution is very viscous, air bubbles will be trapped during the stirring. These bubbles will disappear if the viscous solution is left to stand overnight.

If the needle is plugged, place it in dilute citrate solution for a few minutes. Passing a fine wire through the needle also helps. Resuspending cells in 1% (w/v) sodium alginate solution will dilute the alginate and could give tear-drop shaped beads when the solution is extruded. To solve this problem, increase the concentration of sodium alginate solution to 3 or 4%.

Extrusion of alginate droplets using a 5.7 kV fixed-voltage power supply showed that there is a direct relationship between the electrode distance and the bead diameter. For example, at 10-cm electrode distance, the bead diameter was 1500 µm while at 2 cm it decreased to 800 µm. The greatest effect on bead diameter was observed between 2 and 6 cm electrode distance. While there was overlap in bead sizes between 6, 8 and 10 cm electrode distance, there was a significant difference (*i.e.*, no overlap in SD) between bead sizes at 2- and 6-cm electrode distance. An inverse relationship between needle size and microbead diameter was observed. Aside from the 23 gauge needle there was a significant difference between bead sizes produced by all needles (*i.e.*, no SD overlap). As the needle size decreased from 19 to 26 gauge, the bead size decreased from 1400 to 400 µm, respectively. These results support previous work reported by Bugarski *et al.* [12]. The present investigation showed that the alginate concentration does not appear to be important due to overlapping SD intervals for all data points. The bead diameter was found to be 800 µm at both 1% and 3% alginate concentration.

Looking more closely at the effect of electrode distance on bead diameter, as a function of applied potential (Figure 2) and using a 23 gauge needle, we see that the decrease in microbead size was greatest between 5 and 10 kV for all electrode distances tested. The smallest bead, 200 µm, was produced at an electrode distance of 4 cm and an applied potential of 10 kV.

2.2. EFFECT OF ELECTROSTATIC FIELD ON CELL VIABILITY

To assess the effect of an electrostatic field on animal cell viability, an insect cell suspension was extruded using the electrostatic droplet generator. No detectable change in insect cell viability was observed after extrusion. The initial cell density, 4×10^5 cells/ml, remained essentially unchanged at 3.85×10^5 and 3.8×10^5 cells/ml immediately after passing through the generator at an applied potential difference of 6 and 8 kV, respectively. Prolonged cultivation of these cells did not show any loss of cell density or viability.

2.3. FORCES ACTING ON DROPLET IN AN ELECTRIC FIELD

In the absence of an applied voltage, a liquid drop falls from the end of a capillary tube at a critical drop volume dependent on the surface tension (*i.e.*, the liquid drop would continue to grow until its mass overcomes the surface tension). If gravity were the only force acting on the meniscus of the droplet, large uniformly sized droplets would be produced with a radius *r*. Equating the gravitational force on the droplet to the capillary surface force gives:

$$r = (3r_0\gamma/2\rho g)^{1/3} \tag{1}$$

where r_o is the internal radius of the needle, γ is the surface tension, ρ is the relative density of the polymer solution and g is the acceleration constant.

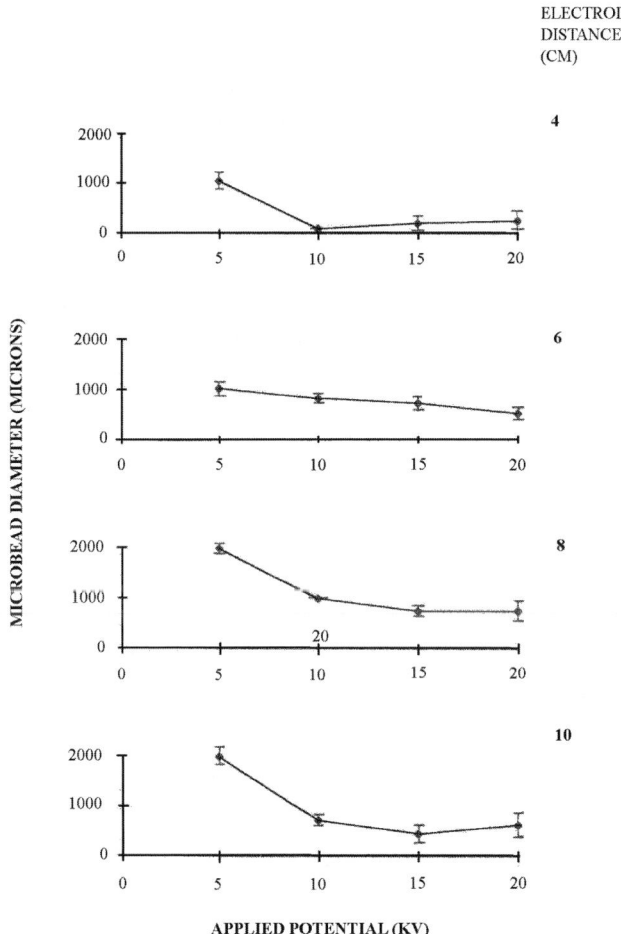

Figure 2. *Effect of electrode distance and applied potential on microbead diameter. Reprinted from Goosen et al. 1997 [21].*

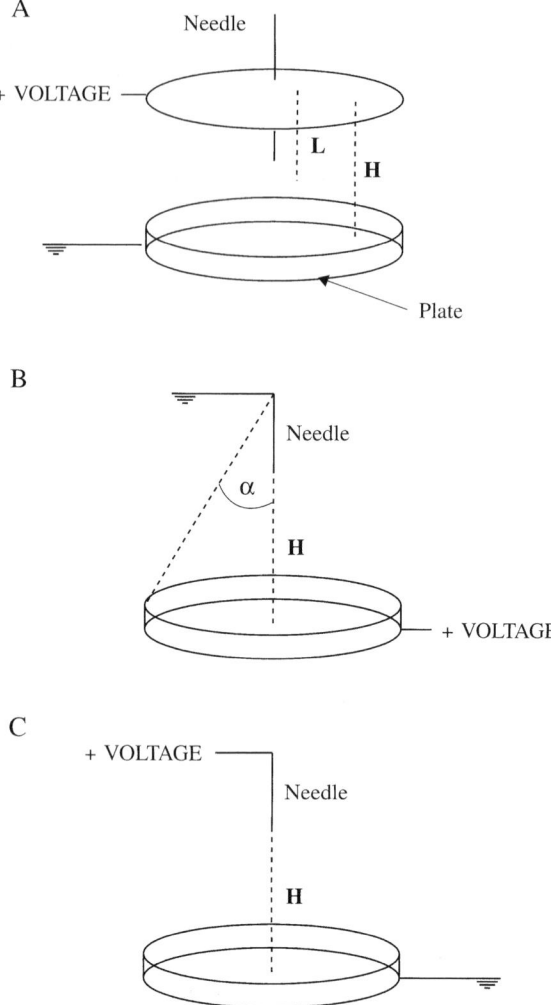

Figure 3. Electrode and charge arrangements; A: Parallel plate set-up with positively charged plate; B: Positively charged collecting plate; C: Positively charged needle [12,26]. Reprinted from Goosen 1996 [26].

Under the action of an electric field, the electric force, F_e, acting along with the gravitational force, F_g, would reduce the critical volume for drop detachment resulting in a smaller droplet diameter. Equating the gravitational and electrical forces on the droplet to the capillary surface force yields:

$$F_\gamma = F_g + F_e \tag{2}$$

Three electrode geometries were considered:
- A parallel plate arrangement in which the charge was applied to a plate held parallel to the collecting solution and through the centre of which the needle protruded (Figure 3A)
- Having the needle alone with the charge applied directly to the collecting solution (Figure 3B)
- Applying the charge to a solitary needle (Figure 3C).

In the case of a parallel plate electrode set up (Figure 3A) the electrostatic force exerted on a needle was found by modifying the expression obtained by Taylor [9,13]:

$$F_e = \pi\varepsilon_0 V^2 L^2 / H^2 [\ln(2L/r_0) - 3/2] \tag{3}$$

in which L is the height of the needle in the electric field, H is the electrode distance, V is the applied voltage, and ε_o is the permittivity of the air.

In the case of a charged needle (Figure 3C), the stress produced by the external field at the needle tip is obtained using a modified expression developed by DeShon and Carlson [14]:

$$F_e = 4\pi\varepsilon_0 \left[\frac{V}{\ln(4H/r_0)}\right]^2 \tag{4}$$

Substituting Equation (3) into Equation (2) results in a relationship describing the effect of the applied potential on the droplet radius for the parallel plate arrangement:

$$r = \left\{[3/4\rho g][r_0\gamma - (\varepsilon_0 V^2 L^2)/(H^2[\ln(2L/r_0) - 3/2])]\right\}^{1/3} \tag{5}$$

Similarly, the effect of applied potential on the droplet radius for the charged needle arrangement can be derived by substituting Equation (4) into (2):

$$r = \left\{[3/2\rho g][r_0\gamma - (2\varepsilon_0 V^2)/(\ln(4H/r_0))]\right\}^{1/3} \tag{6}$$

For the positively charged collecting plate and grounded needle (Figure 5B) an empirical relationship was obtained from the expression given by Hommel et al. [15]:

$$r = \left\{[3/2\rho g][r_0\gamma - (\varepsilon_0 V^2 4\cos(\alpha))/H]\right\}^{1/3} \tag{7}$$

in which α is the angle defined in Figure 3B.

The last three equations can be used to calculate the droplet size as a function of applied voltage for the three electrode geometries studied.

A comparison was made between the measured and calculated droplet/microbead diameters. The general shapes of the calculated curves were similar to those of the experimental curves. Very good agreement between experimental and calculated data was achieved with the 22 gauge needle in all three charge configurations. The calculated droplet diameter, for the range of applied potentials studied (2-12 kV), agreed well with the experimental data with an error of ± 15%. Equally successful was the agreement for the 26 gauge needle, where the calculated and experimental data agreed within an error of ± 10%.

As the voltage increased beyond a critical point (known as the minimum spraying potential), a transition from the dripping mode, where individual droplets could be seen to come off the end of the needle, to a high frequency spraying mode (*i.e.* liquid jet) was observed with all three charge arrangements. The minimum spraying potential for the positively charged needle set-up was observed to be lower than that for the parallel and charged plates. The process of droplet formation suggests that the forces due to the presence of an external electric field and the surface charge are responsible for the instability of the droplets at the needle tip. The charges, which are distributed on the liquid surface, repel each other and cause a force opposing the surface tension. In the case of the positively charged needle, the conducting liquid and needle are in close contact bearing approximately the same potential difference [16]. The electric field (E) at the meniscus is therefore proportional to the applied voltage, V, and radius of the meniscus, r_o. The area over which the surface charge operates varies with $1/r^2$. Due to the small area available for charge distribution (the needle tip) the overall surface charge would be higher for this electrode arrangement, than for the two other set-ups at the same potential difference. For the parallel plate set-up, the uniform electric field between the two electrodes was proportional to the potential difference and electrode spacing, H. Therefore, a much higher potential difference was required to build a sufficiently large charge on the plate to initiate spraying (*i.e.* 8 kV for the parallel plate versus 5 kV for the positive needle set-up).

The positively charged plate, with grounded needle, is the reverse arrangement of the positively charged needle. Results with this geometry show that the reversed polarity had an impact on bead size with the charged plate set-up giving smaller bead sizes. This may be due to the larger area over which the surface charge has to spread (*i.e.* the area of the hardening solution, $CaCl_2$). In this situation the charge density on the forming drop would be lower than in case of a positively charged needle, where the charge area was limited to the liquid meniscus at the needle tip. The net effect being a weaker force pulling the droplet from the tip of the needle.

3. Culture of encapsulated insect cells

Insect cell culture has received an increased amount of attention recently since these cells are hosts for a class of viruses, the baculoviruses, which has been shown to be an excellent vector for genetic engineering [17]. This is largely due to the high expression rate and post-translation processing capabilities of the baculovirus. Such processing includes efficient secretion, proteolytic cleaving, phosphorylation, N-glycosylation, and possibly myristylation and palmitylation.

The main objective of a study by King *et al.* [18], was to investigate whether insect cells (*Spodoptera frugiperda*) infected with a temperature-sensitive mutant of the

Autographa californica Nuclear Polyhedrosis Virus (AcNPV ts-10) could be cultured to a high cell density in alginate/PLL microcapsules and whether the virus could be successfully grown and concentrated within the capsules. Ultimately, nonvirus genes were to be inserted into the temperature-sensitive virus in order to measure their expression and concentration in the microcapsule system.

3.1. GROWTH OF CELLS AND VIRUS IN MICROCAPSULES

The single- and multiple-membrane capsules produced using an initial alginate concentration of 1.4% were spherical in shape and showed no surface irregularities. The cells in the single-membrane capsules remained dispersed throughout the volume of the capsule and, after two days in culture, became enlarged and dark in colour. Recovery of some of the cells from these capsules and subsequent staining with trypan blue indicated that few cells were living. In contrast, the cells in the multiple membrane capsules tended to settle to the bottom of the capsule, indicating that the viscosity of the intracapsular solution was lower. After two days in culture, these cells appeared to be healthy (supported by trypan staining) although little sign of growth could be seen. The doubling time of insect cells is between 17 and 24 h.

The capsules which were produced using a 0.7% alginate/TC100/cell mixture were non-spherical in shape and tended to have pointed ends, tails and creased sides. This non-sphericity, due to the low alginate concentration and hence low viscosity was most prevalent when the alginate concentration was 0.5% and was significantly reduced when increased to 0.6% or 0.7%. The capsules produced with a single, high-molecular-weight cut-off membrane (and either 0.7% or 1.4% alginate) tended to have weak membranes, which broke easily and, consequently, allowed the cells within to escape and proliferate in the growth medium. In those few capsules which did not rupture, the infected cells grew in isolated clumps on the inside surface of the membrane. Intracapsular cell densities and virus titres were not measured in these capsules.

Table 1. Comparison of growth rates of encapsulated and suspended insect cells [18].

Culture system	Specific growth rate (day^{-1})	Doubling time (h)
Suspended cells		
Shaker flasks	0.66	25
Monolayer (T-flasks)	0.78	21
Immobilised cells		
Microcapsules	0.55	30

The capsules that possessed a multiple membrane (initial alginate concentration 0.7%) were much stronger and more flexible than their single-membrane counterparts as judged by pinching the capsules with fine tipped forceps. Consequently, there were fewer ruptured or broken capsules and significantly reduced numbers of cells were found in the supernatant. Cells grew and virtually filled the microcapsules (Figure 4) reaching maximum densities of 8 x 10^7 cells/ml. In comparison, maximum suspension culture densities were at least 10 times lower. The specific growth rate of encapsulated cells was 0.55 day^{-1}, and was comparable (though slightly lower) to the specific cell growth rates in shaker flasks, 0.66 day^{-1}, and monolayer culture, 0.78 day^{-1} (Table 1).

The $TCID_{50}$ assay revealed that the titre of virus (ts10) in the capsules reached $ca.$ 1 x 10^9 infectious units (IFU)/ml ($ca.$ 20 IFU/cell) and that of the supernatant was lower by a factor of 300. This indicates that virtually all of the virus (more than 99%) was retained within the capsules.

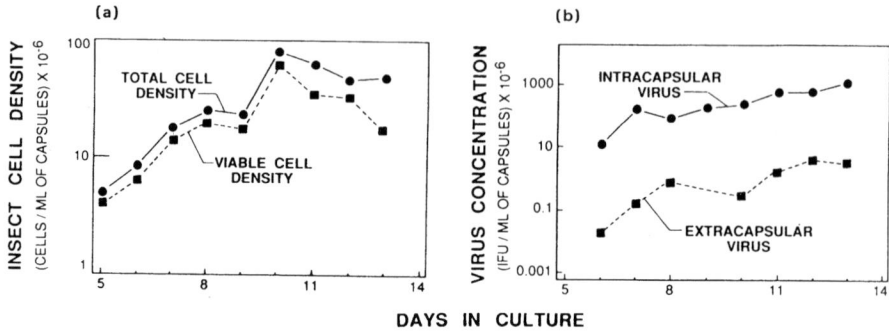

Figure 4. A. Intracapsular insect cell density as a function of time. Batch culture consists of 1-ml capsules in 30 ml medium; B. Intracapsular and extra-capsular virus concentration as a function of culture time. At day 5, the culture temperature was dropped from 33 to 27 °C to initiate virus growth in the infected insect cells. Reprinted from Goosen 1996 [26].

Table 2. Biocompatibility tests: Cells suspended in calcium alginate[18].

Alginate concentration	Cell viability
1.5% (w/v)	All cells enlarged, and dark most dead
1.3% (w/v)	All cells enlarged, and dark most dead
1.0% (w/v)	Some cells dark and granular while some are round and healthy; some cells have multiplied to form small masses
0.75% (w/v)	Good cell growth many cell masses
0.50% (w/v)	Good cell growth many cell masses

3.2. BIOCOMPATIBILITY OF ENCAPSULATION SOLUTIONS

The KCl, CHES, $CaCl_2$ and citrate solutions had no apparent effect on the growth of the cells. However, while exposure of the cells to PLL of $\overline{M}_v = 2.2 \times 10^4$ resulted in virtually no loss of cell viability, the cells exposed to PLL of $\overline{M}_v = 2.7 \times 10^5$ showed a 76% loss of viability. Cells exposed to alginate solutions showed a decrease in viability, which was dependent on the alginate concentration. On mixing the 2X TC100 and the alginate solutions, a gel-like material was produced, due, presumably, to the interaction between the calcium ions in the 2X TC100 (2.64 g/l) and the alginate. The gel tended to be more solid when the alginate concentration was 1.5%, 1.3%, and 1.0%, and was more liquid at 0.75% and 0.5% alginate. Cells suspended in mixtures of sodium alginate and

TC100 [final concentrations of 1.5%, 1.3%, and 1.0% (w/v)] showed little or no growth and usually appeared dark and granular after 2-3 days (Table 2). Trypan blue staining indicated that few if any cells were viable. Cell growth, however, was observed in the lower alginate/ TC100 mixtures [concentrations of 0.75% and 0.5% (w/v)].

Insect cells could not be cultured in either single- or multiple-membrane capsules when the initial intracapsular alginate concentration was 1.4%. The biocompatibility tests (Table 2) support these results. Only at alginate concentrations of 0.75% or less was cell growth observed. In the biocompatibility tests cell density could not be measured directly since the cells were immobilised in calcium alginate gel. It can be postulated that the mechanism of cell growth inhibition may involve electrostatic interaction and/or viscosity effects between alginate and cells.

Cells encapsulated using single membrane capsules (molecular-weight cut-off of *ca.* 6×10^4) grew poorly. In this case, the membrane was relatively impermeable and, thus, only a small fraction of the alginate, with molecular weight of 3.5×10^5, could diffuse through. Cells encapsulated in single membrane capsules (molecular-weight cut-off of *ca.* 3×10^5) grew much better (though in isolated clumps) presumably because there was increased mass transfer and because more of the intracapsular alginate could diffuse out. Similar clumped cell growth has been observed by other workers [5], who cultured hybridoma cells in single membrane microcapsules. The capsule membrane was also weak, causing some of the capsules to collapse and break, allowing the cells to escape into the medium. The capsule membrane molecular-weight cut-off was determined by incubating microcapsules (without cells) in solutions containing a protein standard. The protein concentration in the extracapsular (supernatant) solution was monitored and the membrane was deemed impermeable to the protein if there was 100% rejection of the diffusing protein over a two-hour period.

Multiple-membrane capsules, on the other hand, were significantly stronger than their single-membrane counterparts. Better cell growth was obtained with the former capsules, presumably due to the lower intracapsular alginate content. Why similar cell growth is not obtained in the high-molecular-weight, single-membrane capsules as occurs in the multiple-membrane capsules, is not clear since the intracapsular environment (*i.e.* viscosity) is believed to be virtually identical in both capsule types. One may postulate that the second PLL reaction in the multiple membrane procedure may further reduce the intracapsular alginate concentration in the bulk of the capsule by reacting with the alginate chains near the inside surface of the membrane. It is also unclear whether we have two distinct alginate/PLL membranes or one interacting membrane. The term multiple membrane refers more to the method of capsule preparation rather than to the type of final membrane.

When insect cells were exposed, for one minute, to PLL (0.05% with $\overline{M}_V = 2.7 \times 10^5$), 76% were killed. In contrast, cells exposed to PLL ($\overline{M}_V = 2.2 \times 10^4$) (0.05%) for one minute remained 100% viable. It was possible that some fraction of the PLL may have bound to the glass beaker in which the PLL was dissolved rather than remaining in solution thus lowering the bulk PLL concentration, or perhaps PLLs of different molecular weight may bind to the cells differently.

The ability to control virus replication by lowering the culture temperature has allowed the growth and concentration of virus inside of cell-filled microcapsules. Comparison of the number of IFU per cell obtained in the present study (*ca.* 20 IFU/cell) to that obtained by Knudson and Tinsley [4] (150-500 IFU/cell) suggests that our system is operating far below its optimum. Based on the data of Knudson and Tinsley, intracapsular virus concentrations should, when optimised, be approaching 2.5×10^{10} IFU/ml capsules.

4. Encapsulation and growth of plant tissue in alginate

Animal cell suspensions were successfully extruded using the electrostatic droplet generator. The application of this technology to plant cell immobilisation has only recently been reported [19-21]. A major concern in cell and bioactive agent immobilisation has been the production of very small microbeads so as to minimise the mass-transfer resistance problem associated with large-diameter beads (>1000 µm).

Somatic embryogenesis is a new plant tissue culture technology in which somatic cells (*i.e.*, any cell except a germ or seed cell) are used to produce an embryo (*i.e.*, plant in early state of development) [22]. The technique of somatic embryogenesis in liquid culture, which is believed to be an economical way for future production of artificial seeds, may benefit from cell immobilisation technology. Encapsulation may aid in the germination of somatic embryos by allowing for higher cell densities, protecting cells from shear damage in suspension culture, allowing for surface attachment in the case of anchorage dependent cells, and being very suitable for scale-up in bioreactors.

The long-term objective of the project reported in this section is the development of an economical method for the mass production of artificial seeds using somatic embryogenesis and cell immobilisation technology. The short-term aim was to investigate the production of small alginate microbeads using an electrostatic droplet generator. Callus tissue from Carnation leaves was also immobilised and cultured.

Immobilised callus cells from carnation leaves retained viability as observed by cell growth and plantlet formation [19-21]. In a related study, Shigeta [23] was able to germinate and grow encapsulated somatic embryos of carrot using a 1% sodium alginate solution, as compared to a 2% alginate solution used in the present investigation. The main findings of our experiment, though, indicated that somatic tissue could be electrostatically extruded and aseptically cultured while retaining viability.

Plantlets obtained from 4% alginate beads on agar, originally immobilised at 10 kV, 6 cm distance, were transferred to sterilised potting mixture at two months culture. The plantlets grew well and showed complete leaf and root development by four months [20]. Suspension culture of encapsulated somatic tissue was less successful. Piccioni [24] in a study investigated the growth of plantlets from alginate encapsulated micropropagated buds of M.26 apple rootstocks. He showed that the addition of growth regulators (*e.g.*, indolebutyric acid) to the somatic tissue culture several days prior to the encapsulated, as well as the addition of the same regulators to the encapsulation matrix, improved the production of plantlets in suspension culture from 10% to more than 60%. We can speculate that culturing the Carnation leaf callus tissue in the presence of growth regulators prior and during encapsulation may enhance the production of plantlets from suspension culture.

Electrostatic droplet generation does not appear to have a negative impact on somatic tissue viability since cell growth and plantlet formation was observed. This is in agreement with similar studies reported for insect cells [18] and mammalian cells [25], where it was shown that high electrostatic potentials did not affect cell viability.

In closing, microencapsulation technology is an exciting area to work in. To be able to develop successful and well-understood systems requires close collaboration between scientists with different areas of expertise such as electrical engineering, microbiology, biochemistry, biomaterials and mathematical modelling. Over the next decade we can expect to see many new areas of application of microencapsulated living cell systems such as in the treatment of diseases requiring tissue transplantation, as well as in bioprocess engineering for the production of high value biologicals.

5. References

[1] Goosen, M.F.A. (2002) Microencapsulation methods: Chitosan and alginate. In: Atala, A. and Lanza, R.P. (Eds.) Methods of Tissue Engineering. Academic Press, San Diego. Chapter 76; pp. 857-874.
[2] Inlow, D.; Harano, K. and Maiorella, B. (1987) 194th ACS Meeting, Division of Microbial and Chemical Technology, New Orleans, LA.; abstract 70.
[3] Hink, W.F. (1982) In: E. Kurstak (Ed.) Microbial and Viral Pesticides. Marcel Dekker, New York; pp. 493-506.
[4] Knudson, D.L. and Tinsley, T.W. (1974) Replication of a nuclear polyhedrosis virus in a continuous cell culture of *Spodoptera frugiperda*: purification, assay of infectivity, and growth characteristics of the virus. J. Virol. 14: 934-942.
[5] Posillico, E.G. (1986) Microencapsulation technology for large-scale antibody production. Bio/Technol. 4(2): 114-121.
[6] Balachandran, W. and Bailey, A.G. (1982) The dispersion of liquids using centrifugal and electrostatic forces. In: IEEE Meeting of the Industrial Application Society; pp. 971-980.
[7] Fillimore, G.L. and Van Lokeren, D.C. (1982) Multinozzle drop generator which produces uniform break-up of continuous jets. In: IEEE Meeting of the Industrial Society; pp. 991-998.
[8] Keshavarz, T.; Ramsden, G.; Phillips, P.; Mussenden, P. and Bucke, C. (1992) Application of electric field for production of immobilized biocatalysts. Biotech. Tech. 6: 445-455.
[9] Loscertales, I.G.; Barrero, A.; Guerrero, I.; Cortijo, R.; Marquez, M. and Ganan-Calvo, A.M. (2002) Micro/nano encapsulation via electrified coaxial liquid jets. Science 295: 1695-1698.
[10] Rayleigh, Lord (1945) The theory of sound. Dover Publications, New York, Volume II; p. 372.
[11] Rayleigh, Lord (1882) On the equilibrium of liquid conducting masses charged with electricity. Phil. Mag. 14: 184-188.
[12] Bugarski, B.; Li, Q.; Goosen, M.F.A.; Poncelet, D.; Neufeld R. and Vunjak, G. (1994) Electrostatic droplet generation: Mechanism of polymer droplet formation. AIChE J. 40(6): 1026-1031.
[13] Taylor, G.I. (1966) The force exerted by an electric field on long cylindrical conductors. Proc. R. Soc. A 291: 145-155.
[14] DeShon, E.W. and Carlson, R. (1968) Electric field and model for electrical liquid spraying. J. Colloid. Sci. 28: 161-166.
[15] Hommel, M.; Sun A.M. and Goosen, M.F.A. (1988) Droplet generation. Canadian Patent 1,241,598.
[16] Sample, S.B. and Bollini, R. (1972) Production of liquid aerosols by harmonic electrical spraying. J. Colloid. Sci. 14: 185-193.
[17] Luckow, V.A. and Summers, M.D. (1987) Trends in the development of baculorvirus expression vectors. Bio/Technology 6(1): 47-55.
[18] King, G.A.; Daugulis, A.J.; Faulkner, P.; Bayly, D. and Goosen, M.F.A. (1989) Alginate concentration: A key factor in growth of temperature-sensitive baculovirus-infected insect cells in microcapsules. Biotechnol. and Bioeng. 34: 1085-1091.
[19] Goosen, M.F.A. (1999) Mass transfer in immobilized cell systems. In: Kuhtreiber, W.M.; Lanza, R.P. and Chick, W.L. (Eds.) Handbook of Cell Encapsulated Technology and Therapeutics. Birhauser and Springer-Verlag (Publ.), Chapter 2; pp. 18-28.

[20] Al-Hajry, H.A.; Al-Maskary, S.A.; Al-Kharousi, L.M.; El-Mardi, O.; Shayya, W.H. and Goosen, M.F.A. (1999) Encapsulation and growth of plant cell cultures in alginate. Biotechnol. Progress 15(4): 768-774.
[21] Goosen, M.F.A.; Al-Ghafri, A.S.; El-Mardi, O.; Al-Belushi, M.I.J.; Al-Hajri, H.A.; Mahmoud, E.S.E. and Consolacion, E.C. (1997) Electrostatic droplet generation for encapsulation of somatic tissue: Assessment of high voltage power supply. Biotechnol. Progress 13(4): 497-502.
[22] Teng, W.-L.; Liu, Y.-J.; Tsai, V.-C. and Soong, T.-S. (1994) Somatic embryogenesis of carrot in bioreactor culture systems. Hort. Sci. 29 (11): 1349-1352.
[23] Shigeta J. (1995) Germination and growth of encapsulated somatic embryos of carrot for mass propagation. Biotechnol. Tech. 10(9): 771-776.
[24] Piccioni, E. (1997) Plantlets from encapsulated micropropagated buds of M.26 apple rootstock. Plant Cell, Tissue and Organ Culture 47: 255-260.
[25] Goosen M.F.A.; O'Shea, G.M.; Gharapetian, M.M. and Sun, A.M. (1986) Immobilization of living cells in biocompatible semipermeable microcapsules: Biomedical and potential biochemical engineering applications. In: Chiellini E. (Ed.) Polymers in Medicine. Plenum Publishing, New York; pp. 235-246.
[26] Goosen, M.F.A. (1996) Microencapsulation of living cells. In: Willaert, R.G.; Baron, G.V. and De Backer, L. (Eds.) Immobilized Living Cell Systems: Modelling and Experimental Methods. Wiley, Chichester. Chapter 14; pp. 295-322.

PLANT CELL IMMOBILISATION APPLICATIONS

RYAN SODERQUIST AND JAMES M. LEE
Department of Chemical Engineering, Washington State University Pullman, Washington 99163-2710, USA – Fax: +1-509-3354806 – Email: jmlee@wsu.edu

1. Introduction

For decades plants have played an important role as a source of raw materials for all kinds of products. After Hamberlandt first successfully cultivated single plant cells in 1902, Murashige and Skoog in 1962 capitalised on the findings of several other early researchers to develop a well-defined growth medium for plant cells [1]. The foundation was then in place for the manipulation of plant cells as a machine for the production of several raw materials and valuable secondary metabolites.

Once plant cells were recognized for their ability to produce secondary metabolites, researchers in the 1970's invested large efforts into the cultivation of plant cells for the production of these metabolites and other materials of industrial interest. Soon the applications of immobilisation techniques to enhance the production of secondary metabolites were developed and several reviews [1-4] were published in the 1980's illustrating the potential of plant cell immobilisation techniques.

Parallel to these developments, the era of DNA technology was also emerging in the 1970's and techniques for the manipulation of DNA and subsequent protein production in foreign host cells were blossoming. When early recombinant products such as insulin and human growth factor were successfully manufactured, research efforts and industrial development in this area also greatly increased. The merger of plant cell culture with recombinant protein production techniques was then imminent and the benefits of immobilised plant cells to increase recombinant protein production are now being recognized.

While microbial systems were manipulated for the production of several valuable products on an industrial scale, it was soon found that the production of eukaryotic proteins in prokaryotic cells was limited. In general, microbial systems lacked the machinery to correctly fold, process, and secrete the proteins [5]. Yeast cells posed similar problems, even though they were able to perform some post-translational modifications that were impossible for bacterial cells, they still carried out post-translational modifications in a different manner than 'higher' cells [5]. New efforts, which turned to the use of mammalian and insect host systems for the production of complex eukaryotic proteins, were successful and are the current focus of many research efforts.

Plant cells, however, offer another alternative to the use of mammalian and insect systems and could prove to be an effective vehicle for recombinant protein production in the future. Mammalian systems are expensive, very shear sensitive and must be grown in expensive serum-containing media. Insect cells avert many of the obstacles of mammalian cells; however, the baculovirus expression system leads to the destruction of the insect cells, which ends their reproductive potential and subsequently releases contaminating protein along with the protein of interest into the culture. Plant cells currently express significantly lower levels of protein than either of the above systems, but that difference is mitigated by lower media costs, simpler purification, and the lack of human pathogens in the plant cell system [5].

Immobilisation techniques with plant cells offer the potential of overcoming many of the current obstacles in secondary metabolite and recombinant protein production from plant cells. This review will highlight some of the advances that have been made over the years in secondary metabolite production and explore the more recent advances in the production of recombinant proteins from plant cells that have stemmed from the use of immobilisation techniques.

2. Significance of plant cell immobilisation

Many of the secondary metabolites produced by plants are of industrial interest, and many of the medicinal properties of plants have been recognized for centuries. Subtropical and tropical plants have been used as remedies for thousands of years. The compounds of interest in these plants are often alkanoids or terpenoids that are nearly impossible to artificially synthesize. Other potential secondary metabolites, or non-growth associated chemicals of importance from plants include: flavours, dies, aromas, herbicides and insecticides [6]. It is estimated that approximately 25% of all prescriptions contain some active compound derived from plants [7].

There are many similarities between plant cell cultures that produce the above secondary metabolites and plant cell cultures that express foreign, or recombinant, proteins. Researchers have successfully utilized plant cell culture techniques to secrete biologically active mammalian proteins that are of clinical relevance, such as human interleukin-2 and human interleukin-4 as well as human granulocyte macrophage colony-stimulating factor [8].

It is thus apparent that any technique that could lead to an improvement in secondary metabolite or recombinant protein production from plant cells would lead to significant benefits on a clinical and an industrial scale. There are some inherent problems of plant cells, such as a requirement for dedifferentiation in order to produce some substances, which cannot be necessarily solved by immobilisation techniques. On the other hand, plant cell cultures for the production of secondary metabolites and recombinant proteins both suffer from a common list of obstacles that could be overcome with the aid of immobilisation techniques. Some of the problems that could be reduced, or even mitigated, through the use of immobilisation techniques include: slow growth of cells, low shear resistance of cells, intracellular products, low yield of product, cell aggregation, and genetic instability of the cell line [2].

3. Methods of plant cell immobilisation

The common techniques of plant cell immobilisation include: gel encapsulation (or entrapment), physical entrapment, surface immobilisation, and some other less common techniques. Physical entrapment uses a boundary, such as a hollow fibre tube, to separate the cells from the media. Nilsson *et al.* [9] applied a bead polymerisation technique using hydrophobic phases, such as soy, paraffin, and silicon oil to plant cells that resulted in preserved viability of the immobilised cells. Other reviewed plant cell immobilisation techniques include methods based on adsorption or spontaneous aggregation of the cells [1].

Surface immobilisation involves attaching the cells to a solid support such as glass or polyurethane foam. Surface immobilisation minimizes microbial contamination, because it is a single-step process, and it is a relatively natural method that is harmless to the cells. Unfortunately, surface immobilization can take as long as 10 to 24 days [1]. Lindsey *et al.* [10] gives a detailed description for the use of polyurethane foam particles to entrap *Capsicum frutescens* and *Daucus carota L.* species and found that there was still a 70-80% viability of the cells after 21 days.

Tyler *et al.* [11] found that surface immobilisation promoted the natural tendency for plant cells to aggregate, which might improve the synthesis of secondary metabolites. Ishida [12] successfully increased the total diosgenin production from *Dioscorea deltoida* cell cultures by 40%, with the use of surface immobilisation with polyurethane foam cubes. Archambault *et al.* [13] developed a novel surface immobilisation technique in which the inoculum suspension of plant cells immobilised in the first 24 to 48 hours. This technique involved the attachment of cells to a manmade fibrous material and showed a great deal of promise for the production of secondary metabolites from these plant cells. Alternatively, a method that utilizes loofa, or *Luffa cylindrica* sponge has recently shown the same immobilisation capacity as polyurethane foam for the immobilisation of plant cells [14].

Entrapment in a gel matrix, or gel encapsulation, is the most widely studied method for the immobilisation of plant cells due to its mildness, ease of operation and wide applicability [15]. Plant cells are ideally grown in suspension until they reach the exponential phase and are then mixed into a viscous slurry, with an encapsulating agent, or gel. In general, fine cell suspensions are desirable, and methods to obtain fine cell suspensions can easily be obtained [1]. Various choices for the gel include: agar, alginate, agarose, carrageenan, collagen, pectate, gelatine, polyacrylamide, and combinations of these substances [5,15]. There are two basic types of gels: ionic gels that harden in the presence of a cross-linking agent, and thermal gels that harden when they are cooled. The ionic gels are typically stronger, since they are held together by ionic or covalent bonds, whereas thermal gels are usually only held together by hydrogen bonds. To avoid contamination, the gelling agent should be autoclaved before the addition of a plant cell suspension; however, this can lead to a decrease in the ability of the agent to form a gel. This decrease in the gel formation potential is not always observed and is somewhat dependent on the gelling agent. Researchers have successfully used a lower maximum autoclave temperature of 115°C instead of 121°C to avoid microbial contamination and still obtain firm alginate beads [1].

The viscous slurry is dispersed drop-wise into a gelling agent that causes beads to form. In the case of alginate gels, this is typically a 50 mM solution of calcium chloride. There is some variability in the amount of time that the beads should be allowed to sit in the cross-linking agent, and times anywhere from 5 minutes to 12 hours have been reported. During this hardening procedure, decreases in the gel volume by up to 40% of the original drop volume have been observed [1]. Two-phase solutions, such as an oil/water system, can be used, but in this method the drops must be rapidly dispersed into the solution, which results in a large distribution of various bead diameters.

Once the viscous slurry, or pre-gel solution, has solidified, water-soluble substrates can pass through the beaded gel matrix to provide the immobilised cells with the necessities of life. In general, a film with a higher cell density resides at the wall of the beads [15]. The encapsulating beads may be effectively combined with other strategies, such as inducible promoters or bioreactors [5]. This method is also reversible so that the cells can be investigated in a free suspension subsequent to their confinement in an immobilised state [16]. Some research has been devoted to the mass transfer limitations of this process, especially when the oxygen requirements of the cells are a limiting factor [17-19].

Brodelius and Nilsson [16] reported that alginate, agarose, agar, and carrageenan are the most suitable agents for plant cell entrapment. Alginate gels require Ca^{2+} as a stabilizer, which is frequently bound by chelating agents, especially phosphate that is a component of many types of plant media. It has also been found that decreasing the phosphate level in the media reduces the cell leakage. Another alternative would be to supply the Ca^{2+} ion to the culture at regular intervals, so that the alginate beads retain their stability. Care should be taken when using Ca^{2+} for the immobilisation of plant cells, since it is a trigger for many metabolic processes. The beads can be dissolved by adding EDTA, phosphate or other chelating agents to release the cells without any harm so that they can be used in further investigations [1].

There are several works that report the successful immobilisation of plant cells in a calcium alginate gel [8,16,19-23]. Other gel matrices have been used with varying levels of success [1,8] and reports continue to exhibit that alginate gels are the most compatible for the immobilisation of plant cells [8,16,24]. Since an alginate gel is not a thermal gel, there is little or no 'heat shock' for the plant cells and the added strength of the ionic bonds in the calcium alginate gels imparts a greater deal of stability to the beads. Thermal gels, such as carrageenan or agar, must be heated to temperatures in excess of 50°C. In general any temperature above 40°C can be detrimental to a plant cell. A recent study has shown that thermal gels inhibit the production of recombinant proteins from immobilised plant cells, possibly due to heat inactivation of the genes that code for recombinant protein production [8].

Physical parameters of beads formed from an alginate gel differ, yet these differences do not always lead to changes in productivity. The diameter of the immobilised beads appears to be an inconsequential consideration on a small scale in recombinant protein production from plant cells. Immobilised bead diameters of 4,7 and 8 mm resulted in no significant difference in the recombinant protein production for most of the batch. However, cultures with larger beads of 7 and 8 mm maintained higher protein concentrations near the end of the batch [8]. Thus in scaling up the process larger bead diameters could be more efficient, especially when operating in a continuous mode.

Bodeutsch et al. [8] explored the effects of the initial cell concentration in the beads on the recombinant protein production. The results did not find any significant increase in the recombinant protein production with a higher cell concentration in the beads, but it was concluded that an increase in the volume fraction of beads from a 6% to a 26% (vol/vol) ratio of beads in the solution lead to a higher protein production in the same amount of time, although the peak of the protein production was delayed. Studies on cell immobilisation with alginate gels have indicated that it is possible to alter the parameters of the alginate and Ca^{2+} concentrations in order to minimize compression of the beads, encourage high substrate transfer rates and slow the leakage rate of cells from the beads [19].

4. Important considerations for plant cell immobilisation

As the techniques of plant cell immobilisation advance there are aspects that must be accounted for in future research considerations [1]. The immobilisation procedures have an extra risk for contamination and would require investments on an industrial scale to comply with regulatory requirements. Plant cells exhibit slow growth, a tendency to aggregate, and are currently hindered by low levels of production, thus the economic advantages over mammalian and insect systems would require consideration on a case by case basis. Plants can also show genetic instability over time, which would require special regulatory provisions. Since some plant cell products tend to be stored in the cell, which complicates downstream processing and inhibits continuous processes, products excreted from the cells exhibit the greatest industrial potential.

When beads are used to immobilise plant cells, there are some mass transfer considerations. The diffusion rates of various substrates into an alginate gel, *versus* the substrate consumption rates of the free cells, have been measured in order to analyse the effect of diffusion limitations within the beads [17]. The results have shown that the influence of diffusion resistance within the beads was negligible for most substrates; however, over long periods of time it was shown that the oxygen concentration within the beads decreased considerably. This leads to the conclusion that oxygen mass transfer into the beads could be one of the most important scale up considerations for the cultivation of immobilised plant cells. Many of the techniques used to overcome oxygen mass transfer limitations in the scale up of microbial systems could also find applications in the scale up of immobilised plant cultures. Regardless of the immobilisation method used, the influence of diffusion and convective mass transfer of oxygen to the cells cannot be neglected.

It is also possible that immobilisation of plant cells could alter the physiology of the cells [25]. When cells are grown in a restricted space, such as an immobilisation matrix, they suffer from compressive stresses that could influence the biochemical differentiation and secondary metabolism. It should be noted that secondary metabolite production is usually observed once the cells have reached the stationary phase, thus the 'crowding' effect of these compressive stresses could lead to enhanced secondary metabolite production. On the other hand, recombinant protein production from plant cells tends to occur while the cells are in the exponential phase where growth is relatively unrestricted. This effect is counteracted by the fact that cellular growth within the matrix is much slower than in a free suspension which leads to a prolongation of the exponential phase and hence an enhanced production period. Regardless of whether secondary metabolite or recombinant protein production is the goal, a working knowledge of how the growth kinetics influence the product production can be used to increase the yield from immobilisation techniques.

The immobilisation matrix itself can act as an inducer or a repressor for certain metabolic processes. The matrix can also act as a physical barrier for the formation of plasmodesmata between cells and cause an alteration of the natural age distribution of cells, which exhibit a radial distribution of new cells on the periphery. Unnatural concentration gradients can also be formed within the beads. An effective framework for understanding the above factors is given by Bringi and Shuler [25]. All of these factors could potentially influence the cellular metabolism, which might or might not be the desired outcome of the immobilisation. These potential changes in the cellular metabolism lead to an increase in the probability of unexpected results. This could increase the complexity of regulatory guidelines on an industrial scale for the immobilisation of plant cells.

It is true that the full potential of plant cell immobilisation cannot be realized until all of the biological parameters are clearly understood. Advances in the new biotechnology era are unlocking the doors that lead to a better understanding of the mechanisms involved in the production of secondary metabolites and recombinant proteins from plant cells. As the general understanding of the mechanisms involved in plant physiology under immobilisation conditions advances, and technologies to account for many of the considerations above are developed, plant cell immobilisation will become a useful technique in industrial applications.

5. Potential benefits of plant cell immobilisation

Many benefits of immobilisation have already been realized for increasing the production of secondary metabolites, and more recently recombinant proteins, from plant cells. Earlier in this review the inherent plant cell culture problems including slow growth, low shear resistance, the formation of intracellular products, low yield of product, cell aggregation, and genetic instability of the cell line were mentioned. Immobilised plant cell cultures offer a number of potential advantages that could mitigate or reduce several of the above problems. The potential benefits of immobilisation in overcoming these obstacles will be discussed and some of the more recent improvements recognized from plant cell immobilisation will be mentioned.

Shuler has listed some of these benefits in detail [6]. Immobilisation represses the cell replication, thus problems due to genetic instability can be significantly reduced. It is usually possible to keep the cells in a resting state after immobilisation, which leads to a reduction in the number of cell divisions required for the production of a fixed amount of product and slows the rate of genetic instability [2]. The size of the cell aggregates can also be controlled through immobilisation, which is something that cannot be achieved in a regular suspension culture.

Immobilisation techniques with plant cells have demonstrated a capacity to improve upon low secondary metabolite yields and elicit the excretion of these metabolites into the surrounding medium for many different plant species. As mentioned previously, immobilisation of *Dioscorea deltoidea* cell cultures in polyurethane foam increased the total diogenin production by 40% [12]. Cultures of scopolin-producing *Nicotiana tabacum* plant cell suspensions normally require cell disruption for the recovery of scopolin. Immobilisation of these cultures within calcium alginate gel beads resulted in diffusion of scopolin throughout the gel matrix and a large fraction of scopolin was recovered from the culture medium without disrupting the cells [21]. *Eschscholtzia californica* cells released an increased amount of alkaloids after entrapment in calcium alginate beads [22]. Plant cell cultures of the Mexican species *Solanium chrysotrichum* excreted a desired metabolite into the medium after immobilisation in calcium alginate beads [20]. Navratil and Sturdik [15] have also cited well known applications of immobilised plant cells including embryo production in an immobilised celery callus system, taxol production using immobilised cells of *Taxus cuspidate*, anthraquinone production with immobilized *Cruciata glagra* cells, sanguinarine production by *Papaver somniferum* cells, and the production of scopolin by *Solanum aviculare* immobilised in calcium alginate gel beads.

Increases in the production of certain recombinant proteins from plant cells have also been attained through the use of immobilisation techniques. When *Nicotiana tabacum* cell lines producing human granulocyte-macrophage colony-stimulating factor (GM-CSF) were encapsulated in an alginate gel, the production of GM-CSF in these tobacco cells increased by approximately 50% [8]. It should be noted however that production levels of human interleukin-4 from *Nicotiana tabacum* cell lines in the same report were not improved upon with the use of immobilisation techniques, which suggests that immobilisation of recombinant plant cells will not always result in an increase in recombinant protein production.

The increase in productivity for the recombinant cell line studied by Bodeutsch *et al.* [8] can most likely be attributed to the slower growth of plant cells upon immobilisation. Recombinant protein production from transgenic plant cell cultures is closely linked to the exponential phase of growth and the slower growth that occurs upon immobilisation prolongs all of the growth phases including the exponential phase. They also hypothesized that increases in recombinant protein production upon immobilisation could be attributed to chemical interactions between the cells and the immobilisation matrix or the separation between the cells and the product that occurs as a consequence of immobilisation; however, further studies are necessary to substantiate these hypotheses. The framework proposed by Bringi and Shuler [25], for understanding the effects on cellular physiology and differentiation of plant cells upon immobilisation, could also lead to a better understanding of the mechanisms responsible for productivity increases after immobilisation.

There are also procedures that can induce immobilised cells to release a product without killing the immobilised cells, which can overcome the obstacles created by the formation of intracellular products [2]. Since immobilised cells are frequently entrapped in a polymer gel they can be protected against shear forces. This not only expands the opportunities for the use of simpler bioreactors in the cultivation of these cells [2], but also enables the engineer to choose fluid mixing characteristics that are based solely on mass transfer considerations [6] without a great concern over shear effects.

A significant benefit realized through immobilisation is the decoupling of the growth and production phases for immobilised plant cells. Perhaps the most common example of the potential of immobilised cell cultures is the shikonin process [2], in which there are two stages of growth. In the first stage the conditions are optimised for cellular growth, and in the second stage the conditions are optimised for production. The fact that immobilised plant cells can be transferred from a growth medium to a production medium can be a powerful tool in overcoming the slow growth conditions for many plant cells. Also, since immobilised plant cells can be started with a higher seeding density of cells that are in the proper growth phase, they are more compatible with continuous reactor systems.

Although care must be taken to avoid contamination, the relative ease in transferring the beads from one medium to another is an advantage that should not be underestimated. Now that the techniques to insert an inducible promoter into plant cells are available, there are potential benefits of using inducers to enhance the protein production from plant cells. The goal of inducible protein production is to stimulate a large burst of production, once the cells have reached an optimum biomass concentration or growth phase. Excreted plant proteins are frequently unstable in the harsh and dilute environment of the growth media and are lost over time, yet under induction conditions the total protein production can be achieved in a short amount of time to avoid time dependent protein loss. When a standard suspension culture is used, there are several days of unproductive growth, and frequently a medium exchange is required to rapidly expose the cells to induction conditions [5]. Beads can be made up with concentrations of cells at or near the ideal conditions and can then be easily transferred into an induction medium. Genes for antibiotic resistance can also be incorporated into plant cells in order to decrease the potential for contamination.

Beads of immobilised cells can also be reused several times, which offers great advantages not only for the proposed induction strategy above, but also for production processes in general. Continuous or semi-continuous production strategies are facilitated by the re-use of immobilised cells. The beads offer the potential for high flow rates in continuous reactor systems without concerns for cellular 'washout' [6]. Since the beads have a greater diameter than individual cells they also exhibit a more favourable settling velocity, making them more compatible with a variety of reactor systems [5].

Some noteworthy reviews [1,6,7,24] have also highlighted potential reactor systems for immobilised plant cells that include: packed-bed and fluidised-bed reactors, membrane reactors, hollow-fibre reactors and other reactors. These reactors show potential primarily for products that are secreted into the medium. Beads of encapsulated transgenic plant cells offer the potential for an effective combination with affinity chromatography bioreactors to increase the yield of excreted recombinant proteins [5]. In general, the benefits that are realized from plant cell reactor systems could be greatly increased when effectively coupled with immobilisation techniques.

6. Conclusions

The recent advancements in recombinant DNA technology have provided researchers with the tools to capitalize on the advancements that are being recognized in both of these fields. Perhaps plant cell immobilisation techniques will have their greatest impact in the so-called new product era, such as new bitter flavours and anti-cancer agents [1]. Research continues to show that the benefits of plant cell immobilisation can be used to enhance the yields of natural secondary metabolites and recombinant proteins expressed in plant cell systems. A more in depth understanding of the underlying biological mechanisms responsible for increases in secondary metabolite and recombinant protein production from plant cells upon immobilisation, will enable researchers to fully harness the potential of plant cell immobilisation in order to realize enhanced product yields on an industrial scale.

References

[1] Hulst, A.J and Tramper, J. (1989) Immobilized plant cells: a literature survey. Enzyme Microb. Technol. 11: 546-558.
[2] Brodelius, P. (1985) The potential role of immobilization in plant cell biotechnology. Trends Biotechnol. 3: 280-285.
[3] Brodelius, P. (1984) Immobilized viable plant cells. Ann. N.Y. Acad. Sci. 434: 382-393.
[4] Rosevear, A. and Lambe, C.A. (1983) Immobilized plant and animal cells. In: Wiseman, A. (Ed.) Topics in enzyme and fermentation biotechnology, Vol. 7. John Wiley & Sons, N.Y., USA; pp. 13-37.
[5] James, E. and Lee, J.M. (2001) The production of foreign proteins from genetically modified plant cells. In: Scheper, T. (Ed.) Advances in Biochemical Engineering/Biotechnology, Vol. 72. Springer Verlag, Berlin Heidelberg, Germany; pp. 127-155.
[6] Shuler, M.L.; Hallsby, G.A.; Pyne, J.W. Jr. and Cho, T. (1986) Bioreactors for immobilized plant cell cultures. Ann. N.Y. Adad. Sci. 469: 270-278.
[7] Prenosil, J.E. and Pedersen, H. (1983) Immobilized plant cell reactors. Enzyme Microb. Technol. 5: 323-331.
[8] Bodeutsch, T.; James, E.A. and Lee, J.M. (2001) The effect of immobilization on recombinant protein production in plant cell culture. Plant Cell Rep. 20: 562-566.
[9] Nilsson, K.; Birnbaum, S.; Flygare, S.; Linsefors, L.; Schroder, U.; Jeppsson, U.; Larsson, P.; Mosbach, K. and Brodelius, P. (1983) A general method for the immobilization of cells with preserved viability. Eur. J. Mircobiol. Biotechnol. 17: 319-326.
[10] Lindsey, K.; Yeoman, M.M.; Black, G.M. and Mavituna F. (1983) A novel method for the immobilization and culture of plant cells. FEBS Lett. 155: 143-149.
[11] Tyler, R.T.; Kurz, W.G.W.; Paiva, N.L. and Chavadej S. (1995) Bioreactors for surface-immobilized cells. Plant Cell, Tissue Organ Cult. 42: 81-90.
[12] Ishida, B.K. (1988) Improved diosgenin production in *Dioscorea deltoidea* cell cultures by immobilization in polyurethane foam. Plant Cell Rep. 7: 270-273.
[13] Archambault, J.; Volesky, B. and Kurz, W.G.W. (1988) Surface immobilization of plant cells. Biotechnol. Bioeng. 33: 2933-299.
[14] Kuo, L.Y.; Minoru, S.; Hideo, T. and Shintaro, F. (1998) Characteristics of loofa (*Luffa cylindrical*) sponge as a carrier for plant cell immobilization. J. Ferment. Bioeng. 85: 416-421.
[15] Navratil, M. and Ernest, S. (1999) Bioactive components in productions using immobilized biosystems. Biologia Bratislava 54: 635-648.
[16] Brodelius, P. and Nilsson K. (1980) Entrapment of plant cells in different matrices. FEBS Lett. 122: 312-316.
[17] Furusaki, S.; Nozawa, T.; Isohara, T. and Furuya, T. (1988) Influence of substrate transport on the activity of immobilized Papaver somniferum cells. Appl. Microbiol. Biotechnol. 29: 437-441.

[18] Prince, C.L.; Bringi, V. and Shuler, M.L. (1991) Convective mass transfer in large porous biocatalysts: Plant organ cultures. Biotechnol. Prog. 7:195-199.
[19] Cheetham, P.S.J., Blunt, K.W. and Bucke, C. (1979) Physical studies on cell immobilization using calcium alginate gels. Biotechnol. Bioeng. 21: 2155-2168.
[20] Charlet, S.; Gillet, F.; Villarreal, M.L.; Barbotin, J.N.; Fliniaux, M.A. and Nava-Saucedo J.E. (2000) Immobilisation of *Solanum chrysotrichum* plant cells within Ca-alginate gel beads to produce an antimycotic spirostanol saponin. Plant Physiol. Biochem. 38: 875-880.
[21] Gillet, F.; Roisin, C.; Fliniaux, M.A.; Jacquin-Dubreuil, A.; Barbotin, J.N. and Nava-Saucedo, J.E. (2000) Immobilization of *Nicotiana tabacum* plant cell suspensions within calcium alginate gel beads for the production of enhanced amounts of scopolin. Enzyme Microb. Technol. 26: 229-234.
[22] Villegas, M.; Leon, R.; Brodelius, P.E. (1999) Effects of alginate and immobilization by entrapment in alginate on benzophenanthridine alkaloid production in cell suspension cultures of *Eschscholtzia californica*. Biotechnol. Lett. 21: 49-55.
[23] Brodelius, P.; Deus, B.; Mosbach, K. and Zenk, M.H. (1979) Immobilized plant cells for the production and transformation of natural products. FEBS Lett. 103: 93-97.
[24] Shuler, ML.; Sahai, O.P. and Hallsby, G.A. (1983) Entrapped plant cell tissue cultures. Ann. N.Y. Acad. Sci. 413: 373-382.
[25] Bringi, V. and Shuler, M.L. (1990) A framework for understanding the effects of immobilization on plant cells: differentiation of tracheary elements in tobacco. In: Bont, J.A.M.; Visser, J. Mattiasson, B. and Tramper, J. (Eds.) Physiology of immobilized cells. Elsevier Science Publishers B.V., Amsterdam The Netherlands; pp. 161-172.

PART 5

ENVIRONMENTAL AND AGRICULTURAL APPLICATIONS

WASTEWATER TREATMENT BY IMMOBILISED CELL SYSTEMS

HIROAKI UEMOTO

Bio-science department, Central Research Institute of Electric Power Industry, 1646 Abiko, Abiko-city, Chiba 270-1194, Japan – Fax: 81 471 83 3347 – Email: uemoto@criepi.denken.or.jp

1. Introduction

The immobilisation of whole cells is an important technique in the field of wastewater treatment for purifying water contaminated with various materials, (*e.g.* organic compounds, inorganic compounds, metals, and so on). The processes using immobilised cells have been studied and applied to practical treatment systems. The immobilisation technique simplifies treatment systems, since the immobilised cells as biocatalysts are easily separated from treated water without a settling step. The immobilised cells can be repeatedly used without washing out from the systems. The immobilisation techniques are also effective to concentrate the useful cells in as small a volume as possible. The concentrated cells in the systems can accelerate the treatment of wastewater.

The further advantages of the immobilised cells system have been studied in many fields. Cell protection, which is one of the advantages, is extremely effective in the field of wastewater treatment. Though the useful cells for the treatment are always exposed to the competition with various microorganisms and the inhibition with chemical materials derived from external inflow, high density of microorganisms and bio-polymers surround the useful cells protect them against the competition and inhibition. In case extremely slow-growing cells and sensitive cells are used to treat wastewater, the processes using the immobilised cells are more advantageous than processes with suspended free cells. Thus, immobilisation techniques are often applied to treatment processes using lithoautotrophic bacteria, *e.g.* nitrifying bacteria.

In this chapter, I present small-scale and large-scale reactor systems as environmental applications of immobilised cells, and discuss future development of these systems.

2. Immobilisation techniques

Many immobilisation techniques have been studied and applied to various fields shown in this book. These techniques are classified into four methods: adsorption onto solid inert carriers; entrapment in semi-permeable inert materials such as hydrogels, fibers or membranes; covalent coupling, including treatment with crosslinking agents; and carrierless immobilisation, including the formation of cell aggregates by natural flocculation or flocculation induced by agents [1]. In the field of wastewater treatment the immobilisation is mainly achieved by three methods: adsorption, entrapment, and carrierless immobilisation (Figure 1) [2-5].

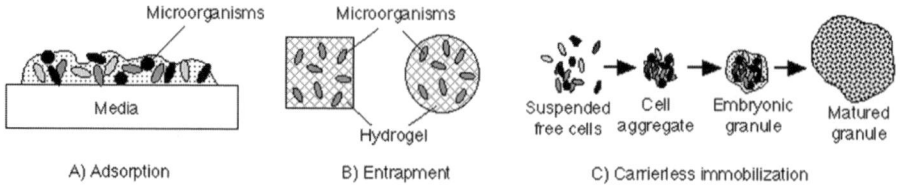

Figure 1. Immobilisation techniques for wastewater treatment.

2.1. ADSORPTION

Adsorption is the simplest and oldest technique for immobilising large quantities of cells. The cells spontaneously attach onto the various media, and form a slime layer called biofilm. Various materials (e.g. porous glass, charcoal, resin, ballast, sand, and wood chips) have been used as media [2,6-8]. Recently, plastic media built in round, square, and other shapes have been used, since they are cheap and have a large surface area. The processes using the biofilm have already been applied to many practical systems (Figure 2A, B). The tricking filter, which is one of the practical systems called fixed-bed reactor, has been used in England since 1893 [2]. Broken stones were first used as a medium in the reactor. These treatment processes using attached biofilms are mainly used to remove organic matter in wastewater, usually measured as biochemical oxygen demand (BOD). They are also used to achieve nitrification (the conversion of ammonia to nitrate via nitrite) [9-17] and metal precipitation (the production of insoluble metal phosphates) [18-19]. The adsorption technique can be used effectively without washing out cells from the system.

2.2. ENTRAPMENT

Entrapment is also a major immobilisation technique, and is applied to practical systems for wastewater treatment. The cells are entrapped with various inert materials of both natural and artificial polymers, which are called hydrogels. The natural polymers (e.g. agar, carrageenan, and alginate), which retain high cell viability, are often used in small-scale experiments [10,15-17]. However, natural polymers are inappropriate for practical applications, since they are susceptible to mechanical stress and are easily bio-degraded by various micro-organisms. The artificial polymers (e.g. polyvinyl alcohol and polyethylene glycol) are often used in practical systems [3,13,14,20]. The entrapment technique protects the useful cells against the fast-growing microorganisms and inhibitors, which are derived from the inflow. Thus, in case the extremely slow-growing cells and sensitive cells are used to treat the wastewater, the entrapment technique is effective to concentrate the useful cells selectively in the system. They are used to achieve nitrification and complex bio-conversion processes (simultaneous aerobic nitrification and anaerobic denitrification) [20-24]. The entrapment technique is rarely used for removing only organic matter by common heterotrophs, since the immobilisation cost of the entrapment technique is the highest among immobilisation techniques.

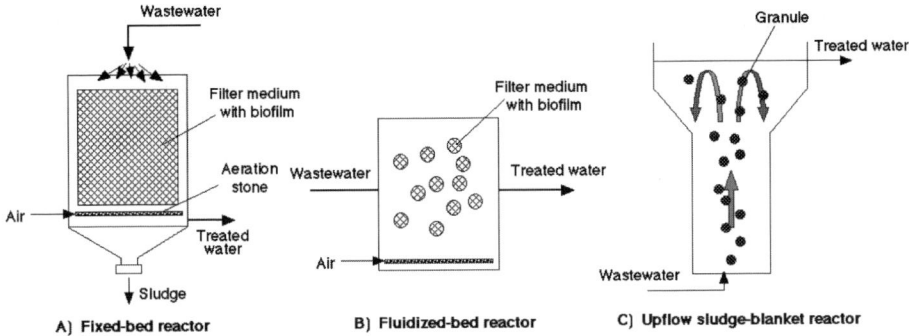

Figure 2. *Immobilised cells bioreactor for wastewater treatment.*

2.3. CARRIERLESS IMMOBILISATION

Carrierless immobilisation is used in the upflow anaerobic sludge-blanket (UASB) process [2,25]. The upward flow of wastewater in the reactor vessel promotes the formation of cell aggregates by natural flocculation (Figure 1C and 2C). The initial aggregates (embryonic granule) grow to matured granules, whose diameter is a few millimeters. The UASB process using the granules is considered desirable in high-strength organic wastewater treatment, since the granules have high biomass concentration and rich microbiol diversity. The granulation is also reported in upflow aerobic sludge-blanket (ASB) processes [25,26]. The granules are formed in both aerobic and anaerobic conditions. However, one major drawback of the carrierless immobilisation technique is its extremely long forming period, which generally requires 2-8 months for the granulation. Though many granulation models are proposed, mechanism of the granulation is poorly understood [25]. Strategies for expediting granular formation are examined to reduce the forming period. The carrierless immobilisation technique mainly applies to high-strength organic wastewater treatment. The removal of nitrogen compounds is also studied in sequencing batch airlift reactors [5].

3. Applications of immobilised cells

3.1. NITRIFICATION

Nitrogen compounds in wastewater accelerate eutrophication in lakes and coasts.

Removal of the nitrogen compounds (especially ammonia removal) is one of the main applications of immobilised cells in the field of wastewater treatment [2]. Ammonia removal from wastewater is mainly achieved by two microbial conversion steps: aerobic nitrification (oxidation of ammonia to nitrate *via* nitrite) and anaerobic denitrification (reduction of nitrate to N_2 gas) (Fig. 3) [27]. The aerobic nitrification is a cascade reaction of two lithoautotrophic microorganisms, *i.e.*, ammonia-oxidising bacteria and nitrite-oxidising bacteria, whose generic term is nitrifying bacteria [28]. Due to the slow growth rate of the nitrifying bacteria and the sensitivity of these organisms to several environmental factors (*e.g.* pH, DO and temperature), the nitrifying bacteria are often washed out from treatment plants using typical activated sludge processes and the plants frequently fail to establish stable nitrification. Thus immobilisation techniques are often applied to stabilise the nitrification, which is the key process of the ammonia removal.

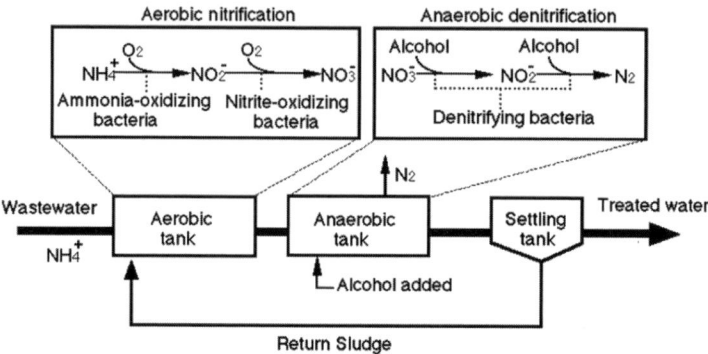

Figure 3. Ammonia removal process using suspended activated sludge.

Adsorption has generally been the preferred method of immobilisation in wastewater treatment, since it is the simplest and lowest-cost technique for immobilizing large quantities of cells. Biofilms formed by adsorption are applied to practical systems for nitrification. Various media, *e.g.* polyurethane foam, polyethylene, and polyvinyl alcohol (PVA), are widely used in various reactors [2,3,30]. The nitrification rates of the reactors were compared in previous reports (Table 1). A typical example of the applications is a cube made of polyurethane foam, which is cut into 6 mm x 6 mm x 6 mm cubes [3]. The cubes are packed into a fluidized-bed reactor (Figure 2B). While air bubbling agitates the cubes in the reactor, nitrifying bacteria attach onto the cubic surface spontaneously. Thus, the slow-growing nitrifying bacteria can be effectively used without washing out from the reactor. However, the adsorption cannot concentrate only nitrifying bacteria selectively into biofilm on the cubic medium. Nitrifying bacteria are competing with the other various microorganisms in biofilms. The population distribution in biofilms is the result of the difference in growth rate of the microorganisms. The slow-growing nitrifiers are located inside the biofilms, the fast-growing heterotrophs are located more in the outer layers of the biofilms. In cases where the heterotrophs are extremely dominant in the biofilms, diffusion limitation of O_2 slows the nitrification process.

The entrapment technique is an effective method to concentrate nitrifying bacteria selectively into treatment plants. Nitrifying bacteria are entrapped with various natural and artificial polymers in small-scale experiments [15-17]. Nitrifying bacteria entrapped with artificial polymers have already been applied at a full-scale plant for nitrification [29]. In this application, a cubic hydrogel made of polyethylene glycol is used. Activated sludge containing nitrifying bacteria (nitrifying biomass) is entrapped with polyethylene glycol, and then the hydrogel is cut into 3 mm x 3 mm x 3 mm cubes.

Table 1. Rates of nitrification and nitrogen removal by immobilised cell systems.

Immobilised cell	Immobilisation technique Medium	Nitrification rate (kg-N m^{-3} d^{-1})	Nitrogen removal rate[a] (kg-N m^{-3} d^{-1})	Reference
Biomass	Adsorption Polyurethane pellet	0.29		[30]
Biomass	Adsorption Polyurethane cube	0.57		[3]
Biomass	Adsorption Polyethylene pasta	0.53		[3]
Biomass	Entrapment Polyvinylalcohol cube	0.70		[3]
N. europaea	Entrapment Ca-alginate beads	0.26		[15]
N. europaea	Entrapment Photocrosslinkable resin beads	0.08		[16]
N. europaea	Entrapment Carrageenan beads	0.99		[17]
N. europaea and Ps. denitrificans	Entrapment Ca-alginate, Carrageenan Double-layer beads		1.54	[21]
N. europaea and P. denitrificans	Entrapment Polyelectro-Ca-alginate beads		0.61	[35]
N. europaea and P. denitrificans	Entrapment Photocrosslinkable resin Packed gel envelope	1.90	1.60	[39]

[a] Nitrogen removal rate means the rate of simultaneous nitrification and denitrification.

The cubic hydrogels containing nitrifying bacteria achieve nitrification in a fluidised-bed reactor (Figure 4). The reactor tank for nitrification is combined with reactor tanks using activated sludge for denitrification and removal of organic matter. The acclimated cubic hydrogels can accelerate nitrification in the treatment system. The rate of nitrogen removal of the system using the cubes was twice higher than that of a typical activated sludge system. PVA particles, which entrapped nitrifying biomass, were also used to accelerate the nitrification process in a fluidised-bed reactor. The volumetric nitrification rate of the PVA particle reactor was higher than those of attached biofilm reactors and suspended biomass reactors [3]. However, the active region in hydrogels is limited from the surface to 0.1-0.2 mm depth of the hydrogel, since diffusion limitation of O_2 restricts the population distribution in hydrogels [31-33]. Cubic media made by adsorption and entrapment techniques showed similar activity (Table 1). It is suggested that most of the inside region in the hydrogel is useless for the nitrification process. Further, when the hydrogel is used to remove both organic matter and nitrogen compounds, fast-growing heterotrophs often attach onto the hydrogel and cover its surface. The covered biofilm limits the diffusion of O_2 into the hydrogel. It results in the elimination of nitrifying bacteria from the hydrogel. Thus, entrapment is an unsuitable immobilisation method in high-strength organic wastewater treatment where heterotrophs are extremely dominant in the reactor.

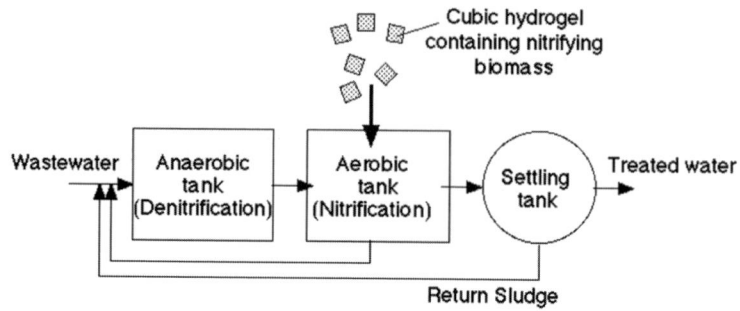

Figure 4. Nitrogen removal system using cubic hydrogels containing nitrifying biomass.

3.2. SIMULTANEOUS NITRIFICATION AND DENITRIFICATION

Ammonia removal from wastewater is mainly achieved by aerobic nitrification and anaerobic denitrification, as mentioned above. Treatment systems for ammonia removal consist of aerobic and anaerobic reactor tanks with complicated operations. The entrapment technique is used to simplify the systems.

A typical example of this application is an immobilisation bead capable of simultaneous aerobic nitrification and anaerobic denitrification [34,35]. Hydrogel beads containing both ammonia-oxidising bacteria, *Nitrosomonas europaea*, and denitrifying bacteria, *Paracoccus denitrificans* were made of polyelectrolyte complex alginate (Figure 5). Nitrification (oxidation of ammonia to nitrite) is accomplished by *N. europaea* in the outer surface of the beads. On the other hand, denitrification (reduction of nitrite to nitrogen gas) is accomplished by *P. denitrificans* in the bead core, where it is anaerobic due to the diffusion limitation of O_2. Thus, the hydrogel beads containing both *N. europaea* and *P. denitrificans* are capable of performing simultaneous nitrification and denitrification by a single step. The hydrogel beads can simplify treatment systems for ammonia removal. Double-layer beads, which consist of an outer layer for nitrification and an inside core for denitrification, were also reported to simplify treatment systems [23,24]. The outer layer containing *N. europaea* and the inside core containing *P. denitrificans* were made by two immobilisation steps. Hydrogel containing *Paracoccus pantotroph*, which has the ability of nitrification and denitrification, was studied to simplify the process for nitrogen removal [36,37]. These hydrogel beads, capable of simultaneous nitrification and denitrification, however, were examined in only small-scale experiments. *P. denitrificans* in the inside core of the hydrogel beads requires an electron donor for denitrification. Organic matter as the electron donor for denitrification stimulates fast-growing heterotrophs in a fluidised-bed reactor. Biofilm covered on the surface of hydogel bead results in diffusion limitation of O_2 into the hydrogel. The O_2 limitation slows down the nitrification process.

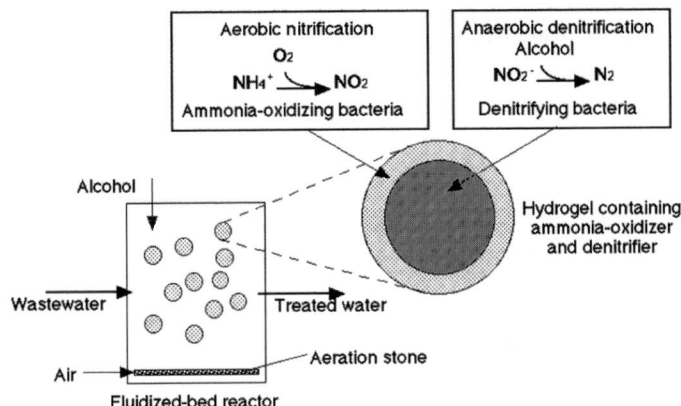

Figure 5. Fluidized-bed reactor using hydrogel beads capable of simultaneous nitrification and denitrification.

Because of the effective supply of an electron donor for denitrification, tubular hydrogel was tried in small-scale experiments [20-22,38]. The tubular hydrogel containing both *N. europaea* and *P. denitrificans* is made of a photo-cross-linkable polymer (PVA-SbQ) (Figure 6).

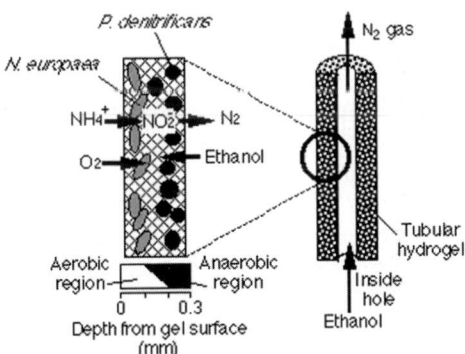

Figure 6. Tubular hydrogel capable of simultaneous nitrification and denitrification.

The tubular hydrogel is capable of simultaneous nitrification and denitrification like the hydrogel bead. Nitrification is accomplished by *N. europaea* in the outer surface of the tube, and denitrification is accomplished by *P. denitrificans* in the inside of the tube. Ethanol as an electron donor for denitrification flows through the inside hole of the tube. Thus, P. denitrificans in the tube can effectively use ethanol supplied from the inside hole of the tube. The advantage of using the tube is that the hydrogel surface remains clean, since the electron donor is supplied to *P. denitrificans* without adding it directly into wastewater. The aerobic region (from gel surface to 0.2 mm in depth) for nitrification and anaerobic region (from 0.2 mm to 0.3 mm in depth) for denitrification were observed in the tubular hydrogel containing both *N. europaea* and *P. denitrificans* [21]. Further, instead of the tubular hydrogel, a packed gel envelope consisting of two plate gels containing both *N. europaea* and *P. denitrificans* has been applied to a practical system for ammonia removal (Figure 7) [39]. The outer surface and the inside space between two plate gels of the packed gel envelope corresponds to the outer surface and the inside hole of the tubular hydrogel, respectively. Ethanol as an electron donor for denitrification is supplied to the inside space without adding it directly into wastewater. A practical system using the packed gel envelopes has been applied for ammonia removal from wastewater in coal power plants.

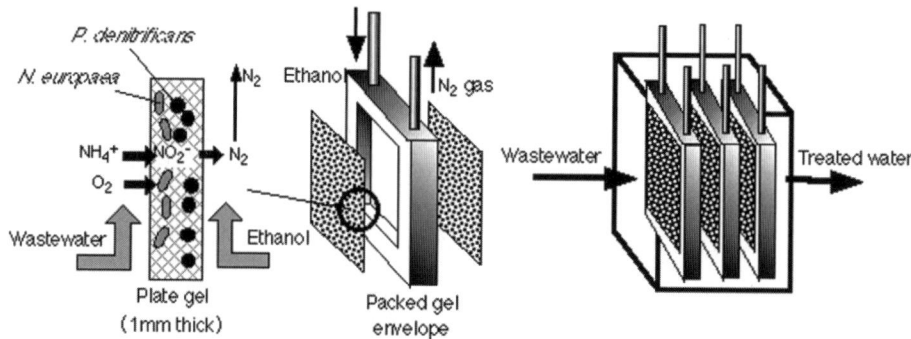

Figure 7. *Packed gel envelope capable of simultaneous nitrification and denitrification.*

Adsorption and carrierless immobilisation techniques are also used in treatment processes of simultaneous nitrification and denitrification. The high density of microorganisms in thick biofilms and matured granules results in the development of both aerobic region and the anaerobic microenvironments. The concentration gradient of O_2 and the presence of anaerobic regions in biofilms and granules were demonstrated by microelectrode measurements [26,40]. Nitrogen removal by simultaneous nitrification and denitrification were observed in appropriate aerobic processes using biofilms and granules [5]. However, the DO concentration has a strong effect on the nitrogen-removal performance of the biofilms and granules. O_2 is needed to accomplish nitrification, and the presence of O_2 slows down the denitrification process. Thus, it is extremely difficult to control the population distributions in biofilms and granules.

3.3. DENITIRIFICATION

Nitrate pollution of drinking water resources, *i.e.* groundwater, has been the focus of intense research due to the potential health risk. Immobilisation techniques are applied to denitrification in the treatment process of nitrate-contaminated groundwater [41-44], though immobilised biomass is also used for denitrification in wastewater treatment processes [2].

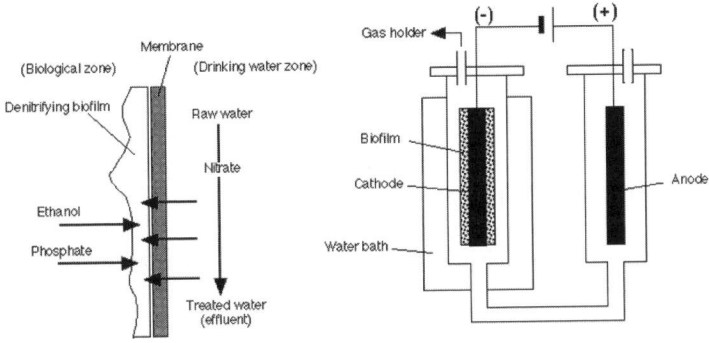

Figure 8. Denitrification reactor using biofilm.

The adsorption technique is used in fluidised-bed reactors using various support media, *e.g.* sand, glass beads, activated carbon particles. In the case of the biological treatment of drinking water, it is important to remove released bacteria and remnant organic matter from treated water. A membrane-fixed biofilm reactor was tested to overcome the disadvantages of the biological denitrification, which were the intensive post-treatment process to remove microorganisms and remnant organic matter [42]. In the membrane-fixed biofilm reactor, the biological reaction zone and carbon supply zone were separated from the raw water stream by a nitrate-permeable membrane (Figure 8A). Denitrification takes place in the biofilm attached on the membrane, *e.g.* regenerated cellulose, polyvinylidene fluoride, and polypropylene. The membrane and the attached biofilm represent a barrier against the passage of 5% ethanol as an electron donor for denitrification and nutrients into the raw water. Autotrophic denitrification using hydrogen instead of organic matter was also studied in a fluidized-bed reactor using various media [41-43]. Hydrogen is an excellent reactant for biological denitrification for drinking water, since it is inherently clean and only very slightly soluble in water. The treatment process using H_2 gas has been already applied to a full-scale plant for drinking water [45,46]. In the case of a small-scale experiment using low-strength nitrate solution, the electro-bioreactor was studied [47]. Biological denitrification was achieved by biofilm attached on a carbon electrode and hydrogen was produced through the electrolysis water (Figure 8B). Previous reports demonstrated that biological denitrification was driven and controlled by an electric current in the electro-bioreactor.

The entrapment technique is also applied to denitrification in treatment process of nitrate-contaminated groundwater. Polymer beads containing denitrifying biomass were studied in small-scale experiments [54]. Tubular gel containing *N. europaea* and *P. denitrificans*, which was also studied for simultaneous nitrification and denitrification, was also used for denitrification in nitrate-contaminated water [22,38]. The tubular gel can use both ethanol and H_2 gas as an electron donor for denitrification effectively, since the biological reaction zone and carbon supply zone are separated from the raw water stream by hydrogel containing microorganisms.

The carrierless immobilisation technique is also used for denitrification of drinking water [49]. An upflow sludge-blanket (USB) reactor, where matured granules containing denitrifying bacteria were formed, was used to treat nitrate-contaminated groundwater. The USB reactor could be operated without problems of channelling and plugging, which were commonly observed in fixed packed-bed reactors.

3.4. METAL ACCUMULATION

Many industrial processes and the mining of metallic ores produce wastewaters containing elevated levels of metals. The concentration of metals in wastewaters is often low and recycling of the metals by chemical extraction uneconomic. An alternative approach has focused on the uptake of metals through biosorption and bioprecipitation mechanisms [50]. However, biosorption systems using non-living biomass are prone to saturation of low levels of metal, and those using live biomass are damaged by metal toxicity. Thus, the immobilisation technique is applied to the bioprecipitation processes [18,19].

A typical example of is an air-lift column reactor containing a biofilm of *Citrobacter* species [18]. A cell-bound and atypical acid phosphatase enzyme is produced by *Citrobacter* species. The enzyme mediates the release of inorganic phosphate (HPO_4^{2-}) from an organic phosphate as a "donor" molecule. It results in the stoichiometric precipitation of metal cations (M) as insoluble metal phosphates ($MHPO_4$). In the case of uranium as a test metal, the air-lift column reactor can remove uranyl ions (UO^{2+}) from wastewater as deposit material (HUO_2PO_4), and Ni^{2+} was also removed into the crystalline $HUO_2PO_4\text{-}4H_2O$ lattice. The bioprecipitation mechanisms, utilising enzymatically mediated metal-accumulation processes, can accumulate metal to high loads without saturation and can function in resting cells decoupled from cell growth and nutrient requirements. Thus, bioprecipitation processes using immobilisation techniques are expected to be applied in large-scale plants and in the removal of other metals.

4. Conclusions

In view of the many applications shown in this chapter, immobilised cell technology will have an increasingly important role to play in the field of wastewater treatment. Immobilisation techniques are especially effective to accelerate the treatment processes using slow-growing microorganisms (*e.g.* nitrifying bacteria), since the useful cells for the treatment are always exposed to competition with various microorganisms and the inhibition with chemical materials derived from external inflow. It was demonstrated by experimental results that the nitrification rate of immobilised nitrifying bacteria was higher than that of activated sludge. Concerning the nitrification rate per unit of surface area of immobilised media, however, the activity of the attached biofilm (adsorption technique) was similar to that of the hydrogel (entrapment technique). It suggests that hydrogels made by the entrapment technique can only use its surface for nitrification, since diffusion limitation of O_2 restricts the population distribution in the hydrogel. Thus, from the viewpoint of immobilisation cost, the adsorption technique would be more effective than the entrapment technique in the nitrification process.

On the other hand, attached biofilms and granules (adsorption and carrierless immobilisation techniques) showed activity of simultaneous nitrification and denitrification. Both aerobic and anaerobic microenvironments were observed in the biofilms and granules. However, it is extremely difficult to control the population distributions in biofilms and granules, since aeration condition and characteristics of wastewater affect bacterial distributions severely. At the moment, the entrapment technique would be effective to control cascade multi-reactions (*e.g.* simultaneous nitrification and denitrification), since the entrapment technique can immobilise only particular microorganisms selectively.

Further investigations with microelectrodes, *in situ* hybridisation, and so on would clarify bacterial consortia and microenvironment in biofilms, granules, and hydrogels. They are certainly effective to improve wastewater treatment processes using immobilisation techniques.

References

[1] Nunez, M.J. and Lema, J.M. (1987) Cell immobilisation: application to alcohol production. Enzyme Microb. Technol. 9: 642-651.
[2] Tchobanoglous, G. and Burton, F.L. (1991) Advanced wastewater treatment. In: Clark, B.J. and Morriss, J.M. (Eds.) Wastewater Engineering, Treatment, Disposal, and Reuse. McGraw-Hill, Inc., Singapore; pp. 711-726.
[3] Rostron, W.M.; Stuckey, D.C. and Young, A.A. (2001) Nitrification of high strength ammonia wastewater: comparative study of immobilisation media. Wat. Res. 35(5): 1169-1178.
[4] Tramper, J.; Suwinska-Borowiec, G. and Klapwijk, A. (1985) Characterization of nitrifying bacteria immobilised in calcium alginate. Enzyme Microb. Technol. 7: 155-160.
[5] Beun, J.J.; Heijnen, J.J. and van Loosdrecht, M.C.M. (2001) N-removal in a granular sludge sequencing batch airlift reactor. Biotechnol. Bioeng. 75(1): 82-92.
[6] Ascon-Cabrera, M.A.; Thomas D. and Lebeault J-M. (1995) Activity of asynchronized cells of a steady-state biofilm recirculated reactor during xenobiotic biodegradation. Appl. Environ. Microbiol. 61: 920-925.
[7] Durham, D.J.; Marshall L.C.; Miller J.G. and Chmurny, A.B. (1994) New composite biocarriers engineered to contain adsorptive and ion-exchange properties improve immobilised-cell bioreactor process dependability. Appl. Environ. Microbiol. 60: 4178-4181.

[8] dos Santos, L.M. and Livingston, A.G. (1995) Biotechnol. Bioeng. 47: 90-95.
[9] Cho, Y.-H. and Knorr, D. (1993) Development of a gel and foam matrix as immobilisation system for cells for microbial denitrification of water. Food Biotechnol. 7: 115-126.
[10] Hunik, J.H.; Bos C.G.; van den Hoogen P.; de Gooijer C.D. and Tramper J. (1994) Co-immobilised *Nitrosomonas europaea* and *Nitrobacter agilis* cells: validation of a dynamic model for simultaneous substrate conversion and growth in κ-carrageenan gel beads. Biotechnol. Bioeng. 43: 1153-1163.
[11] Kuhn, R.H.; Peretti S.W. and Ollis, D.F. (1991) Microfluorimetric analysis of spatial and temporal patterns of immobilised cell growth. Biotechnol. Bioeng. 38: 340-352.
[12] Monbouquette, H.G.; Sayles, G.D. and Ollis, D.F. (1989) Immobilised cell biocatalyst activation and pseudo-steady-state behavior: model and experiment. Biotechnol. Bioeng. 35: 609-629.
[13] Sumino, T.; Nakamura, H. and Mori, N. (1991) Immobilisation of activated sludge by the acrylamide method. J. Ferment. Bioeng. 72: 141-143.
[14] Sumino, T.; Nakamura, H.; Mori, N. and Kawaguchi, Y. (1992) Immobilisation of nitrifying bacteria by polyethylene glycol prepolymer. J. Ferment. Bioeng. 73: 37-42.
[15] van Ginkel, C.G.; Tramper, J.; Luyben, K.Ch.A.M. and Klapwijk, A. (1983) Characterization of *Nitrosomonas europaea* immobilised in calcium alginate. Enzyme Microb. Technol. 5: 297-303.
[16] Itoh, K.; Itadani, T. and Yosimura, H. (1988) Immobilisation of nitrifying bacteria and its application for wastewater treatment. Eisei Kagaku 35: 125-133.
[17] Wijffels, R.H. and Tramper, J. (1989) Performance of growing *Nitrosomonas europaea* cells immobilised in k-carrageenan. Appl. Microbiol. Biotechnol. 32: 108-112.
[18] Finlay, J.A.; Allan, V.J.M.; Conner, A.; Callow, M.E.; Basnakova, G., and Macaskie, L.E. (1998) Phosphate release and heavy metal accumulation by biofilm-immobilised and chemically-caupled cells of a *Citrobacter* sp. Pre-grown in continuous culture. Biotechnol. Bioeng. 35(1): 87-97.
[19] Michel, L.J.; Macaskie, L.E. and Dean, A.C.R. (1985) Cadmium accumulation by immobilised cells of a Citrobacter sp. Using various phosphate donors. Biotechnol. Bioeng. 28: 1358-1365.
[20] Uemoto, H. and Saiki, H. (1996) Nitrogen Removal by Tubular Gel Containing *Nitrosomonas europaea* and *Paracoccus denitrificans*. Appl. Environ. Microbiol. 62: 4224-4228.
[21] Uemoto, H. and Saiki, H. (1996) Behavior of Immobilised *Nitrosomonas europaea* and *Paracoccus denitrificans* in tubular gel for nitrogen removal in wastewater. Prog. Biotechnol. 11: 695-702.
[22] Uemoto, H. and Saiki, H. (2000) Distribution of *Nitrosomonas europaea* and *Paracoccus denitrificans* Immobilised in tubular polymeric gel for nitrogen removal. Appl. Environ. Microbiol. 66: 816-819.
[23] dos Santos, V.A.P.M.; Bruijnse, M.; Tramper, J. and Wijffels, R.H. (1996) The magic-bead concept: an integrated approach to nitrogen removal with co-immobilised micro-organisms. Appl. Microbiol. Biotechnol. 45: 447-453.
[24] dos Santos, V.A.; Tramper, J. and Wijffels, R.H. (1993) Simultaneous nitrification and denitrification using immobilised microorganisms. Biomat. Art. Cells & Immob. Biotech. 21: 317-322.
[25] Liu, Y.; Xu, H-L.; Yang, S-F. and Tay, J-H. (2003) Mechanisms and models for anaerobic granulation in upflow anaerobic sludge blanket reactor. Wat. Res. 37: 661-673.
[26] Satoh, H.; Nakamura, Y.; Ono, H. and Okabe, S. (2003) Effect of oxygen concentration on nitrification and denitrifcation in single activated sludge flocs. Biotechnol. Bioeng. 83(5): 604-607.
[27] Kuenen, J.G. and Robertson, L.A. (1988) Ecology of nitrification and denitrification. In: Cole, J.A. and Ferguson, S.J. (Eds.) The nitrogen and sulphur cycles. Cambridge University Press, Cambridge; pp. 161-218.
[28] Watson, S.W.; Bock, E.; Harms, H.; Koops, H. and Hooper, A.B. (1989) Ammonia-oxidizing bacteria. In: Staley, J.T.; Bryant, M.; Pfenning, N., and Holt, J.G. (Eds.) Bergey's Manual of Systematic Bacteriology. Williams & Wilkins, Baltimore; pp. 1818-1834.
[29] Tanaka, K.; Nakao, M.; Mori, N.; Emori, H.; Sumino, T. and Nakamura, Y. (1994) Application of immobilised nitrifiers gel to removal of high ammonium nitrogen. Water Sci. Technol. 29: 241-250.
[30] Moper, M. and Wildmoser, A. (1990) Improvement of existing wastewater treatment plants effeciencies without enlargement of tankage by application of the linpor process-case studies. Water Sci. Technol. 22: 207-215.
[31] Hunik, J.H.; van den Hoogen, M.P.; de Boer, W.; Smit, M. and Tramper, J. (1993) Quantitative determination of the special distribution of *Nitrosomonas europaea* and *Nitrobacter agilis* cells immobilised in k-Carrageenan gel beads by a specific fluorescent-antibody labelling technique. Appl. Environ. Microbiol. 59: 1951-1954.
[32] Monbouquette, H. G. and Ollis, D. F. (1988) Scanning microfluorimetry of Ca-alginate immobilised *Zymomonas mobilis*. Bio/Technol. 6: 1076-1079.

[33] Wijffels, R.H.; de Gooijer, C.D.; Schepers, A.W.; Beuling, E.E.; Mallee, L.F. and Tramper, J. (1995) Dynamic modeling of immobilised *Nitrosomonas europaea*: implementation of diffusion limitation over expanding microcolonies. Enzyme Microb. Technol. 17: 462-471.

[34] Kokufuta, E.; Shimohashi, M. and Nakamura, I. (1988) Simultaneous occurring nitrification and denitrification under oxygen gradient by polyelectrolyte complex-coimmobilised *Nitrosomonas europaea* and *Paracoccus denitrificans* cells. Biotechnol. Bioeng. 31: 382-384.

[35] Kokufuta, E.; Yukishige, M. and Nakamura, I. (1987) Coimmobilisation of *Nitrosomonas europaea* and *Paracoccus denitrificans* cells using polyelectrolyte complex-stabilized calcium alginate gel. J. Ferment. Technol. 65: 659-664.

[36] Dalsgaard, T.; de Zwart, J.; Robertson, L.A.; Kuenen, J.G. and Revsbech, N.P. (1995) Nitrification, denitrification and growth in artificial *Thiosphaera pantotropha* biofilms as measured with a combined microsensor for oxygen and nitrous oxide. FEMS Microbiol. Ecol. 17: 137-148.

[37] Hooijmans, C.M.; Geraats, S.G.M.; van Neil, E.W.J.; Robertson, L.A.; Heijnen, J.J. and Luyben, K.Ch.A.M. (1990) Determination of growth and coupled nitrification/denitrification by immobilised *Thiosphaera pantotropha* using measurement and modelling of oxygen profiles. Biotechnol. Bioeng. 36: 931-939.

[38] Uemoto, H.; Ando, A. and Saiki, H. (2000) Effect of oxygen concentration on nitrogen removal by *Nitrosomonas europaea* and *Paracoccus denitrificans* immobilised within tubular polymeric gels. J. Biosci. Bioeng. 90: 654-660.

[39] Uemoto, H. and Saiki, H. (2000) Nitrogen Removal Reactor Using Packed Gel Envelopes Containing *Nitrosomonas europaea* and *Paracoccus denitrificans*. Biotechnol. Bioeng. 67: 80-86.

[40] Okabe, S.; Satoh, H. and Watanabe, Y. (1999) *In situ* analysis of nitrifying biofilms as determined by *in situ* hybridization and the use of microelectrodes. App. Emviron. Microbiol. 65(7): 3182-3191.

[41] Kurt, M.; Dunn, J. and Bourne, J. R. (1987) Biological denitrification of drinking water using autotrophic organisms with H_2 in a fluidized-bed biofilm reactor. Biotechnol. Bioeng. 29: 493-501.

[42] Fuchs, W.; Schatzmayr, G. and Braun, R. (1997) Nitrate removal from drinking water using a membrane-fixed biofilm reactor. Appl. Microbiol. Biotechnol. 48: 267-274.

[43] Brosilow, B.J.; Schnitzer, M.; Tarre, S. and Green, M. (1997) A simple model describing nitrate and nitrite reduction in fluidized bed biological reactors. Biotechnol. Bioeng. 54(6): 543-548.

[44] Kurt, M.; Dunn, I.J. and Bourne, J.R. (1987) Biological denitrification of drinking water using autotrophic organisms with H_2 in a fluidized-bed biofilm reactor. Biotechnol. Bioeng. 29: 493-501.

[45] Gros, H.; Schnoor, G. and Rutten, P. (1988) Biological denitrification process with hydrogen-oxidizing bacteria for drinking water treatment. Wat. Supply 6: 193-198.

[46] Vanbrabant, J.; Vos, P.D.; Vancanneyt, M.; Liessens, J.; Verstraete, W. and Kersters, K. (1993) Isolation and identification of autotrophic and heterotrophic bacteria from an autohydrogenotrophic pilot-plant for denitrification of drinking water. System. Appl. Microbiol. 16: 471-482.

[47] Sakakibara, Y. and Kuroda, M. (1993) Electric prompting and control of denitrification. Biotechnol. Bioeng. 42(4): 535-537.

[48] Nelsson, I.; Ohlson, S.; Haggstron, L.; Molin, N. and Mosbach, K. (1980) Denitrification of water using immobilised *Pseudomonas denitrificans* cells. Eur. J. Appl. Microbiol. 10: 261-274.

[49] Tarre, S. and Green, M. (1994) Precipitation potential as a major factor in the formation of granular sludge in an upflow sludge-blanket reactor for denitrification of drinking water. Appl. Microbiol. Biotechnol. 42: 482-486.

[50] Macaskie, L.E. and Dean, A.C.R. (1989) Microbiol metabolism, desolublilisation and deposition of heavy metals: metal uptake by immobilised cells and application to the detoxification of liquid wastes. In: Mizrahi, A. (Ed.) Advances in Biotechnology Processes. Biological waste treatment. New York, Alan R. Liss; pp. 159-201.

IMMOBILISED CELL STRATEGIES FOR THE TREATMENT OF SOIL AND GROUNDWATERS

LUDO DIELS
Dept. of Environmental and Process Technology, Vlaamse Instelling voor Technologisch Onderzoek (VITO), Boeretang 200, B-2400 Mol, Belgium
– Tel: +32 14 33 69 24 - Fax: +32 14 58 05 23 –
Email: ludo.diels@vito.be

1. Introduction

It is well known that soils and groundwaters contain an enormous diversity of microorganisms. Numerous bacteria are involved in the carbon, nitrogen, sulphur, phosphorus and iron cycle and many important bacterial strains have been isolated from contaminated aquifer material during the last few decades. Groundwater pollution consisting of a single pollutant can cover large areas in relatively low concentrations (micrograms per litre or less) or consist of highly complex pollutant mixtures in concentrations up to grams per litre. The former, diffuse pollution, can result from the use of pesticides in agriculture [1] while landfill leachate is an example of the latter source zone pollution. Landfill leachate usually contains both organic pollutants (*e.g.* BTEX (Benzene, Toluene, Ethylbenzene, Xylenes), chlorinated hydrocarbons, pesticides, medical compounds, *etc.*) and inorganic compounds such as heavy metals [2]. Microorganisms play an important role in the degradation, transformation and (im)mobilisation of these contaminants [3]. These processes, leading to a decrease of mass and/or toxicity, are defined as natural attenuation (NA) and if these intrinsic processes are stimulated by man, bioremediation is the right terminology. Whether natural attenuation or bioremediation can be accepted as stand-alone remediation techniques depends on their potential (*i.e.* is the degradation potential intrinsically present), their kinetics (*i.e.* at which rate occurs degradation) and their capacity (*i.e.* is degradation sustainable over long time-scales). When evaluating the potential one can investigate if the right bacteria are present. If so, knowledge about their activity becomes important as well as ways to (further) stimulate these microorganisms. Finally, an estimation on the long-term stability of the population will be required.

Biomass content in the subsurface is low under most natural conditions and microbiological properties can vary by many orders of magnitude, depending on the geological context [4,5]. Culturing-dependent methods have shown that microbial communities in groundwaters can completely differ from those in sediments [6,7]. On the other hand, enumeration of microorganisms by culturing methods only catches a few percent of the total population. Molecular biological methods such as PCR and DGGE therefore reveal the presence of many different and often unculturable bacteria.

At present not much is known about the way microorganisms are present in the soil. First of all, it is generally accepted that microorganisms live preferably attached to aquifer materials (*i.e.* sand particles, clay, organic matter, suspended solids, ...).

It can be stated that many bacteria are present in the aquifer in a kind of dormant way. Once the conditions become favourable, specific types of microorganisms are activated and multiply, drastically increasing their relative proportion within the total microbial community. As fast as one type of organisms can become dominant over other species, as fast can their specific advantage disappear, leading to other community shifts.

Aquifer-associated bacteria show extreme diversity. This means that all kinds of ecological functions are potentially present. Only those bacteria that are uniquely adapted to the specific environmental conditions, however, do emerge and can reach concentrations of up to $10^6 - 10^7$ cfu/g aquifer.

Phylogenetically divergent microorganisms may perform similar processes such as iron reduction [8] and aerobic degradation of aromatic pollutants [9]. In contrast, current knowledge suggests the involvement of a single phylogenetic group for some important subsurface processes such as anaerobic degradation of BTEX by *Azoarcus/Thauera* spp., *Geobacteriaceae* and *Desulfitobacteria* under denitrifying, iron-reducing and sulphate-reducing conditions, respectively [10]. In most cases the lack of specific microorganisms is not among the factors limiting natural biodegradation processes. Instead, specific environmental conditions, such as electron acceptor availability, are critical. The speciation of the contaminants present and their physico-chemical behaviour determine their availability and thus their biodegradation rates. Bioremediation in the water-unsaturated soil compartment will also strongly depend on moisture content since active microbes only occur in aqueous phase.

Phosphorus and nitrogen availability are often key growth-limiting factors controlling the amount and activity of specific pollutant-degrading bacteria. Optimal temperatures for most such bacteria are at 30°C, but biological processes in soil and groundwater will satisfactory occur at much lower temperatures; groundwater temperatures are mostly in the order of 8 to 15°C. The preferable pH ranges between 5.0 and 9.0. Active bacteria are, however, documented that thrive outside this range. Groundwater contaminated with heavy metals for instance, often possesses a pH between 3.5 and 5.0. In such groundwater sulphate-reducing bacteria *Desulfosporosinus* and *Desulfotomaculum* have been identified [11], which seem to have adapted to these rather low pHs. No other sulphate reducing bacteria species so far have been found in acidic aquifers.

2. *In situ* treatment

With the rapid emergence of groundwater and soil contamination as primary environmental issues, the notion has evolved that microbial activity plays a major role in the development and decay of contaminant plumes in the subsurface. Much effort is currently being devoted to designing efficient ways to stimulate *in situ* biodegradation of organic contaminants in aquifers. Since many bacteria are present in a dormant state on the aquifer material, the introduction of ideal conditions by adding nutrients (*e.g.* fertilisers), electron acceptors (most commonly oxygen) or sometimes electron donors (other carbon sources besides the contaminant itself) can cancel the state of dormancy and lead to a fast increase of a specific type of bacteria. Many contaminants can be bioremediated using such a strategy: *e.g.* petroleum hydrocarbons such as mono or polyaromatic compounds and phenols that can be degraded under aerobic conditions or chlorinated aliphatics and aromatics under anaerobic conditions.

Mono-aromatic hydrocarbons benzene, toluene, ethylbenzene and the 3-xylene isomers (all confirmed or suspected carcinogens and classified as environmental priority pollutants) are common groundwater pollutants associated with petroleum product release. Exploiting naturally occurring degradation processes (biodegradation) to remove these priority pollutants from contaminated sites therefore is of utmost importance. It is known that microorganisms, living in thin biofilms, can be activated to degrade these pollutants under aerobic or anaerobic conditions [12]. Aerobic biodegradation is occurring relatively easily with rates that seem to be influenced by biofilm thickness. Babaarslan *et al.* [13] showed that the time required for complete biodegradation of both benzene and toluene decreased from 3 to 2 days with toluene-adapted microorganisms when the initial biomass concentration increased from 0.42 g/l tot 2.34 g/l. The same authors also indicated that the overall specific biodegradation rates of BTEX compounds individually were higher with toluene-adapted microorganisms than with non-adapted and benzene-adapted microorganisms. This is an illustration of the complexity of the bacterial communities in biofilms and of the induction of degradation capacities.

Studies performed by Hendricks *et al.* [14] showed that soil or aquifer material contained about 10 different bacterial species with the capacity to biodegrade BTEX compounds. In the contamination plume, however, only one of these species could be detected that was active and present in strongly increased numbers. This illustrates the complicated way in which bacteria, initially dormant in thin biofilms, can be activated to grow and stimulate biodegradation. Another result of Hendricks *et al.* [14] was that biodegradation of BTEX under anaerobic conditions (denitrifying, sulphate or iron reducing conditions) depends on the local conditions and biofilm composition, which can be highly spatially variable within one contaminated site.

Another study demonstrated that phenols and cresols, which are also widespread contaminants of groundwaters, could be degraded by microorganisms participating in methanogenic consortia. It seems that humus, stable organic matter accumulating in soils, can play an active role in the anaerobic oxidation of various organic compounds. The fact that humus can serve as an electron acceptor for the anaerobic oxidation was demonstrated by stimulating the oxidation through addition of humic acids or the model compound anthraquinone-2,6-disulfonate [15].

In some cases an environmental pollutant does not support microbial growth. However, aerobic cometabolism can offer a biological method for the removal of the pollutant from the contaminated environment. The aerobic cometabolism of chlorinated solvents (*e.g.* trichloroethylene) is such an example often encountered. The physiological role of oxygenases is to initiate the metabolism of growth supporting substrates (*e.g.* methane, propane, butane, toluene, ethylene and ammonia). During cometabolism the growth-supporting and non-growth-supporting substrates can both bind to the oxygenases. Transformation of chlorinated solvents by these enzymes presents the cell with a new set of compounds. Some of these compounds are toxic to the cells, others are stable products that are expelled from the cell, and in a few cases the cells utilise the products. The combined effects of cometabolism can have a profound influence on a cell.

3. Soil and aquifer biofilms

The creation of *in situ* reactive zones with increased biological activity can result in some local clogging due to biofilm formations. This often occurs in injection wells where it may result in increased injection pressures or, alternatively, in decreased injection rates. Significant reductions of the saturated hydraulic conductivity may also occur in the biologically activated aquifer zone itself as a result of the growth and metabolism of microorganisms. This may in turn cause the reactive zone to be by-passed by the injected nutrient solution, thus jeopardising the over-all chances of success of the bioremediation efforts. The decreased conductivity of the porous medium, which is a reduced capacity of the pores to conduct water, is the result of physical or biological processes. Since we are now focusing on biofilm formation, we will discuss the biological processes in more detail.

The addition of carbon or energy sources such as plant residues, monosaccharides, disaccharides, alcohols, *etc.* can accelerate and enhance soil clogging. Very often this is accompanied by an increase in cell number of the aquifer. Attempts to isolate typical clogging bacteria failed as it turned out to be very difficult to isolate specific groups of bacteria that cause the clogging. It was also difficult to establish if aerobic or rather anaerobic conditions mostly contribute to biofilm formation causing clogging. Biofilm formation mostly starts aerobically, depleting the oxygen available, which then leads to an anaerobic biofilm. It is generally accepted that redox conditions and the presence or absence of certain electron donors and acceptors will lead to the appearance and disappearance of certain types of bacteria. It is observed that their numbers increase or decrease in the aquifer, but they are always present in one way or another. To be present in thin biofilms is certainly a way to survive under unfavourable conditions. Biofilms also provide the possibility for several groups of bacteria to co-exist and to live in consortia necessary for the degradation of certain contaminants. It thus turns out that aquifer biofilms are rather thin layers of heterogeneous groups that tend to co-exist. In case of condition changes one or some of these groups can start tot grow relatively fast to reach numbers 10^6 to 10^7 cfu/g aquifer starting from numbers below detection limit.

The formation of exopolymers by microorganisms can play an important role in biofilm formation and subsequent clogging. Exopolymer production and degradation depend on a variety of soil parameters such as soil type, moisture content, presence of surfaces, temperature, redox potential, availability of organic substrates, nitrogen, oxygen and physiological status of the microorganisms involved. In water treatment systems exopolysaccharides are very often involved. At present not very much is known about the occurrence of exopolymers in soil or aquifers. *E.g.* some polyuronides (polymers of uronic acids) appear to be important because they are more resistant to degradation due to the formation of metal salts [16] or complexation with other materials [17]. Slime formation can play an important role in the reduction of hydraulic conductivity. Exopolymers, as slimes, are usually highly hydrated and contain about 99% water [18]. Exopolymers form either gel structures or a highly viscous sol. They affect conductivity by increasing the viscosity of the fluid or by decreasing the volume and size of the conducting pores. It seems that the latter is the only important feature as slimes really tend to prevent water to penetrate. Vandevivere and Baveye [19] observed that the change in the C/N ratio from 39 tot 77 lead to a twenty-fold reduction in hydraulic conductivity. Macleod *et al.* [20] showed that ultramicrobacteria, *i.e.* dwarf cells resulting from prolonged starvation, move through porous media to a much larger extent than their non-starved (normal-sized) counterparts and, as a result, have a much lesser tendency to plug the porous medium.

4. Viruses in soil and groundwater: do they play a role?

It has long been believed that the groundwater was protected from surface contamination because the upper soil mantle removed pollutants during percolation [21]. It was also believed that even if contaminated groundwater would infiltrate, it would be purified through adsorption processes and metabolism of indigenous aquifer microflora [22,23]. Macler [23], however, estimated that approximately 20 to 25% of the United States' groundwater sources are contaminated with microbial pathogens, including more than 100 types of viruses. Viruses are colloidal particles ranging in size from 20 to 350 nm that are negatively charged at neutral pH. The unique feature of viral structure and colloidal properties plays an important role in the transport of viruses in soil and groundwater [24].

It is generally accepted that viruses are transported over greater subsurface distances than bacteria [25] and protozoa, due to their relatively small size with variations depending on their degree of inactivation and adsorption characteristics [26]. A number of pathways are furthermore available for the introduction of viruses to the subsurface (*i.e.* land disposal of untreated and treated wastewater, land deposition of sludge, septic tanks and sewer lines, and landfill leachates), as well as a number of parameters which affect their migration and survival. Adsorption or release of viruses from soil particles is a consequence of the amphoteric nature of the external viral proteins. Both ionic strength and pH strongly affect adsorption and desorption processes [27]. Most viruses adsorb more strongly to the aquifer in acidic water. The role biofilms play in this context is at the moment rather unknown. What is known is that microbial ecology plays an important role in the inactivation of waterborne viruses. Microbial activity can affect viral survival by the action of proteolytic protozoa in destroying the viral capsid protein. It was also shown that enteric viruses survive longer periods of time because they are more resistant to environmental conditions [28].

5. Crossing the bridge between groundwater and surface water

Groundwater and surface water are mostly seen as being physically independent and chemically distinct. The past few years however more research is focusing on the zone where these waters interact beneath the streambed, the so-called hyporeic zone [29], in Europe also called "the interphase". The flux through this zone is spatially and temporally dynamic, governed by physical variables imposed by overlying streamwaters and tempered and hydrogeological characteristics [30]. Even if the surface water contains only limited amounts of carbon sources and one can therefore anticipate low microbial populations, the advection- and diffusion-facilitated cyclings in the hyporeic zone act as a source or sink of nutrients, potentially regulating biotic productivity. Whereas in hyporeic zones interstitial waters tend to be oxidising, zones of microscale anoxy may develop within aggregates of detritus and lithotrophic biofilms [31]. Adjacent aerobic and anaerobic microzones may promote the exchange of redox-sensitive metabolites (*e.g.* Fe(II)/Fe(III)) between diverse microorganisms. This stimulates the microbial metabolism and diversity. In some samples relative large populations of sulphate-, nitrate-, and iron-reducing bacteria were present. Also, in some cases high concentrations of acid-volatile sulphides were measured, indicating the existence of an anoxic zone. In some cases Cr(VI) removal could be observed as a result of microbial activity and chemical reduction (anaerobic microbial metabolism). In other cases the anaerobic dehalogenation of chlorinated compounds was observed in parts of these zones [32]. The hyporeic zone may thus represent both the "last line of defense" in the prevention of river contamination and a factor leading to toxic exposures of sensitive species.

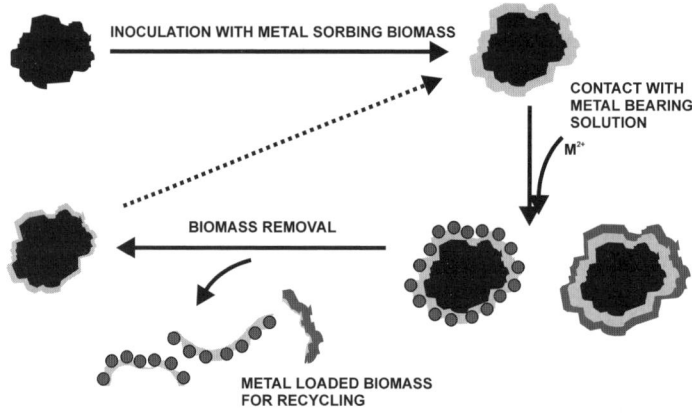

Figure 1. Concept of sand particles inoculated with metal biosorbing or bioprecipitating biofilms.

6. Pump and treat

Bacteria can also be used in immobilised systems above the soil surface. In a "pump and treat" system groundwater is pumped to the surface where it is treated in order to degrade or to remove the contaminants from the water. The water can then be either discharged or reinfiltrated into the aquifer.

A very specific application of using immobilized bacteria is presented in MERESAFIN "MEtal REmoval by SAnd Filters INoculated by bacteria" [33,34]. Bacteria, able to biosorb or bioprecipitate heavy metals, grow in a biofilm on a supporting material (Figure 1). During contact with heavy metal-containing wastewater the biofilm adsorbs the metals. Subsequently the metal-loaded biomass is removed from the supporting material and the biomass residual on the substratum can be reused, after regrowth, for a subsequent treatment cycle.

The supporting material can be sand or other materials retained within a moving bed sand filter which is based on a counterflow principle (Figure 2). The water to be treated is admitted through the inlet distributor (1) in the lower section of the unit and is cleaned as it flows upward through the sand bed, prior to discharge through the filtrate outlet (2) at the top. The sand containing the heavy metals bound to the biofilm is conveyed from the tapered bottom section of the unit (3) by mean of an airlift pump (4) to the sand washer (5) at the top. Cleaning of the sand starts in the pump itself, in which metal-loaded biofilms are separated from the sand grains by the turbulent mixing action.

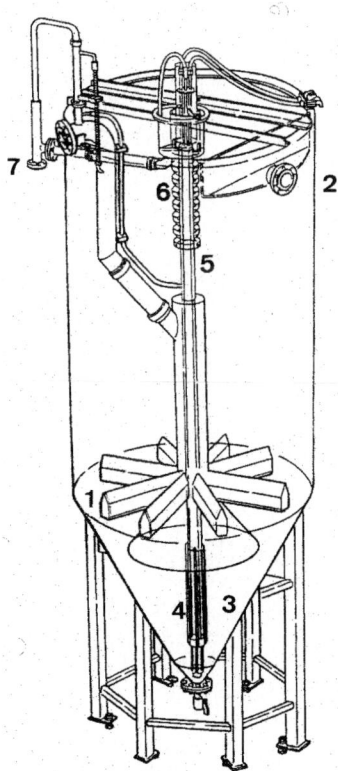

Figure 2. Moving bed sand filter concept: (1) inlet distributor, (2) outlet, (3) dirty sand, (4) air-lift pump, (5) sand washer, (6) washer labyrinth, (7) wash water outlet, (8) cleaned sand.

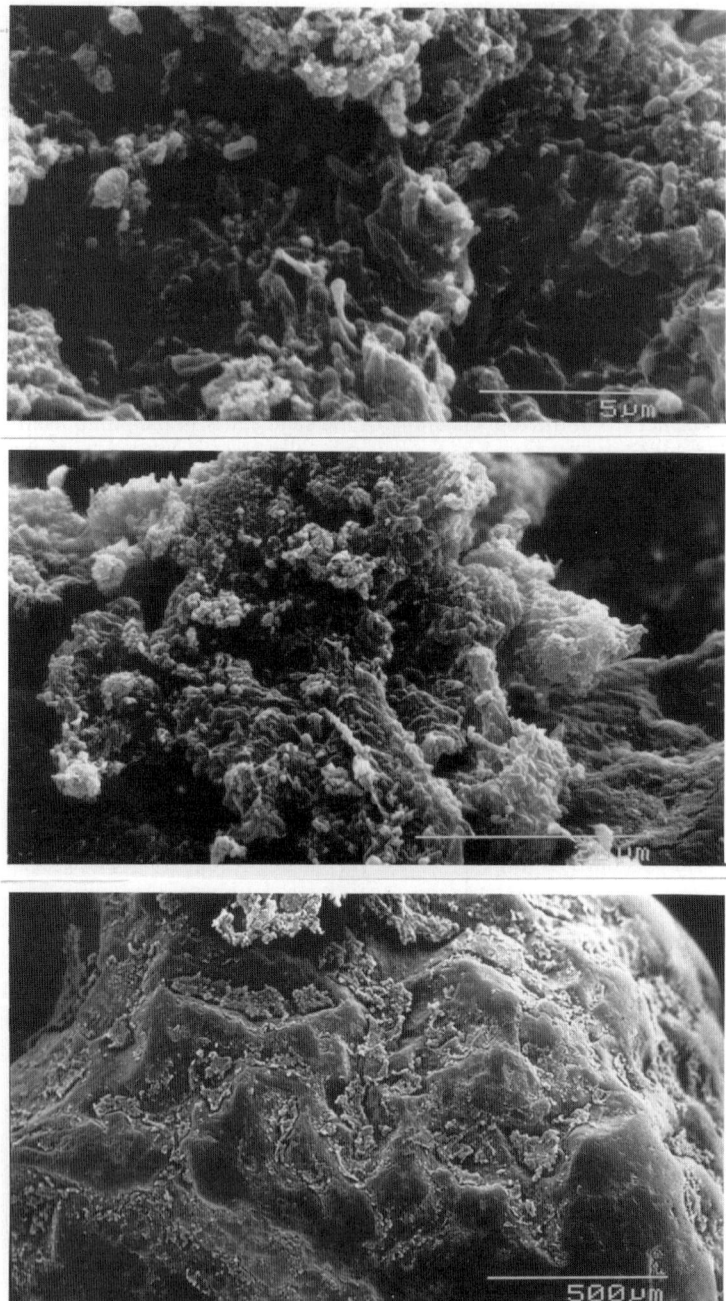

Figure 3. SEM images of sand grain and biofilm from the experimental filter.

The contaminated sand spills from the pump outlet into the washer labyrinth (6), in which it is washed by a small flow of clean water. The metal loaded flocs are discharged through the washwater outlet (7), while the grains of sand with a partly removed biofilm are returned to the sand bed (8). As a result, the bed is in constant downward motion through the unit. In this concept water purification and sand washing both take place continuously, enabling the filter to remain in service without interruption.

The complete water treatment system is shown in figure 2. Wastewater is pumped through the Astrasand filter and purified. The wash water, containing the metal-loaded biomass, is drained to a lamella separator or settling tank. The water, coming from the thickener, is reintroduced in the sand filter. The sludge coming from the thickener is treated further in a filter press or a decanting centrifuge after addition of lime or flocculants respectively. The filtercake obtained in this way, containing the metals, is recycled in a pyrometallurgical treatment facility (shaft furnace) of a non-ferrous company.

Figure 3 shows SEM images of a sand grain of an experimental inoculated sandfilter. Biofilm formation is clearly shown on the sand surface and especially in the crevasses of the particles which allow for biofilms to bind. Molecular biology determination by using PCR based on the specific *czc*-sequence [35] revealed the presence of the metal resistant bacterium *Ralstonia metallidurans* as the main residing bacterium.

7. *In situ* bioprecipitation

Groundwater pollution with (heavy) metals can be treated *in situ* in the aquifer by immobilisation. This can be established by stimulating Sulphate Reducing Bacteria (SRBs) to reduce sulphates into sulphides, which will precipitate the heavy metals as metal sulphides. In order to induce the bacteria, a carbon source (*i.e.* electron donor: molasses, lactate, acetate, Hydrogen Release Compound (HCR®), ...) must be provided to remove the oxygen and to allow fermentative bacteria to transform the energy source into H_2 which will be used as electron donor by the SRBs. It is important to correctly pre-evaluate the feasibility of this technology since low Oxidation Reduction Potentials (< -250 mV) are required and methanogenic activity (depending on the carbon to sulphate ratio) must be avoided. It seems that FeS that is often produced in these processes tends to bind to humus and organic matter and no to the mineral matrix of the soil.

8. Bacterial contamination of aquifers

The application of manure to land is the principal source of pathogens in soil and groundwater. Pathogen numbers will be affected by form, duration and conditions of manure storage, timing of application and method of incorporation into the soil. Pathogens decline in number with storage because of oxygen deprivation, elevated temperatures and pH, and drying conditions. On spreading, pathogens may be exposed to the anti-microbial effect of ultra-violet light. Generally, the risk of pathogen survival and movement to watercourses is smallest with broadcast spreading and greatest with injection spreading. Pathogen survival is also likely to be affected by the season in which spreading occurs.

9. Conclusions

Recent years have witnessed a continued growth in interest in microbially accelerated reclamation of contaminated soils and decontamination of groundwater. Effective techniques will require combined *in situ* and *ex situ* bioremediation, coupled with effective chemical and engineering advances. It cannot be overstressed that the bacterial contribution to this process must be viewed as only one aspect of the developing strategies. These will undoubtedly combine some of the above methodologies with strategies further based on fungal, phytoremediative and physico-chemical clean up.

Fundamental research is providing insight into the natural attenuation of many compounds in a wide variety of subsurface geochemical settings. However, current knowledge of natural attenuation of many compounds and the processes occurring in biofilms on soil particles are still insufficient for the development and implementation of general tools to predict the effects of natural attenuation. Therefore, feasibility testing and detailed monitoring are necessary in most cases. An important new issue arises from the knowledge that probably the hyporeic zones in the transfer from groundwater to surface water will play an important role in the speciation of metals and the biodegradation of organic pollutants.

There is no doubt that the further development of molecular biology tools, allowing to identify bacteria extracted from soil and aquifer at DNA-level, will give more answers on the presence of certain bacterial species and their metabolic activities. The development of primers to identify specific bacteria at species level will allow to understand some of these interactions. On the other hand the isolation of RNA will allow to understand the activation of certain processes.

The transport and survival of pathogens in the subsurface depend on their retention to soil and aquifer materials. Some of the more important factors affecting virus transport include soil water content, temperature, sorption and desorption, pH, salt content, type of virus, and hydraulic stresses. There are indications that the inactivation rate of viruses is the single most important factor governing virus transport and fate in the subsurface.

The "game" going on in biofilms on soil and aquifer particles, the adsorption of bacteria and other microorganisms and the regulation of growth and (de)activation of other bacteria is still largely unknown. We hope to have illustrated that this elucidation will be a big challenge for the future.

References

[1] Ralebitso, T.K.; Senior, E. and Van Verseveld, H.W. (2002) Microbial aspects of atrazine degradation in natural environments. Biodegradation 13: 11-19.

[2] Christensen, T.H.; Kjeldsen, P.; Bjerg, P.L.; Jensen, D.L.; Christensen, J.B.; Baun, A.; Albrechtsen, H.J. and Heron, C. (2001) Biogeochemsitry of landfill leachate plumes. Appl. Geochem. 16: 659-718.

[3] Lovley, D.R. (2001) Reduction of iron and humics in subsurface environments. In: Frederickson, J.K. and Fletcher, M. (Eds.) Subsurface Microbiology and Biochemistry. Wiley-Liss, New York.

[4] Brockman, F.J. and Murray, C.J. (1997) Subsurface microbiological heterogeneity: current knowledge, descriptive approaches and applications. FEMS Microbiol. Rev. 20: 231-247.
[5] Chandler, D.P. and Brockman, F.J. (2001) Nucleic acids analysis of subsurface microbial communities: pitfalls, possibilities and biogeochemical implications. In: Frederickson, J.K. and Fletcher, M. (Eds.) Subsurface Microbiology and Biochemistry. Wiley-Liss, New York; pp. 281-313.
[6] Röling, W.F.M.; van Breukelen, B.M.; Braster, M.; Goeltom, M.T.; Groen, J. and van Verseveld, H.W. (2000) Analysis of microbial communities in a landfill leachate polluted aquifer using a new method for anaerobic physiological profiling and 16S rDNA based fingerprinting. Microbial. Ecol. 40: 177-188.
[7] Röling, W.F.M.; van Breukelen, B.M.; Braster, M.; Lin, B. and van Verseveld, H.W. (2001) Relationships between microbial community structure and hydrochemistry in a landfill leachate polluted aquifer. Appl. Environ. Microbiol. 67: 4619-4629.
[8] Lonergan, D.J.; Jenter, H.L.; Coates, J.D.; Phillips, E.J.P.; Schmidt, T.M. and Lovly, D.R. (1996) Phylogenetic analysis of dissimilatory Fe(III)-reducing bacteria. J. Bacteriol. 178: 2402-2408.
[9] Anderson, R.T. and Lovely, D.R. (1997) Ecology and biogeochemistry of *in situ* groundwater bioremediation. Adv. Microbial Ecol. 15: 289-350.
[10] Spormann, A.M. and Widdel, F. (2000) Metabolism of alkylbenzenes, alkanes, and other hydrocarbons in anaerobic bacteria. Biodegradation 11: 85-105.
[11] Geets, J.; Borremans, B.; Vanbroekhoven, K.; Diels, L.; Van der Lelie, D. and Vangronsveld, J. (2004) Monitoring sulfate-reducing bacteria (SRB) using molecular tools during *in situ* immobilization of heavy metals, European Symposium on Environmental Biotechnology, Ostend April 25 – 28, ESEB 2004, submitted.
[12] Chang, M.K.; Voice, T.C. and Criddle, C.S. (1992) Kinetics of competitive inhibition and cometabolism in the biodegradation of benzene, toluene and p-xylene by two *Pseudomonas* isolates. Biotechnol. Bioeng. 41: 1057-1065.
[13] Babaarslan, C.; Abuhamed, T.; Mehmetoglu, U.; Tekeli, A. and Mehmetoglu, T. (2003) Biodegradation of BTEX compounds by a mixed culture obtained from petroleum formation water. Energy Sources 25: 733-742.
[14] Hendrickx, B.; Faber, F.; Dejonghe, W.; Top, E. M.; Bastiaens, L. and Springael, D. (2004) Diversity of BTEX monooxygenase genes in contaminated sites as assessed by PCR-DGGE, in preparation.
[15] Cervantes, F.J.; van der Velde, S.; Lettinga G. and Field, J.A. (2000) Quinones as terminal electron acceptors for anaerobic microbial oxidation of phenolic compounds. Biodegradation 11: 313-321.
[16] Tollner, E.W.; Hill, D.T. and Busch, C.D. (1983) Manure effects on model lagoons treated with residues for bottom sealing Trans. A.S.A.E. 26: 430-435.
[17] Avnimelech, Y. and Nevo, Z. (1964) Biological clogging of sands. Soil Sci. 98: 222-226.
[18] Rittmann, B.E. and McCarty, P.L. (1980) Evaluation of steady-state-biofilm kinetics. Biotechnol. Bioeng. 22: 2359-2379.
[19] Vandevivere, P. and Baveye, P. (1992) Saturated hydraulic conductivity reduction caused by aerobic bacteria in sand columns. Soil Sci. Amer. J. 56: 1-13.
[20] MacLeod, A.; Lappin-Scott, H.M. and Costerton, J.W. (1988) Plugging of a model rock system by using starved bacteria. Appl. Environ. Microbiol. 54: 1365-1372.
[21] Amundson, D.; Lindholm, C.; Goyal, S.M. and Robinson, R.A. (1988) Microbial pollution of well water in southeastern Minnesota. J. Environ. Sci. Health A 23(5): 453-468.
[22] Dizer, H.; Lopez, J.M. and Nasser, A. (1984) Penetration of different human pathogenic viruses into sand columns percolated with distilled water, groundwater, or wastewater. Appl. Environ. Microbiol. 47: 409-415.
[23] Macler, B.A. (1995) Developing a national drinking water regulation for desinfection of groundwater. Ground Water Monit. Remed. 15: 77-84.
[24] Azadpour-Keeley, A.; Faulkner, B.R. and Chen, J.-S. (2003) Movement and longevity of viruses in the subsurface. EPA/540/S-03/500.
[25] Scheueman, P.R.; Farrah, S.R. and Bitton, G. (1987) Reduction of microbial indicatiors and viruses in a cypress strand. Water Sci. Technol. 19: 539-546.
[26] Herbold-Paschke, K.; Straub, U.; Hahn, T.; Teutsch, G. and Botzenhart, K. (1991) Behaviour of pathogenic bacteria, phages and viruses in groundwater during transport and adsorption. Water Sci. Technol. 24: 301-304.
[27] Duboise, S.M.; Moore, B. E. and Sagik, B.P. (1976) Poliovirus survival and movement in a sandy forest soil. Appl. Environ. Microbiol. 31: 536-543.
[28] Shuval, H.I.; Thompson, A.; Fattal, B.; Cymbalista, S. and Wiener, Y. (1971) Natural virus inactivation processes in seawater. J. San. Eng. Div. Proc. Am. Soc. Civ. Eng. 97: 587-600.

[29] Findlay, S. (1995). Importance of surface-subsurface exchange in stream ecosystems - the hyporheic zone. Limnol. Oceanogr. 40: 159-164.
[30] Moser, D.P.; Fredrickson, J.K.; Geist, D. R.; Arntzen, E. A.; Peacock, A. D.; Li, W.; Spadoni, T. and Mckinley, J.P. (2003) Biogeochemical processes and microbial characteristics across groundwater-surface water boundaries of the Hanford Reach of the Columbia River. Environ. Sci. Technol. 37: 5127-5134.
[31] Eichem, A.C.; Dodds, W.K.; Tate, C.M. and Edler, C. (1993) Microbial decomposition of elm and oak leaves in a karst aquifer. Appl. Environ. Microbiol. 95: 3592-3596.
[32] Middeldorp, P.J.M.; Luijten, M.L.G.C.; Van de Pas, B.A.; Van Eekert, M.H.A.; Kengen, S.W.M; Schraa, G.W.M. and Stams, A.J.M. (1999) Anaerobic microbial reductive dehalogenation of chlorinated ethenes. Bioremediation J. 3: 151-169.
[33] Diels, L.; Spaans, P.H.; Van Roy, S.; Hooyberghs, L.; Wouters, H.; Walter, E.; Winters, J.; Macaskie, L.; Finlay, J.; Pernfuss, B.; Woebking, H. and Pümpel, T. (2004) Heavy metals removal by sand filters inoculated with metal sorbing and precipitating bacteria. Hydrometallurgy, in press.
[34] Pümpel, T.; Ebner, C.; Pernfuss, B.; Schinner, F.; Diels, L.; Keszthelyi, Z.; Macaskie, L.; Tsezos, M. and Wouters, H. (2001) Removal of nickel from plating rinsing water by a moving-bed sandfilter inoculated with metal sorbing and precipitating bacteria. Hydrometallurgy 59: 383-393.
[35] Diels, L.; Dong, Q.; Van der Lelie, N.; Baeyens, W. and Mergeay, M. (1995) The *czc* operan of *Alcaligenes eutrophus* CH34: from resistance mechanism to removal of heavy metals. J. Ind. Microbiol. 14: 142-153.

APPLICATION OF IMMOBILISED CELLS FOR AIR POLLUTION CONTROL

Cleaning air naturally

MARC A. DESHUSSES
Department of Chemical and Environmental Engineering, University of California, Riverside, CA 92521, USA. Fax: 909-787-5696 – Phone: 909-787-2477– Email: mdeshuss@engr.ucr.edu

1. Introduction

The use pollutant-degrading organisms for air pollution control is an important and emerging application of cell immobilisation technology. The principle is relatively simple: a contaminated air stream is passed through a packed bed on which pollutant-degrading organisms are immobilised. Contaminants in the air are transferred to the microorganisms, and are degraded to harmless compounds. Air biotreatment is not a new concept, it has been proposed more than 40 years ago [1]. However, it is only in the past two decades that new environmental regulations have forced engineers to consider alternatives to convention air pollution control methods.

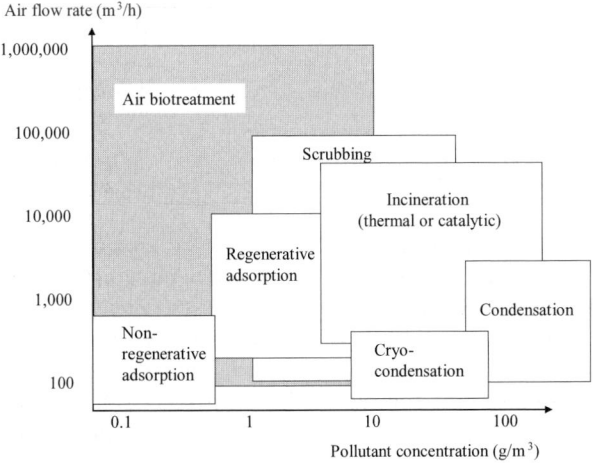

Figure 1. Applicability of various air pollution control technologies.

The most successful applications of biological techniques for air pollution control have been for the treatment of dilute, high flow waste gas streams containing odours or volatile organic compounds (VOCs) (Figure 1) [2-4]. Under optimum conditions, the contaminants are completely degraded to innocuous end-products. The major advantage over conventional treatment technologies is that air biotreatment is accomplished at low temperature, and has lower operating and maintenance costs (see Table 1).

Table 1. Comparison of biotreatment and conventional air pollution control techniques.

Technology	Advantages	Disadvantages
Biotreatment	• Simple and low cost technology • Low to medium capital costs, and low operating costs • Effective removal of odours and low concentrations of contaminants • No production of by-products • Low energy requirement, no fuel needed • Environmentally friendly	• Relatively large footprint requirement • Can not treat non-biodegradable pollutants • Medium replacement every 2-5 years (biofilters only) • Moisture and pH sometimes difficult to control (biofilters only) • Particulate matter may clog the bed • Clogging by growing biomass if too much nutrient is added and high concentrations of VOCs are treated
Wet Scrubbing	• Medium capital costs • Can operate with particulate in gas stream • Relatively small footprint • Ability to handle variable loads • Well proven technology	• Very high operating costs • Reduced performance by scale deposit • Need for complex chemical feed systems • Not effective for most VOCs • Requires toxic/dangerous chemicals
Carbon Adsorption	• Short retention time/small unit • Consistent, reliable operation • Moderate capital costs	• High to extremely operating costs • Carbon life reduced by moist gas • Creates secondary waste streams (spent carbon) • Medium pressure drop
Incineration	• Effective removal of compounds irrespective of nature and concentration • Suitable for very high loads Performance is uniform and reliable • Small footprint	• High operating and capital costs • High flow/ low concentrations not cost-effective • Usually requires additional fuel • Creates a secondary waste (NO_x) • Scrutinised by the public

Biodegradation of the contaminants in bioreactors for air pollution control is usually mediated by mixed cultures or consortia. The primary pollutant-degraders are similar to the organisms found in wastewater treatment processes. They are thriving in a complex and stressful environment that include higher organisms, such as protozoa, rotifers, even larvae, worms, insects and other predators. The nature of the primary degraders and the fate of the pollutant depend on the main pollutant(s) being treated. Further discussion of the microbial ecology of bioreactors for air pollution control can be found in Section 2. Table 2 lists the most important applications and the class of organisms responsible for pollutant removal.

Table 2. Type of air contaminants that can be treated in bioreactors and class of primary degrading organisms.

Pollutant	Primary degraders	End product
VOCs, hydrocarbons	Heterotrophic, use hydrocarbon as carbon and energy source	CO_2, biomass[*]
H_2S	Autotrophic, use CO_2 as carbon source	SO_4^-, biomass
Reduced sulphur compounds (e.g., dimethyl sulphide)	Heterotrophic and autotrophic mixed cultures	CO_2, SO_4^-, biomass
Ammonia	Nitrifying organisms, possibly associated with denitrifiers	NO_3^- (and NO_2^-), possibly N_2 biomass
NO_x (via denitrification)	Heterotrophic or autotrophic (denitrifying)	N_2, oxidised electron donor (organic or inorganic), biomass
NO_x (via nitrification)	Heterotrophic or autotrophic (nitrite oxidisers)	NO_3^-, biomass

[*] Biodegradation of chlorinated VOC will results in chloride, brominated compounds will give bromide, sulphur containing VOC will result in sulphate, nitrogen containing compounds will result in nitrate.

The two most promising bioreactors for air pollution control are biofilters and biotrickling filters. Their principles are explained below and schematically shown in Figures 2-3. A picture of an actual biofilter system in shown in Section 3. Other bioreactor setups (e.g., airlift reactors, bioscrubbers, membrane bioreactors, etc.) have been used in air pollution control, but because they are less relevant for industrial application, this Chapter focuses on biofilters and biotrickling filters.

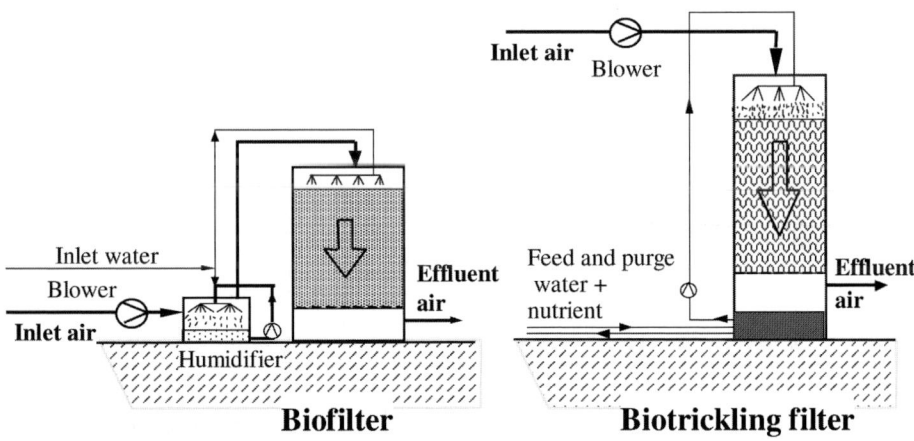

Figure 2. Schematic of biofilter and biotrickling filter setups. In-vessel systems are shown, but open bed design is common for biofilters. The air can be upflow or downflow. The biofilter shown includes sprinklers for additional moisture supply.

In biofilters, a humid stream of contaminated air is passed through a porous packed bed, usually made of a mixture of compost and wood chips or any other bulking agent [2,5,6]. On the packing, pollutant-degrading organisms form a biofilm and degrade the absorbed contaminants. Biofilters are very dry systems with no or little water trickling, hence any metabolite formed during biodegradation will stay in the damp material of the packing. Over time, this may cause inhibition of the process culture. Flushing of the bed is usually effective in removing accumulated metabolites, however it also leaches nutrients and often results in the compaction of the packed bed structure, hence, flushing should be exercised with caution. Biofilters are simple and cost effective. They require only low maintenance and are particularly effective for the treatment of odour and volatile compounds that are easy to biodegrade, and for compounds that do not generate acidic by-products. Biofilters are widely used in industrial applications for either VOC or odour control.

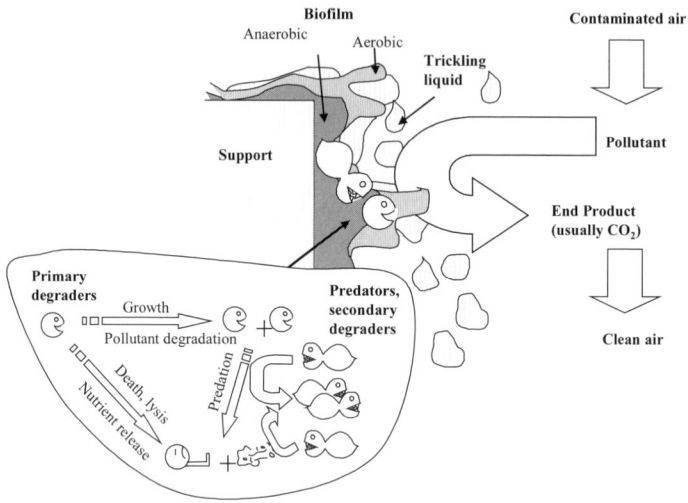

Figure 3. Simplified treatment mechanism in a biotrickling filter. Biofiltration mechanism is similar except that there is no free liquid trickling. Note that in both case, direct gas-biofilm contact exists.

Biotrickling filters work in a similar manner to biofilters, except that an aqueous phase is trickled over the packed bed, and that the packing is usually made of a synthetic or inert material, such as plastic rings, open pore foam, lava rock, *etc.* [7,8]. The trickling solution contains essential inorganic nutrients such as nitrogen, phosphorous, potassium, *etc.* and is either slowly trickled and wasted (one refers often to a trickling bed biofilter in such case), or trickled at a higher rate and partly recycled. Biotrickling filters are more complex to build than biofilters but are usually more effective. They are specially well suited for the treatment of compounds that generate acidic by-products, such as H_2S or methylene chloride, because the free aqueous phase allows for a tight control of the conditions such as pH or ionic strength. Another advantage of biotrickling filters is that they can be built taller than biofilters (2-3 m bed height, *vs.* 1-1.5 m in biofilters) because the packing used for biotrickling filters is usually not subject to compaction. This reduces the required footprint. Biotrickling filters are more recent than biofilters, and have not yet been deployed for industrial applications to the same extent as biofilters.

2. Microbiology of gas phase bioreactors

2.1. MICROFLORA

As shown in a simplified manner in Figure 3, pollutant elimination in biofilters or biotrickling filters is the result of many, interdependent processes that simultaneously take place inside the biofilm. The level of understanding of the detailed mechanisms of biofiltration and biotrickling filtration is still relatively limited, most probably because of the marked differences that exist between organisms in traditional bioprocesses and in biofilters or biotrickling filters (Table 3). However, significant progress has been made in the past decade, and a more detailed discussion of the process microbiology and biofilm architecture is warranted.

Table 3. Differences between traditional biotechnology and biofiltration processes.

Traditional biotechnology process	Biofilter or biotrickling filter
Pure cultures, closed system	Mixed cultures, presence of higher organisms, open system
Plenty of nutrients	Often nutrient limited
Conditions optimised for cell growth, growing cells	Suboptimum growth conditions. Many cells in a resting state, predation, death and lysis
Homogeneous well mixed systems	Mass transfer limitations, presence of significant heterogeneities
Consistent well defined substrate feed	Continuously changing pollutant-substrate nature and concentration

Inside the biofilm, biodegradation of the pollutant is mediated by mixed cultures of bacteria and fungi, thriving in a complex ecosystem. Initial inoculation with competent pollutant-degrading organisms is often unnecessary for biofilters, as most biofilters use compost as packing, which contains a large and diverse microflora suitable for the degradation of common air pollutants [2]. Inoculation is always required for biotrickling filters which packing does not naturally contain microorganisms. The inoculum is usually from enrichment cultures, or simply from activated sludge or another convenient source of pollutant-degrading organisms. For research purposes, some investigators have worked with monocultures and have sterilised air and aqueous feeds [9,10]. However it is practically impossible to maintain such a system free of bacterial contamination for extended periods of time. Almost all biofilters and biotrickling filters are open systems, in which the process culture naturally evolves over time depending on the operating conditions.

Webster *et al.* [11] observed changes in phospholipids fatty acids (PLFA) biomarkers for as long as 6 months after the startup of their biofilters. A common observation by several investigators is that biodiversity decreases over time, as a result of natural enrichment of those organisms for which the conditions are favourable, and the slow disappearance of all other organisms by death, lysis and cryptic growth [12]. Sakano and Kerkhof [13] found that over a period of about 100 days, only 38% of the original species remained in their ammonia treating biofilter.

Usually, effective pollutant removal is observed within a few days after initiating treatment. However, a more careful selection of the inoculum source is required when the pollutant is either very toxic, difficult to degrade, or when the conditions are very stringent (*e.g.*, low pH, or high temperature). Oh and Bartha [14] reported that a 4-week acclimation phase was required for effective treatment of nitrobenzene vapours. During the acclimation, the inlet concentration of nitrobenzene was incrementally increased to avoid toxic shocks. Another example illustrating the effect of the source inoculum is shown in Figure 4 for the treatment of methyl tert-butyl ether (MTBE), a compound which is only degraded by specialised organisms. Effective removal of MTBE was only observed after seven months of operation, when the biotrickling filter was inoculated with contaminated soil and groundwater. This is obviously an unacceptable delay for practical applications. The startup phase could be significantly shortened by inoculating the biotrickling filter with an adapted consortium. Finally, removal was observed only days after inoculating a pure culture capable of degrading MTBE.

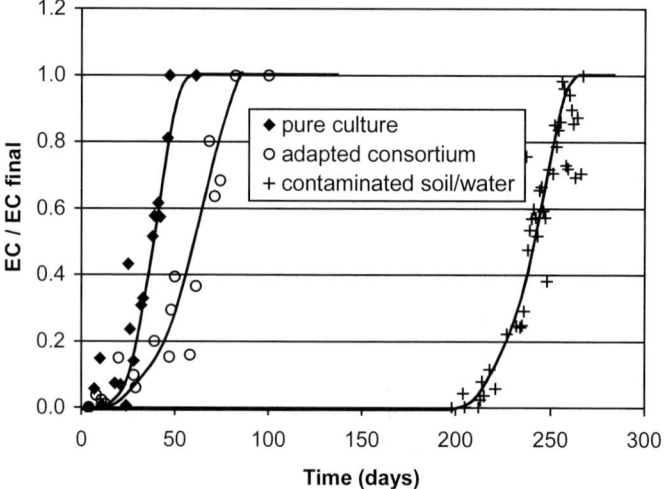

Figure 4. Start-up of MTBE-degrading biotrickling filters inoculated with different sources of microorganisms. EC = pollutant elimination capacity. (Source: unpublished data from Cox and Deshusses for the pure culture and adapted consortium, Fortin and Deshusses [15] for the contaminated soil/water inoculum data).

Many of the organisms that have been identified in biofilters and biotrickling filters are similar to those found in wastewater treatment. They include organisms such as *Pseudomonads, Rhodococcus, Acinetobacter, Corynebacteria, Gordona, Xanthomonas* species, *etc.* [2,16,17,18]. Interestingly, the primary pollutant-degraders are not necessarily the dominant organism in biofiltration systems. Moller *et al.* [19] described the distribution of *Pseudomonas putida* in the biofilm of a toluene degrading biotrickling filter using scanning confocal laser microscopy, 16S rRNA probes, and various staining techniques. Interestingly, *P. putida* constituted only 4% of the total biofilm population, but was responsible for about 65% of the degradation of the toluene vapours. Further, the comparison of rRNA content of *P. putida* in the biofilm and growing under optimum conditions in suspension indicated that toluene degradation activity by *P. putida* in the biofilm was substantially lower than in suspension. This can most probably be attributed to various stresses and limitations imposed on the immobilised culture as outlined earlier in Table 3.

Fungi have also frequently been observed, especially in biofilters, either as primary pollutant-degraders or as secondary degraders living off metabolites or degrading the lignocellulosic material of compost biofilters. Interestingly, white-rot fungi which are very popular in xenobiotic biodegradation studies have not found widespread application in gas phase bioreactors. A possible reason is that they degrade pollutants by means of extracellular peroxidases, a process which results in the formation of intermediate by-products [20] for which the white-rot fungus may not compete effectively. Still, other fungi have been successfully deployed in biofilters. Their resistance to low pH and low water activities makes them particularly well suited for gas-phase bioreactors. It has also been proposed that fungi may "harvest" pollutant *via* their airborne hyphae [21,22], therefore by-passing the mass transfer resistance posed by the liquid film. This may be particularly beneficial for the treatment of hydrophobic compounds. Indeed, effective removal of compounds such as styrene, toluene, alkylbenzenes, and nitric oxide has been reported in fungal based bioreactors [22-26].

2.2. BIOFILM ARCHITECTURE

An important difference between the organisms present in biofilters and biotrickling filters an those discussed in many chapters of this book is that organisms in biofilters and biotrickling filters are present in biofilms rather than entrapped or attached. The biofilms consist mostly of cells and exopolymers and water [27]. The biofilms usually range from 20 micrometer to several millimetres thick, although it has been shown that in most cases, only the section that is close to the gas or liquid interface is biologically active, while the rest of the biofilm is subject to mass transfer limitations [19,27,28].

Early representations of biofilms considered the biofilm to be a flat surface. It is now well known that biofilms are far from being planar but that they are very heterogeneous, with large channels extending from the gas or liquid interface of the biofilm deep into the biofilm, sometimes up to the substratum. Such channels and heterogeneities are thought to contribute to a possible enhancement of pollutant and oxygen mass transfer. de Beer *et al.* [28] evaluated that for submerged biofilms, the supply of oxygen through such voids and channels was roughly 50% of the total oxygen transfer.

Detailed microscopic observation [29] of relatively thick biofilms (2-5 mm) taken from an active H2S degrading biotrickling filters revealed that they include three regions with approximately constant relative thicknesses. The external film (5% of total thickness) had a high bacterial and hyphal tip density and included numerous channels free of cells, as discussed above. While it was not measured, it is reasonable to speculate that this region was responsible for most of the biodegradative activity. The intermediate region (about 20% of total thickness) was characterised by a lower density of bacteria, an increased density of hyphae and an absence of channel-like structures. Also a number of nematodes were detected. Finally, the basal region (75% of total thickness) composed of tightly compacted hyphae against the substratum exhibited little staining with haemato-eosin. This indicates the absence of cytoplasmic material, i.e., the organisms were dead or completely inactive. The heavy presence of presumably heterotrophic fungi throughout the biofilm in a reactor used for hydrogen sulphide and carbon disulfide treatment suggests that the fungi were essentially living off the by-products or metabolites secreted by the primary degraders. A similar observation was made by Woertz et al. [26] who described the presence of heterotrophic denitrifying fungi in toluene degrading biofilters. It was shown that the fungi used metabolites of toluene biodegradation while denitrifying nitric oxide. In fact, heterotrophic denitrification of nitrate (while degrading organic contaminants) could be more widespread in gas phase biotrickling filtration than originally thought. The process would enable biodegradation of organic pollutants in anaerobic pockets resulting from oxygen diffusion limitation, and would be favoured in reactors where plenty of nitrate is available.

2.3. SECONDARY DEGRADERS AND PREDATORS

The current knowledge of the nature and role of secondary degraders and predators in biofilters and biotrickling filters is relatively limited. It is well known that protozoa and other higher organisms are always present in gas phase bioreactors. Their cryptic growth [12] and predation of the primary degraders are expected to play a major role in recycling essential nutrients such as nitrogen, phosphorous and potassium. In fact, applying conventional rules of thumb for carbon to nutrient ratio to operating biofilters and biotrickling filters reveals that those bioreactors are usually moderately to severely limited by the supply nutrients. While nutrient limitation is sometimes done on purpose for limiting the growth of biomass and avoid plugging of the bed [30,31,32], effective VOC biodegradation under severe nutrient limitation can only be explained by the recycling action of the predators and by the fact that the process culture degrades VOCs to satisfy its maintenance requirements.

Higher organisms also impact the rate of biomass accumulation by either physically detaching biofilms through their grazing and tunnelling activity, or by converting part of the biofilm to carbon dioxide as a result of their own metabolism. Cox and Deshusses [33] showed that biomass accumulation was slower in a toluene-degrading biotrickling filter in which selected protozoa were added, compared to a protozoa-free control. The presence of protozoa correlated with a higher rate of toluene degradation, a higher CO_2 yield and lower pressure drop. Interestingly, the startup of the biotrickling filter with protozoa was faster than the control, possibly because of growth factors or recycled nutrients secreted by the protozoa. Cox and Deshusses also observed that the biofilms grown in protozoa-free environment were denser, and would not detach as easily from the substratum. Still, the predation by the protozoa was not sufficient to completely balance the growth of the process culture and the biotrickling filter eventually clogged. Other studies on predation include one by Woertz et al. [34] who observed that biofilters inoculated with the toluene-degrading fungus *Cladophialophora sp.* had a higher performance and a lower pressure drop when *Tyrophagus* mites were added to the system and allowed to graze on the fungus.
In another study, Won et al. [35] described the effect of fly larvae on biotrickling filters for air pollution control. Flies have frequently been associated with biofilters and are usually considered as a nuisance. Here, a small fly identified as *Telmatoscopus albipunctatus*, a *Psychodidae* fly species thriving on decaying material, drain, and

wastewater treatment activities invaded a lab-scale biotrickling filter and larvae rapidly spread throughout the reactor. This resulted in effective removal of biomass from the packing. The wet biomass content of the reactor was reduced from 455 to 28 kg m^{3}reactor in 16 days with 80% of the biomass reduction occurring in 2-4 days. Analysis of the recycle liquid indicated that the major mechanism of biomass removal was detachment of biofilm, while an estimated 2-10% of the removal may have been by consumption by the larvae. It was speculated that larval activity loosened the biofilm structure, thus enhancing biofilm detachment by shear-stress from the trickling liquid.

Overall, these studies conducted with higher organisms illustrate the relatively unexplored potential for controlling biomass and possibly enhance the performance of biofilters and biotrickling filters. However, the complexity of these multispecies systems governed by interconnected relationships still represents a major research challenge.

2.4. BIODEGRADATION AND GROWTH KINETICS

The kinetics of pollutant elimination in the biofilm are influenced by numerous factors including the type and design of the bioreactor, the operating conditions, the pollutant(s) being treated, *etc.* Also, conditions along the height of the reactor are expected to be drastically different. At the micro scale, biodegradation rates are influenced by the environmental conditions such as pH, substrate and nutrient concentrations, *etc.* which are themselves affected by mass transfer limitations. Clearly, the situation is complex. In a first approximation, neglecting heterogeneities and mass transfer effects, Deshusses and Cox [36] proposed that, one could write that the pollutant elimination capacity (*EC*) expressed in g of pollutant degraded per unit bed volume per hour, depended on the intrinsic growth rate of the active fraction of the primary degraders (X_1) and its maintenance requirement, as in Equation 1.

$$EC = \left(\frac{\mu}{Y_{X/S}} + m \right) X_{1 \text{ (active fraction)}} \qquad (1)$$

where μ is the specific growth rate of the primary degraders, $Y_{X/S}$ is the biomass yield, m the maintenance energy requirement, and $X_{1(active\ fraction)}$ is the biomass content of active primary degraders per volume of reactor. Note that the maintenance energy is included in the kinetic relationship, as it is suspected to be an important mechanism of pollutant removal compared to growth associated pollutant degradation. This is because pollutant-substrate concentrations can be very low (see Table 4, below), and because the primary degraders are subject to significant stresses that may result in high maintenance requirements. The shift from a growing to an essentially non-growing metabolism over time was extensively discussed by Cherry and Thomspon [37] and experimentally proven later by Fuerer and Deshusses [38].

The specific growth rate of the active fraction of the primary degraders can be expressed using a modified Monod equation that takes into account all possible limitations.

$$\mu = \frac{\mu_{max} S}{Ks + S} \frac{N}{Ks_N + N} \frac{O}{Ks_O + O} \frac{I}{1 + \frac{I}{K_I}} \qquad (2)$$

where S is the pollutant-substrate concentration, N is a lumped parameter representing the concentration of limiting nutrients, O is the dissolved oxygen concentration, I the concentration of any inhibitor, and Ks, Ks_N, Ks_O, and K_I are the respective half-saturation and inhibition constants.

Interestingly, biofilters and biotrickling filters have been applied for pollutants with an extremely broad spectrum of physico-chemical properties. Of prime importance is Henry's law coefficient of the pollutant undergoing treatment. Table 4 illustrates that the actual concentration seen by the process culture may vary by several orders of magnitude, depending on the application. This will of course affect the biodegradation kinetics (see Equation 2). Pollutants with low Henry coefficients will favourably partition into the biofilm and concentrations will often be above the substrate half-saturation constant Ks. These pollutants may cause oxygen depletion in the biofilm if present at high concentration in the air undergoing treatment. Pollutants with higher Henry coefficients will have concentrations in the biofilm that are comparable or below the Ks value, and pollutant removal will be subject to kinetic limitations. Certainly, the lowest biofilm concentrations indicated in Table 4 raise interesting fundamental questions on the induction of key enzymes in the process culture. This remains a relatively unexplored area.

Table 4. Henry's law coefficient (H) and typical gaseous inlet concentration of selected pollutant and corresponding liquid equilibrium concentration. Note that the actual concentration in the biofilm may be orders of magnitude lower because of mass transfer limitations and/or axial position in the reactor.

Pollutant	Typical application	Typical inlet gas concentration (g m^{-3})	H (-)	Corresponding pollutant equilibrium liquid concentration (mg L^{-1})
Methanol	Wood industry	0.2-1.0	0.0002	1000-5000
Methyl ethyl ketone	Paint spray booth	0.1-0.5	0.0024	40-200
Toluene	Various processing	0.1-0.5	0.275	0.4-2
Toluene	Air exhaust, wastewater treatment plants	0.002	0.275	0.007
H$_2$S	Air exhaust, wastewater treatment plants	0.003-0.15	0.385	0.008-0.1
NO	Combustion gases	0.5	21	0.024
Hexane	Various processing	0.1-0.5	53	0.002-0.01

As far as secondary degraders as concerned, an equation similar to Equation 2 can be written for all the species (or group of species) present in the system. Each will have one or several specific substrates (such as a metabolite, exopolymer, primary degrading organism, *etc.*), specific kinetic constants, and thus a different specific growth rate. The overall rate of biomass accumulation in the biofilter or biotrickling filter system is the sum for all the different species (designated by the index i in Equation 3) of the growth rate minus death and lysis (d_i term), the predation by other species and the wash-out *via* leaching. This is expressed in Equation 3.

$$\text{Rate of biomass accumulation} = \sum_i \left((\mu_i - d_i)X_i - \text{Predation}_i - \text{Wash out}_i\right) \quad (3)$$

As mentioned, Equations 1-3 are highly simplified since they do not take local heterogeneities into account. Still, they define a number of parameters that are impossible to determine. A possible solution is to conceptually split the process culture into large classes of organisms, such as primary degraders, secondary degraders, predators, *etc.* and use lumped kinetic parameters. This is an area of current research. Even so, Equations 1-3 reflect the fact that the pollutant elimination, secondary processes and the observed biomass growth are interconnected in a complex manner. Equation 3 can be further used to support the development of biomass control strategies by minimising or zeroing out the biomass accumulation term. Examination of Equation 3 shows that meaningful biomass control strategies should consider means to reduce the specific growth rate [30-32,39], accelerating death and lysis, stimulating predation [33,34], and mechanically washing out the biomass [40-42]. The challenge is to conduct any of the above biomass control measures in a cost effective manner, without negatively affecting the rate of pollutant elimination by the primary degrading organisms (Equation 1).

3. Biofilter and biotrickling filter applications

3.1. DEFINITIONS AND PERFORMANCE REPORTING

The performance of bioreactors for air pollution control is generally expressed as pollutant removal efficiency, or pollutant elimination capacity and reported as a function of the pollutant inlet concentration, pollutant loading, or gas empty bed retention time (EBRT). These terms are defined below (Equations 4-7).

$$\text{Removal} = RE = \frac{C_{in} - C_{out}}{C_{in}} 100 \quad (\%) \quad (4)$$

$$\text{Pollutant Elimination Capacity} = EC = \frac{(C_{in} - C_{out})}{V} Q \quad (g\ m^{-3}\ h^{-1}) \quad (5)$$

$$\text{Empty Bed Retention Time} = EBRT = \frac{V}{Q} \quad (s\ or\ min) \quad (6)$$

$$\text{Pollutant loading} = L = \frac{C_{in}}{V} Q \quad (g\ m^{-3}\ h^{-1}) \quad (7)$$

where C_{in} and C_{out} are the inlet and outlet pollutant concentrations (usually in $g\ m^{-3}$ or in dilution to threshold for odour removal), respectively, V is the volume of the packed bed (m^3) and Q is the air flow rate ($m^3\ h^{-1}$). It should be stressed that the elimination capacity and the loading are calculated using the volume of the packed bed and not the total volume of the reactor. Such normalisation enables comparison of systems of different sizes operated under different conditions. Depending on the reactor design, the volume of the packed bed will be about 40-90% of the total reactor volume [2]. Also, EBRT is calculated on the basis of the total volume of packed bed (Equation 6) and not the void space in the packing. This is because the porosity of the bed is usually unknown and it may change over time as a result of bed compaction (in biofilters) or biomass growth. Typical porosities range from 40 to 60% of the bed volume, hence the actual gas residence time will be much lower that the EBRT. Typical EBRTs in biofilters and biotrickling filters range from 15-60 seconds. Obviously, compounds that are more difficult to biodegrade require longer contact times.

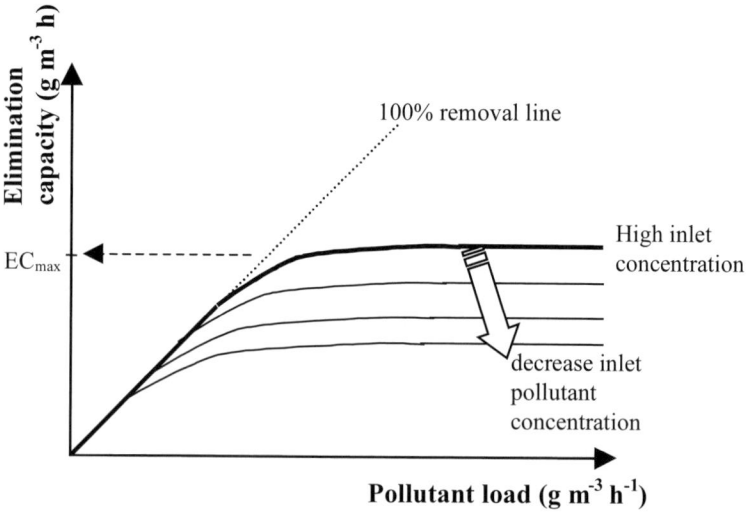

Figure 5. Schematic of a typical elimination capacity vs. load curve for a biofilter or a biotrickling filter.

In the case of a typical VOC, reporting pollutant elimination capacity *vs.* its loading usually results in a curve similar to the one shown in Figure 5. One underlying assumption is that the performance depends only on the pollutant load, hence, that low concentrations-high flow rates conditions lead to similar elimination capacities than high concentrations-low flow rates. This assumption is often valid in laboratory studies because the pollutant concentrations treated in biofilters and biotrickling filters are often high enough for the micro-kinetics to be of zero order. As schematically shown in Figure 5, the assumption is no longer true at low pollutant concentrations, in particular for pollutants with high Henry's law coefficients, because first order kinetics will prevail in the biofilm resulting in a reduction of the maximum elimination capacity.

Examination of Figure 5 reveals that there are essentially three operating regimes:
- Low loading, also called first order regime. The elimination capacity and the loading are identical and the pollutant is completely removed. The bioreactor is operated well below its maximum elimination capacity. The performance increases proportionally with the loading.
- Intermediate range. Breakthrough of the pollutant occurs. With higher inlet concentrations or higher air flow rates, the elimination capacity increases, but to a lesser extent than the loading.
- High loading, also called zero order regime. The biofilter or biotrickling filter is operated at its maximum elimination capacity. Increases in loading do not result in further increases in elimination capacity, the removal efficiency decreases. As indicated on the figure, if the concentration is very low, the maximum elimination capacity will be reduced.

For the evaluation of biofilter or biotrickling filter performance, one should consider both the maximum elimination capacity and the removal efficiency. For practical reasons, academic research is mainly concerned with the maximum elimination capacity or with high performance, which occur at relatively high pollutant concentration and often less than ~ 90% removal efficiency. On the other hand, reactor design for industrial application often needs to meet a certain discharge requirement, or achieve a high removal percentage. Also, lab-scale studies are usually conducted with single pollutant under steady conditions. On the other hand, industrial exhausts usually consist of mixtures of pollutant, with high variability in concentrations over time. Thus there are significant challenges in extrapolating research data for reactor design.

3.2. EXAMPLES OF APPLICATIONS

In the present section, selected examples of application are briefly listed. It should be stressed that research in the area of gas-phase bioreactors has increased exponentially over the past decades, and that numerous papers exist on lab-scale applications.

Comparison of the performance between different studies is made difficult by the differences in the reactor types, packing, operating conditions, *etc*. As a result, the reported performance for a given pollutant often greatly vary depending on the studies, which makes design of large scale biofilters and biotrickling filters a difficult task in absence of pilot testing. Table 5 also lists selected field applications. It should be mentioned that there is a relatively large number of full-scale biofilters operating successfully in the field. Because of environmental regulations, a majority of applications in the USA has been for odour control, whereas in Europe, there is a greater proportion of applications for VOCs control. A picture of an actual biofilter is shown in Figure 6.

Table 5. Examples of biofilter and biotrickling filter applications (laboratory and full-scale). Bf = biofilter; Btf = biotrickling filter.

Pollutant	Type of reactor	Packing	EBRT (s)	EC$_{max}$ (g m^{-3} h^{-1})	Remarks	Ref.
Ethanol	Lab-scale bf	Granular activated carbon	186	200	Acetic acid was formed probably because of oxygen limitation	[43]
Styrene	Lab-scale bf	Perlite	25-90	60	Fungal based system	[22]
Hexane	Lab-scale bf	Compost + perlite	60	21	System performance correlated with nutrient supply	[44]
Diethyl ether	Lab-scale btf	Celite (inert silicate pellets)	25	60	Low trickling rate, one pass	[45]
Methyl ethyl ketone	Lab-scale btf	Wood bars Polyprop. Spheres	88 88	30 (wood) 40 (PP)	Performance correlated with packing area	[46]
Toluene	Lab-scale btf	2.5 cm Polyprop. Pall rings	56	71-83	CO_2 balances are shown	[33]
Toluene	Lab-scale btf	Steel Pall rings	32-160	35	Partial removal at high EC	[47]
Toluene	Lab-scale bf	Celite (inert silicate pellets)	60	270	Fungal based system	[26]
Methanol	Lab-scale btf	Polypropylene	60	85-90	Thermophilic operation at 70°C	[48]
Alpha-pinene	Lab-scale btf	Polypropylene	60	30	At 55°C, no removal at 70°C	[48]
Gasoline vapours	Full-scale bf	Compost + perlite	96-180	Not reached	60-90% removal of 0.4-0.8 g m^{-3}	[49]
CS_2	Full-scale btf	Structured packing, plastic	33-40	220	Simultaneous 99% removal of trace H_2S	[50]
H_2S	Full-scale btf	Open pore polyurethane foam	1.6-2.3	105	Secondary effluent used as nutrient source. pH of btf = 1.8.	[51]
Odours	Full-scale bf	Soil	210	NA	99% odour removal	[2]
Odours	Full-scale btf	Open pore polyurethane foam	11	NA	>90% odour removal	[2]

Figure 6. Picture of an actual biofilter. The in-vessel system treats H_2S and other odours at a wastewater treatment facility. The EBRT is 20 s. (Courtesy of Biorem Technologies, Inc.)

4. Current research, emerging topics

Bioreactors for air pollution control are a major progress in environmental protection. Currently, factors slowing down the deployment of air biotreatment techniques include both regulatory issues, technical issues, and the inherent resistance of the market to new technologies. The latter aspect is also significantly influenced by the preconceived idea that biological systems are unreliable and can not consistently sustain effective treatment over time. With the increasing number of success stories, this element should decrease in the near future. As far as regulations are concerned, there is still a bias towards less environmental friendly techniques such as incineration or carbon adsorption. While, these techniques may be able to achieve higher pollutant removal percentages, they usually do not compare well with biological techniques when the global environmental impact is considered. For example, incineration of low VOC concentrations requires burning additional fossil fuel and generate nitrogen oxides and CO_2. The global impact of these is usually not considered, although it can be significant. Further studies in the area of lifecycle assessment are required to adjust the metrics used in air pollution control regulations, and to inform regulators about the benefits of biotechniques for air pollution control.

The present know-how in air biotreatment is sufficient to deal with a large number of cases, in particular in VOC and odour control. Still, a variety of challenges remain before the complex phenomena that occur during biotreatment in vapour-phase bioreactors are fully understood. Whether such a detailed understanding is absolutely required is debatable. In fact, the situation is not very different from the current situation with wastewater treatment, where biotreatment is widely applied, but fundamental aspects of the treatment are not fully understood. Fundamental research challenges in air biotreatment include understanding the complex microbiology of the process, and mass transfer aspects specific to gas-phase bioreactors. Current research efforts are directed towards developing new vapour-phase bioreactors for air pollution control, new packing supports for microorganisms, optimisation of current applications and finding new ones. Selected exciting recent developments or areas of growing research are briefly summarised below.

As discussed in Section 2, the use of fungi in vapour-phase bioreactors appears to be extremely promising. It has already been shown that fungal based system can often degrade VOCs at faster rate than conventional bacterial based systems. Fungal based systems seem to be less sensitive to pH and relative humidity changes. Still several issues such as the control of the growth of fungi (especially if dimorphic fungi are used), or the possible emissions of spores, remain before fungal systems can be widely applied in the field.

In the recent years, biotreatment has been proposed for the control of trace indoor air contaminants [52,53]. Indoor air is characterised by very low contaminant concentrations and application of biofiltration should consider issues such as the release of microorganisms and of unwanted odours, as well as excessive humidification of the treated air. The biofilter that has been reported for indoor application is somewhat different from those discussed in this chapter. It includes plants, living mosses and is more like a small ecosystem [52]. Presumably, photosynthesis and the resulting plant residues may provide the system with the extra energy it requires to support active bacterial life, so that trace contaminant removal can be accomplished. The reported pollutant removal performance appears to be promising, and if this application matures to a commercial development, the impact could be very large, as indoor air quality is a major concern of industrialised countries.

Recently, effective treatment of low H2S concentrations (2-50 ppmv) has been shown in a high performance biotrickling filter [51]. The originality of the work lies in the fact that an existing chemical scrubber was converted to a biotrickling filter, keeping the original gas contact time of 1.6-2.3 seconds, *i.e.*, a much shorter time than any previous biotrickling filtration studies. The scrubber vessel was reused and the conversion procedure was simple and relatively inexpensive. Reclaimed water was used as nutrient source for the process and for maintaining the pH in the biotrickling filter between 1.5 and 2.2. Under these conditions, successful treatment of H_2S at rates comparable to those of chemical scrubbers was observed. H_2S removal was in excess of 98% for inlet H_2S concentrations as high as 30 to 50 ppm_v. This corresponds to volumetric elimination rates of H_2S of 95 to 105 g H_2S m^{-3} h^{-1}. Such performance is exceptionally high compared with other biofilters or biotrickling filters removing low concentration of H_2S, even at higher gas contact times. The authors speculate that a combination of high pollutant mass transfer rate due to the special polyurethane packing support that was used and optimum operating conditions (nutrient, pH, CO_2) was responsible for the unprecedented performance. Significant removal of reduced sulphur compounds, ammonia and volatile organic compounds present in traces in the air was also simultaneously observed. The demonstration that H_2S can be effectively treated biologically at contact times comparable to those of chemical scrubbers has substantial implications for odour control, as a large number of chemical scrubbers could be similarly converted to biotrickling filters, with substantial cost and environment health and safety benefits.

Highly chlorinated solvents remain some of the most challenging compounds as far as air biotreatment is concerned. This is because of their recalcitrance under aerobic conditions or the requirements of co-substrates for inducing cometabolism. Recently, Schwartz et al. [54] treated simulated landfill gas, consisting of an equimolar mixture of carbon dioxide and methane with traces of containing tetrachloroethylene (PCE) in a biofilter under anaerobic conditions. Sucrose was added as an additional energy and carbon source. Tetrachloroethylene was dechlorinated completely. The final end-product of biodegradation could not be identified, but the usual PCE biodegradation metabolites (vinyl chloride, dichloroethylene and trichloroethylene) were excluded. After an acclimation period the removal efficiency was higher than 98%. Such development could be applied to the remediation of contaminated groundwater, by sparging the aquifer with nitrogen, and treating chlorinated solvent vapours in a biofilter. In another study by Sun and Wood [55] a different approach was used for the treatment of trichloroethylene (TCE) vapours. A pure culture of *Burkholderia cepacia* PR_{123} (TOM_{23C}) that constitutively expresses toluene ortho-monooxygenase (TOM) was used to cometabolize TCE vapours in a biotrickling filter. Usually, aerobic biodegradation of TCE only occurs through cometabolism, and a growth substrate (*e.g.*, toluene, methane, propane, phenol, or ammonia) is required to induce the expression of the appropriate TCE-degrading enzyme. *B. cepacia* PR_{123}, however, expresses TOM constitutively, which avoids the competitive inhibition by the inducer. Sun and Wood used a glucose solution as a carbon and energy source and observed TCE eliminations up to 200 times higher than previously reported. However, rapid inactivation of the TCE-degrading enzyme by TCE breakdown products (TCE epoxide) remained a problem. In both these applications, economical and technological challenges exist before deployment at the industrial scale.

The recent advances in molecular methods for studying microbial communities in complex environments has started to impact the research in vapour-phase bioreactors. These tools include DNA microarrays, 16S rDNA based analyses, fluorescence *in-situ* hybridisation (FISH), 2D gel electrophoresis, specific gene probes, *etc.* [56] While these new tools have only recently been applied to vapour-phase bioreactors, they appear to be extremely promising in describing the complex ecology of biofilters and biotrickling filters [13,18,57,58,59]. A challenge for the application of these tools is to relate the observations made with these advanced tools to the performance of the reactor systems, so that the advanced tools are used to develop a better understanding of the process, or as a diagnostic for troubleshooting, or to support process optimisation.

There are many other research issues in air biotreatment. One important one that has received little attention is this of the biodegradation kinetics of very low concentrations of pollutants in vapour-phase bioreactors. Clearly, a better understanding of the physiology of the process culture, of the induction of key enzymes and of the pollutant metabolism under the complex environmental conditions experienced in vapour-phase bioreactors is desirable. This should enable scientists and engineers to design better reactors that are more effective over a broader range of conditions, and a wider spectrum of pollutants. Another issue is this of pollutant mass transfer. As improvements in the biological kinetics are made, pollutant removal in gas-phase bioreactors may become limited by the rate at which pollutant is transferred to the microorganism. Hence, a better understanding of mass transfer in biofilters and biotrickling filters is required. Advances in this area will support the development of new packings and possible means to improve mass transfer. Undoubtedly, kinetic and mass transfer aspects are closely linked, and an integrated approach will be required.

References

[1] Pomeroy, R.D. (1957) Deodorizing gas streams by the use of microbiological growths. US Patent 2.793.096.
[2] Devinny, J.S.; Deshusses, M.A. and Webster, T.S. (1999) Biofiltration for air pollution control. Lewis Publishers, Boca Raton, FL, USA; 300 pages.
[3] van Groenestijn, J.W. and Hesselink, P.G.M. (1993) Biotechniques for air pollution control. Biodegradation. 4: 283-301.
[4] Leson, G. and Winer, A.M. (1991) Biofiltration - An innovative air pollution control technology for VOC emissions. J. Air Waste Manage. Assoc. 41: 1045-1054.
[5] Kennes, C. and Veiga, M.C. (2001). Conventional Biofilters. In: Kennes, C. and Veiga, M.C. (Eds.) Bioreactors for Waste Gas Treatment. Kluwer Academic Publisher, The Netherlands, 47-98.
[6] Deshusses, M.A. (1997) Biological waste air treatment in biofilters. Curr. Opin. Biotechnol. 8(3): 335-339.
[7] Cox, H.H.J. and Deshusses, M.A. (1998) Biological waste air treatment in biotrickling filters. Curr. Opin. Biotechnol. 9(3): 256-262.
[8] Cox, H.H.J and Deshusses, M.A. (2001) Biotrickling Filters. In: Kennes, C. and Veiga, M.C. (Eds.) Bioreactors for Waste Gas Treatment. Kluwer Academic Publisher, The Netherlands, 99-131.
[9] Kirchner, K.; Wagner, S. and Rehm, H.J. (1992) Exhaust gas purification using biocatalysts (fixed bacteria monocultures) - the influence of biofilm diffusion rate (O2) on the overall reaction rate. Appl. Microbiol. Biotechnol. 37: 277-279.
[10] Zilli, M.; Converti, A.; Lodi, A.; Del Borghi, M. and Ferraiolo, G. (1993) Phenol removal from waste gases with a biological filter by *Pseudomonas putida*. Biotechnol. Bioeng. 41: 693-699.
[11] Webster, T.S.; Devinny, J.S.; Torres, E.M. and Basrai, S.S. (1997) Microbial ecosystems in compost and granular activated carbon biofilters. Biotechnol. Bioeng. 53: 296-303.
[12] Mason C.A.; Hamer, G. and Bryers, J.D. (1986) The death and lysis of microorganisms in environmental processes. FEMS Microbiol. Rev. 39: 373-401.
[13] Sakano, Y. and Kerkhof, L. (1998) Assessment of changes in microbial community structure during operation of an ammonia biofilter with molecular tools. Appl. Environ. Microbiol. 64: 4877-4882.
[14] Oh, Y.S. and Bartha, R. (1997) Removal of nitrobenzene vapors by a trickling air biofilter. J. Ind. Microbiol. Biotechnol. 18: 293-296.
[15] Fortin, N.Y. and Deshusses, M.A. (1999) Treatment of methyl tert-butyl ether vapors in biotrickling filters. 1. Reactor startup, steady-state performance, and culture characteristics. Environ. Sci. Technol. 33: 2980-2986.
[16] Bendinger, B.; Kroppenstedt, R.M.; Klatte, S. and Altendorf, K. (1992) Chemotaxonomic differentiation of coryneform bacteria isolated from biofilters. Int. J. Syst. Bacteriol. 42: 474-486
[17] Juteau, P.; Larocque, R.; Rho, D. and Le Duy, A. (1999) Analysis of the relative abundance of different types of bacteria capable of toluene degradation in a compost biofilter. Appl. Microbiol. Biotechnol. 52(6): 863-868.
[18] Friedrich,U.; Naismith, M.M.; Altendorf, K. and Lipski A. (1999). Community analysis of biofilters using fluorescence in situ hybridization including a new probe for the Xanthomonas branch of the class Proteobacteria. Appl. Environ. Microbiol. 65: 3547-3554.

[19] Moller, S.; Pedersen, A.R.; Poulsen, L.K.; Arvin, E. and Molin, S. (1996) Activity and three-dimensional distribution of toluene-degrading *Pseudomonas putida* in a multispecies biofilm assessed by quantitative in situ hybridization and scanning confocal laser microscopy. Appl. Environ. Microbiol. 12: 4632-4640.

[20] Yadav, J.S. and Reddy, C.A. (1993) Degradation of benzene, toluene, ethylbenzene and xylenes (BTEX) by the lignin-degrading basidiomycete *Phanerochaete chrysosporium*. Appl. Environ. Microbiol. 59: 756-762.

[21] Braun-Lüllemann, A.; Majcherczyk, A.; Tebbe, N. and Hüttermann, A. (1992) Bioluftfilter auf der Basis von Weiβfäulepilzen. In: Dragt, A.J. and van Ham, J. (Eds.) Biotechniques for Air Pollution Abatement and Odour Control Policies. Elsevier, Amsterdam, The Netherlands, 91-95.

[22] Cox, H.H.J.; Moerman, R.E.; van Baalen, S.; van Heiningen, W.N.M.; Doddema, H.J. and Harder, W. (1997) Performance of a styrene-degrading biofilter containing the yeast *Exophiala jeanselmei*. Biotechnol. Bioeng. 53: 259-266.

[23] Garcia-Pena, E. I.; Hernandez, S.; Favela-Torres, E.; Auria, R. and Revah, S. (2001) Toluene biofiltration by the fungus *Scedosporium apiospermum* TB1. Biotechnol. Bioeng. 76(1): 61-69.

[24] Phae, C.G. and Shoda, M. (1991) A new fungus which degrades hydrogen sulphide, methanethiol, dimethyl sulphide and dimethyl disulfide. Biotechnol. Lett. 13: 375-380.

[25] Woertz, J.R.; Kinney, K.A.; McIntosh, N.D.P. and Szaniszlo, P.J. (2001) Removal of toluene in a vapor-phase bioreactor containing a strain of the dimorphic black yeast *Exophiala lecanii-corni*. Biotechnol. Bioeng. 75(5): 550-558.

[26] Woertz, J.R.; Kinney, K.A. and Szaniszlo, P.J. (2001) A fungal vapor-phase bioreactor for the removal of nitric oxide from waste gas streams. J. Air Waste Manage. Assoc. 51(6): 895-902.

[27] Characklis, W.G. and Marshall, K.C. (1990) Biofilms. Wiley & Sons, New York, NY, USA; 796 pages.

[28] de Beer, D.; Stoodley, P.; Roe, F. and Lewandowski, Z. (1994) Effects of biofilm structures on oxygen distribution and mass transport. Biotechnol. Bioeng. 43: 1131-1138.

[29] Hugler, W.C.; Cantu-De la Garza J.G.; and Villa-Garcia, M. (1996) Biofilm analysis from an odor-removing trickling filter. In: Air & Waste Management Association (Ed.) Proc. Annual Meeting and Exhibition of the Air & Waste Management Association. Pittsburgh, PA, USA; paper 96-RA87A.04: 20 pages.

[30] Holubar, P.; Andorfer, C. and Braun, R. (1999) Effects of nitrogen limitation on biofilm formation in a hydrocarbon-degrading trickle-bed filter. Appl. Microbiol. Biotechnol. 51: 536-540.

[31] Wübker, S.M. and Friedrich, C. (1996) Reduction of biomass in a bioscrubber for waste gas treatment by limited supply of phosphate and potassium ions. Appl. Microbiol. Biotechnol. 46: 475-480.

[32] Weber, F.J. and Hartmans, S. (1996) Prevention of clogging in a biological trickle-bed reactor removing toluene from contaminated air. Biotechnol. Bioeng. 50: 91-97.

[33] Cox, H.H.J. and Deshusses, M.A. (1999) Biomass control in waste air biotrickling filters by protozoan predation. Biotechnol. Bioeng. 62: 216-224.

[34] Woertz, J.R.; van Heiningen, W.N.M.; van Eekert, M.H.A.; Kraakman, N.J.R.; Kinney, K.A. and van Groenestijn, J.W. (2002) Dynamic bioreactor operation: Effects of packing material and mite predation on toluene removal from off-gas. Appl. Microbiol. Biotechnol. 58(5): 690-694.

[35] Won, Y.S.; Cox, H.H.J.; Walton, W.E. and Deshusses, M.A. (2002) An environmentally friendly method for controlling biomass in biotrickling filters for air pollution control. In: Air & Waste Management Association (Ed.) Proc. of the Annual Meeting and Exhibition of the Air & Waste Management Association. Pittsburgh, PA, USA; paper #43554: 12 pages.

[36] Deshusses, M.A. and Cox, H.H.J (2002) Biotrickling Filters for Air Pollution Control. In: Bitton, G. (Ed.) The Encyclopedia of Environmental Microbiology, Vol. 2. Wiley & Sons, New York, NY, USA, 782-795.

[37] Cherry, R.S. and Thompson, D.N. (1997) Shift from growth to nutrient-limited maintenance kinetics during biofilter acclimation. Biotechnol. Bioeng. 56(3): 330-339.

[38] Fürer, C. and Deshusses M.A. (2000) Biodegradation in biofilters: Did the microbe inhale the VOC? In: Air & Waste Management Association (Ed.) Proc. of the Annual Meeting and Exhibition of the Air & Waste Management Association. Pittsburgh, PA, USA; paper #799: 13 pages.

[39] Schönduve, P.; Sára, M. and Friedl, A. (1996) Influence of physiologically relevant parameters on biomass formation in a trickle-bed bioreactor used for waste gas cleaning. Appl. Microbiol. Biotechnol. 45: 286-292.

[40] Laurenzis, A.; Heits, H.; Wübker, S.M.; Heinze, U.; Friedrich, C. and Werner, U. (1998) Continuous biological waste gas treatment in stirred trickle-bed reactor with discontinuous removal of biomass. Biotechnol. Bioeng. 57: 497-503.

[41] Smith, F.L.; Sorial, G.A.; Suidan, M.T.; Breen, A.W.; Biswas, P. and Brenner, R.C. (1996) Development of two biomass control strategies for extended, stable operation of highly efficient biofilters with high toluene loadings. Environ. Sci. Technol. 30: 1744-1751.

[42] Cox, H.H.J. and Deshusses, M.A. (1999) Chemical removal of biomass from waste air biotrickling filters: screening of chemicals of potential interest. Wat. Res. 33: 2383-2391.

[43] Devinny, J.S. and Hodge, D.S. (1995) Formation of acidic and toxic intermediates in overloaded ethanol biofilters. J. Air Waste Manage. Assoc. 45: 125-131.

[44] Morgenroth, E.; Schroeder, E.D.; Chang, D.P.Y. and Scow, K.W. (1996) Nutrient limitation in a compost biofilter degrading hexane. J. Air Waste Manage. Assoc. 46: 300-308.

[45] Zhu, X.; Rihn, M.J.; Suidan, M.T.; Kim, B.J. and Kim, B.R. (1996) The effect of nitrate on VOC removal in trickle bed biofilters. Wat. Sci. Technol. 34: 573-581.

[46] Chou, M.S. and Huang, J.J. (1997) Treatment of methyl ethyl ketone in air stream by biotrickling filters. J. Environ. Eng. 123(6): 569-576.

[47] Pedersen, A.R. and Arvin, E. (1995) Removal of toluene in waste gases using a biological trickling filter. Biodegradation 6: 109-118.

[48] Kong, Z.; Farhana, L.; Fulthorpe, R.R. and Allen, D.G. (2001) Treatment of volatile organic compounds in a biotrickling filter under thermophilic conditions. Environ. Sci. Technol. 35(21): 4347-4352.

[49] Wright, W.F.; Schroeder, E.D.; Chang, D.P.Y. and Romstad, K. (1997) Performance of a pilot-scale compost biofilter treating gasoline vapor. J. Environ. Eng. 123(6): 547-555.

[50] Hugler, W.; Acosta, C. and Revah, S. (1999) Biological removal of carbon disulfide from waste air streams. Environ. Prog. 18(3): 173-177.

[51] Gabriel, D. and Deshusses, M.A. (2003) Retrofitting existing chemical scrubbers to biotrickling filters for H_2S emission control. Proc. Natl. Acad. Sci. U.S.A. 100(11): 6308-6312.

[52] Darlington, A.B.; Dat, J.F. and Dixon, M.A.(2001) The biofiltration of indoor air: Air flux and temperature influences the removal of toluene, ethylbenzene, and xylene. Environ. Sci. Technol. 35(1): 240-246.

[53] Darlington, A.; Chan, M.; Malloch, D.; Pilger, C. and Dixon, M.A. (2000) The biofiltration of indoor air: Implications for air quality. Indoor Air-Int. J. Indoor Air Qual. Climate 10(1): 39-46.

[54] Schwarz, B.C.E.; Devinny, J.S. and Tsotsis, T.T. (1999) Degradation of PCE in an anaerobic waste gas by biofiltration. Chem. Eng. Sci. 54(15-16): 3187-3195.

[55] Sun, A.K. and Wood, T.K. (1997) Trichloroethylene mineralization in a fixed-film bioreactor using a pure culture expressing constitutively toluene ortho-monooxygenase. Biotechnol. Bioeng. 55: 674-685.

[56] Wilderer, P.A.; Bungartz, H.J.; Lemmer, H.; Wagner, M.; Keller, J. and Wuertz, S. (2002) Modern scientific methods and their potential in wastewater science and technology. Wat. Res. 36(2): 370-393.

[57] Tresse, O.; Lorrain, M.J. and Rho, D. (2002) Population dynamics of free-floating and attached bacteria in a styrene-degrading biotrickling filter analyzed by denaturing gradient gel electrophoresis. Appl. Microbiol. Biotechnol. 59(4-5): 585-590.

[58] Alexandrino, M.; Knief, C. and Lipski, A. (2001) Stable-isotope-based labeling of styrene-degrading microorganisms in biofilters. Appl. Environ. Microbiol. 67: 4796-4804.

[59] Malhautier, L.; Degrange, V.; Guay, R.; Degorce-Dumar, J.R.; Bardin, R. and Le Cloirec, P. (1998) Estimation size and diversity of nitrifying communities in deodorizing filters using PCR and immunofluorescence. J. Appl. Microbiol. 85: 255-262.

ARTIFICIAL SEEDS

EUGENE KHOR[1] AND CHIANG SHIONG LOH[2]
[1]*Department of Chemistry and* [2]*Department of Biological Sciences, National University of Singapore, 3 Science Drive 3, Singapore 117543 – Fax: 65-779-1691 – Email: chmkhore@nus.edu.sg*

1. Introduction

The idea that basic plant materials can be combined with a nutrient filled environment, preserved and subsequently be capable of revitalizing into a viable living plant connotes endless possibilities for a world where arable agricultural land is diminishing at an alarming rate annually. The case for artificial seeds is an attractive one. Better and clonal plants could be propagated similar to seeds; preservation of rare plant species extending biodiversity could be realised; and more consistent and synchronised harvesting of important agricultural crops would become a reality, among many other possibilities. The concept started inconspicuously, but slowly took hold until today the term artificial seed has matured to acceptance. This is the basis for this chapter, to examine some of the plant materials and the matrix materials used for the making of artificial seeds. No intention is made to provide a comprehensive review of the literature, but rather to briefly highlight some examples and discuss some of the issues related to the use of these materials.

2. Definitions of artificial seed

An artificial seed is often defined as a novel analogue to botanic seed consisting of a somatic embryo surrounded by an artificial coat [1]. This definition is primary based on the similarity of somatic embryo with zygotic embryo in morphology, physiology and biochemistry [2]. Redenbaugh et al. [1] pointed out that the developmental stage of even the most highly matured somatic embryo may at best be equal to that of a zygotic embryo that has reached the stage where it is capable of precocious germination. Such zygotic embryos are not fully developed and lack the vigour of mature embryos [1] as is the case in most artificial seeds containing somatic embryos encapsulated in an artificial seed coat. This implies that the somatic embryo encapsulated in an artificial seed coat is at most equivalent to an immature zygotic embryo, possibly at post-heart stage or early cotyledonary stage. In many plant species such as *Capsella bursa-pastoris* [3] an endosperm is still present in an immature seed-containing embryo of this stage; the endosperm tissue being reduced when the zygotic embryo reaches full maturity. Hence, if an artificial seed is to be defined as an analogue to botanic seed consisting of somatic embryo, the somatic embryo should be surrounded by not only an artificial seed coat but also an artificial endosperm.

Kamada [4] provided an alternative definition to an artificial seed. He defined an artificial seed as "comprises a capsule prepared by coating a cultured matter, a tissue piece or an organ which can grow into a plant body and nutrients with an artificial film". His artificial seed concept comprised of (a) "an external film for strengthening the seed" which possibly implies the seed coat, (b) "an internal film for encapsulating nutrients required for growth of the cultured matter and plant hormones for controlling germination", a layer that possibly simulates the endosperm tissue and (c) "a callus, and adventitious bud or an adventitious embryo which can grow into a plant body". This definition liberates the use of the term "artificial seed" from the requirement of a somatic embryo inside an encapsulated matrix to encompass any cultured materials that can develop into plant bodies whether directly or after further treatment.

Regardless of the definition interpretations, the two main components of an artificial seed remains, namely (1) a plant material that can grow into a plant body and (2) one or more matrix materials (typically polymers) used for the encapsulation of the plant material.

3. Plant materials used for encapsulation

3.1. SOMATIC EMBRYOS AND MICROSPORES-DERIVED EMBRYOS

Somatic embryogenesis is the process whereby either haploid or diploid cells develop into plants through characteristic embryological stages without fusion of the gametes [5]. The process may be direct, with embryogenic cells developing directly from explanted cells or indirect with an intermediate callus phase [6]. Somatic embryos can be produced clonally in large numbers and make them attractive materials for encapsulation. Artificial seeds, consisting of somatic embryos enclosed in a protective coating, have been proposed as a low cost, high volume propagation system [2,7]. The potential and usage of somatic embryos for the making of artificial seeds had been reviewed [1,8,9].

Redenbaugh [7] pointed out that one of the main problems in the use of somatic embryos for the making of artificial seeds is the regeneration of plants from somatic embryos. Although somatic embryogenesis has been achieved in many plant species, in most cases, only a few plants were produced. Unlike the zygotic embryos, many of these somatic embryos required further treatment before complete regeneration. For example, somatic embryos of *Elaeis quineensis* (oil palm) were required to be placed on different culture media for shoot and root regeneration [7]. Similarly, the secondary embryos of *Brassica napus* required transfer to a medium enriched with cytokinin for shoot regeneration and a plant growth regulator-free medium for root formation [10]. Such multiple steps requirement for "embryo conversion" remains one of the hurdles in the use of somatic embryos for field planting of artificial seeds. In addition, an ideal artificial seed made from somatic embryo of non-recalcitrant species should be able to be stored at room temperature under non-aseptic condition without precocious germination, like a natural seed. The somatic embryo, encapsulated in a matrix, should remain dormant, until water is provided for simultaneous shoot and root formation under non-aseptic conditions. This would require further studies in the regulation of growth and development of the somatic embryos and the encapsulation procedure.

Similarly, direct somatic embryogenesis from microspores, either through anther cultures or pollen culture techniques, has been described for many plant species. As microspores are gametophytes, the embryos that derive from microspores are normally haploids. Plants regenerated through androgenesis are therefore mostly haploids and are useful materials in plant breeding programmes. One example for the encapsulation of microspores derived embryos is that of barley [11]. The embryos were encapsulated in calcium alginate beads. For germination, the artificial seeds were placed on a tissue culture medium, presumably under aseptic conditions. One interesting finding was that the percentages of germination in encapsulated embryos were found to be higher than the non-encapsulated controls. The advantage of encapsulation for germination was even more evident when the embryos were stored at low temperature in the dark for 6 months; about 38% of the encapsulated embryos germinated compared to no germination for non-encapsulated embryos. Direct encapsulation of microspore-derived embryos and regeneration will provide a short cut for the efficient production of haploid plants. However, some of the hurdles encountered in the use of microspore-derived embryos would be similar to those of the somatic embryos.

3.2. SHOOT BUDS AND SHOOT-TIPS

Apical and axial shoot buds were used for encapsulation of several plant species ranging from mustard [12], medicinal plant [13] to tree species [14]. Mathur *et al.* [13] encapsulated the apical and axial shoot buds of *Valeriana wallichii* in sodium alginate dissolved in MS medium and calculated that only 200 ml MS medium was required for 350 explants. On the other hand, nearly 4.5 L of MS medium is required for producing the same number of plants. Further, an Erlenmeyer flask of 250 ml capacity can accommodate only 5 non-encapsulated explants whereas the same space was sufficient for placing 30 beads. Hence, encapsulation of apical and axial shoot buds offers easy transportation of a large number of propagules in low bulk. Both apical and axial shoot buds are easy to regenerate into plants provided there is sufficient nutrient and rooting is not a problem. Again, how to desiccate or temporary suspend the growth of apical and axial buds until an appropriate time in an encapsulation matrix needs further investigation.

3.3. SEEDS

Seeds, with viable zygotic embryos are rarely used as plant material for encapsulation. Khor *et al.* [15] and Tan *et al.* [16] encapsulated orchid seeds in an attempt to make artificial seeds of orchids. The rationales for making artificial orchid seeds by encapsulating the zygotic seeds were based on some of the unique features of orchid seeds. Orchid seeds are minute and dust-like, ranging about 0.25-1.20 mm in length and 0.09-0.27 mm in width [17]. Each seed contains an embryo composed of 80-100 cells only [18]. Hence the seeds cannot be handled and propagated like other plant species [16]. The seeds are non-endospermic and the embryo is enclosed with only a membrane-like tissue. Orchid embryos are undifferentiated, and the embryos of most orchid species do not have cotyledons [17]. The minuteness and lack of storage tissues make encapsulation of orchid seeds justifiable. Using a free-flowering tropical orchid species *Spathoglottis plicata* as a model plant, Khor *et al.* [15] and Tan *et al.* [16] developed a method to encapsulate the seeds with two coatings, one simulates the endosperm and the other simulate the seed coat (Figure 1). The method, using coacervation of alginate-chitosan, gave firm, round beads of about 4 mm in diameter, a size comparable to the seeds of many species such as Brassicas. The enlargement of "seed" size theoretically would enable the handling of seeds in an easier manner.

3.4. ORCHID PROTOCORMS AND PROTOCORM-LIKE BODIES

Singh [19] germinated seeds of *Spathoglottis plicata*, an orchid, into protocorms and then encapsulated the protocorms by using a single layer of calcium alginate. The encapsulated protocorms regenerated into complete plantlets in Vacin and Went medium [20]. The germination frequency was found to decrease after long-term storage at 4°C. On the other hand, Khor *et al.* [15] used complex coacervation of alginate-chitosan and alginate-gelatin to develop a two-coat system for encapsulation of *Spathoglottis plicata* portocorms. About 83-100% of the protocorms were found to be viable 4 weeks after encapsulation and able to grow further.

Currently, orchid breeders germinate tens of thousands of seeds in a small container under aseptic conditions. The protocorms tend to become clusters of seedlings that need to be separated manually. Such process is tedious and labour intensive [15]. The development of an artificial seed system with individual seed or protocorm encapsulated in a proper matrix will obviously reduce the difficulty of seedling separation and planting.

Similarly, protocorm-like bodies could be used for encapsulation. In another orchid *Cymbidium giganteum*, the protocorm-like bodies were encapsulated in a single matrix of calcium alginate beads with Murashige and Skoog medium [21,22]. The conversion frequency *in vivo* was 64-88%. Protocorm-like-bodies are multiplied asexually and the artificial seeds made are theoretically clonal seeds, if there is no somaclonal variation among them.

3.5. MYCORRHIZAL FUNGI

Mycorrhiza has a symbiotic association between plant roots and certain soil fungi that play a key role in nutrient cycling in the ecosystem and also protects plants against environmental and cultural stress [23]. This association is usually a mutualistic symbiosis one because of the highly interdependent relationship established between both partners. The host plants normally receive nutrients *via* fungal mycelium while the heterotrophic fungus obtains carbon compounds from the host photosynthesis. Azcon-Anguilar and Barea suggested that the performance of artificial seeds might be greatly enhanced by ensuring a suitable mycorrhizal establishment at out planting.

Tan *et al.* [16] successfully attempted to infect seeds of *Spathoglottis plicata* encapsulated in a two-coat system with *Rhizoctonia*, an orchid mycorrhiza by just placing the encapsulated seeds on a PDA layer containing *Rhizoctonia* mycelium yielded higher percentage of infection (Figure 1, Method 1). About 84% of the encapsulated seeds were infected and the orchid mycorrhizal fungus established a symbiotic relationship with the encapsulated seeds. The addition of sucrose or other nutrients were excluded and hence eliminate the potential of microbial contamination and provided further opportunity for further development of encapsulated materials grown under non-sterile conditions. Alternatively, the *Rhizoctonia* mycelium was encapsulated together with the seeds; about 39% of the encapsulated seeds were infected with the mycorrhiza when the mycelium was encapsulated with a potato-dextrose agar (PDA) block (Figure 1, Method 2).

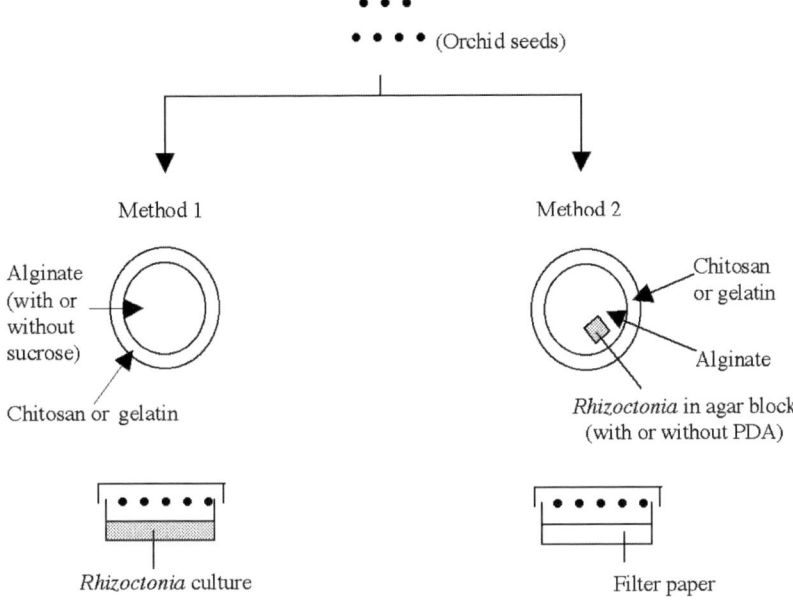

Figure 1. Methods for the infection of Spathoglottis plicata *seeds with* Rhizoctonia *in two-coat encapsulation systems [16].*

The interaction between mycorrhizal fungi and other artificial seed system have not been reported. Previous experience has shown that the earlier the infection of mycorrhizal fungi followed by a symbiotic relationship being established with the host plant at an appropriate stage, the greater the benefit [23]. The incorporation of mycorrhizal fungi also opens the door to developing a non-aseptic artificial seed system for certain plant species such as orchids [16]. Successful infection of protocorm-like-bodies of orchids in an artificial seed system will definitely advance the pragmatic use of the system in orchid micropropagation.

3.6. HAIRY ROOTS-DERIVED MATERIALS

Uozomi *et al.* [24] extended the concept of artificial seeds to include fragments of hairy roots. The infection of plants with *Agrobacterium rhizogenes* causes the transfer-DNA (T_L and T_R) contained in the plasmid to be inserted into the plant genome. Both T_L and

T_R have rhizogenic functions [25] and inoculation of plants with *A. rhizogenes* induces the formation of "hairy roots" that readily give rise to rapidly growing root cultures [26]. These hairy root cultures are considered as useful sources of enzymes [24] and secondary metabolites [25]. Both the root fragments and adventitious shoot primordia derived from horseradish (Armoracia rusticana) hairy roots were used for encapsulation [27]. An efficient method for the production of plantlets from horseradish hairy roots has been developed with the plantlets (0.4-4.0 mm) also being used for encapsulation by Nakashimada *et al.* [28].

4. Matrix material selection for artificial seeds

While the somatic embryo or plant material suitable for propagation into a plant is at the core of the artificial seed concept, the choice of the accompanying matrix materials used for combining with the biological material is equally vital. The matrix is responsible for the immediate surrounding of the plant material. Factors will include mediating the environment such as temperature and humidity to protect the biological material as well as providing a nutrient reservoir. Therefore, the role of matrix materials has an important bearing on the ultimate viability of the artificial seed. The concept of matrix materials has over the years developed into a rather sophisticated interplay that focuses on the germination viability of artificial seeds.

4.1. FLUID DRILLING

The idea of utilising a delivery system began primitively with the use of a fluid gel into which somatic embryos were mixed and sowed directly in the field [1]. The purpose of the gel was to better facilitate the manipulation of small plant materials, possibly in the millions, that was not readily managed because of their small size. This so called "fluid drilling" process was the first realization that a carrier was necessary for the bulk handling of embryos. Magnesium silicate clay and polymeric materials such as potassium starch acrylamide, copolymer of potassium acrylate and acrylamide, starch, and various cellulose-based materials have been tried as the gelling agents in this process [29]. For carrot somatic embryos, the survival rate for this method was a very low 4% and none in the absence of sucrose, attributed to somatic embryos succumbing to desiccation with this process leading to embryo death. In this process, there was really no attempt to encapsulate the biological material. But the realisation that handling and delivery techniques were necessary justifiably credits this work for laying the foundation for artificial seeds that followed.

4.2. POLYMERIC COATING

The next innovation in matrix development chronologically was the concept of a polymeric coating. Somatic embryos or callus were first pre-treated or "hardened" to improve survival rate with by exposure of the plant material to high sucrose concentrations and/or abscisic acid (ABA). Subsequently, commingling of the embryo suspension with water-soluble polyoxyethylene, polyox WSR-N750 [30] and dispensing onto non-adhering surfaces to permit drying under aseptic conditions, gave desiccated artificial seed wafers. These wafers were readily stored and ready for planting. The survival rate for coated seeds with or without ABA for drying times of 5 or 6.5 hours ranged between 10 to 31%. The desiccation step alluded here is another step forward as it simulates what naturally occurs in nature when a seed develops. Presumably, the role of the polymer is as a covering for the plant material where the coverage is most likely sufficient but may not be complete.

4.3. HYDROGELS

What can be considered a true synthetic seed process is credited to Redenbaugh et al. who introduced hydrogels to produce artificial seeds. In the hydrogel approach, somatic embryos are mixed with a polymeric solution that when introduced drop-wise into a separate solution containing divalent metal ions, initiated a crosslinking reaction to form the hydrogel that in the process encapsulates the embryo. With this advancement, the true seed character of an embryo in a simulated endosperm was realised. The most common material used to generate this artificial endosperm is based on sodium alginate that in the presence of divalent calcium ions forms a hydrogel. The nature of the hydrogel being soft ensures that little pressure acts on the embryo therefore minimizing harm to the embryo. Nutrients and other materials such as bactericide and growth hormones for example could be included. Distinct beadlike artificial seeds of good quality could be obtained with an increased high rate of germination in the 80% and above range have been reported.

4.4. TWO-COAT SYSTEM

While the hydrogel systems can be considered as a worthy milestone, limitations of the system soon manifested. Hydrated capsules were tacky unless coated with an additional hydrophobic layer, the hydrogels rapidly dehydrate making storage difficult and their high porosity led to cell and nutrient leaching out. This situation is not difficult to comprehend if one considers that a true seed comprises not only the embryo and endosperm, but also the all-important seed coat.

Towards the resolution of the seed coat stipulation, *Tay et al.* introduced the *in situ* generation of an artificial seed complete with somatic embryo, artificial endosperm and the outermost layer constituting the seed coat [31]. In this process, an additional material – the polycationic chitosan polymer – interacts with the polyanionic alginate to form the "seed-coat" as depicted diagrammatically in Figure 2.

The resulting artificial seeds are less tacky to the touch, easy to handle and store as shown in Figures 3 and 4 using *Spathoglottis plicata* seeds as the plant material. With this final contribution, it can be stated that the puzzle of assembling of the artificial seed has been completed in all its conceptual assembly.

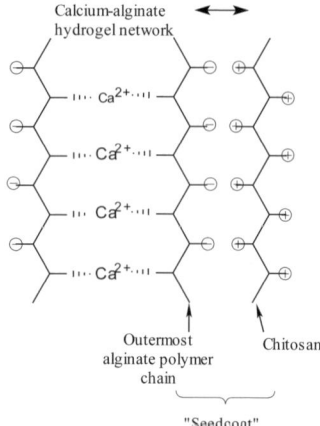

Figure 2. Diagrammatic representation of the two-coat system. The inner core of alginate polymer chains are crosslinked by Ca^{2+} ions forming the hydrogel network while the outermost alginate layer interacts with chitosan to form the polymer seed-coat [31].

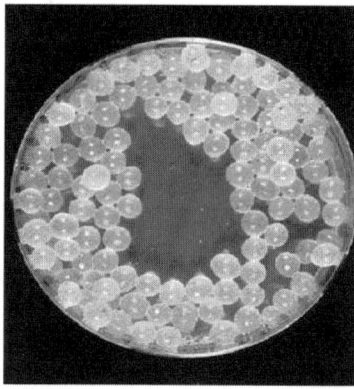

Figure 3. Collection of Spathoglottis plicata *artificial seeds obtained from the two-coat alginate-chitosan encapsulation system [31].*

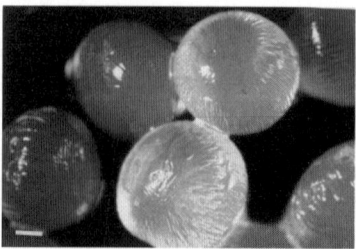

Figure 4. Close up view of Spathoglottis plicata *artificial seeds obtained from the two-coat alginate-chitosan encapsulation system showing the minute plant material in the matrix [31].*

4.5. TAKING STOCK

From the preceding survey, it is evident that the challenge in defining a suitable matrix, carrier or delivery system in artificial seed technology demands a set of requirements that was tricky to fulfil and garnered as the experience with trying various delivery methods grew. The evolved requirements can be conveniently summarised into five major considerations:

- The material must protect the embryo and/or other biological materials.
- The material must be capable of including nutrients and other growth and/or biological factors.
- The material must protect the formed artificial seed during storage and handling.
- The material must incorporate a mechanism for activating "germination".
- And of course the material must be non-toxic, compatible with the biological and chemical systems it comes into contact with and be available in a suitable and simple preparation for use.

In addition, several factors regarding the directions for developing a functional seed delivery system beyond those stated above become apparent:

- The material must have the capability to reversibly imbibe and discharge fluid, primarily water, to simulate desiccation and germination processes.
- The material must be water soluble in the first instance as a mode of introduction.
- The material should be preferably biodegradable.

The materials used to date revolve principally around both natural and synthetic polymers that in one way or another utilise the gelling effect of the polymers. Gelling is based on the ionic and/or polar interactions inherent in the polymers, a result of the individual properties of the functional groups of each monomer repeat unit. In some instances, the use of crosslinking agents are required to obtain the gel as is the case in alginates where calcium ions are used to pair up two anionic groups on the polymers (preferably from different chains) thereby generating the hydrogel. The number and/or extent of these interactions in turn determine the final capsule properties such as shape integrity, burst strength and fluid or media retention.

All these properties have direct bearing on the embryo's or plant material's survival and eventual germination. The parameters involved in determining the capsule properties included the concentration of polymer, the processing time that leads to "hardening" of the capsule and the functional groups distribution especially for natural polymers. It is interesting to note that to date there is no direct interaction *per se* of the polymeric gelling material with either the biological material or the growth and other culture media that is incorporated during the encapsulation process. In other words, a physical entrapment process is at work.

The two-coat system developed by Tay *et al.* presents a more sophisticated "barrier" effect that may be more efficient in retaining encapsulated material but nevertheless still has a physical basis for its action. This leaves a lot of scope to fine tune the encapsulation process where a more systematic consideration of the requirements in achieving the goals of developing a functional seed delivery system can lead to an artificial seed hitherto not realised, provided of course, the cost of such a delivery system is not prohibitive.

5. Outlook

Redenbaugh [7] stated that the main objective for developing artificial seeds of crop plants is to produce "clonal seeds" at a cost comparable to true seeds. The objective is suitable if somatic embryos are used as plant materials for encapsulation. With the extension of the artificial seed concept to include encapsulation of other plant materials, the objective for the making of artificial seeds may be different. We would expect to see more reports on the encapsulation of transgenic plant materials, parasitic plants and the incorporation of various microorganisms in the matrix. We would also expect to see more innovative use of polymers and encapsulation procedures to satisfy the different objectives. Nevertheless, the requirement that the cost of producing artificial seeds should be comparable to that of true seeds remains valid and is a goal that may not be reached in the near future unless there are breakthroughs in the technology, materials used and manpower requirement.

References

[1] Redenbaugh, K.; Fujii, J.A. and Slade, D. (1988) Encapsulated plant embryos. In: Mizrahi, A. (Ed.) Advances in biotechnological processes. Vol. 9. Liss, New York, USA; pp. 225-248.
[2] Redenbaugh, K; Passch, BD; Nichol, JW; Kossler, ME; Viss, PR and Walker, KA (1986) Somatic seeds: encapsulation of asexual plant embryos. Bio/Technol. 4: 797-801.
[3] Esau, K. (1977) Anatomy of Seed Plants. John Wiley & Sons, New York, USA.
[4] Kamada, H (1985) Artificial seed. In: Tanaka, R. (Ed.) Practical technology on the mass production of clonal plants.CMC Publisher, Tokyo, Japan (in Japanese, cited in Redenbaugh et al., 1988).
[5] Williams, E.G. and Maheswaran, G. (1986) Somatic embryogenesis: factors influencing coordinated behaviour of cells as an embryogenic group. Ann. Bot. 57: 443-462.
[6] Merkle, S.A.; Parrott, W.A. and Williams, E.G. (1990) Applications of somatic embryogenesis and embryo cloning. In: Plant tissue culture: applications and limitations. Elsevier Science Publishers, New York, USA; pp. 67-101.
[7] Redenbaugh, K. (1990) Application of artificial seeds to tropical crops. Hort Science 25: 251-255.
[8] Redenbaugh, K. (1992) Synseeds: Applications of Synthetic Seeds to Crop Improvement. CRC Press, Boca Raton, USA.
[9] Bajaj, Y.P.S. (1995) Biotechnology in agriculture and forestry 30: somatic embryogenesis and synthetic seed 1. Springer-Verlag, Berlin, Germany.
[10] Koh, W.L. and Loh, C.S. (2000) Direct somatic embryogenesis, plant regeneration and in vitro flowering in rapid-cycling Brassica napus. Plant Cell Rep. 19: 1177-1183.
[11] Datta, S.K. and Potrykus, I. (1989) Artificial seeds in barley: encapsulation of microspore-derived embryos. Theor. Appl. Genet. 77: 820-824.
[12] Arya, K.R.; Beg, M.U. and Kukreja, A.K. (1998) Microcloning and propagation of endosulfan tolerant genotypes of mustard Brassica campestris through apical shoot bud encapsulation. Indian J. Exp. Biol. 36: 1161-1164.
[13] Mathur, J.; Ahuja, P.S.; Lal, N. and Mathur, A.K. (1989) Propagation of Valeriana wallichii using encapsulated apical and axial shoot buds. Plant Sci. 60: 111-116.
[14] Maruyama, E.; Kinochita, I.; Ishii, K.; Shigenaga, H.; Ohba, K. and Saito, A. (1997) Alginate-encapsulated technology for the production of the tropical forest trees: Cedrela odorata L., Guazuma crinata Mart., and Jacaranda mimosaefolia D. Don. Silvae Genetica 46: 17-23.
[15] Khor, E.; Ng, W.F. and Loh, C.S. (1998) Two-coat systems for encapsulation of Spathoglottis plicata (Orchidaceae) seeds and protocorms. Biotechnol. Bioeng 59: 635-639.

[16] Tan, T.K.; Loon, W.S.; Khor, E. and Loh, C.S. (1998) Infection of Spathoglottis plicata (Orchidaceae) seeds by mycorrhizal fungus. Plant Cell Rep. 18: 14-19.
[17] Arditti, J. (1992) Fundamentals of Orchid Biology. John Wiley & Sons, New York, USA.
[18] Sheehan, T.J. (1983) Recent advances in botany, propagation, and physiology of orchids. In: Janick, J. (Ed.) Horticultural Reviews. Vol. 5. AVI Publishing Company, Westport, Connecticut, USA; pp. 279-315.
[19] Singh, F. (1991) Encapsulation of Spathoglottis plicata protocorms. Lindleyanna 6: 61-63.
[20] Vacin, E. and Went, F. (1949) Some pH changes in nutrient solutions. Bot. Gaz. 110: 605-613.
[21] Murashige, T. and Skoog, F. (1962) A revised medium for rapid growth and bioassay with tobacco cultures. Physiol. Plant. 15: 73-97.
[22] Corrie, S. and Tandon, P. (1993) Propagation of Wall. Through high frequency conversion of encapsulated protocorms under in Cymbidium giganteum vivo and in vitro conditions. Indian J. Exp. Biol. 31: 61-64.
[23] Azcon-Angular, C. and Barea, J.M. (1997) Applying mycorrhiza biotechnology to horticulture: significance and potentials. Sci. Hortic. 68: 1-24.
[24] Uozumi, N.; Asano, Y. and Kobayashi, T. (1992) Production of artificial seed from horseradish hairy root. J. Ferment. Bioeng. 74: 21-26.
[25] Hamill, J.D.; Parr, A.J.; Rhodes, M.J.C.; Robins, R.J. and Walton, N.J. (1987) New routes to plant secondary products. Bio/Technol. 5: 800-804.
[26] Tepfer, M. and Casse-Delbart, F. (1987) Agrobacterium rhizogenes as a vector for transforming higher plants. Microbiol. Sci. 4: 24-28.
[27] Uozumi, N. and Kobayashi, T. (1995) Artificial seed production through encapsulation of hairy root and shoot tips. In: Bajaj, Y.P.S. (Ed.) Biotechnology in Agriculture and Forestry 30: Somatic Embryogenesis and Synthetic Seed I. Springer-Verlag, Berlin, Germany; pp. 170-180.
[28] Nakashimada, Y.; Uozumi, N. and Kobayashi, T. (1995) Production of plantlets for use as artificial seeds from horseradish hairy roots fragmented in a blender. J. Ferment. Bioeng. 79: 458-464.
[29] Gray, D.J.; Compton, M.E.; Harrell, R.C. and Cantliffe, D.J. (1995) Somatic embryogenesis and the technology of synthetic seeds. In: Bajaj, Y.P.S. (Ed.) Biotechnology in Agriculture and Forestry 30. Springer-Verlag, Berlin, Germany; pp. 126-151.
[30] Janick, J. and Kitto, S.L. (1986) Process for encapsulating asexual plant embryos. US Patent 4615141.
[31] Tay, L.F.; Khoh, L.K.; Loh, C.S. and Khor, E. (1993) Alginate-chitosan coacervation in production of artificial seeds. Biotechnol. Bioeng. 42: 449-454.

SPERMATOZOAL MICROENCAPSULATION FOR USE IN ARTIFICIAL INSEMINATION OF FARM ANIMALS

RAYMOND L. NEBEL AND RICHARD G. SAACKE
Department of Dairy Science, Virginia Polytechnic Institute and State University, Blacksburg, Virginia, USA 24061- Fax: 540-231-5014 - Email: rnebel@vt.edu

1. Introduction

In a series of publications, Chang and co-workers reported the development of microcapsules having semi-permeable membranes [1-4]. Chang called these microcapsules "artificial cells". The semi-permeable microcapsules developed by Chang differ from earlier industrial microcapsules in a number of important areas which include: 1) the contents were aqueous solutions or suspensions of biologically active materials, 2) the encapsulated material did not depend on capsule rupture for desired action, 3) the capsule membrane possessed selective permeability, allowing passage of small molecular weight solutes while restricting macromolecules. To simulate biological cell membranes, biological materials were used as the encapsulating membrane: protein [1,2], polysaccharide [5], lipids [6,7] and a complex of lipids and cross-linked protein chains [8].

The concept of using encapsulated sperm for prolonged storage or sustained release of sperm within the reproductive tract of the female mammal has a biological precedent in several classes of animals. In all mammals, sperm remain viable during storage in the *cauda epididymis*; in the male bovine this can be for up to 3 weeks if animals are not sexually active [9]. In concept, the potential role for microencapsulated semen in artificial insemination is to maintain spermatozoal viability while reducing their susceptibility to retrograde action of the uterus and phagocytosis by leukocytes, and allowing their release over an extended period. The potential application of spermatozoal microencapsulation extends beyond that immediately apparent to the artificial insemination industries. Microencapsulation of spermatozoa offers the potential for sequentially exposing discrete populations of sperm to different environments without the trauma of centrifugation and resuspension of the cells (washing). Co-culturing sperm with somatic cells would also be possible.

Encapsulation of live cells began with the successful encapsulation of pancreatic cells [10] and hepatoma cells [11]. Microencapsulated pancreatic islet cells maintained morphology and function for > 4 months *in vitro* and upon implantation into rats with streptozotocin-induced diabetes, neutralised the diabetic state for > 3 weeks. The microencapsulated hepatoma cells grew and multiplied at the same rate as the unencapsulated controls (doubling time of approximately 24 hr). They stretched the capsular membrane with growth in culture until finally rupturing the capsular membrane after 10 days. From such studies it should be clear that much could be learned about sperm physiology from co-culturing spermatozoa with cells of potential reproductive importance, *e.g.*, epididymal cells, oviductal cells, *etc*. In addition, the sequential effects of different environments on physiologically important events in spermatozoa (*e.g.*, capacitation, true acrosome reaction) could be investigated in a population of sperm under more physiological conditions than the harsher centrifugations and resuspensions currently employed.

2. Procedures for microencapsulation of sperm

Microencapsulation procedures developed by Lim and co-workers for pancreatic islet cells of Langerhans [10,11] were modified [12,13] to encapsulate bovine sperm with negligible evidence of injury compared to unencapsulated sperm (based upon *in vitro* evaluation). The following procedures have been reported to be optimal for microencapsulation of bovine sperm. Sperm (up to 150×10^6 sperm/ml) are suspended in a 1.2 to 1.5% (w/v) sodium alginate solution and pipetted into a 10 to 50 ml syringe. Droplets containing sperm are then produced by a syringe pump extrusion technique. Using a syringe pump the aqueous phase to be encapsulated is forced through a 19 or 22 gauge stainless steel hypodermic needle containing in a polyvinyl chamber. Simultaneously, air or nitrogen, at constant pressure, is passed through the air inlet of the holding chamber. The gas pressure is controlled by a flow meter. The needle size and velocity of gas has a profound effect on droplet size. The bottom of the chamber is placed 50 mm above a solution of 1.5% calcium chloride - HEPES buffer. Immediately on contact, the droplets absorb calcium ions, which results in a shape-retaining high viscosity (microgel) for the future microcapsules. Once the microgels settle in the flask, the calcium chloride - HEPES solution is aspirated and the microgels are rinsed three times with physiological saline. The semi-permeable membrane is constructed on the surface of the microgels by interfacial polymerisation.

Various polymers and polyamines have and concentrations of these polyamines have been used to construct the outer membrane, the most successful being: 1% protamine sulfate [14], 0.04 to 1.5% poly-L-lysine [15,16], and 0.01% polyvinylamine [15]. Microgels are suspended in the desired polymer with continuous stirring for 2 to 5 min. Exposure time and polymer concentration influence the membrane thickness by limiting or favouring the "cross-linking" of the alginate and polyamine. Excess polymer is aspirated and microgels rinsed with CHES buffer. The CHES buffer terminates the cross-linking of the polymer and alginate by reacting with the unbound reactive sites of the specific polymer. Membrane enveloped microgels are then rinsed three times with physiological saline and suspended in 3% sodium citrate. This liquefies the gel core of the microcapsules by chelation of calcium ions. Microcapsules containing motile spermatozoa are then transferred to the desired environment, extender, or culture medium.

A novel procedure of reverse encapsulation using 2% (w/v) hydroxypropyl-methylcellulose (HPMC) and either calcium chloride or barium chloride to form a semen suspension and sodium alginate (0.5% w/v) as the hardening solution has been developed for porcine spermatozoa [17-19]. The procedures that produce barium alginate gel capsules are an enhancement of their previous studies and will be briefly described [19]. A saturated barium chloride solution was added to semen to obtain a barium concentration of 20 mM per liter. The sperm suspension was forced with a peristaltic pump through a 2 mm diameter nozzle and a flow rate of 45 droplets per min. The height of the nozzle above the surface of the sodium alginate solution was 120 mm, immediately on contact the barium ions and the alginate chains reacted to form a membrane surrounding each droplet. The thickness of the gel membrane increased during incubation for 1 hr. The gel capsule were collected by sieve filtration, rinsed twice with a glucose bicarbonate buffer and suspended in the same buffer. These barium alginate capsules were then suspended in a 1% (w/v) solution of protamine sulphate and continuously stirred for 20 min, to obtain cross-linked capsules with an outer barium alginate protamine membrane that were then rinsed twice and suspended in the glucose bicarbonate buffer.

3. Capsule size and maintenance of sperm viability *in vitro*

Nebel *et al.* [12] encapsulated bovine sperm at three different concentrations (45, 90 and 180 10^6 sperm/ml) in 0.75 and 1.5 mm (diameter) microcapsules using 0.04% poly-L-lysine, (25,000 to 50,000 kD). Unencapsulated treatments at each of the three concentrations served as controls. For evaluation of viability, treatments were incubated at 37°C for 24 hr in Cornell University Extender [20]. This extender was chosen as the incubation medium because it had been shown to be an effective ambient body-temperature extender. The aim was to maintain the encapsulating temperature and thus, evaluate the effects of encapsulation *per se, i.e.*, without the necessity of prolonging sperm life by cooling. They found that sperm viability (motility and acrosomal integrity) was not influenced by sperm concentration or capsule size and was not different from unencapsulated controls. Only sperm trapped in the capsular membrane appeared to be damaged. Percent motility and percent intact acrosomes for sperm incubated in two capsule sizes (diameters of 0.75 and 1.50 mm) at 45 10^6 sperm/ml are presented in Table 1.

Table 1. *Effect of capsule size on maintenance of percent motility (MOT) and percent intact acrosome (PIA) for semen extended in Cornell University Extender and encapsulated at 45 10^6 sperm/ml (during incubation at 37°C for 24 hours[a]).*

Capsule size, mm diameter	Hours of Incubation					
	2		12		24	
	MOT	PIA	MOT	PIA	MOT	PIA
0.75	46.7	83.0	48.3	81.3	35.0	77.8
1.5	48.3	85.8	48.0	76.2	36.7	74.5
Unencapsulated control	51.7	84.0	46.7	79.3	40.0	73.2

[a]Mean for six replicates, each replicate being a poll of three ejaculates [12].

Similarly, favourable sperm viability results were obtained when sperm were microencapsulated using a CAPROGEN extender [21] and four different concentrations of poly-L-lysine (0.025, 0.05, 0.075 and 0.10% w/v) resulting in capsular membrane thicknesses of 3.2 ± 0.5, 5.3 ± 0.3, 7.1 ± 0.4 and 7.4 ± 0.8 µ respectively (see Table 2, from Nebel et al. [16]). Spermatozoal viability, as assessed by motility estimates at 24 hr intervals during a 120 hr incubation was not influenced by polymer concentration (membrane thickness). Earlier, molecular weight of the poly-L-lysine used in capsular membrane formation was evaluated and found to be unimportant to sperm viability when between 10 and 120 kD [15].

Esbenshade and Nebel [22] reported on the microencapsulation of porcine spermatozoa. They encapsulated sperm in Kiev extender (1:1 v/v with 1.4% Na alginate in saline) and used 25-50 kD poly-L-lysine at 0.1% w/v to form the capsule membrane. Under these conditions, encapsulated sperm motility was similar to unencapsulated controls when examined immediately after encapsulation, but dissipated quickly during a 16 hr incubation at 37°C. This was not the case for acrosomal integrity, which did not differ from unencapsulated controls over the same incubation conditions.

Table 2. *Effect of poly-L-lysine concentration on maintenance for semen extended in CAPROGEN and encapsulated at 20 10^6 sperm/ml (incubated at 37 °C).*

	Motility (%)				
	Hours of incubation[a]				
Treatment	24	48	72	96	120
Unencapsulated	71.1	51.1	33.9	20.5	11.7
Encapsulated Poly-L-lysine (%, w/v)					
0.025	68.3	51.1	38.9	13.3	3.9
0.05	67.2	54.4	46.1	17.8	5.5
0.075	69.4	50.5	37.8	14.4	5.5
0.1	71.7	51.1	42.8	21.1	3.3

[a]Mean of nine replicates, Standard error ± 3.42 [16].

4. Microcapsule behaviour in the female tract

Initial experiments to determine the fate of poly-L-lysine microencapsulated bovine sperm in the female tract were disappointing. Nebel [23], using a heterospermic insemination with marked sperm encapsulated and unmarked sperm unencapsulated (and *vice versa*) differentially counted sperm recovered from the reproductive tract 12 hr following insemination. The data showed that the encapsulated sperm disappeared at a faster rate than the unencapsulated sperm. In a similar subsequent experiment comparing poly-L-lysine encapsulated sperm with polyvinylamine encapsulated sperm, the polyvinylamine capsules improved retention in the female over uncapsulated control sperm while poly-L-lysine encapsulated sperm again disappeared at a faster rate from the uterus than unencapsulated controls (Table 3).
The difference in behaviour of the two capsule types was believed to be due to a "sticky" property of the polyvinylamine relative to the uterine mucosa. The nature of the capsular membrane could therefore be quite important to the availability of encapsulated sperm to the ovum. Since polyvinylamine is non-biodegradable in the female tract, further studies were abandoned using this material in favour of a capsular membrane

formed from protamine sulfate (1% w/v), also having a "sticky" property and which proved to be as efficient as poly-L-lysine with respect to post-encapsulation sperm viability [14]. Since uterine sperm retention alone is not indicative of sperm transport in the female, accessory sperm numbers were utilised to further evaluate the fate of microencapsulated sperm in the female tract.

Table 3. Competitive retention of encapsulated vs. unencapsulated sperm using polylysine or polyvinyl capsule membranes[a] (recovered from uterine flushings 12 hr following insemination).

Poly-L-lysine		Polyvinylamine	
Cow	Difference in % encapsulated vs. unencapsulated sperm retained	Cow	Difference in % encapsulated vs. unencapsulated sperm retained
1	-04.2	7	+20.0
2	-00.8	8	+29.6
3	-13.5	9	+21.0
4	-10.9	10	+24.0
5	-20.8	11	+26.7
6	-24.1	12	+26.9

[a]Competitive retention of sperm; cows were inseminated with equal numbers of encapsulated and uncapsulated sperm. The encapsulated sperm were either marked or unmarked in relation to the unencapsulated sperm. Marked sperm were from a fertile male providing nearly absolute numbers of sperm with a morphologically unique and recognizable trait such that upon recovery from the female tract, the treatment assigned to the unmarked or marked sperm could be distinguished by a differential count (from Nebel [23]).

The presence and quantity of accessory sperm in the *zona pellucida* has been suggested as an indicator of the efficiency of inseminated spermatozoa to be transported to the site of fertilisation and participate in fertilisation [24-27]. Assumptions for the use of accessory sperm as a fertility indicator reside in the understanding that accessory sperm are capable of fertilisation because they have demonstrated the ability to be transported to the oviduct (site of fertilisation), to be capacitated, and to attach to and penetrate the zona pellucida.

Munkittrick et al. [14] conducted a heterospermic study in the bovine to evaluate microgels in which two morphologically distinct sperm types (marked and unmarked) were artificially inseminated with equal sperm numbers under three different conditions: (1) semen encapsulated with protamine sulfate (1% w/v) from a fertile bull providing unmarked sperm cells, and unencapsulated semen from another fertile bull providing marked sperm cells, (2) marked sperm encapsulated in protamine sulfate microcapsules and unmarked sperm unencapsulated, and (3) unencapsulated semen from both marked and unmarked bulls. Microcapsules were 0.75 mm in diameter and contained approximately 1.25×10^5 sperm/capsule. Equal numbers (100×10^6 sperm) of frozen-thawed marked and unmarked sperm were inseminated at the conventional time (approximately 12 hr after the detection of oestrus). Ten embryos were recovered for each treatment. Uterine flushes and embryo recovery was performed on day 7 post inseminations (Results are present in Table 4). Microencapsulated sperm contributed 25.7% of the accessory sperm obtained for inseminates containing microcapsules.

Microencapsulation did not affect the total number of accessory sperm per embryo; however, encapsulated reduced accessory sperm for the specific sperm type encapsulated. Thus, sperm inseminated in protamine sulfate microcapsules qualified functionally as accessory sperm *in vivo*; however, microencapsulation put these sperm at a competitive disadvantage when mixed equally with unencapsulated sperm. It was emphasised by the authors [14] that studies were also needed to compare treatments when insemination time was less favourable than the conventional timing of insemination used in this study.

Table 4. Heterospermic competition of Protamine sulfate encapsulated sperm with unencapsulated sperm[a] (based on accessory sperm).

Inseminate	Marked (n)	Unmarked (n)	% Unmark
Unmarked/marked	100	193	65
Unmarked encap.	142	78	35
Marked encap.	65	271	80

[a]Competitive appearance of accessory sperm; cows were inseminated with equal numbers of encapsulated and unencapsulated sperm. The encapsulated sperm were either marked or unmarked in relation to the unencapsulated sperm. Marked sperm were from a fertile male providing nearly absolute numbers of sperm with a morphologically unique and recognisable trait such that upon digestion of the zona pellucida of 6-d old nonsurgically recovered ova/embryos, the marked and unmarked accessory sperm could be differentially counted. The natural competition of marked sperm without encapsulated was also evaluated as a control (from Munkittrick et al. [14]).

The initial studies of Torre *et al.* [17,18] encapsulated porcine semen in calcium alginate beads. Their results obtained from motility and average velocity tests of treated seminal materials were promising, especially when considering the difficulty of preservation of porcine spermatozoa compared to bovine material. The motility and average velocity of encapsulated spermatozoa were lower than that of the free sperm. It was postulated that the residues of alginate matrix could interfere with sperm kinetic activity, thereby reducing cell velocity. The different concentrations of calcium ions used to prepare the capsules influenced the gel strength and thickness of the alginate membrane and could be used to modulate the delivery of the cells.

The fate of inseminated capsules containing porcine sperm has been reported [22]. The uteri of sows inseminated with 20 ml of encapsulated semen in 60 ml Kiev extender (120 10^6 sperm/ml encapsulated, total of 2.5 10^9 sperm inseminated) were flushed surgically at 3, 6, and 24 hr post-insemination and the flushings searched for capsules and evidence of motility in recovered sperm. Results are presented in Table 5; only at 3 hr did all sows have capsules and an acceptable percentage of motile sperm. By 24 hr there was no evidence of either capsules or motile sperm. There were no unencapsulated controls for comparison.

Table 5. Microcapsules and motility of sperm recovered from uteri after insemination of sows in oestrus with 2.5 10^9 microencapsulated sperm[a].

Hours post insem.	No. of sows	Sows with capsules	Sows with live sperm	Motility (%)
3	3	3	2	40-50
6	3	2	1	25
24	4	0	0	-------

[a]Inseminate was 20 ml of poly-L-lysine microcapsules (0.75 mm diameter) followed by 60 ml of Kiev extender (from Ebenshade and Nebel [22]).

5. Evaluation of encapsulation technology in cattle artificial insemination – field trials

Initial laboratory trials showed that the survival of encapsulated sperm in ambient temperature diluents CUE [20] and CAPROGEN [21] was between 80 to 96 hr at 37°C [12,28]. This prompted the evaluation of encapsulated sperm in the field to determine their fertility potential. In a preliminary trial, oestrus in 23 mixed age cows were synchronised with a progesterone impregnated CIDR-B® (Controlled Internal Drug Release). Insemination was 48 hours after CIDR-B® removal with either poly-L-lysine encapsulated or normally processed semen diluted at ambient temperature (18°C to 22°C). The sperm dose was maintained at 5 10^6 sperm/inseminate. Recovery of eggs three days later from these animals showed that the encapsulated sperm were capable of fertilisation [13]. On this basis, more extensive field trials were undertaken along with consideration of capsule membrane thickness.

Four concentrations of poly-L-lysine were studied (0.025, 0.050, 0.075, and 0.100%, w/v) to investigate the relationship between poly-L-lysine concentration and membrane thickness as well as the effect of these factors on pregnancy rate in 335 oestrous synchronised heifers [16]. Membrane thickness based upon the poly-L-lysine concentration was: 3.2 ± 0.5 μm, (mean $\pm$ SD), 5.3 ± 0.3 μm, 7.1 ± 0.4 μm, and 7.4 ± 0.8 μm, respectively. Semen was extended in CAPROGEN containing a final encapsulated spermatozoal concentration of 20 10^6 sperm/ml. Sixty-five heifers were inseminated with sperm either unencapsulated or encapsulated with one of the four polymer concentrations. Oestrous synchronisation was accomplished with the combination of a progesterone-impregnated CIDR-B® device containing a 10-mg oestradiol benzoate capsule inserted for 10 d with administration of 12.5 mg of prostaglandin F2a on day 6 of CIDR-B® insertion. Heifers were inseminated in the uterine corpus at 24 hr after CIDR-B® removal, which constituted the pro-oestrus stage of the cycle for 95% of the heifers. Inseminate dose was 5 x 10^6 sperm in 0.25 ml. Pregnancy rated was similar for heifers inseminated with encapsulated and unencapsulated spermatozoa (Table 6). From these studies the authors concluded poly-L-lysine concentration and resulting capsule thickness was without effect on fertility of oestrous synchronised heifers.

Table 6. *Effect of Capsular thickness on pregnancy rate in Fresian heifers inseminated 24 hr after CIDR-B® removal (5 10^6 sperm/dose in CAPROGEN extender)* (from Nebel et al. [16]).

Poly-L-lysine (%, w/v)	No. of heifers	Pregnancy rate (%)
0.025	67	53.7
0.050	67	34.3
0.075	65	49.2
0.100	66	47.0
Unencapsulated	70	48.6

The potential effect on fertility of time of insemination using encapsulated semen was investigated in two separate studies. In the first trial [16], 417 synchronised yearling heifers were inseminated with 5 10^6 poly-L-lysine encapsulated or unencapsulated sperm at either 24 or 48 hr after CIDR-B® removal (Table 7). There was a time of insemination by treatment interaction ($P < 10$). Heifers inseminated at 48 hr after CIDR-B® removal had the highest conception rate when bred with encapsulated semen 67.7% vs. 57.0% for the control). However, heifers bred to unencapsulated semen had the best conception when bred at 24 hr after CIDR-B® removal (61.4% vs. 49% for encapsulated semen). The reason for this apparent effect is unclear; however, trends from a second study are in the same direction.

Table 7. *Pregnancy rate in heifers inseminated at 24 and 48 hr after CIDR-B® removal with poly-L-lysine encapsulated sperm; "0.075% poly-L-lysine, 5 10^6 sperm/dose (from Nebel et al. [16]).*

Treatment	Hours after CIDR Removal	No. of heifers	Pregnancy rate (%)
Control	24	101	61.4
	48	107	57.0
Encapsulated	24	104	49.0
	48	105	67.6

In the second study [29] 220 heifers were inseminated at either 12, 24, 36, 48, or 60 hr after CIDR-B® removal with 2 10^6 sperm encapsulated with poly-L-lysine (0.075% w/v) or unencapsulated (control). Overall, 56% of the heifers were pregnant with no difference due to encapsulation. However, pregnancy rates to encapsulated semen appeared to increase at the later insemination times. Together, these two studies suggest that sperm encapsulated in poly-L-lysine are released quickly *in vivo* at which time they compete favourably with unencapsulated sperm (within 18 – 24 hr), whereas, early breedings (more removed from time of ovulation) place encapsulated semen at a disadvantage. More work is necessary to determine the fate of capsules *in vivo* as well as the digestion or breakdown interval of capsular membranes.

6. Conclusions

Microencapsulation of spermatozoa is possible over a relatively wide range of conditions. However, the potential of this technology in artificial insemination in both the bovine and porcine species resides in the need of more intensive research and development in several areas including: 1) the ability to improve capsular retention in the female; 2) improve maintenance of viability of encapsulated sperm *in vivo* and *in vitro*; 3) the ability to control the duration of capsular membrane integrity *in vivo* such that capsule rupture will make sperm available in sufficient numbers during the optimum time for sperm transport and fertilisation in the female. On the positive side we can conclude that current technology in microencapsulation of somatic cells and spermatozoa has reached a point opening many opportunities for advancement of our understanding of cell physiology. This technology provides an opportunity to control the environment or sequential changes in environment of cells. For spermatozoa, sequential effect of environmental factors on such phenomena as capacitation, egg recognition and binding or the true acrosome reaction could be undertaken on discrete cell populations without injurious washing procedures involving centrifugations and resuspensions. The influence of co-culturing spermatozoa with reproductively important cells such as those of the epididymis or oviduct could lead to concepts relevant to prolongation of spermatozoal life *in vitro* or *in vivo*. Thus, microencapsulation of spermatozoa can now be applied to gaining a better understanding of spermatozoal physiology, which in turn could lead to a direct application of microencapsulation in artificial insemination.

References

[1] Chang, T.M.S. (1964) Semipermeable microcapsules. Science 146: 524-525.
[2] Chang, T.M.S. (1966) Semipermeable aqueous microcapsules, "artificial cells": with emphasis on experiments in an extracorporeal shut system. Trans Am. Soc. Artif. Intern. Organs 12: 13-19.
[3] Chang, T.M.S.; MacIntosh, F.C. and Mason, S.G. (1966) Semipermeable aqueous microcapsules I. preparation and properties. Canadian J. Physiol. Pharm. 44: 115-128.
[4] Chang, T.M.S. and Poznansky, M.J. (1968) Semipermeable microcapsules containing catalase for enzyme replacement in acatalasemic mice. Nature (Lond.) 218: 243-245.
[5] Chang, T.M.S. (1967) Microcapsules as artificial cells. Science 3: 63-67.
[6] Mueller, P. and Rudin, D.O. (1968) Resting and action potentials in experimental bimolecular lipid membranes. J. Theor. Biol. 18: 222-258.
[7] Pagano, R. and Thompson, T.E. (1968) Spherical lipid bilayer membranes: electrical and isotopic studies of ion permeability. J. Molec. Bio. 38: 41-58.
[8] Chang, T.M.S. (1969) Lipid-coated spherical ultrathin membranes of polymer or cross-linked protein as possible cell membrane models. Fed. Proc. 28: 461, abstr.
[9] Bedford, J. (1975) Maturation, transport, and fate of spermatozoa in the epididymis. In: Hamilton, D.O. and Greep, O.R. (Eds.) Handbook of Physiology. Section 7: Endocrinology. Vol 5: Male Reproduction System. American Physiology Society, Washington, DC; pp. 303-317.
[10] Lim, F. and Sun, A.M. (1980) Microencapsulated islets as bioartificial endocrine pancreas. Science 210: 908-910.
[11] Lim, F. and Moss, R.D. (1981) Microencapsulation of living cells and tissue. J. Pharm. Sci. 70: 351-354.
[12] Nebel, R.L.; Bame, J.; Saacke, R.G. and Lim, F. (1985) Microencapsulation of bovine spermatozoa. J. Anim. Sci. 60: 1631-1639.
[13] Nebel, R.L.; Vishwanath, R.; McMillan, W.H. and Saacke, R.G. (1993) Microencapsulation of bovine spermatozoa for use in artificial insemination: a review. Reprod. Fertil. Dev. 5: 701-712.

[14] Munkittrick, T.W.; Nebel, R.L. and Saacke, R.G. (1992) Effect of microencapsulation on accessory sperm in the zona pellucida. J. Dairy Sci. 75: 725-731.
[15] Nebel, R.L.; Fultz, S.W. and Saacke, R.G. (1986) A series of studies on the microencapsulation of bovine spermatozoa. In: Chaudry, I.A. and Thies, C. (Eds.) Proc. 13th Int. Symp. Controlled Release of Bioactive Materials. Norfolk, Virginia; pp. 214-215.
[16] Nebel, R.L.; Vishwanath, R.; Pitt, C.J. and McMillan, W.H. (1994) Effects of microcapsule membrane thickness on *in vitro* sperm viability and fertility of estrus synchronised New Zealand Friesian heifers. J. Dairy Sci. 72 (Suppl. 1): 172, abstr.
[17] Torre, M.L.; Maggi, L.; Vigo, D.; Galli, A.; Bornaghi, V.; Maffeo, G. and Conte, U. (1998) Calcium alginate beads containing an hydrophilic polymer for the encapsulation of swine spermatozoa. S.T.P. Pharma Science 8: 233-236.
[18] Torre, M.L.; Maggi, L.; Vigo, D.; Galli, A.; Bornaghi, V.; Maffeo, G. and Conte, U. (2000) Controlled release of swine semen encapsulated in calcium alginate beads. Biomaterials 21: 1493-1498.
[19] Vigo, D.; Asti, A.; Torre, M.L.; Maggi, L.; Conte, U.; Faustini, M.; Munari, E.; Russo, V. and Maffeo, G. (2000) Boar sperm encapsulated in barium alginate membranes: a survey on spermatozoa ultrastructure. In: Proc. 16th International Pig Veterinary Society Congress. Melbourne, Australia; p. 406.
[20] Foote, R.H. and Breton, R. (1960) Survival of bovine spermatozoa stored at 5 and 25°C in extenders containing varying levels of egg yolk, glucose, glycerol, citrate, and other salts. J. Dairy Sci. 43: 1322-1329.
[21] Shannon, P. (1965) Contribution of seminal plasma, sperm numbers and gas phase to dilution effects of bovine spermatozoa. J. Dairy Sci. 48: 1357-1361.
[22] Esbenshade, K.L. and Nebel, R.L. (1990) Encapsulation of porcine spermatozoa in poly-L-lysine microspheres. Theriogenology 33: 499-508.
[23] Nebel, R.L. (1983) Microencapsulation of bovine spermatozoa. Doctoral Thesis, Virginia Polytechnic Inst. and State Univ., Blacksburg, Virginia, USA.
[24] DeJarnette, J.; Saacke, R.G.; Bame, J. and Volger, C.J. (1992) Accessory sperm: Their importance to fertility and embryo quality, and attempts to alter their numbers in artificially inseminated cattle. J. Anim. Sci. 70: 484-491.
[25] Hawk, H.E. and Tanabe, T.Y. (1986) Effect of unilateral cornual insemination upon fertilisation rate in superovulating and single-ovulating cattle. J. Anim. Sci. 63: 551-560.
[26] Hunter, R.H.F. and Wilmut, I. (1983) The rate of functional sperm transport into the oviducts of mated cows. Anim. Reprod. Sci. 5: 167-173.
[27] Nadir, S.; Saacke, R.G.; Bame, J.; Mullins, J. and Degelos, S. (1993) Effect of freezing semen and dosage of sperm on number of accessory sperm, fertility, and embryo quality in artificially inseminated cattle. J. Anim. Sci. 71: 199-204.
[28] Vishwanath, R.; McMillan, W.H. and Curson, B. (1992) Microencapsulation of bovine spermatozoa: impact on semen preservation and artificial breeding. In: Proc. Ruakura Farmers Conf. Hamilton, New Zealand; pp. 116-119.
[29] McMillan, W.H.; Vishwanath, R. and Pitt, C.J. (1994) Pregnancy rate in heifers inseminated with control or encapsulated sperm prior to, during or after the onset of oestrus. In: Bradley, M. and Cummins, J. (Eds.) Proc. 7th Int. Symp. Spermatology. Cairns, Australia. pp. 25, abstr.

BIOSENSORS WITH IMMOBILISED MICROBIAL CELLS USING AMPEROMETRIC AND THERMAL DETECTION PRINCIPLES

JÁN TKÁČ[1], VLADIMÍR ŠTEFUCA[2] AND PETER GEMEINER[1]

[1]*Institute of Chemistry, Slovak Academy of Sciences, Dúbravská cesta 9, SK-842 38 Bratislava, Slovak Republic - Fax: + 421-2-59410222 - E-mail: chempege@savba.sk;* [2]*Department of Chemical and Biochemical Engineering, Faculty of Chemical and Food Technology, Slovak University of Technology, Radlinského 9, SK-812 37 Bratislava, Slovak Republic*

1. Biosensors

Biosensor is a device in which the recognition system utilizes a biological element and this information is further transformed into an analytically useful signal by a transducer (Figure 1). Biosensors can be classified according to the biological element or transducer used. Several transducers were exploited for successful biosensor construction using electrochemical, optical, calorimetric, piezoelectric and others principles of detection [1]. The most common biocomponents that have been used for the preparation of biosensors are enzymes, antibodies, DNA and cells [1]. The basic requirement for effective performance of the biosensor is a close proximity between transducer and a biological element, which is achieved by a suitable immobilisation strategy.

2. Microbial biosensors

Whole cells are very promising for biosensor preparation because of several advantages: the enzyme does not need to be isolated, enzymes are usually more stable in their natural environment in the cell, coenzymes and activators are already present in the system [2,3]. For the preparation of whole cell biosensors, microbial cells and tissues from higher plants and animals can be used. Microbial cells have some advantages over tissue cells such as higher resistance to the physical and chemical changes, easier and cheaper production and handling with higher reproducibility of preparation and microbial cell can be easily modified for construction of biosensors. Moreover, in many cases the response of microbial biosensors is more dynamic making them more attractive for preparation of biosensors. Microbial sensors are well suited for a rapid determination of sum parameters (BOD – Biological Oxygen Demand, toxic agents, assimilable sugars) because of low selectivity [4,5], but they can also be used for selective detection of a single analyte. However, the low selectivity and slow response time have to be overcome. The analyte of interest can be considered as a substrate (carbohydrates, organic acids, amino acids, peptides, alcohols, some gases – CO_2, CH_4, and others), activator (vitamins, cofactors and some metals) or inhibitor (heavy metals, antibiotics, pesticides, herbicides and other toxic agents) for cells [2-6]. Microbial biosensors were successfully used in food and fermentation industry, in environmental control and in the area of clinical analyses [2].

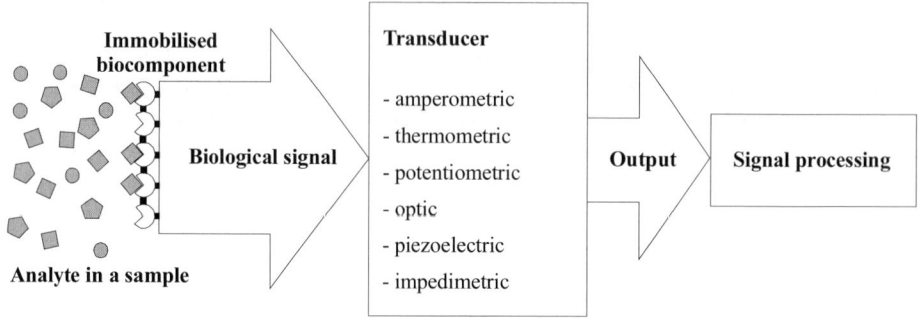

Figure 1. Basic principle of biosensor operation.

3. Immobilisation strategies

When whole and viable microbial cells are involved a gentle immobilisation technique should be used. Covalent binding frequently used for immobilisation of enzymes may result in a lower cell viability or loss of intracellular enzymes negatively affecting biosensor performance. Moreover, a low cell loading is usually achieved [3].

Entrapment into natural polymers prepared from alginate/pectate, κ-carrageenan, collagen, gelatine, chitosan and agar (agarose) is performed under mild conditions with high viability of cells entrapped. On the other hand alginate, pectate and κ-carrageenan matrix can be easily destroyed by chelating agents present in a sample, because the matrix is formed by an ionotropic gelation in the presence of Ca^{2+} or K^+ ions. Gel matrix can be further stabilised using cross-linking agents like glutaraldehyde or glutaraldehyde/polyethyleneimine. Cross-linking using glutaraldehyde is also very often used in combination with proteinacious supports such as gelatine or albumin. Because of similarity with covalent binding, the technique is more useful for immobilisation of non-viable cells [5].

The synthetic polymers including polyvinylalcohol, polyacrylamide, polyurethane and photo cross-linkable resins are more stable than the natural ones, but in many cases loss of viability may happen. The major limitation of entrapment into synthetic or natural polymer matrix is a diffusion barrier slowing down the response of the biosensor. The problem can be solved by choosing a proper gel matrix with larger pores in order to facilitate diffusion of compounds to cells and from cells to the transducer [3]. Adsorption of cells at a solid surface is very simple method of immobilisation and usually no reagents are required with minimum of activation or clean-up steps. On the other hand, the adsorption is susceptible to changes in pH, temperature, ionic strength and other factors and the method is suitable rather for single measurement [4].

The highest cell viability is preserved by a simple trapping of microbial cells into the pores of synthetic or cellulose based membranes. For this purpose filtration of cell suspension is carried out and membrane with cells is attached to the electrode. Another alternative is a coating of cell suspension on the electrode by a membrane, usually using a dialysis membrane. In this case generally lower operational and storage stability is achieved compared to entrapment into gel/polymer matrix [2].

Among other techniques of immobilisation it is worth mentioning biospecific reversible immobilisation using lectins, when adhesion of cells in monolayer is achieved and sensitivity can be further improved by application of several layers of lectin immobilised cells and a fresh portion of biocatalyst can be easily loaded [3]. New techniques including adhesion of cells on a variety of polymeric surfaces using polyethyleneimine with rapid and strong adhesion [3] and biospecific reversible attachment on cellulose particles/membranes using a cellulose binding domain expressed on the cell surface provide an immobilisation with high viability and fast response due to low diffusion resistance [7]. Incorporation of cells into a carbon paste [8] or within a matrix of sol-gel has several advantages [9]. All reagents needed including cells can be packed into carbon paste or sol-gel matrix providing 'reagent less' biosensors. In many cases stability of biomaterial is enhanced with retaining high viability. Carbon paste and sol-gel electrodes can be easily modified and sensitivity restored by a simple polishing.

4. Improvement of microbial biosensor characteristics

The main drawback of microbial biosensors when used for single analyte detection - a low selectivity was improved using several approaches. The selectivity was enhanced by influencing cell physiology (induction of transport/metabolic systems, inhibition of undesired transport/metabolic pathways); by changing cell environment (pH of measurement); by choosing proper transducer or suitable mode of measurement (kinetic *vs.* steady-state); by permeabilisation of the cell membrane (leaching of cofactors from cells) [2,3]. More advanced approaches use recombinant microbial strains lacking specific receptors, fusing of reporter gene for green fluorescent protein or *lux* gene with genes induced by the analyte, by overexpression of natural enzymes in the cells or by improving the analyte affinity [2,3,10]. More general approaches using chemometrics [11] and the use of permselective membranes were recently introduced [12].

The dynamics of the microbial biosensor response can be significantly improved by permeabilisation of cells [3,13] or by engineering of cells to anchor the desired enzyme into the periplasmic space [14]. When the targeted enzyme is localised in the periplasmic space, the situation is simplified and whole cells can be used without a pretreatment.

5. Amperometric and thermal microbial biosensors

The use of two transducers is emphasised in this chapter – amperometric and thermal.

Since more than 100 articles were published to date dealing with microbial biosensors and most of them based on amperometric detection (mainly Clark type oxygen electrode), here a more focused view on amperometric microbial biosensors using one of the most promising microorganism – *Gluconobacter sp.* for preparation of microbial biosensors is introduced. Moreover a chapter dealing with thermal detection using microbial cells is included. This approach offers natural connection between cells and a generally applicable transducer.

5.1. AMPEROMETRIC BIOSENSORS WITH THE USE OF *GLUCONOBACTER OXYDANS*

5.1.1. Gluconobacter oxydans as a prospective biocatalyst

Gluconobacter oxydans possesses an extraordinary unique organisation of metabolic system with surface location of main oxidative enzymes responsible for partial oxidation of carbon substrates (Figure 2) [15,16].

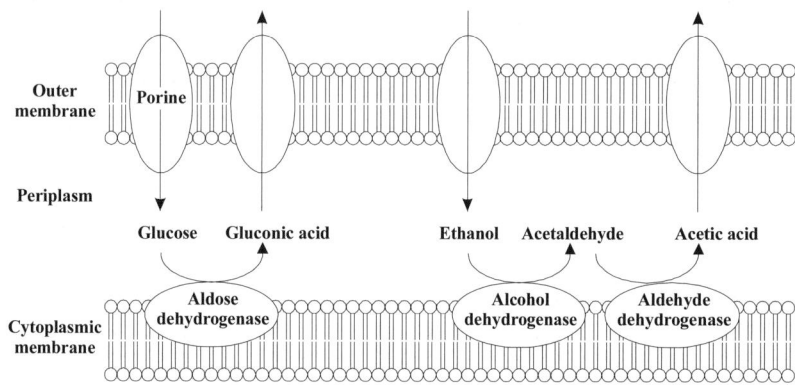

Figure 2. Schematic model of localisation of membrane-bound dehydrogenases in *Gluconobacter sp.*

The genus of Gluconobacter can oxidize a broad range of compounds with high rate and almost quantitatively [15]. The complete genomic sequence of *G. oxydans* strain 621H has been determined by a whole-genome-shotgun approach indicating simple respiration chain with enormous oxidation potential of *G. oxydans* [17]. Fast and high efficiency of oxidation makes this genus ideal for preparation of biosensors based on whole cells [18]. *Gluconobacter sp.* contains many membrane bound enzymes such as alcohol-, aldose-, aldehyde-, sorbitol-, D-gluconate-, fructose- and glycerol dehydrogenases, which were very often used for preparation of enzyme biosensors due to insensitivity of enzymes to oxygen and cofactor does not need to be added to the system. Efficient communication between enzyme and electrode was achieved using a wide range of artificial electron acceptors (mediators) [15,19,20], but also direct electron transfer between electrode and enzyme (containing *haeme c* moiety) is possible [21,22].

5.1.2. Whole cell Gluconobacter oxydans biosensors

Whole cell *Gluconobacter oxydans* biosensors are based on amperometric detection using either oxygen [11,13,[23-25]] or mediators including ferricyanide [12,26-29], p-benzoquinone, dichlorphenolindophenol [[30]] and 2-hydroxy-1,4-naphthoquinone [31]. In case of oxygen detection the biosensor performance can be affected by fluctuation in the concentration of oxygen and by its availability; and slower diffusion of oxygen to the layer with immobilised cells resulting in a long response time of the biosensor. Moreover the sensitivity of mediated microbial biosensors is higher compared to biosensors based on detection of oxygen. Only limited number of immobilisation techniques was used for preparation of whole cell *G. oxydans* biosensors including entrapment with crosslinking, containment behind membranes and adsorption on a filtration paper or a porous nitrocellulose filter (Table 1).

Table 1. G. oxydans biosensors based on whole cells or membranes.

Immobilisation	Analyte	LOD [μM]	RT [s]	Mediator[a]	Ref.
A	Glc	20	120	No	13
B	Glc	5	30	Benzoquinone	26
B	Glc	10	26	Ferricyanide	12
B	Glc, Gal, Ara, Xyl, Man	5-107	120-180	Ferricyanide	27
B	Gly	20	84	Ferricyanide	29
B	EtOH	0.8	13	Ferricyanide	12
C	EtOH	0.9	13	Ferricyanide	12
D	EtOH/Glc	500	-	No	24
D	EtOH/Glc	10	-	No	11
E	EtOH	2	180	No	32

Abbreviation used: A - entrapment and crosslinking; B – containment behind a dialysis membrane; C - containment behind a cellulose acetate membrane; D - adsorption on a filtration paper; E - adsorption of cell membranes on a porous nitrocellulose filter; LOD – limit of detection; RT – response time; Glc – glucose; Gly – glycerol; Gal – galactose; Ara – arabinose; Xyl – xylose; Man – mannose; EtOH – ethanol; [a] – in case when no mediator was used, oxygen electrode was used as a transducer.

Both immobilisation technique and transducer used have a strong influence on the biosensor performance. The best characteristics were achieved using mediator affecting not only response time, but also sensitivity and thus detection limit of assay. The response time of *G. oxydans* biosensors are shorter compared to the response time of microbial biosensors in general (2-10 min) [2]. The sensitivity of determination, when mediated approach was used, is comparable to the sensitivity of enzyme biosensors. Improvement in the selectivity of whole cell *G. oxydans* biosensors was achieved by chemometric approach using either multivariate calibration method [11] or polynomial approximation [24]; by kinetic method of analysis [29]; using a permselective cellulose acetate membrane [12] or membrane fraction from *G. oxydans* [32].

5.1.2.1. Oxygen-based G. oxydans biosensors. Limited availability of dissolved oxygen influences measurable range of a biosensor, which is obvious especially with high cell loading. The problem was effectively solved using a fluorinated organic substance –

perfluorodecalin, when it was possible to have concentration of oxygen in the measurement medium of 37 mg/l compared to 9.3 mg/l for buffer saturated with air [25]. Perfluorodecalin-treated buffer retained high level of oxygen for more than 1 h. However authors claim that this approach can be used for a wide variety of microbial and enzyme biosensors, utilisation of fluorinated compounds due to environmental toxicity is questionable.

Enhanced selectivity of a *G. oxydans* biosensor is of great interest. Two compounds glucose and ethanol are oxidised by *G. oxydans* very efficiently making whole *G. oxydans* biosensor unable to detect one analyte in the presence of the other one. Reshetilov's group developed two strategies how to solve this problem using chemometric tools [11,24]. The first approach is based on the measurement of the sum of ethanol and glucose concentration by an unselective *G. oxydans* biosensor. Glucose oxidase electrode is used for selective detection of glucose and using multivariate calibration it is possible to estimate concentration of glucose and ethanol with error of measurement less than 8%, but measurement range was very narrow (0-0.6 mM) [11]. The second approach uses again unselective G. oxydans cells for assay both glucose and ethanol. Detection of ethanol was performed by a microbial *Pichia pastoris* biosensor. Using chemometric principles of polynomial approximation, data from both of these sensors were processed to provide accurate estimates of glucose and ethanol in the range 1.0 – 8.0 mM [24]. When artificial neural network was used for data processing, glucose and ethanol were accurately estimated in the range 1.0 – 10.0 mM [24]. These two approaches were tested only on mixtures of glucose and ethanol and measurement of real samples will be more problematic.

The possibility to use *G. oxydans* in combination with other microorganisms for preparation of bi-microbial biosensors was recently introduced by our group. *G. oxydans* cells were used for final oxidation of monosaccharides released after hydrolysis of sucrose by co-immobilised *S. cerevisiae* or after hydrolysis of lactose by co-immobilised *Kluyveromyces marxianus*. In the case of sucrose biosensor, cells of *S. cerevisiae* were used without pre-treatment due to periplasmic localisation of invertase. When whole cells of *K. marxianus* were co-immobilised together with *G. oxydans*, no response after addition of lactose was observed. Permeabilisation of *K. marxianus* using mixture of chloroform and ethanol was efficient allowing release of glucose and galactose from *K. marxianus* reaching *G. oxydans*. The response time of *G. oxydans* biosensor was 2 min and the response time of bi-microbial biosensors was 5 min [13]. The presented bi-microbial biosensors are unselective, but when bimicrobial lactose sensor was used for determination of lactose in milk, results obtained were in a good agreement with the reference analytical method [23].

Karube's group developed a method for construction of ethanol biosensor using membrane fraction of *G. oxydans* [32]. The cells were disintegrated using a French press and the cell debris sediment after ultracentrifugation was used for biosensor preparation. Surprisingly, the membrane biosensor was insensitive to glucose addition [32]. This can be explained by disturbed communication between aldose dehydrogenase (AldDH, responsible for oxidation of glucose) and alcohol dehydrogenase in the damaged membrane, when no electrons are coming from AldDH to the respiration chain and finally to oxygen. This is in accordance with the observation that AldDH is not able to transfer electrons to the respiration chain directly, but this communication is mediated only through alcohol dehydrogenase and ubiquinone [33].

5.1.2.2. Mediated Gluconobacter sp. biosensors. As mentioned earlier the use of artificial electron acceptors in combination with microbial cells has some advantages including higher sensitivity, shorter response time and in many cases a higher solubility of a mediator in aqueous solutions compared to oxygen. The most interesting facts that were found using mediated *G. oxydans* biosensors are highlighted here.

Lignocellulose hydrolysate as a waste from pulp industry can be used as a source of sugars for microbial fermentation improving the economy of the process. The concentration of saccharides (glucose, galactose, xylose, arabinose and mannose) in the hydrolysate is quite low, containing many inhibitors making monitoring of the process by the enzyme biosensors difficult. Ferricyanide-mediated *G. oxydans* biosensor was sensitive enough to detect all the saccharides present in the hydrolysate even at low concentrations. Storage stability was better, when the sensor was stored in a phosphate buffer at 4°C compared to the storage in a dry state. Moreover, the use of trehalose as a stabilizer resulted in two times higher storage stability (half-life of 20 days). The measurement was very reproducible and when the response was expressed as arabinose equivalents, the results were in a good agreement to the results obtained by a quantitative paper chromatography [27].

A triglyceride measurement was performed by hydrolysis of triglyceride by a non-specific lipase. Glycerol released from triglyceride was detected by *G. oxydans* biosensor. Two approaches were used: a) separate hydrolysis of triglyceride and detection of glycerol by ferricyanide-mediated biosensor and b) kinetic measurement, when triglyceride hydrolysis and detection of glycerol were performed in the same vessel. The use of surfactants was omitted since it negatively affected the performance of the microbial biosensor. The hydrolysis time was up to 20 min with a total time of triglyceride assay of 25 min. To shorten the assay of triglyceride to 10 min, a kinetic approach was used. The kinetic measurement was not disturbed by the presence of free glycerol in a sample, because the response was read 5 min after the sample was introduced into the system. The basic parameters of our biosensor are comparable with enzyme ones in terms of sensitivity, detection limit, reproducibility of measurement and stability [29].

Selective detection of ethanol in the presence of excess of glucose was performed using a permselective cellulose acetate membrane cast *in situ* on a *G. oxydans* layer. In a preliminary study it was found that glucose does not affect the biosensor performance and the measurement of ethanol in the presence of glucose is possible [28]. The discrimination between analytes is based on the size exclusion effect of the membrane, when compounds with higher molecular weight are excluded from reaching an electrode with immobilised *G. oxydans*. The biosensor allows very fast and sensitive determination of ethanol (Table 1) comparable to that published for enzyme biosensors. The operational stability of the biosensor was excellent; during 8.5 h of continuous use no significant decrease in the sensitivity was observed. Ethanol concentrations measured in the real samples by the microbial biosensor were in a good agreement to the values obtained by a reference analytical method ($R^2=0.998$, Figure 3) [12].

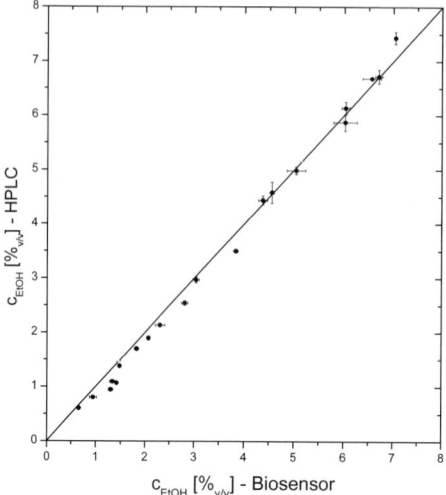

Figure 3. Correlation between G. oxydans biosensor and HPLC assays of ethanol during batch fermentation with initial concentration of glucose of 200 g/l. Bars at each point represent RSD of sample measurement by both methods of analysis.

Ikeda's group examined the effect of various electron acceptors on the magnitude of oxidation of glucose, fructose, glycerol and ethanol by *G. industrius* biosensor [26]. They found that the response time of whole cell mediated glucose biosensor is the same as in the case of enzyme-based electrode with the same configuration, indicating that outer membrane is not a dominating factor influencing response time [26]. Mediated bioelectrochemistry of *G. industrius* was very carefully characterised with some interesting results. It was found that benzoquinone is preferred electron acceptor for oxidation of glucose by *G. industrius* to oxygen, even though the permeability of the bacterial membrane to the mediator is smaller than that of oxygen [30]. According to the kinetic constants obtained, the system consisting of 10 g of dry weight of *G. industrius* cells in 1L can produce 6.1 mol of gluconic acid from glucose *per* day in the presence of benzoquinone, demonstrating extremely high oxidation potential of *Gluconobacter sp.* [30].

Recently the effect of initial carbon source on the response time and efficiency of glucose oxidation by *G. oxydans* mediated by 2-hydroxy-1,4-naphthoquinone was investigated. The study showed that the time needed to reach the maximum current after glucose addition, sensitivity of the response and the Coulombic efficiency of glucose oxidation were dependent on initial carbon source used for cultivation of *G. oxydans*. The highest Coulombic efficiency of glucose oxidation was achieved using galactose as a substrate [31].

5.1.3. Conclusion and future perspectives of G. oxydans biosensors

G. *oxydans* based microbial biosensors were very successfully used in different fields including monitoring of fermentations, food and environmental analyses. The biosensors were efficiently used for determination of sum parameters (utilisable saccharides) or for determination of a single analyte using various approaches. Especially operational stability of G. *oxydans* biosensors is very high, whilst storage stability is low due to the use of artificial electron acceptors. The storage stability can be enhanced using an approach utilised for stabilisation of enzyme biosensors [34]. The biosensor offers very fast response time with high sensitivity of detection and low detection limit. Moreover preparation of cell suspension is very cheap with very low nutritional demands during cultivation.

Due to unspecific and high-efficient oxidation of various types of compounds including sugars, alcohols and polyols the G. *oxydans* cells can be easily used for preparation of BOD biosensors or for construction of biofuel cells.

A new concept for amperometric pH sensing based on the use of a redox dye changing redox properties due to change in pH was introduced by Stredansky et al. [35]. This approach has many advantages compared to potentiometric pH sensing including faster response, lower drift of the baseline, lower susceptibility to the presence of other ions in the sample and lower detection limits. Recently our group developed a method to enhance stability of adhesion of a redox dye by its electropolymerisation on the surface of electrode [36], allowing simultaneous immobilisation of enzymes. By amperometric pH sensing it is possible to use other enzymes (amidohydrolases, esterases, decarboxylases *etc.*) besides oxidoreductases. This concept can be used more generally in combination with immobilised microbial cells, when oxidation of substrates leads to the external change of pH (especially the case of *Gluconobacter sp.*).

Immobilisation of enzymes within a film formed by an electropolymerisation is a well-established procedure for construction of enzyme biosensors with many advantages such as controlled loading of enzymes, controlled spatial distribution of biomolecule with a possibility to modulate the enzyme activity. Moreover, the thickness of the film can be easily adjusted with fast response time and sensitivity of measurement [37]. The question: "Is this immobilisation technique suitable for immobilisation of microbial cells (membranes)?" should be answered.

5.2. THERMAL BIOSENSORS

Metabolism of viable cells in non-immobilised form is often investigated by calorimetry that is preformed mainly in batch calorimetric instruments [38]. The application of this versatile and simple detection principle for immobilised cells was up to this day limited to flow calorimeter representing one type of thermal biosensor. The measurement by flow calorimeters is based on registration of the temperature change induced by the reaction heat released in a column with immobilised cells that is operated as a packed bed reactor. Principles and applications of thermal biosensors with the focus on flow calorimeters were reviewed in the literature [39,40].

Flow calorimeters with immobilised cells are used in two principally different ways: as analytical devices for determination of concentration of metabolites and other compounds, and as typical calorimeters in which properties of immobilised cells are followed. These two application areas imply different criteria on immobilisation techniques. While for analytical applications the most important criteria are specific activity, mechanical and catalytic stability, and sensor-fitted geometry, in the case of investigation of cell properties the most important requirement is to minimise the contribution of immobilisation to the cell behaviour. In addition to that, application of immobilised cells on industrial scale puts in advance economical criteria. Immobilisation methods used in calorimetry copy general trend meaning that the dominant technique is entrapment in hydrogels. This is obvious so from the frequency of the use of this technique according to the published literature [41], as from the development of commercial instruments for preparation of bulk quantities of cell immobilisates [42].

5.2.1. Analytical applications

The basic design of flow-injection analysis (FIA) system with flow calorimeter based on immobilised cells is depicted in Figure 4. The typical analysis is performed by the injection of the sample that is carried by buffer stream to the column where it is metabolised by immobilised cells. The released heat is proportional to the extent of chemical reactions and it is followed through the measurement of the temperature difference between the immobilised cell and reference columns. The temperature difference is measured by thermistors [43] and under ideal conditions (no heat losses cross the column wall) it is directly proportional to the total amount of produced heat.

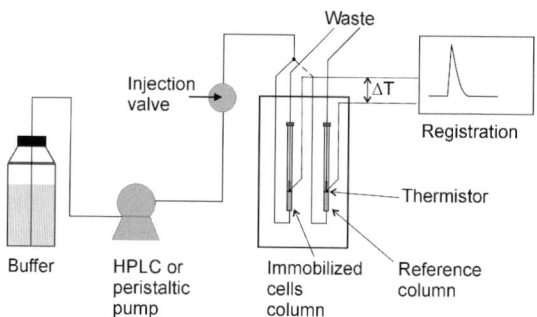

Figure 4. Using flow calorimeter as a detector device in FIA system.

The analyte concentration in samples is determined through the method calibration on standard concentrations.

First instruments known up to this day under the name "Enzyme Thermistor" were designed for analysis based on immobilised enzymes. Advances in cell immobilisation have logically brought idea to replace enzymes by whole cells also in the flow calorimetry. Even though the tests with whole cells are a natural ambition to find new applications of the flow calorimetry, analytical applications with immobilised cells are quite rare. This is mainly because of lack of specificity given that the registered temperature change reflects the overall heat produced by reactions of different compounds in the sample. This disadvantage is only partially compensated by possibly low price of immobilised cells compared to purified enzymes. On the other hand cells can be tailored by permeabilisation and chemical inactivation of undesired enzymes. It may happen that such biocatalysts are more stable than isolated enzymes.

Some promising applications can be expected in environmental control. The inhibitory effect of a pollutant on immobilised cells or the heat of cell metabolic route converting the pollutant can be detected in a flow calorimeter [44]. For example cells of *Pseudomonas cepacia* were immobilised in calcium alginate gel and their metabolic responses when exposed to a range of derivatised aromatic compounds were studied [45]. Metabolic responses *vs.* many derivatised aromatics were obtained. In addition to the direct dose-response studies, it was also possible to use the cell-based sensor to obtain information concerning the cellular physiology.

In certain cases metabolites can be monitored by this way. Cells of *Gluconobacter oxydans* immobilised in calcium alginate were used for determination of glycerol [46]. The calibration curve was linear up to 2 mM glycerol. Immobilised cells of *Enterobacter aerogenes* were used as citrate-sensing biocatalyst [47]. The whole cell sensor was prepared by immobilising cells in barium pectate beads hardened by polyethyleneimine and glutaraldehyde. The sensor provided linear range form 0.2 to 6 mM citrate. The immobilisation of whole cells rather than isolated enzyme, citrate lyase, was found to be better solution, as citrate lyase undergoes inactivation during its reaction due to its deacetylation. In cells the enzyme is reactivated *via* acetylation providing good stability of the sensor that could be used for more than 200 analysis runs.

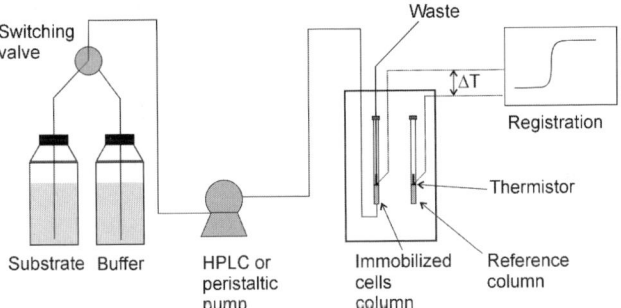

Figure 5. *Using flow calorimeter as a monitoring device for continuous flow reactor operating in steady-state.*

5.2.2. Investigation of properties of immobilised cells

Metabolic or biotransformation activity of cells immobilised in porous particles is often followed in flow reactors as the relation between cell activity and conditions around cells. Hence, the cell activity is estimated from the rate of chemical reactions occurring in the reactor. However, the metabolism of cells is accompanied by production of heat, so that the metabolic activity can be investigated by calorimetric methods [48]. The flow calorimeter systems previously designed for flow injection analysis can easily be adapted for such measurements (Figure 5). Using configuration depicted in Figure 5 the system is first stabilised by pumping buffer solution. After that the system is switched to substrate solution and run until the temperature is stabilised on a new steady state characterised by stabilisation of registered temperature change, ΔT. By choosing the composition of substrate solution different properties of immobilised cells can be followed. The main advantage of the method is linkage of the cell flow reactor and analysis in one step in the calorimeter column equipped with the output temperature measurement.

Thermometric registrations provide useful qualitative information about cell properties. Moreover, kinetic or more quantitative data can be obtained by mathematical treatment of thermometric data based on material and heat balances of processes in the calorimetric column [49]. The heart of the system is a particle with immobilised cells. The mass balance for reacting compounds in spherical particle at steady-state conditions is given by equation

$$D_{ei}\left(\frac{d^2c_i}{dr^2} + \frac{2}{r}\frac{dc_i}{dr}\right) + v_i v = 0 \qquad (1)$$

where D_{ei} is effective diffusion coefficient of the compound, c_i is its concentration in particle, r is radial coordinate in particle, and v is rate of chemical reaction. Stoichiometric coefficient of the compound, v_i, is of negative or positive value depending on whether it concerns one of reaction substrates or products.

Solving Eq. (1) for boundary conditions

$$r = 0: \quad \frac{dc_i}{dr} = 0 \qquad (2)$$

$$r = R: \quad c_i = c_{ib}$$

the observed overall particle reaction rate per unit volume of particle, v_{obs}, can be calculated

$$v_{obs} = -\frac{AD_{ei}v_i}{V}\left(\frac{dc_i}{dr}\right)_{r=R} = -\frac{3D_{ei}v_i}{R}\left(\frac{dc_i}{dr}\right)_{r=R} \qquad (3)$$

where A, V, and R are particle surface area, volume, and radius, respectively.

The calculation of the temperature change in the column necessitates linking Eq. (3) to balance equations for the column. According to the calorimeter configuration the column should be described as a continuous packed bed reactor in which the layer of immobilised cells is sandwiched by two layers of inert glass beads as depicted in Figure 6.

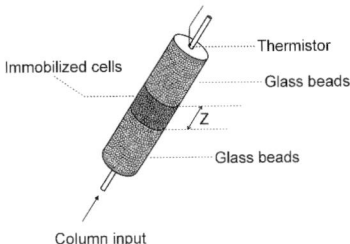

Figure 6. Schematic representation of calorimetric column with immobilised cells.

Balance equations were derived from following assumptions: the reactor is differential, plug flow occurs in the reactor, interstitial velocity of flow is high enough to prevent the effect of external diffusion, heat loss through the reactor wall is negligible so that the reactor is considered to be adiabatic [49].
Mass balances:

$$w\frac{dc_{ib}}{dz} = (1-\varepsilon)v_{obs}v_i \tag{4}$$

Heat balance:

$$w\rho C_p \frac{dT}{dz} = (1-\varepsilon)v_{obs}(-\Delta_r H) \tag{5}$$

Symbols in equations are: c_{ib} – bulk concentration, w – superficial liquid flow rate, z – axial coordinate in the column, ρ and C_p – liquid density and specific heat capacity, $\Delta_r H$ - molar reaction enthalpy, ε – bed void fraction. If the reaction extend in the column is very low enabling to apply a differential reactor conception, it can be written: $dT/dz = \Delta T/\Delta z = \Delta T/Z$ and Eq. (5) can be rearranged

$$\Delta T = \frac{(1-\varepsilon)Z(-\Delta_r H)}{w\rho C_p} v_i v_{obs} \tag{6}$$

$$\Delta T = \alpha v_{obs} \tag{7}$$

Parameter $\alpha = \frac{(1-\varepsilon)Z(-\Delta_r H)}{w\rho C_p}$ regroups quantities that are considered constant for each individual experimental system. Eq. (7) was up to now applied for simple biotransformations $S \rightarrow P$ [50-52]. In that case stoichiometric coefficients $v_s = -1$ and $v_p = 1$. The initial rate of lot of enzyme reactions was well described by substrate inhibition model

$$v = \frac{V_m c_S}{K_m + c_S + c_S^2/K_i} \qquad (8)$$

where c_S is substrate concentration, V_m, K_m, and K_i are kinetic parameters. The implementation of particle mass balance into column balance equations can be simplified by introducing the effectiveness factor [51]:

$$\eta = \frac{v_{kin}}{v_{obs}} \qquad (9)$$

where v_{kin} is the rate v when particle substrate concentration, c_S, is considered to be equal to bulk substrate concentration, c_{Sb}. Then, from Eqs. (7), (8), and (9)

$$\Delta T = \alpha\eta \frac{V_m c_{Sb}}{K_m + c_{Sb} + c_{Sb}^2/K_i} \qquad (10)$$

If cells are immobilised on the particle surface or particles are small enough to prevent internal diffusion limitation, $\eta = 1$. Then, K_m and K_i in Eq. (10) can be estimated by non-linear regression from experimental dependencies between ΔT and c_{Sb} where, according to the differential bed assumption, the bulk substrate concentration is the concentration of the substrate entering the bed of immobilised cells. Although this approach has already been applied for example for studying kinetics [53], inactivation [49], screening [54] of immobilised purified enzymes, applications of immobilised cells were limited to the case when essentially $\eta \neq 1$.

Examples of immobilised cell systems that were investigated by means of flow calorimetry are reviewed in Table 2. According to the general experience that microbial cells are preferred type of sensing cells used for construction of biosensors (*e.g.* electrochemical sensors), exclusively microbial cells were studied by flow calorimetry. So far, the most important part of studies was oriented toward β-lactame antibiotics transforming cells.

Properties of *T. variabilis* cells with D-amino acid oxidase activity were studied under low substrate concentration [50] that was practical for rapid screening of immobilised cell preparations [57]. The low substrate concentration enabled to use first order kinetics approximation that simplified solution of particle mass balance by effectiveness factor calculation *via* explicit mathematical equation [58]. More complex situation was solved later in the case of *E. coli* cells with penicillin G acylase activity [51]. Here higher substrate concentrations were applied and the particle mass balance in form of Eq. (1) had to be solved. By this approach kinetic parameters in Eq. (10), K_m and K_i, as well as the extent of internal mass transfer limitation expressed in values of effectiveness factor values were determined.

Table 2. Investigation of properties of immobilised cells by means of the flow calorimetry.

Cells	Biocatalytic activity	Immobilisation technique	Ref.
Trigonopsis variabilis	D-Amino acid oxidase	Entrapment + cross-linking	[50]
Trigonopsis variabilis	D-Amino acid oxidase	Cross-linking	[50]
Trigonopsis variabilis	D-Amino acid oxidase	Entrapment	[50,55]
Escherichia coli	Penicillin G acylase	Entrapment + cross-linking	[52]
Escherichia coli	Penicillin G acylase	Entrapment	[51]
Nocardia tartaricans	*cis*-Epoxysuccinate hydrolase	Entrapment	[56]

Mathematical treatment of thermometric data is always based on their transformation to more standard kinetic data. It means from values of the registered temperature change the observed reaction rate is calculated *via* Eq. (7). The value of transformation parameter α can by estimated by several ways:
- by calibration on immobilised cell preparations with known activity [50,57,59]
- by calibration applying post-column analysis of reactant concentration [60]
- by autocalibration technique using infinite recirculation of reaction mixture [61] (no post-column analysis is required)
- by direct calculation from its definition

The last approach is rather difficult to apply given that values of some quantities cannot be provided with sufficient precision. Moreover, the calibration techniques enable to compensate for some uncertainties coming from instrument systematic errors not included in the mathematical model.

5.2.3. Conclusions and perspectives

Examples introduced in this chapter indicate that flow calorimetry is a simple and versatile technique for investigation of kinetics of biotransformation reactions catalysed by immobilised microbial cells. Prevalent number of previous applications was oriented toward permeabilised and therefore non-living cells used like biocatalysts. The experience gained during the last period could be extended to other type of cells (animal, plant) and to more complicated systems, *e.g.* viable cells and their metabolic response to various factors. Such type of applications can be advisable with the actual type of instrumentation, but further instrument improvements will be suitable concerning sterility, robustness, easy of use, and automation. Finally, it is clear that each improvement of properties of immobilisation material and techniques will help to promote the introduced method in different research areas.

References

[1] Thevenot, D.R.; Toth, K.; Durst, R.A. and Wilson G.S. (1999) Electrochemical biosensors: Recommended definitions and classification. Pure Appl. Chem. 71: 2333-2348.
[2] Racek, J. (1995) Cell-based biosensors. Technomic Publishing Company, Inc., Basel, Switzerland; ISBN 1-56676-190-5; pp. 1-72.
[3] D'Souza, S.F. (2001) Microbial biosensors. Biosens. Bioelectron. 16: 337-353.

[4] Wittmann, C.; Riedel, K. and Schmid, R.D. (1997) Microbial and enzyme sensors for environmental monitoring. In: Kress-Rogers, E. (Ed.) Handbook of biosensors and electronic noses. Medicine, food and environment. CRC Press, London (UK); ISBN 0-8493-8905-4; pp. 299-332.
[5] Riedel, K. (1998) Microbial biosensors based on oxygen electrodes. In: Mulchandani, A. and Rogers, K.R. (Eds.) Enzyme and microbial biosensors: Techniques and Protocols. Volume 6 from the series Walker, J.M. (Ser.Ed.) Methods in biotechnology. Humana Press Inc., Totowa (USA); ISBN 0-8960-3410-0; pp. 199-224.
[6] Arikawa, Y.; Ikebukuro, K. and Karube, I. (1998) Microbial biosensors based on respiratory inhibition. In Mulchandani, A. and Rogers, K.R. (Eds.) Enzyme and microbial biosensors: Techniques and Protocols. Volume 6 from the series Walker, J.M. (Ser.Ed.) Methods in biotechnology. Humana Press Inc., Totowa (USA); ISBN 0-8960-3410-0; pp. 225-236.
[7] Wang, A. A.; Mulchandani, A. and Chen, W. (2001) Whole-cell immobilization using cell surface-exposed cellulose-binding domain. Biotechnol. Prog. 17: 407-411.
[8] Gorton, L. (1995) Carbon paste electrodes modified with enzymes, tissues and cells. Electroanal. 7: 23-45.
[9] Gill, I. (2001) Bio-doped nanocomposite polymers: Sol-gel bioencapsulates. Chem. Mater. 13: 3404-3421.
[10] Daunert, S.; Barrett, G.; Feliciano, J.S.; Shetty, R.S.; Shrestha, S. and Smith-Spencer, W. (2000) Genetically engineered whole-cell sensing systems: Coupling biological recognition with reporter genes. Chem. Rev. 100: 2705-2738.
[11] Reshetilov, A.N.; Lobanov, A.V.; Morozova, N.O.; Gordon, S.H.; Greene, R.V. and Leathers, T.D. (1998) Detection of ethanol in a two-component glucose/ethanol mixture using a nonselective microbial sensor and a glucose enzyme electrode. Biosens. Bioelectron. 13: 787–793 and references cited therein.
[12] Tkac, J.; Vostiar, I.; Gorton, L.; Gemeiner, P. and Sturdik, E. (2003) Improved selectivity of microbial biosensor using membrane coating. Application to the analysis of ethanol during fermentation. Biosens. Bioelectron. 18: 1125-1134.
[13] Svitel, J.; Curilla, O. and Tkac, J. (1998) Microbial cell-based biosensor for sensing glucose, sucrose or lactose. Biotechnol. Appl. Biochem. 27: 153-158.
[14] Mulchandani, P.; Chen, W.; Mulchandani, A.; Wang, J. and Chen L. (2001) Amperometric microbial biosensor for direct determination of organophosphate pesticides using recombinant microorganism with surface expressed organophosphorus hydrolase. Biosens. Bioelectron. 16: 433-437.
[15] Lusta, K.A. and Reshetilov, A.N. (1998) Physiological and biochemical features of *Gluconobacter oxydans* and prospects of their use in biotechnology and biosensor systems. Appl. Biochem. Microbiol. 34: 307-320 and references cited therein.
[16] Matsushita, K.; Toyama, H.; Yamada, M. and Adachi, O. (2002) Quinoproteins: structure, functions and biotechnological applications. Appl. Microb. Biotechnol. 58: 13-22.
[17] Deppenmeier, U.; Hoffmeister, M. and Prust, C. (2002) Biochemistry and biotechnological applications of *Gluconobacter* strains. Appl. Microbiol. Biotechnol. 60: 233–242.
[18] Macauley, S.; McNeil, B. and Harvey, L.M. (2001) The genus *Gluconobacter* and its applications in biotechnology. Crit. Rev. Biotechnol. 21: 1-25.
[19] Laurinavicius, V.; Razumiene, J.; Kurtinaitiene, B.; Lapenaite, I.; Bachmatova, I.; Marcinkeviciene, L.; Meskys, R. and Ramanavicius, A. (2002) Bioelectrochemical application of some PQQ-dependent enzymes. Bioelectrochem. 55: 29-32.
[20] Tkac, J.; Vostiar, I.; Sturdik, E.; Gemeiner, P.; Mastihuba, V. and Annus, J. (2001) Fructose biosensor based on D-fructose dehydrogenase immobilised on a ferrocene-embedded cellulose acetate membrane. Anal. Chim. Acta 439: 39-46 and references cited therein.
[21] Schuhmann, W.; Zimmermann, H.; Habermüller, K. and Laurinavicius, V. (2000) Electron-transfer pathways between redox enzymes and electrode surfaces: Reagentless biosensors based on thiol-monolayer-bound and polypyrrole-entrapped enzymes. Faraday Discuss. 116: 245-255 and references cited therein.
[22] Ikeda, T. (1997) Direct redox communication between enzymes and electrodes. In: Scheller, F.W.; Schubert, F. and Fedrowitz, J. (Eds.) Frontiers in Biosensorics I, Fundamental aspects. Springer Verlag, Basel (Switzerland); ISBN 3-7643-5475-5; pp. 243-266.
[23] Tkac, J. and Svitel, J. (1997) Determination of glucose and lactose in milk by microbial biosensors. Bull. Food Sci. 36: 113-121 (in Slovak).
[24] Lobanov, A.V.; Borisov, I.A.; Gordon, S.H.; Greene, R.V.; Leathers, T.D. and Reshetilov, A.N. (2001) Analysis of ethanol-glucose mixtures by two microbial sensors: application of chemometrics and artificial neural networks for data processing. Biosens. Bioelectron. 16: 1001-1007.

[25] Reshetilov, A.N.; Efremov, D.A.; Iliasov, P.V.; Boronin, A.M.; Kukushskin, N.I.; Greene, R.V. and Leathers, T.D. (1998) Effects of high oxygen concentrations on microbial biosensor signals. Hyperoxygenation by means of perfluorodecalin. Biosens. Bioelectron. 13: 795-799.
[26] Takayama, K.; Kurosaki, T. and Ikeda, T. (1993) Mediated electrocatalysis at a biocatalyst electrode based on a bacterium *Gluconobacter industrius*. J. Electroanal. Chem. 356: 295-301.
[27] Tkac, J.; Gemeiner, P.; Svitel, J.; Benikovsky, T.; Sturdik, E.; Vala, V.; Petrus, L. and Hrabarova, E. (2000) Determination of total sugars in lignocellulose hydrolysate by mediated *Gluconobacter oxydans* biosensor. Anal. Chim. Acta 420: 1-7.
[28] Tkac, J.; Vostiar, I.; Gemeiner, P. and Sturdik, E. (2002) Monitoring of ethanol during fermentation using a microbial biosensor with enhanced selectivity. Bioelectrochem. 56: 127-129.
[29] Tkac, J.; Svitel, J.; Novak, R. and Sturdik, E. (2000) Triglyceride assay by amperometric microbial biosensor: Sample hydrolysis and kinetic approach. Anal. Lett. 33: 2441-2452.
[30] Ikeda, T.; Kurosaki, T.; Takayama, K.; Kano, K. and Miki, K. (1996) Measurement of oxidoreductase-like activity of intact bacterial cells by an amperometric method using a membrane-coated electrode. Anal. Chem. 68: 192-198.
[31] Lee, S.A.; Choi, Y.; Jung, S. and Kim S. (2002) Effect of initial carbon sources on the electrochemical detection of glucose by *Gluconobacter oxydans*. Bioelectrochem. 57: 173–178.
[32] Kitagawa, Y.; Ameyama, M.; Nakashima, K.; Tamiya, E. and Karube, I. (1987) Amperometric alcohol sensor based on immobilised bacteria cell membrane. Analyst 112: 1747-1749.
[33] Shinagawa, E.; Matsushita, K.; Adachi, O. and Ameyama, M.J. (1990) Evidence for electron transfer *via* ubiquinone between quinoproteins D-glucose dehydrogenase and alcohol dehydrogenase of *Gluconobacter suboxydans*. J. Biochem. 107: 863-867.
[34] Tkac, J.; Navratil, M.; Sturdik, E. and Gemeiner, P. (2001) Monitoring of dihydroxyacetone production during oxidation of glycerol by immobilized *Gluconobacter oxydans* cells with an enzyme biosensor. Enzyme Microb. Technol. 28: 383-388.
[35] Stredansky, M.; Pizzariello, A.; Stredanska S. and Miertus, S. (2000) Amperometric pH-sensing biosensors for urea, penicillin, and oxalacetate. Anal. Chim. Acta 415: 151-157.
[36] Vostiar, I.; Tkac, J.; Sturdik, E. and Gemeiner, P. (2002) Amperometric urea biosensor based on urease and electropolymerized toluidine blue dye as a pH-sensitive redox probe. Bioelectrochem. 56: 23-25.
[37] Gerard, M.; Chaubey, A. and Malhotra, B.D. (2002) Application of conducting polymers to biosensors. Biosens. Bioelectron. 17: 345-359.
[38] Kemp, R.B. and Lamprecht, I. (2000) La vie est donc un feu pour la calorimetrie: half a century of calorimetry - Ingemar Wadsö at 70. Thermochim. Acta 348: 1-17.
[39] Ramanathan, K.; Rank, M.; Svitel, J.; Dzgoev, A. and Danielsson, B. (1999) The development and applications of thermal biosensors for bioprocess monitoring. TIBTECH 17: 499-505.
[40] Ramanathan, K. and Danielsson, B. (2001) Principles and applications of thermal biosensors. Biosens. Bioelectron. 16: 417-423.
[41] Gerbsch, N. and Buchholz, R. (1995) New processes and actual trends in biotechnology. FEMS Microbiol. Revs. 16: 259-269.
[42] Prüsse, U., Jahnz, U., Wittlich, P., Breford, J. and Vorlop, K.-D. (2002) Bead production with JetCutting and rotating disc/nozzle technologies. Landbauforschung Völkenrode SH 241: 1-10 and references cited herein.
[43] Danielsson, B.; Mosbach, K.; Winquist, F. and Lundström, I. (1988) Biosensors based on thermistors and semiconductors and their bioanalytical applications. Sensors and Actuators 13: 139-146.
[44] Mattiasson, B.; Larsson, P.-O. and Mosbach, K. (1977) Microbe thermistor. Nature 268: 519-520.
[45] Thavarungkul, P.; Hakanson, H. and Mattiasson, B. (1991) Comparative study of cell-based biosensors using *Pseudomonas cepacia* for monitoring aromatic compounds. Anal. Chim. Acta 249: 17-23.
[46] Danielsson, B. and Mosbach, K. (1988), Enzyme thermistors. In: Mosbach, K (Ed.) Immobilized Enzymes and Cells. Volume 137 from the series Colowick, S.P. and Kaplan, N.O. (Ser.Eds.) Methods in Enzymology. Academic Press, San Diego (USA); ISBN 0-12-182037-8; pp. 181-197.
[47] Svitel, J.; Vostiar, I.; Gemeiner, P. and Danielsson, B. (1997) Determination of citrate by FIA using immobilized *Enterobacter aerogenes* cells and enzyme thermistor/flow microcalorimeter detection. Biotechnol. Tech. 11: 917-919.
[48] Wadsö, I. (1986) Bio-calorimetry. TIBTECH – February: 45-51.
[49] Stefuca, V. and Gemeiner, P. (1999) Investigation of catalytic properties of immobilized enzymes and cells by flow microcalorimetry. In: Scheper, T. (Ed.) Thermal Biosensors Bioactivity Bioaffinity. Volume 64 from the series Scheper, T. (Ser.Ed.), Advances in Biochemical Engineering/Biotechnology. Springer-Verlag, Berlin-Heidelberg (Germany); ISSN 0724-6145; pp. 71-99.

[50] Gemeiner, P.; Stefuca, V.; Welwardova, A.; Michalkova, E.; Welward, L.; Kurillova, L. and Danielsson, B. (1993) Direct determination of the cephalosporin transforming activity of immobilized cells with use of an enzyme thermistor. 1. Verification of the mathematical model. Enzyme Microb. Technol. 15: 50-56.
[51] Stefuca, V.; Welwardova, A.; Gemeiner, P. and Jakubova, A. (1994) Application of enzyme flow microcalorimetry to the study of microkinetic properties of immobilized biocatalyst. Biotechnol. Tech. 8: 497-502.
[52] Welwardova, A.; Gemeiner, P.; Michalkova, E.; Welward, L. and Jakubova, A. (1993) Gel-entrapped penicilin G acylase optimized by an enzyme thermistor. Biotechnol. Tech. 7: 809-814.
[53] Stefuca, V.; Gemeiner, P.; Kurillova, L.; Danielsson, B. and Bales, V. (1990) Application of the enzyme thermistor to the direct estimation of intrinsic kinetics using the saccharose-immobilized invertase system. Enzyme Microb. Technol. 12: 830-835.
[54] Docolomansky, P.; Gemeiner, P.; Mislovicova, D.; Stefuca, V. and Danielsson, B. (1994) Screening of Concanavalin A bead cellulose conjugates using an enzyme thermistor with immobilized invertase as the reporter catalyst. Biotechnol. Bioeng. 43: 286-292
[55] Vikartovska-Welwardova, A.; Michalkova, E.; Gemeiner, P. and Welward, L. (1999) Stabilization of D-amino acid oxidase from *Trigonopsis variabilis* by manganese dioxide. Folia Microbiol. 44: 380-384.
[56] Vikartovska, A.; Bucko, M.; Gemeiner, P.; Nahalka, J., and Hrabarova, E. (2003) Flow calorimetry - a useful tool for determination of immobilized *cis*-epoxysuccinate hydrolase activity from *Nocardia tartaricans*. Artif. Cells, Blood Substit. & Biotechnol. 32 (1), 000-000, 2004, in press.
[57] Vikartovska-Welwardova, A.; Gemeiner, P.; Stefuca, V.; Vrabel, P.; Michalkova, E. and Welward, L. (1998) Screening of immobilized *Trigonopsis variabilis* strains with cephalosporin C transforming activity by enzyme flow microcalorimetry. Biologia 53: 705–712.
[58] Satterfield, C.N. (1976) Massoperedaca v geterogennom katalize (in Russian). Izdatelstvo Chimija, Moskva, (Russia); 66.015.23:66.097.13; p. 133.
[59] Vikartovska-Welwardova, A.; Michalkova, E. and Gemeiner, P. (1998) Enzyme flow microcalorimetry - a useful tool for screening of immobilized penicillin G acylase. J. Chem. Technol. Biotechnol. 73: 31-36.
[60] Gemeiner, P.; Docolomansky, P.; Nahalka, J.; Stefuca, V. and Danielsson B. (1996) New approaches for verification of kinetic constants of immobilized concanavalin A: Invertase preparations investigated by flow microcalorimetry. Biotechnol. Bioeng. 49: 26-35.
[61] Stefuca, V.; Vikartovska-Welwardova, A. and Gemeiner, P. (1997) Flow microcalorimeter auto-calibration of immobilized enzyme kinetics. Anal. Chim. Acta 355: 63-67.

INDEX

acidified wort 263, 266
adult stem cells 49, 169, 175, 178
agarose gel 23, 104, 110
air pollution control 507, 508, 509, 514, 517, 521, 522
alcoholic fermentation 276, 281, 282, 286, 287, 290
aldehydes 265, 365, 367
alginate 21, 22, 23, 24, 25, 26, 27, 30, 31, 41, 42, 43, 44, 74, 75, 76, 77, 78, 79, 80, 118, 119, 147, 160, 162, 188, 189, 190, 200, 201, 204, 211, 213, 214, 215, 216, 217, 218, 219, 220, 221, 222, 237, 240, 241, 242, 250, 261, 262, 266, 278, 279, 280, 281, 286, 287, 288, 289, 296, 297, 298, 304, 308, 309, 310, 312, 313, 314, 323, 324, 326, 327, 328, 329, 330, 341, 343, 346, 347, 348, 349, 351, 352, 359, 360, 361, 362, 378, 381, 382, 383, 384, 385, 386, 387, 388, 396, 397, 398, 399, 408, 409, 413, 414, 415, 416, 417, 418, 424, 426, 427, 435, 444, 450, 455, 456, 457, 458, 463, 464, 465, 466, 471, 472, 473, 475, 482, 485, 486, 529, 530, 533, 534, 540, 542, 544, 550, 559
alginate beads 11, 22, 41, 43, 77, 214, 217, 218, 221, 222, 261, 262, 266, 278, 279, 281, 286, 313, 314, 323, 328, 346, 347, 349, 360, 361, 381, 383, 384, 396, 397, 398, 408, 417, 427, 435, 456, 466, 472, 475, 485, 529, 530, 544
alginate gel 21, 22, 215, 219, 262, 287, 289, 304, 308, 323, 324, 326, 328, 330, 343, 346, 383, 384, 408, 414, 418, 427, 465, 472, 473, 475, 541, 559
alginate microcapsules 22, 23, 25, 26, 27, 30, 31, 188, 201, 204, 214, 215, 217, 218, 219, 220, 221, 222, 237, 240, 241, 242
amperometric 551, 553, 557
antibiotics 88, 301, 316, 321, 388, 407, 408, 409, 413, 414, 415, 416, 417, 452, 550, 562
aquifer 495, 496, 497, 498, 499, 500, 503, 504, 523
artificial cells 185, 249, 250, 251, 539, 547
artificial insemination 539, 545, 547
bacteriocins 295, 305, 312, 331, 338, 341
bacteriophages 303, 304, 345
beer 259, 260, 261, 262, 263, 264, 265, 266, 268, 275, 278, 279, 287, 321, 324, 330, 331, 358, 359, 360, 365, 366, 367, 513,
beer maturation 265, 365
beer production 259, 260, 261, 263, 268, 279, 365
bioartificial 39, 41, 42, 55, 69, 70, 71, 77, 80, 186, 191, 426, 547
biocompatibility 17, 22, 24, 40, 44, 61, 86, 144, 149, 156, 187, 188, 189, 190, 193, 200, 219, 220, 221, 464, 465
bioconversion by microorganisms 362
bioconversions 321, 322, 331
biodegradation 491, 496, 497, 504, 508, 509, 510, 511, 513, 514, 515, 516, 523
bioencapsulation 205, 249, 452

Index

biofilms 482, 484, 488, 491, 497, 498, 499, 500, 501, 503, 504, 513, 514
biofilter 509, 510, 511, 516, 517, 518, 519, 520, 521, 522, 523
bioflavour 358, 359
biopharmaceuticals 407, 419
biopolymers 94, 117, 298, 379
bioprecipitation 490, 503
bioreactor .. 70, 71, 72, 73, 74, 77, 79, 100, 101, 103, 104, 105, 106, 107, 109, 111, 118, 119, 121, 129, 141, 147, 148, 158, 159, 198, 200, 214, 215, 222, 260, 261, 263, 264, 266, 267, 268, 278, 285, 296, 297, 299, 300, 303, 304, 308, 309, 311, 313, 314, 322, 323, 324, 331, 369, 375, 376, 378, 380, 383, 384, 385, 387, 388, 389, 391, 392, 394, 395, 396, 397, 398, 409, 410, 411, 415, 416, 424, 425, 426, 428, 429, 431, 432, 434, 435, 468, 483, 490, 491, 509, 515, 519
biotransformation 358, 417, 559, 563
biotrickling filter .509, 510, 511, 512, 513, 514, 515, 516, 517, 518, 519, 520, 522, 523, 524, 525, 526
blood substitutes 249, 251, 252
blood vessel 18, 21, 85, 86, 87, 88, 90, 91, 92, 93, 95, 117, 135, 147, 154, 162, 178, 198, 199, 200, 205, 214, 237
bone .. 19, 62, 96, 101, 104, 109, 117, 147, 153, 154, 155, 156, 157, 158, 159, 160, 161, 162, 169, 170, 172, 174, 175, 200, 231
capillary 71, 86, 93, 94, 95, 126, 192, 202, 204, 387, 424, 456, 458, 460
cardiovascular 85, 116, 135, 142, 147, 149
carrier materials 261, 268
carrier scaffolds 154, 155, 156, 160
cartilage 99, 100, 101, 103, 104, 105, 106, 107, 108, 110, 111, 112, 113, 117, 120, 129, 154, 155, 160, 170, 174, 175, 200
cell .. 17, 19, 20, 22, 24, 27, 29, 30, 33, 34, 39, 40, 41, 42, 43, 44, 45, 46, 47, 48, 49, 50, 55, 56, 58, 59, 60, 62, 65, 69, 70, 71, 72, 73, 74, 75, 76, 77, 78, 79, 80, 85, 86, 87, 88, 89, 94, 99, 100, 101, 103, 104, 108, 109, 110, 111, 114, 116, 117, 118, 119, 120, 121, 122, 123, 126, 127, 128, 129, 135, 136, 139, 140, 141, 142, 143, 144, 145, 146, 147, 148, 149, 153, 154, 155, 156, 157, 158, 159, 160, 161, 162, 167, 168, 169, 170, 171, 172, 173, 174, 175, 177, 178, 185, 186, 187, 188, 190, 191, 197, 198, 199, 200, 201, 203, 204, 205, 211, 213, 214, 216, 217, 218, 219, 221, 222, 231, 232, 233, 234, 235, 236, 237, 238, 241, 242, 249, 250, 251, 252, 254, 260, 261, 262, 263, 264, 266, 275, 277, 278, 280, 281, 282, 285, 286, 287, 288, 290, 295, 296, 297, 298, 299, 300, 301, 302, 304, 305, 306, 307, 308, 309, 311, 312, 313, 314, 321, 322, 323, 324, 325, 327, 328, 330, 331, 337, 342, 344, 346, 347, 348, 349, 355, 358, 360, 361, 362, 365, 366, 375, 376, 377, 378, 379, 380, 381, 382, 383, 384, 385, 386, 388, 389, 391, 395, 398, 399, 407, 408, 409, 410, 411, 412, 413, 414, 415, 416, 417, 418, 419, 423, 424, 425, 426, 427, 428, 429, 430, 431, 432, 433, 434, 435, 439, 442, 443, 450, 455, 456, 458, 462, 463, 464, 465, 466, 469, 470, 471, 472, 473, 474, 475, 476, 481, 482, 483, 485, 490, 491, 495, 498, 507, 511, 533, 539, 544, 547, 549, 550, 551, 553, 554, 556, 557, 558, 559, 560, 562, 563
cell encapsulation 42, 71, 74, 185, 187, 191, 193, 198, 200, 201, 204, 211, 215, 218, 231, 233, 238, 242, 250, 297, 313, 402, 424, 426, 427

cell immobilisation .42, 70, 71, 72, 74, 80, 135, 149, 185, 211, 264, 275, 278, 304, 305, 307, 311, 322, 330, 378, 379, 381, 384, 386, 391, 407, 413, 415, 416, 417, 418, 419, 424, 427, 435, 455, 456, 469, 470, 471, 473, 474, 507, 558
cell physiology 261, 281, 301, 349, 407, 547, 551
cell sourcing 39
cell therapy 6, 37, 163, 245, 249, 250
cell transplantation 19, 22, 50, 96, 181, 200
cellobiose............327, 396, 397, 398, 399
cellulose......30, 31, 32, 33, 34, 74, 75, 78, 190, 237, 246, 261, 262, 264, 266, 268, 281, 285, 298, 330, 331, 378, 379, 396, 397, 398, 399, 424, 426, 429, 450, 489, 532, 551, 553, 555
cheese 295, 304, 309, 310, 314, 327, 395, 396
cheese whey............395, 396
cider............275, 276, 277, 278, 280, 281, 282, 289, 321, 330, 360
continuous fermentation 261, 266, 277, 278, 296, 300, 304, 305, 306, 307, 308, 323, 398, 409, 411, 412, 415, 416
de novo biosynthesis 359, 360
drinking water 489, 490
electrified liquid jet 455, 456
electrostatic droplet generation............41, 456, 457
embryonic stem cells 48, 167, 168, 169, 170, 171, 172, 180
encapsulated cells ...22, 28, 75, 77, 78, 80, 187, 190, 193, 198, 201, 203, 211, 213, 214, 218, 219, 220, 221, 222, 229, 230, 231, 233, 234, 235, 236, 237, 238, 239, 240, 241, 242, 249, 313, 346, 463
encapsulation 27, 28, 29, 30, 31, 39, 42, 43, 71, 74, 75, 142, 185, 186, 187, 188, 190, 191, 193, 198, 200, 201, 211, 215, 216, 218, 222, 230, 231, 233, 235, 236, 237, 238, 242, 250, 297, 299, 313, 314, 321, 331,341, 343, 345, 346, 349, 388, 424, 426, 427, 445, 451, 455, 464, 466, 471, 528, 529, 530, 531, 532, 534, 535, 539, 540, 541, 542, 543, 545, 546
enzyme therapy............249
ethanol 266, 276, 277, 278, 281, 286, 287, 289, 290, 321, 324, 325, 326, 330, 359, 366, 369, 370, 371, 375, 376, 380, 381, 383, 384, 385, 386, 387, 388, 389, 391, 395, 396, 397, 398, 399, 4487, 489, 490, 520, 553, 554, 555, 556
exopolymers 498, 499, 513
exopolysaccharides............317, 318, 498
expression vectors 231, 235, 240, 242, 467
extracellular matrix 42, 61, 62, 75, 80, 87, 88, 90, 91, 92, 93, 94, 99, 103, 116, 136, 137, 139, 146, 148, 149, 157, 160, 199, 214
fermentation 259, 260, 261, 263, 264, 265, 266, 267, 268, 275, 276, 277, 278, 279, 280, 281, 282, 285, 286, 287, 288, 289, 290, 296, 299, 300, 302, 303, 306, 307, 308, 309, 311, 312, 322, 323, 324, 325, 326, 330, 331, 338, 339, 341, 342, 344, 345, 346, 349, 355, 358, 359, 360, 361, 362, 365, 366, 367, 375, 376, 378, 382, 383, 385, 386, 389, 391, 395, 396, 397, 398, 399, 408, 409, 410, 411, 412, 413, 414, 415, 416, 417, 439, 442, 444, 451, 550, 555, 556
fibroblasts ...28, 56, 59, 60, 61, 62, 63, 64, 65, 67, 86, 87, 88, 90, 93, 94, 116, 126, 136, 139, 143, 146, 219, 231, 235, 236, 238, 240, 241, 430

flow calorimetry 558, 562, 563
fluid drilling..532
fluidised bed bioreactor..398, 428, 429, 432, 433, 435
foetal stem cells...170, 177, 178
food 162, 197, 220, 229, 259, 260, 261, 263, 282, 295, 296, 298, 303, 305, 310, 311,
  312, 313, 321, 322, 323, 324, 325, 326, 328, 329, 330, 331, 337, 338, 339, 340, 341,
  347, 348, 349, 355, 356, 358, 359, 360, 361, 367, 379, 384, 395, 396, 439, 440, 446,
  447, 448, 449, 452, 550, 557
foreign protein 477
gene therapy149, 171, 172, 177, 187, 190, 193, 198, 199, 202, 222, 229, 230, 231, 232,
  233, 234, 235, 237, 238, 239, 240, 242, 249, 433, 435
gluconobacter oxydans...552, 553, 559
groundwater..............................489, 490, 495, 496, 497, 499, 500, 503, 504, 512, 523
hairy roots-derived materials...531
heart valve 135, 136, 137, 139, 141, 142, 143, 144, 149
heavy metals 40, 44, 188, 287, 495, 496, 501, 503, 550
hepatocytes 70, 71, 72, 73, 74, 77, 79, 80, 168, 172, 250, 426, 434
hollow fibre 21, 71, 72, 73, 74, 77, 80, 263, 360, 361, 371, 388, 394, 415, 416, 426,
  427, 434, 435, 455, 471
hollow fibre bioreactor 72, 388, 415, 427, 434, 435
human ...19, 20, 33, 44, 48, 50, 55, 56, 57, 58, 61, 62, 63, 64, 65, 69, 71, 72, 75, 76, 79,
  80, 85, 86, 87, 88, 89, 90, 92, 93, 94, 95, 100, 104, 106, 118, 119, 126, 131, 135,
  136, 139, 145, 154, 155, 157, 158, 159, 160, 161, 162, 171, 177, 178, 197, 198, 201,
  202, 203, 205, 211, 214, 219, 221, 230, 232, 233, 236, 237, 238, 240, 241, 251, 252,
  311, 341, 407, 416, 424, 425, 432, 434, 439, 449, 452, 469, 470, 475
hydrogels 22, 23, 75, 142, 147, 408, 415, 481, 482, 485, 486, 491, 533, 558
hyporeic zone 499, 504
immobilisation....24, 40, 42, 70, 71, 72, 74, 80, 135, 136, 139, 140, 142, 149, 185, 186,
  189, 190, 195, 211, 217, 260, 261, 262, 263, 264, 266, 268, 269, 271, 275, 277, 278,
  279, 280, 281, 282, 284, 286, 291, 295, 296, 299, 301, 304, 305, 306, 307, 311, 312,
  313, 314, 321, 322, 325, 326, 327, 328, 329,330, 333, 341, 342, 343, 344, 345, 349,
  360, 361, 376, 378, 379, 380, 381, 382, 383, 384, 386, 387, 388, 389, 391, 400, 401,
  402, 403, 404, 407, 408, 409, 410, 412, 413, 414, 415, 416, 417, 418, 419, 420, 423,
  424, 427, 428, 431, 432, 436, 455, 456, 466, 469, 470, 471, 472, 473, 474, 475, 476,
  477, 481, 482, 483, 484, 485, 486, 488, 490, 491, 492, 503, 507, 549, 550, 551, 553,
  557, 558, 559, 563
immobilised cell culture 321, 325, 413, 414, 415, 424, 435, 476
immobilised cell reactors...264
immobilised cells..17, 186, 187, 188, 190, 191, 192, 193, 235, 259, 260, 262, 263, 282,
  286, 287, 288, 289, 290, 295, 296, 311, 322, 323, 324, 325, 326, 327, 328, 329, 330,
  331, 342, 344, 345, 346, 359, 361, 365, 366, 376, 378, 380, 382, 383, 384, 388, 394,
  396, 397, 407, 408, 409, 411, 412, 413, 414, 415, 416, 417, 418, 430, 433, 463, 471,
  472, 475, 476, 481, 483, 484, 492, 493, 507, 551, 553, 557, 558, 559, 560, 561, 562,
  563
immobilised yeast........264, 265, 267, 279, 285, 288, 289, 323, 326, 330, 376, 383, 388
immunoisolation..........21, 22, 23, 25, 35, 40, 42, 74, 185, 186, 187, 238, 242, 244, 254

immunoprotection 23, 40, 42, 43
in vitro cultivation 104, 117, 129, 141
insect cells 462, 463, 464, 465, 467, 470
islet .18, 19, 20, 21, 22, 23, 24, 25, 26, 28, 29, 30, 31, 34, 39, 40, 41, 42, 44, 45, 46, 47, 48, 50, 188, 201, 217, 221, 250, 539, 540
Jerusalem artichokes.. 395, 397
keratinocytes.. 56, 58, 59, 60, 61, 62, 63, 64, 65, 93, 94, 175
lactic acid bacteria 263, 264, 266, 275, 277, 278, 282, 285, 295, 297, 303, 306, 330, 338, 347, 361, 452
liver.18, 19, 28, 39, 42, 45, 47, 50, 69, 70, 71, 72, 73, 74, 75, 77, 79, 80, 167, 168, 169, 173, 178, 191, 200, 235, 240, 249, 250, 426
macroporous microcarriers... 424, 426, 428, 429, 430, 431, 434
malolactic fermentation 275, 276, 277, 280, 282, 286, 290
mammalian cells 41, 230, 236, 455, 470
mathematical modelling 467
meat 321, 337, 338, 339, 340, 341, 342, 343, 344, 345, 346, 347, 348, 349, 356, 360, 442
mesenchymal stem cells 62, 143, 154, 155, 156, 160, 168, 175
metabolites....71, 100, 101, 231, 237, 295, 310, 321, 322, 329, 360, 407, 412, 424, 427, 428, 434, 469, 470, 471, 474, 475, 500, 510, 513, 514, 523, 532, 557, 559
methods ..25, 33, 34, 40, 43, 44, 48, 68, 90, 95, 139, 142, 153, 160, 180, 197, 200, 205, 235, 250, 259, 261, 262, 264, 282, 288, 289, 296, 298, 305, 306, 314, 321, 322, 325, 356, 361, 378, 379, 381, 418, 427, 434, 444, 451, 452, 455, 471, 481, 495, 507, 523, 531, 535, 556, 558, 559
microbes 349, 407, 411, 439, 496
microbial biosensors.. 549, 550, 551, 552, 553, 554, 557
microencapsulation......23, 36, 41, 44, 186, 193, 196, 200, 205, 219, 263, 264, 296, 297, 299, 445, 452, 455, 539, 540, 542, 544, 547
monoclonal antibodies......................... 192, 215, 423, 424, 425, 426, 427, 431, 437, 455
multifunctional microcapsules.. 30, 31
mycorrhizal fungi 530, 531
myocardium..99, 100, 101, 103, 116, 117, 118, 121, 127, 129, 145, 146, 148, 149, 151
non-conventional feedstock.. 395
odour control 510, 519, 521, 523
optimisation 40, 147, 189, 268, 285, 323, 358, 362, 383, 434, 522, 523
organic pollutants 495, 504, 514, 530
osteoblasts 142, 153, 157, 158, 159, 160
oxygen 28, 29, 40, 41, 70, 72, 73, 74, 101, 105, 113, 114, 115, 117, 118, 120, 121, 123, 124, 125, 126, 148, 157, 186, 201, 218, 251, 272, 298, 302, 307, 313, 323, 328, 345, 361, 369, 391, 409, 410, 413, 414, 415, 416, 423, 426, 427, 428, 430, 432, 433, 434, 435, 440, 443, 448, 455, 472, 473, 482, 497, 498, 503, 513, 514, 516, 520, 549, 552, 553, 554, 555, 556
packed bed bioreactor.. 383, 428, 433, 434, 435
perfusion........................ 77, 121, 122, 123, 148, 158, 159, 162, 391, 424, 431, 432, 434
plant cell 355, 356, 358, 378, 455, 466, 469, 470, 471, 472, 473, 474, 475, 476,
plant tissue.. 466

plasmid stability..305, 306, 317, 322
polymeric coating..532
precursor fatty acids 364
preservation 63, 80, 178, 295, 310, 338, 339, 340, 341, 342, 345, 527, 544
primary beer fermentation 264, 268
probiotic ...295, 297, 299, 301, 302, 307, 313, 314, 325, 341, 343, 345, 349, , 439, 440, 441, 442, 444, 445, 446, 447, 449, 450, 451, 452
probiotic microorganisms..439, 440, 444, 446
probiotics..................295, 307, 314, 321, 341, 342, 343, 345, 349, 439, 440, 442, 452
products 47, 59, 60, 61, 85, 100, 156, 157, 162, 185, 186, 187, 190, 191, 193, 199, 218, 219, 220, 225, 229, 236, 238, 262, 276, 286, 289, 295, 297, 298, 299, 302, 303, 306, 308, 311, 312, 313, 323, 325, 327, 331, 337, 338, 339, 340, 341, 343, 345, 348, 355, 356, 358, 359, 375, 378, 379, 381, 382, 385, 388, 391, 399, 407, 417, 419, 423, 424, 425, 426, 431, 433, 435, 439, 440, 442, 445, 448, 449, 450, 451, 452, 455, 469, 471, 473, 474, 475, 476, 498, 508, 510, 513, 514, 523, 560
protein delivery..197, 205
protocorm-like bodies..530
psychrotrophs..303
recombinant cells...196, 230, 235, 236, 237, 238, 305
seeds 466, 527, 528, 529, 530, 531, 532, 533, 534
shoot buds...529
shoot-tips..529
skin 55, 56, 57, 58, 59, 60, 61, 62, 63, 64, 65, 86, 88, 93, 94, 95, 147, 154, 169, 174, 175, 200, 250, 337
sparkling wines...288, 289
Spermatozoal..539, 540, 541,542, 544, 545
stabilisation................205, 298, 439, 440, 442, 443, 444, 446, 448, 449, 452, 557, 559
starters 276, 277, 290, 304, 305, 306, 307, 308, 309, 342, 343, 344, 351,
stem cell differentiation..50
stem cells 35, 37, 44, 45, 46, 47, 48, 49, 50, 53, 54, 62, 143, 153, 154, 155, 156, 159, 160, 167, 168, 169, 170, 171, 172, 173, 174, 175, 176, 177, 178, 233, 237, 246, 250
stirred tank bioreactor...389, 423, 427, 429, 431, 432, 434
sulphate reducing bacteria 496
therapeutic application 169, 191, 197
therapeutic cloning 50, 171
thermal biosensors..557, 565
three dimensional 117, 150, 172
tissue engineering...7, 39, 55, 58, 65, 85, 86, 95, 99, 100, 101, 103, 104, 106, 108, 116, 117, 118, 119, 120, 121, 129, 135, 136, 140, 142, 143, 144, 149, 153, 154, 155, 156, 158, 159, 160, 162, 167, 168, 172, 173, 174, 176, 199
two-coat system..530, 533,534, 535
vanillin..362, 363
vapour-phase bioreactors..522, 523
wastewater..376, 481, 482, 483, 484, 486, 488, 489, 490, 491, 499, 501, 503, 508, 512, 515, 516, 521, 522
wine 264, 277, 278, 280, 285, 287, 288, 289, 290, 321, 330, 358, 361, 362, 367

xylose 361, 396, 399, 553, 555
yeasts ..264, 275, 285, 286, 287, 288, 289, 337, 365, 366, 367, 383, 389, 396, 397, 398
yoghurt 308, 313, 448